Drug Discovery

A History

Walter Sneader

School of Pharmacy
University of Strathclyde, Glasgow, UK

John Wiley & Sons, Ltd

Telephone (+44) 1243 779777

Email (for orders and customer service enquiries): cs-books@wiley.co.uk

Other Wiley Editorial Offices

John Wiley & Sons Inc., 111 River Street, Hoboken, NJ 07030, USA

Jossey-Bass, 989 Market Street, San Francisco, CA 94103-1741, USA

Wiley-VCH Verlag GmbH, Boschstr. 12, D-69469 Weinheim, Germany

John Wiley & Sons Australia Ltd, 33 Park Road, Milton, Queensland 4064, Australia

John Wiley & Sons (Asia) Pte Ltd, 2 Clementi Loop #02-01, Jin Xing Distripark, Singapore 129809

John Wiley & Sons Canada Ltd, 22 Worcester Road, Etobicoke, Ontario, Canada M9W 1L1

Wiley also publishes its books in a variety of electronic formats. Some content that appears in print may not be available in electronic books.

Library of Congress Cataloging-in-Publication Data

Sneader, Walter
Drug discovery : a history / Walter Sneader.
p. ; cm.
Includes bibliographical references and index.
ISBN-10 0-471-89979-8 (hardback : alk. paper) – ISBN-10 0-471-89980-1 (pbk. : alk. paper)
ISBN-13 978-0-471-89979-2 (HB) 978-0-471-89980-8 (PB)
1. Pharmacy–History. 2. Drug development–History. 3. Pharmacognosy–History. I. Title
[DNLM: 1. Pharmaceutical Preparations–history. QV 11.1 S671da2005]
RS61.S637 2005
615′.19′90–dc22

2005041804

British Library Cataloguing in Publication Data

A catalogue record for this book is available from the British Library

ISBN-13 978-0-471-89979-2 (HB) 978-0-471-89980-8 (PB)
ISBN-10 0-471-89979-8 (HB) 0-471-89980-1 (PB)

Typeset by Dobbie Typesetting Ltd, Tavistock, Devon
Printed and bound in Great Britain by Antony Rowe Ltd, Chippenham, Wiltshire
This book is printed on acid-free paper responsibly manufactured from sustainable forestry in which at least two trees are planted for each one used for paper production.

Dedication

The quest for the elixir of life by the mediaeval alchemists may not have resulted in success, but the last one hundred years have witnessed the introduction of a number of drugs that have literally saved countless millions of lives. Society at large is indebted to the men and women in the laboratories and clinics of the pharmaceutical industry, research institutes, hospitals and universities whose commitment to science has brought about what in an earlier era would simply have been called miracles. This book is dedicated to these men and women.

Contents

Preface

The contents of this book range from the faltering attempts of our ancient ancestors to discover herbal remedies to the present quest of scientists to develop safe, efficacious medicines. This will enable readers to understand both why efficacious drugs were not developed until the twentieth century and why progress has been so rapid over the past fifty years.

The book consists of three parts. The first part progresses from the possible use of medicinal herbs by Neanderthals on to the endeavours of ancient civilisations in Mesopotamia, Egypt, Greece and Rome. A subsequent consideration of the impact of Greek medicine on the Arab world and then renaissance Europe is followed by an examination of the heritage of the alchemists whose attempts to transmute base metals into gold provided the chemical techniques that allowed purified metallic compounds to be introduced into medicine. After reviewing how the influence of the enlightenment on drug discovery was impeded by the introduction of unsubstantiated medical systems that were popular from the seventeenth to nineteenth centuries, this part of the book concludes with an account of the twentieth century development of life-saving organic compounds incorporating various metals.

The second part of the book examines the commonest source of drug prototypes from which other medicines are derived – the natural world. Prototypes from vegetable, animal and microbial sources are described in separate chapters, as are the medicinal compounds prepared from them. The problem of relying on nature as a source of drug prototypes is confronted.

Modern drug discovery has consisted of a series of thematic developments that began with the isolation of pure alkaloids and glycosides from plants in the early years of the nineteenth century. One hundred years later, a parallel development witnessed the isolation of pure hormones from mammalian sources. The chapters throughout the book are organised so as to heighten awareness of such thematic developments, but this is particularly relevant in the second part of the book.

The third part considers those synthetic compounds that have served as drug prototypes, followed by a consideration of compounds derived from them. The role of serendipity in providing synthetic prototypes is also investigated. The book ends with a cautionary note concerning the future of drug discovery.

The specialist reader will find that the book contains an extensive list of bibliographic references relating to the discovery of drugs, but the book has been written for the general reader. No particular expertise in chemistry is assumed, but chemical structures have been interspersed with the text to enable those familiar with the subject to obtain a deeper understanding. Their presence should also enable others to begin to appreciate why chemistry has appealed to so many great minds. The chemical structures are presented in such a way as to allow even those with little experience of the subject to understand how prototypes and compounds obtained from them were manipulated in order to provide novel compounds with unprecedented healing properties.

Walter Sneader
December 2004

History of Drugs website compiled by Walter Sneader: http://www.historyofdrugs.net

Foreword

One of the best presents I ever had was a copy of Dr Sneader's 1985 book *Drug Discovery: The Evolution of Modern Medicines*. It was a great introduction to the fascinating world of the way in which many medicines have been discovered, a world of which I had gained only snippets from books on clinical pharmacology. And now Dr Sneader has produced not just an updated version of that book but a completely new work which comprehensively spans the whole subject from the earliest times to the present.

Dr Sneader is exceptionally well qualified to write such a book having trained in both pharmacy and medicinal chemistry. His distinguished academic career at the University of Strathclyde, Glasgow, where he is the Head of the School of Pharmacy, includes the writing of two other acclaimed books, *Drug Development from Laboratory to Clinic* and *Drug Prototypes and Their Exploitation*.

The author takes us right back to the pre-historic period, but he does so critically and with caution as to how to interpret earlier evidence, for example as to when opium poppy was first used. Critical evaluation is a hallmark of Dr Sneader's writing. I like the way that the book has been divided into three major sections, dealing first with previous history, then with drugs developed from natural compounds and finally with synthetic drugs. Phytochemicals are of especial interest and we know that over 30 modern medicines owe their origin to plants in one way or another. Sometimes a folk remedy has stood by itself, but more often it has been scientific investigation that has either found a new use for the plant substance, or has developed a synthetic compound from it. Here as elsewhere one is continually impressed by the details that the author gives as to who were the actual people making the discoveries. Likewise Dr Sneader has a plethora of fascinating stories, for example the one that recounts how the discovery of coal gas for illuminating the home led to the use of antiseptic treatment by Joseph Lister. In another he relates how the Nazis suppressed the fact that aspirin was developed by a Jewish chemist.

Natural remedies or their derivatives are of great importance but so are purely synthetic drugs, sometimes sneered at by the ignorant as "chemical medicine", not realising that chemical molecules are the same whether produced by nature or by a pharmaceutical company. The beta blockers owed everything to Sir James Black's ideas and nothing to the natural world.

Dr Sneader deals with synthetic drugs in one complete section of the book and here as elsewhere our understanding is greatly helped by the chemical formulae he carefully places within the text. This is where his expertise in chemistry is so useful.

A good chapter is the one on screening. Companies are doing mass screening of natural compounds in the hope of finding an interesting looking molecule, and furthermore by combinatorial chemistry they are creating huge numbers of totally new compounds. The author notes that there are 17 million compounds on the chemical register, but will screening find a new treatment for say psoriasis?

It is a pleasure to give a whole hearted commendation to a really fine account of a very important subject, written and written clearly by someone who has a deep understanding of it. There are many references to the literature and the index is excellent.

Arthur Hollman, MD FRCP FLS
Emeritus Consulting Cardiologist, University College Hospital, London
Formerly Chairman of the Advisory Committee of the Chelsea Physic Garden

1

Introduction

The quest to discover healing drugs has always been influenced by prevailing social and cultural factors, one of the most important being the ability to communicate what has been learned to others. The success of modern drug research has in no small measure been dependent upon the rapid and universal publication of scientific results. Before the appearance of writing about five thousand years ago, accumulated knowledge and wisdom could only be passed on by word of mouth from generation to generation, or be preserved in epics and poems recited by bards who memorised their content. In some instances, these verses were eventually committed to writing.

According to some, ancient paintings on the walls of caves may hold clues to the early use of drugs that had a pronounced effect on the mind. However, it is only the written accounts that communicate something of the reasoning that led our ancestors to seek out remedies for treating disease, such as those compiled by the inhabitants of Ancient Egypt, Mesopotamia and particularly Greece.

How our earliest ancestors selected drugs is a matter for speculation. Clues may be found by considering the behaviour of contemporary primitive peoples and the nature of their folk medicines. This, in conjunction with the written record, suggests that most societies initially adopted a straightforward empirical approach to treatment of the minor ailments of everyday life, selecting healing herbs by a process of trial and error. Those living in hot, sandy desert conditions sought out soothing balms and lotions to apply to their dry skin and eyes, while others whose diet was deficient in fibre sought herbs to relieve their constipation. The ache following the eating of contaminated foods was relieved by swallowing other herbs that irritated the wall of the stomach to induce vomiting, while the bleeding of minor cuts or the pain of burns was dealt with by rubbing with leaves or barks rich in astringent tannins.

Serious illness was a different matter. Unable to explain how it was possible to be fit and healthy one moment yet be writhing in agony or even dying a moment later, our ancestors turned to magic and the supernatural. The art of healing became inextricably interwoven with magic and religion.[1] Disease was then seen as a manifestation of the power of demonic characters to enter the body, presumably through one of its orifices. Attempts were made to expel these demons through their supposed route of entry by administering drugs that could produce vomiting, purgation, urination or sneezing. An alternative was to administer foul-tasting or obnoxious substances such as dung or dead flies. Sympathetic magic was also employed, whereby properties associated with plants, animals or other objects were deemed to be transferable to the sick. A patient might then have been treated with a skin newly sloughed from a reptile because the act of sloughing was considered to be a process of life renewal. Whether expurgating demons or transferring life-endowing properties, the objective was to treat the demon-possessed patient rather than a specific disease. Therein lies the origins of holistic medicine.

Witch doctors, sorcerers, shamans and medicine men preceded their administration of herbs by elaborate rituals involving fervent dancing, weird utterances, grotesque facial expressions, use of charms and other devices – all likely to have a powerful effect on the mind of the sick

Drug Discovery. A History. W. Sneader.
©2005 John Wiley & Sons Ltd

person and thereby enhance any benefit to be derived from the treatment. Even when belief in magic was undermined by the arrival of both pantheistic and monotheistic religions, the remedies themselves hardly changed. Treatment was instead administered in conjunction with appropriate benedictions rather than the incantations previously uttered in the hope of driving out the demons. The successes seen in those few patients who recovered despite the ministrations of their attendants were probably produced by the incantations and benedictions rather than the medicaments themselves, for these were invariably naturally occurring substances of which few could produce any effect other than purgation. Failure was explained away by asserting that the medicament had not been properly prepared or administered.

Once a truly effective remedy has been recognised, the quest for alternatives usually dissipates or, in our modern era, becomes limited to a few commercially viable alternatives. It is a telling point that the natural products employed in the past, and still favoured in some quarters, were frequently recommended for a wide variety of disparate medical conditions, a feature that is suggestive of a lack of any clearly defined effect. This is not to say that none of the ancient remedies had any beneficial action; the few that did were the exception rather than the rule.

When the philosophically inclined Greeks arrived on the scene, supernatural beliefs were rejected in favour of rational concepts. The quest to find drugs became based on the belief that they had to correct an imbalance in body humours thought to be the cause of disease. This was to have a major impact on the selection of plants for use as drugs.

The writings of the second century physician Galen so convincingly presented the accumulated knowledge of Greek medicine that they dominated medical thinking until the seventeenth century or later. During the Dark Ages, however, medical knowledge became fossilised in much of Europe. The asceticism of the Christian Church fostered the attitude that treating the sick was in the hands of the Almighty alone. Eventually, it was recognised that the soul had still to be cared for and consequently some monasteries became hospitals. These acquired surviving medical manuscripts which monks laboriously copied by hand. Herbal gardens were also established to supply medicines, yet new treatments were not introduced, or if they were then it was in the hands of local folk healers who did not record their observations for posterity.

The Christian influence did not extend to the Arab world. There, ancient texts were translated and given a new lease of life, especially in ninth century Baghdad. Arab physicians took delight in the legacy from their Greek counterparts and compiled treatises that elaborated at length on this and their own contributions to therapeutics. It was from the Arab world that Europe was to rediscover its own medical heritage three centuries later, when the first European medical school was established in the southern Italian port of Salerno. Its influence and that of its successors quickly developed, largely because the Church took steps to prevent the practice of medicine in its monasteries out of concern that it was in danger of becoming their prime activity.

The secularisation of medicine in Europe helped its practitioners to question the authoritarianism that had crippled it for so long. Another factor in this was the introduction of the printing press in the second half of the fifteenth century, which gradually made available the old Greek herbals and the Arab additions to them. By this time the Renaissance was well under way and there were those prepared to challenge the authoritarian views of Galen and his devotee ibn Sina. Prominent among them was the maverick physician Paracelsus. His writings were complex and at times obscure, but they attracted a band of followers who were to change drug therapy. Among them were chemists who strove to isolate active principles from medicinal preparations rather than attempt to transmute base metals into gold as the alchemists had aspired. Initially, they met with some success when using traditional alchemical techniques such as distillation or sublimation to refine metallic compounds. However, when plants were exposed to these techniques, the heat treatment involved destroyed any active principles. All that could be obtained were odiferous waters and a few volatile oils.

ISOLATION OF ACTIVE PRINCIPLES

For any drug to produce a real as opposed to an imagined response, it must at the very least interact at the molecular level with either a component of the body or an infectious micro-organism. A drug may also interact elsewhere to produce an undesirable effect. As natural products such as plants contain a complex and variable mixture of chemicals, the risk of unwanted side effects occurring is always high. The simplest way to minimise this is to isolate the component that produces the desired response. While this may be obvious today, it was not the reason behind the successful isolation of active principles in the second decade of the nineteenth century. The sole motivation for this was the need to identify adulterated plant products by determining the amount of active principles in them. Only after this could be done was it recognised that the amounts of these principles varied both quantitatively and qualitatively throughout the life cycle of the plant, being influenced by changing climatic conditions. The time of harvesting, adequacy of the drying process, and nature and duration of storage further influenced the composition of these chemicals.

The inability to identify and quantitatively analyse the active substances in herbal medicines before the nineteenth century means that there is uncertainty surrounding the nature of all herbal remedies used before this period. It may also explain why there was no effective challenge to the beliefs of the ancient healers other than when mineral drugs were recommended, which was from the tenth century onwards. Unlike herbs, these could be purified and analysed due to their ability to withstand high temperatures. They could also survive exposure to strong acid or alkali to form salts that could be refined by crystallisation.

The uncertainty about the identity and quality of ancient herbal remedies meant that they could never be properly assessed in the clinic. Added to that was the total failure of reports about their medicinal applications to even consider, let alone apply, the various factors that we now know must be taken into account when evaluating any drug in the clinic, irrespective of its origin. Thus the ancient and mediaeval medical literature that we have inherited reveals no awareness of that classic example of mind over matter, the placebo effect. Nor does it take into account how patients were selected or how the nature of their ailment was assessed prior to treatment. Indeed, that would have been fraught with difficulty for those whose diagnostic skill was restricted to assessing which of the four bodily humours was present in excess or was deficient. In addition, there was never any attempt to eliminate bias on the part of the attendant physician. As to any consideration of the dose, or frequency and duration of the treatment, little was ever written. Side effects of medication were sometimes even considered to be evidence of their efficacy and at other times a sign of their failure. As for defining what could be considered as a successful outcome, again little was ever written. All of this means that claims made for traditional remedies in the past have no validity.

It is not a matter of commenting unfairly on what was done in the past by applying our modern understanding of the subject. The reality is that out of the myriad of plants and minerals found on this planet, remarkably few possess the ability to relieve disease when rigorously evaluated by the criteria of modern, evidence-based medicine which sets high standards for the conduct of clinical trials. Few minerals are able to penetrate across biological membrane barriers within the body, and if they do they are often highly toxic. As for the chemicals in plants, their very role as irritant or toxic materials that provide a defence against foraging predators is singularly inauspicious with regard to healing the sick. Plants did not evolve for the purpose of producing medicines! There are an estimated 320 000 plant species currently growing in this planet, with perhaps as many as 25 000 of these having been utilised in various systems of folk medicine. Many active principles were extracted over the last two centuries, yet only 120 or so are currently in use around the world. For any single country, the number is much lower; in the United Kingdom fewer than 40 are in regular use in mainstream medicine.

Pharmacologically active compounds from plants became available at the beginning of the nineteenth century through the technique of solvent extraction. Access to pure alkaloids and glycosides meant that patients received exactly what the prescriber intended, rather than a product of highly variable composition, even if it was not adulterated. Furthermore, the ability of alkaloids to form water-soluble salts meant that parenteral medication became a reality in the middle of the nineteenth century when the hypodermic syringe was developed.

As more alkaloids and other pure products were isolated, it became obvious that most of these were highly toxic when administered to patients. This is hardly surprising in view of their role as poisons to protect the plants from foraging predators. While the ancient materia medica has certainly been the source of a variety of valuable pharmacologically active principles, there has been a tendency to underestimate their toxicity. Plant-derived substances have a role to play, though they tend to be among the most toxic drugs in the therapeutic armamentarium. It is noteworthy that the herbs promoted by the retailers of herbal nostrums are supplied in formulations containing low doses for precisely that reason. Homeopathy developed in the early nineteenth century by taking this approach to the extreme by diluting toxic plant products to such an extent that patients ingested no active drug at all.

The influence of plant products on drug research has been disproportionate to their therapeutic value. During the twentieth century, both academic and commercial researchers spent time in seeking out novel preparations employed by native tribes or users of folk medicines. The paucity of useful outcomes from this is hardly surprising, for the rationale of treating the sick with such products is inherently unsound since in most cases the products were chosen on account of a perceived property such as taste, smell, colour or shape. Alternatively, plants were selected because of a dramatic effect like purgation, emesis, sneezing, increased urination, salivation, sweating, stupefaction, hallucination or mood alteration. Truly effective modern drugs such as antibiotics, antivirals, statins, hormone derivatives, and the like, are often devoid of any perceptible effect that is related to their action. If research into plant products is to be more successful in the future, rigorous criteria for their selection for investigation will need to be developed.

SYNTHETIC DRUGS

The initial enthusiasm for the use of alkaloids and other active substances from plants in the first half of the nineteenth century was followed by a period of therapeutic nihilism during which many medical practitioners came to doubt the value of drug therapy. That was, however, to change with the rapid development of synthetic organic chemistry to meet the demands of the emergent synthetic dyestuffs industry in the wake of the serendipitous discovery of mauveine by William Perkin in 1856. This meant that for the first time in history it was not necessary to rely on nature to provide new drugs. This did not mean that natural products were no longer important. On the contrary, these have remained the progenitors of many drugs developed since then, but the universal success of the synthetic hypnotic chloral hydrate in 1868 and then sodium salicylate seven years later resulted in the emergence of the industrial production of synthetic drugs. At first these were discovered in universities, but by the end of the nineteenth century the leading pharmaceutical companies were developing new drugs in their own laboratories. A significant event at this time was the arrival on the scene of industrial pharmacologists and chemists specialising in drug discovery.

Initially, chemists simply made minor changes to the chemical structures of natural products by altering functional groups that had been identified. Some of these were converted to esters or ethers, while others were alkylated, hydrolysed, oxidised or reduced. By the end of the nineteenth century, the chemical structures of alkaloids were beginning to be determined. This meant that analogues could be designed in which only essential parts of the molecule were retained. In the second half of the twentieth century, the process of analogue design became highly sophisticated as medicinal chemists wrestled with structural manipulations that both affected transportation of drugs to their intended sites of action and also ensured optimal docking on arrival there. In pursuing their manipulations chemists occasionally stumbled upon molecules that exhibited wholly unexpected activities, thereby extending the available range of pharmacophores. All this activity resulted in many hundreds of valuable new therapeutic agents that generated fortunes for their industrial exploiters. Patients, too, benefited and by the end of the twentieth century drug therapy had become the major mode of bringing succour to the sick. There still remained a strong dependence upon naturally occurring molecules as sources of inspiration for the design of new drugs, thanks to the discovery of mammalian hormones and also antibiotics.

DRUG PROTOTYPES

It was not until demonology, astrology and humoral imbalances had been thoroughly discredited that drug research could make progress. Yet until the late nineteenth century its scope had been limited to testing and evaluating readily available natural products. Once organic chemistry had finally developed sufficiently for a limitless range of new drugs to be synthesised, researchers began to modify the chemical structures of natural alkaloids in the hope of finding better products, or attempted to synthesise substitutes for them in order to overcome shortages and high costs of production.

Many of the successes of drug research have involved the preparation of analogues of natural products or synthetic compounds possessing some form of exploitable pharmacological or chemotherapeutic activity. It is helpful to describe the first pure compound to have been discovered in any series of chemically or developmentally related therapeutic agents as a drug prototype. In certain cases, there has been no perceived need to develop a prototype further. Such prototypes continue to serve as medicinal compounds in their own right, but others have been rendered obsolete by the analogues derived from them. In some instances, both the prototype and its analogue even compete in the market-place. In a recent study, nearly 250 drug prototypes were described, from which 1200 medicinal compounds were derived.[2]

Until the middle of the twentieth century, most drug prototypes were derived from plants. For a period thereafter, the major sources of these were micro-organisms that yielded antibiotics. Throughout the century, however, prototypes were being isolated from biochemical sources and this eventually became the principal route for discovering novel drug prototypes. Spurred by major advances in physiology, biochemistry, microbiology and pathology, chemists were at last provided with the full scope to apply their newly acquired skills. Hormones and antibiotics alike were then altered to limit undesired effects and to permit their administration as medicines that could survive in the human body long enough to exert a useful action. Indeed, the twentieth century was the first in which drug therapy contributed more to the progress of medicine than did any other single factor.

The start of the twentieth century had been marked by the isolation of the first pure hormone, epinephrine (adrenaline), yet analogues of it were not developed until after the introduction of the alkaloid ephedrine in 1926, when it was recognised that it was a chemical analogue of epinephrine with some advantages when used therapeutically. By that time, thyroxine, insulin and acetylcholine had also been isolated, followed by the corticosteroids and sex hormones in the 1930s. This formed the basis for some of the greatest achievements of medicinal chemists in the twentieth century when they later prepared analogues of epinephrine and cortisone in which undesirable effects on the body were largely overcome while leaving desired effects unaltered or even enhanced. This successful exploitation of hormones in drug research had much to do with their natural role, which is to regulate physiological function, not to disrupt it. With plant poisons it is the converse, since their role is to protect plants against foraging predators. However, the hormones had presented problems when they were selected to serve as drugs. Not only did they often occur in only trace amounts, making isolation difficult, but they were also often chemically unstable. Additionally, when employed as drugs, they lacked sufficient specificity of action, and it required considerable effort by medicinal chemists to develop them into selective therapeutic agents.

During the year 2004, the 25 millionth compound was listed in the registry of the Chemical Abstracts Service. The availability of so many compounds prepared for a wide variety of purposes other than drug therapy has provided opportunities for serendipitous discoveries of unexpected pharmacological activity. In addition, the screening of ever expanding collections of synthetic compounds prepared by drug companies for earlier projects has provided another potential source of novel prototypes.

The first major break with reliance upon natural products as drug prototypes was inspired by the development of synthetic dyes. This was the introduction in 1904 by Paul Ehrlich of the antitrypanosomal agent known as Trypan Red. Developed as a result of screening over one hundred dyes in a quest to find one that would bind to and kill the trypanosomes that caused the tropical disease known as sleeping sickness, Trypan Red proved toxic to the optic nerve and was reserved for veterinary use. Five years later, Ehrlich evaluated arsphenamine, an organic arsenical drug that cured syphilis. It was an analogue of Atoxyl®, a highly toxic compound used for treating sleeping sickness and originally introduced as an anticancer drug in the unsubstantiated belief that it would be less toxic than arsenic trioxide as it was an organic compound. Further dyes and arsenicals were developed from those introduced by Ehrlich, leading to effective chemotherapy against sleeping sickness, malaria and a variety of bacterial diseases.

By the middle of the twentieth century the majority of drug prototypes were no longer gleaned from the plant kingdom. Biochemists had come to grips with delicate molecules harboured within cells and vesicles or circulating in body fluids. The animal organism had now become the single greatest provider of drug prototypes. It has retained that role, rivalled only by prototypes obtained from fungi and other microbes. In the 1940s a therapeutic revolution occurred when the first antibiotics were introduced. However, the vast majority of the antibiotics that have been discovered are, like the products from plants, too toxic for therapeutic application. This toxicity is a reflection of their role in the battle for survival of their host organisms in the natural world.

The range of drug prototypes now includes hormones, cellular metabolites and other substances found in the human body, as well as vitamins and antibiotics. Chemists have used their skills to create new drugs by synthesising analogues of those that exhibited desirable characteristics such as greater safety or efficacy, enhanced stability, improved selectivity or duration of action, superior absorption and distribution characteristics, and so forth. In the closing years of the twentieth century, attempts to generate novel prototypes from which new medicines could be developed moved in a new direction with the introduction of combinatorial chemistry. This permitted the automated synthesis of vast numbers of molecular variants that were fed into high-throughput screening systems in the hope of finding active compounds. While the variants can be analogues of compounds with known activity, the possibility of now screening compounds with no resemblance to anything found in nature may further lessen the dependence upon nature as a source of drug prototypes.

ANALOGUES OF DRUG PROTOTYPES

All drugs exhibit unwanted side effects in at least some individuals. For that reason alone, attempts will always be made to find alternatives so long as it is economically viable to do so. Another reason is the need for commercial companies to hold patents on their products. As these have only a limited lifespan, companies will protect their own interests by seeking successor products to patent. Whatever the motivation, the simplest and most frequent way of finding an alternative to an existing drug is to synthesise chemical analogues and test these.

Analogues of existing drugs could only be made once pure compounds had been isolated. Some of the first to be made were analogues of morphine, quinine, atropine and cocaine. In these particular cases, the correct chemical structure of the prototype had not yet been determined when its analogues were synthesised. Chemists merely replaced or modified functional groups believed to be present. At first, simple derivatives were made by esterification, hydrolysis, oxidation and reduction. Analogue formation became more sophisticated once the chemical structures of these and other compounds were correctly established and when recognition of the importance of the physical properties of the drug became widespread, especially in the second half of the twentieth century.

Structural modification was used to change lipophilicity in order to control the ability of drug molecules to penetrate biological membranes by passive diffusion; e.g. alkyl side chains were lengthened or shortened to make compounds less or more polar and thus increase or decrease the likelihood of penetrating into the brain. This was followed by a growing awareness of the relevance of constraints placed on organic molecules through the degree of ionic dissociation, which could render potential drugs useless if the charged molecular species was that mainly present in the body. Careful placing of chemical substituents was found to alter radically the tendency of molecules to ionise, thereby enhancing their pharmacological activity.

Apart from altering the physicochemical properties of drugs to develop novel analogues, chemists have also manipulated the molecular shape or electronic distribution within drugs with a view to modifying molecular interactions at their site of action. Unfortunately, such change has frequently proved to be incompatible with the desired physical properties that affect the distribution and metabolism of drugs within the body. This is the principal reason why the vast number of analogues fail to behave in a satisfactory manner.

It is exceedingly difficult to develop new drugs that are both safe and effective. It may take years of intensive research before structural alterations to a prototype or an existing drug will provide a compound that can satisfy the stringent safety requirements universally applied to all new pharmaceutical products. Even when that much is achieved, sometimes all that is obtained is a novel compound that is no more than a commercial success. Such products have been much maligned for their failure to confer any therapeutic advantage over their predecessors. Nonetheless, there have been several instances in which drugs that appeared to be unnecessary variants of an existing product have been found from long-term studies to be safer, more efficacious or sufficiently different in their mode of action as to constitute important therapeutic developments. This factor alone could justify the marketing of almost any variant that is at least as safe and efficacious as the existing product, but the cost of bringing a new drug to the market is now so high that companies are reluctant to invest in any drug that does not have some significant advantage over its rivals. In the past, this restraint was not so powerful and consequently there are still redundant drugs on the market.

REFERENCES

1. M.A. Powell, Drugs and pharmaceuticals in ancient Mesopotamia, in *The Healing Past. Pharmaceuticals in the Biblical and Ancient World*, eds I. Jacob, W. Jacob. Leiden: E.J. Brill; 1993, p. 54.
2. W. Sneader, *Drug Prototypes and Their Exploitation*. Chichester: Wiley; 1996.

The Prehistoric Period

In 1975, Ralph Solecki of Columbia University described excavations at the Shanidar Cave in the Zagros Mountains of northern Iraq, where he had unearthed 60 thousand year old Neanderthal bones laid out in the foetal position.[1] Alongside were clusters of pollen grains from 28 species of flowers. As seven of these were recognised to be medicinal herbs, it was suggested that this could explain their presence.[2] Solecki dismissed the possibility that the Persian jird, *Meriones persicus*, a rodent that stores seeds and flowers, may have contaminated the site by its burrowing. A later investigation by Jeffrey Sommer of the University of Michigan revealed that the types of pollen grains in the Shanidar grave were also found in jird burrows.[3] While this does not prove contamination had occurred, it does mean that further investigation is required before firm conclusions about the Shanidar Cave can be drawn. Moreover, the presence of a few plants that are reputed to possess medicinal properties does not in itself mean they were used for such purposes. They could simply have been food, especially as ancient medicines often originated as components of the diet that were deemed to have beneficial effects on health.

There is indirect evidence of drug use in the Neolithic period, which began around 11 thousand years ago. Among remains of plants found in the Spirit Cave in the north-west of Thailand were seeds of the mildly psychoactive betel nut (*Areca catechu*), placed there between 7000 and 5500 BC.[4] The earliest direct evidence for human consumption of betel nut comes from the Duyong Cave in the Philippines, where a skeleton from 2680 BC was found buried beside shells containing lime. This is reminiscent of the practice still seen in India, of wrapping the nut in betel leaf (*Piper betel*), adding lime (thus liberating the readily absorbable free base of arecoline), then chewing. As with modern users of betel nut, the teeth of the Duyong Cave skeleton were blackened.[5]

Warming of the Fertile Crescent in the Middle East after the end of the last Ice Age permitted agricultural development. A transition from hunting and gathering to settling in villages developed over a six thousand year period, spreading to Europe, the Indus Valley and onwards, with parallel developments occurring in Central America and Northern China. Following upon the domestication of the horse by the fourth millennium BC, our ancestors acquired the ability to travel great distances with alacrity, a freedom the slow-moving ox had denied them. The consequences of this were profound. Trade developed, cities were established and conquest ensued, all of which created a need for law and order that in turn stimulated the development of writing in Mesopotamia and Egypt. With improved communications and the interchange of ideas, knowledge of herbs that could be used as spices, to treat disease or to offer some other beneficial action was no longer confined to isolated communities, and trading began in those products that could be preserved.

ROCK ART

Even before the appearance of writing there was another form of communication that has preserved information about the lifestyle of our prehistoric ancestors. This was rock art,

Drug Discovery. A History. W. Sneader.
©2005 John Wiley & Sons Ltd

consisting of paintings and inscriptions on rocks either exposed to the elements or protected inside caves. In 1958, the American anthropologist Thomas Campbell drew attention to intriguing parallels between aspects of the ancient American Indian cult of the mescal bean (*Sophora secundiflora*) and drawings on the walls of caves near the confluence of the Pecos River and the Rio Grande in the south-west of Texas.[6] The drawings depicted dancers with bows and arrows, hunting and deer. Campbell also noted that mescal beans had repeatedly been found in nearby caves over the preceding 25 years and in one case had been radiocarbon-dated as originating in the period 7500 BC to AD 200. In another cave were remains of peyote, a mescaline-containing cactus (*Lophophora williamsii*), and Mexican buckeye (*Ungnadia speciosa*), both known to be psychoactive.[7,8]

The Pecos River cave drawings are now believed to be only four thousand years old, but that has not deterred the publication of reports[5] about prehistoric cave drawings involving psychoactive drugs since the appearance of the paper by Campbell.[6] There are many who dismiss such claims as being highly speculative and it should be pointed out that cave art with similar content to that of the Pecos River drawings has been discovered in parts of the world where there have been no findings of mescal beans or other psychoactive materials. Just as controversial are reports of cave paintings that are said to depict shamanic healing rituals in which the medicine man entered a trance-like state that may have been induced by either hyperventilation, ritual music and dancing, or drugs.[9] It has, for example, been proposed that central Saharan Desert cave paintings from the period 7000–5000 BC represent the ritual use of hallucinogenic mushrooms since they portray scenes of harvesting, adoration, dancing and the offering of mushrooms.[10]

DRUGS THAT ACT ON THE MIND

The historical record of drugs consumed by our Neolithic ancestors could have been unduly influenced by those that exhibited an effect on the mind, such as mescal bean, peyotl, betel, the poppy, alcohol, coca, cannabis, tobacco, soma and pituri.[11,12] The social and recreational use of these drugs is associated with artefacts, in much the same way as nicotine and alcohol are today involved with pipes and liquor glasses respectively. Drugs employed for medicinal purposes, by contrast, do not leave enduring evidence of their consumption, particularly if they were also considered to be foods rather than medicines – as is still the case today among North American Indians who cultivate certain plants for consumption as food, yet also employ them for therapeutic purposes.[13] It is hardly surprising therefore that archaeologists have found more evidence of drugs used for social and recreational purposes than those selected as medicinal substances. It is noteworthy that when written records appeared in the third millennium BC onwards, psychoactive drugs did not appear to be dominant.

Not only has it been suggested that the original use of drugs was in religious rituals[14] but it has even been claimed that early religious practices arose from the cultic use of psychoactive drugs.[15] This may be controversial, but the role of wine in religious practice in ancient times is beyond dispute if we consider the Greek Dionysian festivities, the Roman Bacchanalia and the importance of wine in Judaism and Christianity.

Alcohol

The most widely consumed drug is, of course, alcohol. It is not surprising that alcoholic beverages were among the earliest of drugs to have been discovered since their effects must have been readily perceived within a short period of administration. This remained a basic necessity for drug discovery until the isolation and testing of active ingredients from plants in the nineteenth century. When toxic effects also occurred with little delay, the drug was considered to be a poison. However, if these were not manifest until a considerable time had

elapsed, there was little likelihood they would be connected with the drug. This remains a problem for those who consume herbal remedies, but it also applies to modern synthetic drugs. In recent years, thousands of people were prescribed drugs such as thalidomide, benoxaprofen, practolol and others without anyone being aware of the damage these caused because of the delay before it happened. It was left to astute physicians to detect the link between these drugs and their toxic effects.

Alcohol appears to have first been consumed in the form of wine. Evidence of wine being available over seven thousand years ago has been obtained from chemical analysis of yellow residues found in six vessels sunk in the floor of a mud brick building of the Neolithic Hajji Firuz Tepe village unearthed in the Zagros Mountains of western Iran.[16] The capacity of each vessel was nine litres, indicating that organised production had taken place. Infrared, high-performance liquid chromatographic and chemical analysis of the yellow residue indicated that it consisted largely of calcium tartrate together with resin from the terebinth tree. The calcium tartrate could only have come from grape juice as there is no other common natural source. The juice would have fermented into wine on storage, while the resin served as a preservative to prevent despoliation by bacteria that convert wine to vinegar.[17] Excavations further south in the Zagros Mountains uncovered a military centre known as Godin Tepe, where earthenware jars from 3500 to 3000 BC were found. Again, deposits of calcium tartrate were present.[18] By this time wine and beer were being prepared throughout the Aegean.[19] Silver and gold drinking vessels unearthed by Heinrich Schliemann when he was excavating the second city of Troy reveal the significance attached to the consumption of wine around 2500 BC.[20]

It is not clear whether the production of beer preceded that of wine. A 4000 BC seal depicting two people drinking beer with a straw (characteristically used to filter out the mash floating in the beer) was discovered in the 1930s during excavations of the Tepe Gawra site near the ancient city of Nineveh in north-east Iraq, and beer is frequently featured in Sumerian drawings and text of the third millennium BC.[21]

Opium

Alcohol may have been the most frequently consumed drug in history, but the most effective medicinal drug available before the dawn of the twentieth century was opium obtained from the poppy, *Papaver somniferum* L. – one of the oldest cultivated species known. Claims that opium was used in the Neolithic period should be viewed with extreme caution. They are based on the finding of seeds of the closely related *P. somniferum* subsp. *setigerum* in Germany at the Danubian settlements, the location of the first central European farms around 4400–4000 BC. Much greater quantities of poppy seeds have been found in northern France and farming settlements on the shores of lakes in Switzerland and surrounding areas, which date back to 3700–3625 BC.[22] Both *P. somniferum* and its subspecies *setigerum* produce morphine and related narcotic compounds, though the former generates greater amounts.[23] However, the seeds of both plants contain insufficient amounts of active alkaloids to produce any narcotic effect.[24] The greatest concentration of active alkaloids is to be found in the latex exuded from the unripe capsule, which is dried to prepare opium. The seeds from the Neolithic period were probably used for the production of oil, as they certainly were in later times.[25]

Evidence of what appeared to be early opium use came from the Cueva de los Murciélagos (Cave of the Bats), situated 50 metres above a ravine near Granada in Spain. In 1868 Góngora y Martínez reported that he had found poppy capsules in esparto grass bags in this ancient burial site.[26] The capsules were radiocarbon-dated in 1975 as having originated about 4200 BC,[27] but this was later revised to around 2500 BC.[25] It thus appears that there is no reliable evidence for the use of opium prior to the appearance of writing.

REFERENCES

1. R.S. Solecki, Shanidar IV, a Neanderthal flower burial in northern Iraq. *Science*, 1975; **190**: 880–1.
2. J. Lietava, Medicinal plants in a middle paleolithic grave Shanidar IV? *J. Ethnopharmacol.*, 1992; **35**: 263–6.
3. J.D. Sommer, The Shanidar IV flower burial: a re-evaluation of Neanderthal burial ritual. *Camb. Arch. J.*, 1999; **9**: 127–9.
4. C.F. Gorman, Hoabinhian: a pebble-tool complex with early plant associations in Southeast Asia. *Science*, 1969; **163**: 671–73.
5. R. Rudgley, *Lost Civilizations of the Stone Age*. New York: The Free Press; 1998, pp. 126–41.
6. T.N. Campbell, Origin of the mescal bean cult. *Am. Anthropol.*, 1958; **60**: 156–60.
7. R.E. Schultes, Antiquity of the use of New World hallucinogens. *Heffter Rev. Psychedelic Res.*, 1998; **1**: 1–7.
8. W. La Barre, Shamanic origins of religion and medicine. *J. Psychedelic Drugs*, 1979; **11**: 7–11.
9. J.D. Lewis-Williams, T.A. Dowson, The signs of all times – entoptic phenomena in upper paleolithic art. *Current Anthropol.*, 1988; **29**: 201–45.
10. G. Samorini, The oldest representations of hallucinogenic mushrooms in the world (Sahara Desert, 9000–7000 B.P.). *Integration*, 1992; **2–3**: 69–78.
11. R. Rudgley, *Essential Substances: A Cultural History of Intoxicants in Society*. New York: Kodansha International; 1993.
12. J. Goodman, P.E. Lovejoy, A. Sherratt (eds), *Consuming Habits: Drugs in History and Anthropology*. London: Routledge; 1995.
13. D.E. Moerman, North American food as drug plants, in *Eating on the Wild Side*, ed. N.L. Etkin. Tucson: University of Arizona Press; 1994, pp. 166–81.
14. A. Sherratt, Sacred and profane substances: the ritual use of narcotics in later Neolithic Europe, in *Sacred and Profane*, eds P. Garwood, P. Jennings, R. Skeates, J. Toms, Oxford Committee for Archaeology, Monograph No. 32. Oxford: Oxbow; 1991, pp. 50–64.
15. R.G. Wasson, S. Kramrisch, J. Ott, C.A.P. Ruck, *Persephone's Quest: Entheogens and the Origins of Religion*. New Haven: Yale University Press; 1986.
16. M.M. Voigt (ed.), *Hajji Firuz Tepe, Iran: The Neolithic Settlement*. Philadelphia: The University Museum; 1983.
17. P.E. McGovern, D.L. Glusker, L.J. Exner, M.M. Voigt, Neolithic resinated wine. *Nature*, 1996; **381**: 480–1.
18. G. Algaze, The Uruk expansion: cross cultural exchange in Early Mesopotamian civilization. *Current Anthropol.*, 1989; **30**; 571–608.
19. A.H. Joffe, Alcohol and social complexity in ancient western Asia. *Current Anthropol.*, 1998; **39**: 297–322.
20. A. Sherratt, Alcohol and its alternatives. Symbol and substance in pre-industrial cultures, in *Consuming Habits: Drugs in History and Anthropology*, eds J. Goodman, P.E. Lovejoy, A. Sherratt. London: Routledge; 1995, pp. 11–46.
21. S. Katz, M. Voigt, Bread and beer: the early use of cereals in the human diet. *Expedition*, 1986; **28**: 23–34.
22. D. Zohary, M. Hopf, *Domestication of Plants in the Old World: The Origin and Spread of Cultivated Plants in West Asia, Europe, and the Nile Valley*, 2nd edn. Oxford: Clarendon Press; 1994.
23. C.G. Farmilo, H.L.J. Rhodes, H.R.L. Hart, H. Taylor, Detection of morphine in *Papaver setigerum* DC. *Bull. Narcotics*, 1953; **1**: 26–31.
24. R.E. Struempler, Excretion of codeine and morphine following ingestion of poppy seeds. *J. Anal. Toxicol.*, 1987; **11**: 97–9.
25. M. Merlin, *On the Trail of the Ancient Opium Poppy: Natural and Early Cultural History of Papaver somniferum*. East Brunswick: Associated University Presses; 1984.
26. M. Góngora y Martínez, *Antiguedades Prehistoricas de Andulucia*. Madrid: C. Moro; 1868.
27. M. Hopf, A.M. Muñoz, Neolithische Pflanzenreste aus der Höhle Los Murciélagos bei Zuheros (Prov. Córdoba). *Madrider Mitteil.*, 1974; **15**: 9–27.

Pre-Hellenic Civilisations

Writing developed in Mesopotamia and Egypt six thousand years ago in the form of crude pictographs representing everday objects, but gradually these were modified to represent ideas and ultimately sounds. Considerable progress was made in the nineteenth century in transcribing the wedge-shaped cuneiform text introduced by the Sumerians around 3100 BC, as well as the hieroglyphics of the Egyptians. Consequently, there now exists substantial knowledge of the lifestyle of these civilisations, including how they used and administered drugs. However, progress in identifying individual drugs has been more difficult.

A major source of information about drugs used in Mesopotamia is provided by the 660 tablets of clay recovered in the 1920s from the library of the palace in Nineveh of the last king of Assyria, Asshurbanipal (669–626 BC). Most of them appear to have been written around 1700 BC. They were among 20 thousand tablets unwittingly preserved by being baked when invaders set the palace on fire. The tablets were deciphered by R. Campbell Thompson, who established that many were copies of much older texts, with some of the drugs possessing Sumerian names from the third millennium BC.[1]

Thompson painstakingly sought to identify each plant mentioned in the tablets, both from the general context in which its name appeared and by reference to other cuneiform texts. He then adduced philological evidence from other languages before comparing his conclusions with the findings of other scholars, taking fully into account the ancient, mediaeval and modern usage of drugs in the Middle East. His work remains important, despite valid criticism that there were many errors in his identification of the 115 or so drugs.[2] A number of his findings were revised in a posthumous publication,[3] but his mistaken belief that the wild poppy, *Papaver rhoeas* L., produced narcotic substances persisted and has continued to mislead those who are unaware that morphine is not formed in this species. This has resulted in the widespread belief that opium was used 5000 years ago in Sumer. The situation has been further confounded by the suggestion that a third millennium BC ideogram on a clay tablet from the holy city of Nippur depicted *HUL GIL*, the plant of joy, this allegedly being the opium poppy. That opinion was originally expressed in a personal communication from a professor at Yale.[4] There is no evidence to support it.[5]

The herbs cited in his posthumous publication that Campbell Thompson claimed to have identified include aloes, ammi (toothpick plant), anemone, anise, asafoetida, balm of Gilead, beetroot, black cumin, black hellebore, bitter or black nightshade, cannabis, carob, cassia, castor oil, cedar, chamomile, chasteberry, citron, colcynth, cornflower, cress, cumin, darnel ryegrass, date palm, elder, fennel, fig, frankincense, galbanum, garlic, ginger, heliotrope, henbane, hound's tongue, laurel, leek, liquorice, mandrake, meadow saffron, mint, mustard, myrrh, myrtle, nettle, oak, pellitory, pine, pomegranate, poppy, rocket, rosemary, Syrian rue, saffron, squirting cucumber, styrax, sweet flag, sycamore fig, tamarisk, thistle, thorns, thyme, turmeric, white hellebore, winter cherry, willow and wormwood. Caution must be exercised in accepting his conclusions, though many of these plants are used to this day in herbal medicine – albeit with little more evidence of their efficacy than was available to the inhabitants of Mesopotamia. It is interesting to note that the medical texts distinguished between different

Drug Discovery. A History. W. Sneader.
©2005 John Wiley & Sons Ltd

parts of the plant. This indicates an early awareness that this was an important aspect of using herbal medicines.

Many of the plants mentioned by Thompson had multiple uses – a feature of all ancient systems of medicine since they held that all ailments were different manifestations of the state of disease. Thus any remedy that was believed to expel the demons from the body or ameliorate an imbalance of humours would inevitably have a broad spectrum of application. This did not change until the emergence of pathology as an independent medical science. As for the actual uses mentioned by Thompson, they are diverse and include treatment of a wide range of gastrointestinal, urinogenital, pulmonary and abdominal problems, as well as external afflictions such as scorpion stings, bruising and blistering, irritation of the eyes, ears, mouth, feet and so forth.

In addition to the cuneiform tablets from Nineveh, there are many others, including 420 translated by Franz Köcher.[6] From these sources we learn that there were two types of healers, namely sorcerers who belonged to the clergy, and physicians. The former, known as the *ashipu*, believed that disease was caused by demons and employed magic, incantations, charms and exorcism to expel them, while the latter, the *asu*, followed a therapeutic tradition of using animal, vegetable and mineral remedies in a pragmatic manner. There was often interaction between these groups, but it is clear that an independent therapeutic tradition had developed in Ancient Mesopotamia.[7] This was an old tradition that originated in the prehistoric world and it appears likely that it passed to the Egyptians.

There may be problems with the identity of the drugs cited in ancient tablets, yet they can provide valuable information about how drugs were prepared and administered. Samuel Kramer wrote about a four thousand year old clay tablet excavated at Nippur that may be the oldest pharmaceutical text known.[8] He and chemist Martin Levey translated the tablet and found it contained more than a dozen prescriptions with ingredients from animal, vegetable and mineral sources, but there was no indication of their uses. Of the minerals, both salt (sodium chloride) and saltpetre (potassium nitrate) were preferred, while from the animal kingdom milk, snakeskin and tortoise shell were favoured. However, most of the medicaments were of vegetable origin, a situation that persisted until modern times. Amongst these were said to be cassia, myrtle, asafeotida and thyme. There was no mention of the quantities to be incorporated, but all parts and secretions of the plants were used for making the prescriptions. In some cases, several plants were treated with wine and then mixed with cedar or similar oil to prepare soothing salves. The liquid formulations were obtained by decoction of plants in boiled water, followed by addition of alkali or salts that may have enhanced the extraction process. For prescriptions administered by mouth, the pulverized ingredients were taken up in beer. Sometimes the ingredients were specially treated, such as the burning of soda-rich plants to free the alkali ash from vegetable matter. There was even a degree of chemical interaction, as in two prescriptions where fats were combined with alkali to form a soap. The cleansing and antibacterial action of this when applied directly to infected wounds was probably of more benefit than the odoriferous ingredients of other prescriptions in the clay tablet.

EGYPT

The most useful record about drugs used in Ancient Egypt is a document purchased in Thebes in 1872 by the Egyptologist Georg Ebers. Generally known as the *Ebers Papyrus* and dating back to around 1550 BC, this lists more than 800 prescriptions then in vogue, including most of the drugs cited in Thompson's interpretation of the medical tablets recovered from Nineveh.[9] This would be due to the trade routes between Egypt and Mesopotamia, resulting in cultural exchange. Both sources include elements dating back to the third or fourth millennium BC. In the *Ebers Papyrus*, there is once again a problem with the identification of specific herbs, but there are many prescriptions incorporating culinary materials such as dates, figs, grapes, pomegranates, roasted barley, wheat, honey, sweet beer, wine and milk. Spices abound, including absinth, cassia, carraway, coriander, cumin, fennel, gentian, juniper, peppermint and thyme. Fats are also present, albeit from unusual sources such as cats, crocodiles, geese, hippopotamuses, lions, oxen and serpents. Among the oils are cedar wood, linseed, turpentine and other essential oils. There are also locally obtained minerals such as common salt, natron, alum, red and yellow ochres (iron oxides), as well as salts of lead, copper, mercury and antimony. Many of the prescriptions are for skin and eye complaints, as in the tablets from Nineveh, while remedies for gastrointestinal complaints are also common. Once again, the former doubtlessly reflects the hot, sandy environment and presumably the latter is related to the unwholesome diet of those days.

The prescribing of soothing balms for skin problems and purgatives for what appears to be constipation indicates that there was a strong empirical element in both the Mesopotamian and Egyptian systems of medicine, while the preponderance of spices and culinary herbs supports the view that drugs were initially considered to be foods that had some beneficial value. However, what lay behind the selection of the herbs?

Many plants contain substances that will irritate one part or another of the gastrointestinal canal, from the mouth to the rectum. The response to these substances, and consequently their application, will depend on which region is irritated, varying from salivation, or emesis, to purgation. If the elements of magic, superstition or religion are involved rather than pragmatism, the scope for innovation is immense. It does not, however, augur well for efficacy, and therein lies the problem for those who look to herbal remedies as a source of new drugs. It is conceivable that some might be, but much effort will be wasted examining worthless nostrums in the vague hope that something of value will be found.

Whereas in Mesopotamia the treatment of the sick was in the hands of two distinct classes of healers, the priestly *ashipu* and the therapeutic *asu*, most physicians in Egypt were priests. Imhotep, the earliest physician known to historians, lived during the Third Dynasty of the Old Kingdom around 2650 BC and was High Priest of Heliopolis. At that time, elements of religion and magic were closely intertwined with drug use, incantations routinely being uttered prior to administration in order to confer the healing property upon it. Without such a spell being cast, the drugs would in most cases have been ineffective.[10] Modern knowledge of the placebo effect, whereby psychological elements influence the response to therapy, suggests that this greatly contributed to any success of the Egyptian priest–physicians. Although there were some herbs that did have real properties, such as the ability of castor oil to produce purgation, the likelihood of recognising such effective remedies would have been diminished as the outcome of their administration was difficult to distinguish from the many worthless herbs that were introduced.

Medicine acquired such a high status in Egypt that its physicians were sought out by leaders of surrounding nations. There was considerable specialisation in the treatment of different parts of the body, further increasing the demand for physicians. Although there were some who were not priests, they lacked the influence of the priest–physicians. All, nevertheless, were ever ready to administer a wide range of herbs. Regrettably, it was not the tradition of the *asu* of Mesopotamia that was passed to all subsequent generations by the Ancient Egyptian priest–physicians, but rather a heritage of mainly worthless remedies. Most of these are still being sold in our modern shopping malls as 'herbal' or 'natural' medicines.

Finally, it should be noted that claims that traces of morphine had been found in the tomb of the Egyptian chief royal architect Kha, who died in 1405 BC, have been disproved by modern chemical analysis.[11]

INDIA

Recent scholarship has established that not only were there trade links between Mesopotamia and Egypt at the beginning of the second millennium BC but that the Silk Road to China was also being established. Due to the immense distances involved, no single individual was able to travel from the West to the Far East. Although goods were exchanged, the transfer of knowledge and culture was therefore erratic and unreliable.[12] Consequently, there are both similarities and differences in the ancient medical systems of the East and the West.

There are many who believe that the four vedas that constitute the Brahamanic basis of Hinduism could have been compiled by the second millennium BC and that some of the mantras in them trace their origins back a further one or two thousand years. A recent challenge to established concepts of Indian history could further complicate the dating of the compilation of the vedas.[13]

Each of the vedas covers the role of ritual in a different sphere of life. The last, the Atharvaveda, is held by believers to have originally contained the basis of Ayurvedic medicine, which was rearranged and revised perhaps around 800 BC in samhitas (treatises) by the physician Charaka and the surgeon Sushruta. Charms for expelling demons are to be found in their works and there is a god of medicine, Dhanvantari. The *Charaka Samhita* may not have been written in its present form until the first century AD. It refers to over 300 vegetable drugs, the remainder being from animal and mineral sources. The current version of the *Sushruta Samhita* might date to the seventh century AD. Although a text on surgery, it mentioned nearly 400 vegetable drugs. The number of plant medicines cited in these two works is greater than that of any other ancient people who left a record of such matters. However, it does not rest there, for in sutra 26 the *Charaka Samhita* asserts that there is nothing in the world that cannot be used as a medicine, so long as an appropriate use can be found for it!

Ayurveda considers man to be born in a state of humoral balance that is gradually lost through changes in diet, lifestyle and environment. The humours are wind, bile and phlegm. In some traditions there is a fourth humour, blood.[14] To restore a healthy balance in those who are sick requires not merely the use of appropriate drugs but also dietary change, controlled breathing and spiritual meditation. There is an obvious similarity to the Greek system of humoral medicine insofar as both are holistic in aiming to restore a healthy balance of humours. In both there is liberal use of emetics, purgatives, enemas, sneezing powders and the like, not to expel demons but rather to eliminate excess of undesired humours. This would hardly be surprising if the writings upon which Ayurveda is based were compiled hundreds of years after the *Hippocratic Corpus* of the fifth century BC, but if the concept of the humours originated before 2000 BC, a major revision of medical history would be necessary. The matter has not yet been settled, but from the perspective of the history of drug discovery, the debate is of limited concern since humoral medicine was nothing other than an impediment to the discovery of efficacious drugs.

As with several other ancient medical systems, that of India has been associated with the use of a psychoactive drug. In this instance it is soma, a substance featured in the oldest of the vedas, the *Rig Veda*. Many suggestions as to its identity have been made, including the celebrated claim by Gordon Wasson that it was a psychedelic mushroom, *Amanita muscaria*.[15] More recently, it has been suggested that soma consisted of a mixture of herbs including Syrian rye, *Peganum harmala*, the substance originally proposed to be soma in 1794.[16] This new claim, however, depends on the assertion that soma was introduced into Northern India by Aryan invaders. However, the proponents of the antiquity of vedic medicine strongly deny such an invasion ever occurred.[13] From this, the reader may conclude that unravelling the history of early drug use in India is no simple matter. A thorough investigation of the subject is required.

CHINA

During the Han Dynasty in China (206 BC–AD 220) there was a proclivity to fabricate history in order to elevate the reputation of the state.[17] Mythical figures were said to have rescued the world from chaos, such as the inventors of fire, writing and agriculture. The latter was supposed to be a peasant who became the Emperor Shen Nung in the third millennium BC and it was claimed that valuable information about the use of drugs could be gleaned by scouring the pages of his herbal, *Pen Ts'ao Ching* (*Classic of Materia Medica*), in which he described how he tested more than 300 herbs upon himself. Acceptance of this story has persisted in China where Shen Nung is revered as the Father of Medicine, but in fact this herbal was written during the last years of the Han Dynasty and then modified in the sixth and seventh centuries. By this time, it would have been influenced through contacts with India and the Hellenic world and so diminished in value as a guide to the early use of drugs.

The concept attributed to Shen Nung that herbs have properties of inducing warmth or coolness will strike a chord with anyone familiar with the writings of Aristotle (384–322 BC). In the absence of reliable evidence to the contrary, the concept must be considered as coming from Aristotle. Another text that is just as prominent in Chinese medicine is the *Huang Di Neijing* (*The Yellow Emperor's Classic of Internal Medicine*), written around the same time as *Pen Ts'ao Ching* but supposedly authored by a fifth millennium BC ruler.[18] This introduced the concepts of yin and yan, the forces that must be balanced for the maintenance of good health. Yin is a negative or passive force, while yan is positive or active.

The best known of the Chinese herbals is *Pen Ts'ao Kang Mu*, which was compiled during the Ming Dynasty by Li Shih-Chen (1518–1593). Featuring nearly 2000 herbs, as well as minerals and animal products, it summarised all that was known about Chinese medicine at that time. Since then, the number of herbs used has increased to around 5000.

Cannabis

Although native to Asia, it is generally accepted that *Cannabis sativa* L. was first cultivated in China around 4500 BC when its seeds served as a grain crop. Soon after, its decaying stems became the source of hemp fibre for making clothes, ropes and fishing nets. However, reports of early hemp use in China are open to question either because of inadequate identification of fibre remains or because they relate to impressions on pottery and other objects. These are assumed to have been made by hemp since other fibres, except the much finer silk, are not found in the region. Such cord impressions have been seen on material from 4000 to 3000 BC unearthed in Taiwan and also on similarly dated pots found in Henan Province, eastern China, at an archaeological site of the Yang-Shao culture, the first civilisation to manufacture pottery. Outside China, clay shards showing impressions of what are thought to be *Cannabis* seeds from around 3000 BC were discovered during excavations at a Linearbandkeramik Culture site north of the Black Sea. There are also reports of hemp fibres from objects recovered at Adaouste in the south of France (2000 BC), Gordium in Turkey (800 BC) and St Andrews in Scotland (800 BC).[19] While these and many other findings indicate that hemp fibre was of importance as long ago as the third millennium BC, they do not constitute reliable evidence that *Cannabis* was being used as a drug. The frequently repeated assertion that it was then being used in China as a medicine or a euphoriant is based on the mistaken belief that *Pen Ts'ao Ching* was written in the third millennium BC by Shen Nung.

REFERENCES

1. R.C. Thompson, *The Assyrian herbal...A Monograph on the Assyrian Vegetable Drugs*, the subject matter of which was communicated in a paper to the Royal Society, March 20, 1924. London: Luzac; 1924.
2. M.A. Powell, Drugs and pharmaceuticals in ancient Mesopotamia, in *The Healing Past. Pharmaceuticals in the Biblical and Ancient World*, eds I. Jacob, W. Jacob. Leiden: E.J. Brill; 1993, pp. 47–67.
3. R.C. Thompson, *A Dictionary of Assyrian Botany*, ed. C.J. Gadd. London: British Academy; 1949.

4. C.E. Terry, M. Pellens, *The Opium Problem*. Montclair, New Jersey: Patterson Smith; 1928.
5. P.-A. Chouvy, Le pavot à opium et l'homme: origines géographiques et premières diffusions d'un cultivar. *Ann. Géograph.*, 2001; **618**: 182–94.
6. F. Köcher, *Die Babylonisch-Assyrische Medizin in Texten und Untersuchungen*, 6 vols. Berlin: W. de Gruyter; 1963–1980.
7. R. Biggs, Medicine in Ancient Mesopotamia. *Hist. Sci.*, 1969; **8**: 94–105.
8. S.N. Kramer, *History Begins at Sumer*. Philadelphia, Pennsylvania: University of Pennsylvania Press; 1958; pp. 100–4.
9. C.P. Bryan, *The Papyrus Ebers*, translated from the German version. London: G. Bles; 1930.
10. H.E. Sigerist, *A History of Medicine*, Vol. 1. New York: Oxford University Press; 1951, p. 283.
11. N.G. Bisset, J.G. Bruhn, S. Curto, *et al.*, Was opium known in 18th dynasty ancient Egypt? An examination of materials from the tomb of the chief royal architect Kha. *J. Ethnopharmacol.*, 1994; **41**: 99–114.
12. D. Christian. Silk roads or steppe roads? The silk roads in world history. *J. World Hist.*, 2000; **11**: 1–26.
13. G. Feuerstein, S. Kak, D. Frawley, *In Search of the Cradle of Civilization: New Light on Ancient India*. Wheaton, Illinois: Quest Books; 1995.
14. K. Morgan, *Medicine of the Gods: Basic Principles of Ayurvedic Medicine*. Oxford: Mandrake; 1994.
15. R.G. Wasson, *Soma: Divine Mushroom of Immortality*. New York: Harcourt, Brace; 1971.
16. A. Sherratt, Alcohol and its alternatives: symbol and substance in pre-industrial cultures, in *Consuming Habits: Drugs in History and Anthropology*, ed. J. Goodman. London; New York: Routledge; 1995.
17. J. M. Triestman, China at 1000 B.C.: a cultural mosaic. *Science*, 1968; **160**: 853–6.
18. N. Maoshing (trans.), *The Yellow Emperor's Classic of Medicine: A New Translation of the Neijing Suwen with Commentary*. Boston: Shambhala Publications; 1995.
19. M.P. Fleming, R.C. Clarke, Physical evidence for the antiquity of *Cannabis sativa* L. (Cannabaceae). *J. Int. Hemp Ass.*, 1998; **5**: 80–92.

Greece and Rome

There is a reference in the *Odyssey* to nepenthe, a drug that Helen of Troy cast into wine to induce forgetfulness in her guests who were grieving the long disappearance of Odysseus.[1] Earlier in the *Odyssey*, Homer tells us that Helen had learned of this and other drugs during a visit to the land of the Pharaohs. The identity of nepenthe, however, is not revealed. Despite the absence of reliable evidence for the use of either opium or cannabis in Egypt before the *Odyssey* was written in the eighth century BC, both have frequently been suggested as candidates. An alternative proposal is one of the Solanaceae, such as henbane, mandrake or belladonna, all of which contain the centrally depressant alkaloids hyoscine and atropine. However, there are those who maintain that nepenthe was merely a fictional creation by Homer or an allegorical allusion to the charm of Helen.

What is particularly relevant in the present context is the allusion to Helen of Troy learning about nepenthe and other drugs in Egypt. The Greeks revered Imhotep, who was deified by the Egyptians in the sixth century BC, some two thousand years after his death, and they saw him as fulfilling the role of their own god of healing, Asklepios. Temples where the sick could receive succour were dedicated to Asklepios, just as they had been to Imhotep. The Greeks believed Asklepios was the son of the god Apollo and a mortal mother, Coronis. He was said to have learned about healing herbs from the wise centaur Chiron who dwelt at the foot of Mount Pelion in Thessaly. Asklepios then healed the sick and dying, but when he brought a corpse back to life, he incurred the wrath of Pluto who complained to Zeus that Hades was being depopulated. Zeus then struck down Asklepios with a thunderbolt. All was forgiven, however, when Asklepios was elevated to the status of an Olympian god. Whether or not he was originally a healer whose achievements led to the granting of divine status will never be known with certainty, but in the fourth book of Homer's *Iliad*, Asklepios is presented as a prince of Thessaly in northern Greece and a healer whose sons, Machaon and Podaleiris, can cure those wounded in battle.[2] It is noteworthy that the first temple dedicated to Asklepios was built in Thessaly. Over 300 more were constructed in Greece and elsewhere in the Hellenic world. When the sick were brought to them, their treatment consisted of fasting, sleep, hygiene and diet, supplemented by recourse to divination by the oracles. The most illustrious of the temples eventually became centres for the training of physicians. It was from one such temple in the island of Cos that physicians who wrote under the collective name of Hippocrates ensured that the therapeutic tradition of the Asklepiadae would never be forgotten.

THE FOUR HUMOURS

The forerunners of the apothecary in Ancient Greece were the rhizotomoi (root cutters) who recited traditional spells and followed ancient rituals as they plucked plants from the earth, such as tethering a dog to the mandrake to wrench it from the ground. They believed the beast would die as the bifurcate root uttered a shriek on being pulled out. Better a dog should die than a human. The cynic might see this as a cautionary tale intended to scare off the

Drug Discovery. A History. W. Sneader.
©2005 John Wiley & Sons Ltd

unqualified herb collector, but behind their ritual was the conception that the mandrake root had the appearance of a man and hence could be used to treat infertility.

Empedocles

The philosopher and statesman Empedocles who lived in Acragas in the island of Sicily around the period 460 BC was much travelled and was familiar with the lore of the rhizotomoi. This enabled him to combine the Greek mysteries and magic with medical science and natural philosophy. His doctrine of the four roots had far-reaching consequences. He argued that the universe was balanced between love and strife, the former influencing the development of superior species from primaeval monsters by interacting with the four roots from which all matter is derived.[3] Different objects consisted of different proportions of these roots; they have frequently been named 'elements', but Empedocles did not use that term. He identified them as earth, air, fire and water, believing they had a mystical as well as a material dimension, including magicospiritual associations with the gods. One interpretation of his writings suggests that he associated 'earth' with Hera (the goddess of marriage and birth), 'air' with Zeus (the god of gods and men), 'fire' with Hades (god of the underworld) and 'water' with Persephone (goddess of the underworld).[4] There was also a connection with the macrocosm: 'earth' linked to the land, 'air' to the heavens, 'fire' to the sun and 'water' to the sea. Since his views applied to man as well as the natural world, they inevitably led to the idea of the four humours, but these are not mentioned in the extant writings of Empedocles.

Hippocratic Corpus

Hippocrates is one of the most famous figures in the history of medicine, often being called the 'Father of Medicine'.[5] He is said to have been born in Cos around 460 or 450 BC and was referred to by Plato and Aristotle, his younger contemporaries. About 60 medical treatises have been attributed to him, but they reveal the hand of different authors with conflicting opinions, writing over an extended period of time, mainly during the period 420–370 BC. Collectively known as the *Hippocratic Corpus*, these prose works in the Ionian dialect of Athens may have originally been housed in Cos, but they were gathered together in the great library of Alexandria around 280 BC. Even at that time there was uncertainty over their authorship.

The *Hippocratic Corpus* was compiled at the height of the classical period when philosophers were attempting to explain the nature of the world in rational terms that excluded magic and superstition. The views of Empedocles clearly had a strong influence, with the first known account of the four humours, blood, phlegm, yellow bile and black bile, appearing in the *Hippocratic Corpus*. The origins of these may be traceable to the more distant past, but they are manifestly the application of the doctrine of the four roots to the human body. As such, it was to have a negative influence on the development of physiology for over two millennia until Rudolf Virchow proposed the cellular theory of tissue organisation in 1858.

Individual contributors to the *Hippocratic Corpus* frequently contradicted each other, with one even being scathing about the views of Empedocles, but there was a general acceptance of the prevailing view that the body was in a state of balance that was disrupted in disease. Phlegm and yellow bile were considered as prime factors involved in illness, the former being associated with water and the latter with fire. Their presence could be detected when they flowed out of the sick patient. Blood was considered essential for life, although it was also viewed by some of the writers as a cause of disease when present in excess. In stark contrast to blood stood black bile, a vaguely defined malevolent humour of indeterminate composition but related to yellow bile. To complicate matters further, each humour was linked to one of the seasons and to a phase of the life cycle of man.[6] Despite differences in opinion as to their exact nature, the requirement for a healthy balance of humours to return the sick patient to good health provided the physician with a rational approach to therapy. Magical and superstitious factors were at last set aside.

In his writings Plato twice refers to Hippocrates as an Asclepiad. This is not true of those writers of the *Corpus* who travelled extensively around the Greek islands, but it is reflected in the approaches taken by the others who put stress upon first observing the patient with great care in order to determine the nature of the disease and thereby determine its course and means of treatment. There was a firmly held conviction that the body would heal itself if left alone, so treatment was often based on a change of diet as it was believed the imbalance of humours could be related to unhealthy eating and lifestyle, the latter being inherent in the Greek word *dieta*. In those times the quality of food left much to be desired and nothing was known of vitamins, so this belief in a change of diet to prevent or cure disease had some justification. Only when this approach had failed was recourse made to drug therapy. This involved either prescribing mildly acting herbs or else referral of the patient to the rhizotomoi. As it was believed that disease could be caused by one of the humours blocking an internal vessel, drug treatment often involved the administration of a herb that would clear the blockage. Cholagogues, diuretics, emetics, emmenagogues, expectorants, purgatives, sudorifics and a variety of other herbs that irritated tissues to produce muscular contraction had previously been relied upon to drive demons out of the body, so there was no shortage of herbs that could clear the various types of blockage. Remedies once thought to expel demons were now believed to expel humours. However, the emphasis now was on treating the patient rather than the disease. This was an advance on religious, magical or superstitious approaches, yet this holistic method was fundamentally flawed because it dissuaded physicians from investigating the real causes of disease and ultimately finding effective treatments. This was coupled with the typical Greek failure to undertake experimental investigation.

Aristotle

The allure of humoral medicine seems to have been irresistible. Even Aristotle (384–322 BC) was ensnared by it. His wide knowledge of plants and animals led him to recognise the power of the seed to grow into a fully developed plant. He ultimately came to believe that all objects had within them the power to change in a predetermined manner and he wrestled with the nature of that internal power. His conclusion was that the four elements were associated with four powers, or qualities, existing in two opposing sets, such as hot and cold and moist and dry. Thus he held that each of the humours consisted of one of the elements of Empedocles possessing two of the four qualities. Furthermore, these qualities were said to possess spiritual dimensions. The implications of this were far-reaching. Alcmaeon, a contemporary of Empedocles, had already asserted that disease was due to an imbalance, possibly of heat or cold brought about by an inappropriate diet. Aristotle therefore concluded that as drugs possessed elements with sets of qualities, imbalance of the humours could be treated with appropriate medication to restore any deficiencies. So far as drug discovery was concerned, the idea that the qualities of an object were a reflection of its elemental content might seem attractive in the twenty-first century, but all it did in an era when people believed there were only four elements was to muddy the waters.

THE GREEK HERBAL TRADITION

Aristotle had been head of the Academy of Athens, a forerunner of the universities, which Plato had established in 387 BC. Aristotle was succeeded by Theophrastus (c. 370–287 BC), his friend and former student. Theophrastus extended the academy and purchased adjacent ground where he created a large garden in which his pupils could study plants, including those sent by the army of Alexander the Great from the lands conquered in the East, as far away as the River Indus in India. The knowledge thus acquired enabled Theophrastus to write two Greek texts, *Enquiry into Plants* (*Historia Plantarum*), in which he classified and described over 500 plants,[7] and *Growth of Plants* (*De Causis Plantarum*), in which he explored the physiology of plants.[8] These were the first publications of their type and ensured that the concepts about the four qualities developed by Aristotle would never be forgotten. They also perpetuated the myths created by the rhizotomoi that have already been mentioned, including the warning to collect peony only at night lest a woodpecker attacks the digger's eyes.

Alexandria on the Mediterranean coast of Egypt gradually became the centre of Greek culture and the greatest minds were attracted to the city, such as Herophilus who came there from Chalcedon, a town near Byzantium, and established a medical school around 300 BC. As Alexandria was also a major trading centre, knowledge of the herbs in vogue there spread throughout the ancient world, including the emerging city of Rome, where there was an interest in Greek art, literature, philosophy and science. When Rome grew in might and wealth, the best Greek physicians went there to practise.

King Mithridates VI of Pontus (120–63 BC) on the southern coast of the Black Sea was a ruler with many enemies, having incurred the wrath of the Romans by conquering surrounding territories. Not unwisely, he feared being poisoned and sought out an antidote that could offer protection. He turned for assistance to the rhizotomist Crateuas, author of several books including the first illustrated herbal, *Rhizotomikon*. Crateuas then became personal physician to Mithridates and set about compounding a large number of ingredients, possibly over 40, into an eponymous antidote known as mithridatium. According to Galen, this was tested on slaves who had been deliberately poisoned. After Mithridates was defeated by Pompey, the formula came into the hands of the Romans. Nero's physician Andromachus added further ingredients, as did Galen, who brought the total up to 77 and renamed it 'Theriac of Andromachus'. The large number of ingredients, according to Galen's thinking, ensured that the body would select one that was a suitable antidote to the poison. He even incorporated the flesh of a viper in the belief that this would offer protection against snakebites.

Dioscorides

Crateuas is believed to have recommended colchicum corm (autumn crocus) as a cure for gout, but it became unpopular because of its severe toxicity. When pure colchicine was isolated in the nineteenth century it indeed proved effective for gout, especially as it was now free from contamination with the veratrine that had caused much of the toxicity. Colchicine is the sole legacy of Crateuas since none of his books have survived. However, his *Rhizotomikon* may well have formed the basis of some of the text and illustrations of plants later attributed to Dioscorides in the oldest extant copy of his herbal, the sixth century illuminated manuscript in the Vienna State Library known as *Juliana Anicia Codex*.[9] This is a copy of *De Materia Medica*, written by Dioscorides in Greek between AD 60 and 78. This herbal is the major source of information about medicinal herbs used in the Hellenic world, citing over 600 plants that remained the mainstay of herbal medicine through the ages. Dioscorides provided a description of each herb and its habitat, discussing its humoral qualities and its characteristics so that adulteration could be recognised. He took great care to include not only as much information as he could about the medicinal uses of the herbs in man and animals but also reported the side effects. He also warned that the qualities of drugs were not obvious from mere inspection or tasting; they had to be tested. The approach taken by Dioscorides was essentially empirical and he had little time for theoretical speculation, but he did organise his herbal according to the qualitative properties of the drugs, such as warming or cooling. As this made it difficult to comprehend, others reorganised the text for greater clarity after his death.

Dioscorides was an astute observer and claimed to have tested many herbs on himself. It is possible that he was a physician who had travelled widely, as he claimed, for he described drugs from many lands, among which were almond oil, belladonna (deadly nightshade), bramble, caraway, cassia, cherry syrup, cinnamon, cumin, dill, fennel, ginger, henbane, lavender, liquorice, marjoram, mandrake, mint, mustard, olive oil, opium, parsley, pepper, rosemary, rhubarb, rue, sage, sorrel, tarragon, thyme and wormwood. Many of these, of course, were probably listed on the Assyrian clay tablets recovered from Nineveh, revealing how much interchange of medical lore there was in the Ancient World. There were also about 90 minerals, including calamine, mercury and sulfur. Passed down the generations, *De Materia Medica* strongly influenced Arab physicians after the fall of Rome. It was subsequently among the first books to be printed and it continued to be widely accepted as an infallible authority until the eighteenth century.[10,11]

Pliny the Elder

Unlike many of the prominent figures in Greek medicine, Pliny the Elder (23–79) was a Roman. He compiled an encyclopaedic work dealing with around 900 herbs, *The Natural History*. Written with somewhat less rigour and care for detail than was *De Materia Medica*, it reflected the use of folk medicines by those who could not afford the services of a physician and was more readable than the herbal of Dioscorides. Pliny was less selective than Dioscorides and, like Theophrastus, had no hesitation in incorporating magical and superstitious material. The use of Latin later facilitated its being copied in monasteries, ensuring that its magical and superstitious elements would be perpetuated for many generations thereafter. For over 500 years after the fall of Rome it was the principal source of medical knowledge in many parts of Europe.

Galen

The most influential of all the Greek physicians was unquestionably Galen (AD 129–199). Born in Pergamon, he began his study of medicine at nearby Smyrna, completing his training 12 years later in Alexandria before moving to Rome in AD 162. He achieved considerable fame there as a clinician and his amassed wealth allowed him to become a prolific writer whose output dwarfed that of all other ancient writers. His collected work in 20 volumes is entitled *Opera Omnia*. This enshrines literally all the medical knowledge of his day. In it, Galen expressed admiration of Hippocrates and Dioscorides, although distorting their ideas to suit himself. Specifically, he taught that imbalance of the four humours was the cause of disease, but went much further by offering the dogmatic view that it should be possible to correct humoral imbalances, not only by supplementing deficient qualities with herbs possessing similar properties but also by administering other herbs that had opposing qualities. He expressed this succinctly in his maxim '*contraria contrariis curantur*' (opposites cure opposites). This was the origin of a medical system that beguiled physicians for the best part of two millennia. It was the antithesis of what modern medicine strives to do by identifying the causes of disease and addressing them directly by experimental investigation, supported by controlled clinical trials to evaluate efficacy of drugs.

Galen's approach to therapy also involved polypharmacy, numerous plants being compounded in complex formulations that became known as galenicals. His enthusiasm for the Theriac of Andromachus illustrates this. He argued that the body was capable of selecting whichever ingredient would correct the humoral imbalance. Whenever the followers of Galen encountered failure of a remedy to cure, they exercised their minds trying to decide which extra ingredients to incorporate in the hope that they might possess appropriate properties to correct specific qualities that were deemed to be either in excess or else deficient.

Galen claimed that nature acted with perfect wisdom and the body was but the instrument of the soul. This appealed to the Church and reinforced the authority of Galen, making it difficult for anyone to question his teachings. The credence given to his writings exerted a stultifying influence on therapeutic practice, not only while the Roman Empire flourished but right up until comparatively recent times. So pervasive was his influence that until pathology developed into a mature science in the nineteenth century, when it was bolstered by Pasteur's germ theory of disease, it was generally accepted that illness consisted of a disruption of the constitution. This severely impeded the search for external causes of disease and consequently the development of effective drugs.

Many of the extant works of Galen are available on-line from the Bibliothèque Interuniversitaire de Médecine.[12] They are also available in print.[13]

REFERENCES

1. Homer, _Odyssey_, IV.
2. Homer, _Iliad_, XI.
3. M.R. Wright, _Empedocles: The Extant Fragments_. New Haven: Yale University Press; 1981.
4. P. Kingsley, _Ancient Philosophy, Mystery and Magic: Empedocles and Pythagorean Tradition_. Oxford: Oxford University Press; 1995.
5. J. Jacques, _Hippocrates_, translated by M.B. DeBevoise. Baltimore: Johns Hopkins University Press; 1999.
6. V. Nutton, Humoralism, in _Companion Encyclopedia of the History of Medicine_, Vol. 1, eds W.F. Bynum, R. Porter. New York: Routledge; 1993, pp. 281–91.
7. Theophrastus, _Enquiry into Plants_, 2 vols, translated by A.F. Hort. Cambridge: Loeb Classical Library; 1916.
8. Theophrastus, _De Causis Plantarum_, 3 vols, translated by B. Einarson. Cambridge: Loeb Classical Library; 1976–1990.
9. C. Singer, The herbal in antiquity and its transmission to later ages. _J. Hellenic Stud._, 1927; **47**: 1–52.
10. J.M. Riddle, _Dioscorides on Pharmacy and Medicine_, History of Science Series, No. 3. Austin: University of Texas Press; 1986.
11. R. Gunther, _The Greek Herbal of Dioscorides_. New York: Hafner Publishing Co.; 1959.
12. http://www.bium.univ-paris5.fr/histmed/medica/galien_va.htm
13. Claudius Galenus, _Opera Omnia_, Hildesheim: Georg Olms Verlag; 2001.

5

The Arab World

The origins of Baghdad as the leading medical centre of the Muslim world in the ninth century are obscure, but it has often been claimed that in the year 765 Jirjis Bakhtyishu', the head of a prominent medical school in Jundishapur (situated about 300 miles to the south-east of Baghdad), was summoned to the capital to treat Caliph al-Mansur for dyspepsia. The likelihood of Bakhtyishu' covering such a distance over rugged terrain in time to treat such a minor, self-limiting complaint must be questioned, but the successful outcome is said to have resulted in his remaining in the city for many years. His son, Jibra'il ibn Bakhtishu', is then believed to have become one of its leading physicians and founded its first hospital in 805 at the request of the fifth Abbasid Caliph, Harun al-Rashid. There is, however, no convincing evidence that a leading medical school ever existed in Jundishapur. The Bakhtyishu' family, one of two leading Nestorian medical families in Baghdad over a period of nearly three centuries, may have created the story.

A more probable account of the emergence of Baghdad as a medical centre involves Nestorian Christians. Nestorians believed that Jesus was two persons, one divine and one human. Their stronghold had been Edessa, situated in what is now south-eastern Turkey, near the border with Iran. The Nestorian Church spread out from this centre of Syrian culture to form its own traditions. Its scholars translated the Hebrew Bible into their native language, Syriac, with ease as it was similar to Aramaic, itself not greatly unlike Hebrew. They also studied Greek so as to translate the Septuagint, a Greek rendition of the Hebrew Bible compiled by 70 Jewish scholars in Alexandria in 132 BC. Both versions of the Bible were similar, but in order to address theological issues their differences were compared in Syriac, the native language of the Nestorians. When the Byzantine Emperor Zeno expelled the Nestorians from Edessa in 489 for their heretical views, they fled eastwards, taking with them knowledge of the Greek language and literature. After Baghdad was established some settled there and so, too, did Arab, Jewish, Alexandrian and Indian scholars, all contributing to the intellectual milieu of the new city.

In the ninth century, Baghdad became the focus for a revival of humoral medicine, which elsewhere was receding as classical culture declined and medicine became compromised by the Christian belief in disease being a punishment for sin. Several factors can account for the revival, not the least of which was the presence in Baghdad of Nestorians and other immigrants who were able to translate into Arabic the classical medical texts from either Greek or Syriac versions. Arabic had replaced the now neglected Greek as a universal language of scholarship. Also, the prosperity of the city favoured the existence of an educated elite who could afford the teaching and books required to study medicine. Finally, as a result of the Muslim territorial expansion that had occurred there was access to classical texts stored in the monasteries of Egypt, Syria and Iraq that could be borrowed, purchased or simply taken as the spoils of war.[1]

Drug Discovery. A History. W. Sneader.
©2005 John Wiley & Sons Ltd

HUNAYN IBN ISHAQ

While translation of classical texts had already begun during the caliphate of Harun al-Rashid, it was boosted by the desire of his son Ma'mun to support the Mu'tazila clergy who were seeking a rational basis for Islamic belief that could challenge that of Christianity and other religions through the assimilation of acceptable ideas and opinions enunciated elsewhere. A year before his death in 833, Ma'mun established the Bayt al-Hikma (House of Wisdom) in Baghdad, where scholars gathered to collect and translate into Arabic not only classical texts but any others that were considered to contribute to the development of Islamic culture. It was here that the Nestorian translators came to the fore, the most notable being Hunayn ibn Ishaq (known in mediaeval Europe as Johannitius). He had been a pupil of Yuhanna ibn Masawayh (777–857), a member of a prominent Nestorian family who was physician to four Caliphs of Baghdad, and was to become well known in the West through his treatises on various aspects of medicine that were translated into Latin with the name Mesue as the author.

Established as a translator of the Hebrew Bible into Arabic, Hunayn together with his associates tirelessly sought far and wide to recover a vast number of medical and other texts, especially those written by Galen, and then translate them with great care and attention to detail. This hectic activity continued unabated after the death of Ma'mun until Caliph Mutawakkil, an opponent of the Mu'tazila, came to power in 847. A steady decline in the output of the Bayt al-Hikma followed, and eventually lone translators were left to work unassisted in that once great institution. However, the legacy of the Bayt al-Hikma remains with us in the hundreds of texts that were translated there into Arabic and thereby saved. Its collection of so many works by Galen also resulted in the spread of humoral medicine throughout the expanding Muslim world and planted the seed for the further development of Arabian medicine. There were many who took up this cause, resulting in a massive literature being compiled over many centuries by a large number of writers, of whom only a few can be mentioned here. From the tenth century onwards, the complexity of this stimulated the writing of encyclopaedic tomes that charted the path for physicians to follow.[1]

One of the consequences of the revival of Galenic medicine during the caliphate of Ma'mun was a demand for complex formulations that required specialised skills for their manufacture. Initially this resulted in the development of a new profession, that of the apothecary – later to be described as a pharmacist – and soon these licensed apothecaries opened shops in Baghdad which were under the watch of inspectors employed by the state.

ABU BAKR AL-RAZI

Among the greatest of the physicians who practised in Baghdad was Abu Bakr al-Razi, or, simply, al-Razi (known in the West as Rhazes), a Persian who died in the year 925. He was not only the director of a hospital in the city but was also an eminent philosopher and author of around 200 books on a diverse range of subjects. His medical writing included two major works, *Kitab al-Mansuri fi al-tibb* (*The Book of Medicine for Mansur*) and *Kitab al-Hawi fi al-tibb* (*The Comprehensive Book on Medicine*). The former consisted of ten volumes covering a wide range of specialities from both a theoretical and practice-based viewpoint, the use of drugs being covered in the first four volumes. Gerardo de Cremona (1114–1187) translated the entire work into Latin as the *Liber Almansoris* in Spain in the late twelfth century. The second of two major works by al-Razi was compiled by his students after his death from his extensive files and was published in 20 volumes that then constituted the largest collection of medical writings, including material from Greek, Hebrew, Syriac and Arabic sources. It is noteworthy that al-Razi recommended the testing of drugs on animals before exposing humans to them. His caring attitude towards his patients even extended to his introduction of opium to sedate those upon whom he practised surgery.

ABU ALI AL-HUSSAIN IBN ABDALLAH IBN SINA

Al-Razi was surpassed in his influence by yet another Persian philosopher who both practised and wrote prolifically about medicine, namely Abu Ali al-Hussain ibn Abdallah ibn Sina (980–1037), generally known in the West as Avicenna. He was a devout follower of Galen, whose ideas he expounded and developed further. His fame was in large part due to the encyclopaedic *al-Qanun fi at-tibb* (*The Canon of Medicine*), of which the Latin translation by Gerardo de Cremona became adopted as the standard text for medical students throughout Europe until the eighteenth century. In this magnum opus of nearly one million words, ibn Sina merged Greek and Arab traditions to encompass all that he had learned from diligent reading of ancient manuscripts, and then he embellished his book with his own clinical experience. He also adopted an Aristotelian approach that resulted in Galen's complex ideas being lucidly presented for the first time. In the second of the five main sections, ibn Sina described the uses and efficacy of 760 drugs arranged alphabetically, these being his preferred drugs. Thousands of others favoured throughout the Arab world were rejected. Such was the influence of this authoritative book that it ensured the perpetuation of Galenic medicine for many centuries, during which time no one dared question the teachings of ibn Sina. He undoubtedly had a brilliant mind and was one of the greatest physicians of all time, but his influence delayed the demise of an ancient system of medicine that in retrospect can be seen to have ill served those who sought benefit from it. If a genius of the stature of ibn Sina could have embraced humoral medicine so wholeheartedly, it is little wonder that it still persists to this day in the Unani medicine of India and strongly influences the use of contemporary herbal products favoured by proponents of complementary medicine.

ABU AL-QASIM AL-ZAHRAWI

It was not just in the East that Arabic culture thrived after the Muslim conquests. Under the enlightened rule of Abd-er-Rahaman III, Córdoba, and in particular the suburb of the Umayyad royal palace, rivalled Baghdad in its grandeur until its conquest by the Berbers a century later. With a library of 400 000 manuscripts, it became a great centre of learning and scholarship where physicians could study the ancient classics. The most illustrious of them was Abu al-Qasim al-Zahrawi (936–1013), often known in the West through his Latinised name Albucasis. He was the author of one of the greatest encyclopaedic medical texts ever written, *al-Tasrif li-man 'ajaza 'an al-ta'lif* (*The Guide for Him Who Cannot Compose ...*).[1] This masterpiece was written a few years before the *Canon* of ibn Sina at the time when Arabian medicine was approaching its zenith. It records the state of medical and pharmaceutical knowledge 1000 years ago, but unfortunately only parts were ever translated. The last of its 30 volumes, an illustrated manual of surgery, was translated into Latin by Gerardo de Cremona and subsequently printed in Venice in 1497. Under the title *De Chirurgia Parva*, it became widely studied throughout European medical schools during the Middle Ages, resulting in its author becoming best known as a surgeon. The rest of *al-Tasrif*, however, contained a vast amount of chemical, pharmaceutical and therapeutic information. The 28th volume was translated into Latin in 1288 by Simon of Genoa and printed in Venice in 1471, becoming known as *Liber Servitoris de Praeparatione Medicinarum Simplicium*, or simply *Liber Servitoris*. It was relied on by many generations of European apothecaries as a first-hand source of information on pharmaceutical processes. That a physician wrote it might seem surprising, but can readily be explained. The development of an independent profession of apothecaries that took place in Baghdad was not paralleled in Spain; hence physicians there were left to compound their own medicines. As there was no reliable source of written information about such matters, possibly because the apothecaries in Baghdad preferred to keep their knowledge secret, al-Zahrawi set about filling the void in the 28th volume of *al-Tasrif*.

The 28th volume is in three parts. The first deals with how to obtain, store, grind and reduce or oxidise a variety of minerals used in medicine, including alum (aluminium sulfate), copper, galena (lead sulfide), iron, stibnite (antimony sulfide) and vitriols (natural sulfates). The preparation of the oxide and carbonates of lead are fully described together with the purification of ceruse, a white lead carbonate found near Córdoba and which was used as a cosmetic. The preparation of verdigis (basic copper acetate) from copper is likewise discussed. The text also expounds on the roasting (calcinations) of mercury and of arsenic in a furnace to form their oxides, and also details a variety of chemical processes such as the purification of tutty, a sublimate of zinc oxide that adhered to the flues of furnaces in which zinc ores were smelted, and the preparation of sal ammoniac (ammonium chloride) from igneous rocks.

As Arab physicians were strongly influenced by Galen, it might seem surprising to encounter such detailed information on the preparation and uses of minerals, about which Galen was certainly far from enthusiastic. It has been suggested that this happened because of a strong association between Arab and Indian medicine which arose from the presence of Indian physicians in late eighth century Baghdad.[2] In India, links between medicine and alchemy had been developed in the fourth and fifth centuries and subsequently many Arab physicians dabbled in it, thereby ensuring a continuing supply of minerals for medicinal purposes. Troubled by a hot, dry climate and frequent sandstorms, their patients often developed eye and skin diseases for which the application of soothing ointments containing sulfur and mercury became almost routine. So far as the internal administration of minerals was concerned, however, restraint seems to have been exercised.

The second part of the 28th volume, in contrast, is a treasure trove of pharmacognosy that is concerned with the correct manner of handling plant products for medicinal application, with an emphasis on the importance of drying and storage techniques. Examples include acacia, aloes, cardamom, colocynth, fleawort, fumitory, galbanum, lily, liquorice root, lycium, mandrake, opium, scammony, sandalwood, spurge, squill and wormwood. The natural habitat of each herb is described, together with a description of the plant and how to select the appropriate component and at what season. The expression of juices and collection of gums is covered, as is the filtering of decoctions and other liquid preparations. There is information about preparing oils, vinegar, aromatic waters, amber, coral, turpentine and so forth. A wide variety of pharmaceutical techniques familiar to pharmacists of recent generations are also expounded upon in this part of the book.[1]

The last section of the 28th volume deals with products obtained from animal sources. Guidance is provided on the collection of shells, nails, hooves, bones, fat, blood, milk, urine, snake skins, scorpions and other similar necessities of mediaeval pharmacy. As in the other two sections, a variety of pharmaceutical processes are again described, complete with illustrations of the apparatus required.

Throughout the ages there were many who believed that the *Liber Servitoris* represented the complete pharmaceutical writings of al-Zahrawi. That view was entirely mistaken. The 3rd volume of al-Tasrif, for example, gives a detailed account of the compounding and storage of concoctions, while the 10th volume ends with a discussion on preparations of seeds and nuts to treat gastrointestinal complaints. Similarly, the 15th volume encompasses the production of conserves prepared from fresh or dried herbs, even going into the details of incorporating aromatic spices to enhance their acceptance. The 16th volume concentrates on the preparation and storage of powders, whereas the 18th volume considers a more diverse range of products, including astringent and dusting powders for application to wounds, ear drops, effervescent powders, fumigations, gargles and incenses. The 20th volume covers preparations for the eye, including compresses, drops, washes, lotions and ointments, whereas the 21st describes applications for the mouth and throat, such as dentifrices and lozenges. Cough remedies appear in the next volume, while ointments prepared from animal fat are to be found principally in the 24th volume. Here, as elsewhere in the *al-Tasrif*, al-Zahrawi clearly acknowledges his indebtedness to previous authorities such as Galen and al-Razi. The 25th volume deals with the manufacture of fats and oils extracted from plants, which were more than just ancient products as far as al-Zahrawi was concerned; he was adamant that they remained of major importance as medicines. Listed among the sources of what he called 'adhan' are apricot, castor seeds, hazelnuts, henbane, linseed, peach, poppy seeds, radish, sesame nuts, sweet almonds, walnut, wheat and many more. Not only does al-Zahrawi describe the various manufacturing processes of these but he also expands upon their role in medicine, complementing this with detailed formulae and dosage. The 29th volume was of particular value to the apothecary or pharmacist because of the multilingual lists of drug names and synonyms that appeared in it, accompanied by details of their stability.

The *al-Tasrif*, however, was written for physicians, not apothecaries; hence much of the text was concerned with the use of drugs. This was covered strictly within the confines of humoral medicine, with all products evaluated in terms of their qualities, namely hot, cold, dry or wet. As an admirer of Galen, its author did not hesitate to compound a variety of ingredients together in the expectation that the body will select the most appropriate one. This was carried to an extreme in the 4th volume where he elaborated upon the seven-stage preparation and the 84 ingredients of the famous antidote known as the Great Theriac. It incorporated balm, black pepper, cinnamon, ginger, glycyrrhiza, honey, mushroom, opium, psillium, rhubarb, saffron and viper flesh. Similar multi-ingredient concoctions were developed by others to protect against stings and the bite of venomous animals. Under the name 'theriaka' these worthless herbal preparations acquired the reputation of being panaceas against all ills and, as such, were so highly valued that their content was kept secret. From mediaeval times until the nineteenth century, leading European civic authorities were responsible for providing their own locally produced theriacs to their populace. Perhaps this accounts for the continuing faith in herbal remedies among the general population. Consider, for example, garlic – a panacea first proposed as a cheap substitute for theriac in the seventeenth century by Nicholas Culpepper under the name of 'treacle', a corruption of 'theriac'.

There were other products that rivalled the theriacs. The 5th volume of *al-Tasrif* described bitter medicines called 'hieras'. Originally introduced by the Greeks and used in their temples, the most popular of these was hiera picra (the sacred bitters). Containing a variety of bitter herbs such as aloes, cinnamon and colocynth, the hieras gained in reputation what they lacked in efficacy, and at times they, too, acquired the status of panaceas for all ills.

There were more mundane products as well in *al-Tasrif*. For example, the 6th volume discussed purgatives, selecting principally those bitter ones that could expel black bile and yellow bile. This included such powerful agents as aloes, colocynth, euphorbium, scammony and squill. These acted by irritating the bowel, so it is just as well that al-Zahrawi recommended that they should be reserved for healthier patients, with smaller doses being more appropriate for certain cases! More pleasant tasting purgatives were discussed separately in the 8th volume, these being considered suitable for those who vomited if prescribed a bitter agent. The 10th volume was similar, describing a variety of seeds and nuts that could be administered as purgatives. The apparently disproportionate coverage of purgatives is merely a reflection of the limited range of procedures for ridding the body of excess humours.

As would be expected in a system of medicine that claimed all illness was due to humoral imbalance, al-Zahrawi recommended specific herbs on the basis that they could correct this and thereby heal a wide variety of ailments associated with that imbalance. This remains a common feature in contemporary herbal medicine. There was, in addition, the alternative of treating a single disease with a variety of herbs thought to have similar corrective effects on the perceived imbalance. Thus, in his 9th volume, al-Zahrawi advocated hot, aromatic drugs to treat heart disease; this was considered to arise from an excess of phlegm. The discerning observer may, of course, complain that these two approaches gave carte blanche to the incompetent healer to administer just about anything at all. This is evident in the 12th volume, which covers remedies for both obesity and for lack of weight, for either excess or deficiency of semen, for over- or under-production of mother's milk and, inevitably, aphrodisiacs.

The *al-Tasrif* was a remarkable work which al-Zahrawi himself readily acknowledged leant heavily on the contributions of generations of medical writers from the time of Dioscorides and Galen onwards. The *al-Tasrif* now exists only as volumes and fragments scattered around the world in different libraries, and until the publication of the study by Hamarneh and Sonnedecker[1] and subsequent publications by Hamarneh[3] only the *Liber Servitoris* and the surgical sections were generally known in the West. The account given here is an attempt to distil some of the essence of their brilliant work. However much we may admire the humanity, scholarship and surgical skills of al-Zahrawi and his fellow physicians, there can be no hiding from the fact that they remained uncritical of the medicine of Galen and consequently their contribution to history was the perpetuation of an unsound system of medicine that ultimately failed to discover safe, effective medicines. Instead, Arab physicians continued to introduce worthless and sometimes harmful products, which is best illustrated by a text of the thirteenth century Andalusian ibn al-Baytar which listed over 3000 plants, animals and minerals. However, in their development of pharmacy as a highly skilled profession, the apothecaries of Baghdad unquestionably played a decisive role.

ARABIAN MEDICINE IN EUROPE

After the collapse of the Western Roman Empire in the fifth century, the ancient Greek heritage that it had nourished was gradually lost to most of Europe. There was, however, a notable exception in Southern Italy as this territory remained part of the Eastern Empire of Byzantium even after the invasion of Italy by the Lombards in 568. Greek culture still persisted when Benedictine monks established a hospital in the west coast city of Salerno in the late seventh century. A medical school gradually developed there, probably in the ninth or tenth century. It was clearly a secular institution, for both women and Jews taught Greek medicine alongside priests, while some of the staff were even married. Furthermore, the Jewish and Christian practice of one master teaching one pupil was abandoned in favour of the entire staff collaborating in the teaching of students.

The Muslim conquest of Sicily in the year 827 inevitably had an influence on medical practice in Salerno, especially when Arab communities began to settle nearby. However, the arrival in Salerno of a scholar born in Tunis and who had travelled throughout much of the Arab world was the defining point for the introduction of Arabian medicine in Salerno. Generally known as Constantinos Africanus (c. 1020–1087), he translated a vast number of Arabic texts into Latin, including some that were much more reliable translations of Ancient Greek texts than those that had hitherto been available in Salerno – despite these having originally been translated directly into Latin. Most of his translations were made at the nearby Benedictine monastery of Monte Casino where he resided after converting to Christianity. The impact of his work was to strengthen the understanding of Greek medicine through the new insights his translations afforded of the works of its leading proponents. He also introduced a familiarity with early Arabian medicine into the school. There was consequently some movement from a preoccupation with Hippocratic medicine, with its emphasis on diet and hygiene coupled with restraint in the use of drugs and a reliance on preparations of single drugs, towards the interventionist polypharmacy of Galen so much favoured by Arab physicians.

In 1140, Nicolas of Salerno produced his *Antidotarium* of 139 compound formulas that featured Galenical remedies for opposing the excess humours that were believed to cause disease. It also included a table of weights and measures that was later to become known as the apothecaries' system after the book became the first drug formulary to be set in print in a Venice edition of 1471. However, Hippocratic teaching still prevailed at Salerno as a few years later Matthaeus Platearius of Naples wrote his *Circa Instans*, so named on account of its opening words. It was loosely based on a Latin translation of Dioscorides, being quite different to the *Antidotarium* insofar as it described 273 simple herbs that were used in the medical school at Salerno in the Hippocratic manner. These were arranged alphabetically, an arrangement first employed by Galen. The *Circa Instans* strongly influenced *Le Grand Herbier* printed in Paris around 1520, and a translation of it was printed in England in 1526 under the title *The Grete Herball*.

By the eleventh century the reputation of Salerno had spread throughout Europe, and pupils were drawn from far and wide. However, in 1139 the Lateran Council banned Christian clergy and monks from practising medicine for the practical reason that increasing demand for their services was diverting them from their ecclesiastical duties. This eventually led to a greater need for schools such as Salerno; hence the following century saw the establishment of rival institutions at Montpellier, and in the new universities of Paris, Oxford, Padua and Bologna. These were soon to eclipse Salerno, hitherto the bastion of Hippocratic medicine in Europe. For the next three or four centuries it was the polypharmacy of Galen enshrined in Arabic medicine that was to become fashionable. At Salerno, preoccupation with Greek medicine kept its scholars from seizing the opportunity of translating the works of a new generation of Arab physicians. Translating al-Razi, ibn Sina, Al-Zahrawi and their contemporaries was left to others.

TOLEDO

Just as Greek medicine was revived in the Arab world after the fall of the Western Roman Empire, through a twist of history a similar revival of Arabian medicine ensued in Christian Europe after the Arabs were driven out of Spain. Toledo fell to the Christians in 1085, Córdoba in 1236 and finally Granada in 1492.

Reference has already been made to the translation of al-Razi's *Kitab al-Mansuri fi al-tibb*, ibn Sina's *al-Qanun* and the 30th volume of al-Zahrawi's *al-Tasrif* into Latin in the late twelfth century by Gerardo de Cremona. It was largely through the efforts of this prolific translator that Christian Europe obtained access to the major Arabic medical texts that had been neglected at Salerno. He had settled in Toledo, a cosmopolitan centre in Central Spain where Arabs, Christians and Jews freely intermingled. It was here that most of the manuscripts from the great library of Córdoba had been brought for safekeeping after its conquest and looting by the Berbers in 1013. Gerardo was able to translate more than 60 of the major works in this collection before his death in 1185.

LATIN TRANSLATIONS OF ARABIC TEXTS

Christian Europe had lost access to Greek medical texts during the fifth to twelfth centuries due to the supremacy of Latin in the Church, which meant that it became the language of intellectual discourse. Ancient manuscripts were copied and preserved in the monasteries, but were often restricted to those written in Latin as most monks knew no Greek. However, the only major medical text in Latin was the unreliable work of Pliny.

The motivation for looking after the sick may have reflected the highest ideals of Christianity, but lack of access to Greek medical texts meant that the basis of medical practice was considerably restricted until these became available in the twelfth century in the Latin translations of Arabic renditions of the Greek texts. When this finally happened, these were eagerly sought in the rapidly growing number of hospitals then being established throughout Europe by Benedictine monks and others. They were also sought by the newly designated apothecaries, so named in Venice because they were responsible for storage of the many herbs imported from Arab countries and elsewhere. The physicians and students in the new medical schools also required access to the texts. In short, there was an insatiable demand for the translations of Arabic texts, one that could not be met until Gutenberg introduced the printing press at Mainz in Germany in the mid-fifteenth century.

Due to the inferiority of mediaeval education compared with classical times, the clarity and apparent logic of ibn Sina was greatly preferred to the difficult and sometimes convoluted writings of the man who had inspired him, Galen. A thorough knowledge of the *Canon of Medicine* soon became a *sine qua non* for a medical career and Arabian medicine became fashionable throughout Europe. Few challenged its basis, with no other significant medical developments taking place. A notable exception was the Italian poet Francesco Petrarca (1304–1374) who argued that, because medicine was a practical art, those physicians who applied the methods of philosophical enquiry to it failed to cure their patients. He was proved correct by the onset of the bubonic plague against which all the ancient nostrums were to prove utterly worthless.

REFERENCES

1.   S.K. Hamarneh, G. Sonnedecker, *A Pharmaceutical View of Abulcasis al-Zahrawi in Moorish Spain: with Special Reference to the 'Adhan'*. Leiden: E.J. Brill; 1963.
2.   L.J. Goldwater, *Mercury: A History of Quicksilver*. Baltimore: York Press; 1972.
3.   S.K. Hamarneh, Pharmacy and materia medica, in *The Different Aspects of Islamic Culture*, Vol. IV, part II, *Science and Technology in Islam*, ed. A.Y. Al Hassan. Paris: Unesco; 2003, Chapter 5.6.

CHAPTER

6

Herbals

With the development of the printing press by Johann Gutenberg in the mid-fifteenth century, the medicine of Galen and ibn Sina attained a new lease of life, but initially not through the publication of their books. After the collapse of the Western Roman Empire there had been a continuing decline in scholarship throughout Europe. Consequently, major medical texts were beyond the comprehension of many. As a result, there was a demand for simplified herbals and formularies listing drugs and their uses, exemplified in the popularity of texts such as the *Herbal of Apuleius Platonicus*, written by an unknown fourth or fifth century author, originally in Greek. It was little more than a recipe book, full of spells and pagan incantations to the earth goddess, yet its Latin translation circulated widely throughout mediaeval Europe. It was the first herbal to be printed when a version based on a manuscript from the monastery at Monte Casino in Italy was produced in Rome in 1481, bound together with a sixth century Latin translation of Dioscorides.

Gutenberg's partner Peter Schoeffer printed editions of two herbals in 1484 and 1485, respectively. The first was compiled from the works of earlier writers and was crudely illustrated. Like other early printed works it did not have a title page, so it has become known as either *The Latin Herbal* or *Herbarius Moguntinus* (*The Mainz Herbal*). It was intended for local consumption, for it featured only herbs cultivated in Germany. There was still no awareness of the regional variation in plants, but once herbals began being printed in different locations around Europe this gradually became apparent and stimulated interest in botanical investigation among members of the medical profession who recognised that accurate plant identification was essential if their patients were to receive the intended medication. Furthermore, the wide circulation of printed herbals exposed errors and inconsistencies in the identification of plants, not only in the translations of ancient texts but also in the newly produced herbals. The correction of these errors had the unintended merit of introducing a degree of standardisation of medicines for the first time in history.

Encouraged by the success of his first herbal, Schoeffer turned his hand to producing a text in German, *The German Herbal* (again known by a variety of names including *Herbarius zu Teutsch, Gart der Gesundheit*). It may have been compiled by von Cube, the city physician of Frankfurt, and was a far more significant work than Schoeffer's earlier one, covering 435 plants and animal products as well. The crude illustrations, 65 of which were drawn from nature, were woodcuts.[1] For half a century they were frequently copied despite the improbability of the plants being identified. The quality of illustrations did not improve until Otto Brunfels (1488–1534), a Lutheran preacher born in Mainz and who had turned to a career in medicine, published *Herbarium Vivae Eicones* in 1530. This had 135 illustrations prepared from woodcuts by Hans Weiditz, an artist who broke with convention by copying the plants in great detail as he made his engravings. This unique approach allowed physicians to recognise the plants mentioned in the book. Brunfels also encouraged Hieronymus Bock (1498–1554) to publish a similarly illustrated herbal in German in 1546, his *Kreuterbuch*, which included his own observations on plants that had not hitherto been described by others. This was an early stirring of independent enquiry that was now appearing throughout Europe.

Drug Discovery. A History. W. Sneader.
©2005 John Wiley & Sons Ltd

The most influential of the Lutheran herbal writers was Leonhart Fuchs (1501–1566), whose *De Historia Stirpium* was published in 1542. As professor of medicine at Tübingen, he was disturbed by the relative ignorance of herbs among his fellow physicians and produced his herbal to remedy matters. Over 500 plants were described, most of which flourished in Germany. The text was also printed in German the following year. The Flemish herbalist Rembrant Dodoens (1517–1585) was strongly influenced by Fuchs in the writing of his *Cruydeboeck*, published in 1554, and *Stirpium Historia Pemptades Sex*, published in 1583. The latter was plagiarised by John Gerard (1545–1612) and expanded in 1597 as *The Herball* or *Generall Historie of Plantes*. A revised edition compiled by the London apothecary Thomas Johnson in 1633 corrected many of its errors and remained highly popular until the nineteenth century.

Despite the spirit of enquiry that was in the air, the ancient works of Dioscorides were not forgotten. The recognition of the unreliability of versions of his work that had appeared in print encouraged the apothecary Valerius Cordus (1515–1544) to travel widely to collect the herbs described by the Greek master. The herbal that resulted from his endeavours, *Historia Plantarum*, was edited by the Swiss naturalist Conrad Gesner and published 17 years after the early death of Cordus.

During a visit to Nuremberg in 1542, Cordus had received 100 guilders from the City Council for his dispensing notes. The Council published these two years after his death as its *Dispensatorium*. The first official pharmacopoeia to be printed was *Nuovo Receptario*, compiled in Florence by its Guild of Apothecaries and the Medical Society in 1498. Other cities subsequently produced their own pharmacopoeias too, including Basel (1561), Augsburg (1564) and London (1618).

In 1544, the Italian medical botanist Pietro Andrea Gregorio Mattioli (1501–1577) published *Di Pedacio Dioscoride Anazarbeo Libri Cinque*, his celebrated translation of the herbal of Dioscorides, complete with a commentary. He translated it into Latin shortly afterwards as *Compendium de Plantis Omnibus*. Many editions were published, the later ones featuring large woodcuts that greatly assisted identification of the plants.

HERBS FROM THE NEW WORLD

The early herbals were written at the same time as the voyages of discovery were opening up new trades routes. Fernandez de Oviedo y Valdes, a Viceroy of Mexico, sent the first account of medicinal plants from the Americas to King Charles V of Spain. This was published in Toledo in 1525 as *Sumario dela Natural y General Istoria delas Indias*, but it had little impact compared to the contribution from Nicolas Monardes (1493–1588), a physician who practised in Seville, a port with a monopoly in trade with the Americas. He took full advantage of the opportunity of investigating the strange herbs that his patients frequently obtained from the captains of newly returned galleons. It was probably fear of the unknown that drove him to test these on animals before contemplating administering them to his patients. Before long, he had begun cultivating the herbs in his own garden to ensure adequate supplies. His enthusiasm for the new herbs from the Americas led to the publication of the first volume of his three-part illustrated *Historia Medicinal de Indias Occidentales* in 1565. The remaining parts were published in 1571 and 1574 and were so enthusiastically received that expanded editions and numerous translations soon appeared. An English translation was printed in 1577 under the title *Joyfull Newes Out of the Newe Founde Worlde*. The complete work mentioned products such as balsam, china root, coca, guaiac, jalap, passion flower, guava, sarsaparilla, sunflower and tobacco, all of which soon appeared in the premises of apothecaries throughout the continent.

Monardes was not the first to recommend guaic, a resin exuded from the heartwood of the small West Indian tree known now as *Guaiacum officinale* L. This was already in wide use against syphilis, then generally believed to have been imported into Spain on the return from the West Indies of those who sailed with Columbus. Infected Spanish colonists on observing that the indigenous population were treating the similar disease of yaws with guaiac, promptly concluded it would also cure syphilis. Their view was bolstered by the ability of large doses of guaiac to induce diuresis, expectoration, purgation and profuse sweating. For those familiar with humoral medicine, no better indication of efficacy need have been sought. They even went so far as to recommend the resin for the relief of rheumatism and gout. When guaiac failed to cure syphilis, there were other herbs from the same region they could turn to, such as china root (*Smilax china* L.) or sarsaparilla (*S. aristolochia*).

Nor was Monardes the first to describe tobacco, *Nicotiana tabacum* L. The Spaniards became acquainted with tobacco smoking when Columbus landed in Cuba in 1492. Hernandez de Toledo later brought the plant back to Spain because it was thought to have medicinal properties. Joan Nicot subsequently sent the seeds to France about 1560. From then until the middle of the nineteenth century, tobacco was used as an antispasmodic, often administered rectally in the belief that the smoke would rise further up the intestines than any liquid enema and so be more efficacious. Medicinal cigarettes were popular in the nineteenth century, incorporating ingredients such as belladonna, atropine, hyoscyamus, aniseed, amber or nitre. The latter was converted to nitrite on burning, which relaxed the bronchi. Particularly popular were commercial brands such as Grimaud's Cigarettes® containing cannabis resin, Cigarettes de Joy® containing arsenic, Crevoisier's Cigarettes® containing foxglove and Savory & Moore's Cigarettes® containing camphor.

CINCHONA BARK

Two genuinely efficacious herbs from South America were introduced into Europe in the sixteenth century. The first of these was cinchona bark, *Cinchona officinalis* L. There are two accounts of how it was introduced into Europe. A fictitious version was included in a book known as the *Anastasis Corticis Peruviae*, published in 1663 by a Genoese physician called Sebastianus Badus. He claimed that 30 or 40 years earlier, during an epidemic in the city of Lima in Peru, the bark had cured the Countess of Cinchon, wife of the viceroy of the province, of a severe attack of tertian fever (malaria). In gratitude, the countess distributed the bark to sick people in the town and they were also cured. After the countess returned to Spain, the bark was used there and then taken to Rome for testing by Cardinal de Lugo in 1649. Romantic though this tale may be, a fascinating book written by Saul Jarcho has thoroughly undermined it.[2]

Jarcho also examined the other account of the introduction of cinchona. He found several sources for it, the principal one being a book by Gaspar Caldera de Heredia, entitled *Tribunalis Medici Illustrationes et Observationes Practicae*, published in Antwerp in 1663. In this, it was recounted that in a remote part of the Peruvian province of Quito, the Indians had to cross a river up to their necks in water and thereby became intensely cold. To stop shivering, they would drink a decoction of the bark in hot water. When Jesuit priests observed the striking effects of the bark, they asked for some of it to investigate. Reasoning by analogy, they tried the bark to relieve their own bouts of coldness and shivering caused by the intermittent fevers of malaria. Before long, this revealed to them the efficacy of the bark in relieving the disease.

Jarcho also cited the 2nd edition of the *Opera Medica* by the chief physician to Philip IV of Spain, Pedro Miguel de Heredia, who died about 1661. This 2nd edition of his book, which was not published until 1689, described the use of powdered 'quarango' bark by the Indians of Quito after exposure to coldness and how the Jesuits successfully exploited this in treating malaria. The details given in this account are similar to those published by Caldera de Heredia, but the author goes on to say that in Madrid he had been shown the bark by the son of the late Count of Cinchon.

The recurring account of the use of the bark by the Indians to relieve their coldness and shivering was addressed by Jarcho, who noted that a paper on hypothermia in animals, published in 1977, described how quinine (the major active principle in the bark) stopped their shivering. This was consistent with its ability to decrease the excitability of the motor end-plate region of skeletal muscle to repetitive nerve stimulation.

Jarcho also noted that Caldera de Heredia went on to relate in his book how the news of the discovery of the bark spread throughout Peru and was carried back to Spain by Dr Juan de Vega, physician to the Count of Cinchon. Trials in Spain, coupled with further reports from Jesuits returning there from Peru, ensured the general adoption of the bark. Other similar reports of the introduction of the bark into Spain corroborate this version, including a reference to it by Badus in his *Anastasis*.

Jarcho, who cited evidence relating to dates and persons which appears to confirm the validity of the account given by Caldera de Heredia, suggested that the bark first became known to Europeans around 1630, and samples were taken to Rome by Jesuits in 1631 and 1632. Jarcho also mentioned the work of others who considered it probable that Cardinal Juan de Lugo learned of the bark from Bartolome Tafur, a Jesuit from Peru who visited Rome in 1645. The Cardinal had the bark tested, and subsequently used his own funds to obtain large amounts of it for distribution to the sick. This led to cinchona being described as pulvis cardinalis or pulvis de Lugo.

Following its acceptance in Rome, the bark was sought throughout Europe. However, it was not without its dangers, and disputes broke out over its merits. The fact that it was distributed by the Jesuits led to an element of religious bigotry adding to the confusion, and it was only the efforts of an English quack, Robert Talbor, that finally ensured Jesuits' bark a role in medicine. Talbor was an apprentice to an apothecary in Cambridge, where the Professor of Physic, Robert Brady, had begun to prescribe the bark in 1658 during a serious outbreak of malaria. Talbor moved to London, where he treated many cases of malaria with a secret remedy. In 1672, he published a small book called *A Rational Account of the Cause and Cure of Agues*. In this he warned the public of the dangers surrounding the administration of Jesuits' bark by the unskilled. Cunningly, he suggested that this remedy should not altogether be condemned, but nowhere did he even hint that he himself had used it! The book boosted his reputation, and King Charles II awarded him a knighthood and issued letters patent appointing Talbor as his physician in ordinary. This was done despite fierce criticism from the College of Physicians, angered at the appointment of an unqualified practitioner. The physicians of Paris and Madrid were likewise outraged when, later that same year, Talbor cured the Dauphin of France and the Queen of Spain. The following year, King Charles' judgement was vindicated when he fell ill with tertian fever at Windsor and was cured by Talbor's secret remedy.

Grateful for the curing of his son, Louis XIV persuaded Talbor to part with the secret of his remedy for a sum of 2000 louis d'or plus an annual pension of 2000 livres. Talbor agreed on condition that the formula would not be revealed during his lifetime. When he died in his fortieth year in 1681, Louis arranged for this to be divulged in a small volume that appeared a year later, disclosing that it consisted of large doses of Jesuits' bark infused in wine. An English translation was published in 1682 under the title *The English Remedy: or Talbor's Wonderful Secret for Curing Agues and Fevers*.

The revelation of the nature of Talbor's remedy was followed by the publication in 1712 of a book known as *Therapeutice Specialis* by Francesco Torti, professor of medicine at Modena. It was written to counter the prejudiced arguments raised against the use of the bark, especially those claiming that as the drug had no purgative properties it could not be expected to cure disease! Torti provided an elaborate taxonomy of fevers in order to explain which ones would be cured by the use of cinchona bark. He distinguished the continued fevers from the intermittent fevers. The latter, often called the ague, were milder and occurred in regular cycles that were described as tertian or quartan, depending on whether the attacks took place every third or fourth day. These were the fevers that responded to the Peruvian bark. *Therapeutice Specialis* did much to establish the efficacy of cinchona bark, particularly as it dealt with the question of which types of fever would respond. The problem here was that until the general acceptance of the germ theory of disease in the latter part of the nineteenth century, physicians considered fevers to be diseases in their own right.

It is a measure of how deeply entrenched the humoral system of medicine was that it caused so many physicians to discount the value of the first curative drug they encountered. Many people had clearly been cured by it, yet there remained widespread resistance to its use. It took the efforts of a quack to overcome this. What a terrible indictment of the physicians of that period.

As demand for the Peruvian bark grew, confusion arose over its botanical origins, especially as the colour varied in different samples. The situation was only clarified after Charles-Marie de la Condamine, while engaged in an astronomical study near Quito in 1737, investigated the origins of the barks and obtained specimens. He learned the true identity of the cinchona tree from a botanist, Joseph de Jussieu. This enabled the Swedish botanist Linnaeus to classify the genus of trees from which the barks were obtained as Cinchona, a term he invented in 1742.

IPECACUANHA ROOT

The second effective herb introduced from South America in the seventeenth century was ipecacuanha root, *Cephaelis ipecacuanha* Willdenow. A report about its ability to cure what is now recognised as amoebic dysentery appeared in the *Historia Naturalis Brasiliae*, published in 1648 by William Piso, a Dutch physician who had returned from Brazil. Supplies did not reach Europe until 1672, but when he recommended it to his patients in Paris, they experienced toxic effects. His partner, Jean-Adrian Helvetius, was then urged by a merchant to try freshly imported root. After conducting suitable tests, Helvetius placed placards all over Paris advertising the virtues of his secret remedy for dysentery. Louis XIV heard of the remedy and arranged for it to be tested at the Hôtel-Dieu. On the successful outcome of this trial, the king paid Helvetius 1000 louis d'or for its formula. The merchant who had urged Helvetius to test the fresh root received nothing. Ipecacuanha root remained highly popular while amoebic dysentery was a regular occurrence and was one of the first herbs from which an active principle was extracted in the early nineteenth century, in Paris.

The impact of the new herbs from the Americas was significant because they had never been described by Galen or ibn Sina, yet they were believed to be powerful drugs. For some, this began to raise doubts about the opinions of these old authorities about the discovery of new drugs.

THE DOCTRINE OF SIGNATURES

A form of sympathetic magic which ultimately came to be known as the doctrine of signatures was originally expressed in earlier times by actions such as wearing the skin of powerful animals in order to acquire their strength or the adoption of horns as an aid to fertility because their crescent shape is similar to either the moon, which undergoes a cycle of birth and renewal, or else to the plough, which was so important in the fertilisation of the soil. To this day, rhinoceros horn is prized in the East as an aphrodisiac. A variant of this is the account in the *Book of Genesis* which relates how Rachel, the infertile wife of the patriarch Jacob, gave birth to a son after obtaining mandrakes. As mandrake root (*Mandragora officinale*) is bifurcate, it is not difficult to envisage it resembling the shape of a human. Those who sought to profit from this herb frequently strengthened this perception by appropriate carving of the root. Ginseng was an alternative to mandrake for the same reason. Many more examples can be found in the works of the historic medical writers, but it was especially in the writings of Paracelsus that the concept was fully developed. However, Paracelsus was not an author who was easy to comprehend. It was left to one of his admirers, the German master shoemaker and theologian Jacob Böhme (1575–1624), to formalise the ideas of Paracelsus as the doctrine of signatures in his *Signatura Rerum* (*The Signature of all Things*), published posthumously in 1651.[3] In this devotional text, Böhme asserted that the Almighty had specially marked every object to reveal its purpose. This led him to believe that there were signs in herbs that could reveal their intended medicinal use. Thereafter, all who were gullible enough to believe in this doctrine eagerly sought such signs. These could take the form of colour or shape, so that iris petals were ideal for treating bruises on account of their blue–black colouring, goldenrod could relieve jaundice because of its yellow hue, blood diseases could be treated with red herbs, lung complaints would respond to lung wort with its lung-shaped leaves, likewise liverwort for liver disease, pilewort for haemorrhoids and toothwort for toothache, while dropsy (oedema), which was diagnosed by swelling of the ankles, would respond to byrony root as this resembled a swollen foot. There were other types of signs that could be identified, such as the quacking aspen for those afflicted with palsy, plants with rough surfaces for rough skin, while plants or animals that survived for many years were evidently to be considered as aids to longevity.

The doctrine of signatures lay at the heart of the writings of Nicolas Culpeper, a healer who refused to join the London College of Physicians as he held its members in utter disdain. Influenced by the belief of Paracelsus that within every object there was some sort of volatile spirit that had counterparts both in the stars and in creations of the natural world such as plants and minerals, Culpeper added an astrological dimension to Böhme's writings. In his *Judgement of Diseases*, Culpeper asserted that the stars had imprinted signatures upon plants. When, in 1649, he had the impunity to translate the College of Physicians' *Pharmacopoeia Londinensis* from Latin into English, Culpeper intertwined astrological concepts with the original text. At that time, astrology was universally held to be valid, its foundations having been firmly laid in many of the admired Arabic texts. It was employed in diagnosing disease, determining the most appropriate time to administer medicines and even in the preparation of medicines.

Culpeper wrote the best-known English herbal of his time, published in 1653 under the title *The English Physician*. In it he discussed 369 medicinal plants, and for each one he described the astral influence that governed its virtues. For example, on cowslip he wrote, 'Venus lays claim to this herb as her own, and it is under the influence of Aries', dandelion '...is under the dominion of Jupiter' and fenugreek '...is under the influence of Mercury'. The impact of this book gradually waned during the next hundred years or so, but did not ever completely disappear. In an 1826 edition, the editor wrote: 'Hereby you see what reason may be given for medicines, and also what necessity there is for every physician to be an astrologian.' Many modern herbal publications for the mass market are still littered with astrological material lifted straight from Culpeper. The more serious herbals have somehow managed to extricate themselves from it, but in the public mind there still lingers the fond notion that herbs exist for the benefit of mankind and, as such, are providers of medicines to cure our ills.

SCIENTIFIC INVESTIGATION

The popularity of herbals from the late fifteenth century onwards resulted in their becoming the prime source of information about drug therapy rather than the longer treatises by Galen and ibn Sina. This not only contributed to a reduction in the influence of these and other ancient writers but encouraged physicians and apothecaries to conduct and publish their own investigations. However, it was not until the late eighteenth century that these began to adopt what would now be recognised as a scientific approach.

James Lind

The Scottish naval surgeon James Lind is remembered as the first man in medical history to conduct a controlled clinical trial when he proved that drinking lemon juice could cure scurvy.[4] In May 1747, he tested a variety of reputed remedies on 12 scorbutic sailors quartered in the sick bay of HMS *Salisbury*. Two of them were restricted to a control diet, but each of the others was additionally given one of the substances under trial. Both sailors who had been provided with two oranges and a lemon every day made a speedy recovery, one of them being fit for shipboard duties within six days. The only others to show any signs of recovery were those who had been given cider. Lind observed no improvement in the condition of the sailors who had received oil of vitriol (dilute sulfuric acid), vinegar, an electuary of garlic, myrrh, mustard and other ingredients, sea water or solely the control diet.[5] He drew the obvious conclusions, which were duly acted upon by Captain James Cook on his second voyage round the world. Cook was at sea for three years, yet not a single member of his crew died from scurvy thanks to adequate provision of lemon juice, as well as fresh fruit and vegetables. Disappointingly, it was not until 1795 that the Admiralty finally agreed to a regular issue of lemon juice on British ships. The effects of this action were dramatic; whereas in 1780 there had been 1457 cases of scurvy admitted to Haslar naval hospital, only two admissions took place between 1806 and 1810. The situation then deteriorated for over a century until it was discovered that cheaper lime juice which had replaced lemon juice had only about a quarter of its antiscorbutic activity.

Anton von Störck

The prohibitive cost of the services of medical practitioners meant that folk and domestic remedies continued to be relied upon throughout the eighteenth century, the reputation of the more popular ones spreading by word of mouth. The first to conduct an examination of such medicines in a scientific manner was Baron Anton von Störck (1731–1803), a physician who began life as an orphan in the poorhouse, yet through his efforts as a physician rose to the ranks of the aristocracy in Vienna and became a confidant of the Empress Maria Theresa. He had been a pupil of Gerard van Swieten (1700–1772), who had studied under Boerhaave in Leiden before establishing the Vienna medical school.

In order to determine the safe dose of folk medicines that had previously been considered too poisonous for therapeutic application, Störck first conducted animal experiments and then tested the herbs on himself. He finally administered small doses to patients, which he gradually increased on a sliding scale until he observed beneficial effects. Störck began by investigating the deadly poison hemlock (*Conium maculatum* L.), the Athenian state poison that claimed the life of Socrates. He reported that it had a beneficial effect in cancer but this claim remained controversial until rejected at the end of the nineteenth century. His findings were published in 1761 in Latin[6] and translated into English the following year as *Essay on Hemlock*. He next investigated thorn apple (*Datura stramonium* L.), henbane (*Hyoscyamus niger* L.) and monkshood (*Aconitum napellus* L.), which is also known as aconite.[7] The latter he established as having both diaphoretic and diuretic properties, whereas the first two were said to be of value in mental disorders. As a result of these and other investigations by Störck, several neglected herbs were accepted into mainstream medicine, including monkshood, thorn apple, henbane and also colchicum corm (*Colchicum autumnale* L.) as a remedy for dropsy (oedema) and pulmonary effusions.[8,9] Colchicum was originally found growing in the Colchis Region of Asia Minor, being known today as the meadow saffron. It was classified by Dioscorides as a highly poisonous plant which caused death by choking, yet the most influential of the Byzantine physicians, Alexander of Tralles, recommended it in the sixth century for 'podagra', an opinion endorsed by ibn Sina. The German botanist Hieronymous Tragus mentioned in the 1552 edition of his *Kreuterbuch* that Arabian physicians employed this root in cases of gout and rheumatism. Despite this, colchicum went out of fashion among physicians, but it remained in favour among country folk. Its misuse earned it the nickname 'colchicum perniciosum' within the medical profession.

Nicolas Husson, an officer in the French army, in 1783 introduced a secret remedy known as Eau Médicinale, which acquired a considerable reputation for, among many other things, alleviating the pain and inflammation of gout.[10] Eau Médicinale was brought to England in 1808, soon being rivalled by similar formulations such as Wilson's Tincture and Reynold's Specific. The formula of Wilson's Tincture was a closely kept secret until it was revealed in 1818 by William Henry Williams who, by then recommending the administration of colchicum for gout, ensured the place of the plant in the treatment of that painful ailment.[11] In the middle of the nineteenth century, a classic work on the administration of colchicum for gout was published.[12]

William Withering

Foxglove (*Digitalis purpurea* L.) was another folk remedy introduced into medicine in the eighteenth century. It grew throughout Europe, originally being known as the folksglove on account of the shape of its flowers. Both this and the colour are revealed by the systematic name, which is derived from one given to it in 1539 by Hieronymus Bock. He explained that 'digitalis' was an allusion to the German word *fingerhut* (finger stall), since the blossoms resembled the fingers of a glove. Foxglove has a long history of folk use, being known to Ancient Celtic tribes.[13] The Myddvai, the family of thirteenth century Welsh healers, applied it to the body by inunction to relieve headache, abscesses and cancerous growths.[14] It was also listed among herbs used by Edward III of England (1327–1377).

In 1775, the English physician and botanist William Withering (1741–1799) was asked his opinion of a herbal tea for the relief of dropsy, a condition marked by accumulation of body fluid due to right heart failure. The cause was then unknown, so the aim of treatment was to expel the excess body fluid with herbs. Withering realised that among the 20 or so ingredients in the herbal tea, it was the foxglove that was most likely to be causing the violent vomiting and purging. During the next nine years, Withering treated 158 patients with it, of whom about two-thirds responded favourably. In 1785, he wrote his treatise entitled *An Account of the Foxglove, and Some of its Medical Uses: with Practical Remarks on Dropsy, and Other Diseases*.[15] This described how to determine the correct dosage, which was highly relevant since foxglove was a potent poison that was ineffective unless administered at near the toxic dose level. Like Störck, Withering took care to standardise the doses he used, going even further by determining the correct dosage for each patient. Withering also discussed different ways of preparing foxglove, preferring the use of powdered leaves. However, he did not realise that its stimulant action on the heart was responsible for its beneficial role in dropsy. Indeed, it was not until 1799, the year of his death, that Ferriar suggested that the increased urinary output was of secondary importance compared with the power of foxglove to reduce the pulse rate.[16] Because of the lack of a clear understanding of how it was able to affect dropsy, foxglove was rarely used as a cardiac stimulant during the nineteenth century. Although the German pharmacologist Ludwig Traube revealed the stimulating effect of foxglove on heart muscle in 1850, it was not until 1901 that a clear understanding of the effects emerged through the development of the polygraph by the Scottish physician James Mackenzie and the electrocardiograph by the Dutch physician Willem Einthoven. Subsequent investigations elucidated the correct indications for the use of digitalis, namely in atrial fibrillation, where the co-ordination of the contraction of individual muscle fibres is so disorganised that the heart beats become weak and irregular, or in certain forms of heart failure in sinus rhythm.

Through the efforts of Störck and Withering, experimental medicine was now able to exploit the single greatest breakthrough in the history of drug discovery when alkaloids were first extracted from plants early in the nineteenth century. Progress from prehistoric times until then had been pitiful because two essential factors were missing, namely access to chemical compounds with consistent potency and effective methods of clinical investigation. Access to these during the past two hundred years has made it possible to identify and exploit beneficial drugs, free from the dogmatic teachings that were the legacy of the past and a major obstacle in the path of drug discovery.

REFERENCES

1. A. Arber, *Herbals. Their Origin and Evolution: A Chapter in the History of Botany, 1470–1670*, 2nd edn. Cambridge: Cambridge University Press; 1953.
2. S. Jarcho, *Quinine's Predecessor: Francesco Torti and the Early History of Cinchona*. Baltimore: Johns Hopkins University Press; 1993.
3. J. Böhme, *Signature of All Things; of the Supersensual Life; of Heaven and Hell; Discourse between Two Souls*. Whitefish, Massachusetts: Kessinger; 1997.
4. L.H. Roddin, *James Lind, Founder of Nautical Medicine*. New York: Schuman; 1950.
5. J. Lind, *Treatise on Scurvy* [containing a reprint of the 1st edn of *A Treatise of the Scurvy*, with additional notes], eds C.P. Stewart, D. Guthrie. Edinburgh: University Press; 1953.
6. A. Störck, *Libellus, quo Demonstratur: Cicutam non Solum usu Interno Tutissime Exhiberi, sed et Esse Simul Remedium Valde Utile in Multis Morbis, qui Hucusque Curatu Impossibiles Dicebantur*. Vienna: 1761.
7. A. Störck, *Libellus quo Demonstratur: Stramonium, Hyoscyamum, Aconitum non Solum tuto posse Exhiberi usu Interno Hominibus*. Vienna: 1762.
8. A. Störck, *Libellus de radice Colchici Autumnalis*. Vienna:1763.
9. A. Störck, *An Essay on the Use and Effects of the Root of the* Colchicum autumnale, *or Meadow Saffron*. London: Becket and De Hondt; 1764.
10. N. Husson, *Collection de Faits et Recueil d'Experiences sur le Spécifique et les Effets de l'Eau Médicinale*. Paris: Brasseur, Bouillon; 1783.
11. W.H. Williams, *Observations Proving that Dr. Wilson's Tincture for the Cure of Gout and Rheumatism is Similar . . . to . . . the Eau Medicinale*. London: J. Callow; 1818.
12. A.B. Garrod, *The Nature and Treatment of Gout and Rheumatic Gout*. London: Watson and Maberly; 1859.
13. J.H.G. Grattan, C. Singer, *Anglo-Saxon Magic and Medicine*. London: Oxford University Press (for the Wellcome Historical Medical Museum); 1952, p. 23.
14. J. Pughe, *The Physicians of Myddvai: Meddygon Myddfai*, ed. J. Williams ab Ithel. London: Llandovery-Longman; 1861.
15. J.K. Aronson (ed.), *An Account of the Foxglove and Its Medical Uses, 1785–1985*. London: Oxford University Press; 1985 (includes a facsimile of Withering's original monograph).
16. J. Ferriar, *An Essay on the Medical Properties of the* Digitalis purpurea, *or Foxglove*. Manchester: Sowler & Russell; 1799.

Chemical Medicines

The first significant challenge to the polypharmacy of Galen and ibn Sina came from a man whose writings seethed with alchemical mysticism. Theophrastus Philippus Aureolus Bombast von Hohenheim (1493–1541), who gave himself the name Paracelsus, undermined conventional medical practice after his controversial writings were published posthumously by his admirers. A similar attack on the anatomical principles espoused by Galen came from Andreas Vesalius of the University of Padua in 1543 with the publication of *De Humani Corporis Fabrica* (*On the Structure of the Human Body*).

Paracelsus was born in Einsiedeln, near Zurich, the son of a physician who moved to a mining area in the south of Austria. It was there that he acquired an extensive knowledge of minerals and alchemy that was to later influence his medical practice. Paracelsus attended classes at several European universities, but was unimpressed with his teachers and their theorising about Greek science and medicine. He came to the conclusion that there was more to learn outside the universities than in them and he may never have qualified in medicine. Taverns were, for him, ideal venues for acquiring local knowledge about what folk remedies might be effective. With his singularly uncouth manner, taverns would have been ideal for him to mix with people, especially during the ten years in which he travelled the length and breadth of Europe and into Arab lands in search of what he believed were the latent forces of nature.

By the time Paracelsus had returned from his travels, his fame had spread across Europe and as a result of successfully treating the influential publisher Johannes Frobenius, he was appointed as municipal physician and professor of medicine in Basle in 1527. In the face of strong opposition from the academic community, he was denied an official position in the university. Their objection can readily be understood, for he would neither produce his medical qualifications nor take the oath of Hippocrates. A few weeks later, he publicly burned copies of ibn Sina's *Canon of Medicine* during St John's Day celebrations. He attracted students from all over the continent to his lectures, but nearly a year later he had to flee the town after denouncing a magistrate who found against him over a disputed fee for treatment. During the remaining years of his life he travelled around Europe and wrote his *Der Grossen Wundartzney* (*The Great Surgery Book*), which was published in 1536.

With his knowledge of alchemy and the theory of transmutation, Paracelsus believed that there were three elements, mercury, sulfur and salt. These were not the materials with which we are familiar; rather they were characteristics of these, namely volatility, combustibility and cohesiveness, respectively. Though this was wrong, it encouraged Paracelsus and his followers to introduce chemical remedies that would never have been considered under the humoral system as they lacked the properties of being moist, dry, hot or cold. He also thought that within every object there was some sort of volatile spirit, the quintessence, which had counterparts both in the stars and in creations of the natural world such as plants and minerals. For him, this explained why certain herbs and mineral salts could influence specific types of disease. His upbringing led him to turn to alchemy to find the quintessences from herbs and minerals in their uncontaminated state, as he supposed they existed at the time of the creation of the universe. He believed that, with the passage of time, impurities had accrued,

Drug Discovery. A History. W. Sneader.
©2005 John Wiley & Sons Ltd

but if these were removed then minerals could safely be given by mouth as medicines. This approach led Paracelsus to adopt monotherapy with purified compounds and was, in the long term, of the utmost importance in the development of drug discovery. Any therapeutic benefit could now be directly attributed to a specific drug, whereas with the polypharmacy of Galen and ibn Sina there was little likelihood of this. The ultimate acceptance of the benefits of cinchona bark and ipecacuanha roots in the century after his death was a direct result of their being administered on their own.

Much that Paracelsus wrote was quite absurd, for he would consider almost any report or claim with equanimity on the grounds that it should be put to the test before judgement could be passed on it. The outcome of this was that magic, superstition and the occult, together with obscure alchemical investigations and philosophical enquiries, often dominated his writings. Few of these were published during his lifetime, but many of his manuscripts were recovered some years after his death and then printed. Their obscurity and mysticism encouraged others to expand upon and develop his ideas with greater clarity, resulting in their wider acceptance. This set in motion a sequence of events that ultimately led to the isolation and identification of therapeutically effective chemicals.

THE PARACELSIANS

Konrad Gesner (1516–1565) was one of the first to follow in the footsteps of Paracelsus by favouring chemical medicines. A Swiss polymath who is better known as the father of bibliography on account of his famous *Bibliotheca Universalis*, he also wrote *De Secretis Remediis*, published in 1552 (translated into English under the title *The Treasure of Euonymus*), in which he recommended the use of chemical remedies. He also prescribed mercury, lead and antimony. Like Paracelsus, he tried to remove impurities from his compounds. One of his favoured methods for doing so involved distillation which '...corrected the malignitie or venimous qualities thereof, as in oyles of quicksilver, of oyle of vitrioll, antimonie...'. Crude forms of distillation had been used by the Greeks and were improved by Arab alchemists, but it was not until around the twelfth or thirteenth century that an efficient glass still-head was developed, enabling fractional distillation to be achieved. This allowed Arnald of Vilanova (c. 1240–1311), the physician who established Montpellier as the leading European medical school, to prepare a variety of distilled waters from plants. In the following century, the Spanish Franciscan monk Joannes de Rupescissa distilled from wine what he called *aqua vitae*, the water of life, believing it came from the heavens as the *quinta essentia*, the fifth essence, and as such was a panacea for all ills.[1] This was the first recorded preparation of concentrated alcohol. Rupescissa's text was circulated in manuscript until a printed version appeared in 1514, entitled *Liber de Consideratione Quintae Essentiae*. A few years before this, in 1500, the illustrated *Liber de Arte Distillandi, de Simplicibus* by Hieronymus Brunschwig (1450–1512) was published in Strasbourg. As a result of the appearance of these two books, distillation now became important in the manufacture of medicinal substances, and hence its use by Gesner.

The first of the so-called Paracelsians to express strong support for the views of Paracelsus was Petrus Severinus (1542–1602), physician to the King of Denmark. His *Idea Medicinae Philosophicae* was published in Basle in 1571. It exposed the inadequacies of the Galenic approach, described what was later to become known as the doctrine of signatures and urged its readers to burn their textbooks, then experiment and travel so as to gain first-hand knowledge of the natural world. Oswald Croll (c.1560–1609), who studied orthodox medicine at Heidelberg, Strasbourg and Geneva, became another early Paracelsian. His *Basilica Chymica*, published in Frankfurt in the year of his death, ran to many editions and included formulations still to be found in pharmacopoeias during the nineteenth century. His clarity of writing was in marked contrast to Paracelsus and the many editions of the *Basilica* ensured that any physician or apothecary wishing to prepare a mineral medicine would find detailed instructions.

The views of the Paracelsians were disputed for many years, but refined mineral preparations were included in the *Pharmacopoeia Londinensis,* published by the London College of Physicians in 1618. As this was the most influential pharmacopoeia yet to have appeared, this was confirmation that these chemical remedies had finally become accepted in orthodox medical circles. From then on, their place in medicine was assured.

The earliest chemical remedies consisted of naturally occurring metals and their salts. As with plant- and animal-derived medicines, it is impossible to ascertain when they were first used as medicines, but some are referred to in ancient literature. Copper, gold, iron, lead, mercury, silver and tin were all known by Roman times. The writings of al-Zahrawi, especially the 28th volume of his *al-Tasrif,* provide detailed information on the preparation and purification of a range of mineral drugs. The range widened thereafter as alchemists conducted chemical reactions in their futile quest to convert base metals into gold. As a result of the rapid development of organic chemistry in the last third of the nineteenth century due to the growth of the dye industry, it became possible to manufacture organometallic compounds. Some of these were introduced into medicine, principally as chemotherapeutic drugs for treatment of infectious diseases. As such, they sometimes proved to be safer and more effective than their inorganic predecessors.

IRON

Iron appears to be the first mineral known to have been administered by mouth for therapeutic purposes. One of the oldest Greek legends tells how Melampus, a shepherd endowed with magical powers, cured the infertility of Iphiclus, King of Phylacea, by giving him wine in which rust scrapings from a knife had been steeped. Ferric oxide occurs naturally as a mineral, red haematite, and in the form of rust. Throughout recorded history, the astringent properties of ferric oxide have been exploited in various ways, such as in preparations for soothing the eye. It is mentioned in the *Ebers Papyrus* as an ingredient of a paste for the treatment of eye disease,[2] while haematite in ox fat was applied to the eye by Babylonian physicians to alleviate photobia.[3] The Hippocratic writings described the use of iron salts to stem bleeding, possibly resulting from the belief that battle wounds were best healed by application of the agent that caused them. Dioscorides noted that rust can stem uterine haemorrhage and also prevent conception,[4] while evidence of iron therapy for debility during Roman times has been obtained from excavations in Gloucestershire, England.[5] A detailed account of the use of iron by the Greeks and Romans was provided by Nicolas Monardes in the 2nd volume of his *Historia Medicinal de las Cosas que se Traen de Nuestras Indias Occidentales . . .* , where he cited them for conditions such as the bloody flux (amoebic dysentery), the flux of women (menstruation) and piles.

Lazarus Riverius, a physician to Louis XIV, in 1640 published the 15th volume of his *Praxis Medica,* a work that was translated into English 15 years later by Culpepper, Cole and Rowland as *The Practice of Physic.* This recommended steel dissolved in wine for chlorosis, a disease of young women sometimes described as love sickness, but now recognised as iron deficiency anaemia. The selection of an acid wine would have facilitated the oxidation of the iron. In 1681, Thomas Sydenham specified iron as the cure for chlorosis.[5] He considered this to be a manifestation of hysteria caused by a weakness of the animal spirits, which could be cured by strengthening the blood as this was the origin of these spirits. He advised that the patient should be bled before starting a 30 day course of pills containing steel filings and extract of wormwood, each dose being followed by a draught of wormwood wine.

The Parisian apothecary Pierre Pomet (1658–1699) wrote his *Histoire Generale des Drogues,* which was translated into English in 1712 as the *Compleat History of Drugs,* with additions by the Apothecary at the Jardin du Roi in Paris, Nicolas Lémery, and the medical botanist, Joseph Pitton de Tounefort. This recommended *crocus martis*, which is ferric oxide, for chlorosis and dropsy. A year after the publication of the English edition, Lémery and Geoffy provided a rational basis for iron therapy by demonstrating the presence of iron in ash prepared from blood.[6]

The first use of ferrous sulfate was by the French physician Pierre Blaud (1774–1858).[7] This reduced form of iron occurs naturally as a hydrate in several minerals, but can readily be prepared by dissolving iron in sulfuric acid. In 1832, Blaud reported that he had cured 30 cases of chlorosis within 10 to 30 days by administering pills containing 320 mg each of ferrous sulfate and potassium carbonate. He believed earlier failures of iron therapy were due to the underdosing of patients. The amount of ferrous sulfate he administered was gradually increased until his patients were taking 12 pills daily, a dose that would now be considered excessive as 200 mg two or three times a day is recommended. Blaud's nephew, a pharmacist, sold the 'veritable pills of Doctor Blaud' throughout the world.[8] Many similar products appeared on the market and other ferrous salts were introduced, such as the carbonate, citrate, fumarate, gluconate, iodide, lactate, phosphate and succinate.

MERCURY

Since around 500 BC, the mines at Sisapo in Almadén (Arabic for the mine) in Southern Spain have been the principal source of Iberian cinnabar, red mercuric sulfide, which possesses a bright scarlet red colour that ensured its use throughout the ages as a pigment. Cinnabar was also the source of quicksilver, metallic mercury, which was of major importance to the alchemists. A scholarly account of the history of cinnabar and mercury by Goldwater has provided much of the information given here.[9] He concludes that there is inadequate support for claims that either cinnabar or mercury were used medicinally in the ancient world by the Egyptians, Assyrians, Chinese or Indians. While mercury was prized above all other metals by the Indians, the date of its introduction into Indian medicine cannot be ascertained. Goldwater accepts that Hippocrates could have made use of cinnabar, but considers evidence of this is lacking.

Cinnabar was mined by the Etruscans until banned by the Roman Senate, possibly late in the second century BC, due to the damage caused to vineyards, olive groves and streams.[10] Pliny the Elder (AD 23–79), who gave a description of how mercury was obtained by roasting cinnabar, pointed out that the importation of the equivalent of 1000 kg of cinnabar from Spain each year had become so important to the Roman economy that the price had to be fixed by law.[11] Several of the prescriptions given by Celsus in his *De Medicina*, which was written around AD 30, contain cinnabar. In the applications listed it was employed as a purgative, which indicates internal administration. Cinnabar was also incorporated in salves for treatment of the eye disease trachoma and of genital ulceration, which could have been a venereal infection. In the following century, however, Galen warned that cinnabar was so poisonous that he would not even sanction its external application.

The appearance of syphilis in Europe at the end of the fifteenth century saw desperate measures being taken to cure this new plague. The old medicines failed miserably and healers and afflicted alike were prepared to turn in any direction to find a cure. It was against this background of near panic that mineral medicines such as mercury salves became established as a standard treatment for syphilis. The first book about syphilis, variously known as *Consillium in Morbum Gallicum, Consillium in Pustulas Malas* or *De Morbo Gallico*, was written by Conrad Schellig, a professor at Heidelberg, perhaps as early as 1488 and certainly no later than 1496.[9] The last page of his book warned that great caution must be observed when administering mercury by inunction. At least eight others before the year 1500 recommended the use of mercury in the treatment of syphilis, but its strongest advocate was Paracelsus. He wrote 11 books on syphilis in which he rejected the use of guaiac in favour of mercury.

The term 'syphilis' first appeared in a revised version of a poem written in 1530 by the Italian physician Girolamo Fracastoro (1478–1553), in which the shepherd Syphilus is striken with the disease. The victim is advised to use mercury ointment: 'Without hesitation, spread this mixture on your body and cover with it your entire skin. . . . Ten days in succession renew this treatment . . . very soon, you will feel the ferments of the disease dissolve themselves in your mouth in a disgusting flow of saliva. . . .'

This allusion to the flow of saliva reveals the rationale behind mercury inunction. Salivation was one of the many ways in which to expel harmful humours. This was clearly stated by the French surgeon, Ambroise Paré (1510–1590): 'Verily we must so long use frictions and inunctions, until the virulent humours be perfectly evacuated, by spitting and salivation, by stools, urine, sweat or insensible perspiration.'

The influential seventeenth century London physician Thomas Sydenham (1624–1689) was similarly convinced that mercury cured syphilis through its ability to induce a profuse flow of saliva. This was still held to be true even in the middle of the nineteenth century, when Jonathan Pereira, another renowned London physician, described it as '. . . the great value of mercurials'. It was only after Fritz Schaudinn and Erich Hoffmann in 1905 isolated the spirochaete which caused syphilis, *Treponema pallidum*, that the capacity of mercury inunction to induce salivation was finally recognised to be nothing more than an indication of mercurial intoxication. It is, however, quite plausible that if mercurial treatment had an effect, it may have required a toxic dose to produce any benefit, hence the observation was made that salivation was necessary before a cure could be effected. As to whether mercury inunction ever did cure syphilis, it is the opinion of Goldwater[9] that we shall never know since modern standards of clinical investigation cannot accept the reliability of the many records of patients who may have had syphilis and who may have been cured, especially since the spontaneous remission of the disease even without treatment was commonplace.

The influential *Pharmacopoeia Augustana*, published in Augsburg in 1564, included monographs on Unguentum Mercurium cum Theriaca, an ointment in which mercury was mixed with fat and other ingredients. However, in 1582 the Augsburg Senate included mercury and all its preparations in a list of substances that apothecaries were not permitted to dispense. Nevertheless, mercury later appeared in the pharmacopoeias of London in 1618, Brussels in 1671, Lille in 1694 and Valencia in 1698, reflecting its widespread acceptance in the treatment of syphilis by inunction. In the following century Thomas Dover acquired the name of 'the quicksilver doctor' after he published a monograph that hailed mercury as a panacea for all ailments. His lack of restraint produced an inevitable backlash, with controversy raging throughout the second half of the century. However, at the end of the century mercury was still being frequently prescribed, with Johann Friedrich Gmelin devoting an entire volume of his *Apparatus Medicaminum* to mercurial preparations. The nineteenth century saw even greater enthusiasm for mercury as it was prescribed not only for syphilis but also for testicular disorders, hepatic disease, plague, pleurisy, purpura, peritoneal inflammation, erysipelas, cholera, chronic diarrhoea, amenorrhoea, eczema, hydrophobia, chronic brain diseases, typhoid fever, amaurosis and inflammation. During the first half of the twentieth century alternatives became available to deal with all the established conditions for which mercury had previously been prescribed. Notwithstanding, ointments containing mercury appeared in all the major pharmacopoeias until the 1960s. The *British Pharmaceutical Codex* of 1959, for example, included strong mercury ointment containing 30% metallic mercury in a base of wool fat, white soft paraffin and beeswax. This kind of formulation had been the standard treatment for syphilis since mediaeval times, with patients smearing large areas of their body with it in the hope of a cure. Even after the introduction of the effective antisyphilitic drug arsphenamine in 1910, mercury ointments were widely used. That they remained in existence after the introduction of penicillin is disturbing.

In 1530, Paracelsus advised the city council of Nürnberg that mercury compounds should be taken in this way to treat syphilis. What became known as 'blue pills' were later widely used, especially after their composition was revealed by the pirate Hairedin Barbarossa to François I of France, who then published it. These pills remained popular until the end of the nineteenth century, and the *British Pharmaceutical Codex* of 1954 still included mercury with chalk tablets as a purgative for children. It was omitted from the next edition as a consequence of the belated recognition of the harmful effects of mercurous chloride in infants.

Paracelsus recommended as a purgative a preparation in which mercury coagulated with egg albumin was mixed with alum water and distilled to form a powder which was actually red mercuric oxide. This oxide is less chemically reactive than the more finely divided yellow oxide, which is the one that has generally been used for medicinal purposes. A popular eye ointment known as Pagenstecher's ointment was introduced in 1865, which contained 12.5% yellow mercuric oxide in spermaceti ointment. Variants on this appeared in most pharmacopoeias and were used to treat conjunctivitis, but they are no longer considered either safe or effective.

Mercurous chloride was discovered in the quest for a mercurial salt that would be less reactive than mercuric chloride (corrosive sublimate). Multhauf has suggested that this might represent the first product of deliberate chemical research.[1] The earliest written account of its preparation appeared in 1608 by Joannes Renodaeus (Jean de Renou) in his *Dispensatorium Medicum*, who precipitated it from a solution of mercury in nitric acid by the addition of a solution of common salt. A year later, Oswald Croll in his *Basilica Chymica* recommended its administration by mouth to produce a catharsis that would expunge harmful humours from the body. It was included in the *Pharmacopoeia Londinensis* of 1618 as mercurius dulcis. Later, it became known as calomel, derived from *kalos* (Greek for beautiful) and *melas* (Greek for black) on account of the formation of a black powder on the addition of caustic alkali. The term is believed to have been introduced by the royal physician Sir Theodore Turquet de Mayerne (1573–1655), an enthusiastic advocate for the use of chemical remedies. Calomel is a white, tasteless powder that is practically insoluble in water. It was considered to be an 'alterative' for the relief of a diverse range of diseases and, like mercury itself, was among the most frequently prescribed drugs in the nineteenth century.

Many thousands were poisoned by calomel before infant teething powders containing it were withdrawn from the British market in 1954 after it was realised they caused pink disease, characterised by pinkish colouration of the palms and soles, fever, insomnia, lethargy, loss of appetite and pain in the extremities. The mortality rate was about one in ten. It was previously believed the symptoms were caused by teething, a complaint for which calomel was recommended. One such product was Steedman's Powders®, which contained 26.3% mercurous chloride. In 1953, Joseph Warkany, a Cincinnati physician collaborating with Donald Hubbard of the nearby Kettering Laboratory, produced convincing evidence that it was the mercury in teething powders that caused pink disease and it was promptly withdrawn from the market.[12]

In the middle of the eighteenth century, Gerard Van Swieten (1700–1772) introduced the highly toxic mercuric chloride as a remedy for syphilis, employing a very dilute solution (0.1%) so that it could be taken by mouth. He had been curious as to whether the salivation produced by mercury treatment was a requisite part of the cure of syphilis. To settle the matter, he sought a mercurial preparation that could be diluted at will to allow administration of very small doses. After Robert Koch reported in 1881 that low concentrations of mercuric chloride were effective in destroying most of the micro-organisms that he had examined *in vitro*, Van Swieten's solution and other similar formulations became popular as disinfectants. Their activity was due to the capability of the mercuric ions to form insoluble complexes with proteins.

A variety of other inorganic water-soluble salts of mercury were also introduced into medicine. Ointments containing mercuric nitrate were used until the 1960s to treat skin conditions, including eczema, psoriasis and syphilitic warts. Dilute solutions of mercuric cyanide were considered to be less irritant than those of mercuric chloride, although there may well have been a parallel reduction in efficacy as a disinfectant. Another popular salt was red mercuric iodide, a naturally occurring mineral that can be rendered soluble by the addition of an equivalent weight of potassium iodide to form mercuric potassium iodide (K_2HgI_4). Gilbert's syrup contained 0.05% w/v mercuric iodide rendered soluble with potassium iodide for the treatment of syphilis. It remained in the *Pharmacopoiea Français* until 1949. Ointments incorporating up to 2% mercuric iodide were used in treating ringworm. Until the middle of the twentieth century, dilute solutions of mercuric oxycyanide were used to bathe the eye for conjunctivitis as they were considered to be less irritant than solutions of mercuric chloride.

The first organomercurials appeared in the late 1880s. Mercury benzoate, which was slightly soluble in water, was one of several such compounds that were introduced in an attempt to reduce the irritancy and toxicity of inorganic mercurials by obtaining a slow, sustained release of mercuric ions from the organic complex. This was one of a few approaches to drug design at that time, as typified by the introduction of drugs such as chloral hydrate and urethane, which had been designed to release chloroform and ethanol respectively. A number of injectable organomercurials were subsequently marketed, including merbaphen, a double salt of sodium mercurichlorophenyloxyacetate with barbitone that was introduced in 1912 by F. Bayer & Company for the treatment of syphilis. Seven years later a novel use was to be found for it after Arthur Vogl, a third year medical student at the Wenckebach Clinic in Vienna's First Medical University, had ordered a 10% solution of mercury salicylate to be prepared by the hospital pharmacy.[13] When told that such a solution could only be prepared as an oily injection, Vogl accepted a colleague's offer of merbaphen injection and duly proceeded with the treatment. As the nursing staff were meticulous in maintaining records of their patients' progress, the amount of urine voided by every patient was routinely entered on a chart. Vogl saw at once that after receiving her first injection of merbaphen, his patient's 24 hour urine output increased to 1200 ml from a mere 200–500 ml noted previously. After the third daily injection, this increased to 2 litres! On interruption of the treatment for a few days, the urine outflow decreased, only to rise on resumption of the injections. Vogl then administered merbaphen to another syphilitic patient, in whom the disease had damaged the heart, causing advanced congestive heart failure. Conventional diuretics had been tried on this man without success. Vogl found that after receiving the injection, his patient passed a massive amount of almost colourless urine. The flow continued throughout the day and the night and by the next morning he had eliminated over 10 litres. A similar profound diuresis was produced in other patients when injected with merbaphen, but not when they were given other antisyphilitic mercurials which were found to lack the potent activity of merbaphen. A thorough clinical evaluation was then carried out by Paul Saxl.[14] This transformed the treatment of the severe oedema of congestive heart failure. By removing so much fluid, the pressure on the heart was relieved, allowing it to recover normal function. None of the many substances previously used as diuretics had anything like comparable activity. Nevertheless, it was soon recognised that merbaphen injections posed a risk of severe kidney damage or even fatal colitis. It was quickly superseded by another antisyphilitic agent, mersalyl, which was administered on an intermittent schedule to minimise toxicity.

mersalyl

The diuretic action of mercury was originally observed long before its rediscovery in the Wenckebach Clinic. It had been employed by Paracelsus in his treatment of dropsy, but it was not until the late nineteenth century that mercury and mercurous chloride were introduced as oral diuretics. Tablets combining finely dispersed metallic mercury with digitalis (a cardiac stimulant) remained on the market until rendered obsolete by the introduction of the thiazide diuretics in the 1950s.

Ethacrynic acid

In the 1950s, a novel approach to the problem of designing a non-toxic, potent diuretic was taken by a team of researchers from Sharp & Dohme, under the direction of Karl Beyer. Recent developments in renal physiology had convinced him that the moment was opportune for a fresh attempt to design a safe, effective diuretic. By then, it was known that reabsorption of water from the glomerular filtrate inside the lumen of the kidney tubules depended largely on the efflux of sodium ions across the tubule wall. It was believed that mercurial diuretics probably interfered with this movement of ions by inhibiting dehydrogenase enzymes inside the cells of the tubules. The ability of mercury compounds to bind to dehydrogenase enzymes (presumed to involve reaction with thiol groups) was readily demonstrable. This convinced the Sharp & Dohme scientists that it would be worth trying to design mercury-free inhibitors of dehydrogenases in order to avoid the toxic effects of mersalyl. It proved to be a daunting task that took many years of intensive research. The turning point was the realisation that both merbaphen and mersalyl possessed a phenoxyacetic acid moiety. When an unsaturated ketone was attached to the 4-position of the benzene ring, potent hydrogenase inhibitors were obtained. Analysis of the influence of additional substituents revealed that chlorine or methyl groups attached to the benzene ring further enhanced potency. Ultimately, ethacrynic acid emerged as a safe, orally active diuretic in 1962, five years after a separate group of Sharp & Dohme chemists had announced their discovery of the thiazide diuretics with similar properties.[15]

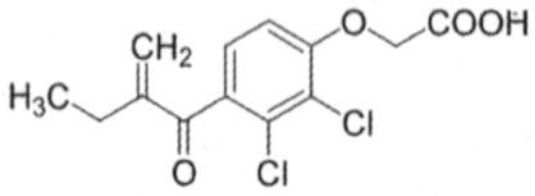

ethacrynic acid

Ethacrynic acid can trace its medicinal chemical ancestry all the way back to the introduction of pure mercury in the sixteenth century. Using the terminology previously promoted by the present writer, mercury can be considered as one of the oldest of all drug prototypes.[16] The majority of drug prototypes come from the animal and vegetable kingdoms; mercury belongs to a minority that derive from the mineral kingdom. The development of many more therapeutic drugs from a single prototype is a theme that appears repeatedly in this book.

ARSENIC

Arsenic trisulfide (Greek: *arsenikon* = potent), orpiment, is a yellow ore that has been used since ancient times as a pigment. Another pigment was arsenic disulfide, red orpiment or realgar, which was used to dye hair, but it was arsenic trioxide that was the first arsenic compound to become established for medicinal use. It was originally obtained 5000 years ago as a by-product from the smelting of copper. There is, however, no conclusive evidence of the medicinal use of arsenic trioxide before the eleventh century, when ibn Sina recommended it for cancer, both internally and topically.[17] From then onwards, salves containing arsenic formed the basis of popular remedies for cancer. While Paracelsus believed it to be effective in the treatment of cancer as well as ulcers and wounds, he warned that it was too toxic for anything other than external use.[18] He also fused arsenic with saltpetre to form potassium arsenate, which was dissolved in alcoholic solution to form Liquor Arsenici Fixi, a preparation later recommended by Jan Baptista Van Helmont in his *Ortus Medicinae* published in 1648. Van Helmont warned that this should only be used for external application. A similar preparation was introduced in the following century by Thomas Fowler, a medical practitioner in Stafford who published his *Medical Report of the Effects of Arsenic in Cases of Agues, Remittant Fevers, and Periodical Headaches* in 1786. He had carried out an investigation into the value of arsenic in the treatment of malaria as a consequence of his favourable results with a medicine popular in Lincolnshire, known as the Tasteless Ague Drops. Throughout the nineteenth century, Fowler's Solution was widely accepted as a less reliable alternative to quinine for malaria, as well as for skin diseases, chorea, dropsy, hydrophobia and glandular obstructions.[19] In the middle of the century, exaggerated reports that the inhabitants of Upper Styria in the Austrian Tyrrol kept themselves in good health by sprinkling their food with small amounts of arsenic compounds led to the widespread use of it by the Victorians as a tonic.[20] Even in the 1940s, Fowler's Solution was being prescribed as a tonic to treat pernicious anaemia. This was probably inspired by the heightened colouring of the cheeks arising from an increased fragility of the blood capillaries consequent upon chronic arsenic poisoning!

In 1865, Heinrich Lissauer of Breslau reported in the *Berliner Klinische Wochenschrift* that he had ameliorated the condition of a young woman with acute leukaemia.[21] She had presented with enlarged spleen and liver, together with a large number of white cells in her blood. Her disease was clearly advanced. Acting on the recommendation of a colleague, Lissauer administered Fowler's Solution as a tonic, hoping it would improve his patient's appearance. Her general condition improved dramatically, with the result that she was discharged from hospital five months later, only to die soon after. Lissauer disappeared from the scene, but the use of Fowler's Solution in leukaemia persisted until after the introduction of the first modern cytotoxic drugs in the 1940s. In 1931, its use in chronic myeloid leukaemia was described.[22] Reports from China of its intravenous administration to achieve remissions in patients with acute promyelocytic leukaemia appeared in the 1990s and were followed up by clinical investigations in the United States.[23] In 2001, FDA approval was granted for a proprietary formulation of Fowler's Solution by injection in promyelocytic leukaemia.[24]

Arsphenamine

In 1858 David Livingstone, the Scottish medical missionary who explored much of Central Africa, recommended Fowler's Solution for the alleviation of the symptoms of sleeping sickness.[25] This was to take on greater significance after David Bruce demonstrated in 1896 that trypanosomes were present in the blood of animals afflicted with nagana, a form of sleeping sickness, and could be temporarily eliminated by administering Fowler's Solution. Trypanosomes were protozoa somewhat larger than red blood corpuscles. Several varieties were subsequently shown to be responsible for what had previously been classified as 80 different tropical diseases affecting both men and animals. The most devastating of these in humans was sleeping sickness, caused by *Trypanosoma brucei* infection carried by blood-sucking tse-tse flies. In one recorded instance, the mortality had risen from 13 to 73% between the years 1896 and 1900. There were even fears that the disease could depopulate the whole of Central Africa, where it was being transmitted by the tse-tse fly, especially after an epidemic in Uganda killed 250 000 people in the first two decades of the twentieth century. Attempts to prepare vaccines proved futile.

In 1899, Alfred Lingard at Muktesar in India tried unsuccessfully to cure horses infected with the form of trypanosomiasis known as 'surra' by administering Fowler's Solution. Three years later, Alphonse Laveran and Felix Mesnil at the Pasteur Institute announced that they had been able to infect laboratory mice and rats with two varieties of the causative organism of trypanosomiasis. Subcutaneous injections of the sodium salt of arsenious acid into the infected rodents resulted in a rapid disappearance of trypanosomes from their blood, but the parasites reappeared within a few days to cause the death of the animals.[26]

Antoine Béchamp synthesised the sodium salt of the *meta*-anilide of phenylarsonic acid in 1863. At the beginning of the twentieth century it was marketed by the Vereinigte Chemische Werke in Charlottenburg with the unsubstantiated claim that it was 40–50 times less toxic than any of the inorganic arsenicals previously used. On this basis it was given the overly optimistic name of Atoxyl®.[28] The director of the factory at Charlottenburg, Ludwig Darmstädter, had been particularly interested in finding a cure for cancer and had sponsored some of Paul Ehrlich's work in this field at the Institute for Experimental Therapy in Frankfurt. Whether Darmstädter sent Ehrlich a sample of Atoxyl® with a view to its being evaluated in cancer or in trypanosomiasis is not clear. Atoxyl® was, however, the first substance Ehrlich investigated after receiving cultures of trypanosomes from Edmond Nocard of the Pasteur Institute. Unfortunately, it proved to be inactive.[29] After this, Ehrlich began his investigation of chemotherapeutic dyes, which preoccupied him until 1905, when he was surprised to read a paper by Wolferstan Thomas, of the Liverpool School of Hygiene and Tropical Medicine, describing the success of Atoxyl® in the treatment of animals experimentally infected with trypanosomiasis.[30] Ehrlich at once reopened his investigation of Atoxyl® and confirmed the validity of the results obtained by Thomas. He then realised that his earlier study had been misleading because it had been carried out on isolated cultures of trypanosomes, rather than on infected animals. This implied that Atoxyl® either stimulated immunity or else had to be metabolically converted to an active form by the animal before any trypanocidal effect could be exerted. Meantime, the bacteriologist Robert Koch was asked by the German Sleeping Sickness Commission to evaluate Atoxyl® in East Africa, where the mortality from the disease was reaching alarming proportions. Koch established that a 500 mg injection of the drug could cause the disappearance of trypanosomes from the blood for up to eight hours. His recommendation was that effective treatment would require constant medication for six months, a protocol that would present the risk of blindness through damage to the optic nerve in up to 2% of patients. This convinced Ehrlich that it would be worth his while examining analogues of Atoxyl® to see whether the ratio of efficacy to toxicity could be improved. Thanks to the generosity of the widow of Frankfurt banker George Speyer, who had donated a million marks, and also to John D. Rockefeller, who supported Ehrlich's research, a chemotherapy institute was built alongside the existing Institute for Experimental Therapy. Called the Georg Speyer-Haus, it was officially opened in September 1906. There were excellent facilities for chemists to synthesise the analogues of Atoxyl® required by Ehrlich.

Ehrlich was fond of coining Latin aphorisms to express the complex ideas underlying his researches. To describe the principle guiding the design of the new chemotherapeutic agents he proposed to synthesise, he used the phrase *corpora non agunt nisi fixata*. This was his way of explaining that infecting organisms are not killed unless the chemotherapeutic agent has a high affinity for them. Ehrlich termed such drugs parasitotropic. Their toxic effects he attributed to their also being, to varying extents, organotropic. His ideal agent would have a high therapeutic index because it exhibited a high parasitotropic activity and a low organotropic activity. The therapeutic index could be measured simply by comparing both the curative and lethal doses of the test compounds in mice. Around the time he moved into the Georg Speyer-Haus, he began to accept the ideas on the existence of drug receptors newly propounded by the Cambridge physiologist John Langley. Ehrlich then advanced the hypothesis that the parasitotropic action of arsenicals was due to their binding to arsenoceptors on the surface of the parasites, these being specific types of chemoceptors. It was his fervent hope that he could develop drugs that, like the antibodies discovered by his former colleague Emil von Behring, would act like 'magic bullets' insofar as they would bind only to receptors in the parasite and not the patient.

Initially, Ehrlich believed the scope for molecular modification would be limited since Atoxyl® was known to be the chemically unreactive anilide of arsenic acid. His unexpected discovery that the drug reacted with nitrous acid to form a diazonium salt led him to the realisation that the structure originally assigned to it by Béchamp was incorrect. Ehrlich correctly concluded that Atoxyl® was the sodium salt of 4-aminophenylarsonic acid, which meant that methods existed for preparing a wide range of its analogues.[31]

sodium salt of the *meta*-anilide of phenylarsonic acid

Atoxyl®

In the spring of 1907, Ehrlich signed a formal contract with the nearby Cassella Dye Works, granting them exclusive rights to the commercial exploitation of his arsenic compounds in return for their sponsorship of his research. The arrangement continued after the Hoechst Dyeworks took control of Cassella in 1908. Ehrlich's chief chemist, Alfred Bertheim, then synthesised a range of Atoxyl® derivatives with substituents on the amino group. One of these was the *N*-acetyl derivative of Atoxyl®, known as arsacetin. It was less toxic to mice, but when administered at the high dose level required to effect a cure it caused the mice to rotate in the manner of Japanese waltzing mice. This indicated damage to the vestibular nerve (associated with balance) similar to that produced by Atoxyl®, and it suggested that arsacetin would likewise cause blindness. Recalling his earlier work on lead poisoning, Ehrlich became convinced that this nerve damage was due to chronic arsenic poisoning caused by the large amount of arsenic that had been administered. In seeking a more potent series of arsenicals he hoped to avoid this particular problem. To this end chloro, hydroxy, cyano, sulfonic acid and amino groups were introduced into the benzene ring, but all exhibited their own particular types of toxicity over and above neurotoxicity. None of them exhibited any enhancement of potency.

The discovery that higher concentrations of phenylarsonic acids were required to kill test tube cultures of trypanosomes than could possibly be achieved when curing infected mice led Ehrlich to conclude that they were undergoing metabolic activation, presumably through a process of chemical reduction. To test this hypothesis, he asked Bertheim to prepare the two possible types of reduction products from phenylarsonic acids. One type, the arsenoxides, were highly toxic to trypanosomes, but their general toxicity to the tissues of the host was also high. They were also very irritant when injected, due to the presence of impurities. The second series, the arsenobenzenes, were not as potent, although still more potent than the phenylarsonic acids, and they were less toxic to the host. This meant that small doses could be given to avoid neurotoxicity from chronic arsenic poisoning. Ehrlich subsequently restricted his investigations to arsenobenzenes, unaware that the neurotoxicity of his phenylarsonic acids was mainly due to impurities.

Ehrlich and his contemporaries believed that the arsenobenzenes consisted of two fragments linked by a double bond between their respective arsenic atoms. It was only years later that this was shown to be incorrect by Kraft and his colleagues in Russia,[32] who established that the arsenobenzenes are polymeric mixtures formed from several molecules joined by single bonds between their arsenic atoms. Polymers cannot penetrate mammalian cells, which explains why the toxicity of arsenobenzenes was much weaker than that of the arsenoxides formed from only two molecules joined via arsenic–oxygen–arsenic linkages. The chemotherapeutic action of both arsenobenzenes and arsenoxides was shown in the 1920s to be due to the release of arsenite molecules which had only a single arsenic atom that reacted with thiol groups in receptors on the surface of the parasites.

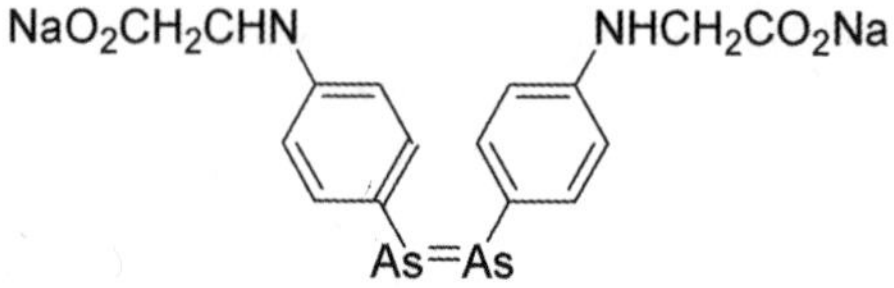

Ehrlich's structure for arsenophenylglycine

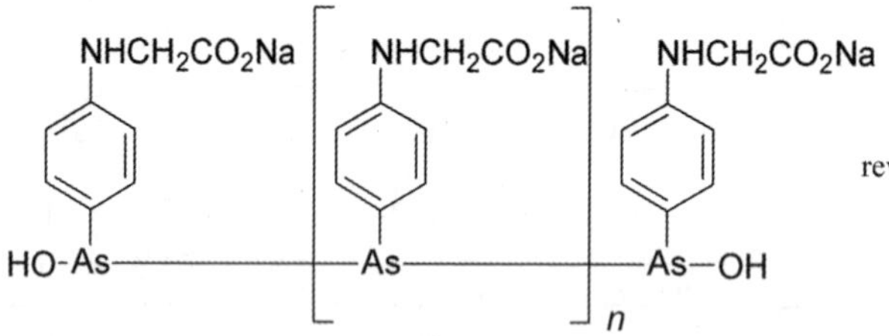

revised structure for arsenophenylglycine

Arsenophenylglycine (compound 418) emerged as a promising trypanocidal agent. It was administered to patients in 1907 and for many it proved safe and effective. Unfortunately, in a small number severe and often fatal hypersensitivity to it was exhibited. Despite this, some physicians deemed it acceptable to continue to use it for those forms of trypanosomiasis with a high mortality rate. Ehrlich strongly disagreed.

Insertion of a hydroxyl group on the 4-position of the benzene ring was readily achieved by warming diazotised Atoxyl® in water. Reduction of the product yielded arsenophenol. This was highly effective against trypanosomes, but was prone to oxidation to the severely irritant arsenoxide and exceedingly difficult to purify. However, Ehrlich's prior experience with chemotherapeutic dyes (see below) convinced him that the introduction of a substituent adjacent to a phenolic hydroxyl group enhanced chemotherapeutic activity. An arsenophenol substituted in this manner was synthesised by Bertheim in 1907 as potential trypanocidal agent 606.[31] The assistant who tested it told Ehrlich that it was ineffective, and consequently it remained on a shelf in the laboratory for over a year before being re-examined.

The isolation of the organism that caused syphilis, *Treponema pallidum*, was reported from the Reichsgesundheitsamt in Berlin by Fritz Schaudinn and Erich Hoffmann in 1905. The following year, Hoffmann visited the Georg Speyer-Haus and told Ehrlich that this spirochaete was in many ways similar to trypanosomes. He asked for samples of Ehrlich's new compounds for testing on syphilitic patients in his clinic at Bonn, where he had just been appointed to the chair of dermatology. Ehrlich agreed, urging great care when using the arsenicals. Shortly after this, Uhlenhuth found that Atoxyl® had a curative effect against syphilis in chickens, but was inferior to mercurials and could cause blindness through optic nerve damage.[33] Ehrlich, meantime, arranged for his friend Albert Neisser, director of the dermatology clinic at Breslau, to take arsacetin and arsenophenylglycine to Java, where he had managed to inoculate apes with *T. pallidum*. The results were published in 1908 and confirmed the efficacy of arsenophenylglycine.[34] This was the only way in which Ehrlich was then able to have his compounds tested on animals. The arrangement lasted until the spring of 1909, when Ehrlich was joined by Sacachiro Hata who had developed a method of infecting rabbits with syphilis while working in the Kitasato Institute in Tokyo. On his arrival, Hata was asked to test every arsenical that had been synthesised by Bertheim and his colleagues over the preceding three years. Working swiftly and with precision, he discovered that compound 606 had outstanding curative properties in the rabbits infected with syphilis! As a result, Ehrlich arranged for the Farbwerke Hoechst to apply for a patent on '606'; the application was submitted on 10 June 1909. Further details of the synthesis were published later.[35]

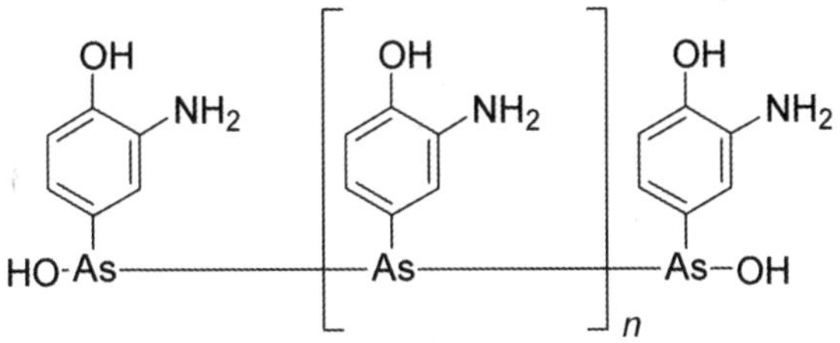

arsphenamine

Disappointment over the hypersensitivity reactions among patients receiving arseno-phenylglycine led Ehrlich to exercise great caution over the assessment of '606'. Once the results of animal studies confirmed that it was likely to be safe and effective, samples of '606' were sent to Iversen at the Obuchow Hospital for Men in St Petersburg and Alt at Uchtspringe (Altmark). Iversen was the first to administer the drug to patients when he successfully treated relapsing fever, a disease transmitted by lice. The causative organism, _Borrelia recurrentis_, was similar to the _T. pallidum_. Alt, on the other hand, spent three months evaluating the safety of '606' in dogs before he would let two of his assistants volunteer to be injected with the new arsenical. Only after this was the drug administered to several syphilitic patients suffering from general paralysis of the insane. Once its relative safety had thus been confirmed, samples of '606' were given to Schreiber at Magdeburg, where trials were initiated on large numbers of patients with primary syphilis. By January 1910, several other trusted investigators had received '606' for trials.[36] Ehrlich insisted on personally checking the records of every patient who received the drug, and it was not until the following April that he was prepared to make any public announcement about '606'.[37]

On 19 April 1910, Ehrlich told the Congress for Internal Medicine at Weisbaden of his work leading up to the discovery of '606'. Hata then described his experiments and Schreiber explained how he had been able to cure syphilis with '606'. These announcements received enthusiastic applause from those present, and in the following days the press throughout the world carried headline reports of the new drug that could cure syphilis. The immediate result of this publicity was that the Georg Speyer-Haus was besieged by physicians and patients anxious to obtain samples of '606'. Letters also poured in from around the world. Under this unprecedented pressure, Ehrlich modified his policy of supplying the drug only to a small circle of trusted colleagues. He agreed to provide the physicians, at no charge, with a small number of vials of the drug that had been prepared under his supervision, on the condition that he received full reports of every case treated. Between April and December 1910, some 65 000 vials of '606' were freely supplied in this manner while the necessary plant for its commercial production was being installed in the Farbwerke Hoechst. Production began there at the end of the year. '606' was then marketed under the proprietary name of Salvarsan®, the approved name given to it later being arsphenamine. To the general public it remained known as either '606' or 'Ehrlich–Hata 606'.

Many syphilitic patients were cured by a single intravenous 900 mg dose of arsphenamine. This injection was prepared from powdered arsphenamine hydrochloride in a sealed ampoule. Prior to being injected, this had to be treated with the correct amount of dilute alkali to form a solution containing the highly unstable sodium salt, a procedure that brought many complaints from physicians. As more experience was gained, Ehrlich realised repeated dosing was necessary.[38] Typically, 600 mg doses were repeated every few days until a total of 5 g of arsphenamine had been administered. Four to six weeks later, this was repeated and then another one or two similar courses of treatment were required. In all, each patient would receive 25–30 injections.

What had happened in the Georg Speyer-Hause in the spring and summer of 1909 was the culmination of a truly remarkable effort. Ehrlich's secretary, Martha Marquardt, has commented as follows:[39]

> "No outsider can ever realise the amount of work involved in these long hours of animal experiments, with treatments that had to be repeated and repeated for months on end. No one can grasp what meticulous care, what expenditure and amount of time were involved. To get some idea of it we must bear in mind that arsenophenylglycine had the number 418, Salvarsan® the number 606. This means that these two substances were the 418th and 606th of the preparations which Ehrlich worked out. People often, when writing or speaking about Ehrlich's work, refer to 606 as the 606th experiment that Ehrlich made. This is not correct, for 606 is the number of the substance with which, as with all the previous ones, very numerous animal experiments were made. The amount of detailed work which all this involved is beyond imagination."

Despite early setbacks mainly because of careless administration, arsphenamine analogues remained the standard treatment for syphilis until the end of the Second World War, when adequate supplies of penicillin became available. There can be no question whatsoever that arsphenamine was the first major chemotherapeutic agent and that Ehrlich rightly deserves to be remembered as the founder of modern chemotherapy.

Shortly after starting to produce arsphenamine, the Hoechst Dyeworks patented neoarsphenamine (Neosalvarsan®), a water-soluble derivative that was made less sensitive to oxidation by condensing some of the amino groups on the polymer with sodium sulfoxylate, an antioxidant widely used in the dyestuffs industry.[40] Trials revealed this to be less potent than arsphenamine. This worried Ehrlich, but he was reluctantly persuaded that it should be made available because of the preference of the medical profession for a preparation that required only the addition of water to the ampoule immediately before injection. Neo-arsphenamine ultimately superseded arsphenamine, usually being co-administered with either mercury ointment or oral mercury. It also became the standard treatment for anthrax until the introduction of antibiotics.[41]

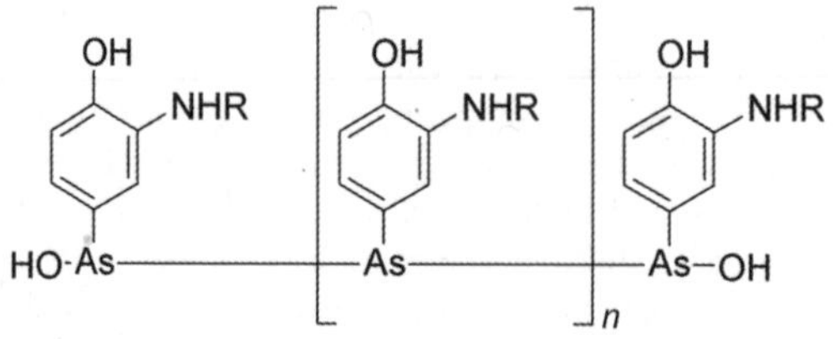

neoarsphenamine R=H or CH₂SO₂Na
sulfarsphenamine R=CH₂SO₂Na

Intramuscular injections of arsphenamine or neoarsphenamine produced pain. This could be avoided by the use of sulfarsphenamine, obtained by treating arsphenamine with an excess of sodium sulfoxylate.[42]

In 1919, Walter Jacobs and Michael Heidelberger at the Rockefeller Institute in New York introduced tryparsamide, an arsonic acid analogue of Atoxyl®.[43] It had the valuable property of being able to penetrate the central nervous system and combat the central spread of the trypanosomiasis. Unfortunately, it also caused optic nerve damage in some patients, but until the early 1960s it was the drug of choice for cases of human trypanosomiasis. The problem of resistance led to its abandonment apart from when used in the treatment of Gambian sleeping sickness.

The Rockefeller Foundation patented tryparsamide, but they issued licenses free of charge to manufacturers who wished to produce it. It was truly fitting that the Rockefeller Institute, founded in 1901 with an endowment of $200 000, should succeed in fulfilling Ehrlich's original objective only four years after his death. John D. Rockefeller had been one of the first backers of Ehrlich's work on arsenicals.

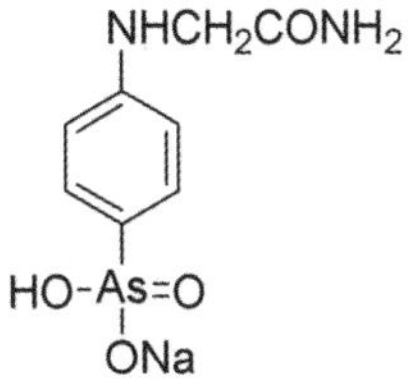

tryparsamide

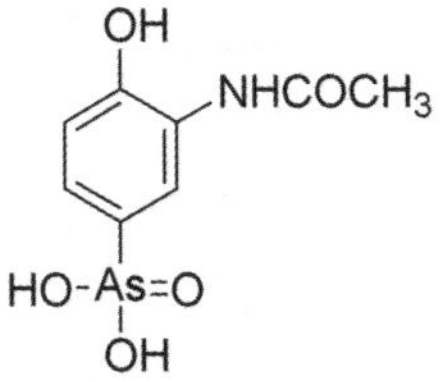

acetarsol

Ernest Fourneau at the Pasteur Institute questioned Ehrlich's belief that the neurotoxicity of phenylarsonic acids was due to the large doses required. Fourneau was convinced that the blindness caused by Atoxyl® was due to impurities, and this led him to examine thousands of phenylarsonic acids, from which acetarsol emerged as a valuable antisyphilitic and amoebicide.[44] Acetarsol had been used in 1911 as an intermediate in the preparation of some of Ehrlich's arsphenamine analogues. Fourneau arranged for it to be manufactured by Poulenc Freres under the proprietary name of Stovarsol®, a pun on his own name (French: *fourneau* = stove).

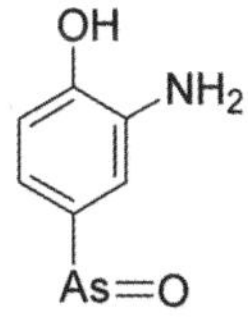

oxophenarsine

The product formed when solutions of arsphenamine were exposed to the air was found to be a highly potent antisyphilitic agent and possibly a safer drug than earlier arsenicals.[45] It was introduced as oxophenarsine and was welcomed by physicians in the United States because it overcame the problems associated with the administration of earlier arsenicals with regard to solubility, stability and local irritancy on injection. The reason for this can now be appreciated, namely that oxophenarsine was a pure compound rather than a polymeric mixture. Another advantage was that it could be chemically assayed instead of having to undergo bioassay for standardisation. One indication of its superiority over earlier arsenicals is that it was still included even in the 1963 edition of the *British Pharmacopoeia*.

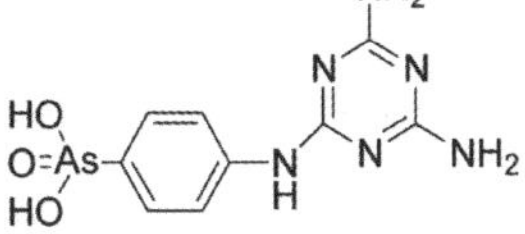

melarsen

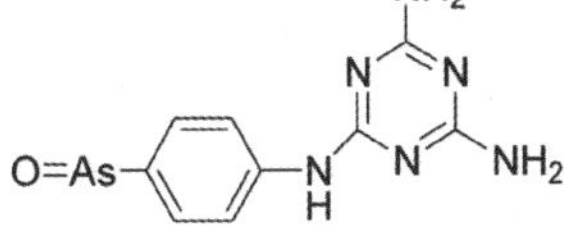

melarsen oxide

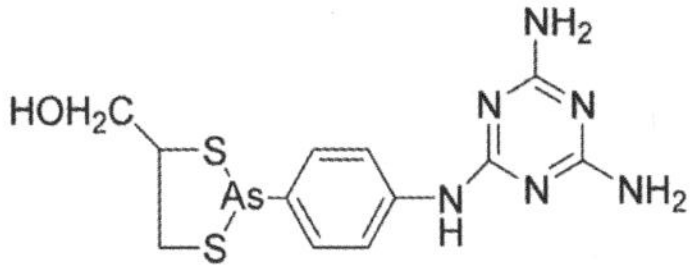

melarsoprol

Working in his private laboratory (originally his kitchen!), Swiss chemist E.A.H. Friedheim attempted to find a trypanocide that would not damage the optic nerve. Concluding that the most potent of the existing drugs all contained nitrogen atoms, he introduced a melamine moiety because of its high nitrogen content, leading to the introduction of melarsen in 1938.[46] It proved to be less toxic than existing drugs, but was much more expensive. Friedheim next prepared the trivalent analogue of melarsen, viz. melarsen oxide. This was more effective than melarsen, but also more toxic. Freidheim then tried to reduce its toxicity by reacting it with various thiols to mask the chemical reactivity of the trivalent arsenic moiety. Only after he received a supply of the arsenical poisoning antidote dimercaprol from the British Ministry of Supply was he able to prepare a safe analogue. This was melarsoprol (also known as Mel B), which was introduced in 1949 for the treatment of human African trypanosomiasis. It had a therapeutic index in mice infected with *Trypanosoma equiperdum* 40 times greater than that found for melarsen oxide.[47] Melarsoprol fulfilled its early promise and is still recommended by the World Health Organization for the treatment of human trypanosomiasis in combination with other drugs. Until recently, it was the only effective drug for advanced trypanosomiasis caused by *T. brucei gambiense* or *T. brucei rhodesiense*.

During the twentieth century, more than 12 500 arsenicals were synthesised in the quest for a safe trypanocide. Paracelsus would have been impressed. However, not all were made for the benefit of mankind. W. Lee Lewis of the US Chemical Warfare Service prepared a highly vesicant arsenical in 1918, known as Lewisite®, which produced its destructive effects by penetrating the skin and then hydrolysing to form an arsenite.[48]

Lewisite®   dimercaprol (BAL)

Fears that Lewisite® would be used by the enemy during the Second World War led the British Ministry of Supply to set up a team of scientists charged with the responsibility of finding an antidote. Based in the department of biochemistry at Oxford and led by Rudolf Peters, they established that arsenite reacted with two adjacent thiol groups on the vital pyruvate oxidase enzyme system. The discovery that this reaction could be competitively inhibited by the presence of simple sacrificial molecules that also contained two adjacent thiol groups ultimately resulted in the preparation of a highly effective antidote, namely dimercaprol (BAL, or British Anti-Lewisite).[49,50] It was a viscous liquid that could be applied to the skin in an ointment, or else be injected. This wartime episode constitutes one of the finest examples of rational drug design, and dimercaprol is still used as an antidote to poisoning by arsenic, mercury and other metallic poisons that react with thiol groups in the pyruvate oxidase system.

ANTIMONY

Red antimony trisulfide occurs naturally in stibnite, a mineral mentioned by Dioscorides and found in Hungary, France, Cornwall and Borneo. It appears in Arabic writings about eye diseases as well as in the *Liber de Gradibus* compiled in the eleventh century by Constantinus Africanus,[51] who introduced the term 'antimony'. The medicinal use of stibnite was, however, limited by its insolubility in water. Paracelsus was largely responsible for the popularity of antimony trisulfide and other antimonial preparations from the sixteenth century onwards. He distilled the trisulfide from stibnite and recommended ointments and balms containing it for treatment of wounds, ulcers, leprosy and other skin ailments.[52] Shortly before his death, the metallurgical text by Vannoccio Biringuccio, entitled *De la Pirotechnia*, mentioned that a metal could also be formed from stibnite, although instructions for making this did not appear until 1571 in *De Secretis Antimonii* by Alexander von Suchten. When finely divided, metallic antimony formed a slightly water-soluble oxide that could be taken by mouth as a purgative. It caused a number of deaths in Paris, provoking the Faculty of Paris in 1566 to persuade the Parlement to ban its use.[53] The University of Heidelberg did likewise, but those who sought to evade these bans drank from cups of metallic antimony which, if filled with acidic white wine for 24 hours, could deliver a fatal dose of antimony tartrate.[54]

Antimony and its compounds were hailed as a valuable remedy for syphilis, melancholy, chest pains, fevers and the plague in the *Triumphal Chariot of Antimony*, published in 1604 under the alias of Basil Valentine, supposedly a fourteenth century Benedictine monk but probably Johann Tholde (fl. 1595–1625). It had considerable influence, providing details of how to make a variety of antimony preparations. A gathering swell of opinion in favour of antimony ensued, which made the ban on it increasingly irrelevant. In the first half of the seventeenth century, Charles de Lorme grew rich and famous by prescribing antimony preparations for Henry IV, Louis XIII, Cardinal Mazarin and Madame de Sévigné.[54] Eventually, the Parlement of Paris not only withdrew the ban in 1665, but voted by a large majority to recommend antimony potassium tartrate and other compounds as valuable medicines.[55]

Antimony potassium tartrate was described by the German physician Adrian von Mynsicht in his *Medico-Chemical Treatise and Arsenal* of 1631. It became known as tartar emetic since a dose of 65 mg produced violent emesis and sweating. It was prescribed for fever on the basis that humoral imbalances were being corrected, reflecting the views of those who accepted chemical remedies without abandoning humoral medicine. It was used for this purpose until the introduction of modern antipyretic drugs in the late nineteenth century. A particularly successful product introduced in the mid-eighteenth century was known as 'Dr James' Fever Powders', which combined the alleged virtues of antimony with those of mercury and remained on the market for over one hundred years. Another popular use of tartar emetic was as an external application when mixed with lard or in a plaster. This produced a burning sensation, followed by the appearance of a characteristic pustular eruption. It was thought to be beneficial in the treatment of severe ulcers. There was also a wider application, as can be seen from a letter written by Jenner (of vaccination fame) in 1822 which advocated the inunction of the shaven head with a tartar emetic ointment as a cure for insanity.[56]

During the nineteenth century the popularity of antimonial drugs declined, but a reversal of this occurred at the start of the twentieth century after Alphonse Laveran demonstrated the ability of Fowler's Solution (dilute arsenic trioxide) to kill trypanosomes in experimentally infected mice, indicating a possible role in the treatment of sleeping sickness. His colleagues at the Pasteur Institute, Mesnil and Brimont, went on to demonstrate the temporary disappearance of trypanosomes from the blood of experimentally infected mice.[57] They proceeded to show that when cattle infected with trypanosomiasis were injected intravenously with antimony potassium tartrate, a cure could be achieved if large doses were given. Patrick Manson, the Scottish physician who demonstrated that some tropical diseases were transmitted by insects, then cured several patients with repeated injections of antimony potassium tartrate. There were severe side effects which were later attributed to the large quantity of saline administered with the intravenous injections.[58] The need for prolonged treatment was confirmed when Jean Kérandel, a French physician in the Congo who had become infected, started to treat himself.[59] The consensus that emerged was that antimony potassium tartrate was too toxic to be administered to patients for the prolonged schedule of injections that would be required. The subsequent development of safer, less irritant organic antimonial drugs also failed to provide anything to rival suramin or the organic arsenicals.

After detecting a hitherto unrecognised protozoon, *Leishmania braziliensis*, in patients afflicted with a skin disease known locally as *espundia*, a young Brazilian physician, Gaspar de Oliveira Vianna, tested antimony potassium tartrate. In 1912, he reported that the injections cured the disease.[60] Antimony potassium tartrate remained the standard treatment for cutaneous leishmaniasis until superseded by sodium stibogluconate, another antimonial, but leishmaniasis remains one of the most widespread of all communicable diseases in the world. A common form of it is kala-azar, a visceral disease that affects millions of children in Africa, Central and South America, China, India, the Middle East and parts of Russia. Without treatment the mortality rate is in the order of 90%. In 1915, di Cristina and Caronia successfully treated children in Sicily by injecting antimony potassium tartrate.[61] The intravenous injections caused irritancy and, on occasion, sudden death, but they have led to a reduction of the mortality rate from kala-azar to around 10%, saving the lives of millions of children. Few other drugs have ever made such impact.

In 1915, James McDonagh found schistosomiasis (bilharzia) also responded to antimony potassium tartrate therapy.[62] His findings were overlooked until three years later when John Christorpherson in the Civil Hospital at Khartoum noticed that tartar emetic acted against schistosomiasis in a patient being treated for leishmaniasis.[63] McDonagh then published a paper in the *Lancet* shortly after that by Christorpherson which had been published in the same journal.[64] Treatment with antimony potassium tartrate on a large scale began in Egypt in 1919 and antimonials became the standard therapy for schistosomiasis until the introduction of praziquantel. Schistosomiasis still affects the lives of 200 million people, mainly in Egypt. The parasite is transmitted by fresh-water snails.

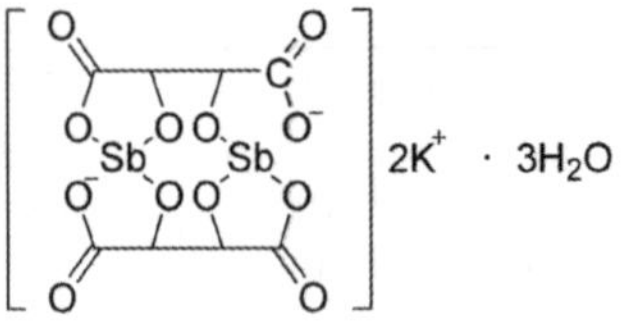

antimony potassium tartrate

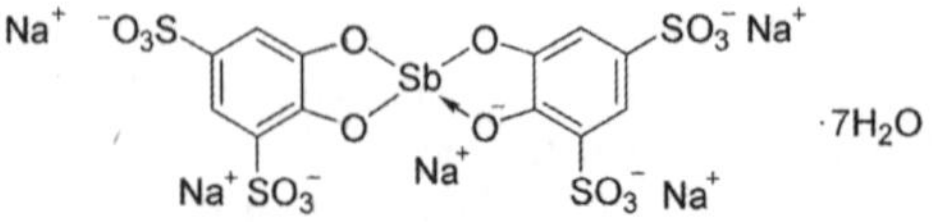

stibophen

Antimony potassium tartrate had major drawbacks, especially the need for prolonged treatment and its serious side effects. This drove forward research into the development of safer, less-irritant analogues by Hans Schmidt at the von Heyden Chemical Works in Radebeul, Germany, for testing at the University of Freiberg by Paul Uhlenhuth who had previously found sodium 4-aminobenzenestibonate, the antimony analogue of Atoxyl®, to be active *in vitro* but chemically unstable.[65] The first successful product of this collaboration between Uhlenhuth and Schmidt was stibenyl, the antimonial equivalent of Ehrlich's arsacetin. Stibenyl was introduced into the clinic in 1915 as it appeared to be less toxic than antimony potassium tartrate. It removed trypanosomes from the blood of infected patients for several months, but relapses invariably occurred. However, it was successfully used in treating kala-azar in children until superseded by stibophen, which had been developed by Uhlenhuth and Schmidt because they believed that the severe irritancy of antimony potassium tartrate was due to breakdown of the complex to release antimony oxide.[66] They had set about preparing stronger complexes, resulting in the marketing in 1924 of the potassium salt of stibophen. Stibophen was re-formulated five years later as the less irritant sodium salt. It was astutely marketed under the brand name of Fouadin® to impress King Fouad of Egypt, a country where schistosomiasis was, and remains, a major health problem causing debility in millions of people. Schmidt also synthesised sodium stibogluconate, a pentavalent antimonial in which gluconic acid replaced tartaric acid, which was evaluated in the IG Farben laboratories at Elberfeld in 1937.[67] It is now considered to be the drug of choice in all types of leishmaniasis, being rivalled by only meglumine antimonate, introduced by Rhône-Poulenc in 1946.[68] This is a pentavalent antimonial in which tartaric acid has been replaced by an aminosugar derived from glucose. It is one of the most effective drugs currently available for the treatment of kala-azar and other forms of leishmaniasis.

BISMUTH

Until the seventeenth century, bismuth was confused with antimony, lead and especially tin.[69] Claude Geoffrey the Younger, a French nobleman, demonstrated that it was a unique substance in 1753. Several salts were then introduced into medicine, their structures being uncertain as they vary with the method of preparation.

Bismuth subnitrate was introduced into medicine by Louis Odier, a Geneva physician (1748–1817). Administered for the relief of irritated mucous membranes and also as an antacid, it was thought not be absorbed from the gut. Bismuth subcarbonate[70] was introduced in 1857 and the subsalicylate followed in the first decade of the twentieth century.[71] The latter decomposed in the stomach, forming salicylic acid and insoluble bismuth salts. Such salts are now known to form a protective barrier to attack by gastric acid on ulcers.

Tripotassium dicitratobismuthate is a colloidal mixture of components.[72] The mixture is soluble in water, but precipitates below pH 5 to form bismuth oxide and bismuth citrate. When administered by mouth, it precipitates in the stomach to form a complex with glycoprotein that protects exposed ulcers from attack by gastric acid and pepsin. More recently, ranitidine bismuth citrate has been introduced for the treatment of ulcers, combining the effects of ranitidine with bismuth.

Bismuth salts were tested during the eighteenth century in patients with syphilis and gonorrhoea, and again in 1889 by Balzer.[73] After they had been found to exert bactericidal activity against *Spirochaeta gallinarum* in chickens,[74] Constantin Levaditi and Robert Sazerac at the Pasteur Institute then introduced parenteral bismuth therapy for the treatment of syphilis in the 1920s, but found this less potent than arsenicals and more toxic.[75] The best results were obtained with lipophilic preparations formulated in oil, such as bismuth ethylcamphorate.[76] Oily injections provided effective depot medication by giving a slow, prolonged release of bismuth from the site of injection. Such formulations were frequently prescribed in combination with arsenicals, until they were rendered obsolete by penicillin.

GOLD

Gold is mentioned as a medicine in both the *Ebers Papyrus* and the Hippocratic writings,[77] but it was only after the mineral acids were introduced in the fourteenth century that it became possible to prepare potable (drinkable) gold, *aurum potabile*, for internal administration by using soluble gold trichloride. Despite the inability to be absorbed from the gut into the circulation, it remained in frequent use as a medicine well into the twentieth century! While Paracelsus believed in the therapeutic value of potable gold, recommending it for paralysis, fevers, palpitations and various acute diseases,[78,79] it was Johann Glauber (1604–1668) who was responsible for its lasting popularity. His book, called *De Auri Tinctura, sive Auro Potabili Vero, quid Sit et Quommodo Differat ab auro Potabili Falso et Sophistico; quomodo Spagyrice Præparandum et Quomodo in Medicina Usurpandum*, was published in Amsterdam in 1651. This recommended gold trichloride for the relief of melancholy, epilepsy, sterility and uterine disorders.[80] Two centuries later, Jean André Chrestien at the University of Montpellier likewise recommended gold sodium chloride for syphilis[81] and, in addition, tuberculosis.[82] His views on the latter seemed vindicated when Robert Koch told an international congress in

Berlin in 1890 that gold–cyanide complexes were the most effective of all known antiseptics against the tuberculosis bacillus, *Mycobacterium tuberculosis*, being active *in vitro* at a dilution of one part in two million.[83] Unfortunately, the complexes could not cure infected animals.

At the Bacteriology Institute of the Hoechst Dyeworks in 1913, Adolf Feldt confirmed that it was the gold component rather than the cyanide that had been responsible for the toxicity of potassium gold–cyanide to mycobacteria.[84] This led to the introduction in 1917 of Krysolgan®, a complex of gold and the sodium salt of 4-amino-2-thiophenolcarbonic acid. A few years later Hoechst marketed the water-soluble sodium salt of aurothiobenzimidazole carbonic acid (Triphal®) for injection in the treatment of tuberculosis.[85] This was followed by another product from the same company in 1924, namely gold sodium thiosulfate (Sanocrysin®), a compound originally prepared in 1845.[86] It was found by Holger Molegaard in Copenhagen to have beneficial effects in cattle with tuberculosis[87]. Unfortunately, it was rather toxic in humans, but its use became firmly established until 1931, when a randomised, blind clinical trial using a matched control group was carried out in Detroit. It was the first such trial to be conducted in the United States and it revealed that in two groups of 12 patients the control group given injections of distilled water responded more favourably than those receiving the gold injections![88] This effectively put an end to the use of gold therapy for tuberculosis.

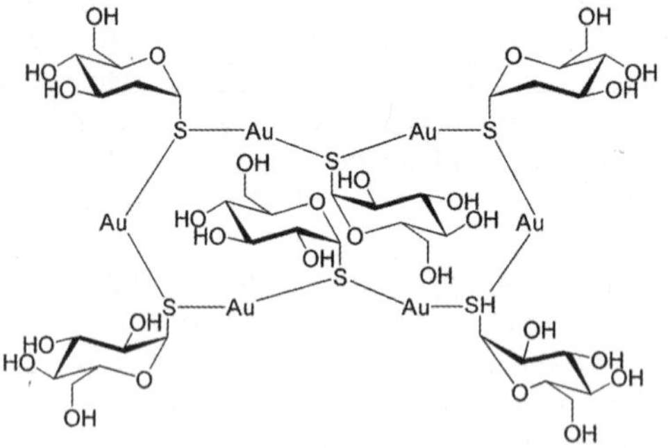

aurothioglucose

Most gold thiolates were foul smelling liquids that oxidised in air, but an exception was aurothioglucose (Solganal®), introduced by the Schering Corporation of Berlin. It was later shown to be hexameric in aqueous solution. Unlike other gold thiolates, it was formulated as a suspension in oil. The resultant slow release of the gold complex may account for its apparent reduced tendency to produce side effects when compared with other gold complexes. Landé in Germany administered it to 39 patients suffering from tuberculosis, bacterial endocarditis or a variety of complaints attributed to bacterial infection, but in most cases probably due to rheumatic fever.[89] He believed that gold complexes had a general antiseptic effect in addition to any possible antituberculosis activity. The most notable outcome of his ill-defined clinical trial was that joint pain was relieved in some of his patients who happened to have arthritis. This serendipitous observation led Landé to recommend a trial of the drug for arthritis. Jacques Forestier in Paris acted on this recommendation and began to administer once-weekly intramuscular injections of gold-thiopropanol sodium sulfonate (Allocrysine®) to patients with rheumatoid arthritis. Six years later he published a detailed report revealing that 70–80% of patients gained some benefit from the treatment, despite a risk of serious toxicity.[90] Forestier advised that a large-scale clinical trial of gold therapy for rheumatoid arthritis should be conducted, but the outbreak of war prevented this. It was not until 1945 that the results of a properly controlled, large-scale, clinical trial of gold therapy for arthritis were published. Fraser reported from the Western Infirmary in Glasgow that improvement occurred in 82% of 57 rheumatic patients given intramuscular injections of sodium aurothiomalate (Myochrysin®), as opposed to only 45% of 46 patients who were given dummy injections.[91] Such placebo responses to blank injections is commonplace, making evaluation of many drugs difficult. In the Glasgow trial neither patients nor their attendants knew if the injection contained the drug or not, making it a 'double-blind trial'. Only after the Glasgow trial was its value in therapeutics generally accepted. As a consequence of the pioneering work of Forestier and the subsequent clinical trials, gold therapy is now firmly established, with sodium aurothiomalate being the most frequently administered injectable drug.

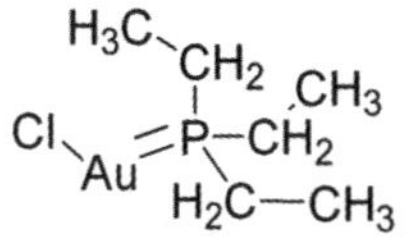

triethylphosphine gold chloride

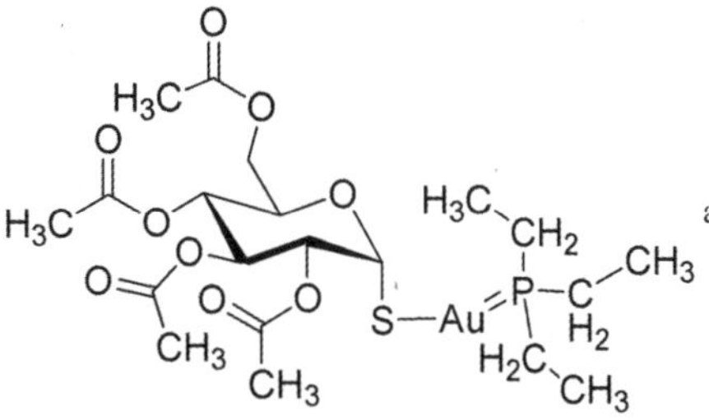

auranofin

The gold compounds that have been described have low lipophilicity and high molecular weights due to their polymeric nature; consequently they are not well absorbed from the gut. The lack of an orally active gold preparation imposed a major limitation on gold treatment for arthritis. Blaine Sutton, however, found that highly lipophilic alkylphosphine gold complexes exhibited anti-inflammatory activity in rats when orally administered. The most potent of these was triethylphosphine gold chloride, which was as potent as injected sodium aurothiomalate.[92] Unfortunately, it was highly toxic in humans, whereas auranofin, an acetylated glucose derivative of it, proved to be safe. Although somewhat less effective than injected gold thiolates, it was better tolerated by most patients. About 25% of an oral dose was absorbed from the gut, a reflection not only of its lipophilicity but also of its monomeric nature. The other gold complexes used in the treatment of rheumatoid arthritis are polymers, and as such cannot be absorbed from the gut.

SILVER

Metallic silver has seen little use as a medicine, although it was mentioned in both the *Ebers Papyrus* and in the Hippocratic writings.[93] Paracelsus recommended it for treating what he described as complaints of the cerebrum, spleen and liver.[94] The Dutch physician Franciscus de le Böe (1614–1672) applied silver nitrate (*crystallis lunae*) to wounds, sores and ulcers,[95] but it was in the following century that the topical use of fused silver nitrate, *lapis infernalis*, became established for this.[96] Throughout the nineteenth century silver nitrate was applied to fresh burns and to remove granulation tissue from burns. Solutions, ointments or application by a stick of fused silver nitrate were all in use at that time. In 1887, Emil von Behring established that silver ions had antibacterial properties and recommended a dilute solution for treatment of gonorrhoea by irrigation.[97] Six or seven years before this, the Leipzig obstetrician Carl Credé had begun instilling 2% silver nitrate eye drops to prevent gonococcal infection in new-born infants.[98] This became established practice until the introduction of antibiotics, as did the application of silver foil to wounds, a procedure recommended by the American surgeon William Halstead.

There was revived interest in the use of silver nitrate solution following the publication of a report in 1965 on its value in the treatment of extensive burns, by Carl Moyer and his colleagues at the Department of Surgery in the University of Washington. Meanwhile, Charles Fox at Columbia University College of Physicians and Surgeons had been testing scores of silver compounds over a nine year period before concluding that an ointment containing the silver salt of sulfadiazine was superior to silver nitrate in the treatment of burns infected with *Pseudomonas aeruginosa*.[99] He licensed the product to Marion Laboratories in 1969.

ZINC SULFATE

Zinc occurs in the form of various minerals and the metal may have been obtained in ancient times from naturally occurring carbonates or sulfides. The water-soluble sulfate was prepared by dissolving zinc in dilute sulfuric acid. It was administered in solution as an emetic by Franciscus de le Böe in the seventeenth century.[95] Two hundred years later it was considered to be the best emetic available for dealing with poisoning by narcotics as its action was rapid.[99] In recent years, the use of emetics to treat poisoning has declined due to lack of evidence of any real benefit.

Zinc oxide occurs in nature as the mineral zincite. As it is insoluble in water, it was initially only applied externally in ointments for its absorbent and dessicant properties. It is still used for this purpose. However, in 1771, it was reported that in Amsterdam the German astrologist and quack Johan Christoffel Ludeman (1683–1757) had been administering a secret panacea called *Luna fixata*.[100] Hieronumus David Gaubius, Boerhaave's successor as professor of chemistry at Leiden, witnessed Ludeman using it to achieve apparently miraculous cures of convulsions in children. Consequently, he also began to prescribe zinc oxide, but cautioned that further evaluation of the treatment was necessary. It was still being prescribed for epilepsy in the middle of the nineteenth century, but doses of 5 or 6 g per day were used instead of the 20–30 mg used by Glaubius. This ill-founded treatment for epilepsy eventually became obsolete once effective drugs were introduced.

The medicinal use of oral zinc sulfate to reduce tissue copper levels in patients with Wilson's disease was reported by G. Schouwink in 1961 in a thesis for the University of Amsterdam.[101] He stated that the zinc sulfate had induced excretion of copper in the stools, but his findings were never published. His work was later followed up in The Netherlands by Tjaard Hoogenraad and his associates.[102] However, it was a serendipitous observation by George Brewer and his colleagues at the University of Michigan Medical School that ultimately led to zinc sulfate becoming licensed for this purpose. On treating sickle cell anaemia patients with zinc it was discovered that this caused copper deficiency. They then began to treat patients with Wilson's disease with zinc sulfate, but it was not until 1997 that the FDA approved this.[103]

In patients with the potentially lethal genetic disorder known as acrodermatitis enteropathica, zinc absorption is defective and zinc-containing enzymes are inadequately synthesised. Zinc sulfate tablets are successfully given to treat this condition.[104]

LITHIUM CARBONATE

John Cade conducted an experiment in a psychiatric hospital in Melbourne in 1948 with the aim of obtaining evidence to support his conjectures as to the nature of manic-depressive illness.[105] He had observed that patients afflicted with thyroid disorders sometimes behaved in a similar manner to those suffering from manic-depression. Extreme hyperactivity of the thyroid seemed to cause a form of mania, while a marked absence of thyroid function could be correlated with depression. The issue addressed was whether manic-depression could be due to either over- or underproduction of a hormone. Cade then collected urine from manic patients, schizophrenics, melancholics, as well as from normal individuals. To test whether an active substance was present, the urines were concentrated and then injected into guinea pigs. If sufficient urine was injected, the animals convulsed, fell unconscious and died. Cade noted, however, that the urines from manic patients were up to three times as toxic as any others. He established that the toxicity was caused by urea, but as the amount of this in the manic urines did not differ much from that in normal urines it seemed possible that there might be something present that augmented the toxicity of urea. The first substance considered was uric acid, but Cade ran into difficulties when preparing solutions of this highly insoluble compound. He overcame this problem by using its most soluble salt, lithium urate. On injecting a saturated solution of this to the urea that had been added to test his hypothesis, Cade found the toxicity was unexpectedly low. A protective factor appeared to be present. Subsequent investigation confirmed that the lithium afforded protective action against the expected convulsions.

Cade next examined the effects of lithium carbonate injections. He found that after a 2 h delay, the guinea pigs became lethargic and unresponsive. Two hours or so later, they reverted to normal behaviour, unharmed by the injections. As lithium salts had been used during the nineteenth century to treat epilepsy, gout and cancer, Cade considered it safe to test both lithium citrate and carbonate on himself. Finding no harmful effects, he administered 1200 mg of the citrate thrice daily to a 51 year old patient who had been in a state of manic excitement for five years. After five days, there was an improvement in his condition, and by the time three weeks had elapsed he was well enough to be transferred to a convalescence ward. During his fourth month of continuous treatment, he was released from hospital with instructions to take 300 mg of lithium carbonate daily, this causing less nausea than the citrate. Such was the improvement that the patient returned to his former occupation. Unfortunately, he became negligent about taking his medication and after six weeks without treatment he was readmitted to hospital on account of his irritability. Cade achieved similar success with nine other cases, the most impressive results being with excited patients.[106] Cade's pioneering work was not widely recognised until 20 years later after it was put on a firm clinical foundation by the Danish investigators Mogens Schou and Poul Christian Baastrup.[107] Today, lithium is considered to be a valuable agent for the prophylaxis of manic-depressive illness. Since it is a rather toxic drug, careful monitoring of blood levels is required, especially as the amount of lithium ion that enters the circulation varies markedly between different brands containing the same nominal quantity of lithium carbonate.

PLATINUM

Platinous chloride ($PtCl_2$), which had been prepared by dissolving platinum in a mixture of nitric and hydrochloric acids, was tried in the treatment of secondary syphilis in the nineteenth century.[108] However, it was well over a hundred years later before platinum compounds became established as therapeutic agents.

In 1964, Barnett Rosenberg and his colleagues in the biophysics department at Michigan State University began to examine the effects of an electric current on a suspension of bacteria.[109] They detected no change in the growth rate, but did observe abnormal elongation of the *Escherichia coli* cells, suggestive of interference with cell division. After extensive investigations, they came to the surprising conclusion that this must have been due to an electrolysis product being formed in the culture medium. It was then established that the action of ammonium and chloride ions on the platinum electrode was responsible for the formation of platinum ions. When similar platinum salts with the *cis* configuration were added to bacterial cultures, they produced the same effects on the bacterial cells. Several platinum compounds were then tested against transplanted tumours and were found to inhibit growth.[110] The structural requirements for this activity were confirmed to be either a neutral bisamine or a primary amine complex with a *cis* configuration.[111] Clinical trials of a compound known as Peyronie's salt confirmed its activity against human tumours, with the outcome that this received the approved name of cisplatin and was licensed for use in 1978.

NH₃
Cl—Pt-NH₃   cisplatin
Cl

carboplatin

oxaliplatin

Cisplatin has become established in the treatment of malignant teratoma of the testis and ovarian cancer. Tumours of the bladder, head and neck also respond. Cisplatin acts as an alkylating agent, forming cross-links both between and within the strands of DNA in the nucleus of dividing cells.[112] A major problem has been its toxic effects on the kidneys and also the nausea and vomiting experienced by patients. The former has been ameliorated by administering cisplatin by slow intravenous infusion over 24 hours and fully hydrating the patient before and during treatment.[113]

Platinum complexes prepared at the Johnson Matthey Technology Centre in Reading, England, were evaluated by the Institute for Cancer Research in Sutton, England, and Bristol-Myers and the National Cancer Institute in the United States. The discovery by Michael Cleare and James Hoeschele in Reading that malonate–amine platinum complexes had reduced general toxicity led to a variety of these being examined for both their effects on the kidneys of small animals, as evidenced by measurement of blood urea nitrogen levels, and on the white cell counts.[114] From this it was apparent that the lack of chemical reactivity of the malonate ester complexes reduced their toxicity to the kidneys, albeit while increasing their activity against white cells. A total of about 1000 platinum complexes of all types had been screened by 1978, when seven were selected for detailed pharmacological evaluation on rats and parallel studies on their action against transplanted human lung carcinomas in mice. From these studies, carboplatin emerged as the only compound that was non-toxic at effective antitumour dose levels.[115] Clinical trials confirmed that carboplatin was not only less nephrotoxic than cisplatin but also less likely to cause nausea and vomiting.[116] It is administered by infusion for the treatment of advanced ovarian cancer and small-cell carcinoma of the lung. Oxaliplatin was synthesised in Nagoya City University, Japan, and evaluated at the Roger Bellon Laboratories in Neuilly, France, and then by Debiopharm in Lausanne, Switzerland.[117] It is used in colorectal cancer in combination with fluorouracil.

MINERAL SALTS

Magnesium sulfate was isolated from the medicinal waters at Epsom in the south of England by Nehemiah Grew at the end of the seventeenth century.[118] A factory was subsequently established at Shooter's Hill, London, ensuring a cheap price for the salts. By the time supplies had run out in the middle of the nineteenth century, Epsom salts had become established as the most widely used purgative in England.[119] Magnesium carbonate, which also occurs naturally, was commercially introduced at the beginning of the eighteenth century as an antacid and purgative. After Joseph Black distinguished magnesium oxide from lime in 1755, it also became established as an antacid and purgative.[120]

A rival product was developed by Johann Glauber who became ill in Vienna in 1624 and then recovered after drinking water from a local spring. Believing that this was due to a constituent in the spring water, he extracted a salt which he called *sal mirabile*. It consisted of hydrated sodium sulfate and was later called Glauber's salt and became a popular saline purgative.[121] A later product to enter the market was Seidlitz water (*aqua sedlitziana*), from a spring in Bohemia. In 1724, Professor Friedrich Hoffmann found it to have a similar composition to Epsom waters. The product was first marketed under the name of Seidlitz Powders by the London pharmacist Thomas Field Savory and did not even contain magnesium sulfate; nor was it prepared from Seidlitz waters. It consisted of two packets of salts, one containing tartaric acid and the other a mixture of sodium bicarbonate and sodium potassium tartrate. Upon mixing the contents of both in water, effervescence occurred. It was recommended for the treatment of a diverse range of complaints.[122]

ORAL REHYDRATION THERAPY

Fundamental advances in the understanding of intestinal transport took place in the 1960s. One of these was the demonstration by Stanley Schultz and Ralph Zalusky at the US Air Force School of Aerospace Medicine in Texas that sodium ions were co-transported when sugars were absorbed from the small intestine into the circulation.[123] This became the basis for the reintroduction of oral rehydration therapy (ORT) to replace body fluids lost by infants with diarrhoea. Harold Harrison had pioneered this treatment in Baltimore in 1945, but commercial products based on his work contained excess carbohydrate and caused too much sodium to be absorbed, with consequent complications.[124,125] ORT then fell out of use until after various formulations were evaluated in the 1960s. In 1978, the World Health Organization adopted ORT as its principal strategy for preventing the deaths of more than 4.5 million children from diarrhoea every year. The treatment was typically provided in sachets containing sodium chloride (3.5 g), trisodium citrate (2.9 g), potassium chloride (1.5 g) and glucose (20 g). The contents of a sachet were dissolved in 1 litre of clean water. Variants of this formula that could be readily prepared in remote villages were also devised. As a consequence of the introduction of ORT by United Nations agencies such as UNICEF, the annual deaths are estimated to have fallen to around 1.5 million in 1999. No other form of drug therapy has ever made such a major impact.

CHLORINATED DISINFECTANTS

Claude Berthollet, the famous French physician who taught Napoleon chemistry, having noted that an aqueous solution of the recently discovered chlorine gas prevented the foul odours of decaying organic matter, in 1789 recommended this as a disinfectant solution.[126] Later that year, a similar chlorine preparation, *Eau de Javelle* (named after the Paris suburb of Javelle), was marketed. In 1825, Antoine Labarraque, a Parisian apothecary, introduced the similar *Eau de Labarraque* for wound disinfection.[127] Two years later, Thomas Alcock did likewise for the purification of drinking water in the United Kingdom.[128] His solution was included in the *London Pharmacopoeia* of 1836 under the name *Liquor Sodae Chlorinatae*. By the middle of the century, solutions of both sodium and calcium hypochlorites were firmly established as disinfectants.[129]

One of the first attempts to control an outbreak of the frequently fatal childbirth fever was that by the Dublin obstetrician Robert Collins in 1829 – one year before Oliver Wendell Holmes, in America, asserted that this disease was contagious, being transmitted by midwives.[130] Holmes eventually was successful in stemming puerperal fever by insisting on hand disinfection with chloride of lime solution. When he reported his conclusions to the Boston Society of Medical Improvement in 1843, he met considerable hostility from many of its members who felt his views to be insulting.[131] Similar opposition was experienced four years later by Ignaz Semmelweis, an assistant at the Lying-in Hospital in Vienna, when he showed that the mortality from puerperal fever in the First Division maternity ward, where medical students were trained, was higher than that occurring in the Second Division ward, where nurses and midwives received instruction. He argued that this was due to the students infecting the patients with putrid particles adhering to their fingers after post-mortem examinations. By insisting that students wash their hands in chloride of lime solution, Semmelweis ultimately reduced the mortality rate in his wards from 12% to just over 1%. However, this did not satisfy the conservatism of some of his senior colleagues. They resented his interference and he became embroiled in bitter arguments, with the result that he was not reappointed when his post in the hospital was due to be renewed. He then had to move to Budapest, where his classical study, *Die Aetiologie, der Begriff und die Prophylaxis des Kindbettfiebers*, was published in 1861.[132] It included striking mortality statistics that should have convinced all but the most biased, yet it failed to win over some of his critics and he became insane through the aggravation this caused. In 1865, the very year that Lister first used carbolic acid to prevent wound sepsis, Semmelweis died from a septic wound on his finger.

Wider use of sodium hypochlorite solution as a wound disinfectant resulted from investigations by Henry Dakin, begun at the University of Leeds at the outset of the First World War.[133] For application to infected, open wounds, he recommended a 0.5% solution prepared by mixing chlorinated lime with sodium carbonate and boric acid. When this was reported to be too irritating for wounds, he replaced the boric acid with sodium bicarbonate. This formulation became known as Dakin's solution. Its effectiveness was never in doubt, but there were problems when it was applied to wounds and delicate tissues. Unless special precautions were taken, both free alkali and chlorine were present, resulting in tissue irritation. This problem was overcome after further work was done by Dakin. He was aware that hypochlorite solution reacted with free amino groups such as are found in proteins. On examining the products, known as chloramines, he found these to have antiseptic properties. Their instability precluded clinical use, but Dakin and Cohen were able to exploit certain aromatic chloramines previously prepared by Chattaway at St Bartholomew's Hospital, London, as these formed stable water-soluble sodium salts.[134] From the Berter Laboratory at the Rockefeller Institute in New York, Dakin reported that chloramine-T was a powerful antiseptic that possessed all the advantages of hypochlorite solutions, but with lower irritancy.[135] It acted by slowly releasing hypochlorous acid over a prolonged period, this rendering it highly efficacious in wound disinfection.

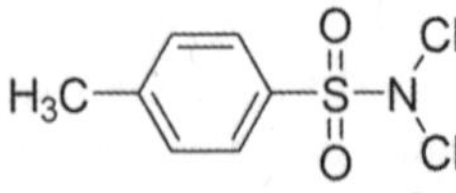

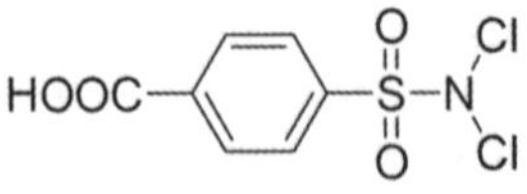

In 1917, Abbott Laboratories in Chicago marketed halazone, which Dakin had developed to meet the wartime demand for tablets to sterilise drinking water.

IODINE COMPOUNDS

Iodine occurs in seaweed as well as various minerals. It was isolated in 1811 by Bernard Courtois, a saltpetre manufacturer in Paris.[136] He observed the violet fumes of iodine when treating seaweed (a substitute for willow which had become unavailable during the naval blockade during the Napoleonic Wars) with sulfuric acid while attempting to isolate nitrates for the manufacture of explosives. Joseph Gay-Lussac then proved that iodine was an element.[137]

Iodine was introduced into medicine in 1820 in Geneva by Jean François Coindet for patients with goitre after he had found it to be the only ingredient likely to account for any beneficial action of burnt sea-sponge, an old remedy for goitre that had been used since the middle ages.[138] Coindet treated over 150 patients, but toxicity due to overdosing became a problem which, coupled with the general ineffectiveness of the therapy in patients with long-standing goitre, led to its eventual abandonment. In 1846, his townsman, Prevost, claimed that a local dietary deficiency of iodine might be the cause of goitre and cretinism. The obvious way to settle this matter was to carry out an analysis of iodine in food and water, but suitable analytical techniques did not become available until the middle of the century. Even then, these lacked refinement and the issue was not settled until after the First World War, when David Marine in the United States finally confirmed Prevost's suspicions.[139] This led to the introduction of iodine supplements for inhabitants of goitrogenic areas.

It was not until 1896, when Eugen Baumann at the University of Freiburg detected iodine in the thyroid gland, that its role in the body became evident.[140] Prior to this, there was no clear view as to what limit there was to the effects it had on the body. It had been assumed that the alleviation of goitre was due to a general stimulant action on all the glands of the body, which explains how in the nineteenth century iodine came to be used both in glandular diseases and as an 'alterative' to treat a wide variety of other diseases. It was as frequently administered to treat tuberculosis as it was for goitre.[141] To cater for all these uses, the French physician Jean Lugol introduced an aqueous solution of iodine in 1829, rendering the iodine sufficiently soluble by the addition of two parts of potassium iodide for each of iodine. He used this preparation (subsequently known as Lugol's iodine) during his clinical studies of 1829–1832. It became universally accepted as a way of administering iodine by mouth as an alternative to tincture of iodine, which was an irritant and caused a stinging sensation.[142] Iodoform, which had been synthesised by G.S. Srullas in 1822, was recommended as an acceptable form of iodine for internal administration in goitre by a Newcastle physician, Robert Glover, in 1847.[143] Another palatable alternative, especially if large doses were to be administered, was developed in Glasgow by Buchanan, who combined iodine with starch in a preparation that was introduced into the pharmacopoeias under the name of iodised starch.[144]

In 1839, Davies, a physician in Hertford, used an iodine solution to disinfect wounds, a purpose for which it was later employed in the American Civil War.[145] Nevertheless, iodine disinfection failed to gain general acceptance until the 1870s, when the French bacteriologist Casimir Davaine confirmed that iodine solutions could kill a wide variety of micro-organisms.[146] In 1878, iodoform was introduced as a hospital disinfectant by Apollinaire Bouchardat, the chief pharmacist at the Hôtel-Dieu in Paris.[147] The mild disinfectant action, as well as any toxicity after absorption from open wounds, was due to small amounts of iodine being released on contact with the tissues, but it is doubtful whether iodoform had any value in the treatment of wounds.

Detailed studies on the value of tincture of iodine as a skin disinfectant in surgery appeared from 1905 onwards. Although it was widely used until the 1930s, it was superseded by the iodophors, aromatic compounds that release iodine. These are less irritant, odourless and do not stain clothing to the same extent as iodine. Examples include iodol (1886), soziodol (1887), thioform (1894), sanoform (1896), clioquinol (1898) and many more introduced in the twentieth century.

chiniofon

clioquinol

Chiniofon was marketed by F. Bayer & Company in 1893 and was strongly promoted for a diversity of conditions.[148] In 1921, it was recognised to have value in the treatment of intestinal amoebiasis. A similar drug, clioquinol, was introduced in 1899 by Ciba.[149] It became popular with travellers for the prevention of diarrhoea, but there was little evidence of its efficacy. It was withdrawn during the early 1980s after repeated reports from Japan of toxicity to the optic nerve.[150]

Povidone (polyvinylpyrolidone) is a water-soluble polymer of widely varying chain length that was developed by IG Farbenindustrie just before the Second World War as a plasma volume expander for the emergency treatment of shock, both on account of its resemblance to bilirubin and its colloidal properties.[151] It was superseded by dextrans produced by *Leuconostoc mesenteroides*, as these do not accumulate in reticulo-endothelial tissue.[152] Following reports that povidone bound and detoxified various drugs in a manner similar to that of plasma proteins, Shelanski and Shelanski at the Industrial Toxicological Laboratories in Philadelphia investigated its effect on toxic inorganic materials. In the course of preparing a formulation with tincture of iodine, they found that a unique complex was formed between iodine and povidone and that it was no longer necessary to add solvent or solubilisers in order to dissolve the iodine. *In vitro* tests for antibacterial activity were done and it was noticed that the povidone-treated iodine was less toxic to animals than was tincture of iodine. Skin tests on volunteers and then tests involving application to mucous membranes were followed by clinical trials, all of which confirmed the superiority of the product over other iodine formulations.[153] It was marketed in 1955 with the approved name of povidone-iodine, since when it has become the universally preferred iodine disinfectant. It was originally believed that the iodine covalently bound to the povidone, but spectroscopic studies have since revealed that the iodine binds through hydrogen bonding of the hydriodic acid (HI_3) to carbonyl oxygen atoms near the end of the polymer chain.[154]

POTASSIUM BROMIDE

Bromine was discovered in sea-water in 1826 by Antoine-Jérôme Balard, a pharmacy professor in Montpellier.[155] He used it to prepare potassium bromide. Once the chemical similarity between bromine and iodine had been recognised, physicians attempted to use it as an alternative to iodine in the treatment of a wide variety of diseases, including syphilis and goitre.[156]

The highly corrosive nature of bromine water led physicians to experiment with the potassium salt, which they then prescribed in massive quantities. In London, Williams claimed success in a case of enlargement of the spleen by administering potassium bromide given by mouth. This led directly to its introduction into the *British Pharmacopoeia* of 1836.[157]

On 11 May 1857, the obstetrician Charles Locock presided at a meeting of the Royal Medical and Chirurgical Society of London at which the topic for discussion was epilepsy.[158] He expressed the view, prevalent at the time, that many cases of the disease were due to masturbation, particularly in young patients. He also considered one type of epilepsy to be associated with menstruation. Locock told the meeting that he had met with little success in treating this until he chanced to read an interesting report by a German who became impotent after taking nearly 2 g of potassium bromide daily for two weeks. On withdrawal of the bromide, his normal sexual function had returned. Locock then administered bromide to several young women suffering from hysteria. Large doses were given with good results, so he proceeded to treat a patient who had suffered from 'hysterical epilepsy' for more than nine years. The result was striking. The epilepsy ceased as long as medication was continued, and all but one of 15 subsequent cases were said by Locock to be cured! Despite Locock's eminence, little attention was paid to his comments. He did not publish any further details, and it was not until 1868 that the use of bromide in epilepsy came to be accepted when Thomas Clouston conducted a clinical trial to establish the correct dosage.[159] The trial lasted for several months and demonstrated that patients experienced a reduction in the number of fits while receiving bromide. It was the first drug to be of proven value, despite its obvious drawback of only providing benefit at almost incapacitating dose levels. Potassium bromide remained the standard drug used in the treatment of epilepsy until after the introduction of phenobarbitone. It remained in use as a sedative until the introduction of the benzodiazepines rendered it obsolete. In some countries, potassium bromide is still incorporated in non-prescription sedatives and hypnotics. Needless to say, the efficacy of bromide in epilepsy had nothing whatsoever to do with its effect on libido or sexual function!

INORGANIC COMPOUNDS AS DRUG PROTOTYPES

Inorganic drugs account for only a small proportion of all those in the modern therapeutic armamentarium. Their historical importance in general should not be overlooked, but in particular they deserve recognition for being the earliest purified products that were introduced into therapeutic medicine. They were also the first drug prototypes and, as such, have provided a legacy from the past that cannot be matched by the countless number of animal and vegetable materials that littered medical practice for thousands of years.

As has been seen, the first synthetic drug capable of curing a major disease was an organoarsenical called arsphenamine. Another organometallic drug, antimony potassium tartrate, has saved millions of lives. These two were introduced early in the second decade of the twentieth century. Since then, there have been many advances in the field of organometallic chemistry, especially in recent years. It would not be unreasonable to expect further therapeutic developments in this area in the future.

REFERENCES

1. R.P. Multhauf, John of Rupescissa and the origin of medical chemistry. *Isis*, 1954; **45**: 359–67.
2. C.P. Bryan, *The Papyrus Ebers*. New York: Appleton Century Crofts; 1931, p. 156.
3. A.C. Krause, Babylonian ophthalmology. *Ann. Med. Hist.*, 1934; **6**: 42–55.
4. A.C. Wootton, *Chronicles of Pharmacy*, Vol. 1. London: Macmillan; 1910, p. 13.
5. G.D. Hart, Descriptions of blood and blood disorders before the advent of laboratory studies. *Br. J. Haematol.*, 2001; **115**: 719–28.
6. H.A. Christian, A sketch of the history of the treatment of chlorosis with iron. *Med. Lib. Hist. J.*, 1903; **1**: 176–80.
7. P. Blaud, Sur les maladies chlorotiques et sur un mode de traitement spécifique dans ses affections. *Rev. Méd. Franç. Étrang.*, 1832; **45**: 341–67.
8. J. Humphrey, The origin of Blaud's pills. *Pharm. J.*, 1903; **16**: 643–4.
9. L.J. Goldwater, *Mercury: A History of Quicksilver*. Baltimore: York Press; 1972.
10. J.U. Nef, Mining and metallurgy in medieval civilization, in *Cambridge Economic History of Europe*, Vol. 2, eds M. Postan, E.E. Rich. Cambridge: Cambridge University Press; 1952.
11 C. Plinius Secundus, *Historia Naturalis*, Book 33, translated by H. Rackman, W.H.S. Jones, D.E. Eicholz, 10 vols. London: Heinemann; 1938–1963.
12. J. Warkany, C. Hubbard, Medical progress: acrodynia and mercury. *J. Pediatr.*, 1953; **42**: 365–86.
13. A. Vogl, The discovery of the organic mercurial diuretics. *Am. Heart J.*, 1959; **39**: 881–3.
14. P. Saxl, R. Heilig, Ueber die diuretische Wirkung von Novasurol und aderen Quecksilberpraparaten. *Wein. Klin. Wochenschr.*, 1920; **33**: 943.
15. E.M. Schultes, E.J. Cragoe Jr, J.B. Bicking, *et al.*, α,β-Unsaturated ketone derivatives of aryloxyacetic acids, a new class of diuretics. *J. Med. Pharm. Chem.*, 1962; **5**: 660–2.
16. W. Sneader, *Drug Prototypes and Their Exploitation*. Chichester: Wiley; 1996.
17. J.M. Riddle, Ancient and medieval chemotherapy for cancer. *Isis*, 1985; **76**: 319–30.
18. A.E. Waite (ed.), *The Hermetic and Alchemical Writings of Aureolus Philippus Theophrastus Bombast, of Hohenheim, Called Paracelsus the Great*, Vol 2. London: James Elliott; 1894, p. 210.
19. J. Pereira, *The Elements of Materia Medica and Therapeutics*, Vol. 1, 4th edn. London: Longman, Brown, Green, and Longmans; 1854, p. 710.
20. G. Przygoda, J. Feldmann, W.R. Cullen, The arsenic eaters of Styria: a different picture of people who were chronically exposed to arsenic. *Appl. Organomet. Chem.*, 2001; **15**: 457–62.
21. H. Lissauer, Zwei Fälle von Leucaemie. *Berl. Klin. Wochenschr.*, 1865; **2**: 403–5.
22. C. Forkner, T.F. Scott, Arsenic as a therapeutic agent in chronic myeloid leukemia. *J. Am. Med. Ass.*, 1931; **97**: 3.
23. S.L. Soignet, P. Maslak, Z.G. Wang, *et al.*, Complete remission after treatment of acute promyelocytic leukemia with arsenic trioxide. *N. Engl. J. Med.*, 1998; **339**: 1341–8.
24. J. Zhu, Z. Chen, V. Lallemand-Breitenbach, H. de Thé, How acute promyelocytic leukaemia revived arsenic. *Nature Rev. Cancer*, 2002; **2**: 705–14.
25. D. Livingstone, Arsenic as a remedy for the tse-tse bite. *Br. Med. J.*, 1858; **1**: 360–1.
26. C.L.A. Laveran, F.E.P. Mesnil, *Trypanosomes et Trypanosomiases*. Paris: Masson et Cie; 1904.

27. M.A. Béchamp, *Compt. Rend.*, 1863; **56**: 1172.
28. P. Ehrlich, S. Hata, in *The Collected Papers of Paul Ehrlich*, Vol. 3, ed. F. Himmelweit, London: Pergamon; 1960, p. 282.
29. P. Ehrlich, K. Shiga, Farbentherapeutische Versuche bei Trypanosomenerkrankung. *Berlin Klin. Wochenschr.*, 1904; **41**: 329–62.
30. H.W. Thomas, A. Breinl, The experimental treatment of trypanosomiasis in animals. *Proc. Roy. Soc., Ser. B.*, 1905, **76**: 513.
31. P. Ehrlich, A. Bertheim, Zur Geschichte der Atoxylformel. *Med. Klin. Berlin*, 1907; **3**: 1298.
32. M.Y. Kraft, E.B. Agracheva, Structure of arsphenamine and the magnitude of its molecular weight. *Doklady Akad. Nauk SSR*, 1955; **100**: 279.
33. P. Uhlenhuth, E. Hoffmann, K. Roscher, Untersuchungen über die Wirkung des Atoxyls auf die Syphilis. *Deut. Med. Wochenschr.*, 1907; **33**: 873–6.
34. A. Neisser, Ueber die Verwendung des Arsacetins (Ehrlich) bei der Syphilis-behandlung. *Deut. Med. Wochenschr.*, 1908; **34**: 1500–4.
35. P. Ehrlich, A. Bertheim, Uber das salzsaure 3.3′-Diamino-4.4′-dioxyarsenobenzol und seine nächsten Verwa ndten (Darstellung und Eigenschaflen des Salvarsans). *Ber.*, 1912; **45**: 756.
36. P. Ehrlich, S. Hata, *Die Experimentelle Chemotherapie der Spirillosen*. Berlin: Springer; 1910. Translated in part in *The Collected Papers of Paul Ehrlich*, Vol. 3, ed. F. Himmelweit, London: Pergamon; 1960, pp. 235, 240, 282.
37. P. Ehrlich, Pro und Contra Salvarsan. *Wien. Med. Wochenschr.*, 1910; **61**: 14–20.
38. P. Ehrlich, *Theorie und Praxis der Chemotherapie*. Leipzig: Klinkhardt; 1911.
39. M. Marquardt, *Paul Erlich*. London: Heinemann Medical Books; 1949.
40. Ger. Pat., 1911: 245 756 (to Meister, Lucius and Brüning).
41. A.B.M. Thomson, Treatment of anthrax by N.A.B. *Br. Med. J.*, 1931; **2**: 921.
42. C. Voegtlin, J.M. Johnson, The preparation of sulfarsphenamine. *J. Am. Chem. Soc.*, 1922; 44: 2573–7.
43. W.A. Jacobs, M. Heidelberger, Aromatic arsenic compounds v. *n*-substituted glycylarsanilic acids. *J. Am. Chem. Soc.*, 1919; **41**: 1809–21.
44. E. Fourneau, A. Navarro-Martin, J. Tréfouel, J. Tréfouel, Les dérivés de l'acide phénylarsinique (arsenic pentavalent) dans le traitement des trypanosomiases et des spirilloses expérimentales. Relation entre l'action thérapeutique des acides arsiniques aromatiques et leur constitution. *Ann. Inst. Pasteur*, 1923; **37**: 551–617.
45. J.P. Swann, Arthur Tatum, Parke-Davis, and the discovery of Mapharsen as an antisyphilitic agent. *J. Hist. Med.*, 1985; **40**: 167–87.
46. E.A.H. Friedheim, L'acide triazine-arsinique dans le traitment de la maladie du sommeil. *Ann. Inst. Pasteur*, 1940; **65**: 108.
47. E.A.H. Friedheim, Mel B in the treatment of human trypanosomiasis. *Am. J. Trop. Med. Hyg.*, 1949; **29**: 173–80.
48. W.L. Lewis, G.A. Perkins, The beta-chlorovinyl chloroarsines. *Ind. Eng. Chem.*, 1923; **15**: 290–5.
49. R.A. Peters, L.A. Stocken, R.H.S. Thompson, British Anti-Lewisite (BAL). *Nature*, 1945; **156**: 616.
50. J. Vilensky, British anti-lewisite (dimercaprol): an amazing history. *Ann. Emergency Med.*, 2003; **41**: 378–83
51. R.P. Multhauf, *The Origins of Chemistry*, London: Oldbourne; 1966, p. 202.
52. A.E. Waite (ed.), *The Hermetic and Alchemical Writings of Aureolus Philippus Theophrastus Bombast, of Hohenheim, Called Paracelsus the Great*, Vol 2. London: James Elliott; 1894, p. 199.
53. J. Lévy-Valensi, *La Médecin et les Médecins Français au XVIIe Siècle*, Paris: J.B. Baillière; 1933, p. 132.
54. P. Manson-Bahr, *Advances in the Therapeutics of Antimony*. Leipzig: Thieme; 1938.
55. T. Thomson, *History of Chemistry*, Vol. 1. London: Colburn and Bentley; 1830, p. 200.
56. R. Hunter, I. Macalpine, *Three Hundred Years of Psychiatry 1553–1860*. London: Oxford University Press; 1963, p. 293.
57. F. Mesnil, E. Brimont, Sur l'action de l'émétique dans les trypanosomiases. *Bull. Soc. Path. Exot. (Paris)*, 1908; **1**: 44–8.
58. P. Manson-Bahr, *Advances in the Therapeutics of Antimony*. Leipzig: Thieme; 1938, p. viii.
59. J. Kérandel, *Bull. Soc. Path. Exot. (Paris)*, 1910; **3**: 642.
60. G.O. Vianna, *Anais do 7° Congresso Brasileiro de Medicina e Cirurgia 1912, Vol. 4*, p. 426.
61. G. di Cristina, G. Caronia, Ueber die Therapie der inneren Leishmaniosis. *Deut. Arch. Klin. Med.*, 1915; **117**: 263.
62. J.E.R. McDonagh, *Biology and Treatment of Venereal Diseases*. London: Harrison; 1915.
63. J.B. Christopherson, The successful use of antimony in bilharziosis: administered as intravenous injections of antimonium tartaratum (tartar emetic). *Lancet*, 1918; **2**: 325.
64. J.E.R. McDonagh, Antimony in bilharzias. *Lancet*, 1918; **2**: 371.

65. H. Schmidt, F.M. Peter, *Advances in the Therapeutics of Antimony*. Leipzig: Thieme; 1938.
66. P. Uhlenhuth, P. Kuhn, E. Schmidt, New trypanocidal antimony compound (Heyden 661). *Deutsch. Med. Wochenschr.*, 1924; **50**: 1288–9.
67. H. Schmidt, *Die Entwicklung der Kala azar Mittel. Festschrift Prof. Nocht.* Hamburg: De Gruyter; 1937.
68. P. Durand, S. Benmussa, M. Carnana, Traitment du kala-azar par un nouveau composé stibié: le 2. 168 R.P. (premiers résultants). *Bull. Soc. Méd. Hôp. Paris*, 1946; **62**: 399–409.
69. J.R. Partington, *A History of Chemistry*, Vol. 2. London: Macmillan; 1961, p. 148.
70. M. Hannon, *Ann. de Thér.*, 1857: 214.
71. D.W.S. Bierer, Bismuth subsalicylate: history, chemistry, and safety. *Rev. Infect. Dis.*, 1990; **12**: Suppl. 1, S3–S8.
72. Ger. Pat., 1975: 2501787 (to Gist-Brocades).
73. F. Balzer, Expériences sur la toxicité du bismuth. *C. R. Soc. Biol.*, 1889; **41**: 537–44.
74. A.E. Robert, B. Sauton, Action du bismuth sur la spirillose des poules. *Ann. Inst. Pasteur*, 1916; **30**: 26.
75. C. Levaditi, *Le Bismuth dans le Traitement de la Syphilis.* Paris: Masson; 1924.
76. W.M. Lauter, H.A. Braun, Preparation and toxicity of bismuth salts on camphoric acid esters. *J. Am. Pharm. Ass.*, 1936; **25**: 394–7.
77. J.R. Partington, *A History of Chemistry*, Vol. 1. London: Macmillan; 1970, p. 32.
78. J.R. Partington, *A History of Chemistry*, Vol. 2. London: Macmillan; 1961, p. 146.
79. A.E. Waite (ed.), *The Hermetic and Alchemical Writings of Aureolus Philippus Theophrastus Bombast, of Hohenheim, Called Paracelsus the Great*, Vol 2. London: James Elliott; 1894, p. 225.
80. S.P. Fricker, Medical uses of gold compounds: past, present and future. *Gold Bull.*, 1996; **29**: 53–60.
81. J.G. Niel, J.A. Chrestien, *Recherches et Observations sur les Effets des Préparations d'or du Dr. Chrestien dans le Traitement de Plusieurs Maladies, et Notamment dans Celui des Maladies Syphilitiques.* Paris: Gabon; 1821.
82. J. Pereira, *The Elements of Materia Medica and Therapeutics*, 4th edn, Vol. 1. London: Longmans, Brown, Green and Longmans; 1854, p. 976.
83. R. Koch, *Deut. Med. Wochenschr.*, 1890; **16**: 757.
84. A. Feldt, Zur Chemotherapie der Tuberkulose mit Gold. *Deut. Med. Wochenschr.*, 1913; **39**: 549–51.
85. N. Galatzer, A. Sachs, Erfahrungen mit dem neuen Goldpräparat 'Triphol' bei interner Tuberkulose. *Wiener Klin. Wochenschr.*, 1925; **38**: 583–5.
86. M.J. Fordos, A. Gélis, *Ann. Chim. Phys.*, 1845; **13**: 394.
87. H. Mollgaard, *Chemotherapy of Tuberculosis.* Copenhagen: NYT Nordisk Forlag, Arnold Busck; 1924.
88. J.B. Amberson, B.T. McMahon, M. Pinner, A clinical trial of sanocrysin in pulmonary tuberculosis. *Am. Rev. Tuberculosis*, 1931; **24**: 401–35.
89. K. Landé, Die günstig Beeinflüssung schleichender Dauerinfekte durch Solganal. *Munch. Med. Wochenschr.*, 1927; **74**: 1132–4.
90. J. Forestier, Rheumatoid arthritis and its treatment by gold salts: results of six years' experiments. *J. Lab. Clin. Med.*, 1935; **20**: 827–31.
91. T.N. Fraser, Gold treatment in rheumatoid arthritis. *Ann. Rheum. Dis.*, 1945; **4**: 71–5.
92. B.M. Sutton, E. McGusty, D.T. Waltz, M.J. Di Martino, Oral gold. Antiarthritic properties of alkylphosphine gold coordination complexes. *J. Med. Chem.*, 1972, **15**: 1095–8.
93. J.R. Partington, *A History of Chemistry*, Vol. 1. London: Macmillan; 1970, p. 32.
94. A.E. Waite (ed.), *The Hermetic and Alchemical Writings of Aureolus Philippus Theophrastus Bombast, of Hohenheim, Called Paracelsus the Great*, Vol 2. London: James Elliott; 1894, p. 226.
95. J.R. Partington, *A History of Chemistry*, Vol. 2. London: Macmillan; 1961, p. 288.
96. H.J. Klasen, Historical review of the use of silver in the treatment of burns. I. Early uses. *Burns*, 2000; **26**: 117–30.
97. E. Behring, Der antiseptische Werth der Silberlösungen und Behandlung von Milzbrand mit Silberlösungen. *Deutsch. Med. Wochenschr.*, 1887; **13**: 805–7, 830–4.
98. C.L. Fox Jr, Silver sulfadiazine – a new topical therapy for Pseudomonas in Burns. *Arch. Surg.*, 1968; **96**: 184.
99. J. Pereira, *The Elements of Materia Medica and Therapeutics*, Vol. 1, 4th edn. London: Longman, Brown, Green, and Longmans; 1854, p. 771.
100. T.U. Hoogenraad, *Trace Elements in Medicine*, 1984; **1**: 47.
101. G. Schouwink, De Hepatocerebrale Degeneratie, mer een Onderzoek naar de Zincstofwisseling. Thesis. Amsterdam, 1961.
102. T.U. Hoogenraad, R.Koevoet, E.G.W.M. De Ruyter Korver. Oral zinc sulfate as long-term treatment in Wilson's disease (hepatolenticular degeneration). *Eur. Neurol.*, 1979; **18**: 205–11.

103.   G.J.Brewer, E.B. Schoomaker, D.A. Leichtman, *et al.* The use of pharmacological doses of zinc in the treatment of sickle cell anemia, in *Zinc Metabolism: Current Aspects in Health and Disease*, eds G.J. Brewer, A.S. Prasad. New York: Alan R. Liss; 1987, pp. 241–54.

104.   E.J. Moynahan, Acrodermatitis enteropathica: a lethal inherited human zinc-deficiency disorder. *Lancet*, 1974; **2**: 399–400.

105.   J.F.J. Cade, The story of lithium, in *Discoveries in Biological Psychiatry*, eds F.J. Ayd, B. Blackwell. Philadelphia, Pennsylvania: Lippincott; 1970, p. 218.

106.   J.F.J. Cade, Lithium salts in the treatment of psychotic excitement. *Med. J. Austral.*, 1949; **36**: 349–52.

107.   P.C. Baastrup, M. Schou, Lithium as a prophylactic agent: its effect against recurrent depressions and manic-depressive psychosis. *Archives of General Psychiatry*, 1967; **16**: 162–72.

108.   Hofer, *Observations et Recherches Expérimentales sur la Platine*. Paris: 1841.

109.   B. Rosenberg, L. Van Camp, T. Krigas, Inhibition of cell division in *Escherichia coli* by electrolysis products from a platinum electrode. *Nature*, 1965; **205**: 698–9.

110.   B. Rosenberg, L. Van Camp, J.E. Trosko, V.H. Mansour, Platinum compounds: a new class of potent antitumour agents. *Nature*, 1969; **222**: 385–6.

111.   M.J. Cleare, J.D. Hoeschele, Studies on the antitumor activity of group VIII transition metal complexes. Part I. Platinum (II) complexes. *Bioinorg. Chem.*, 1973; **2**: 187.

112.   J.J. Roberts, J.M. Pascoe, Cross-linking of complementary strands of DNA in mammalian cells by antitumour platinum compounds. *Nature*, 1972; **235**: 282–4.

113.   E. Cvitkovic, J. Spaulding, V. Bethune, *et al.*, Improvement of cis-dichlorodiammineplatinum (NSC 119875): therapeutic index in an animal model. *Cancer*, 1977; **39**: 1357–62.

114.   M.J. Cleare, J.D. Hoeschele, Studies on the antitumor activity of group VIII transition metal complexes. Part I. Platinum (II) complexes. *Bioinorg. Chem.*, 1973; **2**: 187–210.

115.   C.F.J. Barnard, M.J. Cleare, P.C. Hydes, Second generation anticancer platinum compounds. *Chem. in Britain*, 1986; **22**: 1001–4.

116.   S. Mong, C.H. Huang, A.W. Prestakyo, S.T. Crooke, Effects of second-generation platinum analogs on isolated PM-2 DNA and their cytotoxicity *in vitro* and *in vivo*. *Cancer Res.*, 1980; **40**: 3318–24.

117.   A.G. Mathe, Y. Kidani, M. Noji, *et al.*, Antitumor activity of 1-OHP in mice. *Cancer Lett.*, 1985, **27**: 135–43.

118.   N. Grew, *A Treatise on the Nature and Use of the Bitter Purging Salt*. London: Joseph Bridges; 1697.

119.   J. Pereira, *The Elements of Materia Medica and Therapeutics*, Vol. 1, 4th edn, London: Longman, Brown, Green, and Longmans; 1854, p. 658.

120.   J. Pereira, *The Elements of Materia Medica and Therapeutics*, Vol. 1, 4th edn, London: Longman, Brown, Green, and Longmans; 1854, p. 646.

121.   J.R. Partington, *A History of Chemistry*, Vol. 2. London: Macmillan; 1961, p. 340.

122.   P. Holmes, Seidlitz – the morning-after powder. *Pharm. J.*, 2001; **267**: 911.

123.   S.G. Schultz, R. Zalusky. Ion transport in isolated rabbit ileum. II. The interaction between active sodium and active sugar transport. *J. Gen. Physiol.*, 1964; **47**: 1043–59.

124.   H.E. Harrison, The treatment of diarrhea in infancy. *Pediatr. Clin. N. Am.* 1954; **1**: 335–48.

125.   M. Santosham, E.M. Keenan, J. Tulloch, D. Broun, Oral rehydration therapy for diarrhea. *Pediatrics*, 1997; **100**: 10–13.

126.   C.L. Berthollet, *Ann. Chim.*, 1789; **2**: 151.

127.   A.G. Labarraque, *De l'Emploi des Chlorures d'Oxide de Sodium et de Chaux*. Paris: Labarraque and Huzard; 1825.

128.   T. Alcock, *Essays on the Use of the Chlorurets*. London: Burgess and Hill; 1827.

129.   J. Pereira, *The Elements of Materia Medica and Therapeutics*, Vol. 1, 4th edn, London: Longman, Brown, Green, and Longmans; 1854, p. 638.

130.   R. Collins, *A Practical Treatise on Midwifery*. London: Longman, 1835.

131.   O.W. Holmes, The contagiousness of puerperal fever. *N. Engl. Quart. J. Med. Surg.*, 1843; **1**: 503–30.

132.   I. Semmelweis, *The Etiology, Concept, and Prophylaxis of Childbed Fever*. Translated and edited, with an introduction, by K.C. Carter. Madison: University of Wisconsin Press; 1983.

133.   H.D. Dakin, The antiseptic action of hypochlorites: the ancient history of the 'new antiseptic'. *Br. Med. J.*, 1915; **2**: 809–10.

134.   F. Chattaway, Nitrogen halogen derivatives of the sulphonamides. *J. Chem. Soc.*, 1905; **87**: 145–71.

135.   H.D. Dakin, J.B. Cohen, On the use of certain antiseptic substances. *Br. Med. J.*, 1915; **2**: 318–20.

136.   B. Courtois, *Ann. Chim. Paris*, 1813; **88**: 304.

137.   L.J. Gay-Lussac, *Mémoires d'Iodide*. Paris: 1814.

138.   J.F. Coindet, Découverte d'un nouveau remède contre le goître. *Ann. Chim. Phys.*, 1820; **14**: 190.

139. D. Marine, O.P. Kimball, The prevention of simple goiter. *Am. J. Med. Sci.*, **163**: 634–9.
140. E. Baumann, Ueber das normale Vorkommen von Jod in Thierkorper. *Z. Physiol. Chem.*, 1896; **21**: 319–30 and 481–93.
141. J. Pereira, *The Elements of Materia Medica and Therapeutics*, Vol. 1, 4th edn, London: Longman, Brown, Green, and Longmans; 1854, p. 406.
142. G.B. Wood, F. Bache, *The Dispensatory of the United States of America*, 16th edn. Philadelphia: J.B. Lippincott; 1890, p. 834.
143. R.M. Glover, On the physiological and medicinal properties of iodoform. *Monthly J. Med. Sci.*, 1847; **8**: 184.
144. G.B. Wood, F. Bache, *The Dispensatory of the United States of America*, 16th edn. Philadelphia: J.B. Lippincott; 1890, p. 195.
145. J. Davies, *Selections in Pathology and Surgery*, Part II. London: Longman, Brown, Green, and Longmans; 1839.
146. C.J Davaine, *B. Acad. Méd. Paris*, 1875; **4**: 581.
147. A. Bouchardat, *Annalen*, 1838; **22**: 225.
148. A. Claus, *Arch. Pharm.*, 1893; **231**: 704.
149. Ger. Pat. 1899: 117796 (to Ciba).
150. T. Tsubaki, Y. Honma, M. Hoshi, Neurological syndrome associated with clioquinol. *Lancet*, 1971; **1**: 696–7.
151. G. Hecht, H. Weese, Perriston, Ein Neuer Bluttflüssig-Keitersatz. *Munch. Med. Wochenschr.*, 1943; **90**: 11–15.
152. W.B. Neely, Dextran: structure and synthesis. *Adv. Carbohyd. Res.*, 1960; **15**: 341–69.
153. H.A. Shelanski, M.V. Shelanski, PVP–iodine: history, toxicity and therapeutic uses. *J. Int. Coll. Surg.*, 1956; **25**: 727–34.
154. H.U. Schenck, P. Simak, E. Haedicke, Structure of polyvinylpyrrolidone–iodine (povidone–iodine). *J. Pharm. Sci.*, 1979; **68**: 1505–9.
155. A.-J. Balard, Sur une substance particuliere contenue dans l'eau de la mer. *Ann. Chim. Phys.*, 1826; **32**: 337.
156. J. Pereira, *The Elements of Materia Medica and Therapeutics*, Vol. 1, 4th edn. London: Longman, Brown, Green, and Longmans; 1854.
157. J. Pereira, *The Elements of Materia Medica and Therapeutics*, Vol. 1, 4th edn. London: Longman, Brown, Green, and Longmans; 1854, p. 527.
158. C. Locock, Contributions to discussion of a paper by EH Sieveking. Analyses of 52 cases of epilepsy observed by the author. *Lancet*, 1857; **2**: 527.
159. T.S. Clouston, *J. Ment. Sci.*, 1868; **14**: 305.

8

Systematic Medicine

The publication of the first textbook on human anatomy, *De Humanis Corporis Fabrica* by Andreas Vesalius (1514–1564), was a landmark in the history of medicine. It set in motion a continuing investigation into the structure of the human body that was to have far-reaching consequences. As an increasing number of dissections of human cadavers took place, major findings began to alter the understanding of the body, such as when the English physician William Harvey (1578–1657) discovered the circulation of the blood, which he reported in 1628 in his *Exercitatio Anatomica de Motu Cordis et Sanguinis in Animalibus* (*Anatomical Essay on the Motion of the Heart and the Blood in Animals*). This book challenged the teachings of Galen by disproving the claim that the body consisted of three linked systems – the brain and nervous system, the venous system and the arterial system. Galen had believed that the brain was the source of an animal spirit that facilitated sensation and thinking, whereas the liver was the centre of the circulatory system, from whence blood formed from food was transported around the body by the veins to provide nourishment. Blood that was consumed in this process was believed to be replenished as more food from the intestines entered the liver via the portal vein. The heart was not recognised as a pump, being considered merely as part of Galen's third system that provided energy by carrying blood and vital spirit round the body through pulsating arteries. Again, it was thought that blood was consumed in the process; there was no conception of it being recycled.[1]

After the Italian anatomist Marcello Malpighi (1628–1694) had used the microscope in 1661 to reveal the capilliary network linking arteries and veins, the claims of Harvey seemed incontrovertible. During the next 150 years or so there were several attempts to reconcile the new anatomical findings with the views of Galen by formulating alternatives to the classical humoral system, which still retained the concept of disease arising from a disturbance of the equilibrium of the body. It may seem strange that in an age when leading thinkers were rejecting authoritarian religious doctrines, many physicians still wished to cling to the teachings of Galen, but this was often driven by their need to make a diagnosis and then take any necessary corrective action. There were also those who viewed the body as nothing more than a complex machine that obeyed the laws of physics, but they were challenged by others who believed the body was controlled by the soul or the mind.

It might be thought that with the various investigations into the nature of the body there would have been an opportunity to introduce innovative treatments, yet the drugs preferred by their proponents were neither novel nor selected through objective clinical testing; they were simply chosen on the basis of uncontrolled self-experimentation or theorising.[2] Once again, the process of drug discovery was to gain little from theoretical speculation about the nature of the body, and it was not until after active principles were isolated from plants in the nineteenth century that any significant progress was made. That particular breakthrough was not driven by theory but rather by the practical necessity of finding a way to detect adulterated herbal products.

Drug Discovery. A History. W. Sneader.
©2005 John Wiley & Sons Ltd

SOLIDISM

The physiological text *L'Homme, et un Traité de la Formation du Foetus* (*Man, and a Treatise on the Formation of the Foetus*) by Réne Descartes was published in Paris in 1664. It was in this work that Descartes professed his belief that all bodies, including the human body, are basically machines that operate according to mechanical principles. Among those who accepted this view were influential medical teachers such as Hermann Boerhaave (1668–1738), Professor of Medicine at the University of Leiden, which was situated near where Descartes had written his opus. Boerhaave tried to reconcile the new anatomical observations with Hippocratic medicine, in particular wrestling with the nature of the solid and fluid parts of the body in his *Institutiones Medicae*, published in 1708. This portrayed the solids as composed of fibres that could also form the membranes of the vessels through which the fluids could flow. Boerhaave's system of medicine portrayed disease as arising from a disturbance of the fibres, a rationalisation that his many students who came to Leiden from all over the world admired as an explanation of the diverse collection of observations on health and disease that their master had compiled. Boerhaave and others who thought that disease arose out of a malfunction in the solid fibres of the body rather than through humoral imbalance collectively became known as solidists. In reality, they had not abandoned humoralism; they had merely modified it in keeping with the times. Friedrich Hoffmann (1660–1742), the Professor of Medicine at the newly established University of Halle, thought that disease arose from alteration of the flow of fluid through the vessels, heart or muscles, resulting in alteration of the 'tonus', or tension, of the fibres, thereby inducing spasm or atony. In his *Medicina Rationalis Systematica* (1728–1740), he laid emphasis on the control that the nervous system had on movement, which was so important in the maintenance of good health. He accounted for the rapidity of the movement of animal spirits from the brain through nerves to the voluntary muscles by arguing that as the body operated like a hydraulic machine such fluids are speedily pushed along the tubes or vessels between its chambers. In disease, there is a disruption of this flow. However, he made no clear distinction between the nature of the fibres from which the tubes of the nerves, muscles and blood vessels were formed. It was left to Albrecht von Haller (1708–1777), a Swiss pupil of Boerhaave who became Professor of Medicine at Göttingen, to conduct animal experiments that showed muscles contract when deliberately stimulated, whereas nerves transmit pain or sensation. Von Haller had performed these experiments during his investigation into the irritability of different tissues, i.e. their ability to contract upon stimulation. His views on irritability appeared in his *Primae Lineae Physiologiae* (*First Lines of Physiology*), published in 1747. They so strongly influenced William Cullen (1710–1790), co-founder of the Glasgow School of Medicine and subsequently Professor of Medicine at Edinburgh, that he similarly named his work *First Lines of the Practice of Physic*, published in 1777, in which he taught his students that disease arose when nerves did not receive sufficient stimulation to provide the tonus that would maintain the flow of fluids needed for the normal function of organs. For many years, this remained the leading text on the treatment of disease.

The opinions of solidists such as Hoffmann, Boerhaave, von Haller and Cullen influenced the selection and use of drugs in the eighteenth and nineteenth centuries. The belief that disease arose from altered tone of the fibres in the nerves or blood vessels in affected organs led to attempts to increase the tone by using tonics (a term introduced by Hoffmann) or to reduce it with depletive drugs.[2] A major benefit that arose out of this was a return to single-drug therapy, though polypharmacy did not disappear.

Once the concept of irritability had gained hold, diagnosis of hyperirritability was often made by the detection of an increased pulse rate, as this was a measure of the heart rate. Depletive drugs were then prescribed in order to reduce this, principally agents such as emetics, purgatives, diuretics, diaphoretics, antispasmodics, sedatives or narcotics. Since fevers were judged to arise from hyperirritability, the use of such agents became widely accepted.

However, when disease was considered to arise from decreased tone of the heart, blood vessels or other organs, it was deemed necessary to prescribe a tonic.[2] The problem with this approach was that it was as open to abuse as humoral theory had been by those who wished to justify the use of any favoured remedy. Coupled with the unreliability of assessing whether a patient required depletive or tonic therapy, opportunities to justify therapeutic intervention were not missed by those so inclined, resulting in the introduction of a vast range of tonics for the heart, liver, kidney, brain, hair, skin and other organs.

It was in the hands of John Brown (1735–1788), a fellow Scot and former student of Cullen, that solidist theory took its most extreme form. His Brunonian system was expounded in *Elementa Medicinae*, published in 1780, in which he argued that the unique characteristic of animal and human life was 'excitability', particularly of the nerves and muscles. Excitability was steadily maintained by bodily and environmental stimuli which he termed 'exciting powers'. Disease occurred when there was either a surfeit or deficit of excitability, the former condition being described by Brown as 'sthenia' and the latter as 'asthenia'. He instituted an 80 point scale for indicating the degree of excitability of patients, with the normal state being mid-scale. Treatment was based on restoring the level of excitability to mid-scale by administering remedies that served as 'stimuli'. In asthenic patients, this required stimulant or sthenic remedies supposedly acting with more energy than is normally appropriate for maintenance of good health, notably animal-derived food, wine, spirits, musk, camphor, ether and opium, in order of increasing strength. For sthenic patients, it was deemed necessary to administer remedies acting with less energy than that required for maintenance of good health, described by Brown as 'antisthenic' or 'debilitating'. In addition to bleeding, exposure to cold and dietary restriction, this included drugs that produced vomiting, purging, sweating and all the other effects previously associated with depletive drugs.[3] Here was yet another medical system that contrived to explain disease in terms of an imbalance or disturbance of the status quo of the body and then restore the balance with agents believed to have opposing properties. Again there was no attempt to assess the efficacy of the selected agents in an objective manner.

The absurdity of the Brunonian system can be seen in its use of cinchona bark. Believing that intermittent fevers were preceded by a state of debility, Brown had concluded that those afflicted with it were asthenic. This meant that they should be given a sthenic, stimulant drug to effect a cure. Since cinchona bark frequently cured patients, this was then taken as proof that it was a sthenic drug! By extension of this argument, cinchona was prescribed for a diverse range of diseases arbitrarily designated as asthenic. Throughout the nineteenth century, the bark or its active principle quinine was hailed as a tonic and was prescribed for '. . . debility with atony and laxity of the solids, and profuse charges from the secreting organs'.[4] Misuse of medication in this manner was not confined to cinchona bark as enthusiastic practitioners of the Brunonian system, both in Scotland and in continental Europe, poisoned their patients with a diverse range of stimulants, including alcohol, a substance that in no small part contributed to the early death of John Brown.

The Brunonian system was modified around 1800 by the Italians Rasori and Borda. The two classes of drugs were now described as stimulants or hypersthenics, and contrastimulants or hyposthenics. This approach rejected the Brunonian concept that all drugs were stimulants and instead accepted that some had the ability to depress excitability. Once again arbitrary arguments abounded. For example, it was claimed that since wine was a stimulant, anything that relieved intoxication from it must be a contrastimulant. Purgatives were no longer considered to be stimulants since they allegedly induced depression. However, opium was still considered to be a stimulant. This variation on the Brunonian system cannot escape the criticism that was applied to both systems by Pereira in the middle of the nineteenth century, when he pointed out that nine-tenths of the drugs then known possessed neither stimulant nor depressant activity. They were merely 'alterative' in their action. Pereira clearly recognised that the failure of these systems lay in their classifying drugs by their secondary effects rather than their primary action.[3]

The various medical systems that were in vogue during the eighteenth century encouraged physicians to prescribe from a constantly increasing range of drugs made available by the expansion of international trade. This all played a part in provoking Samuel Hahnemann (1755–1843) of Leipzig to introduce his homeopathic system in 1796, the first systematic account of which was published 14 years later as his *Organon der Rationellen Heilkunde*. Hahnemann prescribed infinitesimally small doses of drugs, the selection of which was based on the principle *similia similibus curantur* (like cures like). This was the opposite of that elaborated by Galen 16 centuries earlier, namely *contraria contrariis curantur*, symbolising Hahnemann's rejection of Galen's antipathy, the use of drugs held to possess effects opposite to those caused by the disease itself. Hahnemann also rejected what he designated as allopathy, namely the use of drugs with effects that differed from those produced by the disease.

Hahnemann developed his system from a misunderstanding of what happened to him after he had taken a large dose of cinchona bark. It had induced nausea, vomiting and coldness of the limbs, probably caused by severe gastric irritation, but Hahnemann believed it was identical to the malarial paroxysm he had experienced on an earlier occasion. This led him to the conclusion that cinchona relieved the intermittent fever of malaria because large doses produced the nausea, vomiting and coldness associated with the disease.

Hahnemann also diluted his preparations to such an extent that no active material was left! At each stage of dilution, the container was vigorously shaken to increase potency through the release of spiritual powers. The year after he published his *Organon*, Avogadro postulated that equal volumes of gases, at equivalent temperatures and pressures, contain the same number of molecules, but the number of molecules in a mole ($6.022\,1367 \times 10^{23}$) was not calculated until later in the century. Hahnemann therefore remained blissfully unaware that for his popular 12C dilutions (i.e. containing 1 part of drug in 10^{24} of water) there was less than a one-in-two chance that a single molecule of any drug would be present.

In favour of homeopathy, it can certainly be said that it delivered many patients from exposure to the drugs favoured by adherents of the Brunonian and similar medical systems. There is some substance to his allegation that medicine as commonly practised at that time merely sought to evacuate the contents of the stomach and sweep the intestines clear of the materials assumed to cause disease. The contrast between the Brunonian and homeopathic systems was highlighted by the use of one of the most popular drugs of the day, namely cinchona bark – the one system grossly overdosed and the other hopelessly underdosed patients.

PNEUMATIC CHEMISTRY

In 1786, shortly after his breakthrough in understanding the role of oxygen in respiration, Lavoisier was visited in Paris by Thomas Beddoes, a young doctor who was keen to learn of the latest developments in pneumatic chemistry. Following upon his appointment as Reader in Chemistry at Oxford two years later, Beddoes acquired the reputation of being the leading exponent of pneumatic medicine in England. He was an adherent of the Brunonian system and believed that oxygen was the principle of irritability.[5] He therefore administered air enriched with oxygen to his asthenic patients, while his sthenic patients were persuaded to inhale air deficient in oxygen. Beddoes continued with these experiments at Oxford until 1792, when his outspoken views on the merits of the French Revolution finally forced him to resign his post and establish an institution where patients could be treated with specially manufactured 'factitious airs'.

Beddoes' father-in-law and several acquaintances were members of the elite Lunar Society, quaintly named because it met monthly in Birmingham when the moon was full and so its members could ride home safely.[6] When Joseph Priestley had lived in Birmingham during the 1780s, he greatly influenced the philosophical activities of this small group. Other distinguished members included Josiah Wedgwood, the pottery magnate, William Withering, the physician who introduced digitalis into medicine, his rival Erasmus Darwin and the famous engineer James Watt, who had been persuaded by members of the Lunar Society to settle in Birmingham. It was to the Society, then, that Beddoes turned for patronage in 1793 after he published a pamphlet on the treatment of consumption (tuberculosis) by inhalation of 'factitious airs'. Several members of the Lunar Society had good cause to support his efforts, for members of their own families were dying of consumption, including young Gregory Watt, the son of the engineer. Intrigued by the imaginative nature of Beddoes' plans and their potential, the Society members backed him financially. Watt collaborated with Beddoes in designing a machine to manufacture factitious airs in 1794.[7,8] Beddoes forged ahead with his plans to open what was to become known as the Pneumatic Institution for Relieving Diseases by Medical Airs. This opened in Clifton, Bristol, in 1798. In the basement was the massive machine built by Watt for the production of gases under the supervision of a young Humphry Davy, who was encouraged to experiment with new gases for patients to inhale. He examined nitrous oxide as an American, Samuel Mitchill, had claimed it was highly toxic. After establishing that small animals could be immersed in a jar of nitrous oxide without any apparent harm, Davy inhaled the gas himself, experiencing what he described as '... the most vivid sensation of pleasure accompanied by a rapid succession of highly excited ideas'.[9] Later, he wrote in his laboratory notebook that, 'Sensible pain is not perceived after the powerful action of nitrous oxide because it produces for the time a momentary condition of other parts of the nerve connected with pleasure'. Davy was a scientist, not a physician, and saw no therapeutic application in this. For Beddoes, the physician, his obsession with the Brunonian system convinced him only that Davy had found a stimulant for asthenic patients that did not have the drawbacks of alcohol. Davy never agreed with this interpretation. Neither of these historical figures recognised the significance of what had been found, reflecting the attitude of their time that pain was an inevitable aspect of surgery. Later, when chloroform anaesthesia was introduced into obstetric practice by Simpson, religious critics considered this deliberate removal of pain to be nothing less than a challenge by man against divine providence, quoting the words in Genesis III, '... in pain thou shalt bring forth children...'. Simpson ably brushed off this challenge by citing the creation of Eve in Genesis II, 21, 'And the Lord God caused a deep sleep to fall upon the man, and he slept; and He took one of his ribs, and closed up the place with flesh instead thereof.'

$$^-N=\overset{+}{N}=O \quad \text{nitrous oxide}$$

Despite the neglect of its ability to numb pain, its euphoriant properties ensured that nitrous oxide soon became widely known, receiving the popular name of 'laughing gas'. It was the most celebrated product of the Pneumatic Institution, which itself developed into a nursing home once physicians and patients alike came to realise that 'factitious airs' were not the panacea for all ills.

SURGICAL ANAESTHESIA

Nitrous oxide

Early in December 1844, Horace Wells, a dentist in Hertford, Connecticut, attended a public lecture at which Gardner Quincy Colton demonstrated the effects of laughing gas. Such lecture demonstrations of popular science were then in vogue, and men like the ex-medical student

Colton were able to earn a living touring towns around the United States. On this occasion, one of the audience who had volunteered to inhale the nitrous oxide accidentally gashed his leg while staggering around under the influence of the gas. Noticing that his injury did not seem to hurt the man, Wells pondered whether it might be possible to exploit nitrous oxide to ease the pain of dental extraction, a problem that he was particularly concerned with at that time. Two years earlier, he and his former partner in Boston, William Morton, had discovered the importance of the removal of entire decayed teeth rather than just the crowns. Although this was clearly in their patients' best interest, such was the pain involved that few ever returned, and the dental partnership had to be dissolved.

After the lecture, Wells asked Colton if he would administer nitrous oxide to him while a troublesome molar tooth was removed. Colton duly appeared on the morning of 11 December 1844 and made history when he administered the first dental anaesthetic. Wells summed up the occasion succinctly, 'A new era in tooth-pulling!' Excited by his success, Wells went to Boston, where he contacted his former partner, Morton, and asked him to arrange a demonstration of anaesthesia in the presence of America's leading surgeons in the Massachusetts General Hospital. Sadly, this was to prove his downfall. The demonstration was a miserable failure due to insufficient anaesthetic being administered. Wells was jeered out of the operating theatre in disgrace.[10,11]

Ether

The next substance to be investigated as a potential anaesthetic was ether, a volatile liquid which was synthesised in 1540 by Valerius Cordus through treating spirit of wine with oil of vitriol (sulfuric acid). Details of his synthesis of the oleum vitrioli dulce (sweet oil of vitriol) appeared in 1561 in *De Artificiosis Extractionibus*, a posthumous collection of his writings edited by Konrad Gesner that was later translated into English.[12] Cordus thought ether had medicinal properties similar to those of sulfur, with greater penetrability because of its volatility. He recommended taking it by mouth to relieve pneumonia, pleurisy or a hacking cough. Sigismund August Frobenius, a German chemist working in Robert Boyle's laboratory in London, described the preparation of 'sulphuric aether' to the Royal Society in 1730.[13] Six years later, Friedrich Hoffmann described a mixture of ether dissolved in twice its volume of alcohol in his *Observationum Physico-Chemicarum Selectiorum*, published in 1736. This *liquor anodynus minerali Hoffmanni* became popularly known as Hoffmann's drops, being taken by mouth to relieve stomach, abdominal and pelvic pain. The preparation was often abused because of its alcohol content, which probably explains its continued popularity two hundred years later!

$$H_3C \diagup O \diagup CH_3 \quad \text{ether}$$

A claim that inhalation of the vapour of ether produced much the same effects as nitrous oxide appeared in an anonymous article in the *Quarterly Journal of Science and the Arts* in 1818, which may have been written by Michael Faraday, an assistant to Humphry Davy, Director of the Royal Institution. As ether was obtainable from apothecaries, anyone who wished to experience its effects upon inhalation had little difficulty in obtaining supplies. This was to have far-reaching consequences in the 1840s when 'ether frolics' became popular in the United States and led to the discovery that ether inhalation could remove pain in much the same way as nitrous oxide.

Crawford Long, a young doctor in Jefferson, Georgia, was asked by friends to obtain nitrous oxide for inhalation at a party they were planning in December 1841. Long supplied ether instead of nitrous oxide and it proved a great success with the partygoers. So high-spirited were the frolics that next morning Long and his friends found they had collected several bruises, none of them recollecting any pain during the party. Fortunately, Long had by now regained his wits sufficiently to realise the full significance of what had happened.[14]

There had, at that time, been much interest in medical circles in Mesmer's claims that hypnosis could remove pain during surgery. Long had not been impressed by these claims, believing that mesmerism had only limited uses. Nevertheless, the idea of painless surgery now came back to him and he made up his mind to pursue the matter further if an opportunity arose. His chance came when, on 30 March 1842, he placed a towel soaked in ether over the face of James Venables, a young man who had frequently inhaled ether. As soon as his patient was insensible, Long proceeded to excise a large growth on his neck. The youth felt no pain. History had been made, but the country practitioner did not bother to report his success until 1849, by which time he considered he had acquired enough experience to tell the medical profession what could be achieved with ether anaesthesia.[15]

There was an even earlier use of ether a few weeks before Long anaesthetised James Venables, an event that was not reported until 1881 in *Artificial Anesthesia and Anesthetics* by Henry M. Lyman. In January 1842, William E. Clarke, a medical student in Rochester, New York, anaesthetised a woman undergoing a dental operation. Clarke also had experience of ether frolics.[16] This isolated case, like the experiments by Long, failed to influence the practice of surgery. However, the failed demonstration by Wells in Boston did have consequential effects through its influence on William Morton, his former partner in dental practice. Morton had remained dubious about the value of nitrous oxide for removing pain during surgery. He visited Wells and discussed its use before starting to experiment on his household pets in the summer of 1846.[11] He mentioned his doubts about nitrous oxide to Charles Jackson, a chemist tutoring him for his medical studies. Jackson proposed ether as an alternative, a suggestion which was later to cause an unseemly row over who deserved the credit and financial rewards for the discovery of surgical anaesthesia. In retrospect, it has to be said that many chemists and physicians were at that time aware of ether as an alternative to nitrous oxide. Had not Long offered to provide it when his friends had been seeking laughing gas? Most of the credit must, therefore, go to Morton, whose endeavours were to ensure the general acceptance of ether anaesthesia.

On 30 September 1846 Morton successfully maintained anaesthesia with ether for three-quarters of a minute while he extracted a tooth. For the next three weeks he used it on all his patients, observing that their exhaled breath contaminated the ether, which was held in a rubber bag. This resulted in a lack of uniform response. To avoid this, Morton designed an inhaler with a valve that vented exhaled breath into the atmosphere. The rubber bag was also replaced by a sponge-filled bottle, from which the patient inhaled ether through a tube held between the lips while the nostrils were pinched shut. This new inhaler was ready just in time to be used in a public demonstration before a large gathering in the domed amphitheatre of the Massachusetts General Hospital on 16 October 1846. Morton administered the anaesthetic, whose nature he unsuccessfully tried to disguise by colouring it and labelling it 'Letheon'. Gilbert Abbott, a 20 year old youth, was then presented to surgeon John Warren, who proceeded to make a deep incision several inches long near the lower jaw. Working with the swiftness and dexterity then required of a master surgeon, Warren removed a tumour without his patient feeling anything. The youth remained quite still most of the time. The following morning, the Boston newspapers carried reports of the operation. A month later, a paper on ether anaesthesia presented by Warren to the Boston Society of Medical Improvement appeared in the *Boston Medical and Surgical Journal*. No sooner did the world learn of the remarkable achievement in Boston than a row over priority of discovery of surgical anaesthesia broke out. Not only did Jackson challenge Morton's claims, but so too did Wells, without whose investigations on nitrous oxide Morton would never have become involved in the quest to develop a safe anaesthetic.[17] This dispute forced Morton to protect his own interests by patenting ether, but this had little effect. Boston surgeons, however, followed up a suggestion in the patent by abandoning the use of the inhaler and instead anaesthetised their patients simply by pouring a few drops of ether at a time on to a sponge held over the mouth and nose. When London surgeon Robert Liston received reports of Morton's demonstration, he had the eminent pharmacist Peter Squire fabricate an inhaler similar to Morton's for use in an operation the following December. Subsequently English anaesthetists relied on inhalers, which they were forever improving to overcome a variety of design shortcomings. In Edinburgh, however, the influence of the American surgeons was stronger, with the open-mask technique prevailing. This continued to be used in Scotland, and on the Continent, well into the following century. As it was simple to use, there was little demand for specialist anaesthetists in Scotland when the more dangerous chloroform was introduced, whereas the English preference for inhalers necessitated specialist training.

Chloroform

Chloroform was first prepared in 1831 by three investigators working independently. Samuel Guthrie at Sacketts Harbor in the north of New York State prepared an alcoholic solution of chloroform by heating alcohol with chloride of lime in a copper still, naming the distillate chloric ether as he believed it to be the ethylene chloride (i.e. 1,2-dichloroethane) that he had been attempting to obtain.[18] Around the same time, chloroform was prepared in France by Eugene Souberian and in Germany by Augustus von Liebig. The French chemist Jean-Baptiste André Dumas revealed the true nature of the dense liquid and both gave it the name of chloroform and elucidated its chemical structure.[19]

chloroform

In 1842, the English physician Robert Glover demonstrated that dogs could be rendered unconscious by injections of chloroform, but he failed to recognise the significance this could have for surgery.[20] It was the introduction of ether anaesthesia four years later that led to the discovery of chloroform as an anaesthetic. James Young Simpson, Professor of Midwifery at the University of Edinburgh, witnessed ether anaesthesia in January 1847, and became the first to use it routinely in obstetric practice. Two problems became apparent. Edinburgh was then the largest city in Scotland, with many of its inhabitants residing in high, crowded tenement blocks. Climbing up several flights of stairs on his domiciliary visits was difficult enough for Simpson, but having to carry the large bottles of ether required by the inefficient open-mask method was too much! This technique permitted wasteful evaporation of the ether from the mask into the atmosphere, and it was this that also caused Simpson's second, more serious, problem. Gas lighting was by now installed in all the tenements; as Simpson proceeded to carry out his work, the concentration of ether in the air soon reached a hazardous level, posing the risk of a fire or even an explosion. Consequently, Simpson sought an alternative to ether. He was unaware that a solution of chloroform in alcohol had already been investigated at the Middlesex Hospital, only to be rejected on account of its cost. However, at St Bartholomew's Hospital, also in London, William Lawrence found this so-called chloric ether to be a satisfactory alternative to ether. In Paris, Pierre Flourens, a former student of the physiologist François Magendie, had already tested both chloroform and ethyl chloride on animals. The latter was subsequently used in Ferdinand Heyfelder's clinic in Erlangen, Germany, but it also was ultimately rejected because of its high cost.

Simpson began his search by testing a variety of volatile fluids on himself and his friends throughout the summer months. These fluids included Dutch liquid (1,2-dichloroethane), acetone, ethyl nitrate, benzene, and iodoform vapour. In October he met David Waldie, an old friend from the days when both had lived in Linlithgow. Waldie was now a pharmacist in Liverpool whose company supplied chloric ether to local physicians for inhalation by patients with respiratory disorders. This was made by dissolving chloroform in alcohol, so Waldie knew that chloroform was safe to inhale. Unfortunately, he was unable to supply Simpson with chloroform as his laboratory had just been destroyed by fire. Instead, Simpson acquired some from the Edinburgh druggists, Duncan, Flockhart & Company. In the company of a circle of friends gathered in the dining room of his home at 52 Queen Street on the evening of 4 November 1847, Simpson and two colleagues proceeded to inhale chloroform from glass tumblers and quickly fell asleep. A few minutes later, all three had recovered unharmed by the experience. His niece then tried it and others present also did likewise.[21] Simpson at once recognised that chloroform met all his requirements, and duly reported to the Medical-Chirurgical Society of Edinburgh that it had many advantages over ether: it was more potent, requiring only a small volume of liquid; the onset of anaesthesia was quicker; it was also pleasant to inhale, and it was simple to administer by impregnating a cloth held over the mouth. Moreover, it was also non-inflammable. On 12 November he published his findings in a pamphlet that sold over 4000 copies within a few days.[22] A paper appeared in the *Lancet* the following week.[23]

When London surgeons used chloroform with inhalers they experienced difficulties not previously encountered with ether, and several deaths from cardiac irregularities were reported. The problem lay in the design of the inhalers, which tended to exclude oxygen-containing air. The first specialist anaesthetist, John Snow, carried out animal and clinical experiments which established the importance of controlling the proportion of anaesthetic in the mixture of air and anaesthetic being inhaled. To this end, he designed improved apparatus, and sought out a safer volatile liquid from the increasing range of organic solvents then becoming available.[26] Snow determined both the boiling point and the concentration of solvent when air was saturated with it, using these physical properties to compare one substance with another prior to animal experimentation. He sought a solvent with the physical properties of chloroform combined with the clinical safety of ether, but his early death denied him of his goal, although not before he had achieved worldwide fame by anaesthetising Queen Victoria with chloroform when she gave birth to Prince Leopold in 1853.

Chloroform was not the ideal anaesthetic, nor was it entirely safe. Only 11 weeks after its first use by Simpson, a death occurred in Newcastle during an operation to remove an ingrown toenail. Many more deaths followed, although there were no problems for the majority of patients; during the decade 1870–1880, the *British Medical Journal* reported 120 deaths from chloroform anaesthesia. Official enquiries held throughout the world during the second half of the century led to no solutions. It was not until 1911 that an explanation was forthcoming after Goodman Levy at Guy's Hospital in London found in cats that chloroform produced death by causing sudden cardiac failure in a manner not previously observed. He found three cats in which fibrillation had produced complete absence of regular cardiac pulsation and he went on to show that exactly the same effect was produced by injecting adrenaline into cats lightly anaesthetised with chloroform.[25] The following year, the American Medical Association recommended that chloroform no longer be used as an anaesthetic, advice that was ignored by many anaesthetists. Chloroform anaesthesia only began to diminish in frequency after it was demonstrated in 1937 that chloroform sensitised the pacemaker (controller of the cardiac conduction system) to the action of both endogenous and injected adrenaline, shortening the refractory period between electrical impulse propagation and thereby causing ventricular fibrillation.[26]

Nitrous oxide

It was not until 1862, after others had successfully demonstrated the value of ether and chloroform, that Colton again tried nitrous oxide for dental anaesthesia. On this occasion, he so impressed the dentist concerned, Dr Dunham, that he was asked for information on how to manufacture the gas. On returning to Connecticut a year later, he learned that Dunham had successfully administered nitrous oxide to more than 600 patients. Colton then persuaded him and a colleague to join him in partnership in New Haven. There, in a period of only 23 days, they painlessly extracted over 3000 teeth. Colton then moved to New York, where he built up a highly profitable dental practice. He administered nitrous oxide to thousands of patients without mishap. Word of this soon spread throughout the United States and Europe, and by 1868, nitrous oxide was regularly used in London hospitals. Not long after, it became available as a compressed gas in cylinders.[27]

Ethylene

Ethylene was the first new anaesthetic to be introduced in the twentieth century. The sequence of events leading up to this began when William Crocker and Lee Knight, at the botany laboratory in the University of Chicago, were asked by carnation growers who shipped their products into the city to find out why their flowers closed up when stored in greenhouses and the buds also failed to open. Heavy financial losses had been incurred through this in the 1908 season, so Crocker and Knight began work at once. They traced the cause to ethylene in the gas used to illuminate greenhouses, confirming that as little as 1 part of ethylene in 2 million parts of air caused already-open flowers to close! Further investigations revealed that ethylene damaged a large variety of plants.[28]

$$HC \equiv CH \qquad \text{ethylene}$$

Concern that ethylene might also prove toxic to humans led Arno Luckhardt, a colleague of Crocker and Knight, to expose frogs, rats and a dog to the concentration of ethylene (4% v/v) present in illuminating gas. Although the gas rapidly anaesthetised the animals, no results were published until a full investigation into the possibility of using ethylene clinically was completed five years later in 1923.[29] Luckhardt had by then established that ethylene induced anaesthesia very quickly and pleasantly, while recovery was also rapid and less eventful than with existing anaesthetics. Just days before Luckhardt was ready to publish his work, Easson Brown, an anaesthetist at the Toronto General Hospital, read a paper to the Academy of Medicine in Toronto describing animal experiments that indicated ethylene might be a promising anaesthetic.[30] It was subsequently used until the early 1950s, when there was mounting concern about the risk of explosion with anaesthetics, especially those that had no major advantage over newer agents.

Cyclopropane

Work at Toronto on the clinical evaluation of ethylene was carried out both at the General Hospital and in the department of pharmacology where Velyien Henderson extended the investigations to include the chemically related gas, propylene. This was also a promising anaesthetic when freshly prepared, but after storage in pressurised steel tanks a toxic impurity was generated, causing patients to experience nausea and even cardiac sequelae. The matter was not pursued and propylene was no longer used. Several years later, however, after a chemist, George Lucas, joined the research team, Henderson asked if he could hazard a guess as to the likely nature of the contaminant. When Lucas suggested it might be cyclopropane Henderson asked that some be synthesised and tested on animals. In November 1928, two kittens inside a bell jar were exposed to low concentrations of cyclopropane mixed with oxygen. To the amazement of all present the kittens quickly fell asleep, and on removal from the jar recovered with no hint of any toxic effects![31]

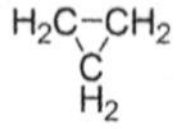

$$H_2C-CH_2$$
$$\underset{H_2}{C} \qquad \text{cyclopropane}$$

Extensive investigations began at once, with the result that a year later Brown safely anaesthetised Henderson, confirming the safety of cyclopropane in humans. These observations were reported to a meeting of the Canadian Medical Association, but the hospital authorities then announced that they would not permit patients to be anaesthetised with cyclopropane. Brown had no choice but to ask Ralph Waters at the University of Wisconsin to carry out the first clinical trials. These were completed by 1934, confirming the status of cyclopropane as a valuable and potent anaesthetic.[32] It is still in use 50 years later. The toxic contaminant in propylene was eventually traced to the presence of hexenes.

Vinyl ether (divinyl ether)

In the twentieth century it was no longer necessary to rely on serendipity to discover new anaesthetics. Chemists and pharmacologists had now become adept at designing stucturally related derivatives of existing drugs, as was the case when Chauncey Leake at the University of California Medical School realised that ethylene lacked some of the desirable properties of ether as an anaesthetic. Believing it would be reasonable to synthesise molecules containing both ether and ethylenic functions, he had the physical properties of five such compounds examined. It was found that only vinyl ether, a novel substance synthesised for Leake by Randolph Major at Merck & Company, had a boiling point and partition coefficient similar to those of ether.[33] Predicting that vinyl ether would be an anaesthetic, Leake and Mei Yu Chen then administered it by inhalation to mice.[34] Their experiment was a success, perhaps too much so, for when vinyl ether was introduced clinically in 1930, anaesthetists found its main drawback was a tendency for it to produce too deep a level of anaesthesia very rapidly.

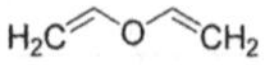

$$H_2C \diagup O \diagdown CH_2 \quad \text{vinyl ether}$$

Reports of liver damage following prolonged administration may possibly have been exaggerated, but they certainly undermined the reputation of vinyl ether. On the credit side, it did not irritate the respiratory tract to the extent that ether did and recovery was quicker than with ether. Although eclipsed by the post-war development of the fluorocarbons, vinyl ether was an important landmark in drug design.

Halothane

Early in 1930, Thomas Midgley was asked by the management of General Motors to develop a non-toxic, non-flammable, low boiling point refrigerant for the company's Frigidaire division. Together with his colleague Albert Henne, he reasoned that the atoms on the right-hand side of the periodic table of chemical elements were the only ones that could permit sufficient volatility for a refrigerant, yet most of those were likely to prove toxic. The inert gases, on the other hand, were already known to be highly volatile. By a process of elimination, Midgley concluded that fluorine should be incorporated into hydrocarbons. He rejected the assumption that such compounds would necessarily be toxic, believing that the stability of the carbon–fluorine bond would be enough to prevent release of either fluorine or hydrogen fluoride. Once the first compound, dichloromonofluoromethane, was prepared, a guinea pig was immediately exposed to it in a bell jar and experienced no ill effects. Thorough toxicological studies then confirmed that fluorocarbon refrigerants were non-toxic under normal conditions of use.[35]

After the Second World War ended, a wider range of compounds became available as a consequence of the wartime development of stable fluorocarbons as solvents for uranium hexafluoride during the separation of uranium isotopes by diffusion. A total of 46 of these were synthesised at Purdue University by Earl McBee and then sent for pharmacological evaluation to B.H. Robbins at Vanderbilt University, Tennessee. It was found that the more volatile compounds were convulsants, whereas those containing heavy atoms such as bromine had high boiling points and were apparently safer anaesthetics than ether or chloroform.[36] The most promising compound then tested would today be recognised as the analogue of halothane (see below) in which the chlorine is replaced by a bromine atom.

Scientists at the laboratories of Ohio Medical Products followed up Robbins' suggestion that more fluoro-analogues should be examined. By this time there was considerable concern over the explosive nature of oxygen–anaesthetic mixtures, but later the British Ministry of Health reported that during the years 1947–1954 there had been a total of only around 40 explosions in the course of 7 million operations carried out in the United Kingdom.

The risk of explosions caused considerable apprehension among operating theatre staff, so much so that it made good commercial sense for Ohio Medical Products to develop a non-explosive anaesthetic. This proved to be a difficult task for their chemists since the first non-inflammable fluorocarbon to be tested lacked anaesthetic activity, whereas the first promising anaesthetic (which was the 24th fluorocarbon to be synthesised) turned out to be flammable, albeit less so than ether. This was fluroxene. Fluroxene was considered by John Krantz at the University of Maryland to compare so favourably with ether that, after extensive animal testing, he proceeded to administer it to a volunteer in April 1953.[37] It was released for general use three years later, but its flammability rendered it inferior to halothane, which was by then undergoing its first clinical trials.

$F_3C\diagdown O\diagup CH_2$ fluroxene   halothane: F_3C—C(Cl)(H)—Br

James Ferguson, the director of the general chemicals division of ICI, was unaware of the developments with fluorocarbon anaesthesia in the United States when he was reviewing his company's range of fluorocarbons in 1950. Their total lack of chemical reactivity convinced him that they should be examined as anaesthetics. Responsibility for the project was given to Charles Suckling, who selected fluorocarbons that satisfied a range of critical requirements drawn up by himself and James Raventos, a pharmacologist from the medicinals section of the company's Dyestuffs Division at Manchester. Their prime objective was to attain low toxicity. The other priorities were that the anaesthetic must be non-explosive, have the right degree of volatility, induce anaesthesia both rapidly and uneventfully, not cause respiratory irritation and be highly potent. The exacting nature of these requirements was to pay handsome dividends because, unlike the American development of fluroxene, the product eventually introduced remained unchallenged for many years – even though it was only the sixth compound synthesised by Suckling. It was first prepared in January 1953 and came quickly to the fore by meeting all the physicochemical and animal criteria.[38,39] After three years of exhaustive testing it was finally put on clinical trial at Crumpsall Hospital in Manchester. Patients were smoothly anaesthetised, their circulatory systems functioned normally and recovery was free from nausea in most cases. The new fluorocarbon was given the approved name of halothane. It quickly became the most widely used inhalational anaesthetic throughout the world, a pre-eminence that remained unchallenged until the 1990s, when concern deepened over the fact that patients who were again anaesthetised with it could suffer life-threatening liver toxicity. This led to the introduction of similar fluorocarbon compounds, such as desflurane, enflurane, isoflurane and sevoflurane. These do not produce the toxic metabolite that is suspected of causing the liver toxicity seen with halothane.

REFERENCES

1. R Porter, *The Greatest Benefit to Mankind: A Medical History of Humanity from Antiquity to the Present*. London: Fontana; 1999.
2. See J.W. Estes, The therapeutic crisis of the eighteenth century, in *The Inside Story of Medicines: a Symposium* eds G.J. Higby, E.C. Stroud. Madison: American Institute of the History of Pharmacy; 1997, pp. 31–49.
3. J. Pereira, *The Elements of Materia Medica and Therapeutics*, Vol. 1, 4th edn. London: Longman, Brown, Green, Longmans and Roberts; 1857, pp. 99–102.
4. J. Pereira, *The Elements of Materia Medica and Therapeutics*, Vol. 2, 4th edn, Part 2. London: Longman, Brown, Green, Longmans and Roberts; 1857, p. 138.
5. E.F. Cartwright, *The English Pioneers of Anaesthesia (Beddoes; Davy; Hickmann)*. Bristol: John Wright; 1952, p. 108.
6. R.E. Schofield, *The Lunar Society of Birmingham*. Oxford: Clarendon Press; 1963.
7. T. Beddoes, *Considerations on the Medicinal Use and on the Production of Factitious Airs*, Part I. Bristol: Bulgin and Rosser; 1794.
8. J. Watt, *Considerations on the Medicinal Use and on the Production of Factitious Airs*, Part II. Bristol: Bulgin and Rosser; 1794.
9   H. Davy, *Researches, Chemical and Philosophical, Chiefly Concerning Nitrous Oxide*. London: Butterworths; reprinted 1972.
10 R.J. Wolfe, L.F. Menczer, I awaken to glory. Essays celebrating the sesquicentennial of the discovery of anesthesia by Horace Wells. December 11, 1844–December 11, 1994. Boston: Boston Medical Library; 1994.
11. B.M. Duncum, *The Development of Inhalational Anaesthesia*. London: Oxford University Press; 1947.
12. G.K. Tallmadge, The third part of the de Extractione of Valerius Cordus. *Isis*, 1925; **7**: 394–420.
13. J.A.S. Frobenius, *Phil. Trans. Roy. Soc. Lond.*, 1730; **36**: 283.
14. F.L. Taylor, *Crawford W. Long and the Discovery of Ether Anesthesia*. New York: Hoeber; 1928.
115. C.W. Long, An account of the first use of sulphuric ether. *Southern Med. J.*, 1849; **5**: 705–13.
16. R.J. Wolfe, *Tarnished Idol: William Thomas Green Morton and the Introduction of Surgical Anesthesia, A Chronicle of the Ether Controversy*. San Anselmo, California: Norman Publishing; 2001.
17. J.F. Fulton, M.E. Stanton, *The Centennial of Surgical Anaesthesia. An Annotated Catalog*. New York: Henry Schuman; 1946.
18. S. Guthrie, New mode of preparing a spiritous solution of chloric ether. *Am. J. Arts Sci. (Silliman's)*, 1831; **21**: 64.
19. J.B. Dumas, *Ann. Chim. Phys.*, 1834; **56**: 120 and 1840; **71**: 353.
20. R.J. Defalque, A.J. Wright. The short, tragic life of Robert M. Glover. *Anaesthesia*, 2004; **59**: 394–400.
21. H. Laing Gordon, *Sir James Young Simpson and Chloroform (1811–1870)*. London: Fisher Unwin; 1897, p. 106.
22. J.Y. Simpson, *Account of a New Anaesthetic Agent, as a Substitute for Sulphuric Ether*. Edinburgh: Sutherland and Knox; 1847.
23. J.Y. Simpson, A new anesthetic agent, more efficient than sulphuric ether. *Lancet*, 1847; **2**: 549–50.
24. J. Snow, *On Chloroform and Other Anaesthetics*. London: Churchill; 1858.
25. A.G. Levy, T. Lewis, Heart irregularities resulting from the inhalation of low percentages of chloroform vapour, and the relationship to ventricular fibrillation. *Heart*, 1911; **3**: 99.
26. W.J. Meek, H.R. Hathaway, O.S. Orth, Effects of ether, chloroform, and cyclopropane on cardiac automaticity. *J. Pharm. Exp. Ther.*, 1937; **61**: 240.
27. W.D.A. Smith, *Under the Influence: History of Nitrous Oxide and Oxygen Anaesthesia*. London: Macmillan; 1982.

28. W. Crocker, L. Night, Effect of illuminating gas and ethylene upon flowering carnations. *Bot. Gaz.*, 1908; **46**: 259.
29. A.B. Luckhardt, J.B. Carter, Ethylene as a gas anesthetic; preliminary communication. *J. Am. Med. Ass.*, 1923; **80**: 1440.
30. W. Easson Brown, Preliminary report on experiments with ethylene as a general anaesthetic. *Can. Med. Ass. J.*, 1923; **13**: 210.
31. G.H. Lucas, The discovery and pharmacology of cyclopropane. *Can. Anesth. Soc. J.*, 1960; **7**: 237–56.
32. R.M. Waters, E.R. Schmidt, Cyclopropane anesthesia. *J. Am. Med. Ass.*, 1934; **103**: 975.
33. W.L. Ruigh, R. Major, The preparation and properties of pure divinyl ether. *J. Am. Chem. Soc.*, 1931; **53**: 2662–71.
34. C.D. Leake, M.Y. Chen, Anesthetic properties of certain unsaturated ethers. *Proc. Soc. Exp. Biol. Med.*, 1930; **28**: 151–4.
35. T. Midgley, A.L. Henne, From the periodic table to production. *J. Ind. Eng. Chem.*, 1937; **29**: 241–4.
36. B.H. Robbins, Preliminary studies of anesthetic activity of fluorenated hydrocarbons. *J. Pharm. Exp. Ther.*, 1946; **86**: 197–204.
37. J.C. Krantz Jr, C.J. Carr, G. Lu, F.K. Bell, Anesthesia. XL. The anesthetic action of trifluoroethyl vinyl ether. *J. Pharm. Exp. Ther.*, 1953; **108**: 488–95.
38. J. Raventos, The action of fluothane; a new volatile anaesthetic. *J. Pharm. Exp. Ther.*, 1956; **11**: 394–410.
39. C.W. Suckling, Some chemical and physical factors in the development of fluothane. *Br. J. Anaesth.*, 1957; **29**: 466–72.

Alkaloids

The greatest advance in the process of drug discovery was the isolation of pharmacologically active substances from plants in the early years of the nineteenth century. The first to be isolated was morphine and the work was carried out in Germany, but had this not come to the attention of the leading French chemist of the day, Joseph Gay-Lussac, it could have remained an obscure piece of scientific investigation. Instead, French chemists were able to exploit the discovery and lead the way in isolating more active principles. The seeds of the French dominance in this area had been planted some years earlier by Antoine Fourcroy (1755–1809).[1] The son of an apothecary, he had been supported financially while he studied medicine by influential members of the Société Royale de Médecine. Such was his talent for chemistry that the Société permitted him to participate in its analysis of mineral waters for the government even before he graduated. He soon rejected the approach of evaporating the waters to dryness, and instead used specific chemical reagents to determine which minerals were present. Fourcroy went on to work with Lavoisier and then devoted much of his career to the application of chemistry to medicine. In 1792 he published an analysis of St Lucia and St Domingo barks, which had been recommended as substitutes for cinchona bark in the treatment of malaria.[2] For many years his study was considered a classic of vegetable analysis. Significantly, it stimulated others to examine cinchona and opium, the two most important vegetable drugs then in common use, and which were often adulterated by unscrupulous dealers.

New institutions of higher learning were established in France following the decision of the Convention in 1793 to suppress academic and professional bodies that had enjoyed privilege under the monarchy. Responsibility for medicinal analysis passed to the Société de Pharmacie and the Ecole Supérieure de Pharmacie, which opened in Paris in 1803. The first director of the Ecole Supérieure was a protégé of Fourcroy, Nicolas Louis Vauquelin, the discoverer of chromium and beryllium.[3] As the leading analytical chemist in Paris, he responded to growing concern in medical circles over the variable quality of plant products such as opium by encouraging his faculty members and others to isolate the active principles from plants so that reliable chemical assays could be established.

It is not merely that the appropriate institutions were in place that accounts for the ensuing breakthrough in the isolation of active principles. The methodology of solvent extraction had also been developed, albeit in a different field of chemistry, with its origins in 1747 when the German chemist Andreas Marggraf observed crystals forming after beetroot had been extracted with brandy. The crystals tasted very sweet and Marggraf was able to show they were identical to those obtained from sugar-cane. Margraaf tried to develop the isolation process, but his pupil, Franz Achard, realised that the yield of sugar extracted was low and set about breeding sweeter crops of beet. Progress was initially very slow, but in 1774 he began small-scale production. Finally, in 1801 he opened a factory in Silesia. This never made a profit, but the solvent extraction technique was now firmly established. When cane sugar supplies from the West Indies could no longer enter mainland Europe due to Napoleon's ban on trade with Britain, Benjamin Delessert opened a factory in Paris to produce beet sugar in

Drug Discovery. A History. W. Sneader.
©2005 John Wiley & Sons Ltd

1813, much to the delight of the Emperor. An acute awareness of the importance of solvent extraction now existed among chemists in the city.

The Swedish chemist Carl Wilhelm Scheele (1742–1786) had been the first to apply the process of solvent extraction in the isolation of acids from plant juices.[4] He also exploited the solubility of substances of varying polarity in polar and non-polar solvents whereby polar impurities were left behind in water when it and an immiscible organic solvent such as ether were shaken together in a vessel from which the aqueous layer could be removed after the immiscible layers separated on standing. Once the remaining solvent had been evaporated, the crude residue was purified by crystallisation. This enabled Scheele, during the 1780s, to isolate citric, malic, oxalic and gallic acids. He only attempted to isolate organic acids, possibly because of the prevailing impression that active principles from plants would be acids.

So successful was solvent extraction that it still remains a powerful tool in the hands of the natural product chemist. However, since the quest to obtain penicillin during the Second World War, it has been complemented by chromatographic techniques that exploit the divergent abilities of different molecules to adhere to materials such as silica gel or dried alumina in the presence of solvents of varying polarity. Columns and thin layers of adsorbent materials have been used for both qualitative and quantitative separation of natural products, supplemented in recent times by preparative high-performance liquid chromatography.

Once a product had been isolated it had to be subjected to testing in order to ensure that it was indeed an active principle. For Scheele, this presented no problem since he isolated fruit acids with instantly identifiable taste and which were not harmful, but when later experimenters chose to test the material on themselves it was not without risk! Chemists recognised they were isolating potent poisons and soon switched to testing their extracts on animals. It is no coincidence that within 30 years of the initial isolation of active substances from plants, experimental pharmacology had begun to develop as a distinct discipline in the hands of Rudolf Buchheim (1820–1879). He was appointed to the first-ever university Chair of Pharmacology at Dorpat in Estonia when he was only 27 years old. Although it was then an Imperial Russian university, the language of instruction and most of the staff were German. Buchheim then built and paid for the first pharmacology laboratory in the world in the basement of his own home. By 1860 his reputation was such that the university finally provided him with a large purpose-built institute.[5] He was succeeded in 1869 by his former student Oswald Schmiedeberg, a son of a local forester, who had obtained his doctorate three years earlier. Schmiedeberg moved to the newly re-established University of Strassburg (as it was then known), which opened when the city was returned to Germany after the Franco-Prussian War of 1870–1871. No expense was spared by the Reichsland of Elsass-Lothringen (now Alsace-Lorraine) in turning its university into one of the finest in Germany. Schmiedeberg was able to build a magnificent Institute of Pharmacology which took until 1887 to complete. His research proceeded unhindered, enabling Schmiedeberg to persuade the best students to work with him at a time when elsewhere pharmacology was held in low esteem. During the 46 years Schmiedeberg occupied his chair, he trained most of those who were to become professors of pharmacology at German and foreign universities.[6] His endeavours led directly to the overwhelming pre-eminence of the German pharmaceutical industry from the 1880s until the outbreak of the Second World War in 1939.

OPIUM ALKALOIDS

Opium (Greek: *opos* = juice) is obtained by drying the latex that exudes from the capsule of the poppy, *Papaver somniferum* L. The suggestion that the *Ebers Papyrus* refers to a mixture of poppy pods and flies as a sedative for children must be questioned in the light of a recent chemical investigation of material from the tomb of an Egyptian royal architect, which has thrown doubt on claims that the opium poppy was known in Ancient Egypt around the time of the compilation of the *Ebers Papyrus*.[7] In the *Iliad*, Homer mentions poppy growing in gardens, implying that their cultivation had been established by the eighth century BC. Later, Aristotle stated that the poppy was narcotic since its vapours rose to the brain.

The first clear description of the poppy appeared in the *Historia Plantarum* by Theophrastus. The collection of opium from the plant is mentioned, but the earliest accurate account of the procedure is to be found in the *Compositiones* written by Scribonius Largus around AD 43. Pliny the Elder (23–79) also mentioned opium, in his 37-volume work on natural history. Sharing his fellow citizens' contempt for Greek physicians who monopolised medical practice in Rome, he did not lose the opportunity to warn of its dangers. Nonetheless, its pain-relieving properties were clearly established and opium was prescribed by Galen (129–199). Alexander of Tralles also mentioned it repeatedly in his *De Re Medica*, written in the middle of the sixth century. Macht has written an account of the subsequent history of opium.[8]

In 1803, Charles Louis Derosne, owner of a fashionable Parisian pharmacy in the Rue St Honoré, delivered a memoir to the Société de Pharmacie, in which he reported that in the course of devising an assay for opium he had isolated a novel crystalline salt.[9] A variety of tests revealed that it had alkaline properties, which Derosne attributed to contamination by the potash used to precipitate it from acid solution. He appreciated that he had been handling a peculiar substance that was not a plant acid, but he described it as a salt simply because it readily crystallised. He added the rider that this was a circumlocution to compensate for his inability to assign it to any known class of chemical compounds. The matter rested there.

Armand Séguin, director of a successful tannery outside Paris, reported to the Institut de France in December 1804 that he had isolated a novel plant acid and also a crystalline narcotic from opium. His findings were presented in a paper that he submitted to the Académie des Sciences, but it was laid aside. Séguin did not continue with his investigations, and by the time his paper finally appeared in the *Annales de Chimie* in 1814[10] similar observations had been reported elsewhere.

Friedrich Wilhelm Sertürner (1783–1841), the son of an Austrian engineer in the service of Prince Friedrich Wilhelm of Paderborn and Hildesheim, completed his apprenticeship with the court apothecary in 1803 when he was 20 years old. He remained at the Adlerapotheke in Paderborn for a further 30 months, during which period he was able to carry out a variety of chemical experiments. Examining opium, he was quickly able to extract an organic acid that had not been reported in the literature. He named it meconic acid (Greek: *mekon* = poppy). When tested on dogs, it was inactive. However, alkalinisation of the mother liquors with ammonia caused precipitation of a substance that he crystallised from alcohol. When he administered this to a dog, it induced drowsiness. He published a preliminary report in 1805 in Johann Trommsdorff's *Journal der Pharmacie*. A detailed account of his isolation of the 'principium somniferum' appeared the following year.[11] Sertürner included in his paper a footnote stating that he had not been aware of Derosne's work until after his own had been completed. That he also wrote of the 'almost alkali-like character' of this principle from a plant source should have caught the attention of chemists, for all plant principles isolated prior to this were acidic in nature. That his work failed to arouse interest was due to his isolation and the fact that he published his findings in a journal read mainly by practising apothecaries.

Sertürner had explained that his 'principium somniferum' could neutralise free acid, yet he did not recognise the significance of this aspect of his work. It was another three years before he briefly renewed his investigations after opening his own apothecary in Einbeck, Westphalia. In 1811, he published further papers in Trommsdorff's *Journal der Pharmacie*, in which he confirmed that the narcotic principle of opium was an alkaline substance, or base, that formed salts with acids. However, it was not until 1815 that he carried out a more detailed investigation on it in the Einbeck Ratsapotheke and identified the presence of carbon, hydrogen, oxygen and possibly nitrogen. In 1817, he published yet another paper, but this time it appeared in a prominent journal, Gilbert's *Annalen der Physik*.[12] Sertürner drew attention to the particular ease with which what he called morphium reacted with acids to form readily crystallisable salts. He also described how he and three youths swallowed doses of about 100 mg and experienced the symptoms of severe opium poisoning for several days, despite recourse to strong vinegar to induce vomiting when these symptoms first appeared! This time, Sertürner was not ignored. Joseph Gay-Lussac, the doyen of French chemists, read the paper and immediately had it translated and republished in the prestigious *Annales de Chimie*, the journal founded by Lavoisier in 1789 and which Gay-Lussac now edited.

Gay-Lussac wrote an editorial to accompany the translation of Sertürner's paper. In this, he expressed surprise that Sertürner's work had been ignored for so many years, but not simply because the isolation of the active principle of opium was important. Of much greater significance, according to Gay-Lussac, was the discovery of a salt-forming organic plant alkali analogous to the familiar organic acids. He predicted that many other organic alkalis would be found in plants, for there was already some evidence to suggest that the few crude active principles isolated in the previous decade contained nitrogen and had alkaline properties. To ensure a degree of conformity in the naming of plant bases, Gay-Lussac proposed that their names should end with the suffix '-ine'. This was the first time such standardisation of nomenclature had been suggested in organic chemistry. For this reason, Gay-Lussac altered Sertürner's term 'morphium' to 'morphine'. In 1818 the German chemist Wilhelm Meissner introduced the term 'alkaloid' to describe plant alkalis, though several years passed before it was generally accepted.

Gay-Lussac asked Professor Robiquet at the Ecole Supérieure de Pharmacie to check Sertürner's experimental work. Robiquet noted the differences in the properties of the salts isolated by Derosne and Sertürner, and concluded that they were different plant alkalis. He then purified the base isolated by Derosne and gave it the name 'narcotine', but despite this it had no narcotic properties. It was later found to retain the cough suppressant properties of morphine, and is still prescribed for this purpose under the name of 'noscapine'.

It subsequently transpired that even the purest samples of morphine had been contaminated with narcotine and it was not until 1831 that William Gregory, Professor of Chemistry at the University of Edinburgh, devised an economic process for obtaining the alkaloid in a high state of purity. The following year he prepared pure morphine hydrochloride, which a local laudanum (i.e. tincture of opium) producer, John Macfarlan, immediately began to manufacture in bulk. Morphine hydrochloride gained considerable importance after Alexander Wood, also of Edinburgh, developed the hypodermic syringe in 1853 for administering subcutaneous injections. The quality of pain relief obtained by patients was thereby greatly enhanced.

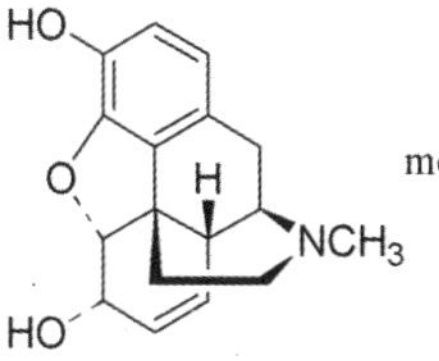

morphine

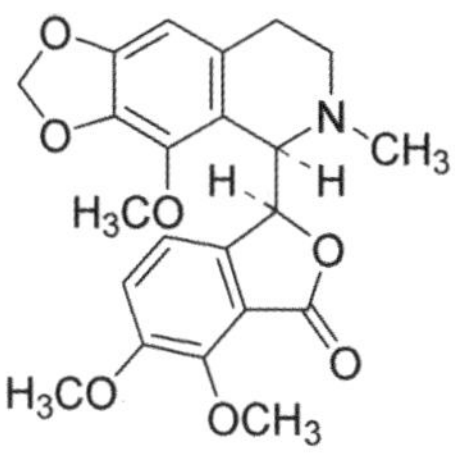

noscapine (narcotine)

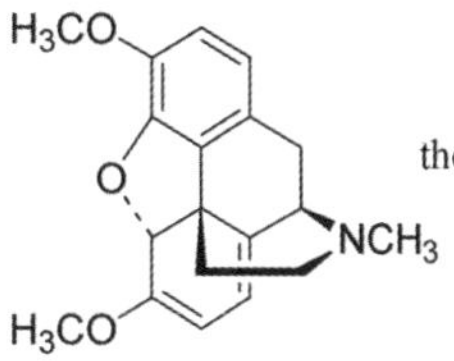

thebaine

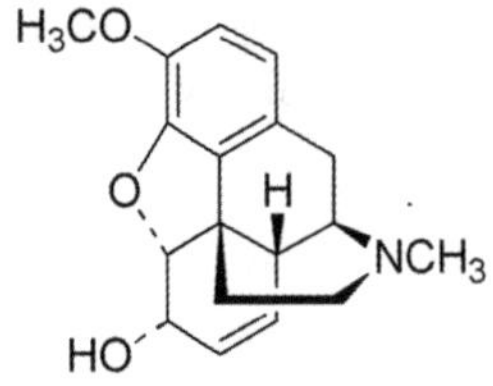

The chemical structure of morphine was not elucidated until 1923, when J. Masson Gulland and Robert Robinson at the University of St Andrews in Scotland succeeded in establishing it correctly.[13] Morphine was synthesised in 1950 by Marshall Gates and Gilg Tschudi at the University of Rochester, New York, but their complex synthesis was purely of academic interest.[14] This has proved to be the case with most alkaloids. Despite its addictive liability and hazardous depressive effect on the respiratory centre plus a tendency to cause nausea and constipation, morphine still remains the most effective drug available for the alleviation of severe pain. Its analgesic power is enhanced by the associated state of euphoria and detachment.

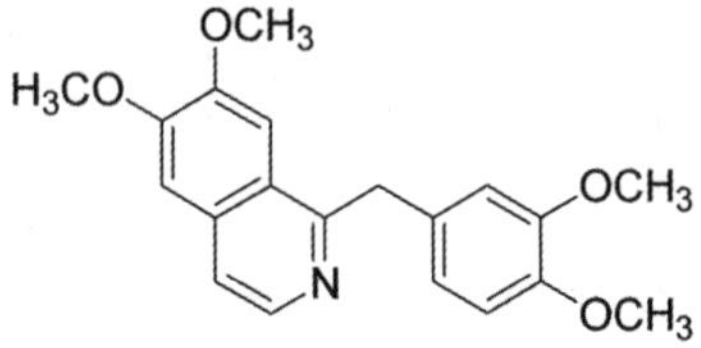

Three more opium alkaloids were eventually isolated from the mother liquors remaining after the extraction of morphine from opium. Thebaine was isolated in 1832 by Thibouméry, the manager of the factory built by Pelletier to produce alkaloids. It was of no medicinal value, yet it eventually turned out to be a key compound for chemists since the reactivity of its conjugated diene system permitted the synthesis of novel codeine (and morphine) analogues from it.

In the same year, Pierre-Jean Robiquet stumbled upon another new alkaloid while examining an alternative morphine extraction procedure.[15] He named it codeine (Greek: codeia = poppy capsule). Although its narcotic properties were mild, it retained the full capacity of morphine as a cough suppressant. This property of opium had long been recognised, being cited in the fourteenth century by Actuarius of Constantinople.[16] Magendie considered the main value of codeine was as a cough suppressant, especially in whooping cough. To this day, codeine linctus remains a popular remedy for coughs, but it is also a popular ingredient in combination with aspirin or paracetamol for the relief of mild to moderate pain.

It was not until 1848 that George Merck, the son of an alkaloid manufacturer in Darmstadt, Germany, isolated papaverine from the mother liquors remaining after the extraction of morphine. As it exhibited only feeble analgesic activity it was ignored until after the American pharmacologist David Macht, in 1916, described its spasmolytic action on smooth muscle.[17] From then until the introduction of synthetic atropine analogues in the 1930s, it was frequently prescribed as a vasodilator and antispasmodic.

EMETINE

Shortly before the French translation of Sertürner's 1817 paper on morphine was published, one of the professors at the Ecole Supérieure de Pharmacie, Pierre-Joseph Pelletier, turned his attention to the isolation of plant alkalis. Impure crystals that he obtained from ipecacuanha root were given the name 'emetine'[18] and were tested on animals by François Magendie, who subsequently recommended their use in the treatment of amoebic dysentery. Pelletier prepared purer crystals in 1823, but these were later found to consist of a mixture of several related alkaloids, all of which were less effective.

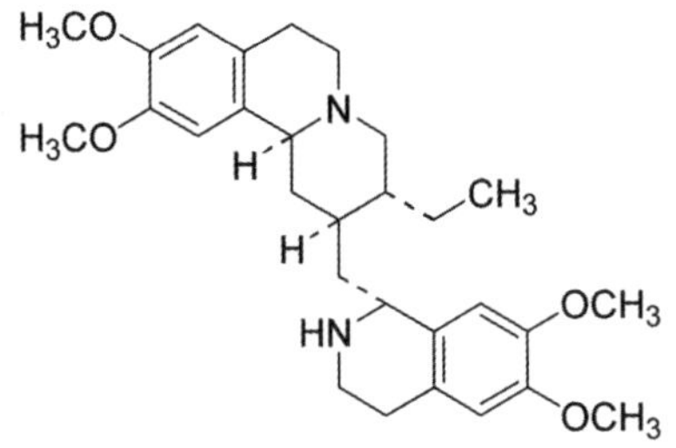

emetine

The pure alkaloid was not obtained until 1887.[19] Robert Robinson[20] determined the structure of emetine in 1948 and the alkaloid was synthesised in 1952 by R.P. Evstigneeva and his colleagues at the M.V. Lomonssov Chemical Technical Institute in Russia.[21] A chiral synthesis was independently achieved later that decade.

QUININE

Concern about the variable responses of his patients to treatment with cinchona bark led the Portuguese surgeon Bernardino Gomez to attempt to isolate the active principle. In 1811, he managed to obtain silvery crystals by first extracting the bark with dilute acid and then neutralising it with alkali. He named the crystals 'cinchonin' (later changed to cinchonine), and assumed they were the sole active principle. They did not possess the bitter taste of cinchona. Gomez noted that the chemical properties of cinchonin were unlike those of any known plant product, but he did not have any reason to believe that it was a plant alkali. It was only after Gay-Lussac had drawn attention to the existence of such alkalis that cinchona bark was carefully examined for their presence.

In 1820, Pelletier and Joseph-Bienaimé Caventou repeated Gomez's experiments on cinchona. By modifying the procedure somewhat, they isolated not only cinchonine from samples of grey bark, but also a more potent alkaloid, quinine, from yellow bark (*Cinchona cordifolia*).[22] In their paper of 1820 announcing their isolation of quinine, Pelletier and Caventou urged medical practitioners to study the therapeutic properties of the pure cinchona alkaloids. This marked a new departure, for previously the active principles of plants had been isolated for analytical rather than therapeutic purposes. By urging their medical colleagues to study pure plant principles, Pelletier and Caventou set the course of drug discovery in a new direction.

Among the first to experiment with quinine was Magendie. He evaluated it in dogs and then went on to treat patients. By including it in his widely consulted *Formulaire pour la Préparation et l'Emploi de Plusieurs Nouveaux Médicamens*, he ensured that knowledge of quinine would quickly spread around the world.[23] This small volume, of which there were to be many editions, listed pure chemicals rather than the hitherto standard mixtures and plant products found in similar works. This was a significant advance, for it enabled Magendie's successors to consolidate the practice of experimental medicine on a firm basis, with the assurance that the drugs they studied were of unchanging constitution.

Magendie was so impressed with quinine and the other alkaloids that he wrote in an English translation[24] of his *Formulaire*:

"We know how advantageous it is in the treatment of disease to be certain of the precise dose of all active remedies; this advantage especially applies to the present case, because the quantity of the alkalis contained in the cinchonas varies prodigiously, according to the nature

and quality of the bark which is employed. It is often also very desirable to administer this medicine in a small volume, and in an agreeable form. Patients often die of malignant fevers because they cannot swallow the necessary quantity of the bark in powder. Some throw it up after having taken it; and in others superpurgation arises, so that the powder passes through the intestinal canal without producing any effect; even in the most favourable cases it is necessary the patient's stomach should, as it were, chemically analyse the bark with which it is filled, and extract the febrifuge principle. A process like this will be always difficult and fatiguing even for the strongest stomach. Chemistry, therefore, has done a great service to medicine, by shewing how this separation may be accomplished beforehand."

Other Parisian physicians confirmed that quinine was able to arrest their patients' fevers and could be used instead of the unpalatable, nauseating powdered cinchona bark. To meet the immediate demand for supplies of quinine, Pelletier and Caventou manufactured it in their own pharmacies in the Rue Jacob and Rue Gaillon. The following year, Pelletier rushed supplies of quinine to Barcelona where an epidemic had broken out. Soon, factories had to be opened, and by 1826 Pelletier's factory was processing more than 150 000 kg of cinchona bark yearly, yielding around 3600 kg of quinine sulfate. Production elsewhere was also expanding, thanks to Pelletier and Caventou openly publishing the full details of how to obtain quinine from cinchona bark. Several manufacturers in Germany also prepared quinine. These and others quickly recognised the economies obtained through large-scale production of quinine and morphine, and also acquired specialised knowledge that enabled adulterated or inferior plant material to be rejected. In this way, the modern pharmaceutical industry gradually became established.

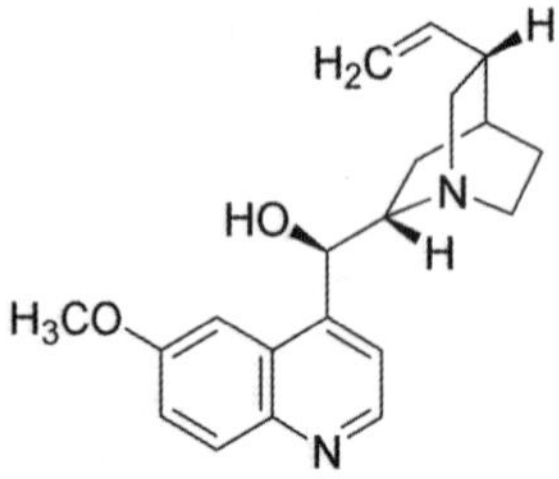

quinine 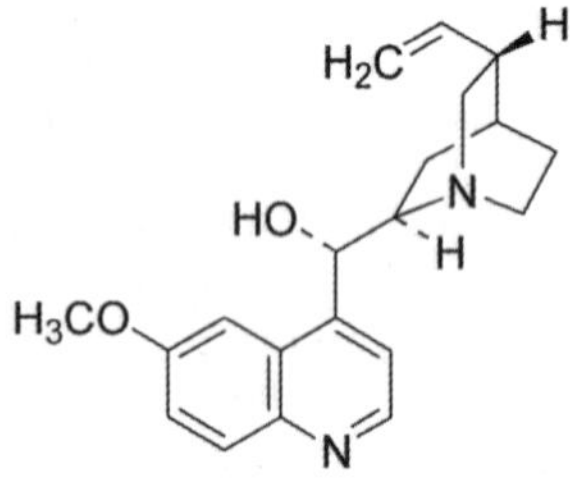   quinidine

Soon after its introduction, quinine began to replace the vegetable products described as 'antiperiodics', which had been commonly used to relieve malarial paroxysms, still the commonest febrile condition at that time. This drew attention to its antipyretic properties, with the result that until the 1880s quinine became the most frequently prescribed drug for lowering temperature in any febrile condition. It was not until the end of the nineteenth century that the mode of action of quinine in malaria was revealed. At St Petersburg in 1891, Yuri Romanovsky applied a stain of eosin and methylene blue to reveal that the malarial parasites in blood samples taken from patients treated with quinine were damaged.[25] These observations led him to the conclusion that the action of quinine was directly on the parasites rather than on the patients. This was a revolutionary concept, yet it drew little attention.

Despite its toxicity, quinine was unrivalled in the treatment of malaria until the introduction of safer synthetic drugs. These displaced it, but the problem of widespread drug resistance has resulted in the reintroduction of quinine for treatment of malaria caused by resistant strains of *Plasmodium falciparum*.

The chemical structure of quinine was confirmed by its synthesis at Harvard University in 1944 by Robert Woodward and William von Eggers Doering.[26]

QUINIDINE

Soon after the introduction of cinchona bark, the German physician Georg Stahl had recognised its action on the heart and subsequently sporadic attempts were made to exploit this. In the second half of the nineteenth century, some physicians prescribed quinine on account of its depressant action on heart muscle. In 1912, a patient told the Dutch cardiologist Karl Wenckebach that he took quinine to bring his atrial fibrillation under control. When Wenckebach tried quinine on other patients, only one responded favourably. In his book on cardiac arrhythmias, published in 1914, Wenckebach referred to the matter in passing. Four years later, a leading Viennese medical journal carried a report from Berlin by Walter Frey, which stated that quinidine was the most effective of the four principal cinchona alkaloids in controlling atrial arrhythmias. Quinidine was also the least toxic, but was still a hazardous drug requiring skilled administration. It was an optical isomer of quinine isolated in 1833 by L. Henry and Auguste Delondre and was found to be inferior to quinine as an antimalarial.[29] It is still used for suppression of supraventricular and ventricular arrhythmias.

XANTHINE ALKALOIDS

The stimulating effects of coffee, *Coffea arabica* L., have long been recognised in the Middle East. Coffee was introduced into England and France from Constantinople in the middle of the sixteenth century and became a popular beverage. Ferdinand Friedlieb Runge isolated a stimulant alkaloid from mocha beans given to him by the German poet and chemist Goethe in 1821 and named it caffeine.[30] The chemical structure was established by Emil Fischer at the University of Berlin in 1882. He also synthesised it three years later.[31]

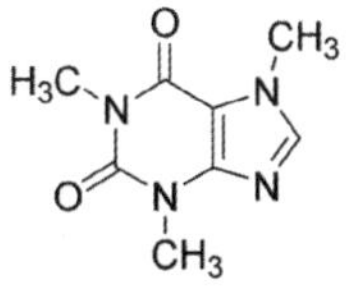

caffeine 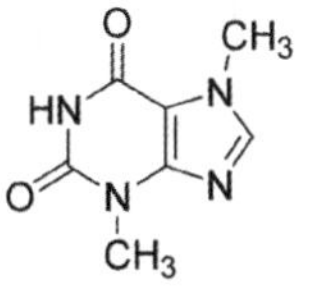   theobromine 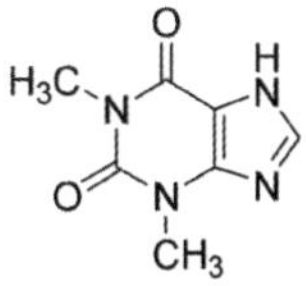   theophylline

Caffeine soon became popular as a stimulant. Until the introduction of the synthetic diuretics, it and the other xanthine alkaloids present in tea, cocoa and coffee were frequently prescribed as diuretics until the introduction of the thiazides. Theobromine, isolated in 1878 from seeds of the cacao tree, *Theobroma cacao* L. (Sterculiaceae), was found to be more effective than caffeine as a diuretic, while less stimulating.[32] Theophylline from the tea plant, *Camellia sinensi* L. (Theaceae), that grows in India and China was isolated in 1888.[35] Its chemical structure was elucidated in 1895.[31] As the yield of alkaloid was only 0.1% it was not evaluated clinically until after its synthesis. In 1902, Oskar Minkowski established that it was around three times as potent a diuretic as caffeine.

In a widely read treatise on the treatment of asthma published in 1860, the London physician Henry Salter noted that strong black coffee could relieve the breathing problems of asthmatics.[34] It was later shown that theophylline was a bronchodilator. However, the safety margin of theophylline in the treatment of asthma is quite narrow and it has been largely superseded by modern bronchodilators.

ATROPINE AND HYOSCINE

Country people have long been aware that belladonna, *Atropa belladonna* L. (Solanaceae), can dilate the pupil, but it was a study by Karl Himly in Paris in 1802 that encouraged ophthalmologists to use belladonna to assist examination of the eye.[35] An impure alkaloid was obtained in 1822 by Rudolph Brandes, pure atropine being isolated from the roots by Philipp Geiger in 1833.[36] Atropine is an artefact formed during the extraction of the alkaloid from the plant. The naturally occurring material is hyoscyamine, present in plants as a laevorotatory stereoisomer. During isolation, hyoscyamine transforms into a mixture of equal parts of its two optical isomers. This mixture is atropine, and is the form in which the alkaloid is used clinically, even though most of the activity resides in the *laevo*-isomer.

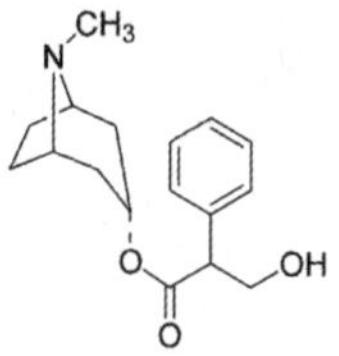

atropine 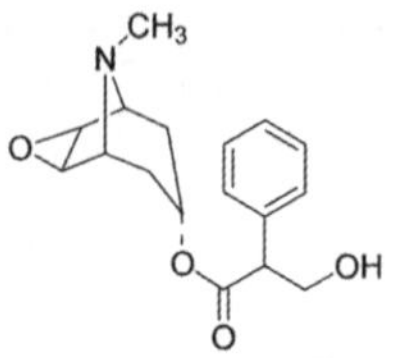   hyoscine

Several plants of the family Solanaceae contain both hyoscyamine and a structurally similar alkaloid known as hyoscine. Both alkaloids block the access of the neurohormone acetylcholine to the receptor site where it acts once it is released from cholinergic nerve endings. In the United States, hyoscine is known as scopolamine, reflecting the use of *Scopola carniolica* as a source of this mydriatic alkaloid. This plant is so named because it was studied at the University of Pavia by the eighteenth century botanist Johann Scopoli.

There is a long history in Ayurvedic medicine of the inhalation through a hukka of the vapour produced by warming *Datura ferox* L. leaves as a means of relieving asthma. Dr James Anderson, Fellow of the Royal Society and Physician General at the Madras Hospital in India, was an enthusiastic botanist who benefited from this treatment in 1802 and duly reported it to his friend Dr Sims in Edinburgh. Following the publication in 1812 of a report in the *Edinburgh Medical and Surgical Journal* by Dr Sims, cigarettes containing the leaf of *D. stramonium* became widely used by asthmatics.[37] After a report by Henry Salter in 1869 describing the successful administration of belladonna and also atropine in asthma, the latter became widely used for this purpose.[38] Other preparations containing the leaf were also used in the nineteenth century, including the highly popular Potter's Asthma Cure®.

Rudolf Buchheim believed that the sedative properties of henbane might be due to the presence of a second alkaloid besides hyoscyamine, but it was Alfred Ladenburg at the University of Kiel who isolated hyoscine in 1881 and showed that it was structurally similar to atropine.[39]

Workers in the Hoffmann–LaRoche laboratories in Switzerland overcame a major impediment to the clinical acceptibility of hyoscine preparations when they discovered that transformation of the alkaloid took place in solution to form equal amounts of its two optical isomers. The dextrorotatory isomer was shown to be a central nervous system stimulant, in contrast to the laevorotatory isomer, which was a sedative. This explained why the racemic combination produced unpredictable effects. It became possible to use the laevorotatory isomer of hyoscine for pre-anaesthetic medication in formulations such as papaveretum and hyoscine injection, introduced in 1910.[40]

PHYSOSTIGMINE

William Daniell, a British Army surgeon, reported to the Ethnological Society of Edinburgh in 1846 that he had observed the use of the Calabar bean, *Physostigma venenosum* Balf. (Leguminosae), as an ordeal poison in Old Calabar, Nigeria.[41] Criminals convicted of capital offences were forced to swallow a watery extract of the beans. It has been suggested that those with a clear conscience swallowed the drug, which either caused them to vomit or resulted in hydrolysis of the active principle, whereas those who were guilty retained it in their mouth and thus facilitated buccal absorption.

Daniell's report interested Robert Christison, Professor of Materia Medica at Edinburgh, who arranged for a missionary to send him a Calabar bean, which he then had cultivated. Animal experiments showed that extracts of the bean killed by stopping the heart. Uncertain about the authenticity of the bean he had received, Christison experimented on himself. He experienced extreme weakness and was fortunate to survive, for his heart had probably gone into atrial fibrillation.[42] His distinguished student Thomas Fraser isolated the active principle as an amorphous powder. Jobst and Hesse in Germany obtained pure crystals in 1864 and named their product 'physostigmine'.[43]

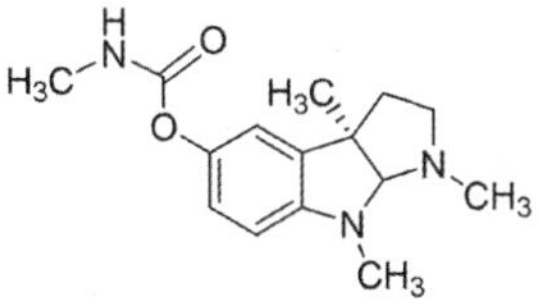

physostigmine

The value of physostigmine in preventing blindness caused by glaucoma was revealed in 1875 by Ludwig Laqueur.[44] It remains in use for the treatment of chronic simple glaucoma, drops being applied to the eye several times a day to relax the ciliary muscles and improve outflow of aqueous humour with consequent reduction in the elevated intraocular pressure.

COCAINE

The earliest written account of the coca bush, _Erythoxylon coca_ L. (Erythoxylaceae), appeared in Nicolas Monardes' _Historia Medicinal de Indias Occidentales_, published in Seville in 1565. To the Indians of Peru it was the incarnation of a god, with the fields where it grew being revered for their holiness. Its use, however, was restricted to only the most noble of the Incas until the Spanish authorities introduced legislation at the second Council of Lima, in 1569, making the plant freely available to all. The subsequent popularity of the coca leaf among the Indians was described by William Prescott in 1847 in his _History of the Conquest of Peru_. This stimulated interest in coca leaves among organic chemists, culminating in the isolation of pure crystals of cocaine in 1860 by Albert Niemann at Göttingen University.[45]

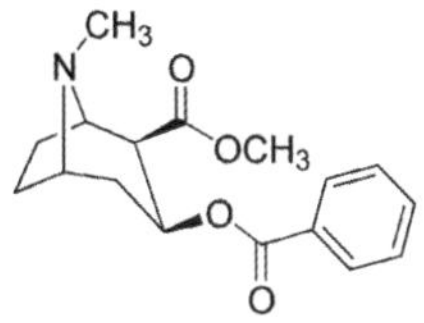

cocaine

During the next two decades, cocaine was considered to be a mild stimulant similar to caffeine. Parisian society imbibed Vin Mariani, a medicinal wine first prepared from an infusion of coca leaves by Alberto Mariani in 1863. The popularity of this and similar beverages spread to the United States, where even the medical profession lauded Mariani Elixir, Mariani Lozenges and Mariani Tea! One of the ingredients when Coca-Cola® was introduced in the United States in 1886 was coca leaf, until the formulation was changed in 1903.[46]

A report that cocaine had helped Bavarian soldiers to withstand hunger and fatigue during taxing military manoeuvres led Sigmund Freud to investigate whether cocaine could counter fatigue, his findings later being published as _The Cocaine Papers_.[47] He was assisted by Carl

Koller, a trainee ophthalmologist. Koller knew that anaesthesia with ether or chloroform was hazardous in eye surgery as patients could vomit or be restless on awakening. He had experimented with various drugs such as chloral hydrate, sodium bromide and morphine as potential topical anaesthetics for the eye, but met with no success. It was therefore fortuitous that in the course of one of Freud's experiments a volunteer told Koller that cocaine numbed his mouth. Although a recognised characteristic of the alkaloid, Koller now appreciated its significance for his work. Returning to the Vienna General Hospital, he was met by his professor's assistant, who later published his recollection of what happened after Koller produced a small flask containing cocaine hydrochloride[48]:

> "*A few grains of the powder were dissolved in a small quantity of distilled water, a large, lively frog was selected from the aquarium and held immobile in a cloth, and now a drop of the solution was trickled into one of the protruding eyes. At intervals of a few seconds the reflex of the cornea was tested by the touching of the eye with a needle. . . . After about a minute came the great historic moment. I do not hesitate to describe it as such. The frog permitted his cornea to be touched and even injured without a trace of reflex action or attempt to protect himself – whereas the other eye responded with the usual reflex action to the slightest touch. With the greatest, and surely considering the implications, most justifiable excitement, the experiment continued. The same tests were performed on a rabbit and on a dog with equally good results.*
>
> *Now it was necessary to go one step further and repeat the experiment upon a human being. We trickled the solution under the upraised lids of each other's eyes. Then we put a mirror before us, took a pin in hand, and tried to touch the cornea with its head. Almost simultaneously we could joyously assure ourselves, 'I can't feel a thing!' We could make a dent in the cornea without the slightest awareness of the touch, let alone any unpleasant sensation or reaction. With that, the discovery of local anaesthesia was completed.*"

Koller's findings were presented to a meeting of the Ophthalmological Society in Heidelberg on 15 September 1884. Within weeks, cocaine anaesthesia was introduced in Europe and the United States and had a far-reaching impact on surgical practice. Because of its widespread abuse, it is now used medicinally only as a topical anaesthetic for the nose, throat and larynx. For other purposes it has been superseded by synthetic local anaesthetics.

PILOCARPINE

A drug frequently employed to mimic the action of acetylcholine on the eye is pilocarpine, an alkaloid obtained from the South American shrub *Pilocarpus jaborandi* Holmes (Rutaceae). Symphronio Coutinhou, a Brazilian physician, observed that the leaves caused excessive salivation when chewed by natives and took samples with him on a visit to Europe in 1873. Pilocarpine was isolated the following year by A.W. Gerrard in London[49] and, independently, by E. Hardy in France.[50]

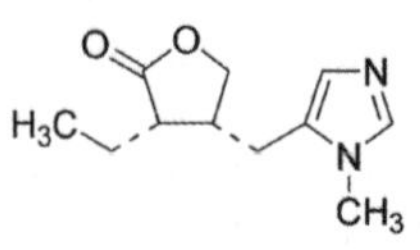

pilocarpine

Pilocarpine was so effective as a diaphoretic in the treatment of oedema that it extended the scope of this type of therapy until the introduction of potent diuretics in the 1920s. In 1877, Adolf Weber discovered its value in decreasing intraocular pressure in glaucoma. Nowadays, it is used solely for this purpose. Pilocarpine is one of the few examples of a natural product

from which analogues have not been developed, principally because of its inherent potency
and acceptability for the use to which it was put.

CURARE

Pietro Martire d'Anghiera, Royal Chronicler to Charles V of Spain, published *De Orbe Novo*
in 1516. Describing the experiences of the first travellers to return from the New World, he
wrote of a soldier being mortally wounded by an arrow tipped with poison. Further details of
this poison emerged in over 50 other reports by visitors to South America during the sixteenth
and seventeenth centuries.[51] One written by Laurence Keynes, who served with Sir Walter
Raleigh on his expedition to the region now known as Venezuela, featured a list of poisons
that included a herb known by the natives as 'ourari'. From this term arose the word 'curare'
to describe South American arrow poisons that cause paralysis.

In 1735, the French Academy of Sciences sponsored an expedition to Ecuador in which
Charles Marie de la Condamine was to conduct topographical studies that would determine
the length of a degree of meridian at the equator and thus establish how much the planet
flattens towards the poles. During the eight years he remained there he collected samples of
curare that could kill chickens when injected. In 1850, Claude Bernard reported that even
though curare prevented the ligated limb of a frog from responding to nerve stimulation, both
the nerve and muscle retained their ability to function.[52] He concluded that curare acted at the
neuromuscular junction to interrupt the stimulation of the muscle by the nerve impulse. In
1936, Dale, Feldberg and Vogt at the National Institute for Medical Research in London
showed that acetylcholine was the chemotransmitter at the neuromuscular junction and that
curare owed its effects to antagonism of it.[53]

Sporadic attempts to employ curare in the treatment of rabies, tetanus, chorea and epilepsy
met with little success during the nineteenth century. The extracts were of dubious quality and
variable potency due to the uncertainty that surrounded the source of the plant from which
they were made. It was not until 1939 that a firm identification of the source of a
physiologically active curare preparation was made by Richard Gill, a coffee and cacao
plantation owner in Ecuador who spent five months collecting 26 varieties of lianas (vines),
from which he prepared over 50 kg of curare.[54] Gill supplied a large quantity of his curare to
Abram Bennett, a neuropsychiatrist practising in Nebraska, who had been trying to obtain
samples of curare to prevent vertebral fractures during the newly developed technique of
convulsive shock therapy.[55] The curare was given to A.R. McIntyre of the Department of
Pharmacology at the University of Nebraska for standardisation.[56] Bennett then carried out a
clinical trial of the curare on 12 spastic children at the Nebraska Orthopedic Hospital. The
relaxing action of curare proved to be transient, but the results in shock therapy were much
more encouraging. It rendered this otherwise dangerous therapy for depressive patients safe
enough for routine application.

Gill signed a contract with E.R. Squibb & Sons, who purchased his remaining supply of
curare. Horace Holaday then simplified the standardisation procedure and Squibb made
curare available free of charge to clinical researchers as Intocostrin®, which they described as
'unauthenticated extract of curare'.

The use of curare to permit pelvic examinations on disturbed female patients gave Lewis
Wright, a medical consultant to Squibb, the idea that it might be possible to administer curare
to overcome the lack of adequate muscle relaxation associated with recently introduced
anaesthetics such as ethylene and cyclopropane. He believed it might also permit adequate
relaxation with lighter levels of older anaesthetics such as ether and chloroform.
Unfortunately, when Intocostrin® was administered to cats and dogs anaesthetised with
ether, the cats died and the dogs became deeply distressed. Wright persuaded Harold Griffith,
a part-time anaesthetist at the Homeopathic Hospital in Montreal, to test Intocostrin® on

patients anaesthetised with cyclopropane. Since Griffith kept his patients on artificial respiration when administering cyclopropane because of the risk of respiratory depression, he unwittingly overcame the two problems with Intocostrin® in anaesthesia. Firstly, ether had neuromuscular blocking activity in its own right, thereby potentiating the effects of the Intocostrin®. Secondly, paralysis of the respiratory musculature by Intocostrin® made artificial respiration a necessity. The outcome of a test of the drug in a youth undergoing an appendectomy encouraged Griffith to inject curare in a further 25 patients. This and other clinical trials confirmed the role of the drug in making anaesthesia safer by permitting low doses of anaesthetic while still attaining muscular flaccidity.

Gill's particular sample, which had provided the highest yield of curare for Intocostrin® production, was identified as *Chondodendron tomentosum* Ruiz et Pavon (Menispermaceae) at the New York Botanical Gardens. In 1943, Squibb chemists Oskar Wintersteiner and James Dutcher[57] isolated crystals from this which proved to be identical to those obtained eight years earlier by Harold King at the National Institute for Medical Research in London from a preparation of curare obtained from the British Museum.[58] The botanical source of that preparation was unknown, but since the material had been packed in a bamboo tube, King had named it tubocurarine.

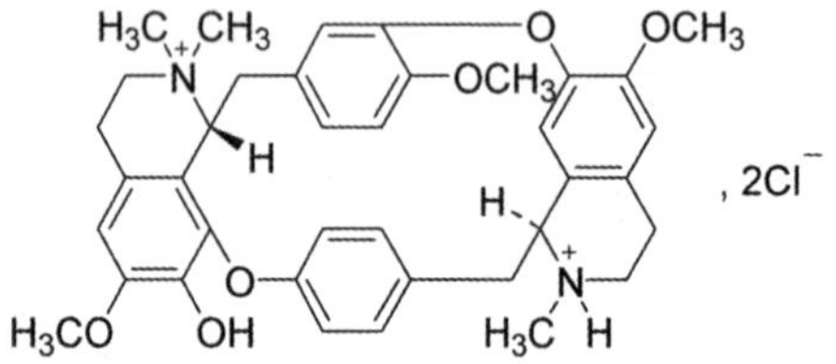

tubocurarine chloride

The isolation of tubocurarine from *C. tomentosum* coincided with the completion of the clinical trials on Intocostrin®, enabling pure tubocurarine chloride to be introduced into anaesthetic practice. As it paralysed respiratory as well as all voluntary muscles, anaesthetised patients had to be artificially ventilated until its long-lasting effects wore off. Newer analogues have generally superseded tubocurarine since they have a shorter duration of action.

EPHEDRINE

Ma Huang, a mixture of *Ephedra sinica*, *E. equisetina* and *E. gerardiana* (Ephedracea), is well known in Chinese medicine. Li Shih-Chen in the late sixteenth century herbal *Pen Ts'ao Kang Mu* claimed that it could improve the circulation, cause sweating, ease coughing and reduce fever. An impure alkaloid was extracted from it in 1885 by G. Yamanashi at the Osada Experimental Station. His work was continued by Nagajosi Nagai who isolated it at Tokyo University and named it as 'ephedrine'.[59] Nagai also elucidated its structure and asked a colleague to examine its physiological properties. These were reported to be comparable to those of atropine, with the advantage that the mydriatic effect lasted only two hours. It was employed in Western medicine to dilate the pupil, but was considered too toxic for other applications. Ephedrine was synthesised by Späth in 1920.[60]

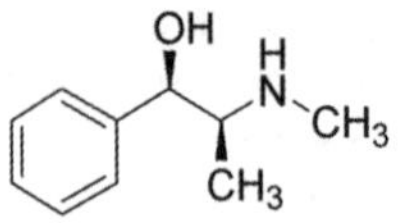

ephedrine

Unaware of these earlier studies, Ku Kuei Chen and Carl Schmidt investigated extracts of Ma Huang at Peking Union Medical College in 1923.[61] A decoction injected into a dog caused a sustained rise in blood pressure, accompanied by an accelerated heartbeat and constriction of the blood vessels in the kidney. These effects were similar to those of adrenaline and tyramine. Chen then used solvent extraction techniques to isolate crystals of the alkaloid previously obtained by Nagai. Further tests demonstrated that, unlike adrenaline, ephedrine was effective by mouth and long-acting. Recognising the clinical potential of this, Chen and Schmidt supplied some ephedrine to Thomas G. Miller at the University of Pennsylvania, who investigated its potential in asthma, and L.G. Rowntree at the Mayo Clinic. Their clinical investigations culminated in the approval of the drug by the Council of Pharmacy and Chemistry of the American Medical Association in 1926. It rapidly became one of the most widely used drugs in the pharmacopoeia because of its value in asthma, where the prolonged action on the bronchi helped to ward off attacks of bronchospasm. Ephedrine has been superseded by drugs such as salbutamol and terbutaline, but is still prescribed as a nasal decongestant. Pseudoephedrine is a naturally occurring stereoisomer of ephedrine with similar properties.[62] It is used as a decongestant in proprietary cough mixtures and nasal drops.

RAUWOLFIA ALKALOIDS

In the early years of the twentieth century, the British colonial authorities in India were concerned about the failure of Western medicine to make inroads into local practice. Committees of Inquiry investigated Indian herbal remedies such as the snakeroot plant, so called on account of its appearance having led to its use as an antidote to snakebite. The earliest mention of snakeroot appeared in the *Charaka Samhita*, a revered Ayurvedic treatise probably written around the first century AD. In 1563, it was described as 'the foremost and most praiseworthy Indian medicine' by the Portuguese physician Garcia de Orta (1500–1569) in his work about Hindu medicine, *Cloquios dos Simples e Drogas e Consas Medicinais da India*. The French botanist Charles Plumier in 1703 named the plant *Rauwolfia serpentina* (Apocynaceae) after the German explorer and botanist Leonhard Rauwolf, who had written a description of it in 1582. More than 30 applications were known early that century in India, yet no serious investigation of it took place until 1931.

In 1931, Salimuzzaman Siddiqui and Rafar Hussein Siddiqui at the Research Institute for Unani–Ayurvedic Medicine in Delhi isolated an alkaloid from snakeroot which they named 'ajmaline', after the physician–wizard Hakim Azmal Khan who founded the Institute.[63] Another five alkaloids were later isolated, but none of them exhibited any properties of interest when tested on frogs. Meanwhile, Gananath Sen and Kartick Bose in Calcutta reported that rauwolfia not only reduced blood pressure if administered for a few weeks, but it also produced sedation.[64] This vindicated the folk use in Bihar where it was given to put babies to sleep and to calm violent lunatics.

Ram Nath Chopra and his colleagues at the School of Tropical Medicine in Calcutta spent more than ten years investigating the pharmacology of rauwolfia and its alkaloids.[65] They confirmed its ability to reduce blood pressure and, although they also demonstrated that crude extracts of the root had powerful sedative properties, they were unable to isolate an alkaloid with sedative activity. Nevertheless, around one million Indians received rauwolfia for their high blood pressure during the 1940s.

There was no interest in rauwolfia in the West until 1949, when Rustom Jal Vakil published the findings of a controlled clinical trial of the effects in 50 patients treated over a five year period at the King Edward Memorial Hospital in Bombay.[66] His results indicated the beneficial effects of the therapy and caught the attention of Robert Wilkins, director of the hypertension clinic at Massachusetts General Hospital. He persuaded E.R. Squibb & Sons to obtain rauwolfia for him to test on patients and then reported his findings in 1952, confirming the mild hypotensive activity of the powdered root and drawing attention to its atypical sedative effect, which caused patients to feel relaxed rather than drowsy.[67] By a fortunate coincidence, the isolation of the long sought active principle responsible for these therapeutic effects was reported around the same time.

In 1949, Emil Schlittler and Johannes Müller of the Ciba research division in Basle investigated the sedative principle of rauwolfia believed to be present in the resinous material remaining in the mother liquors of extracts from which large amounts of ajmaline had already been harvested. Their task was complicated by the presence of a wide variety of pharmacologically active compounds, some of which antagonised the action of the compound they were attempting to isolate. A further complication was the slow onset of the sedative action. Hugo Bein overcame this difficulty by selecting from the varied physiological effects of the extracts one that seemed to be present in most active fractions, namely the miotic effect on the eye of a hare. This was used to identify the best method for recovering a sparingly soluble alkaloid, which was eventually isolated in 1951.[68] This was named 'reserpine', and it was soon confirmed as responsible for most of the hypotensive and sedative activity of rauwolfia root. In November 1953, Ciba began to market reserpine. It was synthesised in 1956 by R.B. Woodward at Harvard.[69]

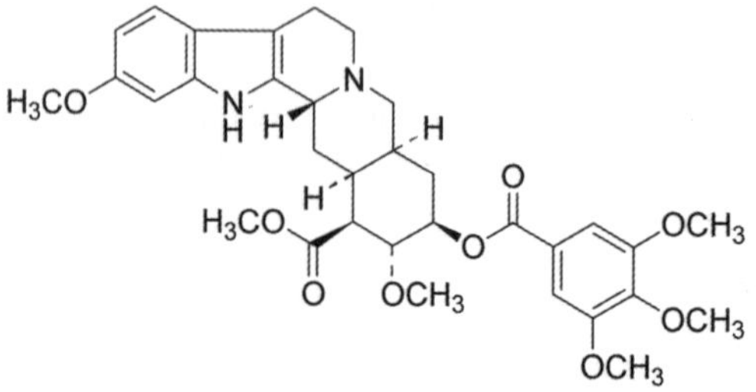

reserpine

For several years reserpine was widely prescribed both as a hypotensive agent and as a tranquilliser, but its popularity waned. The effect on blood pressure when given by mouth proved inferior to that of new synthetic drugs, while there was a risk of precipitating a severe depression that could lead to suicide in some patients. Consequently, neither reserpine nor its semi-synthetic analogues are prescribed any longer in the United Kingdom.

VINCA ALKALOIDS

The species of periwinkle known as *Catharanthus roseus* (L.) G. Don. (Apocynaceae), formerly described as *Vinca rosea*, was indigenous to tropical zones but its attractive appearance led to its cultivation as an ornamental plant. There have been many reports of its use in folk medicine, especially in the treatment of diabetes. In 1949, Ralph Noble at the University of Western Ontario found extracts had no effect on blood sugar levels in rats either by mouth or by injection, but in the latter case the rats succumbed to infections. He then examined the blood of one rat and found that the number of white cells was dramatically reduced. He established that this was due to bone marrow damage, a phenomenon previously associated with the antileukaemic drugs then being introduced.

Noble and Charles Beer separated a pure alkaloid, vinblastine (originally known as 'vincaleukoblastine'), in 1958.[70] Gordon Svoboda of Eli Lilly then informed them that his company had followed a similar path after examining *C. roseus* for antidiabetic activity. On submitting extracts of the plant to a general pharmacological screening programme, the activity against a transplanted acute lymphocytic leukaemia in the mouse was revealed, resulting in its use in lung and breast cancer.[71]

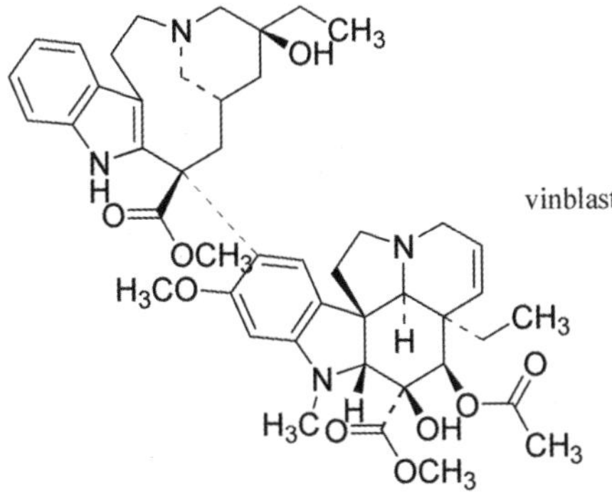

vinblastine

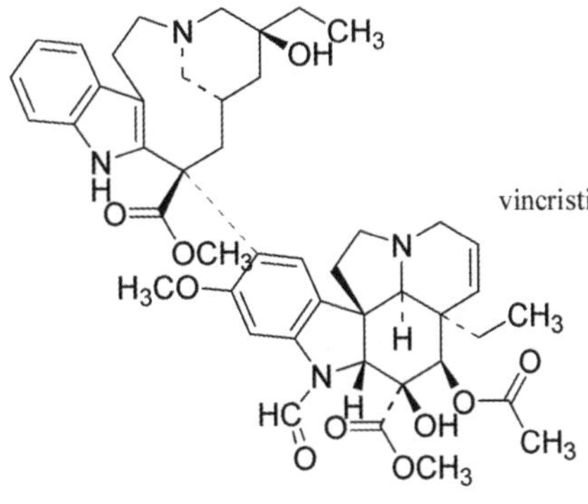

vincristine

In 1961, Svoboda announced the isolation of vincristine, another alkaloid with promising activity.[72] Although vincristine differed from vinblastine solely in having a formyl group instead of a methyl group attached to the nitrogen of one of the indole rings, its toxicity and range of clinical applications was different. It is used mainly to treat leukaemias and lymphomas, especially for the induction of remissions. After the introduction of these two alkaloids, the Lilly chemists isolated many more alkaloids, but none of these attained clinical acceptability. However, two semi-synthetic analogues, vindesine and vinorelbine, are used in certain cases of advanced breast cancer. Like the natural vinca alkaloids, they act by binding to the microtubules that form the spindles to which chromosomes are temporarily attached during cell division. This results in mitosis (cell division) being blocked during metaphase.

CAMPTOTHECIN

While seeking plant sterols that could be converted to cortisone, Monroe Wall of the US Department of Agriculture extracted material from the Chinese tree *Camptotheca acuminata* Decsne in 1958. As a matter of routine, this was sent to the National Cancer Institute where it was screened by Jonathan Hartwell and found to have antitumour activity. Wall and Mansukh Wani then isolated the active compound, camptothecin.[73] Unfortunately, it turned out to be too toxic when tested in the clinic in the early 1970s.[74] Subsequent investigations confirmed that the cytotoxic action arose from damage to both DNA and RNA.

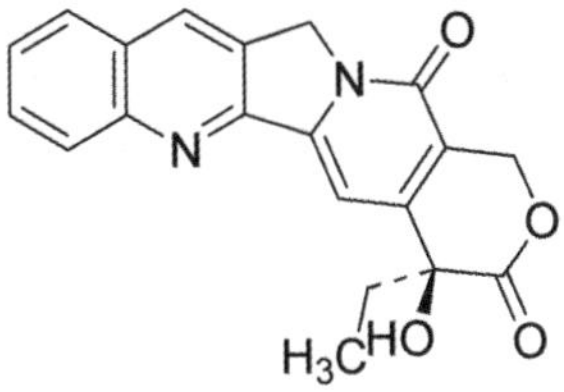

camptothecin

Although camptothecin found no place in the clinic, its analogues topotecan and irinotecan are now routinely used.

REFERENCES

1.   W.A. Smeaton, Fourcroy, Antoine François de, *Dictionary of Scientific Biography*, 1976; **XIII**: 596.

2. A.F. Fourcroy, *Philosophie Chimique, ou Vérités Fondamentales de la Chimie Moderne, Disposées dans un Nouvel Ordre.* Paris: Levrault, Schoell; 1792.
3. W.A. Smeaton, Vauquelin, Nicolas Louis, *Dictionary of Scientific Biography*, 1976; **XIII**: 596.
4. H. Cassebaum, [The work of Carl Wilhelm Scheele]. *Pharmazie*, 1986; **41**: 878–82.
5. G. Kuschinsky, The influence of Dorpat on the emergence of pharmacology as a distinct discipline. *J. Hist. Med.*, 1968; **23**: 258–71.
6. J. Koch-Weser, P. Schechter, Schmiedeberg in Strassburg 1872–1918: the making of modern pharmacology. *Life Sci.*, 1978; **22**: 1361–71.
7. N.G.Bisset, J.G. Bruhn, S. Curto, *et al.*, Was opium known in 18th dynasty ancient Egypt? An examination of materials from the tomb of the chief royal architect Kha. *J. Ethnopharmacol.*, 1994; **41**: 99–114.
8. D.I. Macht, The History of Opium and some of its preparations and alkaloids. *J. Am. Med. Ass.*, 1915; **64**: 477–81.
9. C.L. Derosne, Mémoire sur l'opium. *Ann.Chim.*, 1802; **45**: 257–85.
10. M.A. Séguin, Premier memoire sur l'opium. *Annales de Chimie* 1814; **92**: 225–45.
11. F.W. Sertürner, Darstellung der reinen Mohnsäure (Opiumsäure) nebst einer Chemischen Untersuchung des Opiums mit vorzüglicher Hinsicht auf einen darin neu entdeckten Stoff und die dahin gehörigen Bemerkungen. *J. Pharmacie für Aerzte und Apotheker*, 1806; **14**: 47–93.
12. F.W. Sertürner, Ueber das Morphium, eine neue salzfähige Grundlage, und die Mekonsäure, als Hauptbestandtheile des Opiums. *Annalen der Physik*, 1817; **55**: 56–89.
13. J.M. Gulland, R. Robinson, CXII. The morphine group. Part I. A discussion of the constitutional problem. *J. Chem. Soc.*, 1923; **123**: 980–98.
14. M. Gates, G. Tschudi, The synthesis of morphine. *J. Am. Chem. Soc.*, 1952; **74**: 1109.
15. P.J. Robiquet, *Ann. Chim.*, 1832; **51**: 259.
16. J.R. Partington, *A History of Chemistry*, Vol. 1. London: Macmillan; 1970, p. 204.
17. D.I. Macht, Pharmacological investigation of papaverine. *Arch. Int. Med.*, 1916; **17**: 786–806.
18. P. Pelletier, F. Magendie, Mémoire sur l'émétine et sur les trois espèces d'ipécacuanha. *J. Pharm.*, 1817; **3**: 145–64.
19. H. Kunz, *Jahresb. der Pharm.*, 1887: 416.
20. R. Robinson, *Nature*, 1948; **162**: 524.
21. R.P. Evstigneeva, *et al.*, *Zh. Obshch. Khim.*, 1952; **22**: 1467.
22. J. Pelletier, B. Caventou, Recherches chimiques sur les quinquinas. *Ann. Chim.*, 1820; **15**; 289.
23. F. Magendie, *Formulaire pour la Preparation et l'Emploi de Plusieres Nouveaux Medicaments, tels que la Noix Vomique, la Morphine, etc.* Paris: Mequignon-Marvis; 1822.
24. F. Magendie, *Formulaire pour la Preparation et l'Emploi de Plusieres Nouveaux Medicaments, tels que la Noix Vomique, la Morphine, etc.*, translated by J. Houlton. London: T and G. Underwood; 1828.
25. D.L. Romanovsky, Thesis, St Petersburg, 1891; reprinted in D. Zasuchin, *Outstanding Investigations of Native Scientists on the Causative Organisms of Malaria.* Moscow: 1951.
26. R. B. Woodward, W. E. Doering, The total synthesis of quinine. *J. Am. Chem. Soc.*, 1944; **66**: 849.
27. K.F. Wenckebach, Cinchona derivatives in the treatment of heart disorder. *J. Am. Med. Ass.*, 1923; **81**: 472.
28. W. Frey, Weitere Erfährungen mit Chinidin bei absoluter Herzunregelmässigkeit. *Wien. Klin. Wochenschr.*, 1918; **55**: 849–53.
29. L. Henry, A. Delondre, *J. Pharm.*, 1833; **19**: 623.
30. F. Runge, *Neueste Phytochemische Entdeckungen (Berlin)*, 1820; **1**: 144.
31. E. Fischer, Synthese des Caffeines. *L. Ach. Ber.*, 1895; **28**: 3135–42.
32. W. von Schroeder, Ueber die Wirkung des Caffeins als Diureticum. *Centralb. F.d. mer Wissensches Berl., Arch.*, 1886; **24**: 465–7.
33. A. Kossel, Ueber eine neue Base aus dem Pflanzenreich. *Ber.*, 1888; **21**: 2164.
34. H. Salter, *Asthma. Its Pathology and Treatment.* London: Churchill; 1860, p. 181.
35. K. Himly, *De la Paralysie de l'Iris, Occasionee par une Application Locale de la Belladonna, etc.* Paris: 1802.
36. P.L. Geiger, H. Hesse, *Annalen*, 1833; **5**: 43.
37. J. Sims, Communications relative to the *Datura stramonium*, or thorn-apple as a cure or relief of asthma. *Edin. Med. Surg. J.*, 1812; **8**: 364.
38. H.H., Salter, On the treatment of asthma by belladonna. *Lancet* 1869; **1**: 152.
39. A. Ladenburg, *Annalen*, 1881; **206**: 274.
40. H. King, The resolution of hyoscine and its components, tropic acid and oscine. *J. Chem. Soc.*, 1919; **37**: 476–508.
41. W.F. Daniell, On the natives of Old Callebar. *Proc. Ethnolog. Soc. Edin.*, 1846; **40**: 313–27.
42. R. Christison, On the properties of the ordeal bean of Old Calabar. *Monthly J. Med. Sci. (London)*, 1855; **20**: 193–204.

43. J.O. Jobst, O. Hesse, Über die Bohne von Calabar. *Annalen*, 1864; **129**: 115–21.
44. L. Laqueur, History of my glaucomatous illness. *Am. J. Ophthalmol.*, 1929; **12**: 984–9.
45. A. Niemann, *Ueber eine Neue Organische Base in den Cocablattern.* Gottingen: Huth, 1860.
46. M. Pendergrast, *For God, Country and Coca-Cola.* London: Weidenfeld; 1993.
47. S. Freud, *The Cocaine Papers*, ed. R. Byck. New York: Stonehill; 1974.
48. H. Koller-Becker, Carl Koller and Cocaine. *Psychoanal. Quart.*, 1963; **32**: 309.
49. A.W. Gerrard, The alkaloid and active principle of Jaborandi. *Pharm. J.*, 1875; **5**: 865.
50. E. Hardy, Sur le jaborandi (*Polycarpus pinnatus*). *Bull. Soc. Chim. Fr.*, 1875; **24**: 497–501
51. P. Smith, *Arrows of Mercy.* New York: Doubleday; 1969.
52. C. Bernard, Action du curare et de la nicotine sur le systeme nerveux et sur le systeme musculaire. *C.R. Soc. Biol. Paris*, 1850; **2**: 195.
53. H.H. Dale, W. Feldberg, M.Vogt, Release of acetylcholine at voluntary motor nerve endings. *J. Physiol.*, 1936; **86**: 353–80.
54. R.C. Gill, *White Water and Black Magic.* New York: Henry Holt & Company; 1940.
55. A.E. Bennett, Preventing traumatic complications in convulsive shock therapy by curare. *J. Am. Med. Ass.*, 1940, **114**: 322–4.
56. A.R. McIntyre, *Curare. Its History, Nature and Clinical Use.* Chicago: The University of Chicago Press; 1947.
57. O. Wintersteiner, J.D. Dutcher, Curare alkaloids from *Chondodendron tomentosum. Science*, 1943; **97**: 467–70.
58. H. King, Curare alkaloids. Part I. Tubocurarine. *J. Chem. Soc.*, 1935; 1381–9.
59. N. Nagai, *Ephedrin. Pharm. Ztg.*, 1887; **32**: 700.
60. E. Späth, R. Göhring, Die Synthesen des Ephedrins, des Pseudoephedrins, ihrer optischen Antipoden und Razemkörper. *Monatsch. Monatsch.*, 1920; **41**: 319–38.
61. K.K. Chen, C.F. Schmidt, The action of ephedrine, an alkaloid from Ma Huang. *Proc. Soc. Exp. Biol. Med.*, 1924; **21**: 351–4.
62. A. De Vriesse, L'éphédrine et la pseudo ephedrine, nouveau mydriatiques. *Annals d'Oculist*, 1889; **101**: 182.
63. S. Siddiqui, H. Siddiqui, *J. Indian Chem. Soc.*, 1931; **8**: 667–80.
64. G. Sen, K. Bose, *Rauwolfia serpentina*, a new Indian drug for insanity and blood pressure. *Indian Med. Wld.*, 1931; **2**: 194.
65. R.N. Chopra, J.C. Gupta, B. Mukerjee, The pharmacological action of an alkaloid obtained from *Rauwolfia serpentina* Benth. *Indian J. Med. Res.*, 1933; **21**: 261–71.
66. R.J. Vakil, A clinical trial of *Rauwolfia serpentina* in essential hypertension. *Br. Heart J.*, 1949; **11**: 350–9.
67. R.W. Wilkins, New drug therapies in arterial hypertension. *Ann. Intern. Med.*, 1952; **37**: 1144–55.
68. J.M. Müller, E. Schlittler, H.J. Bein, Reserpin, the sedative principle from *Rauwolfia serpentina* B. *Experientia*, 1952; **8**: 338.
69. R.B. Woodward, F.E. Bader, H. Bickel, *et al.*, The total synthesis of reserpine. *J. Am. Chem. Soc.*, 1956; **78**: 2023–5.
70. R.L. Noble, C.T. Beer, J.H. Cutts, Role of chance observations in chemotherapy: *Vinca rosea. Ann. N.Y. Acad. Sci.*, 1958; **76**: 882–94.
71. N. Neuss, M. Gorman, G.H. Svoboda, *et al.*, Vinca alkaloids. III.1 Characterization of leurosine and vincaleukoblastine, new alkaloids from *Vinca rosea* Linn. *J. Am. Chem. Soc.*, 1959; **81**: 4754–5.
72. G.H. Svoboda, I.S. Johnson, M. Gorman, N. Neuss, Current status of research on the alkaloids of *Vinca rosea* Linn. (*Catharanthus roseus* G. Don). *J. Pharm. Sci.*, 1962; **51**: 707–20.
73. M.E. Wall, M. C. Wani, C. E. Cook, *et al.*, Plant antitumor agents. I. The isolation and structure of camptothecin, a novel alkaloidal leukemia and tumor inhibitor from *Camptotheca acuminate. J. Am. Chem. Soc.*, 1966; **88**: 3888–90.
74. J.A. Gottlieb, A.M. Guarino, J.B. Call, *et al.*, Preliminary pharmacologic and clinical evaluation of camptothecin sodium (NSC-100880). *Cancer Chemother. Rep.*, 1970; **54**: 461–70.

Non-alkaloidal Plant Products

In 1763, Revd Edward Stone of Chipping Norton in Oxfordshire sent a letter to the Royal Society reporting that had found a locally available substitute for the expensive, imported cinchona bark.[1] He wrote,

> *"About six years ago, I accidentally tasted willow bark and was surprised at its extraordinary bitterness; which immediately raised me the suspicion of its having the properties of the Peruvian bark. As this tree delights in a moist or wet soil, where agues chiefly abound, the general maxim, that many natural maladies carry their cures along with them, or that their remedies lie not far from their causes, was so very apposite to this particular case, that I could not help applying it. . . ."*

Stone had obviously been influenced by the doctrine of signatures in this instance, but his subsequent investigation was more rational. He told the Royal Society that during the previous six years he had successfully treated 50 people with agues (malarial fevers) by giving them dried willow bark to consume.

Others had written about the uses of the bark before Stone, but he was the first to note its antipyretic properties.[2] In 1792, Samuel James, a surgeon in Hertfordshire, evaluated different species of willow barks against fevers by assessing their astringent effect on the tongue.[3] He found that willow bark could reduce fevers unaffected by cinchona bark. This was taken further by William White, an apothecary from Bath, who made extracts of the bark.[4] There was little further interest in the use of willow as a substitute for cinchona until the English naval blockade of continental Europe during the war with Napoleon, when it became important to obtain substitutes for imported medicinal herbs.

SALICIN

A pure glycoside, salicin, was isolated in 1829 from meadowsweet, *Filipendula ulmaria* L. (Rosaceae), formerly known as *Spirea ulmaria*, by Leroux, a French pharmacist.[5] Soon after, it was introduced in France as a substitute for quinine. Raffaele Piria later showed that salicin could be decomposed into salicylic acid.[6] Salicin was synthesised in 1879.[7]

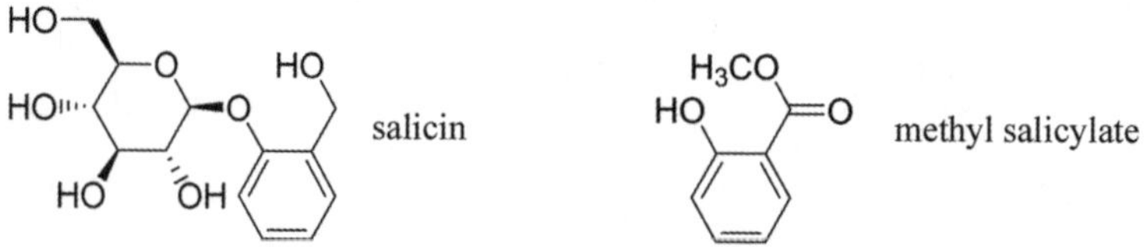

salicin   methyl salicylate

Thomas Maclagan, a Scottish physician, first used salicin in 1874 to treat a patient afflicted with rheumatic fever.[8] A year and a half later, on reporting to the *Lancet* that he had attained successful results in around 100 patients, he also declared his conviction that there was a similarity between intermittent fever (malaria) and acute rheumatism.[9] Like Stone a century before, he noted that cinchona bark was to be found in tropical areas where the ravages of malaria were most severe, and believed that a remedy for acute rheumatism would be found in a low-lying damp locality with a cold climate. Maclagan felt that the plants whose haunts best corresponded to his requirements were the various forms of willow. This may have been unscientific, but his patients benefited. He was to learn that Marcellus von Nencki of Basle had shown in 1870 that salicin is converted in the body into salicylic acid. When the first reports of the antipyretic action of salicylic acid appeared, Maclagan also tried this. Again, he eased the affliction of his patients. Salicin was more expensive than synthetic salicylic acid, so its use was gradually abandoned. Nevertheless, it remained in the *British Pharmaceutical Codex* until 1954.

Methyl salicylate was isolated in 1843 by William Procter from oil of wintergreen, *Gaultheria procumbens* L. (Ericaceae), a traditional American remedy for the ague.[10] It has been incorporated in topical anti-inflammatory and counter-irritant formulations, as well as in numerous cosmetic products.

DIGITALIS GLYCOSIDES

Attempts were made in the 1820s to isolate an active principle from digitalis. To stimulate research, the Société de Pharmacie in Paris offered a prize of 500 francs for the isolation of a pure principle from the plant, the sum having to be doubled after five years had passed without any claimant coming forward. In 1841, the French pharmacists E. Homolle and Theodore Quevenne won the award for their isolation of an impure crystalline material which consisted mainly of impure digitoxin.[11] They called this 'digitalin', a name also applied to various products obtained by other workers.

Digitoxin, the principal cardiotonic glycoside present in the leaves of *Digitalis purpurea*, was isolated in 1875 by Schmiedeberg at the University of Strassburg (now Strasbourg).[21] He obtained crystals from digitalis leaves by modifying a method employed by the pharmacist Claude Nativelle six years before for the isolation of a similar material called 'digitalin crystalisée'. Adolf Windaus finally obtained pure digitoxin in 1925, at the University of Gottingen.[13]

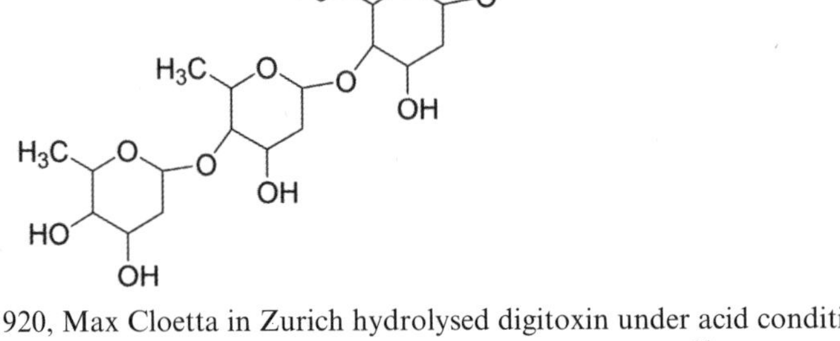

digitoxin

In 1920, Max Cloetta in Zurich hydrolysed digitoxin under acid conditions and isolated the aglycone, digitoxigenin, which had weak cardiotonic activity.[14] The structure of digitoxigenin was determined by Walter Jacobs and his colleagues at the Rockefeller Institute in New York,[15] but it was not until 1962 that chemists at Sandoz in Basle elucidated the structure of the sugar residue and hence that of the entire molecule of digitoxin.[16]

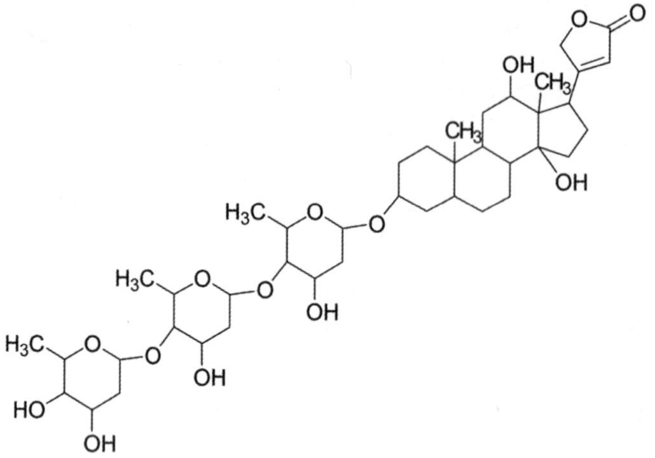

digoxin

In the late 1920s, it was discovered that the powdered leaves of *Digitalis lanata*, once popularly known as 'woolly foxglove', had greater physiological activity than those of *Digitalis purpurea*. This led Sydney Smith of Burroughs Wellcome in London to isolate digoxin.[17] This is now used more than either powdered digitalis leaves or digitoxin since it does not bind as strongly to proteins in the tissues and plasma, resulting in less delay before a therapeutic concentration of unbound drug can build up. Clearance from the body is also faster as only unbound drug is filtered by the kidneys; consequently digoxin is less cumulative and thus safer to use. It has become the standard digitalis preparation in current usage.

GUAIACOL

Ascani Sobrero was an Italian chemist best known for his discovery of nitroglycerin, but he also isolated guaiacol from dry distillation of guaiac resin in 1843.[18] After Auguste Béhal and Eugène Choay extracted crystalline guaiacol from beechwood tar in 1887, it quickly became popular as it was now free of any unpleasant smell and could even be cheaply isolated from coal tar creosote. In Berne, Hermann Sahli recommended it for the treatment of tuberculosis.

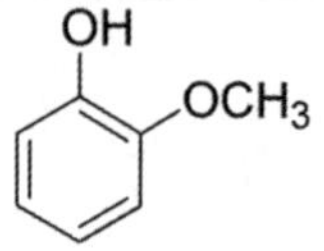

guaiacol

The industrial chemist Karl von Reichenbach distilled creosote from beechwood tar in 1830. Later, its physical similarity to the heavy oil distilled from coal tar led him to examine its antiseptic properties. Over the next few years numerous articles in medical journals extolled the virtues of creosote in the treatment of gangrene, cancerous lesions, eczema, impetigo and foul-smelling wounds. Pieces of cotton soaked in creosote were also applied to tooth cavities to afford genuine relief. Despite its irritant nature, beechwood creosote was even taken internally. It acquired an undeserved reputation for having extraordinary powers to arrest vomiting! Responsible medical opinion questioned this, and also claims that it could relieve diarrhoea. Creosote was also inhaled, apparently to relieve excessive bronchial secretion. Pulmonary tuberculosis was treated by inhalation of creosote combined with administration by mouth of as much as 1 g daily.

All these applications resulted in a demand for a cheap substitute for the expensive genuine beechwood creosote. This demand was often met by the supply of coal tar creosote at around one-fifth of the price. As this was contaminated with phenol it was much more irritant than genuine creosote, especially if inhaled or taken internally.

PODOPHYLLOTOXIN

The May Apple (or American mandrake), *Podophyllum peltatum* L. (Berberiaceae), is a perennial found throughout the eastern side of North America. The juice expressed from its root was employed by the Wyandotte Indians as a laxative and anthelmintic and by the Cherokees for deafness. Podophyllotoxin was isolated from the rhizome in 1880 by Podwyssotski.[19] Its structure was established in 1951 by Anthony Schrecker and Jonathan Hartwell at the National Cancer Institute,[20] and a total synthesis was achieved by Walter Gensler and Christos Gatsonis at Boston University.[21]

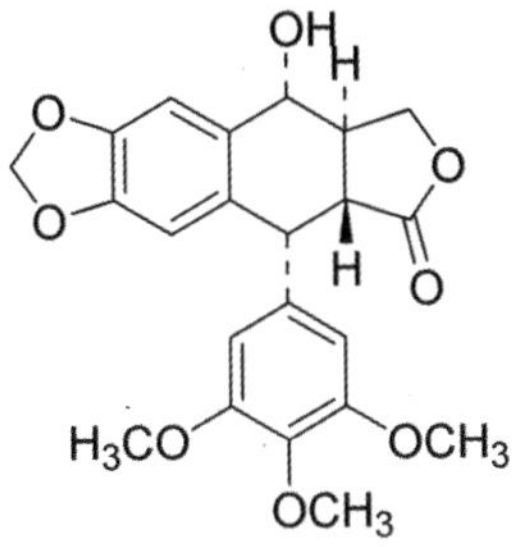

podophyllotoxin

John King, a physician in Ohio, pioneered the use of the highly potent podophyllum resin obtained from the juice, persuading William Merrell, a Cincinnati pharmacist, to include it in 1847 in the range of products his company supplied to the medical and pharmaceutical professions.[22] It was sold as Podophyllin®, under which name it quickly became established as one of the most popular laxatives throughout the world. The name podophyllin was accepted into pharmacopoeias, e.g. the *British Pharmacopoeia* of 1864. Today, the use of podophyllin as a laxative is discouraged since this effect is induced by severe irritation of the intestinal mucosa.

In the *Indian Doctor's Dispensatory*, published in Cincinnati in 1818, Peter Smith recommended the topical application of powdered podophyllum root as an escharotic to deal with warts because of its obvious irritant nature. This remains the principal application of both the resin and podophyllotoxin.

Podophyllotoxin was shown to be a mitotic poison, inhibiting cell division during metaphase in a similar manner to colchicine by binding to the protein fibres known as tubulin, which constitute the mitotic spindle that holds the chromosomes during the metaphase of cell division.[23] Although this finding led to various clinical trials, podophyllotoxin has proved to be too toxic for use as a cancer chemotherapeutic agent.

It is of interest to note that Dioscorides referred to the role of oil of juniper for treating warts, carbuncles and leprosy, while Pliny the Elder mentioned that some smaller species of juniper could be used to treat tumours. The tenth century English *Leech Book of Bald* similarly recommended the root of cow parsley, *Anthriscus sylvestris* (L.) Hoffm., for cancer. Juniper contains podophyllotoxin and deoxypodophyllotoxin, while *A. sylvestris* contains the latter.[24]

KHELLIN

Ammi visnaga (L.) Lam. (Umbelliferae) is a herbaceous plant that grows wild in Egypt, where a decoction from it acquired a reputation in folk medicine as a diuretic that could relieve renal colic. This was a common problem in Egypt arising from renal stone formation associated with schistosomiasis. A study on a refined extract by K. Samaan at the University of Cairo in 1930 revealed its ability to relax smooth muscle, including that of the ureter and coronary arteries, in a variety of animal species.[25] This formed the basis of a clinical investigation that confirmed that, in addition to the diuretic action, it could relieve spasm of the ureter.

Crystals isolated from the seeds by Ibrahim Mustapha at the Ecole Supérieure de Pharmacie at Montpellier in 1879 were given the name 'khellin' (from *El Kellah*, the Moorish name of the plant).[26] In 1938, Ernst Späth obtained pure crystals at the University of Vienna and elucidated the structure.[27]

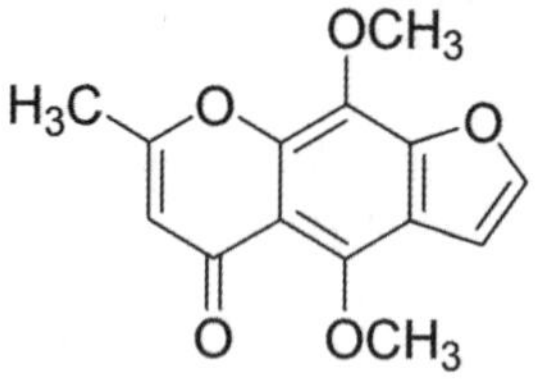

khellin

Much of the interest in khellin was directed at its action on the heart, following a serendipitous observation by Gleb von Anrep, the professor of pharmacology at Cairo, who noticed that after his laboratory technician had self-medicated his renal colic with it his angina disappeared. Anrep then conducted tests of khellin on the coronary blood flow in dogs, followed by a clinical trial in patients with angina. This and further studies by Anrep suggested the drug had beneficial effects when given by mouth.[28] Subsequent studies were generally in favour of the use of khellin for angina as it had a longer duration of action than had glyceryl trinitrate. It became widely prescribed until a study in the United States showed that angina patients who benefited from khellin treatment all experienced some toxic symptoms.[29] As a result of this and similar studies, efforts were made to find less toxic analogues.

DICOUMAROL

In 1922, cattle in North Dakota and Alberta were found to be bleeding to death. A veterinary pathologist, Francis Schofield, discovered that this was caused by the eating of mouldy hay prepared from sweet clovers *Melitotus alba* Desv. and *M. officinalis* (Leguminosae). Sweet clovers had been introduced from Europe earlier in the century since they could withstand harsh winters and dry summers. After Schofield had established that the clotting time was prolonged in the sick cattle, Lee Roderick found that an unknown haemorrhagic factor caused the disease by reducing the activity of prothrombin, the precursor of thrombin. However, it was not until 1938 that progress in isolating the factor was made by Karl Link and Harold Campbell, at the Experimental Station run by the University of Wisconsin College of Agriculture. They modified a new method developed by Armand Quick for one-stage quantitative measurement of prothrombin activity in patients afflicted with various blood diseases. In 1939, Campbell isolated crystals of a potent anticoagulant. The following year, Charles Huebner confirmed by synthesis that the anticoagulant was 3,3'-methylenebis(4-hydroxycoumarin).[30,31] It was formed from coumarin (which causes the smell of new-mown hay) by fungal oxidation to 4-hydroxycoumarin, followed by coupling with formaldehyde.

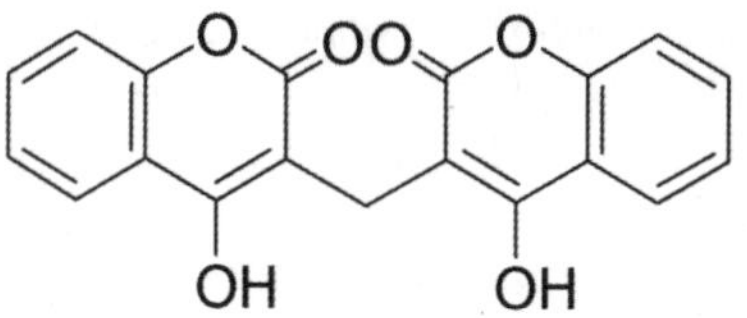

dicoumarol

In 1942, the Abbott and Lilly companies marketed the new anticoagulant, which had been given the approved name of 'dicoumarol'.[32] Since then, numerous studies concerning the merits or otherwise of this and related anticoagulants in thromboembolic disease have been published. At present, the main uses of these drugs are in the prevention and treatment of deep vein thrombosis in the legs and to prevent thrombi forming on prosthetic heart valves. Hundreds of coumarins have been found to exhibit anticoagulant activity, the principal differences between them concerning potency, the rate of onset of action and duration. Since their mode of action is to interfere with vitamin K in the hepatic biosynthesis of clotting factors, there is a time lag of 1–3 days before reserves are depleted. Coupled with a degree of unpredictability of response, this led to dicoumarol being superseded by analogues such as warfarin and phenindione.

TETRAHYDROCANNABINOL

Preparations of _Cannabis sativa_ L. (Cannabidaceae) are known by many names, including Indian hemp, marihuana, bhang, ganja, charas, kif, hashish and pot. These reflect the widespread usage of this euphoriant plant, which occurs naturally in Asia, the Middle East, India, and also Central and North America. Cannabis was known to the Ancient Greeks. Herodotus described how the Scythians made clothes from its fibres and medicated their vapour baths with its seeds. Lloyd provided a detailed list of the many references to the use of cannabis in ancient Indian and Arabian literature.[22]

The recreational use of cannabis in the West can trace its origins to Le Club des Haschischins, which met in the Hôtel Pimodan on the Isle St Louis in Paris in the 1840s. The club was regularly visited by leading intellectuals, including Honoré de Balzac, Alexandre Dumas and Charles Baudelaire. Legal restrictions on cannabis first appeared after the International Opium Convention met in 1925.

Early investigations using cannabis in any of its numerous local formulations had little scientific value as the content of different active principles varied widely. Clinical studies were put on a more secure footing after one of the main ones, Δ^1-3,4-_trans_-tetrahydrocannabinol, was isolated in 1964 at the Hebrew University in Jerusalem by Raphael Mechoulam and Yehiel Gaoni.[33] Following up reports that leukaemia patients were less nauseated if they smoked marihuana before receiving cytotoxic drugs, Stephen Sallan conducted a double-blind trial comparing tetrahydrocannibinol against placebo, confirming its antiemetic activity when given by mouth.[34] Several studies have supported this finding, but others reported a high incidence of cardiovascular and central side effects. Other effects of smoking cannabis have been considered as the basis of potential clinical applications. One of these is decreased lacrimation on exposure to peeling of onions, which was found on further investigation to be associated with reduced intraocular pressure. This raised the possibility of tetrahydrocannabinol being able to reduce intraocular pressure in patients with glaucoma, which was subsequently confirmed.[35]

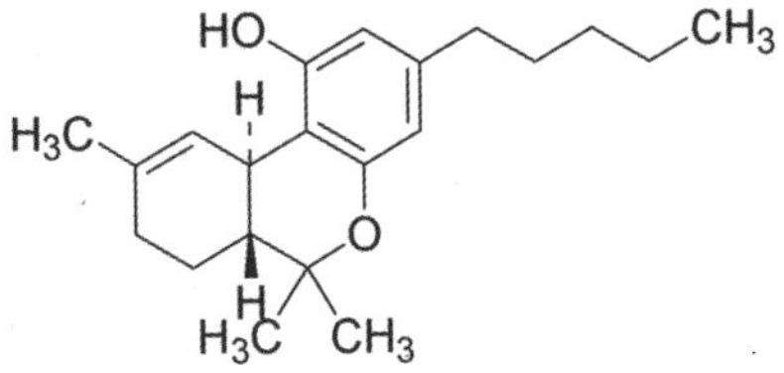

tetrahydrocannabinol

Disappointing results followed investigations into the use of tetrahydrocannabinol as an anti-epileptic drug. Cannabis extract was tested in epilepsy as long ago as 1838[36] and other reports subsequently appeared. A pharmacological basis for these was provided in 1947 when marihuana extract was shown to protect animals against electroshock-induced convulsions.[37] Several more recent investigations with tetrahydrocannabinol have confirmed this in animals and man, but this has not yet resulted in the licensing of any cannabinoid for clinical use.[38]

PACLITAXEL (TAXOL)

A screening programme was established in 1960 by the US National Cancer Institute to detect antitumour activity in extracts of plants gathered by the US Department of Agriculture. Four years later, samples collected in the State of Washington were given to the Research Triangle Institute in North Carolina for investigation. These included the bark, twigs, leaves and fruit of the Pacific yew tree, *Taxus brevifolia* Nutt. (Taxaceae), extracts of which had been found by the US Cancer Chemotherapy National Service Center to exhibit *in vitro* cytotoxic activity as measured by inhibition of the growth of KB cells. Utilising both conventional solvent extraction and Craig countercurrent distribution techniques, with monitoring of the activity of extracts by their ability to inhibit Walker WM solid tumour growth, Monroe Wall and his colleagues in 1966 isolated 0.5 g of a compound they named 'taxol'. It was moderately active against murine L1210, P388 and P1534 leukaemias, Walker 256 carcinosarcoma, sarcoma 180 and Lewis lung carcinoma, and was cytotoxic in the KB cell culture system. This was reported in a paper which also revealed the chemical structure of taxol.[39] Taxol was synthesised in 1994 by two independent groups from Florida State University and the Scripps Research Institute.[40,41]

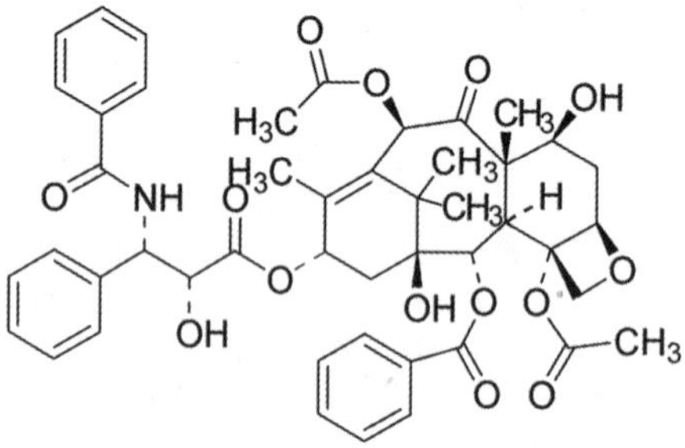

paclitaxel (taxol)

Wall has written of his frustration at failing to interest National Cancer Institute administrators in obtaining large quantities of the bark and wood of *T. brevifolia*.[42] They were understandably apprehensive about the complex extraction procedure which recovered only minute quantities of taxol from a tree in scarce supply. To make matters worse, the substance was insoluble in water and hence difficult to inject. Years passed before two dramatic reports transformed the situation. Firstly, the National Cancer Institute found that taxol was highly active against the relatively resistant murine B16 melanoma. Secondly, Susan Horwitz and her colleagues at the Albert Einstein Medical Center in New York discovered that taxol inhibited cell division at concentrations that had no effect on the synthesis of nucleic acids or protein.[43] Horwitz went on to establish that taxol acted by binding with the tubulin protein fibres to which chromosomes attach during the metaphase period of cell division. Unlike other drugs that did this, taxol caused realignment of the microtubules and consequent metaphase arrest at concentrations that had no effect on nucleic acid or protein synthesis. These new findings stimulated renewed interest in taxol, with the National Cancer Institute taking the initiative for organising the collection of sufficient bark and wood for the extraction of enough taxol to permit clinical trials. Bristol–Myers Squibb was now brought in to assist with production of the drug and the term 'taxol' was dropped in favour of the approved name of 'paclitaxel'.

A superior source of the drug was also found by converting a product from the Himalayan yew tree (*Taxus bacatta*) into paclitaxel.

Early clinical investigations had encouraging results in ovarian cancer, when paclitaxel sometimes produced complete remissions. It is now approved for use in patients with recurrent ovarian cancer and in patients with metastatic breast cancer, often in combination with cisplatin. There is also evidence of its value in lung and prostate cancer.

REFERENCES

1. E. Stone, An account of the success of the bark of the willow tree in the cure of agues. *Phil. Trans.*, 1763; **53**: 195–200.
2. W. Sneader, The discovery of aspirin. *Pharm. J.*, 1997; **259**: 614–17.
3. S. James, *Observations on the Bark of a Particular Species of Willow Showing its Superiority to the Peruvian Bark*. London: J. Johnson; 1792.
4. W. White, *Observations and Experiments on the Broad-Leaved Willow Bark*. Bath: Hazard; 1798.
5. M. Leroux, [Discovery of salicine], *J. Chim. Méd.*, 1830; **6**: 341.
6. R. Piria, Recherches sur la Salicine et les produits qui en dérivent. *C. R. Acad. Sci. (Paris)*, 1839; **8**: 479–85.
7. A. Michael, On the synthesis of helicin and phenolglucoside. *Am. Chem. J.*, 1879; **1**: 305–12.
8. T. Maclagan, The treatment of rheumatism by salicin and salicylic acid. *Br. Med. J.*, 1876; **1**: 627.
9. T. Maclagan, The treatment of acute rheumatism by salicin. *Lancet*, 1876; **110**: 342–3.
10. W. Procter, *Am. J. Pharm.*, 1843; **15**: 241.
11. E. Homolle, T.E. Quevenne, *Mémoire sur la Digitaline et la Digitale*. Paris: Germer Baillière; 1854.
12. J.E.O. Schmiedeberg, *Arch. Exp. Path. Pharmacol.*, 1875; **3**: 27.
13. A. Windaus, *J. Freeze, Ber.*, 1925; **58**: 2503.
14. M. Cloetta, *Arch. Exp. Path. Pharmakol.*, 1920; **88**: 113.
15. W. Jacobs, R. Elderfield, The structure of the cardiac aglucones. *J. Biol. Chem.*, 1935; **108**: 497–513.
16. H. Lichti, M. Kuhn, A. von Wartburg, Zur Struktur der Zuckerkomponente des Digitoxins. *Helv. Chim. Acta.*, 1962; **45**: 868.
17. S. Smith, Digoxin, a new digitalis glucoside. *J. Chem. Soc.*, 1930; 508–10.
18. A. Sobrero, *Annalen*, 1843; **48**: 19.
19. V. Podwyssotski, Pharmakologische Studien über *Podophyllum peltatum*. *Arch. Exp. Pathol. Pharmakol.*, 1880; **13**: 29–52.
20. J.L. Hartwell, A.W. Schrecker, Components of podophyllin. V. The constitution of podophyllotoxin. *J. Am. Chem. Soc.*, 1951, **73**, 2909–16.
21. W.J. Gensler, C.D. Gatsonis, Synthesis of podophyllotoxin. *J. Am. Chem. Soc.*, 1962; **84**: 1748–9.
22. J.U. Lloyd, *Origin and History of all the Pharmacopoeial Vegetable Drugs, Chemicals and Preparations*, Vol. 1. Cincinnati: Caxton Press; 1921.
23. L.S. King, M. Sullivan, The similarity of the effect of podophyllin and colchicine and their use in the treatment of condylomata acuminata. *Science*, 1946; **104**: 244–5.
24. A. Koulman, Podophyllotoxin. A study of the biosynthesis, evolution, function and use of podophyllotoxin and related lignans. Thesis. University of Groningen, 2003.
25. K. Samaan, *Quart. J. Pharm. Pharmacol.*, 1930; **4**: 25.
26. I. Mustapha, *C. R. Soc. Biol.*, 1879; **89**: 442.
27. E. Späth, F. Dengel, W.Gruber, *Ber*, 1938; **71**: 106.
28. G. von Anrep, M.R. Kenaway, G.S. Barsoum, Coronary vasodilator action of khellin. *Am. Heart J.*, 1948; **37**: 531–42.
29. H.N. Hultgren, Clinical and experimental study of use of khellin in treatment of angina pectoris. *J. Am. Med. Ass.*, 1952; **148**: 465–9.
30. K.P. Link, The anticoagulant from spoiled sweet clover hay. *Harvey Lectures*, 1944; **39**: 162–97.
31. M.A. Stahmann, C.F. Huebner, K.P. Link, Studies on the hemorrhagic sweet clover disease. V. Identification and synthesis of the hemorhagic agent. *J. Biol. Chem.*, 1941; **138**: 513–27.
32. K.P. Link, The discovery of dicumarol and its sequels. *Circulation*, 1959; **19**: 97–107.
33. Y. Gaoni, R. Mechoulam, Isolation, structure, and partial synthesis of an active constituent of hashish. *J. Am Chem. Soc.*, 1964; **86**: 1646–7.
34. S.E. Sallan, N.E. Linberg, E. Frei, Antiemetic effect of delta-9-tetrahydrocannabinol in patients receiving cancer chemotherapy. *N. Engl. J. Med.*, 1975; **293**: 795–7.
35. R.S. Hepler, I.M. Frank, J.T. Ungerleider, Pupillary constriction after marijuana smoking. *Am. J. Ophthal.*, 1972; **74**: 1185–90.
36. W.B. O'Shaughnessy, *Trans. Med. Phys. Soc. Bengal*, 1838; 71.

37. S.Loewe, L.S. Goodman, *Fed. Proc.*, 1947; **6**: 352.
38. R.K. Razdan, J.F. Howes, Drugs related to tetrahydrocannabinol. *Med. Res. Rev.*, 1983; **3**: 119–46.
39. M.C. Wani, H.L. Taylor, M.E. Wall, et al., Plant antitumor agents. VI. Isolation and structure of taxol, a novel antileukemic and antitumor agent from *Taxus brevifolia. J. Am. Chem. Soc.*, 1971; **93**: 2325–7.
40. R.A. Holton, C. Somoza, H.B. Kim, *et al.*, First total synthesis of taxol. 1. Functionalization of the B ring. *J. Am. Chem. Soc.*, 1994; **116**: 1597–8.
41. K.C. Nicolaou, Z. Yang, J.J. Liu, *et al.*, Total synthesis of taxol. *Nature*, 1994, **367**: 630–4.
42. M.E. Wall, in *Chronicles of Drug Discovery*, Vol. 3, ed. D. Lednicer. Washington: American Chemical Society; 1993, p. 327.
43. P.B. Schiff, J. Fant, S.B. Horwitz, Promotion of microtubule assembly *in vitro* by taxol. *Nature*, 1979; **277**: 665–7.

CHAPTER

11

Plant Product Analogues and Compounds Derived from Them

Morphine was not only the first active principle to be extracted from a plant but was also the first to have analogues prepared from it. The only way any analogues could be prepared before its chemical structure was established was by modifying one or more of its functional groups. The first semi-synthetic analogue formed in this manner was made by Henry How in 1853 at the Department of Chemistry in the University of Glasgow. In an attempt to elucidate the relationship between codeine and morphine, he heated morphine with several alkyl iodides in alcoholic solution. He had hoped to convert the morphine into codeine, but instead obtained a novel substance that he correctly identified as a quaternary ammonium salt of morphine.[1] How did not take matters further, but 15 years later Alexander Crum Brown, the Professor of Chemistry at the University of Edinburgh, prepared the same quaternary salt together with similar salts of several other alkaloids. He gave these for pharmacological examination to his colleague Thomas Fraser, the Professor of Materia Medica. This resulted in a classic investigation that was one of the first attempts to correlate chemical structure with biological activity. Fraser found that the quaternary ammonium salts of strychnine, brucine, thebaine, codeine, morphine, nicotine, atropine and coniine all exhibited curare-like paralysing activity, despite the pharmacological diversity of their parent alkaloids.[2] This proved that the quaternary ammonium function conferred curariform activity, thereby establishing, for the first time, that certain clusters of atoms could confer specific types of biological activity upon molecules and that the pharmacological properties of natural products could be modified by chemical manipulation. The consequences of this were far-reaching.

A year later, the severely disabled chemist Augustus Matthiessen obtained a post at St Mary's Hospital Medical School in London, where he conducted research on opium alkaloids. In 1869 he and Charles Alder Wright heated morphine with concentrated hydrochloric acid in a sealed glass tube and obtained a degradation product which they named 'apomorphia'.[3] Although this was devoid of morphine-like activity, it proved to be a powerful emetic. It was later named 'apomorphine' and its structure was elucidated by Robert Pschorr in 1902.[4] Apomorphine has been used in the treatment of impotence and in Parkinson's disease as it is a potent stimulator of dopamine D_1 and D_2 receptors.

apomorphine

acetylcodeine

diamorphine

Drug Discovery. A History. W. Sneader.

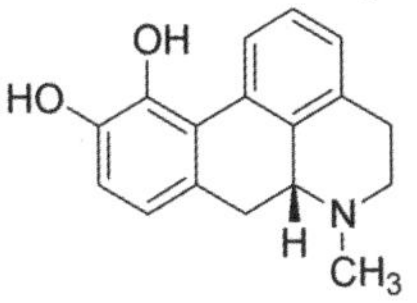

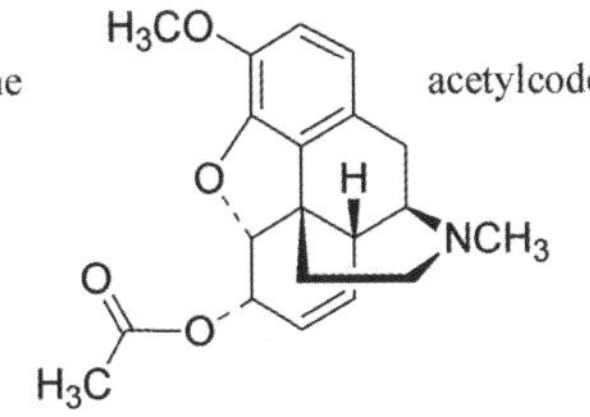

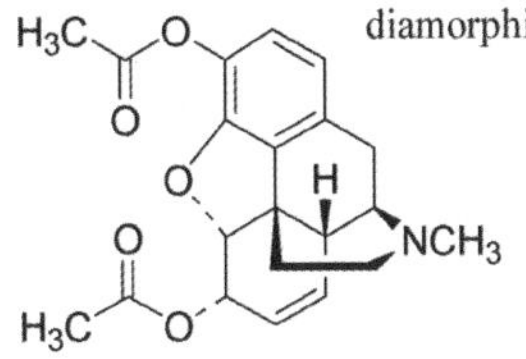

ethylmorphine

After the suicide of Matthiessen, Wright continued the research on opiates by preparing morphine and codeine esters, including hydrochloride salts of acetylcodeine, monoacetylmorphine and diamorphine (diacetylmorphine). These were tested on a dog and rabbit by a physician, F.M. Pierce, but his findings were inconclusive because he failed to compare the new compounds with morphine.[5,6] However, further developments were to take place elsewhere.

The French chemist Edouard Grimaux synthesised ethylmorphine from morphine in 1881.[7] Ralph Stockman, a colleague of Fraser at the University of Edinburgh, and David Dott, a pharmacist, then tested both it and pure diacetylmorphine on frogs and rabbits, observing that ethylmorphine was similar to codeine in its activity.[8] They later reported that while diacetylmorphine had a stronger depressant action than morphine on the spinal cord and respiratory centre, its narcotic action was weaker.[9] Stockman and Dott did not take matters further since their prime interest was on the relationship between chemical structure and biological activity, but the alkaloid manufacturer E. Merck of Darmstadt invited Joseph von Mering to examine 18 esters and ethers of morphine. So far as diacetylmorphine was concerned, he not only confirmed the observations of the Edinburgh researchers, but also tested it on his patients suffering from pulmonary tuberculosis. He reported that it was not only somewhat less potent as a cough suppressant but was also much less potent as an analgesic.[10] However, he found ethylmorphine to be slightly more potent and longer-acting than codeine. Merck then marketed this in January 1898.[11] It was the first semi-synthetic morphine derivative to be made commercially available, being promoted principally as a cough suppressant for use in preference to codeine or other opiates. It is now rarely prescribed.

In 1895, a team of researchers under the direction of Arthur Eichengrün at Friedrich Bayer & Company in Elberfeld, Germany, began preparing a number of esters of phenolic drugs that were known to cause gastric irritation. The expectation was that the formation of an ester would mask the phenolic hydroxyl, which was assumed to be the cause of the irritation. Two years later, Felix Hoffmann synthesised diacetylmorphine.[12] Heinrich Dreser tested it on rabbits and found it both slowed and deepened respiration. This latter aspect encouraged him to examine it in patients as an alternative to codeine for suppressing the agonising coughing of patients with pulmonary tuberculosis while at the same time assisting breathing. Considering it to be a heroic drug with the ability to act as a cough suppressant, yet safer because of its stimulating effects on respiration, he enthusiastically encouraged his company to introduce the drug without delay. Bayer marketed the drug in September 1898 for relief of respiratory disease, using the proprietary name of Heroin®. Dreser then published the most detailed pharmacological study yet to have emerged from any pharmaceutical company, though he characteristically failed to acknowledge either the prior communications from Edinburgh and London or the work of his colleagues Eichengrün and Hoffmann. Under the approved name of diamorphine it became popular throughout the world, with many physicians being convinced it was a respiratory stimulant. It was not until 1911 that this opinion was finally shown to be wrong.[13] Higby has described how from increasing popularity as a cough suppressant, diamorphine went on to become widely prescribed as an alternative to morphine for analgesia.[14]

Being unaware that diamorphine is only active after it is rapidly converted in the body to morphine and monoacetylmorphine, Dreser had assumed that Heroin® was as non-addictive as the earlier morphine ether derivatives. It is, in reality, one of the most addictive drugs ever introduced into medicine. This was not recognised at first since diamorphine was either administered orally to patients who took it for a short period during an acute respiratory complaint and did not become addicted, or else it was given to patients with chronic tuberculosis and the like in whom treatment was continued and so no withdrawal symptoms occurred.[14] What finally led to the demise of the popularity of diamorphine as a medicinal compound was a report in the *Journal of the American Medical Association* in 1912 warning of its abuse by addicts and calling for legislation to prevent its sale over the counter in drugstores at a price of less than $1 for one hundred 30 mg tablets.[15] The press took up this theme and public opinion began to swing against the ready availability of diamorphine, not only in the United States but around the world. Governments were then compelled to introduce laws controlling the supply of diamorphine and other dangerous drugs.

In 1929, the Rockefeller Foundation, the Committee on Drug Addiction of the US National Research Council, the Public Health Department and the Bureau of Narcotics jointly sponsored a research programme on non-addictive analgesics. For ten years they supported a team of chemists led by Lyndon Small at the University of Virginia and one of pharmacologists led by Nathan Eddy at the University of Michigan. Promising agents were tried on volunteers at the Narcotic Prison Hospital in Lexington, Kentucky. Despite testing about 150 compounds at a cost of millions of dollars, Small and Eddy failed to find a non-addictive agent. The most promising compound to come out of their research was metopon (methyldihydromorphinone), which was about three times as potent as morphine but less likely to cause drowsiness or nausea.[16] Unfortunately, it could not be manufactured at an economical price.

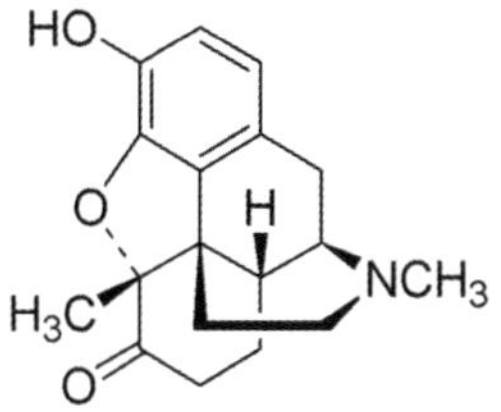

metopon

One of the principal clinical problems presented by morphine is its pronounced inhibitory effect on the respiratory centre. In 1942, Chauncey Leake, who had developed vinyl ether as an anaesthetic, suggested to researchers at the Merck Institute in Rahway, New Jersey, that, as some allyl ethers stimulated respiration, an allyl group should be attached to the morphine ring. This was duly done to form nalorphine, the methyl ether of which had previously been synthesised by Julius Pohl in 1915 and reported then as being able to counter the respiratory depression of morphine.[17] This had never been exploited in the clinic. However, Klaus Unna at Merck found that nalorphine was a potent antagonist of morphine and was of considerable value for treating opiate overdosage since it could restore normal respiration within seconds of being injected.[18] Louis Lasagna and Henry Beecher at the Harvard Medical School subsequently discovered that it was also a non-addictive analgesic when given to patients in the absence of morphine.[19] Unfortunately, it caused disturbing hallucinations, ruling out its use as an analgesic.

The establishment of the chemical structure of morphine by Gulland and Robinson in 1922 permitted the synthesis of a wider range of its analogues than had previously been available when it was subjected to simple oxidation, reduction or addition reactions. One of the most significant developments that now ensued was the synthesis of *N*-methylmorphinan, in which the oxygen bridge had been completely removed from the morphine ring system. This structurally simplified analogue of morphine prepared by Rudolph Grewe at Hoffmann–LaRoche in Switzerland had one-fifth the potency of morphine.[20] Grewe next synthesised racemorphan, its hydroxy derivative. Practically all of its analgesic activity was found to reside in the laevorotatory isomer, levorphanol, which was as potent as morphine. It retained addictive liability.[21]

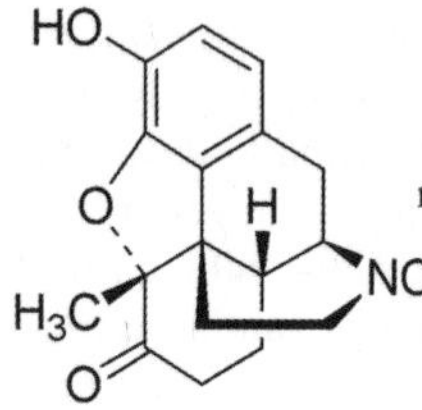

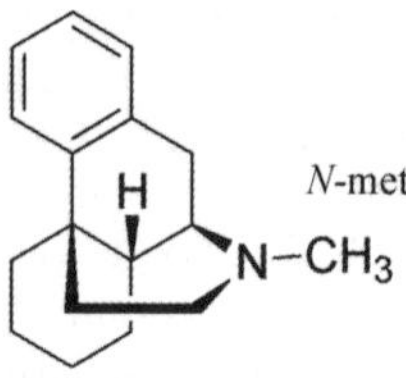

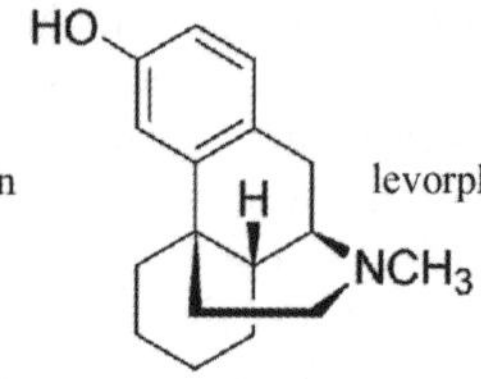

An unanticipated benefit arising from the removal of the oxygen bridge in morphine was an increase in lipophilicity (i.e. the tendency to partition into fat rather than water), as seen in levorphanol. This produced a longer duration of action as the increased lipophilicity ensured extensive distribution throughout the body, which in turn delayed return of the drug to the vascular system prior to its elimination via the kidneys. Additionally, reabsorption of the lipophilic drug from the kidney tubules back into the blood vessels surrounding them also delayed its ultimate excretion. Consequently, levorphanol had a longer elimination half-life than had morphine. This meant less frequent dosing was required, which was especially of value for providing overnight pain relief. The *N*-allyl analogue of levorphanol, levallorphan, was subsequently introduced by Hoffmann–LaRoche in 1951.[22] It had more than ten times the potency of nalorphine as a morphine antagonist because its greater lipophilicity favoured entry into the central nervous system.

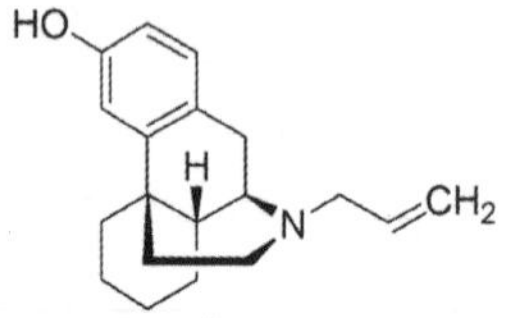

Cyclorphan was synthesised by Marshall Gates at Rochester University, New York, in 1964.[23] It was a potent morphine antagonist, yet had 20 times the potency of morphine as an analgesic. It could not be used clinically because of psychotomimetic side effects, a problem that persisted with its 14-hydroxy analogue, oxilorphan. However, the cyclobutyl analogue of oxilorphan, synthesised at Bristol Laboratories by Irwin Pachter in 1973, was free of psychotomimetic side effects.[24] It was introduced into the clinic with the approved name of butorphanol. Although about 4 times the potency of morphine, it had limited application as it was only effective when administered parenterally.

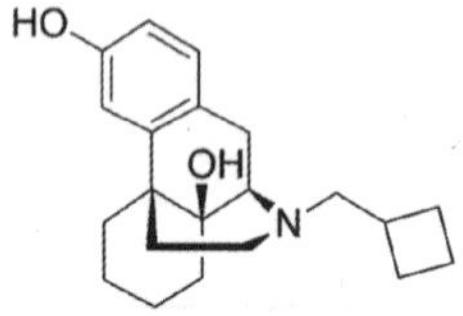

BENZOMORPHANS

Metazocine, the first benzomorphan, was synthesised in 1957 by Everette May at the National Institutes of Health in Bethesda, Maryland, but was not used in the clinic.[25] The related phenazocine was prepared at the Smith, Kline & French laboratories using May's methodology. It was highly lipophilic and, unlike morphine, was well absorbed from the gastrointestinal tract.[26] Tablets could be given for the control of severe pain. Although it had a similar addictive potential to morphine in man, it had the advantage of being about three times as potent an analgesic. It is no longer available in the United Kingdom.

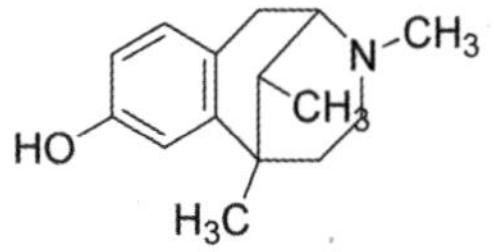

metazocine

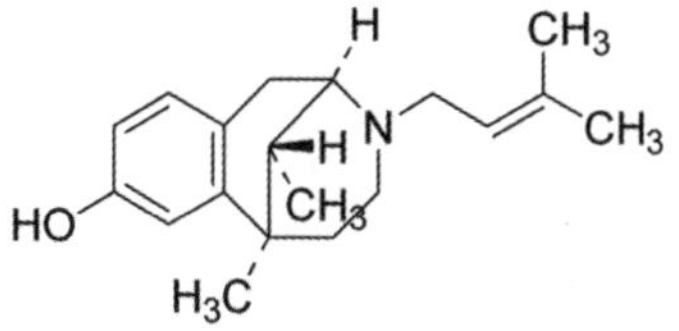

pentazocine

Cyclazocine was patented in 1962 by Sydney Archer of the Sterling–Winthrop Research Institute.[27] It has been used as a morphine antagonist in the treatment of narcotic dependence. Although about 20 times as potent as morphine as an analgesic, it could not be used clinically since it caused severe malaise in patients. The search for an analogue free of this problem resulted in the development of pentazocine by Archer.[28] It was little better than codeine when taken by mouth and had only about one-third of the potency of morphine when injected. Because of its similarity to codeine, its addictive liability was low, but it could induce hallucinations when injected. It is prescribed by mouth for moderate pain.

THEBAINE ANALOGUES

Thebaine has served as a naturally occurring synthetic intermediate rather than as a drug prototype. Compounds were initially synthesised from it without any obvious pharmacological or therapeutic objective, but introduction of a 14-hydroxyl group with appropriate stereochemistry increased the analgesic potency of morphine and codeine analogues. Oxycodone was synthesised in 1916 from thebaine by Martin Freund and Edmund Speyer at the University of Frankfurt.[29] An alternative synthetic route was then developed by Knoll, then the biggest commercial producer of codeine in Germany. Oxycodone is still used in suppositories for patients receiving terminal care. The related oxymorphone, synthesised at the National Institute of Health by Ulrich Weiss in 1955, has ten times the potency of morphine when injected and is an alternative to morphine.[30]

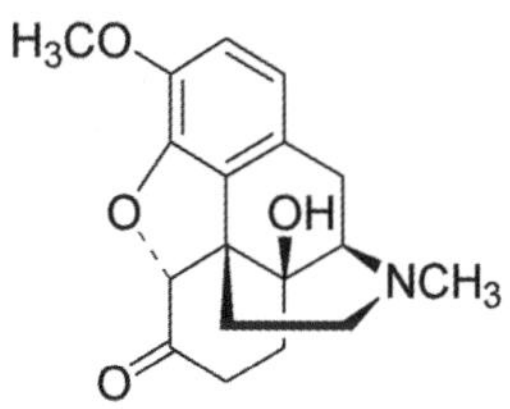

oxycodone

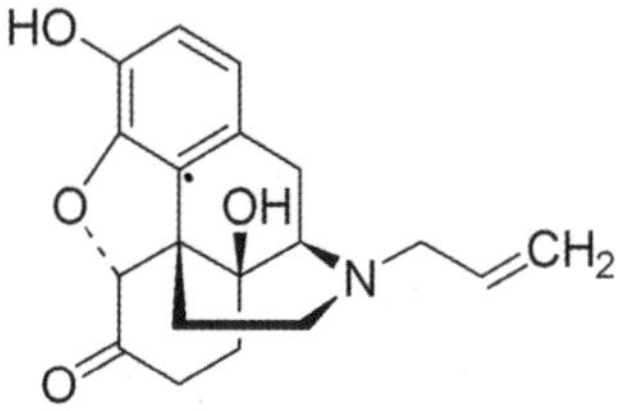

naloxone

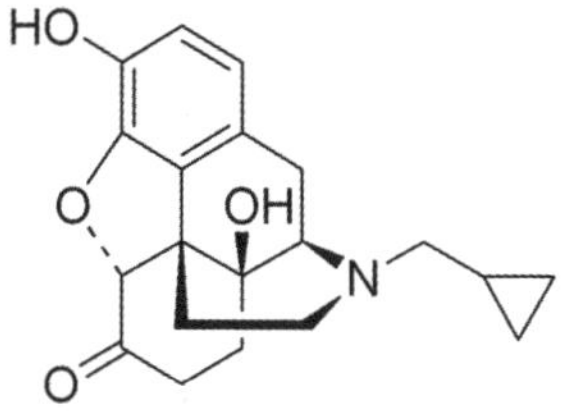

naltrexone

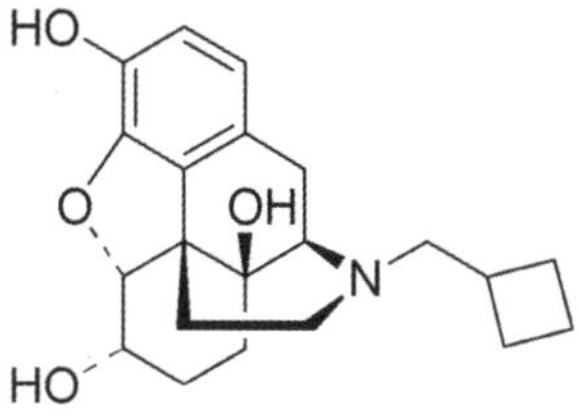

nalbuphine

The oxymorphone analogue naloxone was synthesised by Jack Fishman when he was working as a part-time assistant in the private laboratory of Mozes Lewenstein, who was also the head of the narcotics division of Endo Laboratories in New York. This was intended to be a potent analgesic, but on testing at the Endo Laboratories (to whom it was licensed by Lewenstein), naloxone was found to be solely a potent morphine antagonist. A patent was applied for in March 1961, but it was another five years before it was compared with other morphine antagonists and found to be not only the most potent of them all but also of low toxicity and devoid of the tendency to produce the respiratory depression seen with nalorphine.[31] Naloxone was the first pure morphine antagonist and proved invaluable for investigating drug interactions with opiate receptors. It remains the safest antagonist available for the treatment of opiate overdose or poisoning. Naltrexone has similar activity to naloxone, but with a duration of action extending up to three days, and is thus more effective for treating opiate dependence.[32]

Non-addictive analgesics devoid of hallucinatory properties were not available until 1968. Nalbuphine, which was made by Pachter, was three to four times as potent as pentazocine when injected and almost as potent as morphine.[33] It appeared to cause less nausea and vomiting than morphine, hence it was introduced into the clinic for the parenteral treatment of moderate to severe pain. However, when given by mouth it was no more active than codeine.

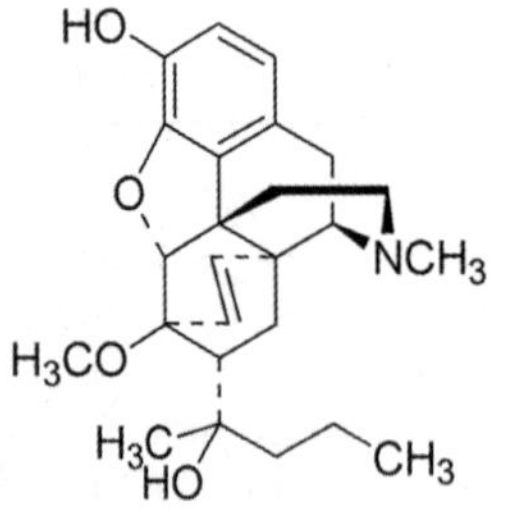

etorphine

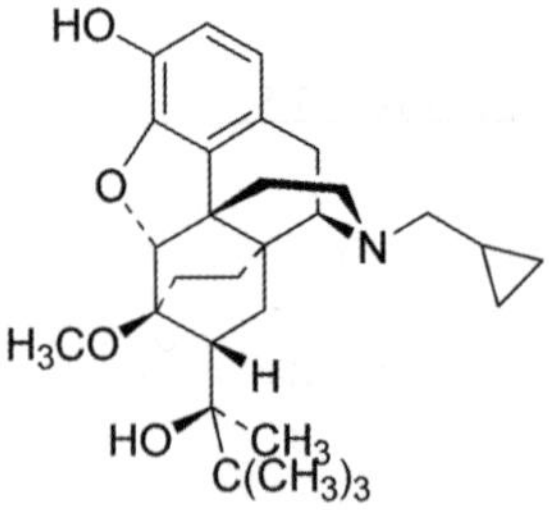

buprenorphine

Taking advantage of the capacity of the conjugated diene in thebaine to undergo the Diels–Alder cyclo-addition reaction, Kenneth Bentley synthesised etorphine in 1963 at the research laboratories in Edinburgh of the alkaloid manufacturer J.F. Macfarlane & Company who were collaborating with the consumer products company Reckitt and Colman to find a replacement for codeine.[34] Etorphine turned out to be about 1000 times more potent than morphine, partly because its higher lipophilicity permitted greater penetration into the central nervous system. The enhanced potency offered no clinical advantage over morphine since it was not accompanied by greater safety, but etorphine was valuable in the immobilisation of large animals with a dart gun. Buprenorphine was synthesised at the same time as etorphine and was later found to be a partial agonist of the *mu* opiate receptors that was free from significant activity at *kappa* receptors.[35] It was 100 times as potent as morphine when injected. It also had a longer duration of action than other opiates, but when given by mouth it underwent rapid hepatic metabolism. This was avoided by formulating it for sublingual administration, avoiding unwanted hepatic metabolism. Buprenorphine is widely used for pain relief as it has a lower dependence liability than even codeine, but it is also of value in the treatment of opiate addiction. Thus the original objective of the Reckitt research programme was ultimately achieved.

ATROPINE ANALOGUES

More drugs have been derived directly or indirectly from atropine than from any other drug prototype. The process began after the publication of reports by K. Kraut and Wilhelm

Lossen[36] that atropine could be split into tropic acid and a base called tropine. Albert Ladenburg prepared synthetic atropine at the University of Kiel in 1884 by heating these two components in dilute hydrochloric acid.[37] Later that year, he reacted tropine with a variety of other aromatic acids to produce a series of physiologically active esters which he called 'tropeines'. Homatropine, the tropeine derived from mandelic acid, proved to act more quickly on the eye than did atropine itself, even though it was not as potent. It also had the advantage of not paralysing the eye muscles for several days, an annoying feature of atropine that prevented patients from focusing in order to read. The manufacturer E. Merck in Darmstadt marketed the new tropeine. It was the first synthetic drug introduced commercially that genuinely improved upon nature.

In 1902, the Bayer Company introduced atropine methonitrate, a quaternary ammonium salt of atropine, as a mydriatic for dilation of the pupil during ophthalmic examination. Because of its highly polar nature it penetrated less readily into the central nervous system than did atropine; hence it was introduced for relieving pyloric spasm in infants. Likewise, ipratropium bromide was introduced by C.H. Boehringer Sohn in 1968 for use by asthmatics, reputedly having fewer side effects than atropine.[38]

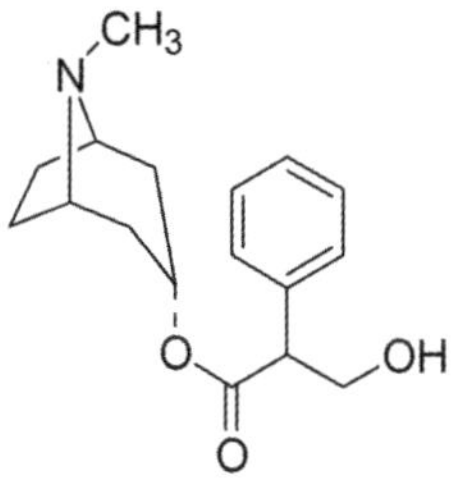

homatropine

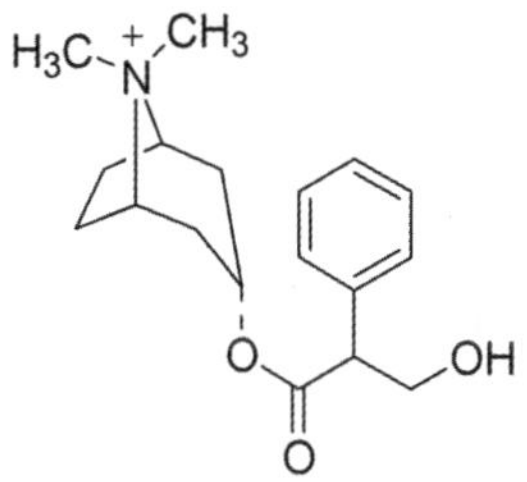

atropine methonitrate

An important advance arose from work initiated at the Landwirtschaftlichen Hochschule in Berlin and continued at the University of Frankfurt by Julius von Braun.[39] He was interested in the effects of transposing functional groups within drug molecules. In the case of atropine, he found that the tropate ester moiety could be moved to a different position on the tropine ring system without loss of mydriatic activity. On further investigation, it became evident that the presence of the intact ring was by no means essential for activity; all that was required was a tertiary amine function situated two or three carbons distant from the tropate ester function. In view of the success achieved in the early years of this century by conducting a similar exercise with the chemically related cocaine molecule to yield novel synthetic local anaesthetics, it is surprising that it took until 1922 before von Braun was able to demonstrate that not only could tropic acid be replaced with retention of mydriatic activity but so, too, could the tropane ring. Once the difficulty in obtaining good yields with the synthetic method had been overcome, the way was clear for a plethora of analogues of atropine to be introduced.

In 1929, the acetyl ester of one of the compounds prepared by von Braun was synthesised by K. Toda and marketed by the Swiss company Hoffmann–LaRoche as a remedy for sea-sickness under the proprietary name of Navigan®.[40] The acetyl function had been introduced as a protecting group during the synthesis to prevent dialkylation of the tropic acid. It was claimed that the activity was comparable to that of atropine. The next development also came from Hoffmann–LaRoche and was based on the reasoning that, as there was a close structural similarity between atropine and cocaine, it would be worth incorporating into an analogue of the former an amino-alcohol which the company already used in the manufacture of its cocaine analogue, known as 'dimethocaine'. This led to the introduction of amprotropine in 1933.[41] It had a fraction of the spasmolytic activity of atropine in the gut, but, being still less potent with regard to antisecretory activity, it had a somewhat diminished tendency to cause a dry mouth when effective antispasmodic doses were administered.

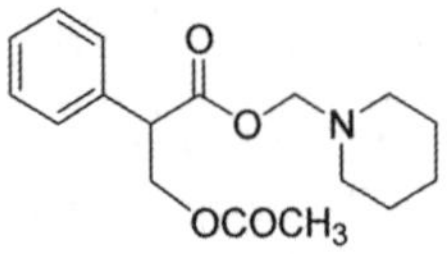

Navigan®

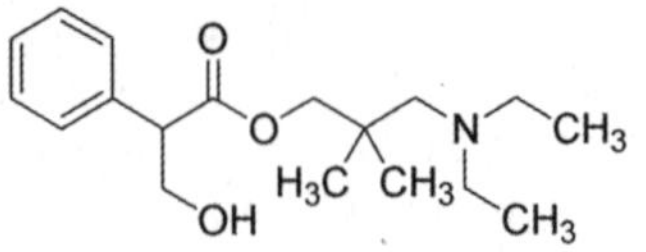

amprotropine

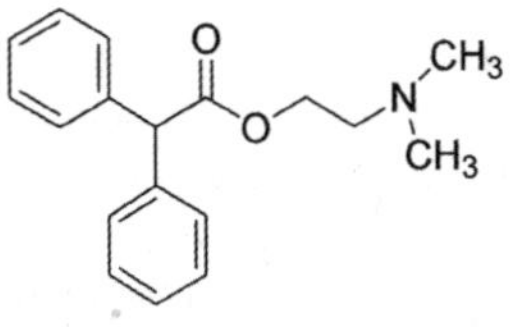

adiphenine

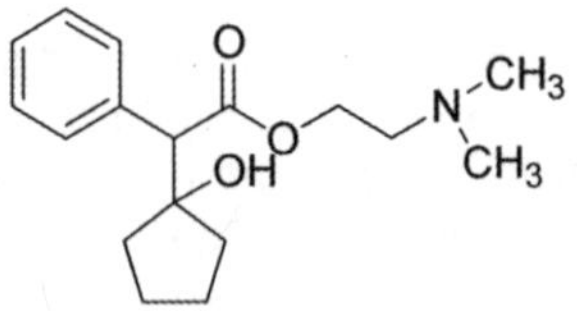

cyclopentolate

Although amprotropine only attained a modest degree of separation of antisecretory from antispasmodic activity, it was the forerunner of the synthetic selective hormone antagonists such as the beta-blockers and histamine H_2 blockers developed by medicinal chemists in the 1960s and 1970s. Its development encouraged Swiss and German pharmaceutical companies to synthesise a vast range of anticholinergic drugs by combining assorted organic acids with amino-alcohols.[42] Among these were drugs like adiphenine, introduced by Ciba and once popular as an antispasmodic.[42] Cyclopentolate was an alternative to atropine as a cycloplegic agent used during eye examination of children and in iridocyclitis to prevent formation of adhesions through its ability to immobilise the iris and ciliary muscles.[44]

Pethidine (Meperidine)

An antispasmodic synthesised in 1930 by Otto Eisleb at the Hoechst laboratories of I.G. Farbenindustrie (formed in 1926 by a merger of eight leading German chemical manufacturers, including Hoechst and Bayer) was pharmacologically examined several years later by Otto Schaumann. To his surprise, it caused the tail of a mouse to flip backwards in a sigmoid shape. Schaumann recognised this as the Straub reaction, characteristic of narcotic analgesics. Further tests confirmed the drug to have about one-tenth of the potency of morphine, but with rapid onset of analgesia and a shorter duration of action.[45] In the mistaken belief that since it bore no chemical resemblance to morphine it would be devoid of addictive properties, it was marketed in Germany for the relief of moderate pain. Eventually, it was found to induce tolerance and habituation. In the United Kingdom, it received the approved name of pethidine, while in the United States it became known as meperidine. It has been of great value in relieving pain during childbirth since it does not depress foetal respiration in anything like the manner of more potent analgesics. It is used as an alternative to morphine if a short duration of analgesia is acceptable.

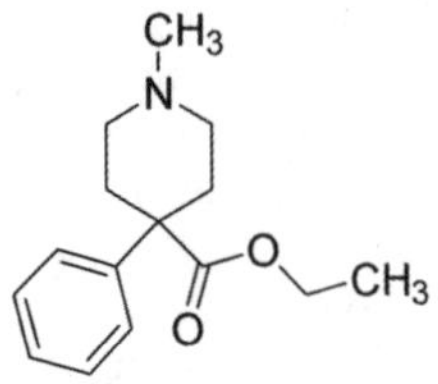

pethidine

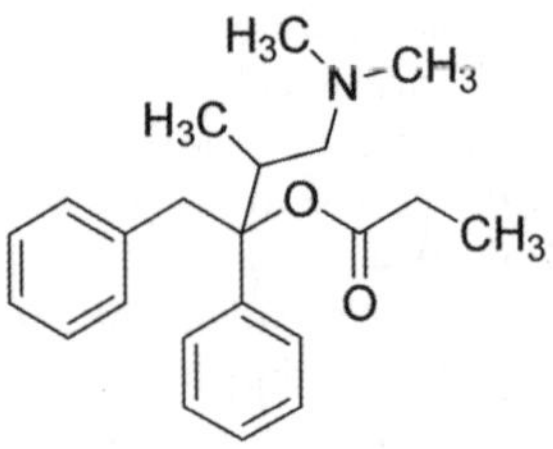

propoxyphene

Propoxyphene is an analogue of pethidine in which the piperidine ring has been opened.[46] Its dextrorotatory isomer, dextropropoxyphene, was found to have less than half the analgesic potency of codeine and has been incorporated in analgesic products.

Methadone

Analogues of pethidine developed by I.G. Farbenindustrie just before the Second World War included a series of compounds prepared by Max Bockmühl and Gustav Ehrhart. One of these had strong analgesic activity and was given the name Amidon®. It was not used clinically at that time, probably because of side effects from the high doses being used.[47]

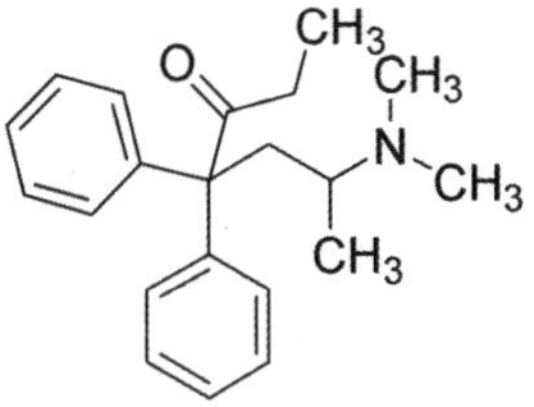

methadone

When the US State Department investigated the wartime activities of the German chemical industry at the end of the war, reports about Amidon® were examined. Information was passed to the Committee on Drug Addiction and Narcotics of the National Research Council, resulting in details being published by Eddy in 1947.[48] Amidon® (approved name methadone) was found to be an alternative to morphine in most types of pain, causing less sedation and respiratory depression. It was still addictive. Slow metabolic deactivation made it longer-acting, a feature that made it useful as a morphine substitute for addicts.

Haloperidol

An analogue of methadone called dextromoramide was one of the early drugs developed in the mid-1950s by Paul Janssen for his newly established Belgian company then known as NV Research Laboratorium Dr C Janssen but later to be called Janssen Pharmaceutica after it merged with Johnson & Johnson in 1961.[49] Janssen was a chemistry graduate who went on to qualify in both medicine and pharmacology. He became one of the most prolific drug discoverers of the twentieth century. His dextromoramide proved to be a potent addictive analgesic that was active by mouth, with a rapid onset of analgesic action and a brief duration of action. It has been used to supplement the action of other analgesics at times when intense pain is experienced. Following this success, Janssen turned his attention to pethidine and discovered a straightforward method of replacing the methyl group attached to its nitrogen atom. Compound R951 proved to be a potent analgesic, so he lengthened the chain to a butyrophenone in R1187, which turned out to be an analgesic.

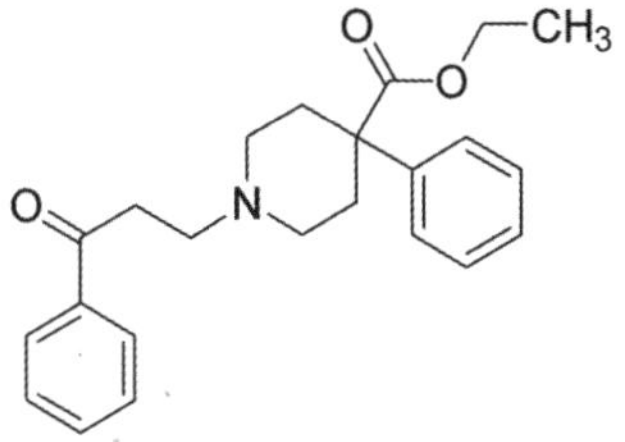

R951

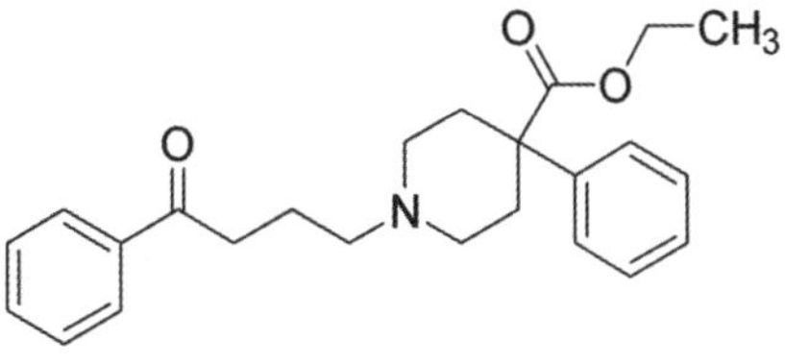

R1187

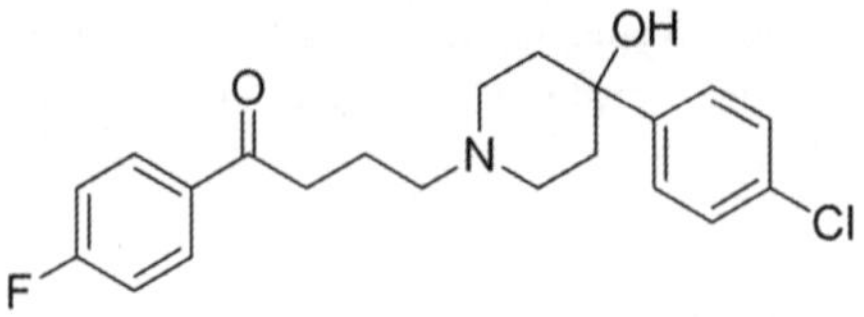

haloperidol

It was noted that the mice injected with R1187 became progressively calm and sedated after initially exhibiting typical pethidine-induced excitement, mydriasis and insensitivity to noxious stimuli. The resemblance of the sedation to that produced by the newly introduced tranquilliser chlorpromazine encouraged Janssen to synthesise analogues of R1187 in the hope of finding a tranquilliser devoid of analgesic activity. This was achieved by replacing the ester function with a hydroxyl group. The next step was to prepare literally hundreds of analogues with different substituents in the benzene rings, as this was already known to enhance potency in similar compounds. From this effort haloperidol emerged, in 1958, as the most potent tranquilliser yet to have been discovered.[50] It was 50–100 times as potent as chlorpromazine, with fewer side effects. It remains in use as an antipsychotic drug despite its tendency to produce extrapyramidal side effects such as tremor.

Another of the pethidine variants synthesised by Janssen in the 1950s was diphenoxylate.[51] It lacked analgesic activity despite being well absorbed from the gut. This was due to rapid metabolism in the liver after the drug was absorbed from the gut. Making a virtue out of necessity, Janssen exploited the fact that diphenoxylate retained a morphine-like constipating effect. This meant it could be used in the control of acute diarrhoea, without the risk of causing morphine addiction. Its effect was enhanced by formulating it together with small amounts of atropine to delay its absorption from the gut.

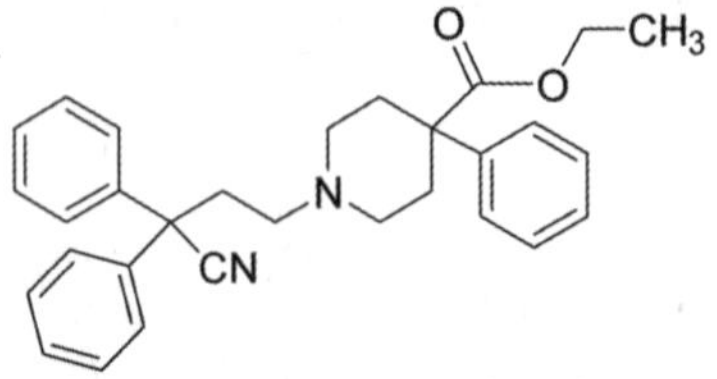

diphenoxylate

In the decade following the discovery of haloperidol and diphenoxylate, over 5000 similar compounds were synthesised and tested, of which 4000 were made in the Janssen laboratories. A dozen or so of these were introduced clinically and, like haloperidol itself, many of them blocked dopamine receptors. The analogues fentanyl[52] and alfentanil[53] are particularly useful as injectable analgesics because they act within 1–2 minutes. Fentanyl is about 500 times as potent as morphine due to its greater lipophilicity, but the anaesthetic effect is of brief duration because of rapid redistribution from the brain to muscle and fatty tissues. It is used for minor surgical procedures.

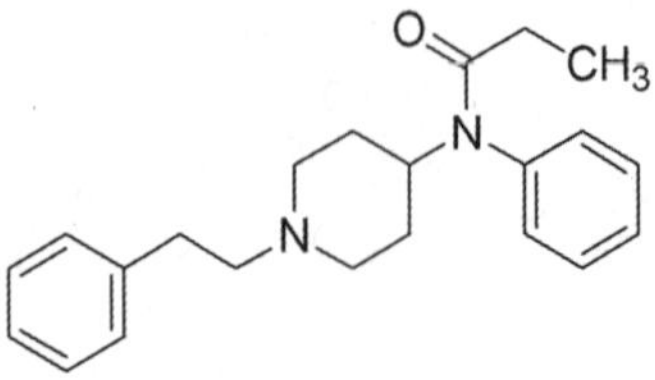

fentanyl

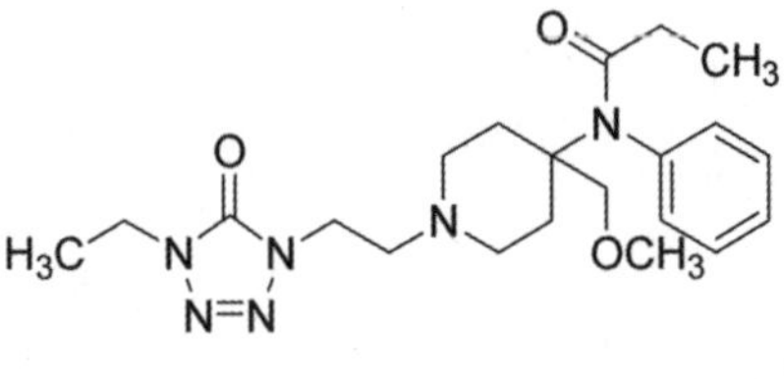

alfentanil

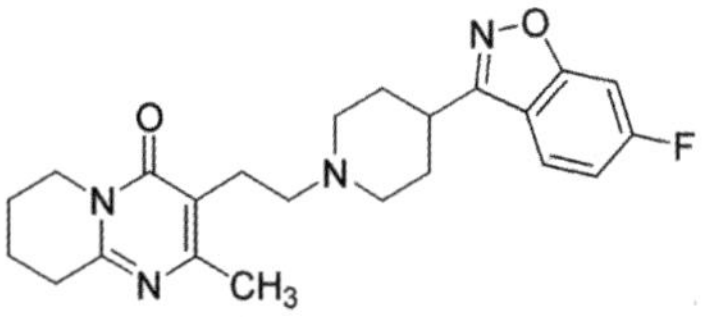

risperidone

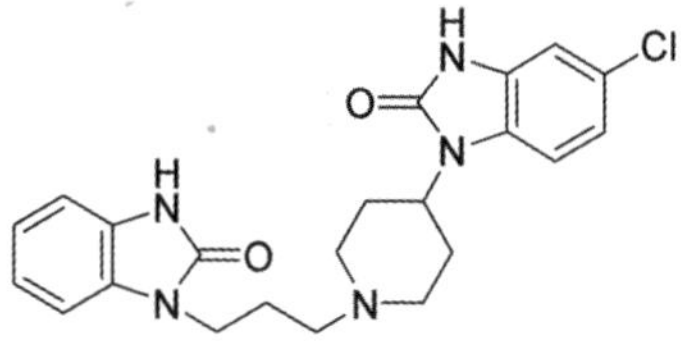

domperidone

While examining the antipsychotic drugs their company had developed, Janssen pharmacologists discovered that some had a greater effect on dopamine receptors in the central chemoreceptor trigger zone that regulated vomiting. This lies outside the blood–brain barrier in the floor of the fourth ventricle of the brain, so a search was made for a dopamine antagonist that would not pass the blood–brain barrier, thereby being free of the extrapyramidal side effects that were associated with drugs of this type. This led to the discovery of domperidone as a strong anti-emetic with minimal central effects.[54] It was marketed in 1982 for prevention of the nausea and vomiting associated with cancer chemotherapy.

Risperidone was introduced by Janssen[55] and turned out to be one of the last antipsychotic drugs to be marketed because of concern among pharmaceutical companies over the risk of being taken to court because of side effects from centrally acting drugs. It appears to cause fewer extrapyramidal side effects than earlier agents.

QUININE ANALOGUES

The original objective for attempting to synthesise quinine was to overcome its high cost brought about through excessive demand during much of the nineteenth century. Various attempts to prepare synthetic analogues of quinine were made even before its correct structure was established in 1944. The first of these was in London in 1856 when William Perkin tried unsuccessfully to synthesise the alkaloid by oxidising allyltoluidine with potassium dichromate. Instead, he fortuitously obtained a dark sludge, which on examination turned out to be exploitable as a purple dye that he was able to produce on a commercial scale under the name mauveine.[56] This serendipitous discovery launched the synthetic dyestuffs industry, which was to become the single greatest stimulus for the development of organic chemistry.

Phenazone

While working in Adolph von Baeyer's laboratory at the University of Strassburg in the spring of 1875, Emil Fischer had unintentionally synthesised phenylhydrazine. In the summer of that year, von Baeyer moved to the University of Munich where Fischer joined him. Fischer then initiated his celebrated research into the reactions of phenylhydrazine, which were to result in the discovery of a variety of novel cyclic compounds incorporating nitrogen atoms in their structure. Working under his guidance, his cousin Otto and doctoral student Wilhelm Koenigs synthesised several tetrahydroquinolines in the mistaken belief that quinine was a tetrahydroquinoline. This was a landmark in the history of drug discovery because it was the first deliberate attempt to synthesise a substitute for an alkaloid. When Wilhelm Filehne examined the compounds at the University of Erlangen, he found they exhibited antipyretic activity at doses that did not produce the severe side effects associated with quinine.[57] One of the tetrahydroquinolines was subsequently marketed in late 1882 under the proprietary name of Kairin® by Farbwerke, vorm. Meister, Lucius und Brüning, a dyestuffs manufacturer based in Höchst am Main near Frankfurt and which was later to adopt the name of the town as its company name. Kairin® was the first medicinal compound any dyestuff company had ever manufactured. A new era had dawned. However, reports that Kairin® was toxic soon appeared.

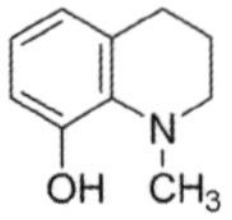

Fischer continued his work on the reactions of phenylhydrazine after transferring to the University of Erlangen in 1882, where he asked his assistant Ludwig Knorr to investigate the reaction between phenylhydrazine and acetoacetic ester. Knorr, having expected the product of this reaction to be a tetrahydroquinoline, gave it to Filehne who was testing novel tetrahydroquinolines that might replace Kairin®. Filehne found Knorr's compound to be devoid of antipyretic activity, but suggested its structure should be modified by attaching a methyl group to the ring nitrogen, as this had previously proved beneficial with compounds related to Kairin®. Knorr then discovered that the *N*-methylated derivative no longer possessed the acidic properties of its parent compound, yet he failed to realise that this change in properties should not have occurred if it had indeed been a tetrahydroquinoline.[58] The new compound turned out to be an antipyretic that was more palatable than either quinine or salicylic acid. Fischer then magnanimously allowed Knorr to take out a patent in his own name. The patent rights were sold in 1884 to Farbwerke, vorm. Meister, Lucius und Brüning. It was marketed under a proprietary name chosen by Knorr while on his honeymoon in Venice, viz. Antipyrin® (the approved name is phenazone). By the time it reached the market, Knorr had established that it was a pyrazolone derivative.

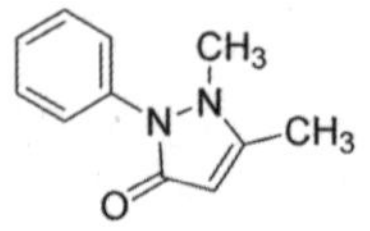

Phenazone was recommended for use in acute rheumatic fever as early as 1885, when its ability to ease joint pain was specifically commented upon. Two years later, a report of its value in relieving headache appeared in the *British Medical Journal*. Its greatest triumph, however, was in 1889 when Antipyrin® became a household name as an antipyretic after a major influenza epidemic swept through Europe. It then remained the most widely used drug in the world until aspirin began to outsell it in the early years of the twentieth century. If any single product can be said to have established the commercial viability of synthetic drugs, it was phenazone. Today, however, it is no longer used because of the risk of the serious blood condition known as granulocytopenia.

After moving to the Institute of Physiological Chemistry at Strassburg University, Filehne suggested to Friedrich Stolz, the chief chemist at Farbwerke, vorm. Meister, Lucius und Brüning, that certain phenazone analogues should be prepared as potential superior antipyretics. When one of these that had previously been synthesised by Knorr was given to him, Filehne found it to be three times as potent as phenazone.[59] This was marketed in 1896 under the name Pyramidon® (the approved name is amidopyrine). Whether it really possessed any significant advantage over the less-potent phenazone is a moot point, but it remained one of the top selling drugs in Europe until, in 1934, it was incriminated in several cases of granulocytopenia.

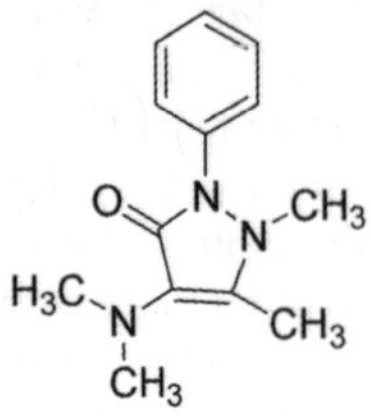

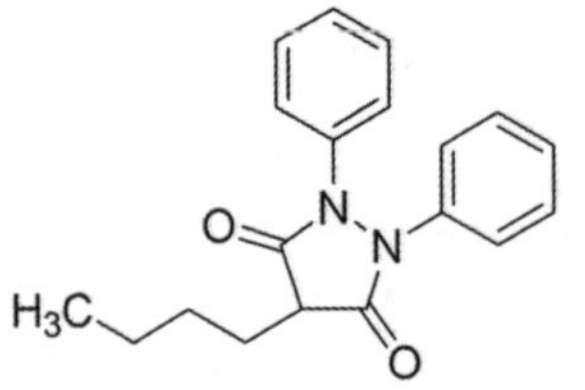

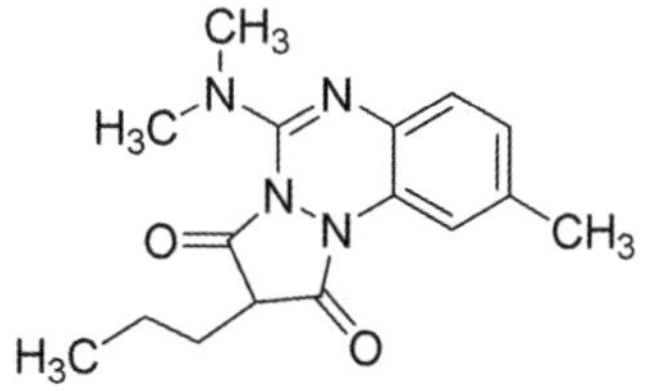

azapropazone

One way to minimise the risk of granulocytopenia was to administer smaller amounts by injecting it directly into the circulation, but to make this practical the insolubility of amidopyrine had to be overcome. The Geigy Company in Switzerland tackled this by formulating basic amidopyrine as a soluble complex with an acidic analogue of it in a formulation marketed as Irgapyrine®. When it was found that blood levels of the acidic analogue were much higher than those of the amidopyrine, this analogue was marketed in 1952 as phenylbutazone.[60] As an analogue of amidopyrine, it still caused granulocytopenia as well as another serious blood condition known as aplastic anaemia. Notwithstanding, it retained a significant market share for nearly 30 years until regulatory authorities finally demanded its withdrawal. Some analogues of phenylbutazone are still available, such as azapropazone, although this has restrictions on its use.[61]

COCAINE ANALOGUES

In 1892, Albert Einhorn of the University of Munich proposed a chemical structure for cocaine based on an erroneous one that had been suggested for tropine by Georg Merling at the Technischen Hochschule in Hannover. After taking up an appointment with Chemischen Fabrik vorm. E. Schering in Berlin, Merling synthesised an analogue of cocaine containing only the piperidine ring so that he could then determine whether both rings in the cocaine molecule were essential for anaesthetic activity.[62] This proved to be a potent local anaesthetic and was marketed in 1896 by Schering as the first cocaine substitute, under the proprietary name alpha-Eucaine®.

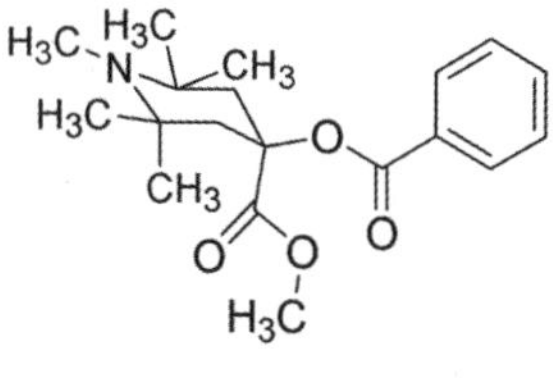

alpha-Eucaine®

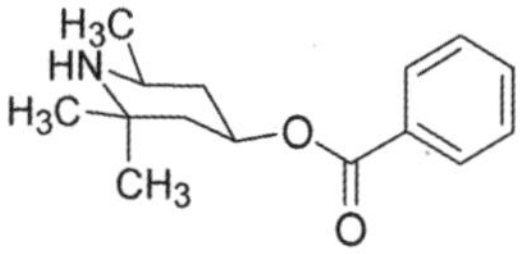

beta-Eucaine®

The sole advantage of alpha-Eucaine® over cocaine was that it was more resistant to hydrolysis, hence injections could readily be sterilised. However, it had to be replaced by the more-soluble beta-Eucaine® after patients complained that it had caused a burning sensation on application to the eye or on injection. This also achieved only a modest degree of clinical acceptability since the problem of irritancy had not been eliminated.

Einhorn now decided to prepare analogues based on the other ring of cocaine, still relying on an incorrect structure for the alkaloid. He synthesised several aromatic compounds containing both the carboxylate ester and benzoyl ester functional groups of cocaine, but was unsuccessful in his attempt to reduce the benzene ring in order to obtain his desired cyclohexane derivatives. Rather than abandon the aromatic compounds he was left with, Einhorn reasoned that as cocaine and some of its active analogues were benzoyl esters, it would be worth while sending his novel benzoyl esters for testing by his former colleague Robert Heinz, who had moved to the University of Erlangen. Several of the esters turned out to be effective, but activity was also unexpectedly found in the phenolic compounds obtained on removal of their benzoyl function by hydrolysis. This was the first ever indication of useful local anaesthetic activity in synthetic compounds that did not contain a benzoate ester function.[63] Einhorn immediately sent one of his phenolic compounds, orthocaine, to Farbwerke Höchst in 1896. As it was a derivative of phenol, they marketed it as a local anaesthetic with antiseptic activity for application to wounds. Its poor water solubility was turned into a virtue by claiming that this ensured the drug remained on wounds to provide pain relief for many hours.

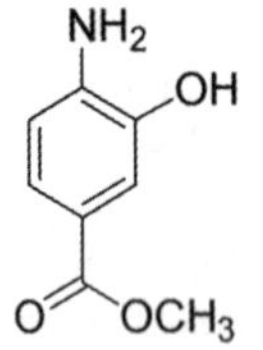

orthocaine

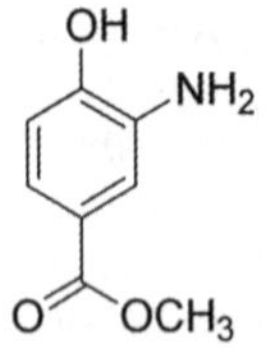

orthocaine new

Because orthocaine tended to form lumps, it was expensive to process. It was replaced the following year by its isomer orthocaine new, a compound prepared by chemists at the University of Heidelberg.[64] This gave fine, smooth-flowing crystals, thereby reducing manufacturing costs.

Einhorn next tried to prepare a water-soluble derivative of orthocaine that would be suitable for injection. As he had already established that it was possible to substitute alkyl chains containing up to five carbon atoms on the amino group of his phenolic compounds, he now examined the effect of introducing chains containing an aliphatic amino group in the hope that this would enable formulation of the resulting compounds as water-soluble hydrochlorides that did not produce highly acidic solutions. This had been the problem with the weakly basic orthocaines. In 1898, Heinz confirmed that one of these amino-compounds was effective as an anaesthetic, although its potency was low. Notwithstanding, it was introduced into the clinic as Nirvanin®, principally because it was less toxic than either cocaine or the eucaines.[65] It was soon rendered obsolete because it was still irritant.

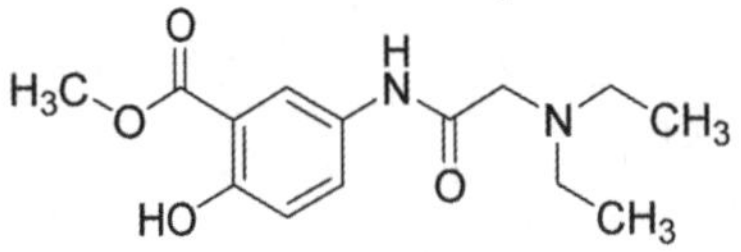

Nirvanin

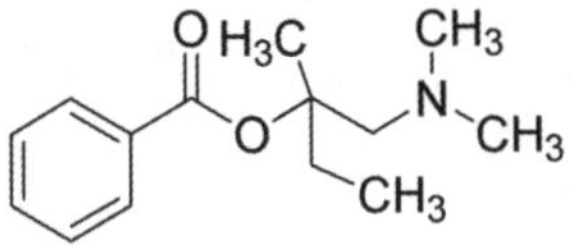

amylocaine

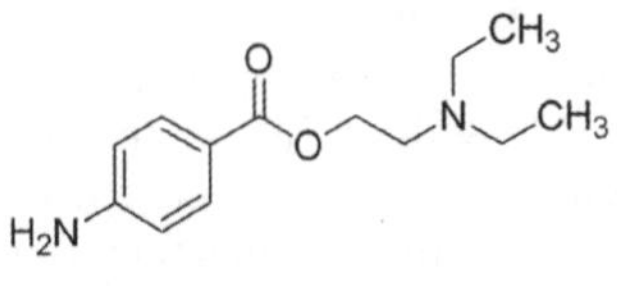

procaine

In 1902, Ernest Fourneau returned to France after having studied with leading German chemists, including Emil Fischer and Richard Willstätter. He believed that the French should not need to import drugs from Germany as they could easily make their own. He persuaded the brothers Camille, Gaston and Emile Poulenc to establish a pharmaceutical chemistry research laboratory in their factory at Ivry-sur-Seine, with himself as its director. Within a year he had developed a new local anaesthetic, amylocaine, which Poulenc Frères marketed with the proprietary name of Stovaine®, an amusing bilingual pun on Fourneau's own name![66] Like Nirvanin®, it was capable of providing water-soluble salts through the incorporation of an aliphatic amino group. Amylocaine was the first drug to achieve the long sought-after goal of being a safe, non-irritant local anaesthetic that could be injected as a substitute for cocaine. It incorporated all the groups present in the cocaine molecule, except for the piperidine ring, as Fourneau had considered this ring would cause toxicity. After all, Socrates had killed himself by taking hemlock, the active principle of which was the piperidine alkaloid coniine!

Einhorn had meanwhile been making esters from both aminobenzoic acids and amino-alcohols for initial screening by Biberfield at the Pharmacological Institute in Breslau. The most promising of these, later to be named procaine, was sent to Heinreich Braun at Leipzig for clinical evaluation.[67] Braun immediately recognised that it was far less irritant than any existing agent, but its action was brief unless high concentrations were injected. Doing this produced as much toxicity as when using cocaine. Only four years before this, however, Braun had improved New York surgeon Leonard Corning's method of prolonging cocaine anaesthesia by using a tourniquet to prevent the drug from being washed away from the injection site. Instead of applying a tourniquet, Braun had mixed cocaine with an adrenal extract to obtain similar results. Brown now combined Einhorn's local anaesthetic with the newly introduced adrenaline and overcame the problem of its short duration of action.[68] It was then marketed by Farbwerke Höchst as Novocaine®. The safety, lack of irritancy and overall efficacy of procaine with adrenaline ensured that it dominated the field of local anaesthesia for half a century.

Procainamide and Its Analogues

Advances in cardiac surgery during the 1930s led Frederick Mautz of Cleveland, Ohio, to investigate several drugs with a view to finding one that could be applied directly to the heart to prevent arrhythmias during the operative or postoperative period.[69] He felt that disturbance of cardiac rhythm might be prevented by local anaesthetics, especially as procaine had already been shown to have some effect on heart muscle. In his experiments on animals, procaine did indeed prove highly effective and superior to cocaine. Mautz then proposed that if a heart proved exceedingly irritable during surgery, or if the beat became irregular, cautious intrapericardial injection of procaine could correct these disturbances.

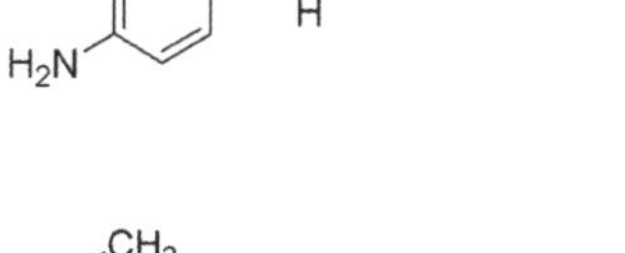

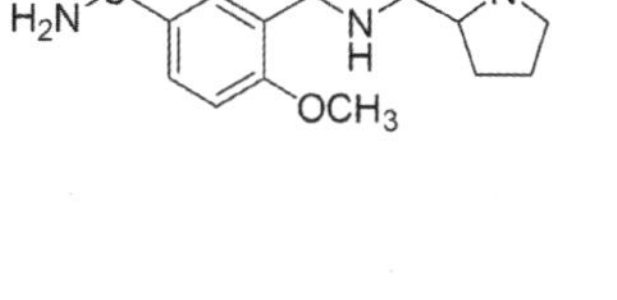

It was only after the Second World War that procaine was generally used clinically as an alternative to quinidine, the action of which it resembled. Unfortunately, procaine was rapidly metabolised by esterase enzymes in the blood, severely limiting its clinical application. There was also a high incidence of central side effects. For these reasons, cardiologists welcomed the introduction of procainamide in the 1950s.[70] It was resistant to esterases and had a lower incidence of central nervous system disturbance as it was less lipid-soluble than procaine. Procainamide is still administered intravenously to control ventricular arrhythmias, but is no longer used for chronic administration because of the risk of inducing systemic lupus erythematosus, which manifests itself as a butterfly type of skin rash. This syndrome can progress to cause granulocytopenia if treatment is not withdrawn.

While examining 2-chloroprocainamide as an anti-arrhythmic agent at the Laboratoires Delagrange in France in 1957, Laville and Justin-Besançon discovered that it also had anti-emetic properties when given by mouth. Analogues were synthesised, leading to the introduction of metoclopramide as an anti-emetic.[71] Because the tranquilliser chlorpromazine was also an anti-emetic, several hundred derivatives of metoclopramide were then examined for tranquillising activity, with sulpiride eventually emerging as a potential antipsychotic agent.[72] Conflicting reports were initially received from French psychiatrists, but it ultimately transpired that sulpiride lacked the sedating properties of drugs such as chlorpromazine and haloperidol, making it a valuable agent in the treatment of schizophrenia. Although available in France for many years, sulpiride was not marketed in the United Kingdom until 1984.

EPHEDRINE ANALOGUES

Monopolisation of supplies of *Ephedra vulgaris* in 1927 led to a scarcity of ephedrine and consequent high prices. Gordon Alles, a chemist working for an allergy specialist in Los Angeles, synthesised amphetamine as a cheap substitute for ephedrine. He found it produced a prolonged rise in blood pressure when given by mouth.[73] The drug was then passed to Smith Kline & French Laboratories in Philadelphia, where it had already been examining it as a potential nasal decongestant. The company was aware that it had been synthesised at the University of Berlin in 1887[74] and examined by George Barger and Henry Dale at the Wellcome Physiological Research Laboratories in London during a major study of sympathomimetic amines in 1910.[75]

Smith Kline & French decided to exploit the volatility of the free base, an oily liquid, by formulating it in a plastic inhaler for insertion into the nostril as a nasal decongestant. This was marketed in 1932 as the Benzedrine Inhaler®, a product that achieved considerable commercial success until its withdrawal in the 1960s because the amphetamine was being extracted from the inhaler and being abused. Early clinical experience with amphetamine as an oral decongestant in hay fever had revealed its stimulant properties, leading to its introduction in 1935 for the treatment of the abnormal sleepiness in the rare condition known as narcolepsy. Alles found that the stimulant properties of amphetamine were principally associated with the dextrorotatory stereoisomer, dexamphetamine. Amphetamine also proved to be an anorectic for curbing appetite, but this is now considered an inappropriate use as it causes drug dependence and can induce psychotic episodes. Another controversial application was in the treatment of hyperactivity in children.

Analogues of amphetamine have reflected the diverse actions of the drug. Some were initially intended to be pressor agents for use in shock, while others were developed as stimulants. The latter had enhanced lipophilicity to encourage penetration into the central nervous system. Numerous analogues were introduced to treat obesity, usually being selected on the basis of their claimed relative freedom from stimulant effects, but they are of no benefit. Methylphenidate is used in children and adolescents with severe attention deficit/hyperactivity disorder.[76]

PHYSOSTIGMINE ANALOGUES

The structure of physostigmine was elucidated in 1925 by Edgar Stedman and George Barger at the University of Edinburgh.[77] Utilising the miotic action on the eye as a bioassay procedure, Edgar and Eleanor Stedman went on to prepare several structurally simpler analogues, the most potent being miotine.[78] John Aeschlimann of the Roche Research Laboratories in Basle then exploited the stability of dimethyl carbamates, overcoming the extreme sensitivity of physostigmine, a monomethyl carbamate, to hydrolytic cleavage.[79] Neostigmine, introduced in 1931, proved superior to miotine in the treatment of the muscular disease myasthenia gravis, a rare form of paralysis in which defective neuromuscular transmission is ameliorated by the anticholinesterase activity of neostigmine. Physostigmine also exhibited this action, but the effects were too brief.

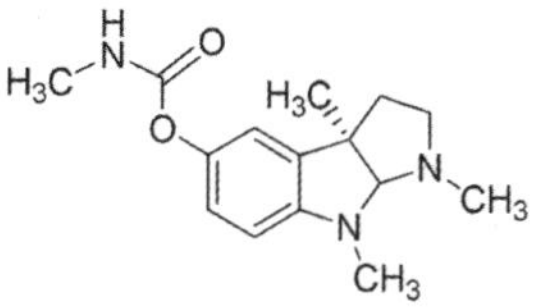

physostigmine

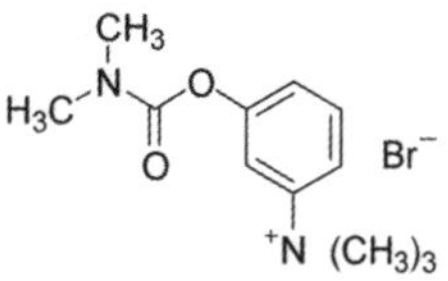

neostigmine

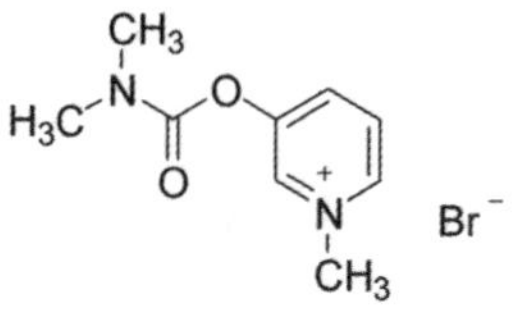

pyridostigmine

At St Alphege's Hospital in London in 1934, Mary Walker noted a resemblance between myasthenia gravis and curare poisoning.[80] Since physostigmine was an antidote to curare, she tried it on her patients. The results were so striking that, after confirmation by others, they made headlines in the national press in March of the following year. Neostigmine was also used and subsequently found to be superior. It was not without its disadvantages, including unwanted cholinergic actions. Attempts to avoid these resulted in the introduction of pyridostigmine.[81]

PAPAVERINE ANALOGUES

An efficient synthesis by Carl Mannich at the University of Berlin in 1927 facilitated the preparation of papaverine analogues in which either or both of the two pairs of methoxyl groups could be replaced by methylenedioxy functions.[82] Within two years, Otto Wolfes of Merck and Company of Darmstadt had obtained a patent for a series of papaverine analogues similar to Mannich's compounds. Merck claimed that one of these, Eupaverin® (a name later given to a different benzylisoquinoline alkaloid), was less toxic than papaverine. The Chinoin

Company of Budapest marketed several of the early analogues of papaverine, including ethaverine, which was said to be three times as potent as papaverine with reduced toxicity. Ethaverine and Eupaverin® were widely prescribed as antispasmodics in the 1930s.

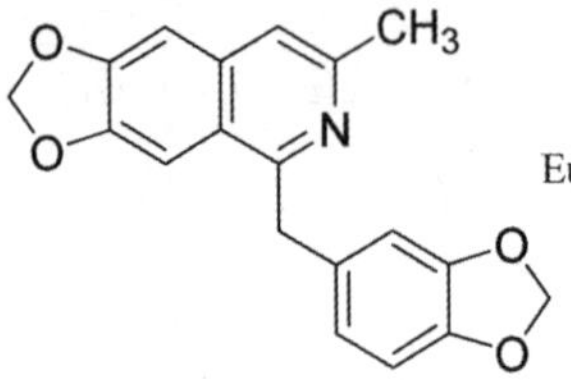

Eupaverin®

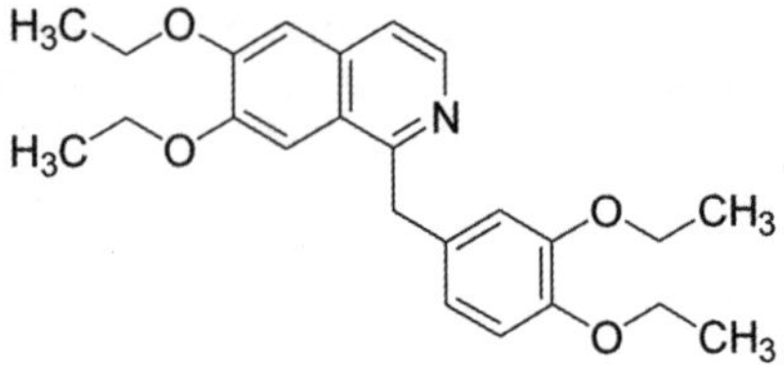

ethaverine

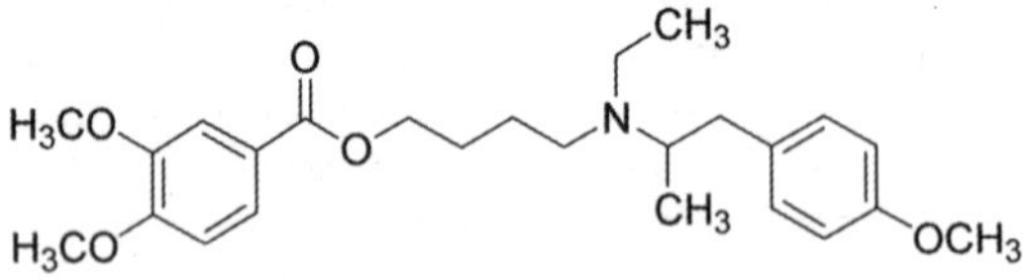

mebeverine

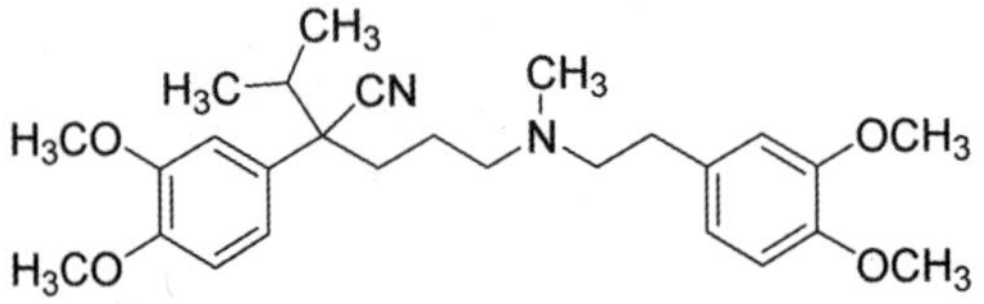

verapamil

The early papaverine analogues proved inferior to the atropine analogues that were introduced as antispasmodics in the 1930s. Open chain analogues were subsequently developed, such as mebeverine and verapamil. The former acts directly on smooth muscle and is still used in a wide variety of gastrointestinal complaints.[83] Verapamil was synthesised by Ferdinand Dengel of Knoll AG as a potential coronary muscle dilator for possible use in patients suffering from angina, and was introduced clinically in 1968.[84] As it caused cardiac depression, it was passed for further investigation to Albrecht Fleckenstein at the University of Freiburg. He established that it prevented contraction of cardiac muscle as well as that of smooth muscle. Observing that this could be reversed by addition of calcium, Fleckenstein proposed that verapamil was acting as a calcium antagonist. Prenylamine, an analogue of methadone developed by Hoechst AG, was also found to have a similar action. Both compounds interfered with the movement of calcium ions into cardiac cells and so were the first substances to be categorised as calcium antagonists.[85] The significance of this became apparent in the 1970s when verapamil was shown to be an anti-arrhythmic agent[86] and then to be capable of reducing high blood pressure.[87]

Prazosin

Pharmaceutical scientists became interested in finding inhibitors of cyclic nucleotide phosphodiesterase enzymes after Earl Sutherland at Western Reserve University in Cleveland, Ohio, had reported in 1962 that caffeine, theophylline and other methylxanthines inhibited the phosphodiesterase that catalysed cyclic adenosine monophosphate (cAMP) hydrolysis.[88] Sutherland had identified cyclic AMP in tissues, shown that its formation was catalysed by adenyl cyclase and then elucidated its role as an intracellular second messenger in the propagation of signals initiated by extracellular hormones. He received the Nobel Prize for Medicine in 1971 for his discoveries. Subsequently, it was found that cyclic guanosine monophosphate (cGMP) also acted as a second messenger. The inactivation of both cyclic nucleotides through hydrolysis of their phosphate ring was catalysed by a large variety of phosphodiesterase (PDE) isoenzymes. Eleven families of PDEs are now known, each produced by different genes. Further subdivisions of isoenzymes exist within each family. The realisation that the level of PDEs varies in different organs opened up the possibility of developing specific inhibitors to control their activity. Although several existing drugs were found to inhibit PDEs and a number of inhibitors were discovered, it was not until the 1990s that any useful drug emerged from the extensive research carried out both in industry and academia. One of the main reasons for this was the difficulty in establishing assay procedures that could identify compounds specifically inhibiting one PDE rather than several.

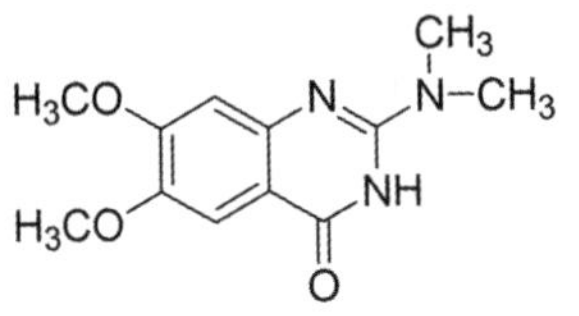

dimethylaminoquinazoline

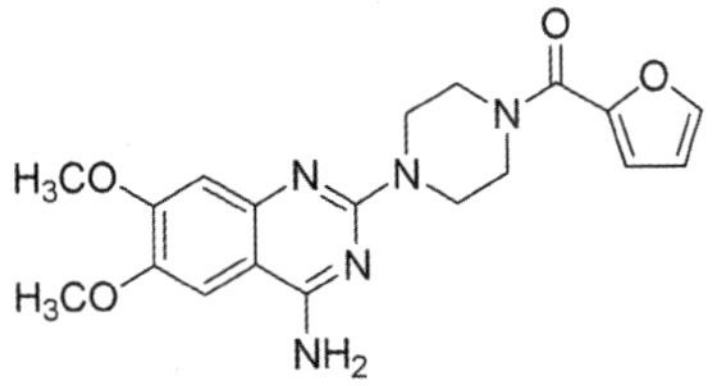

prazosin

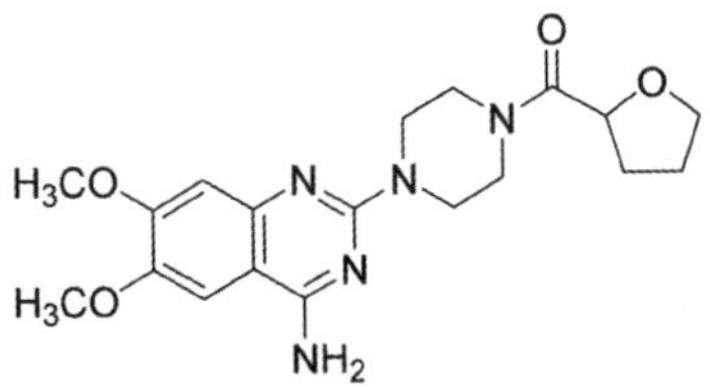

terazosin

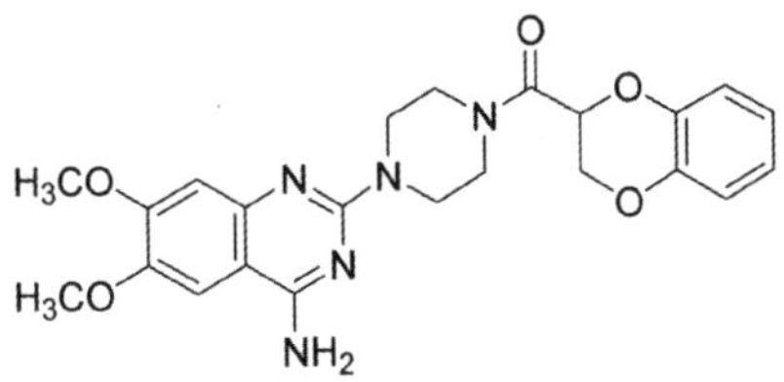

doxazosin

The Pfizer chemist Hans-Jürgen Hess believed that aminoquinazoline analogues of papaverine would resemble the aminopyrimidine moieties of cAMP and cGMP and so might act as phosphodiesterase inhibitors that act as vasodilators with a direct action on vascular smooth muscle. He and his colleagues developed a synthetic route for the preparation of 2-amino-4-oxoquinazolines containing methoxyl groups placed appropriately to resemble papaverine.[89] The most active compound was dimethylaminoquinazoline. The oxo group was then replaced with an amino group. This was followed by the synthesis of analogues in which the dimethylamino group was replaced by other amines. The introduction of a substituted piperazine ring led to a compound that did not slow the heart rate, as did other hypotensive drugs then in use.[90] Although it had been designed to inhibit phosphodiesterases, this did not occur unless the drug was given at higher doses than used in the clinic. It was then shown to act as an alpha-adrenoceptor blocker, which caused dilation of both veins and arterioles, resulting in reduction of peripheral resistance in the vascular system and a consequent drop in blood pressure. When first introduced into the clinic under the name prazosin, patients complained of side effects. It was later recognised that if lower doses were used the incidence of side effects declined. Prazosin is now prescribed in combination with other drugs for hypertension. Terazosin[91] and doxazosin[92] are similar to prazosin in both their pharmacological and clinical profiles.

THEOPHYLLINE ANALOGUES

Theophylline was difficult to formulate because of its lack of solubility in water. This was partially overcome in 1908 by the introduction of aminophylline, a double compound with ethylenediamine that was manufactured in Dr H. Byke's chemical factory in Charlottenburg.[93]

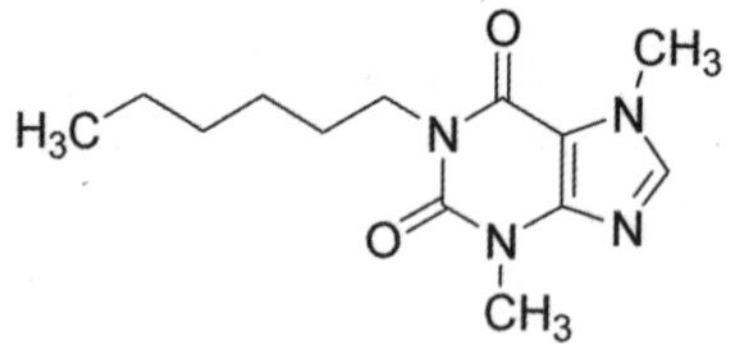

pentifylline

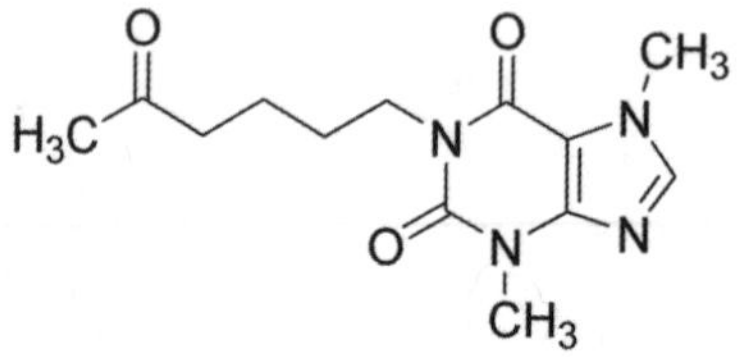

oxypentifylline

· Other double compounds have been marketed, as well as analogues of theophylline substituted on one of its nitrogen atoms, e.g. pentifylline[94] and its metabolite oxypentifylline[95], both introduced as vasodilators for peripheral vascular insufficiency.

Dipyridamole

Dipyridamole was one of a series of compounds described as homopurines in a patent granted to the Karl Thomae Company in 1959.[96] Many of the compounds dilated slices of coronary arteries. Detailed reports on the pharmacology of one of the homopurines, dipyramidole, considered it to be a promising coronary dilator since it produced a persistent effect without raising the blood pressure or the workload of the heart.[97] Dipyridamole was subsequently introduced into the clinic in the early 1960s as a coronary vasodilator.

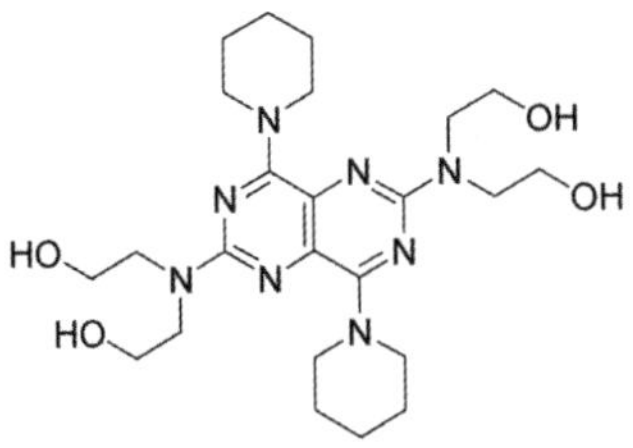

dipyridamole

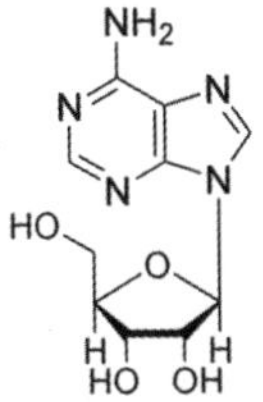

adenosine

Gustav Born of the MRC Thrombosis Research Group in London discovered that adenosine modified the behaviour of platelet thrombi in experimentally injured arteries.[98] When it was examined at the Radcliffe Infirmary in the hope of developing a treatment for vascular thrombosis, it produced a pronounced fall in blood pressure due to its vasodilatory action. Since an investigation into the mode of action of dipyramidole had previously found that it slowed the rate of disappearance of exogenous adenosine from whole blood, an investigation was conducted to see whether dipyramidole could permit lower doses of adenosine that would not be toxic. In the course of this study, it was unexpectedly discovered that dipyramidole itself was a powerful inhibitor of thrombus formation in injured rabbit arteries.[99] Samples of plasma were then collected from volunteers who had taken dipyridamole and it was found that it prevented platelet clumping when the plasma was stirred.[100] Subsequent clinical studies confirmed the ability of dipyridamole to inhibit thrombus formation in the arterial circulation by reducing platelet adhesiveness. Dipyridamole is now combined with oral anticoagulants to prevent thrombosis in patients with prosthetic heart valves. The vasodilating effect for which it was originally introduced is considered of little value.

Sildenafil

A programme to develop a compound that could lower blood pressure by enhancing the activity of atrial natriuretic peptide (ANP) was initiated in 1985 by Simon Campbell and David Roberts of Pfizer UK in Sandwich, Kent. This particular peptide was of interest since it was a vasodilator that also caused the kidneys to excrete sodium and water, hence elevating its level could possibly reduce blood pressure. As it was known that ANP worked by stimulating guanylate cyclase to increase the synthesis of cGMP, the Pfizer researchers decided to seek an inhibitor for the PDE in kidney tissue that destroyed cGMP. They then isolated the isoenzyme free from contamination with others, enabling progress to be made in testing compounds for inhibitory action. As no potent, selective inhibitors of the PDE were known, Nicholas Terrett and his colleagues sought a lead compound from the literature. They chose zaprinast, an unmarketed drug developed by the UK-based May and Baker division of Rhône–Poulenc.[101] It was a xanthine analogue designed as an anti-allergy drug in the wake of reports about the anti-anaphylactic and anti-inflammatory effects of theophylline and other xanthines, but instead lowered blood pressure *in vitro* and turned out to be a weak inhibitor of the PDE that was involved in the hydrolysis of cGMP. Before the synthesis of analogues began, a detailed comparison was made between zaprinast and cGMP, the natural substrate for the PDE. The electronic distribution in both compounds was found to be similar, while computer graphics demonstrated that the preferred shape of cGMP could still be retained if there was appropriate replacement of the phosphate ring by another moiety. Consequently, a variety of compounds with variations in the heterocyclic ring system (i.e. the one containing nitrogen atoms) of zaprinast was prepared. The most promising of these was a pyrazolopyrimidinone with ten times the potency of zaprinast as an inhibitor of the enzyme, which was by now being described as PDE5. Encouraged by this, the researchers focused their attention on the benzene ring, seeking to enhance its similarity to the phosphate ring of cGMP. Two key variations were introduced to increase the similarity to cGMP and thereby enhance fit to the active site of PDE5. Firstly, a 3-methyl group in the pyrazolopyrimidinone ring was enlarged. Secondly, a sulfonamide was introduced on the benzene ring to mimic the cyclic phosphate of cGMP. These changes led to the preparation in 1989 of sildenafil as a PDE5 inhibitor, which was 100 times as potent as zaprinast and highly specific in its action.[102,103]

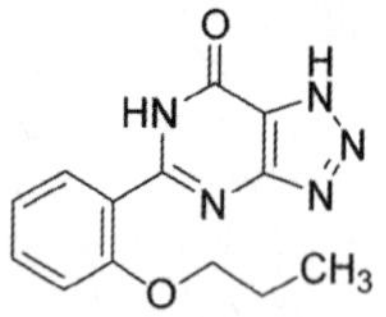

By the time sildenafil had been prepared over 1600 compounds had been synthesised and the objective of the Pfizer programme had changed to seeking a vasodilator of coronary vessels for the relief of angina. Despite encouraging results in the laboratory, sildenafil proved disappointing when tested on patients with coronary heart disease. One of several side effects was only revealed when participants in a trial of sildenafil on 30 men in the Welsh town of Merthyr Tydfil in 1992 were questioned about their reluctance to return unused tablets when the trial was stopped. On questioning by the physician in charge, it emerged that they had discovered that the drug was inducing an increase in erectile function. Further investigations not only confirmed this effect in healthy volunteers, but also revealed that it was of benefit to those suffering from erectile dysfunction. An explanation of this was forthcoming, thanks to concurrent advances in understanding the role of nitric oxide (NO) as a signalling molecule in the cardiovascular system which stimulated guanylate cyclase and consequent formation of cGMP. It transpired that on sexual stimulation NO was released from nerves in the penis and thereby raised cGMP levels in the corpus cavernosum. This, in turn, caused smooth muscle relaxation which permitted blood to fill the area and cause an erection.[104] Sildenafil was eventually launched by Pfizer under the proprietary name Viagra® as a treatment for erectile dysfunction in 1998. Several similar compounds have been introduced since then by other companies.

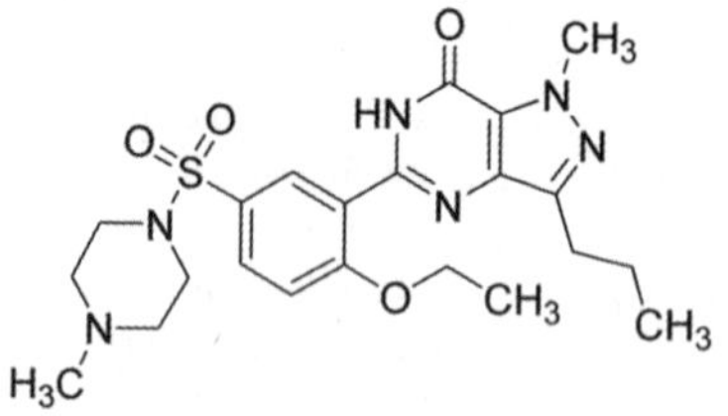

DICOUMAROL ANALOGUES

By 1944, more than 100 hydroxycoumarin analogues of dicoumarol had been synthesised in Paul Link's laboratory at the University of Wisconsin College of Agriculture, the principal differences between them concerning potency, rate of onset of action and duration. Ethyl biscoumacetate proved superior to dicoumarol because it had a quicker onset of action, but this was offset by a shorter duration.[105] As the mode of action of the oral anticoagulants is to interfere with the role of vitamin K in the hepatic biosynthesis of clotting factors, there is a time lag before reserves are depleted. Not only does it take 1–3 days for dicoumarol to exert its action, but after discontinuing treatment its effect may persist for four days or more. Coupled with a degree of unpredictability of response, this led to dicoumarol being superseded by drugs such as warfarin and phenindione.

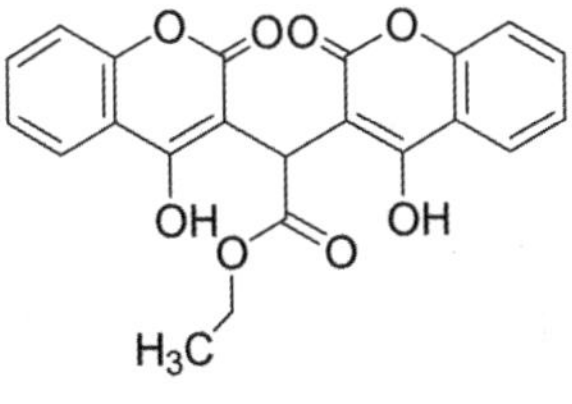

ethyl biscoumacetate

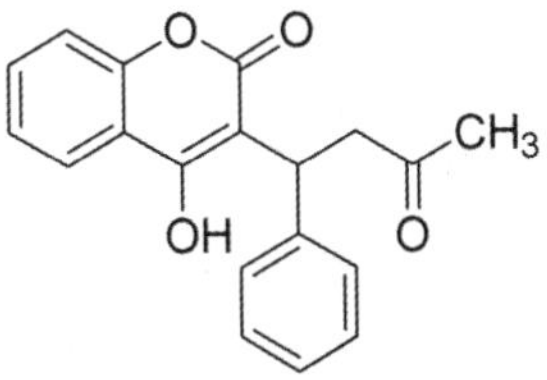

warfarin

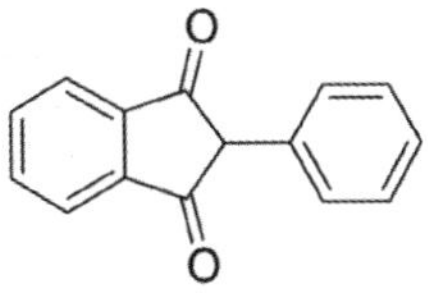

phenindione

Warfarin was longer-acting[106] and 5–10 times more potent than dicoumarol, hence it was selected in 1948 as a rat poison from the hydroxycoumarins synthesised in Link's laboratory. The royalties from its extensive sales were vested in the Wisconsin Alumni Research Foundation, hence its name. Despite the urging of Link, warfarin was not accepted as a clinical anticoagulant until after an attempt by a US army recruit to commit suicide by taking rat poison proved unsuccessful. This indicated that warfarin was safer than other anticoagulants. It was marketed by Endo Laboratories of New York and soon became the most commonly used anticoagulant.

Phenindione was a non-coumarin anticoagulant that was more likely than warfarin to cause hypersensitivity reactions.[107] A report of its anti-inflammatory activity initiated a 15-year-long quest by Pfizer researchers to separate the anti-inflammatory from the anticoagulant activity.[108] Screening of derivatives revealed that the anticoagulant activity (as measured by the prothrombin time) was diminished by large substituents at the meta position on the 2-aryl ring. Control of acidity was critical, since weakly acid compounds with a pK_a value above 5.9 were devoid of anti-inflammatory activity. Despite the recognition of this, all the phenindiones investigated had too brief a plasma half-life to offer prospects of clinical usefulness.

A step forward came with a series of 1,3-dioxoisoquinoline-4-carboxanilides, of which the most promising was tesicam, which had an extended plasma half-life but was not considered to have sufficient potency.

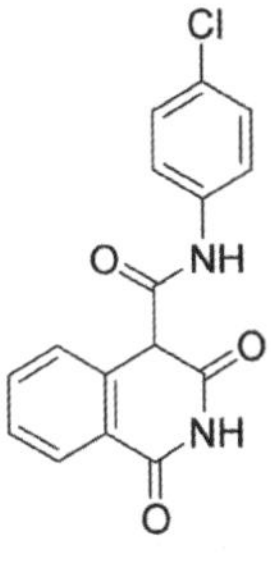

tesicam

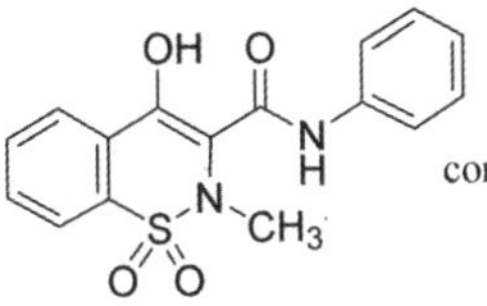

compound A

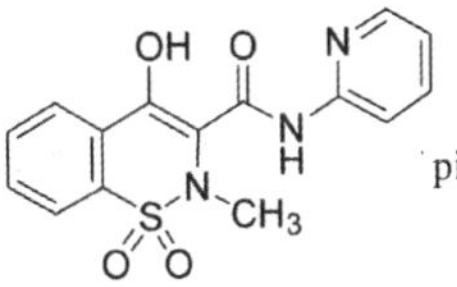

piroxicam

The next step involved the preparation of more than 50 benzothiazine carboxamides, the best of which was compound A. It had a long half-life and double the potency of phenylbutazone. The replacement of the benzene ring by a variety of heterocycles enhanced the acidity and increased anti-inflammatory potency. Piroxicam emerged as the most effective variant.[109] In humans, it had a half-life of about 45 hours and was about ten times as potent as phenylbutazone.

TUBOCURARINE CHLORIDE ANALOGUES

In 1946, Daniel Bovet and his colleagues at the Pasteur Institute in Paris initiated a project to synthesise tubocurarine analogues.[110] Since the structure of the alkaloid proposed by Harold King had two quinoline rings and two quaternary ammonium functions, Bovet's group synthesised a series of phenolic ethers containing quaternary ammonium functions.

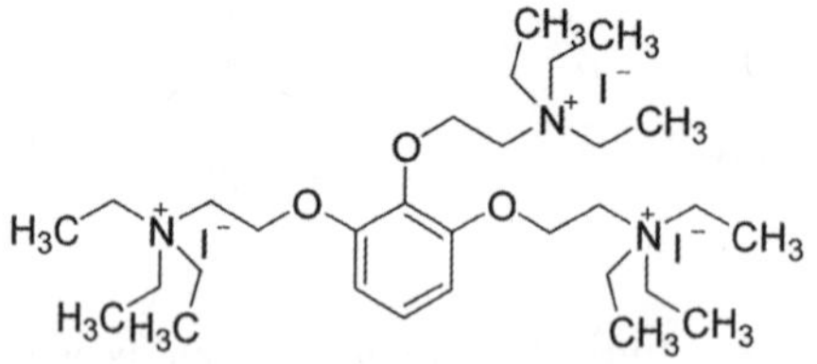

gallamine iodide

This led to the synthesis of gallamine in 1947.[111] Its onset of action was quicker than the 3–5 minutes required for tubocurarine, while recovery was briefer than the 30 minutes or more for the natural alkaloid. As it caused acceleration of the heart, gallamine could not be safely used in patients with kidney or cardiac disease.

Suxamethonium

Around the same time that Bovet began his study of tubocurarine analogues in Paris, a similar investigation was being undertaken at Oxford by Richard Barlow and Harold Ing.[112] Their approach was to establish whether varying the distance between the two quaternary ammonium functions in King's proposed structure for the alkaloid would increase potency. To this end they examined the simplest possible series of compounds containing two quaternary groups at opposite ends of straight carbon chains, the so-called polymethylene bisonium compounds. It was found that the compound with a 10-carbon chain was three times as potent as tubocurarine when assayed by the rabbit head drop test. By a remarkable coincidence, an identical discovery was made simultaneously by William Paton and Eleanor Zaimis at the National Institute for Medical Research in London.[113] While investigating the ability of the octamethylene bisonium compound to liberate histamine in a cat, they discovered that the animal was being asphyxiated without exhibiting the obvious signs of respiratory distress such as gasping or convulsions. This suggested that the drug was acting as a neuromuscular blocking agent. Contact was made with Barlow and Ing for agreement on simultaneous publications early in 1948. Paton and Zaimis confirmed that the 10-carbon compound was the most potent member of the bisonium series. They also obtained evidence that it acted by depolarising the muscle end plate so as to prevent it responding to acetylcholine. Unlike tubocurarine and gallamine, it did not behave as a competitive antagonist of acetylcholine. The 10-carbon compound subsequently underwent satisfactory clinical trials and received the approved name of decamethonium.

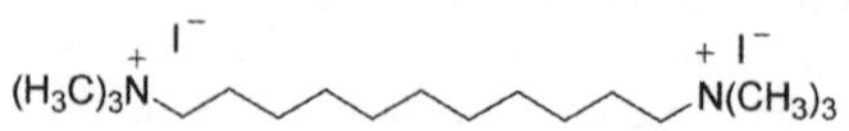

decamethonium iodide

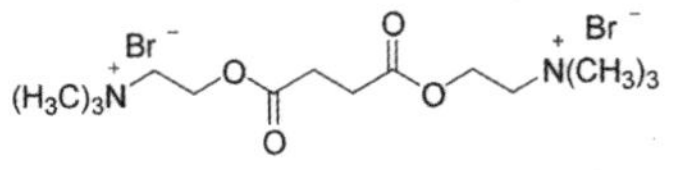

suxamethonium bromide

Decamethonium was marketed for only a few years. Due to its resistance to metabolic deactivation, paralysis was unnecessarily prolonged and could not be reversed with physostigmine. This meant that artificial respiration had to be maintained after completion of surgery until the effects eventually wore off. Seeking a solution to this problem, Bovet examined succinylcholine, a bisonium analogue of acetylcholine.[110] Reid Hunt had studied this compound 40 years earlier, but he failed to observe its neuromuscular blocking activity because his animals had been curarised.[114] Bovet, however, found it to produce very brief neuromuscular blockade; its mode of action was the same as that of decamethonium. It was introduced clinically in 1949, with the approved name of suxamethonium (also known as succinylcholine chloride). With a duration of action of only 5 minutes, it proved ideal for short procedures such as passage of a tracheal tube to assist breathing during general anaesthesia or electroconvulsive shock therapy. Alternatively, it could be administered by intravenous drip for longer operations, with the advantage that paralysis could be terminated by turning off the drip. However, prolonged respiratory paralysis occurred in a small proportion of patients due to abnormally low levels of the pseudo-cholinesterase enzyme responsible for rapid metabolism of the drug. This could not be reversed by physostigmine. Another disadvantage was that the inherent similarity to acetylcholine caused muscle twitching prior to paralysis, which often resulted in postoperative cramp.

Pancuronium

John Lewis, Michael Martin-Smith, Thomas Muir and Helen Ross at the University of Glasgow synthesised and examined a series of monoquaternary analogues of several aminosteroids obtained from Christopher Hewett and David Savage of the nearby Organon Laboratories in Newhouse. A compound that incorporated an acetylcholine-like residue in ring A of androstane exhibited one-sixteenth of the potency of tubocurarine.[115] Consideration of the nature of decamethonium, together with a comparison of this monoquaternary steroid with the naturally occurring investigational bisquaternary neuromuscular blocking agent known as malouetine, led Hewett, Savage and Roger Buckett to duplicate the acetylcholine-like moiety at both ends of the steroid to form an analogue of suxamethonium. This led on to the development of pancuronium bromide, the pharmacological properties of which were first reported in 1966.[116]

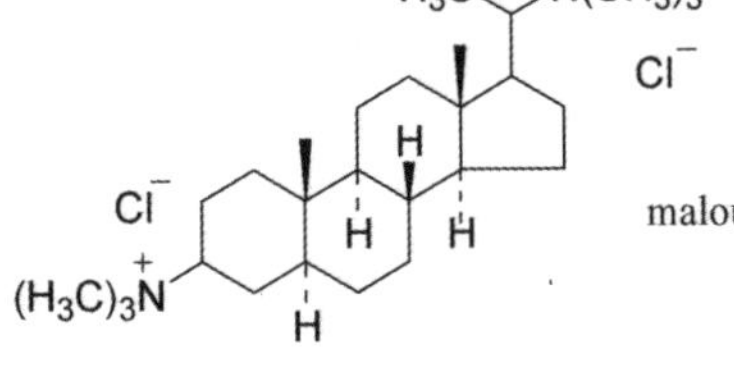

malouetine

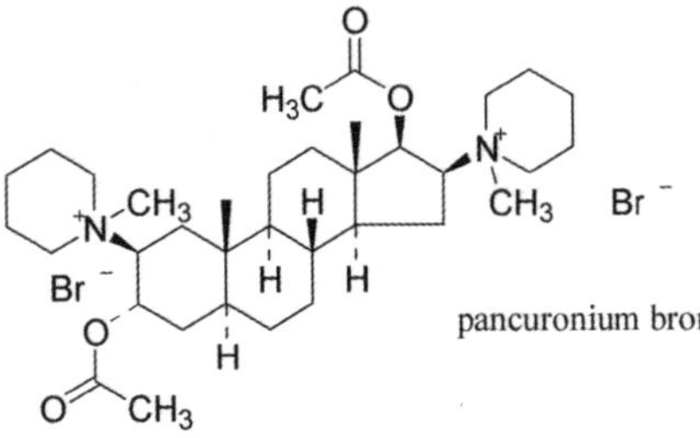

pancuronium bromide

Due to its more rapid onset of action and relative freedom from effects on blood pressure, pancuronium replaced tubocurarine as the relaxant of choice in major surgery. Its onset of action was still slower than that of suxamethonium, while its duration was in the order of 45 minutes. In patients with liver damage, the duration of action could be prolonged because of slow metabolism of pancuronium. The same occurred in those with renal impairment since the rate of elimination of the drug was reduced. Until the introduction of atracurium and vecuronium (see below), pancuronium was the market leader among surgical muscle relaxants.

At the same time as they synthesised pancuronium, Buckett, Hewett and Savage prepared several closely related compounds. Among these was the monoquaternary analogue of pancuronium. At the time of its synthesis, this had not seemed of sufficient interest to warrant detailed study. However, in 1970, Everett, Lowe and Wilkinson of the Wellcome Research Laboratories discovered that tubocurarine was not a bisquaternary compound as previously believed, but instead was a monoquaternary alkaloid.[117] One of the amino groups was tertiary.

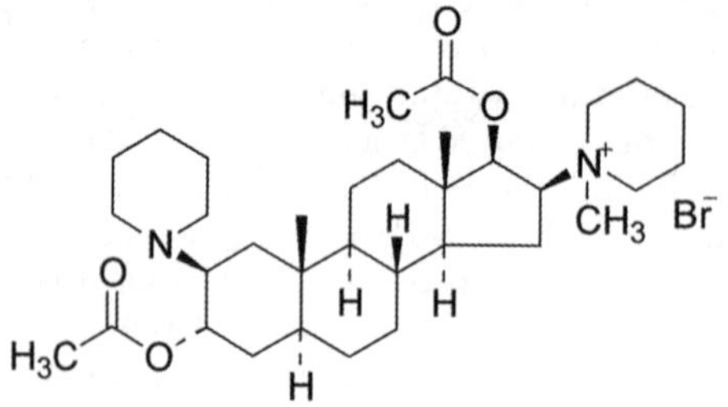

vecuronium bromide

This meant that when Organon submitted more than 100 compounds to Ian Marshall in the Department of Pharmacology at the University of Strathclyde in Glasgow for assessment as neuromuscular blocking agents, the monoquaternary analogue of pancuronium was included among them. When the Organon compounds were subjected by Marshall to pharmacological tests measuring muscle relaxing potency and vagal inhibition, this analogue emerged as having the best selectivity ratio between these properties, and only slightly less potent than pancuronium itself.[118] It was introduced into the clinic with the approved name of vecuronium. It was subsequently found to cause fewer side effects than any other neuromuscular blocking agent.

Atracurium

In the late 1950s, John Stenlake in the Department of Pharmacy at the University of Strathclyde began preparing esters related to suxamethonium in an attempt to find an improved analogue of suxamethonium. A change in the direction of his research occurred in 1964 following the extraction and identification of petaline from *Leontice leontopetalum* L., a Lebanese plant used as a folk medicine to treat epilepsy.[119] Petaline not only had a structure resembling half of the tubocurarine molecule but was also unstable. The facility with which it underwent a rapid Hofmann elimination reaction, normally only seen in quaternary ammonium salts under strongly basic conditions at high temperatures, gave Martin-Smith and Stenlake the idea of synthesising tubocurarine analogues based on petaline, with a view to obtaining a rapidly hydrolysed muscle relaxant that would not persist in the body.

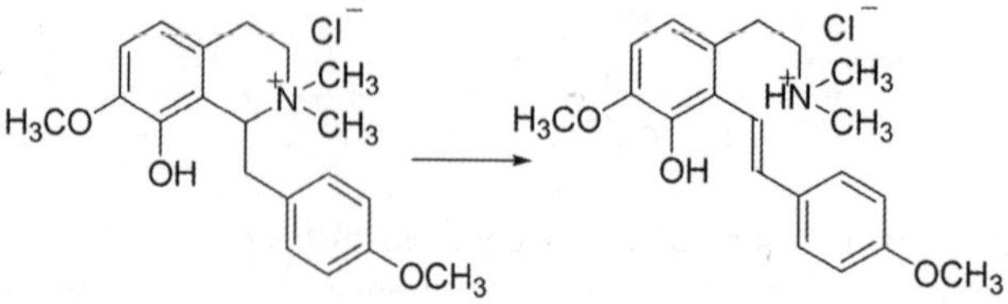

Decomposition of petaline by Hofmann elimination

Unfortunately, the decomposition of petaline was too slow at physiological pH for this to be achieved. Stenlake, Roger Waigh and George Dewar then sought to obviate this difficulty by conducting a systematic investigation of isoquinoline esters and ketones in which the electron-withdrawing properties of the carbonyl group in different positions within the molecule were correlated with the ease of Hofmann elimination.[120] As a result of this, they were able to obtain a series of muscle relaxants that at room temperature underwent Hofmann elimination at neutral pH. These compounds, however, lacked adequate potency. A further series provided potent compounds, but produced an unacceptable level of vagal blockade. It was then discovered that incorporation of ester functions in the central chain, coupled with the use of tetrahydropapaverine derived end-groups, abolished all major side effects. Furthermore, when the carbon chain between the esters function was lengthened, muscle relaxant potency was enhanced. The most promising compound thus obtained was atracurium mesylate.[121] On investigation at the Wellcome Research Laboratories, it was found to exhibit a pharmacological profile that met all expectations. It underwent both Hofmann elimination and esterase hydrolysis _in vivo_ and was a potent muscle relaxant. The only problem that remained was the hygroscopic nature of atracurium mesylate. This was overcome by formulating atracurium as its besylate salt.

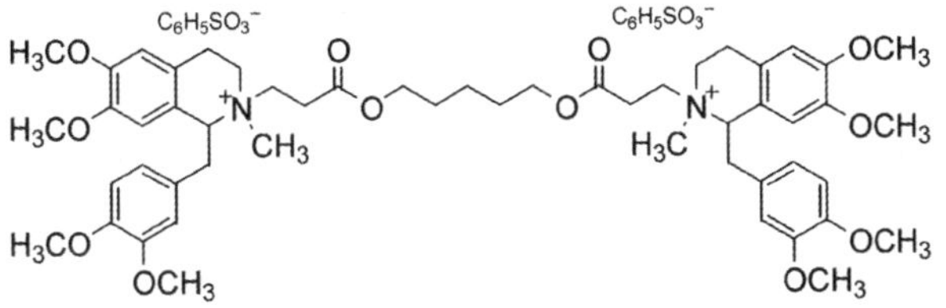

atracurium besylate

Atracurium was patented in 1977 by Wellcome after the signing of an agreement whereby royalties would be paid to the University of Strathclyde. It was subjected to its first clinical trials by James Payne at the Royal College of Surgeons of England, who confirmed its freedom from unwanted cardiac effects, as well as the speedy recovery of all patients irrespective of the status of their overall health.[122] This latter feature was a reflection of the chemical degradation through combined Hofmann elimination and esterase hydrolysis rather than solely metabolic breakdown. It ensured a short duration of muscle paralysis, even in patients with hepatic or renal disease. Apart from causing some histamine release, there were few side effects from atracurium. It could be used in a wide variety of situations that were previously considered hazardous. For all these reasons, it became, with vecuronium, one of the most frequently used muscle relaxants in surgery.

Atracurium consisted of a mixture of three groups of geometric isomers, constituting a total of ten stereoisomers. The _R, cis, R', cis_ isomer, which accounted for 15% of the isomers, was found to be more potent and slightly longer acting.[123] It had the advantage of being free from the risk of causing cardiovascular complications from the release of histamine and was then introduced as cisatracurium.

KHELLIN ANALOGUES

In the mid-1950s in Belgium, Camille Beaudet and Bun Hoi Phuc at the laboratories of the Société Belge de l'Azote et des Produits Chimiques du Marly synthesised and screened hundreds of analogues of khellin.[124] Benzarone was then marketed as an antispasmodic, followed by benziodarone, which was introduced in Europe as a coronary dilator for angina. The latter was withdrawn after the UK Committee on Safety of Drugs noted 11 cases of jaundice associated with its use. Analogues of benziodarone were then prepared in which the phenolic group was masked. This led to the development of amiodarone as a coronary dilator, but concern about corneal deposits, discolouration of skin exposed to sunlight and thyroid disorders also led to its withdrawal, in 1967. Seven years later, amiodarone was discovered to be highly effective in the treatment of a rare type of arrhythmia known as the Wolff–Parkinson–White syndrome.[125] It acted by prolonging the action potential in the cardiac muscle and conducting system, thus prolonging the refractory period. It also proved valuable in the management of arrhythmias resistant to other drugs and was reintroduced in 1980.

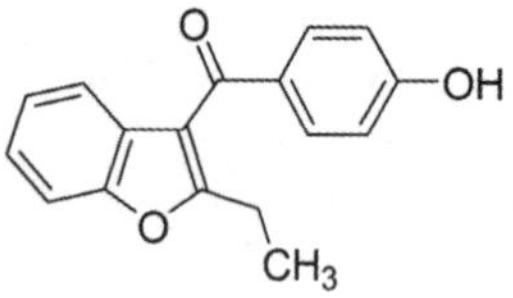

benzarone

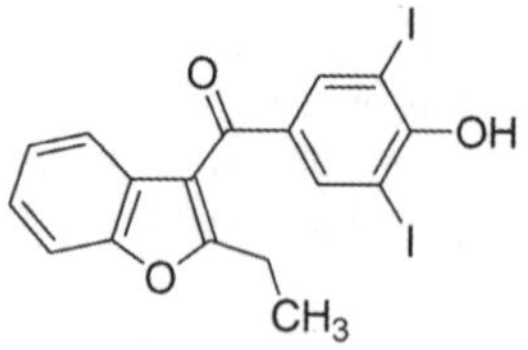

benziodarone

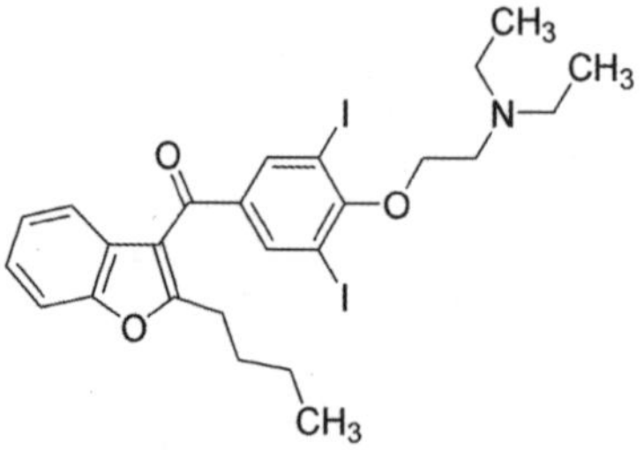

amiodarone

In 1955, chemists at the Benger Laboratories in Cheshire began to synthesise khellin analogues as potential bronchodilators. Compounds were screened for the ability to relieve bronchospasm induced in guinea pigs by histamine or carbachol. Active compounds were further examined in sensitised guinea pigs exposed to aerosols of egg white in order to induce a laboratory model of an asthmatic attack. Roger Altounyan, a physician working with the company, had reservations about the validity of these tests. He took the unprecedented step of evaluating the compounds on himself by inhaling a preparation of guinea pig hair knowing this would induce an asthmatic attack. By 1958, he had found that several carboxylated analogues of khellin were effective, but one in particular had unusual properties. Although it had been reported as having no bronchodilating activity, it somehow protected Altounyan from the effects of guinea pig hair. The compound was only effective if inhaled, but this irritated the lungs. This led the company to abandon the project. Nevertheless, those who had been involved were determined to see it to a successful conclusion, even though only two compounds could be tested on Altounyan each week. They surreptitiously continued with the programme.

A breakthrough came in 1963, when it was found that one of their compounds had been contaminated with highly active material. Brian Lee believed the most likely contaminant was that formed by a side reaction between two chromone molecules to form a bischromone. This was put to the test, progress now being faster due to the takeover of Benger by Fisons and the attitude of the new management being positive. A series of bischromones were then prepared for Altounyan to inhale. In January 1965, he reported that sodium cromoglycate, a bischromone prepared by Colin Fitzmaurice, was ideal as a protective inhalant.[126] Because the effective dose was somewhat inconvenient to administer, Altounyan and a company engineer developed a special device that enabled it to be inhaled as a dry powder liberated from a pierced capsule. The results of the first clinical trial were reported in 1967, with the favourable findings confirmed by subsequent clinical experience throughout the world.[127]

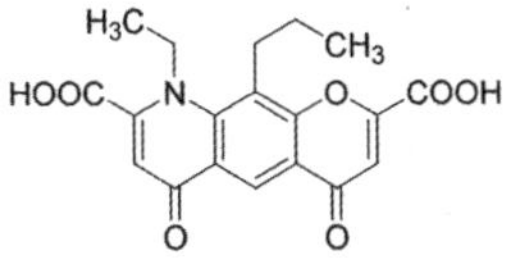

nedocromil

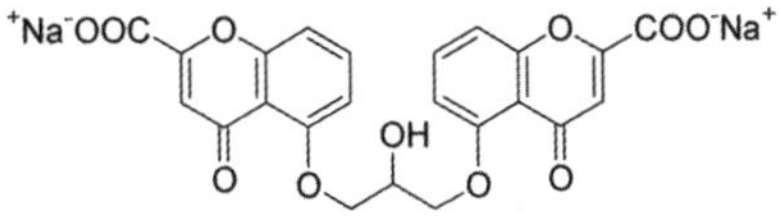

sodium cromoglycate

The new drug was given the approved name of sodium cromoglycate. It was subsequently shown to act by stabilising cell membranes in the lung so as to prevent the allergen-induced release of the substances that would otherwise cause bronchoconstriction in allergic asthma patients. Nedocromil was later found to be more potent than cromoglycate in animals and was effective in some patients who did not respond to cromoglycate.[128]

PODOPHYLLOTOXIN ANALOGUES

Aware that many plant toxins existed in plants as glycosides, which were more potent drugs than their corresponding aglycones, Sandoz chemists led initially by Arthur Stoll began to search for glycosides in _Podophyllum emodi_ and _P. peltatum_, which may have been lost during the extraction of podophyllotoxin. In order to prevent any glycosides being degraded during isolation, precautions were taken to inhibit enzymic activity in the freshly dried rhizomes. This resulted in the isolation of podophyllotoxin glucoside and similar glucosides, which were less toxic than podophyllotoxin, but also less potent.[129]

The Sandoz researchers next prepared a large series of compounds in which either the glycosidic moiety or the podophyllotoxin residue was modified. After ethers and esters of the glucosides were found to offer no advantage, the crude isolation fraction of podophyllotoxin glucoside was treated with benzaldehyde to form a product designated by Sandoz as SP-G, consisting mainly of podophyllotoxin benzylidene glucoside. This was active as an antimitotic agent in animals at a dose level that did not produce depression of the bone marrow, often the limiting factor in the use of anticancer chemotherapy. SP-G was released in 1963 for experimental investigation in the clinic.[130]

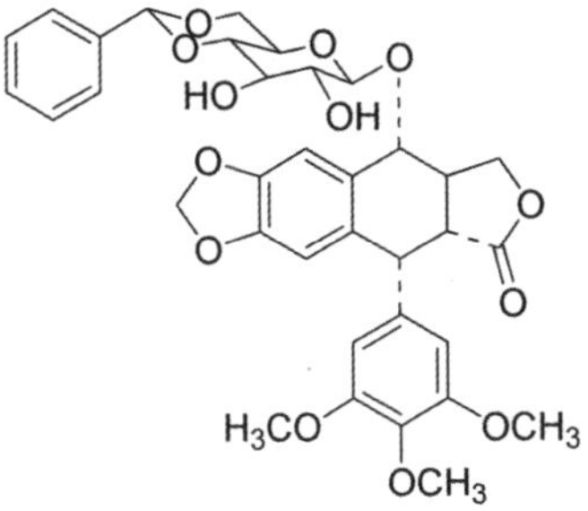

podophyllotoxin benzylidene glucoside

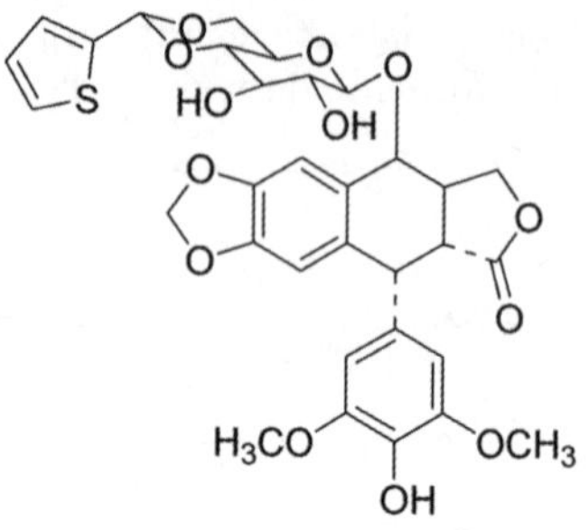

teniposide

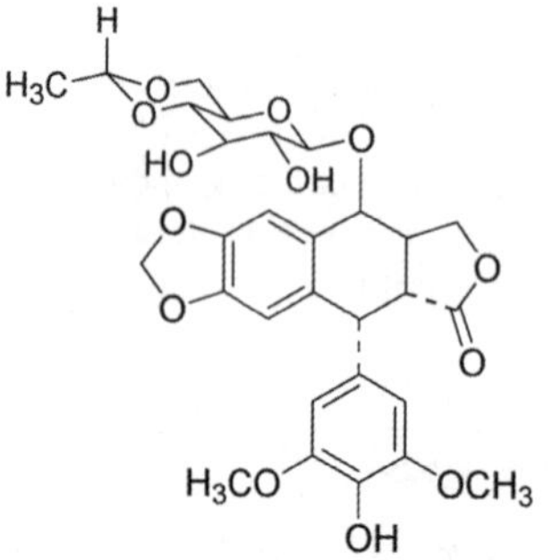

etoposide

During an examination of the cytotoxic activity of SP-G, it became apparent that its known components could not account for all the biological activity. Furthermore, SP-G significantly lengthened the lifespan of mice inoculated with L-1210 leukaemia, yet none of its known components could do so. It was concluded that an unidentified, highly potent antileukaemic factor must be present. Careful chromatographic fractionation of SP-G resulted in the isolation of this factor, which was identified in 1964 as 4′-demethylepipodophyllotoxin benzylidene glucoside. The Sandoz team then prepared and examined about 60 analogues of it, concentrating especially on condensation products of it with carbonyl compounds. The condensation product with thiophene-2-aldehyde, teniposide, was synthesised in 1965 and soon emerged as a promising candidate for therapeutic evaluation.[131] Though it was inactive by mouth, teniposide was introduced into the clinic in 1967 on an experimental basis. After it had been shown to be effective in reticulosarcoma, Hodgkin's disease and bladder carcinoma, teniposide was marketed by Sandoz in some European countries in 1976.

The condensation product between 4′-demethylepipodophyllotoxin glucoside and acetaldehyde was synthesised nine months after teniposide.[132] Known as etoposide, it was only about one-tenth as potent as teniposide when tested for *in vitro* cytotoxic activity, but it was superior in prolonging the lifespan of mice inoculated with L-1210 leukaemia. Unlike teniposide, it was orally active. When tested in the clinic in 1971, it was found of value in the treatment of non-small-cell lung cancer, testicular cancer and lymphomas. It was also the first drug to be effective against acute monocytic leukaemias. Sandoz licensed both teniposide and etoposide to Bristol–Myers in 1978 as they had expertise in the field of cancer chemotherapy. Five years later, etoposide was approved for use in the United States.

The antitumour action of etoposide, teniposide and other condensed-aldehyde derivatives of 4′-demethylepipodophyllotoxin glucoside is different to that of podophyllotoxin. While at high dose levels they act in the same manner as podophyllotoxin, at normal levels they prevent target cells from entering mitosis. This latter mechanism was responsible for the clinical success of etoposide.

PACLITAXEL ANALOGUES

In 1981, the Centre National de la Recherche Sciéntifique (CNRS) in Paris signed an agreement with Rhône–Poulenc for the investigation of compounds related to paclitaxel. The research was carried out by Pierre Potier and his colleagues at the Institut de Chimie des Substances Naturelles. Taking into account the extensive studies on taxoid chemistry that had been conducted elsewhere, they prepared derivatives of two inactive precursors of paclitaxel, namely 10-deacetylbaccatin and baccatin. Both were present in the needles and heartwood of the European yew (*Taxus baccata*).[133]

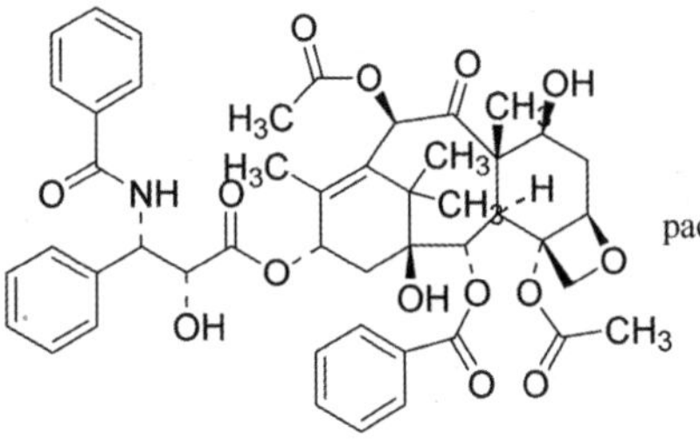

paclitaxel

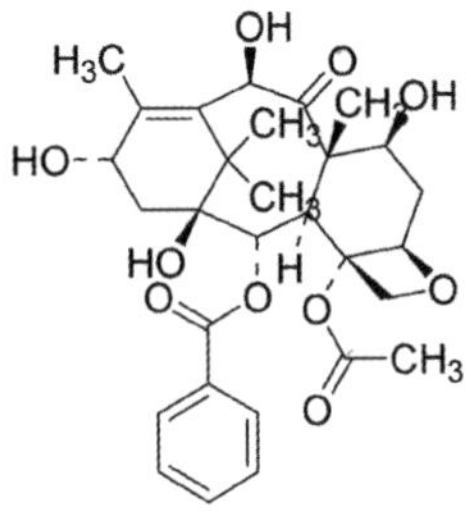

10-deacetylbaccatin

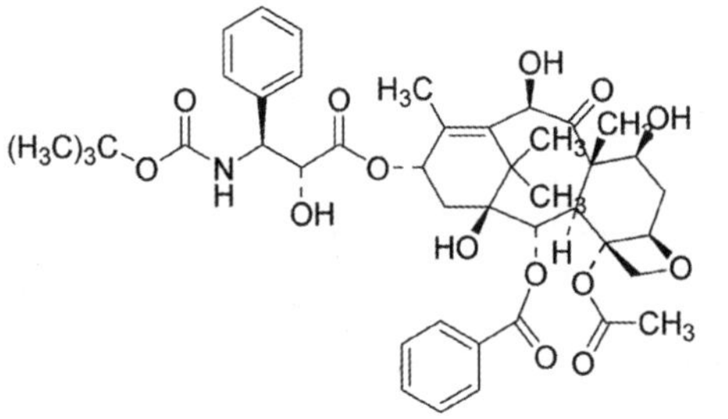

docetaxel

In all, some 40 compounds were synthesised and tested for their activity to inhibit tubulin assembly *in vitro*. The most active compound to emerge from this was an ester of the hydroxyl at position 13, which was made in 1986 and was given the name docetaxel. Further tests in the laboratory showed it to be more cytotoxic than paclitaxel and so it was put on clinical trial by Rhône–Poulenc Rorer. It was licensed in various countries from 1994 onwards for use in advanced or metastatic breast cancer and non-small-cell lung cancer.

CAMPTOTHECIN ANALOGUES

Following the failure of camptothecin in the clinic due to its toxicity and also disappointing results with several analogues,[134] attempts were made to determine the exact nature of its action on tumour cells. This led to the discovery by researchers from Johns Hopkins University collaborating with Smith Kline and French Laboratories that camptothecin acted by inhibiting the enzyme topoisomerase I.[135]

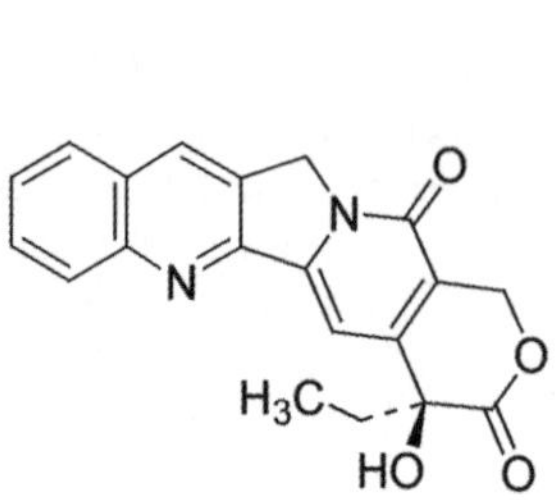

camptothecin

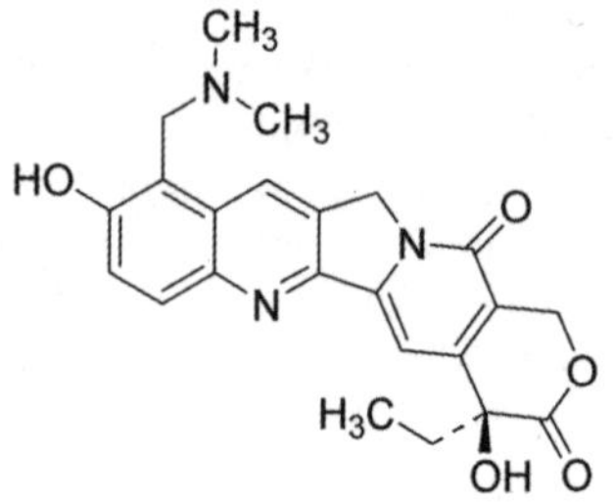

topotecan

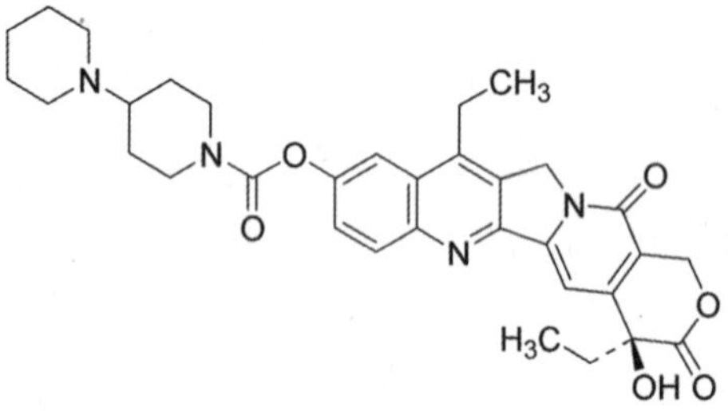

irinotecan

As the mode of action of camptothecin was novel, a renewed effort was made to find a superior analogue, resulting in the preparation of topotecan by William Kingsbury and his Smith Kline colleagues in 1990.[136] It is used when other drugs have failed in the treatment of metastatic ovarian cancer. Topotecan was followed by irinotecan, prepared at the Yakult Institute for Microbiological Research, Tokyo.[137] Irinotecan is a prodrug that is metabolised to form the active metabolite of camptothecin. It is given in combination with fluorouracil in the treatment of metastatic colorectal cancer.

REFERENCES

1. H. How, On some new basic products obtained by the decomposition of vegetable alkaloids. *Quart J. Chem. Soc.*, 1853; **6**: 125–9.
2. A. Crum Brown, T.R. Fraser, On the physiological action of the salts of the ammonium bases, derived from strychnia, brucia, thebaia, codeia, morphia, and nicotia. *Trans. Roy. Soc. Edin.*, 1869; **25**: 151–203.
3. A. Matthiessen, C.R.A. Wright, Researches into the chemical constitution of the opium bases, Part 1. On the action of hydrochloric acid on morphia. *Proc. Roy. Soc. Lond.*, 1869; **17**: 455–60.
4. R. Pschorr, *Ber.*, 1907; **40**: 1984.
5. F.M. Pierce. On the physiological action of the above morphine and codeine derivatives. *J. Chem. Soc.*, 1874; **27**: 1043–4.
6. F.M. Pierce, On the physiological action of some new morphine and codeine derivatives. *Practitioner* 1975; **14**: 437–42.
7. E. Grimaux, Sur la transformation de la morphine en codéine et en bases homologues. *C. R. Acad. Sci.*, Paris, 1881; **92**: 1140–3.
8. D.B.Dott, R. Stockman, The chemistry and pharmacology of some of the morphine derivatives, in *Year Book of Pharmacy*. London: Churchill; 1888, p. 349–55.
9. R. Stockman, D.B. Dott, Report on the pharmacology of morphine and its derivatives. *Br. Med. J.*, 1890; **2**: 189–92.
10. J. Mering, Physiological and therapeutical investigations on the action of some morphia derivatives. *Merck's Jahresbuch*, 1898; **1**: 5.
11. J. von Mering, Physiological and therapeutic investigations on the action of some morphia derivatives. *The Merck Report* 1898; **7**: 5–13.
12. W. Sneader, The discovery of heroin. *Lancet*, 1998; **352**: 1697–9.

13. B. von Issekutz, Über die Wirkung des Codeins, Dionins und Heroins auf die Atmung. *Arch. Ges. Physiol.*, 1911; **142**: 255–67.
14. G.J. Higby, Heroin and medical reasoning: the power of analogy. *New York State J. Med.*, 1986; **86**: 137–42.
15. J. Phillips, *J. Am. Med. Ass.*, 1912; **59**: 2146.
16. L. Small, H.M. Fitch, W.E. Smith, The addition of organomagnesium halides to pseudocodeine types. II. Preparation of nuclear alkylated morphine derivatives. *J. Am. Chem. Soc.*, 1936; **58**: 1457–63.
17. J. Pohl, Ueber das N-Allylnorcodein, einen Antagnisten des Morphins. *Zeitschr. Exp. Pathol. Therap.*, 1915; **17**: 370–82.
18. K. Unna, *J. Pharmacol. Exp. Ther.*, 1943; **79**: 27.
19. L. Lasagna, H.K. Beecher, The analgesic effectiveness of nalorphine and nalorphine–morphine combinations in man. *J. Pharmacol. Exp. Ther.*, 1954; **112**: 356–63.
20. R. Grewe, *Naturwiss.*, 1946; **33**: 333.
21. R. Grewe, Synthetic drugs with morphine action. *Angew. Chem.*, 1947; **59**: 194–9.
22. O. Schnider, A. Grüssner, Oxy-morphinane. *Helv. Chim. Acta*, 1951; **34**: 2211–17.
23. M. Gates, Montzka, some potent morphine antagonists possessing high analgesic activity. *J. Med. Chem.*, 1964; **7**: 127–31.
24. US Pat. 1974: 3819635 (to Bristol-Myers).
25. E. May, E.M. Fry, Structures related to morphine. VIII. Further syntheses in the benzomorphan series. *J. Org. Chem.*, 1957; **22**: 1366–9.
26. E. May, N.B. Eddy, Communications – a new potent synthetic analgesic. *J. Org. Chem.*, 1959; **24**: 294–5.
27. US Pat. 1966: 3250678 (to Sterling Drug).
28. S. Archer, N.F. Alberts, L.S. Harris, *et al.*, Pentazocine. Strong analgesics and analgesic antagonists in the benzomorphan series. *J. Med. Chem.*, 1964; **7**: 123–7.
29. M. Freund, E. Speyer, Transformation of thebaine into hydroxycodeineone and its derivatives. *J. Prakt. Chem.*, 1916; **94**: 135–78.
30. U. Weiss, Derivatives of morphine. I. 14-Hydroxydihydromorphinone. *J. Am. Chem. Soc.*, 1955; **77**: 5891–2.
31. E. Garfield, *Essays of an Information Scientist*, 1983; **6**: 121–30.
32. US Pat. 1967: 3332950 (to Endo).
33. US Pat. 1968: 3393197 (to Endo).
34. K.W. Bentley, D.G. Hardy, Discovery of etorphine. *Proc. Chem. Soc.*, 1963: 220.
35. K.W. Bentley, D.G. Hardy, Novel analgesics and molecular rearrangements in the morphine–thebaine group. III. Alcohols of the 6,14-*endo*-ethenotetrahydrooripavine series and derived analogs of *N*-allylnormorphine and -norcodeine. *J. Am. Chem. Soc.*, 1967; **89**: 3281–92.
36. J. Schmidt, *Organic Chemistry*, 4th edn. Edinburgh: Gurney and Jackson; 1943.
37. A. Ladenburg, L. Rugheimer, Kunstliche Bildung der Tropasaure. *Ber.*, 1880; **13**: 373–9.
38. S. Afr. Pat. 1968: 6707766 (to Boehringer).
39. J. von Braun, O. Braunsdorff, K. Räth, *Ber.*, 1922; **55**: 1666.
40. K. Toda, *Arch. Exp. Path. Pharmakol.*, 1929; **146**: 313.
41. K. Fromherz, *Arch. Exp. Path. Pharmakol.*, 1933; **173**: 86.
42. R.R. Burtner, in *Medicinal Chemistry*, Vol. 1, ed. C.M. Suter. New York: John Wiley; 1951, p. 151.
43. K. Miescher, K. Hoffmann, Uber die Darstellung einiger basischer Ester substituierter Essigsäuren. *Helv. Chim. Acta*, 1941; **24**: 458.
44. US Pat. 1951: 2554511 (to Schieffelin).
45. O. Eisleb and, O. Schaumann, Dolantin, ein neuartiges spasmolytikum und analgetikum (Chemisches und Pharmakologisches). *Deut. Med. Wochenschr.*, 1939; **63**: 967.
46. A. Pohland, H.R. Sullivan, Analgesics: esters of 4-dialkylamino-1,2-diphenyl-2-butanols. *J. Am. Chem. Soc.*, 1953; **75**: 4458–61.
47. K.K. Chen, Pharmacology of methadone and related compounds. *Annals: New York Academy of Sciences*, 1948; **51**: 83–4.
48. N..B. Eddy, A new morphine-like analgesic. *J. Am. Pharmaceut. Ass., Pract. Pharm. Ed.*, 1947; **8**: 536.
49. P.A.J. Janssen, A new series of potent analgesics. *J. Am. Chem. Soc.*, 1956; **78**: 3862–6.
50. P.A.J. Janssen, C. Van de Westeringh, A.W.M. Jagneau, *et al.*, Chemistry and pharmacology of CNS depressants related to 4-(4-hydroxy-4 phenylpiperidino) butyrophenone. Part 1. Synthesis and screening data in mice. *J. Med. Pharm. Chem.*, 1959; **1**: 281–97.
51. US Pat. 1959: 2898340 (to P.A.J. Janssen).
52. US Pat. 1964: 3141823 (to P.A.J. Janssen).

53. J. de Castro, A Van de Water, L. Wouters, *et al.*, Comparative study of cardiovascular, neurological and metabolic side effects of 8 narcotics in dogs. Pethidine, piritramide, morphine, phenoperidine, fentanyl, R 39 209, sufentanil, R 34 995. *Acta Anaesthesiol. Belg.*, 1979; **30**: 55–69.

54. US Pat., 1978: 4066772 (to P.A.J. Janssen Pharmaceutica).

55. P.A.J. Janssen, C.J.E. Niemegeers, F. Awouters, *et al.*, Pharmacology of risperidone (R 64 766), a new antipsychotic with serotonin-S2 and dopamine-D2 antagonistic properties. *J. Pharmacol. Exp. Ther.*, 1988; **244**: 685–93.

56. S. Garfield, *Mauve: How One Man Invented a Colour that Changed the World*. London: Faber & Faber; 2000.

57. W. Filehne, Ueber neue Mittel, welche die fieberhafte Temperatur zur Norm bringen. *Berl. Klin. Wochenschr.*, 1882; **19**: 681.

58. L. Knorr, Ueber die Constitution der Chinizin-derivative. *Ber.*, 1884; **17**: 2032–8.

59. W. Filehne, Ueber das Pyramidon, ein Antipyrinderivat. *Berl. Klin. Wochenschr.*, 1896; **33**: 1061–3.

60. H. Stenzl, 3-Amino-5-pyrazolones. *Helv. Chim. Acta*, 1950; **33**: 1183.

61. G. Mixich, Zum chemischen Verhalten des Antiphlogisticums «Azapyrazone» (Mi 85). *Helv. Chim. Acta*, 1968; **51**: 532.

62. G. Merling, Eucaine. *Ber.*, 1896; **6**: 173–6.

63. A. Einhorn, R. Heinz, Orthoform. Ein Lokalanaestheticum fur Wundschmerz, Brandwunden, Geschwure, etc. *Munch. Med. Wochenschr.*, 1897; **44**: 931–4.

64. K. Auwers, H. Röhrig, *Ber.*, 1897; **30**: 991.

65. A. Einhorn, M. Oppenheimer, Ueber die Glycocollverbindindungen der Ester aromatisiher Amido- und Amidooxysauren. *Annalen*, 1900; **311**: 154–78.

66. E. Fourneau, *Organic Medicaments and Their Preparation*. London: Churchill; 1925, p. 61.

67. A. Einhorn, Ueber neue Arzneimittel. *Annalen*, 1910; **371**: 125–31.

68. H. Braun, Ueber eine neue ortliche Anaesthetica (Stovain, Alypin, Novocain). *Deut. Med. Wochenschr.*, 1905; **31**: 1667–71.

69. F.R. Mautz, Reduction of cardiac irritability by the epicardial and systemic administration of drugs as a protection in cardiac surgery. *J. Thoracic Surg.*, 1936; **5**: 612–28.

70. M. Yamazaki, Y. Kitagawa, S. Hiraki, Y. Tsukamoto, *J. Pharm. Soc. Japan*, 1953; **73**: 294.

71. L. Justin-Besançon, C. Laville, M. Thominet, Le métoclopramide et ses homologues: introduction à leu étude biologique. *C.R. Acad. Sci.*, 1964; **258**: 4384–6.

72. L. Justin-Besançon, M. Thominet, C. Laville, J. Margarit, [Chemical constitution and biological properties of sulpiride]. *C.R. Acad. Sci., Ser D*, 1967; **265**: 1253.

73. G. Piness, H. Miller, G. A. Alles, Clinical observations on phenylaminoethanol sulfate. *J. Am. Med. Assoc.*, 1930; **94**: 790–1.

74. L. Edeleano, Ueber einige Derivate der Phenylmethylacrylsäure under der Phenylisobuttersäure. *Ber.*, 1887; **20**: 616–22.

75. G. Barger, H.H. Dale, Chemical structure and sympathomimetic action of amines. *J. Physiol.* (Lond.) 1910; **41**: 19–59.

76. L. Panizzon, La preprazione di piridil-e-piridilarilace-tonitr ilie di alcuni prodotti di transformazione (part la). *Helv. Chim. Acta*, 1944; **27**: 1748–56.

77. E. Stedman, G. Barger, XLII. Physostigmine (eserine). Part III. *J. Chem. Soc.*, 1925; **127**: 247–58.

78. E. Stedman, E. Stedman, Methylurethans of the isomeric alpha-hydroxyphenylethyldimethylamines and their miotic activity. *J. Chem. Soc.*, 1929: 609–17.

79. J.A. Aeschlimann, M. Reinert, Pharmacological action of some analogues of physostigmine. *J. Pharmacol. Exp. Ther.*, 1931; **43**: 413–44.

80. M.B. Walker, Treatment of myasthenia gravis with physostigmine. *Lancet*, 1934; **1**: 1200–1.

81. US Pat.1951: 2572579 (to Hoffmann–La Roche).

82. C. Mannich, O. Walther, Synthesis of papaverine and related compounds. *Arch. Pharm.*, 1927; **265**: 1–11.

83. US Pat. 1962: 3265577 (to N. V. Philips).

84. G. Sandler, G.A. Clayton, S. Thorncroft, Clinical evaluation of verapamil in angina pectoris. *Br. Med. J.*, 1968; **3**: 224–7.

85. A. Fleckenstein, H. Kammermeier, H.J. Döring, *et al.*, Zum Wirkungsmechanismus neuartiger Koronardilatatoren mit gleichzeitig Sauerstoff-einsparenden Myokard-Effekten, Prenylamin und Iproveratril. *Z. Kreislaufforsch*, 1967; **56**: 716–44, 839–58.

86. L. Schamroth, D.M. Krikler, C. Garrett, Immediate effects of intravenous verapamil in cardiac arrhythmias. *Br. Med. J.*, 1972; **1**: 660–2.

87. G.R.J. Lewis, K.D. Morley, B.M. Lewis, P.J. Bones, The treatment of hypertension with verapamil. *N.Z. Med. J.*, 1978; **87**: 351–63.

88. R.W. Butcher, E.W. Sutherland. Adenosine 3′,5′-phosphate in biological materials. I. Purification and properties of cyclic 3′,5′-nucleotide phosphodiesterase and use of this enzyme to characterize adenosine 3′,5′-phosphate in human urine. *J. Biol. Chem.*, 1962; **237**: 1244–50.

89. H.-J. Hess, T.H. Cronin, A.J. Scriabine, Antihypertensive 2-amino-4(3H)-quinazolinones, *J. Med. Chem.*, 1968; **11**: 130–6.

90. A. Scriabine, J.W. Constantine, H.-J. Hess, W.K. McShane, Pharmacological studies with some new antihypertensive aminoquinazolines. *Experientia*, 1968; **24**: 1150–1.

91. US Pat. 1977: 4026894 (to Abbott).

92. US Pat. 1979: 4188390 (to Pfizer).

93. P. Dessauer, *Therap. Monatsch.*, 1908; **22**: 401.

94. Ger. Pat. 1952: 860217 (to Chem. Werke Albert).

95. W. Mohler, A. Söder, [Chemistry and synthesis of 3,7-dimethyl-1-(5-oxo-hexyl)-xanthine]. *Arzneimittel.-Forsch.*, 1971; **21**: 1159–60.

96. Br. Pat. 1959: 807826 (to Karl Thomae).

97. R. Kadatz, [Pharmacological properties of a new coronary dilator substance 2, 6-bis(diethano-lamino)-4, 8-dipiperidino-pyrimido[5,4-d]pyrimidine]. *Arzneimittel.-Forsch.*, 1959; **9**: 39–45.

98. G.V.R. Born, A.J. Honour, J.R.A. Mitchill, Inhibition by adenosine and by 2-chloroadenosine of the formation and embolization of platelet thrombi. *Nature*, 1964; **202**: 761–5.

99. P.R. Emmons, J.R.A. Mitchill, Postoperative changes in platelet-clumping activity. *Lancet*, 1965; **1**: 71–5.

100. P.R. Emmons, M.J.G. Harrison, A.J. Honour, J.R.A. Mitchill, Effect of dipyridamole on human platelet behaviour. *Lancet*, 1965; **2**: 603–6.

101. S.J. Lee, Y. Konishi, Yoshitaka; D.T. Yu, T.A. Miskowski, C.M. Riviello,, *et al.*, Discovery of potent cyclic GMP phosphodiesterase inhibitors. 2-Pyridyl- and 2-imidazolylquinazolines possessing cyclic GMP phosphodiesterase and thromboxane synthesis inhibitory activities. *J. Med. Chem.*, 1995, **38**(18): 3547–57.

102. N.K. Terrett, A.S. Bell, D. Brown, P. Ellis, Sildenafil (Viagra®), a potent and selective inhibitor of type 5 cGMP phosphodiesterase with utility for the treatment of male erectile dysfunction. *Bioorg. Med. Chem. Lett.*, 1996; **6**: 1819–24.

103. S.F. Campbell, Science, art and drug discovery: a personal perspective. *Clin. Sci.*, 2000; **99**: 255–60.

104. J. Rajfer, W.J. Aronson, P.A. Bush, *et al.*, Nitric oxide as a mediator of relaxation of the corpus cavernosum response to nonadrenergic, noncholinergic neurotransmission. *N. Engl. J. Med.*, 1992; **326**: 90–4.

105. M.A. Stahmann, I. Wolff, K.P. Link, Studies on 4-hydroxycoumarins. I. The synthesis of 4-hydroxycoumarins. *J. Am. Chem. Soc.*, 1943; **65**: 2285–7.

106. M. Seidman, D.N. Robertson, K.P. Link, Studies on 4-hydroxycoumarins. X. Acylation of 3-(α-phenyl-β-acetylethyl)-4-hydroxycoumarin. *J. Am. Chem. Soc.*, 1950; **72**: 5193–5.

107. M. Kabat, L.F. Stohlman, M.I. Smith, *J. Pharmacol. Exp. Ther.*, 1944; **80**: 160.

108. E.H. Wiseman, J.G Lombardino, Piroxicam, in *Chronicles of Drug Discovery*, ed. J.S. Bindra, D. Lednicer. Chichester: Wiley; 1982, p. 173.

109. J.G. Lombardino, E.H. Wiseman, Sudoxicam and related *N*-heterocyclic carboxamides of 4-hydroxy-2H-1,2-benzothiazine 1,1-dioxide. Potent nonsteroidal antiinflammatory agents. *J. Med. Chem.*, 1972; **15**: 848–9.

110. D. Bovet, F. Bovet-Nitti, G.B. Marini-Bettolo (eds), *Curare and Curare-like Agents*. London: Elsevier; 1959.

111. D. Bovet, F. Depierre, S. Courvoisier, Y. de Lestrange, *Arch. Int. Pharmacodyn.*, 1949; **80**: 172.

112. R.B. Barlow, H.R. Ing, Curare-like action of polymethylene bis-quaternary ammonium salts. *Br. J. Pharmacol.*, 1948; **3**: 298.

113. W.D.M. Paton, E.J. Zaimis, The pharmacological actions of polymethylene bistrimethylammo-nium salts. *Br. J. Pharacol.*, 1949; **4**: 381.

114. R. Hunt, R. Taveau, On the physiological action of certain choline derivatives and a new method for determining choline. *Br. Med. J.*, 1906; **2**: 1788–91.

115. J.J. Lewis, M. Martin-Smith, T.C. Muir, H. Ross, Steroidal monoquaternary ammonium salts with non-depolarizing neuromuscular blocking activity. *J. Pharm. Pharmacol.*, 1967; **19**: 502–8.

116. W.R. Buckett, C.R. Hewett, D.S. Savage, Pancuronium bromide and other steroidal neuromuscular blocking agents containing acetylcholine fragments. *J. Med. Chem.*, 1973; **16**: 1116–24.

117. A.J. Everett, L.A. Lowe, S. Wilkinson, Revision of the structures of (+)-tubocurarine chloride and (+)-chondrocurine. *J. Chem. Soc.*, 1970; **D**: 1020–1.

118. I.G. Marshall, S. Agoston, L.H.D.J. Booij, *et al.*, Pharmacology of ORG NC 45 compared with other non-depolarizing neuromuscular blocking drugs. *Br. J. Anaesth.*, 1980; **52**: 11S–19S.

119. J. McShefferty, P.F. Nelson, J.L. Paterson, *et al.*, Studies on *Leontice leontopetalum* Linn. I. The isolation of the chemical constituents of *Leontice leontopetalum*, etc. *J. Pharm. Pharmacol.*, 1956; **8**: 1117–33.

120. J.B. Stenlake, R.D. Waigh, J. Urwin, *et al.*, Atracurium: conception and inception. *Br. J. Anaesth.*, 1983; **55**: 3S–10S.

121. J.B. Stenlake, R.D. Waigh, G.H. Dewar, *et al.*, Atracurium besylate and related polyalkylene di-esters. *Eur. J. Med. Chem.*, 1981; **16**: 515–24.

122. J.P. Payne, R. Hughes, Evaluation of atracurium in anaesthetized man. *Br. J. Anaesth.*, 1981; **53**: 45–54.

123. J.B. Stenlake, R.D. Waigh, G.H. Dewar, *et al.*, Biodegradable neuromuscular blocking agents. Part 6. Stereochemical studies on atracurium and related polyalkylamine di-esters. *Eur. J. Med. Chem.*, 1984; **19**: 441–50.

124. US Pat. 1961: 3012042 (to Société Belge de l'Azote et des Products Chimiques du Marly).

125. M.B. Rosenbaum, P.A. Chiale, D. Ryba, M.V. Elizari, Control of tachyarrhythmias associated with Wolff–Parkinson–White syndrome by amiodarone hydrochloride. *Am. J. Cardiol.*, 1974; **34**: 215–23.

126. Br. Pat. 1969: 1144 906 (to Fisons).

127. J.B.L. Howell, R.E.C. Altounyan, A double-blind trial of disodium cromoglycate in the treatment of allergic bronchial asthma. *Lancet*, 1967; **2**: 539–42.

128. H. Cairns, D. Cox, K.J. Gould, *et al.*, New antiallergic pyrano[3,2-g]quinoline-2,8-dicarboxylic acids with potential for the topical treatment of asthma. *J. Med. Chem.*, 1985; **28**: 1832–42.

129. A. Cerletti, H. Emmenegger, H. Stähelin, [Research on the antimitotic properties of substances derived from podophyllotoxin]. *Actual. Pharmacol. (Paris)*, 1959; **12**: 103–28.

130. H. Emmenegger, H. Stähelin, J. Rutschmann, J. Renz, A. von Wartburg, [On the chemistry and pharmacology of *Podophyllum glucoside* and its derivatives.] *Drug. Res.*, 1961; **11**: 327–33, 459–69.

131. H. Stahelin, 4'-Demethyl-epipodophyllotoxin thenylidene glucoside (VM 26), a podophyllum compound with a new mechanism of action. *Eur. J. Cancer*, 1970; **6**: 303–11.

132. C. Keller-Julsén, M. Kuhn, A. von Wartburg, H. Stähelin, Mitosis-inhibiting natural products. 24. Synthesis and antimitotic activity of glycosidic lignan derivatives related to podophyllotoxin. *J. Med. Chem.*, 1971; **14**: 936–40.

133. F. Gueritte-Voegelein, D. Guenard, F. Lavelle, *et al.*, Relationships between the structure of taxol analogues and their antimitotic activity. *J. Med. Chem.*, 1991; **34**: 992–8.

134. J.A. Beisler, Potential antitumor agents. 1. Analogs of camptothecin. *J. Med. Chem.*, 1971; **14**: 1116–18.

135. Y.H. Hsiang, R. Hertzberg, S. Hecht, L.F. Liu, Camptothecin induces protein-linked DNA breaks via mammalian DNA topoisomerase I. *J. Biol. Chem.*, 1985; **260**: 14873–8.

136. W.D. Kingsbury, J.C. Boehm, D.R. Jakas, *et al.*, Synthesis of water-soluble (aminoalkyl)camptothecin analogs: inhibition of topoisomerase I and antitumor activity. *J. Med. Chem.*, 1991; **34**: 98–107.

137. S. Sawada, S. Okajima, R. Aiyama, *et al.*, Synthesis and antitumor activity of 20(*S*)-camptothecin derivatives: carbamate-linked, water-soluble derivatives of 7-ethyl-10-hydroxycamptothecin. *Chem. Pharm. Bull.*, 1991; **39**: 1446–50.

12

The Origins of Hormone Therapy

The German physiologist Arnold Berthold was the first to find evidence that a gland could secrete a physiologically active substance directly into the blood rather than through a duct. In 1848, he showed that the shrinking of the comb in castrated roosters did not occur if the excised testes were transplanted into the abdominal cavity. As there was no longer any nervous connection to the testes, he concluded that they were secreting some substance into the bloodstream.[1]

As with the isolation of active principles from plants, the city of Paris figured prominently in early developments relating to mammalian hormones. The common factor here was the influence of the protégés of François Magendie on the development of experimental physiology. The most illustrious of them was his research assistant Claude Bernard, generally considered to be the founder of experimental medicine. Shortly after becoming Magendie's successor at the Collège de France in 1855, he reported that his investigations on glycogenesis in the liver had established that an internal secretion is released directly into the blood, thereby helping to maintain the *milieu intérieur* (internal environment) of the body.[2] The following year, Charles-Edouard Brown-Séquard, who was Bernard's friend and later his successor at the Collège, reported to the Académie des Sciences that removal of both adrenal glands from small animals was rapidly followed by muscular weakness, irregularity of breathing and heart beat, lowering of body temperature, vertigo and convulsions, and finally death within 12 hours of removing the adrenal glands. However, if blood from a healthy animal was intravenously injected, survival was prolonged by several hours.[3] Brown-Séquard had clearly been influenced by a book written the year before by Thomas Addison, a physician at Guy's Hospital in London, entitled *On the Constitutional and Local Effects of Disease of the Suprarenal Capsules*. This described a condition in which patients presented with bronzed skin and anaemia. After increasing weakness and general illness, these patients died. Post-mortem examination revealed that the suprarenal (i.e. adrenal) glands were diseased. Since then, the condition has been known as Addison's disease.

There were further developments in Paris in 1856. Though Gabriel Constant Colin, the founder of veterinary physiology, is often remembered for his controversies with Bernard and Pasteur, he also proved that the medulla was a distinct and special part of the adrenal gland by showing that it turned blue when treated with ferric sulfate.[4] Edme Vulpian similarly observed that it became green if ferric chloride solution was used, but he went further than Colin by also showing that venous blood leaving the gland turned green, whereas there was no colour reaction with the arterial blood entering it.[5] Vulpian suggested that the explanation for this was that the medulla synthesised a substance that was secreted into the circulation. It was to take another 40 years before progress was made towards determining the nature of this internal secretion.

ORGANOTHERAPY

At the age of 72, Brown-Séquard self-administered eight injections of an extract of guinea pig testicles. A month later, in June 1889, he created a sensation by telling the Société de Biologie

Drug Discovery. A History. W. Sneader.
©2005 John Wiley & Sons Ltd

that animal testes contained an invigorating principle that might be capable of rejuvenating elderly men.[6] His claim was also published in the *Lancet* and caught the popular imagination throughout the world. Debilitated and senescent patients rushed to visit physicians of renown who, in turn, proclaimed that a new era of organotherapy had dawned.[7,8] Laboratories were established to exploit this novel type of therapy, offering the public a variety of specially prepared glandular extracts at high prices. It seemed to many that the elixir of life had at last been found.

Brown-Séquard had prepared his extract by grinding either a guinea pig or dog testicle in a small volume of water and then filtering it through paper to obtain a clear solution. A company in Geneva subsequently introduced a proprietary preparation called Séquardine®, which was manufactured in exactly the same manner.[9] Since Ernst Laqueur later required more than a ton of bovine testicles in order to isolate enough testosterone for a single course of injections, it seems likely that any invigoration experienced by Brown-Séquard was due to autosuggestion.

Charlatans who sought to enrich themselves through offering organotherapy to gullible clients soon gave organotherapy a bad name, but the publicity was not entirely without benefit. In 1873 at Guy's Hospital in London, William Gull had concluded that some of his female patients were suffering from a form of adult cretinism.[10] This disease, later to become known as myxoedema, was manifested by various mental and physical changes such as fatigue, extreme sensitivity to cold, dryness of the skin and anaemia. Gull had reported atrophy of the thyroid in his patients, but his suggestion that this might be the cause of the disease was not well received. His views were finally vindicated in 1891 in Newcastle-upon-Tyne after George Redmayne Murray had been asked by his superior to implant sheep thyroid into patients suffering from myxoedema. Murray instead followed in the footsteps of Brown-Séquard by subcutaneously injecting the fluid expressed from a freshly removed sheep thyroid gland into a patient gravely ill with myxoedema.[11] Encouraged by the response, he then made aqueous glycerol extracts from sheep thyroid, following a procedure used to extract pepsin (a digestive enzyme) from the stomach lining of pigs. Despite being desperately ill at the outset, his patient made a complete recovery. She continued to receive regular doses of thyroid extract by mouth and remained in good health for 28 years until her death from other causes.[12] Many more patients benefited from this pioneering work by Murray before the thyroid hormones were isolated. The taint of organotherapy had now been removed from the developing field of endocrinology. Still more importantly, the scene was set for a massive endeavour to understand and exploit physiological and biological processes in order to develop new drugs. No longer would reliance have to be placed on products extracted from plants in order to find therapeutic agents.

THYROXINE

Eugen Baumann, who worked at the University of Freiburg, was particularly interested in the biochemistry of the thyroid gland because there was a high incidence of endemic goitre in that locality. Goitre is a deformity of the neck due to swelling of the thyroid. F. Bayer & Company supplied Baumann with crude thyroid extract obtained from more than 1000 sheep glands by boiling them in dilute sulfuric acid, cooling and then collecting the flocculent precipitate that was deposited. This was extracted with alcohol by Baumann and purified.[13] When his colleague Paul Kraske suggested that it should be examined for the presence of iodine, Baumann found that there was 2.9% iodine present. He soon became convinced that there was a correlation between the iodine content and potency. By 1896 he had isolated a fraction that contained around 10% iodine. When administered to patients, its beneficial effects were similar to those previously obtained with the whole gland.[14] Believing he might have obtained the active principle, Baumann called it 'iodothyrin'. F. Bayer & Company then marketed it as

Thyroiodin®. The subsequent isolation of still more potent fractions soon made it evident that it had not been pure.

Kraske had made his suggestion about the possible presence of iodine in the extracts after reading of recent successes in Germany when thyroid had been fed to patients with goitre. This reminded him of the former controversial use of iodine for the same condition, a remedy introduced in Geneva in 1820 after Jean François Coindet had found it to be the only ingredient likely to account for any beneficial action of burnt sea-sponge in goitre.[15] This old remedy had been widely employed since the Middle Ages. Coindet treated over 150 patients with iodine, but toxicity due to overdosing became a problem which, coupled with the general ineffectiveness of the therapy in patients with long-standing goitre, brought about its eventual abandonment. Prevost, his townsman, claimed in 1846 that a local dietary deficiency of iodine might be the cause of goitre and cretinism. The obvious way to settle this matter was to carry out an analysis of iodine content in food and water, but suitable analytical techniques did not become available until after the middle of the century. Even then, these lacked refinement and the issue was not settled until after the First World War, when David Marine in Ohio finally confirmed Prevost's suggestion.[16] This led to the introduction of iodine supplements for inhabitants of goitrogenic locales.

Further developments were slow until Edward Kendall joined Parke, Davis & Company in 1910 and was assigned the task of isolating the thyroid hormone. He subsequently moved to the Mayo Clinic in Rochester, Minnesota, where he isolated material containing 23% iodine, double the highest amount previously recorded. By careful attention to details of the initial hydrolysis of the gland, for which he introduced repeated exposure to alkali instead of dilute sulfuric acid, he obtained crystals of thyroid hormone on Christmas Day 1914.[17] When the process was scaled up, however, Kendall was unable to recover active material. It took 14 months before the cause of this setback was traced to decomposition of the hormone by the galvanised iron vessels in which the hydrolysis of the gland had been conducted. These had to be replaced by enamel vessels. By 1917, Kendall had amassed about 7 g of crystals and was able to start clinical studies. He also put forward a structure for thyroid hormone.[18] Believing it was an oxindole, he coined the name 'thyroxin', a term that was later altered to thyroxine (this name has been retained despite the fact that the hormone is not an oxindole). Kendall studiously avoided giving the hormone any name suggestive of the discredited idea that iodine was responsible for thyroid activity. Thyroxine became available commercially shortly after this when it appeared in the catalogue of E.R. Squibb & Sons of Brooklyn, New York. The price was so high that it could only be used for laboratory investigations. This meant that when Charles Harington of University College Hospital in London had doubts concerning the proposed chemical structure, he had to devise a more efficient isolation process that would furnish enough material for his own investigations. Supported by the Rockefeller Foundation, he collaborated with Francis Carr at British Drug Houses, managing in 1924 to increase the yield of hormone twenty-fivefold. This enabled the company to reduce the price to one-tenth of what it had been.

Harington proposed that thyroxine could be formed from two molecules of an amino acid, diiodotyrosine. Collaborating with George Barger at the National Institute for Medical Research, Harington synthesised the proposed condensation product in 1927 and found it to be identical to thyroxine extracted from thyroid gland; both samples were optically inactive.[19] This confirmed that the chemical structure proposed by Harington was correct. However, as this structure possessed a centre of asymmetry, it meant that during the alkaline hydrolysis in the first part of the extraction process, thyroxine had undergone racemisation into a mixture of optical isomers. Harington and Barger therefore separated the isomers and established that the laevorotatory isomer was three times more potent than the dextrotatory one. In 1930, Harington isolated this isomer from the gland in a modified extraction procedure that relied upon enzymatic rather than alkaline hydrolysis.[20]

Thyroxine did not become available at a price that competed with that of dried thyroid gland until Benjamin Hems led a team of chemists at Glaxo Research Laboratories to a synthesis of it in 1949.[21] Even though eight stages were involved, the overall yield of optically active hormone was 26%.

thyroxine   liothyronine

A more potent hormone than thyroxine was isolated in 1952. Jack Gross and Rosalind Pitt-Rivers detected its presence while carrying out experiments on mice fed with radioactive iodine, which became incorporated into their thyroid glands.[22] This new hormone was identical to a trace contaminant previously found in some samples of synthetic thyroxine. It contained three iodine atoms instead of the four found in thyroxine. Tests at University College Hospital revealed that the new hormone was about twice as potent as thyroxine, and it was suggested that it was the active species formed in the body from thyroxine. The hormone, which was given the approved name of liothyronine, is also obtained by the Glaxo synthesis.

REFERENCES

1. A. Berthold, Transplantation der Hoden. *Arch. Anat. Physiol. Wissensch.*, 1849; **16**: 42–6.
2. C. Bernard, Sur le mécanisme de la formation du sucre dans le foie. *C. R. Acad. Sci.*, 1855; **41**: 461–9.
3. C.-E. Brown-Séquard, Recherches expérimentales sur la physiologie et la pathologie des capsules surrénales. *C. R. Acad. Sci.*, 1856; **43**: 422–5.
4. G. Colin, *Traité de Physiologic Comparée des Animaux Domestiques*. Paris: J.-B. Baillière; 1856.
5. E.F.A. Vulpian. Note sur quelques réactions propres à la substance des capsules surrénales. *C. R. Acad. Sci.*, 1856, **43**: 663–5.
6. C.-E. Brown-Séquard, Expérience démonstrant la puissance dynamogénique chez l'homme d'un liquide extrait de testicules d'animaux. *Archives de Physiologie Normale et Pathologique*, 1889; **21**: 651–8.
7. C.-E. Brown-Séquard, Note on the effects produced on man by subcutaneous injections of a liquid obtained from the testicles of animals. *Lancet*, 1889; **2**: 105–7.
8. M. Borell, Brown-Séquard's organotherapy and its appearance in America at the end of the nineteenth century. *Bull. Hist. Med.*, 1976; **50**: 309–20.
9. M. Tausk, The emergence of endocrinology. In *Discoveries in Pharmacology: Haemodynamica, Hormones and Inflammation*, eds M.J. Parnham, J. Bruinvels, Vol. 2. Elsevier; 1984: pp. 219–49.
10. W.W. Gull, On a cretinoid state supervening in adult life in women. *Trans. Clin. Soc. London*, 1874; **7**: 180.
11. G.R. Murray, Notes on the treatment of myxoedema by hypodermic injections of an extract of the thyroid of a sheep. *Br. Med. J.*, 1891; **7**: 796–7.
12. G.R. Murray, The life history of the first case of myxoedema treated by thyroid extract. *Br. Med. J.*, 1920; **13**: 359–60.
13. E. Baumann, Ueber das normale Vorkommen von Jod in Tierkörper. *Z. Physiol. Chem.*, 1895; **21**: 319–30.
14. E. Baumann, E.Z. Roos, *Z. Physiol. Chem.*, 1895; **21**: 481.
15. J.F. Coindet, *Ann. Chim. Phys.*, 1820; **5**: 49–59.
16. D. Marine, O.P. Kimball, The prevention of simple goitre. *Am. J. Med. Sci.*, 1922; **163**: 634.
17. E.C. Kendall, The isolation in crystalline form of the compound containing iodine which occurs in the thyroid: its chemical nature and physiological activity. *Trans. Ass. Am. Physicians*, 1915; **30**: 420–49.
18. E.C. Kendall, The crystalline compound containing iodine which occurs in the thyroid. *Endocrinology*, 1917; **1**: 153–69.
19. C.R. Harington, G. Barger, Chemistry of thyroxine. III. Constitution and synthesis of thyroxine. *Biochem. J.*, 1927; **21**: 169.
20. C.R. Harington, *The Thyroid Gland: Its Chemistry and Physiology*. London: Oxford University Press; 1933.
21. J.R. Chalmers, G.T. Dickson, J. Elks, B.A. Hems, The synthesis of thyroxine and related substances. Part V. A synthesis of L-thyroxine from L-tyrosine. *J. Chem. Soc.*, **1949**: 3424–33.
22. J. Gross, R. Pitt-Rivers, 3: 5:3′-Triiodothyronine. *Biochem. J.*, 1953; **53**: 652–7.

13

Neurohormones

George Redmayne Murray's work on thyroid extract for the treatment of myxedoema had been carried out in the north of England. It is no coincidence that the development which led to the isolation of the first hormone in a pure state also took place there, in the popular spa town of Harrogate. It was here that George Oliver, a physician, dabbled in physiological experiments using instruments he himself made. Oliver spent his spare time in the autumn of 1893 investigating the action of glycerol extracts of various glands upon the diameter of the radial artery in the wrist, using a new instrument which he had recently designed. He found that when a volunteer (his son!) swallowed an extract of sheep adrenal glands, constriction of the artery could be detected.[1] It says much for the sensitivity of his haemodynamometer that he was able to detect this, since around 90% of the active principle present in the extract would have been deactivated as it passed through the gut wall.

Oliver then collaborated with Edward Schäfer at the physiological laboratory of University College in London, with detailed studies being carried out on adrenal extracts. The following March, Oliver and Schäfer told the Physiological Society that water, alcohol or glycerin could extract from the adrenal gland a substance with a powerful action upon the blood vessels, the heart and the skeletal muscles of a variety of animals.[2] Within a year of this report appearing in print, an adrenal liquid extract was on sale in Germany for topical application to stop bleeding. There was also a paper from Cracow making the same claims as Oliver and Schäfer.[3]

An attempt by Oliver and Schäfer to isolate and determine the nature of the substance in their extract was thwarted when a colleague of Schäfer, Benjamin Moore, mistakenly concluded that it was a pyridine or piperidine compound.[4] However, Sigmund Fränkel in Vienna correctly deduced that the molecule featured both amine and catechol functions.[5]

The presence of the catechol accounted for the sensitivity of the active compound to oxidation. In 1897 John Abel isolated the grey sulfate of the active principle in impure form at Johns Hopkins University in Baltimore.[6] He went on to obtain its inactive but stable benzoate derivative in an almost pure state, now naming the active compound as 'epinephrine'.[7] Otto von Fürth at the University of Strassburg meanwhile developed a crude iron complex of the hormone, which was stable enough to be marketed by Farbwerke Höchst under the name Suprarenin® for use as a haemostatic agent.[8]

While continuing with his efforts to purify epinephrine, Abel received a visit in the autumn of 1900 from Jokichi Takamine, a Japanese industrial chemist based in New Jersey, who had established a close link with Parke, Davis & Company. Several weeks before this visit, Takamine had overcome the stumbling block to the final purification of epinephrine.[9] This involved using the same procedure adopted by Sertürner a century earlier when isolating morphine from an aqueous extract of opium by addition of ammonia, thus avoiding exposure of the alkaloid to strongly alkaline solutions with which it would form a salt that could not be isolated. The new process worked admirably. Takamine obtained 4 g of crystalline base, making this the first hormone ever isolated in a pure state.[10] Before publishing his findings in January 1901, Takamine patented his process for isolating the adrenal hormone, arranging for it to be marketed by Parke, Davis & Company with the proprietary name of Adrenalin®. In

Drug Discovery. A History. W. Sneader.
©2005 John Wiley & Sons Ltd

the US, the hormone was given the United States Approved Name (USAN) of 'epinephrine', but in the United Kingdom and British Commonwealth the term 'adrenaline' was introduced as the British Approved Name (BAN). In order to avoid the confusion caused by this type of situation, the World Health Organization introduced a Recommended International Nonproprietary Name (rINN) for all drugs. The use of rINNs is now mandatory in the European Union, the sole exceptions being adrenaline and noradrenaline. These both retain their BAN on grounds of safety, rather than their rINNs of epinephrine and norepinephrine.

Thomas Aldrich, who had worked with Abel, moved from Baltimore to Detroit to join Parke, Davis & Company in 1899. He also isolated small amounts of adrenaline in the summer of 1900, but Takamine held the patent. Abel subsequently pointed out that the published elemental analysis revealed that neither preparation of adrenaline was pure.[11] He was correct, for it was eventually recognised that these products were contaminated with a significant amount of noradrenaline.

Friedrich Stolz, director of chemical research at Farbwerke Höchst, began his attempts to synthesise adrenaline in 1903. At that time, two possible structures were being proposed. Stolz prepared a ketonic precursor of one of these, and sent it and several related compounds to Hans Meyer and Otto Loewi at the University of Marburg for evaluation.[12] They found that these ketonic compounds retained many of the physiological properties of adrenaline, but only to a much lesser degree. The compound most closely resembling adrenaline was given the name 'adrenalone' and marketed by Höchst as a nasal decongestant and topical haemostat. Meanwhile, at the University of Leeds, Henry Dakin also synthesised adrenalone as one of a series of adrenaline analogues required for a study of the influence of chemical structure on hormonal activity.[13] He obtained similar pharmacological results. It turned out to be anything but a straightforward matter to convert adrenolone to adrenaline. It took until 1906 before Stolz and Fritz Flächer solved the problems involved in large-scale production, the same year that the correct structure was published.

adrenalone   adrenaline

Synthetic adrenaline turned out to have about half the potency of adrenaline extracted from ox adrenals because natural adrenaline consisted solely of the laevorotatory stereoisomer, while the synthetic material contained equal parts of this and the feebly active dextrorotatory isomer. Flächer solved this problem by his discovery that the acid tartrate salt of the unwanted isomer readily dissolved in methyl alcohol, leaving behind the laevorotatory adrenaline acid tartrate to be filtered off. Its activity was identical to that of natural adrenaline.[14] Höchst replaced the earlier synthetic material with the pure isomer.

The most important use of adrenaline was in the treatment of asthma as an alternative to atropine. In 1900, Solomon Solis-Cohen in Philadelphia injected crude adrenal extract in patients with asthma and hay-fever.[15] Three years later, Bullowa and Kaplan reported their successful use in New York of adrenaline injection in asthma, and this became established as the standard therapy for relief of severe asthmatic attacks.[16] Within a few years a nasal spray containing adrenaline was in use for asthma and also as a decongestant for hay-fever and rhinitis. The first report of adrenaline being administered by an inhaler came from a London medical practitioner in 1929.[17]

Adrenaline was also marketed as a haemostatic to control bleeding in surgery and post-partum haemorrhage. It was even incorporated into suppositories for treatment of haemorrhoids. Intravenous adrenaline was administered in sudden cardiac arrest and was also widely used as a pressor agent to reverse the severe drop in blood pressure associated with surgical shock. This latter use is now deprecated because peripheral resistance is already high in shock and any further increase is dangerous.

NORADRENALINE (NOREPINEPHRINE)

Stolz[18] and Dakin[19] independently synthesised noradrenaline (the prefix 'nor' was an acronym derived from *nitrogen öhne radikal*, indicating the absence of a methyl group). Noradrenaline was not used clinically until after the discovery that it was the principal chemical transmitter mediating electrical impulses in the sympathetic nervous system, rather than adrenaline. The long delay before the role of noradrenaline as a neurotransmitter was established was principally due to the difficulty in separating noradrenaline from adrenaline in sympathin.

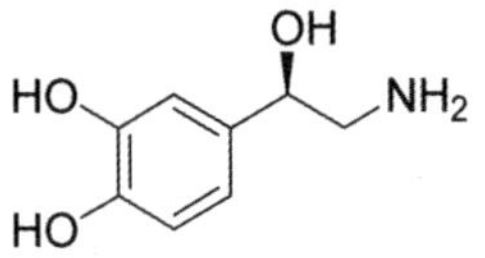

noradrenaline

Ulf von Euler at the Karolinska Institute in Stockholm proved noradrenaline to be the principal sympathomimetic neurotransmitter in humans in 1949.[20] Sterling Winthrop then marketed the laevorotatory isomer for treating clinical shock.

TYRAMINE

Walter Dixon of King's College in London and Frank Taylor of Cambridge University reported to the Royal Society of Medicine in 1907 that an alcoholic extract of human placenta had adrenaline-like effects when intravenously injected.[21] This extract was examined at King's by Otto Rosenheim, who established that adrenaline was not present.[22] He then discovered that the pressor agent was only present in material that had undergone bacterial decomposition. When the extract was investigated at the Wellcome Laboratories by George Barger and G.S. Walpole, the active substance was identified as tyramine, an analogue of adrenaline.[23]

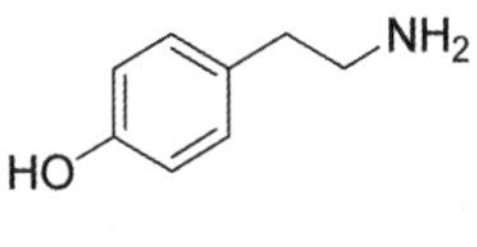

tyramine

Tyramine was formed from the amino acid tyrosine, which was present when protein decomposed. The Wellcome researchers also isolated tyramine from putrified meat extract. Their colleague Henry Dale confirmed that as a pressor agent it had only a fraction of the potency of adrenaline. However, as its duration of action was much longer than that of adrenaline, it was put to clinical use in the treatment of shock, for which purpose it was synthesised and marketed for some years by Burroughs Wellcome.

DOPAMINE

Dopamine was synthesised in 1910 by both Carl Mannich[24] and Barger and Arthur Ewins.[25] Barger and Dale found it to be very much less potent than adrenaline as a sympathomimetic agent.[36]

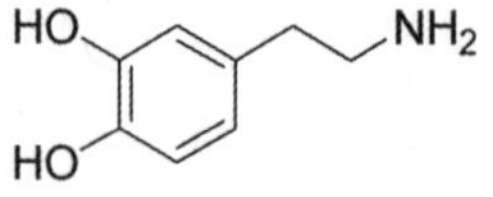

dopamine

There was no further interest in dopamine until the 1950s, when it was finally shown to be present in the adrenal medulla and sympathetic nerves. After it had been reported that dopamine might also be present in the brain, Arvid Carlsson and his colleagues at the University of Lund employed a sensitive assay that confirmed that the levels of dopamine in brain tissue were similar to those of noradrenaline.[26] The subsequent discovery of high concentrations in the basal ganglia of animal brains was followed by similar findings in human brain tissue. This was a crucial piece of evidence for Carlsson since he had already discovered that the tranquillising action of reserpine in animals could be overcome by administration of levodopa. As levodopa was the biosynthetic precursor of dopamine, it thus appeared likely that reserpine acted by depleting dopamine reserves in the brain. Carlsson then proposed that dopamine was a neurohormone in its own right, subsequently demonstrating that dopamine was present in central neurones.[27] Much evidence has accrued since then to confirm the validity of Carlsson's hypothesis.

It is now established that in addition to an agonist action at dopamine receptors, dopamine acts at α- and β-adrenoceptors. This makes dopamine a valuable drug in cardiac failure from infarction or surgery, where inadequate cardiac output reduces the circulating blood volume. One of the major consequences of this can be irreversible renal damage, unless adequate blood supply to the kidneys is restored. Dopamine, injected or infused, immediately produces an increase in the contractility of heart muscle (i.e. a positive inotropic effect), resulting in a rapid increase in cardiac output. This is due to its agonist action at β_1-adrenoceptors. As dopamine also stimulates dopamine (DA_1) receptors in the kidney, the glomerular filtration rate and renal blood flow increases and protects the kidneys from damage.

LEVODOPA

The Italian physician Torquato Torquati in 1913 found a catecholamine present in the broad bean, *Vicia faba* L.[28] The Hoffmann–La Roche biochemist Marcus Guggenheim then acquired some Windsor beans from the garden of Fritz Hoffmann, the founder of his company, and extracted L-dihydroxyphenylalanine.[29] This was found to be inactive and so was not used in medicine until the discovery of the role of dopamine as a neurohormone, after which it became known by the approved name of 'levodopa'.

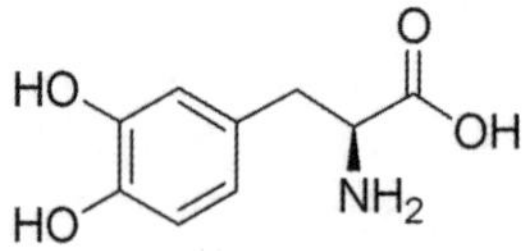

levodopa

An enzyme known as 'levodopa decarboxylase' was discovered in 1939, which degraded any levodopa present in mammalian tissues, so hindering its detection. Both Peter Holtz in Germany and Hermann Blaschko at Oxford then proposed a multistage biosynthetic pathway for the synthesis of adrenaline from L-tyrosine.[30,31] In this, L-tyrosine is converted to levodopa, which decarboxylates to form dopamine. Dopamine is the immediate precursor of noradrenaline, which is converted to adrenaline. Consequently, levodopa can be administered to correct the deficiency of dopamine in Parkinson's disease. It behaves as a prodrug, undergoing metabolic conversion to dopamine after entering the brain.

The introduction of racemic dihydroxyphenylalanine into therapy took place at the University of Vienna, after Oleh Hornykiewicz had acquired evidence that pointed to depletion of reserves of dopamine in the brains of patients with Parkinson's disease. Hornykiewicz attempted to alleviate the disease by administering 50–150 mg of dihydroxyphenylalanine intravenously to 20 patients, using this metabolic precursor of dopamine since the neurohormone itself could not cross into the brain from the general circulation.[32] His results seemed favourable, as were those reported around the same time by André Barbeau at the University of Montreal, who gave dihydroxyphenylalanine by mouth to six patients.[33] However, these findings were disputed by others. It was not until 1967 that the treatment protocols were perfected by George Cotzias and his colleagues at the Medical Research Centre in Brookhaven National Laboratory, who demonstrated that oral doses of up to 16 g each day consistently improved the general clinical condition of more than 50% of patients.[34] This improvement only lasted while treatment continued. Because of the expense involved, racemic dihydroxyphenylalanine had been used in the early trials. Since then, levodopa (the optically active isomer that is the metabolic precursor of dopamine) has become a universal treatment for Parkinson's disease.

HISTAMINE

At the International Physiological Congress in 1907, Henry Dale watched a demonstration of the action of an ergot extract on a uterus removed from a cat.[35] He thought that the extract contained a potent substance different from the ergot alkaloids that he had previously encountered because it had an immediate stimulant effect on the uterus. The extract, known as 'ergotinum dialysatum', had been prepared by a method that allowed putrefaction to occur. To Dale this was especially relevant as his colleagues at the Wellcome Physiological Research Laboratories in London were at that time investigating the action of tyramine, formed by putrefaction of tissue protein, on the uterus and blood vessels.

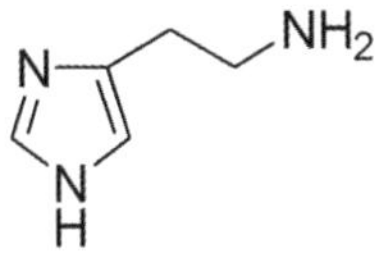

histamine

When Dale returned to London, he asked George Barger to isolate the putative uterine stimulant. This was duly done, but shortly before Barger completed the final identification of this material,[36] Ackerman at the University of Würzburg published an account of the decomposition of histidine, an amino acid, by deliberate putrefaction induced by micro-organisms.[37] The product formed was identical with the compound isolated by Barger. Because of its formation from histidine, Dale called the uterine stimulant 'histamine'. It was not a new substance, having being synthesised for purely chemical reasons by Windaus and Vogt three years earlier.[38] It was unsuitable for therapeutic use due to its wide-ranging physiological effects, though it later found limited use as a vasodilator in the management of peripheral vascular disease, migraine and Ménière's disease. Side effects due to the stimulation of histamine H_1-receptors included flushing, weakness and headache.

The pharmacological actions of histamine were studied by Dale and Patrick Laidlaw.[39] They noted that it caused smooth muscles to contract and lowered blood pressure. They also observed that in various species of animals its effects bore a striking resemblance to what had recently been described as 'anaphylactic shock', ensuring that continuing attention would be paid to it. Dale expanded his views on this later.

For many years, it appeared that histamine was only present in various tissue extracts as a result of bacterial action. Then, in 1926, claims were made in the United States and Canada that liver extracts could reduce blood pressure. Charles Best, visiting Dale to work on the crystallisation of insulin, brought a sample of a liver extract that had been used in Toronto. It was examined physiologically by Best and Dale and chemically by Harold Dudley and Thorpe at the National Institute for Medical Research.[40] They isolated substantial amounts of histamine and then repeated the extraction process under conditions that ruled out any possibility of formation of the histamine from histidine by putrefaction. This proved that the vasodilator present in liver was histamine. It now became apparent that this also accounted for the activity of gastric and pituitary extracts previously reported to reduce blood pressure by vasodilation.

In 1927, Thomas Lewis delivered a lecture in which he presented evidence that a substance with properties that resembled those of histamine was released when skin cells were damaged by various causes, including immune reactions.[41] Dale then gave his opinion that the substance was histamine or some combination owing its action to histamine.[42] This triggered intense interest in the role of histamine in the body, culminating in the development of the antihistamines. The release of histamine in the body was also eventually demonstrated.[43]

ACETYLCHOLINE

In 1898, Reid Hunt joined Abel's department at Johns Hopkins Medical School, where he became particularly interested in the nature of the blood pressure lowering principle that remained in adrenal extracts after the adrenaline had been removed. He found that choline was responsible for some of this activity, although it was certainly not the principal hypotensive substance present. Reporting his findings to the American Physiological Society in 1900, Reid Hunt suggested that either a precursor or derivative of choline was likely to be the main active principle. He never succeeded in isolating the choline derivative, but in 1906 he and Rene Taveau demonstrated that acetylcholine, a substance synthesised by Adolf von Baeyer 40 years earlier, was more than 100 000 times as potent a hypotensive agent as choline itself.[44]

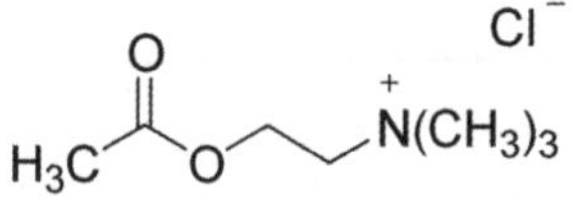

acetylcholine chloride

Although Hunt subsequently established that several other choline derivatives were also active, he could not exploit this clinically since their effects were of only fleeting duration. Nor could Henry Dale, who examined acetylcholine after Arthur Ewins had isolated it from ergot extracts in 1914.[45] However, the similarity of its effects to those resulting from stimulation of the parasympathetic nervous system led Dale to describe it as a parasympathomimetic agent.

In 1921, Otto Loewi demonstrated that repetitive stimulation of the vagus nerve supplying an isolated heart immersed in saline solution caused the release of a chemical that slowed down the rate of a second heart with no nerve supply. Loewi finally isolated this 'vagusstoff' in 1926 after discovering that its metabolic destruction, which Dale and Ewins had proposed was due to the action of an esterase enzyme, could be prevented by the alkaloid physostigmine, a drug known to produce cholinergic effects.[46] Loewi thus showed the vagusstoff to be acetylcholine. The Nobel Prize for physiology or medicine was awarded to Dale and Loewi in 1936 for their work on chemotransmission.

5-HYDROXYTRYPTAMINE

A vasoconstrictor in blood serum was isolated in 1948 by Maurice Rapport, Arda Green and Irving Page at the Cleveland Clinic in Ohio.[47] Initially it was known as 'serotonin'. The structure was proposed as 5-hydroxytryptamine (5-HT) by Rapport[48] and this was confirmed by its synthesis in 1951.[49] 5-HT is stored within the blood platelets and produces vasoconstriction only when these rupture during the first part of the blood clotting process that follows wounding. Vittorio Erspamer in Rome was able to confirm that enteramine, a compound he had extracted from the enterochromaffin cells lining the gastrointestinal tract, was identical to serotonin.[50] In fact, it was the enterochromaffin cells in which the highest concentration of 5-HT in the body was to be found.

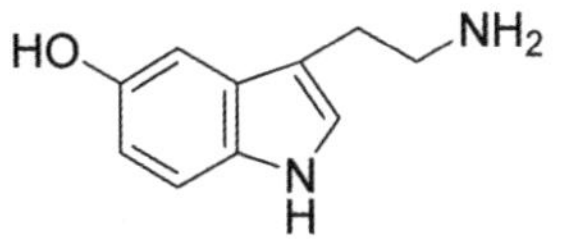

5-HT

Ernst Florey was the first to detect the presence of 5-HT in nervous tissue, namely the stellate ganglia of clams.[51] This was rapidly followed by Page revealing that 5-HT was present in the brains of mammals.[52] At Oxford in 1957 Gaddum established that there were at least two types of 5-HT receptors, describing them as the musculotropic 'D' receptors and the neurotropic 'M' receptors.[53] There has been much progress since then, with radio-labelled ligand-binding studies revealing at least seven families of 5-HT receptors and further subtypes within these.

Once 5-HT had been implicated in both inflammatory processes and as a chemotransmitter in the central nervous system, pharmaceutical companies showed considerable interest in its potential for therapeutic exploitation. They focused their efforts on the development of agonists and antagonists which display a high degree of receptor selectivity. Several major drugs have been developed as a consequence of this approach.

GABA

During the 1960s, the presence of γ-aminobutyric acid, or GABA, in the central nervous system was discovered and its role as an inhibitory neurotransmitter was established.[54] It is now accepted that GABA is the main inhibitory neurotransmitter in the brain. There are three families of GABA receptors, with further subtypes within these.

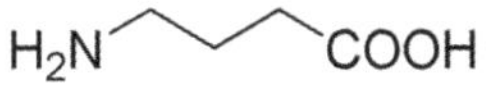

GABA

REFERENCES

1. H. Barcroft, J.F. Talbot, Oliver and Schäfer's discovery of the cardiovascular action of suprarenal extract. *Postgrad. Med. J.*, 1964; **44**: 6–8.
2. G. Oliver, E.A. Schäfer, On the physiological action of extract of the suprarenal capsules. *J. Physiol. (London)*, 1894; **16**: 1–4.
3. L. Szymonowicz, Die Function der Nebenniere. *Arch. Ges. Physiol.*, 1896; **64**: 97–164.
4. B. Moore, The active constituent of the suprarenal capsules. *J. Physiol. (London)*, **1897**; 382–9.
5. S. Fränkel, Physiological action of the suprarenal capsules. *Wien. Med. Blätter*, 1896; **1**: 14–16.
6. J.J. Abel, A.C. Crawford, On the blood-pressure raising constituent of the suprarenal capsule. *Bull. Johns Hopkins Hosp.*, 1897; **8**: 151–7.

7. J.J. Abel, On epinephrin, the active constituent of the suprarenal capsule and its compounds. *Proc. Am. Physiol. Soc.*, 1898; **3–4**: 3–5.
8. O. von Fürth, *Z. Physiol. Chem.*, 1897; **24**: 142–58.
9. J.J. Abel, Chemistry in relation to biology and medicine with especial reference to insulin and other hormones. *Science*, 1927; **66**: 307–19.
10. J. Takamine. Adrenalin; the active principle of the suprarenal gland. *Am. J. Pharm.*, 1901; **73**: 523–31.
11. J.J. Abel, On epinephrine and its compounds, with especial reference to epinephrine hydrate. *Am. J. Pharm.*, 1903; **75**: 301–25.
12. F. Stolz, Ueber adrenalin und alkylaminoacetobrenzcatechin. *Ber. Dtsch Chem. Ges.*, 1904; **37**: 4149–54.
13. H.D. Dakin, Synthesis of a substance allied to adrenaline. *Proc. Roy. Soc. London, Ser. B.*, 1905; **76**: 491–7.
14. F. Flächer, *Z. Physiol. Chem.*, 1908; **58**: 189.
15. S. Solis-Cohen, The uses of adrenal substance in the treatment of acute asthma. *J. Am. Med. Ass.*, 1900; **34**: 1164.
16. J.G.M. Bullowa, D.M. Kaplan, On the hyperdermatic use of adrenalin chloride in the treatment of asthmatic attacks. *Med. News*, 1903; **83**: 787.
17. P.W.L. Camps, A note on the inhalation treatment of asthma. *Guy's Hosp. Rep.*, 1929; **79**: 496–8.
18. F. Stolz, Ueber adrenalin und alkylaminoacetobrenzcatechin. *Ber.*, 1904; **37**: 4149–54.
19. H.D. Dakin, Synthesis of a substance allied to adrenaline. *Proc. Roy. Soc. (London)*, 1905; **B76**: 491–7.
20. U.S. von Euler, U. Harnberg, l-Noradrenaline in the suprarenal medulla. *Nature*, 1949; **163**: 642–63.
21. W.E. Dixon, F. Taylor, Physiological action of the placenta. *Br. Med. J.*, 1907; **2**: 1150.
22. O. Rosenheim, Pressor substances in placental extracts. *J. Physiol. (London)*, 1909; **38**: 337–42.
23. G. Barger, G.S. Walpole, Pressor substances in putrid meat. *J. Physiol. (London)*, 1909; **38**: 343–52.
24. C. Mannich, W. Jacobsohn, Über Oxyphenyl-Alkylamine und Dioxyphenylakylamine. *Ber.*, 1910; **43**: 189–97.
25. G. Barger, A.J. Ewins, Some phenolic derivatives of β-phenylethylamine. *J. Chem. Soc. (London)*, 1910; **97**: 2253–61.
26. A. Carlsson, M. Lindquist, T. Magnusson, B. Waldeck, On the presence of 3-hydroxytyramine in brain. *Science*, 1958; **137**: 471–3.
27. A. Carlsson, B. Falck, N.-Å. Hillarp, Cellular localization of brain monoamines. *Acta Physiol. Scand.*, 1962; **56** (Suppl. 196): 1–28.
28. T. Torquati, Sulla presenza di una sostanza azotata nel baccello verde dei frutti di 'vicia faba'. *Arch. Famacol. Sper. Sci. Affini.*, 1913; **15**: 308–12.
29. M. Guggenheim, Dioxyphenylalanin, eine neue Aminosäure aus Vicia faba. *Z. Physiol. Chem.*, 1913; **88**: 276–84.
30. P. Holtz, Dopadecarboxylase. *Naturwissen*, 1939; **27**: 724–5.
31. H. Blaschko, The specific action of l-dopa decarboxylase. *J. Physiol.*, 1939; **96**: 50–1.
32. W. Birkmayer, O. Hornykiewicz, Der L-3,4-dioxyphenylalanin (= DOPA)-Effekt bei der Parkinson-Akinese. *Wien. Klin. Wochenschr.*, 1961; **73**: 787–8.
33. A. Barbeau, T.L. Sourkes, G.F. Murphy, Les Catécholamines dans la maladie de Parkinson, in *Monoamines et Système Nerveux Centrale*, ed. J. de Ajuriaguerra. Paris: George, Genève and Masson; 1962; pp. 247–62.
34. G.C. Cotzias, M.H. van Woert, L.M. Schiffer, Aromatic amino acids and modification of Parkinsonism. *N. Engl. J. Med.*, 1967; **276**: 374–9.
35. H. Dale, The pharmacology of histamine. *Ann. N.Y. Acad. Sci.*, 1950; **50**: 1017.
36. G. Barger, H.H. Dale, Chemical structure and sympathomimetic action of amines. *J. Physiol. (London)*, 1910; **41**: 19–59.
37. D. Ackerman, *Z. Physiol. Chem.*, 1910; **65**: 504.
38. A. Windaus, W. Vogt, Synthese des imidazolyl-äthylamins. *Ber.*, 1907; **40**: 3691–5.
39. H.H. Dale, P.P Laidlaw, The physiological action of β-imidazolylethylamine. *J. Physiol. (London)*, 1910; **41**: 318–44.
40. C.H. Best, H.H. Dale, H.W. Dudley, W.V. Thorpe, The nature of the vasodilator constituents of certain tissue extracts. *J. Physiol. (London)*, 1927; **397**: 417.
41. T. Lewis, *The Blood Vessels of the Human Skin and Their Respones*. London: Shaw; 1927.
42. H.H. Dale, Some chemical factors in the control of the circulation. *Lancet*, 1929; **1**: 1233–7.
43. E. Gebauer-Fuelnegg, C.A. Dragstedt, R.B. Mullenix, Observations on a physiologically active substance appearing during anaphylactic shock. *Proc. Soc. Exp. Biol. Med.*, 1932; **29**: 1084–6.
44. R. Hunt, R. Taveau, On the physiological action of certain esters and ethers of choline and their relation to muscarine. *Br. Med. J.*, 1906; **2**: 1788–91.

45. H.H. Dale, The action of certain esters and ethers of choline, and their relation to muscarine. *J. Pharm. Exp. Ther.*, 1914; **6**: 147–90.
46. O. Loewi, E. Navratil, Über das Schicksal des Vagusstoffs. *Arch. Ges. Physiol.*, 1926; **214**: 678–88.
47. M.M. Rapport, A.A. Green, I.H. Page, Serum vasoconstrictor (serotonin). IV. Isolation and characterization. *J. Biol. Chem.*, 1948, **176**: 1237, 1243–51.
48. M.M. Rapport, Serum vasoconstrictor (Serotonin): IV, *J. Biol. Chem.*, 1949; **180**: 961–9.
49. K.E. Hamlin, F.E. Fischer, The synthesis of 5-hydroxytryptamine. *J. Am. Chem. Soc.*, 1951; **73**: 5007–8.
50. V. Erspamer, B. Asero, Identification of enteramine, the specific hormone of the enterochromaffin cell system, as 5-hydroxytryptamine. *Nature*, 1952; **169**: 800–1.
51. E. Florey, E. Florey, *Naturwissen.*, 1953; **40**: 413.
52. B.M. Twarog, I.H. Page, Serotonin content of some mammalian tissues and urine and a method for its determination. *Am. J. Physiol.*, 1953; **175**: 157.
53. J.H. Gaddum, Z.P. Picarelli, Two kinds of tryptamine receptors. *Br. J. Pharmacol.*, 1957; **12**: 323–8.
54. P. Krogsgaard-Larsen, Amino acid receptors, in *Comprehensive Medicinal Chemistry*, Vol. 3, eds P.G. Sammes, J.B. Taylor, J.C. Emmett. Oxford: Pergamon Press; 1990, p. 493.

Peptide Hormones

At the University of Strassburg in 1889, Oskar Minkowski suggested to Josef von Mering that his studies on the role of the pancreas in the metabolism of fat might be advanced by investigation of the effects of surgical removal of the pancreas. After the proposed experiment had been carried out on a dog, Minkowski noticed that the animal was urinating excessively. As this was suggestive of diabetes, he tested the urine for sugar and confirmed its presence.[1] Minkowski then removed the pancreas from several more dogs and all of them rapidly developed diabetes.

In 1900, Leonid Ssobolew in St Petersburg demonstrated that ligation of the pancreatic duct did not cause diabetes even though atrophy of the gland occurred.[2] This suggested that the absence of an internal secretion when the gland was removed was the cause of the diabetes that ensued. Ssobolew also noted that after feeding animals with carbohydrate there was an increase in the number of granules in that part of the pancreas described as the islets of Langerhans. This led him to conclude that the islets were the site at which the gland exerted its effect on carbohydrate metabolism. Six years after this, the German pathologist Julius Cohnheim discovered that attempts to isolate a pancreatic hormone had failed because enzymes in the gland were destroying it. He found the problem could be overcome by deactivating these enzymes by boiling the gland prior to alcoholic extraction. This enabled George Zuelzer, a physician in Berlin, to administer an alcoholic extract that maintained the urinary sugar levels of a pancreatectomised dog under control.[3] Three years later, Zuelzer injected five diabetic patients with a pancreatic extract. He reported an improvement in all cases, but supplies ran out and impurities caused fever. This drew the criticism that elevation of temperature might have lowered sugar output, so he spent a further two years purifying his extracts before Farbwerke Hoechst agreed to produce it as Acomatol®. In 1912 a patent was issued and a contract was signed, but Zuelzer never overcame the difficulties in supplying adequate quantities.

In 1908, Ernest Scott decided that his research for his MSc degree in the department of physiology at the University of Chicago would be on pancreatic secretion. He noted at the outset of his investigation that Cohnheim in 1906 had utilised extracts prepared from pancreas in which the digestive enzymatic activity had first been destroyed by boiling in water prior to alcoholic extraction. In 1910, the Berlin physician Erich Leschke had also referred to this, and he pointed out that the enzyme could possibly destroy the hormone during extraction. Scott therefore resolved to inhibit pancreatic enzyme at the start of his extraction procedures, albeit by using less drastic methods than boiling the gland in water. The first method he tried was based on earlier studies by Minkowski. This involved tying off the glandular ducts some time before removal of the pancreas, thereby allowing the enzyme-producing cells of the gland to atrophy. Scott abandoned this procedure because it was not possible to ensure total atrophy. Among the alternative approaches that Scott then tried was an extraction technique in which the fresh, moist gland was first treated with alcohol. This effectively served to inactivate the digestive enzymes. In this manner, he obtained material that caused a temporary diminution in the amount of sugar excreted by pancreatectomised dogs.[4] This was the first safe and potent pancreatic extract to have been successfully employed in the treatment of experimentally induced diabetes. Scott did not have an opportunity to pursue the problem after completing his thesis.

Among those who tried to isolate the pancreatic hormone in the 1890s was Nicolas Paulesco at the University of Bucharest, who renewed his former interest in pancreatic extract following the publication of Zuelzer's results. In 1916, he prepared an aqueous extract of the gland. This proved active when injected into a diabetic dog. Enemy occupation of Bucharest prevented him from taking the investigation further until, in July 1921, he reported his successful isolation of the antidiabetic hormone. He called it 'pancreine' and described how it lowered blood sugar levels in normal as well as diabetic dogs when administered intravenously.[5] Only a small quantity of a soluble powder was isolated and it was irritant when injected.[6] Before the large-scale production could be organised, a similar material became commercially available as a result of Canadian endeavours.

INSULIN

After returning from service in the armed forces during the First World War, Frederick Banting practised in London, Ontario, as a surgeon. In October 1920, he read a review article which postulated that a substance secreted by the pancreas might be capable of alleviating diabetes. He came to the conclusion that the failure to isolate this could be due to its destruction by the digestive enzymes in the gland; he was unaware that this same suggestion had been made 10 years earlier. On the advice of colleagues, Banting approached Professor John Macleod at the University of Toronto, who explained that an investigation into this would require several months of full-time work. Macleod told Banting that facilities could be provided in his department. This was a generous offer, particularly in view of the fact that Banting was not proposing any radical new approach to a problem that had defeated the efforts of experts in the field for over two decades. Later, Banting was to understate the importance of what Macleod did for him, both by way of instruction on how to proceed with the investigation and by providing the services of Charles Best, a final-year biochemistry student. Best had been instructed in the latest microchemical method which permitted sugar concentrations to be measured using only fractions of a millilitre of blood. It was application of this new technique that enabled the two Canadians to succeed where others had failed. They began their experimental work in the middle of May 1921 in a small laboratory within the university medical building.

By late July, the degenerated pancreas (caused by ligation of the duct) had been removed from a dog, and on the 30th of the month 4 ml of a ground-up aqueous suspension of it was injected into a pancreatectomised dog. Blood sugar levels quickly dropped from 0.20 to 0.12%, subsequently being held at this level by a second injection. This proof that their extraction technique was satisfactory encouraged Banting and Best who, from early August until November, worked night and day injecting dogs with extracts prepared by a variety of methods. Studies of the stability of the pancreatic hormone were also carried out, revealing that it was sensitive to alkali and to the protein-digesting enzyme trypsin. This raised the suspicion that the hormone was a protein. An improvement in the isolation procedure was obtained when Macleod's original suggestion to Best was finally acted upon, and Scott's alcoholic extraction method was employed.

The culmination of their endeavours was reached during the Christmas vacation period, when they injected each other with their extract. Other than a little redness at the injection site, there were no untoward effects. Tests on patients now began, using the same extract. It was given the name 'insulin' so as to conform with that suggested for the hormone some 12 years earlier.[7] The first to receive insulin, in January 1922, was 12 year old Leonard Thompson, a dangerously ill diabetic who was being treated at the Toronto General Hospital. He was not expected to live much longer. Unfortunately, the first injections of insulin caused so much local irritation in the sick boy that treatment had to be stopped.

James Collip had been working in Macleod's department as a Rockefeller Fellow when he agreed to assist with the purification of insulin. He made the important breakthrough of precipitating insulin from aqueous alcoholic extract of the pancreas by pouring this into several volumes of pure alcohol. This precipitate of insulin was then reconstituted for injection, and was given to Leonard Thompson only a few days after the original injections had been withdrawn. This time the effect of the injections was dramatic. The boy was rapidly restored to excellent health in only a few days. He remained in the best of health for several years, receiving daily injections until his untimely death in a motor cycle accident. When attempts were made to scale-up Collip's manufacturing process to provide insulin for clinical trials, the yield dwindled and supplies ran out. Tragically, this led to the death of several patients who had been responding well. Facilities to produce insulin had, fortunately, been made available at the Connaught Laboratories, established by the University of Toronto to provide antitoxins and vaccines during the First World War when supplies from Europe were unavailable.

By May 1922, with Best in charge, production was established in the Connaught Laboratories at a satisfactory level, a remarkable achievement when it is realised that there was no previous expertise in this type of production available in Canada. Banting and Best had agreed to apply for patents on the process on the understanding that the University of Toronto would accept and administer these. This it did by setting up an Insulin Committee, which at once put all information it had at the disposal of the Eli Lilly Company of Indianapolis. This public-spirited gesture ensured that mass production of insulin could be rapidly put in hand, thus saving the lives of as many diabetics as possible. Similar arrangements were subsequently made with other responsible agencies throughout the world.

The Lilly group itself consisted of about 100 workers, headed by George Walden. He introduced the technique of isoelectric fractionation, in which advantage was taken of the insolubility of different proteins at critical levels of acidity. Adjustment of acidity allowed insulin to be separated from protein contaminants. Within a year, the Eli Lilly Company was providing enough pure insulin to meet the need of all diabetics in the United States. This established the company as a leading American pharmaceutical manufacturer.

In 1923, a Nobel Prize was awarded to Banting and Macleod. Banting shared his portion of the prize with Best to express his dismay at his colleague being overlooked. Macleod, in response, shared his award with Collip, whom he considered had been largely responsible for turning the experimental findings into a practical reality. Exactly 30 years later at Cambridge, Sanger completed a 10 year investigation which won the 1958 Nobel Prize for chemistry for his elucidation of the chemical structure of bovine insulin.[8] It was the first protein to have its amino acid sequence determined.

```
Gly
 |
Ile
 |
Val
 |
Glu ┌─────────────────────────────┐
 |  |                             |
Gln─Cys─Cys─Thr─Ser─Ile─Cys─Ser─Leu─Tyr─Gln─Leu─Glu─Asn─Tyr─Cys─Asn
          \                                                     |
           His─Leu─Cys─Gly─Ser─His─Leu─Val─Glu─Ala─Leu─Tyr─Leu─Val─Cys      human insulin
            |                                                 |
           Gln                                               Gly
            |                                                 |
           Asn                                               Glu
            |                                                 |
           Val                                               Arg
            |                                                 |
           Phe          Thr─Lys─Pro─Thr─Tyr─Phe─Phe─Gly
```

Pure, crystalline insulin was prepared in 1926 by John Abel of Johns Hopkins University.[9] It caused less local irritation than amorphous insulin, but turned out to be shorter-acting. In Copenhagen, Hans Hagedorn at the Nordisk Insulin Laboratory assumed that the presence of a protein contaminant had accounted for the longer duration of the earlier products. He searched for an acceptable protein that could be added to crystalline insulin, eventually selecting a small protein known as protamine, obtained from fish sperm. Adding this gave a longer-acting insulin preparation that was introduced in 1936.[10] Two years later, Scott at the Connaught Laboratories found that the duration of action of protamine insulin was no longer than that of ordinary insulin after the removal of zinc, an element he had previously found present in crystalline insulin. This led to the introduction of the more reliable protamine zinc insulin for once-daily injection.

Just after the Second World War, Knud Hallas-Moller of Novo Industrie in Denmark realised that the phosphate buffer used in the preparation of insulins was removing zinc and replaced it with acetate buffer. This led to the marketing of the long-acting Lente® range of insulins. These did not require addition of protamine to ensure low solubility for slow dissolution in body fluids.

The discovery by Sanger of minor differences in the amino acid sequences of insulins from oxen, cows and humans[11] stimulated Eli Lilly to use recombinant DNA technology to prepare human insulin from bacteria. A similar product was also made by Novo Industrie by changing one of the amino acids in pig insulin to convert it into human insulin. As a result of these developments, patients now receive human insulin.

Glucagon

Shortly after the introduction of insulin, John Murlin at the University of Rochester discovered a second hormone in pancreatic extracts.[12] This exhibited the opposite effect to insulin when injected into laboratory animals, raising their blood sugar levels. It was later given the name 'glucagon' and shown to be secreted by the α-cells of the islets of Langerhans, whereas insulin was secreted by the β-cells. Shortly after Sanger had determined the chemical structure of insulin, William Bromer and his colleagues at the Eli Lilly laboratories purified glucagon and elucidated its structure.[13] It was synthesised in 1967.

```
His–Ser–Gln–Gly–Thr–Phe–Thr–Ser–Asp–Tyr–Ser–Lys–Tyr–Leu–Asp–Ser–Arg
                                                                    |          glucagon
            Thr–Asn–Met–Leu–Trp–Gln–Val–Phe–Asp–Gln–Ala–Arg
```

Glucagon regulates glucose metabolism by promoting the conversion of glycogen in the liver to glucose. It can be injected to counteract hypoglycaemia arising from excessive dosage with insulin or oral hypoglycaemic drugs.

PITUITARY HORMONES

The pituitary gland, situated at the base of the brain, was one of several glands from which Oliver and Schäfer prepared glycerine extracts in 1895.[14] They found that pituitary extract could increase blood pressure, although the effect was much less marked than with adrenal extracts. William Howell[15] showed that this pressor activity was only present in extracts prepared from the posterior lobe of the gland. In 1908, Thomas Aldrich established that almost all the activity of the entire pituitary gland could be removed by extraction with 0.1–1.0% aqueous acetic acid. This formed the basis of preparing standardised posterior pituitary extracts that met with pharmacopoeial approval.

While working on the pharmacology of ergot extracts, Henry Dale discovered that posterior pituitary extracts had a powerful action on the smooth muscles of the uterus.[16] By 1909 such extracts were being used to hasten labour. Once the risk of rupture of the uterus had been recognised, the administration of extracts was generally restricted until after expulsion of the placenta so as to produce prompt contraction of the uterus in order to limit post-partum haemorrhage.

In 1913, von den Velden in Düsseldorf[17] and Farini in Venice[18] independently discovered that the principal effect of pituitary extract was to inhibit urine production. This led to its use in the treatment of diabetes insipidus, a disease in which copious amounts of watery urine are voided.

A most important advance came from the research laboratories of Parke, Davis & Company in Detroit in 1928, where Oliver Kamm[19] and his colleagues extracted posterior pituitary lobes with 0.25% aqueous acetic acid (to maintain pH in the range 3.8–4.4), followed by salting out with ammonium sulfate to obtain a precipitate that was separated by further manipulation into two active hormones. These became known as oxytocin and vasopressin, the latter of which increased blood pressure. Oxytocin found widespread use in obstetric practice.

Further purification of these hormones was achieved at Cornell University Medical College by Vincent du Vigneaud, who had previously worked with Abel on the crystallisation of insulin. Hundreds of thousands of hog and beef glands were used during the course of his investigation, which relied mainly on electrophoretic separation. Progress could only be monitored by testing the purified fractions on animals. During the Second World War du Vigneaud turned to working on penicillin. This introduced him to Craig's newly developed technique of countercurrent distribution. When he renewed his studies on the pituitary after the war, he combined this technique with starch-column chromatography to purify enough oxytocin for analysis. This showed that it was a peptide (i.e. a small protein), consisting of eight amino acids. At this point it was recognised that the oxytocin could be no more than 50% pure, and another two years passed before a reasonably pure crystalline derivative was obtained in 1953.[20] Du Vigneaud and his colleagues then determined the structure, confirming it by synthesis.[21] It was the first peptide hormone to be synthesised. For his persistence over a quarter of a century, du Vigneaud finally had the satisfaction of being awarded the Nobel Prize for Chemistry in 1955 for his work on the posterior pituitary hormones.

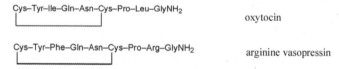

Cys–Tyr–Ile–Gln–Asn–Cys–Pro–Leu–GlyNH$_2$ oxytocin

Cys–Tyr–Phe–Gln–Asn–Cys–Pro–Arg–GlyNH$_2$ arginine vasopressin

Du Vigneaud isolated vasopressin as well as oxytocin in 1953. He established that it was a peptide hormone in which two of the amino acids differed from those in oxytocin, at the third and eighth positions.[22] Conclusive confirmation of the proposed structure was obtained by its synthesis in 1958.[23]

There are two types of vasopressin. The hormone in humans, cattle, sheep, horses and chickens has an arginine residue at position 8, hence it is called 'argipressin'. That found in pigs has a lysine residue in position 8 and is known as 'lypressin'. Vasopressin is also the antidiuretic hormone and as such is injected to treat diabetes insipidus arising either from pituitary disease or as a complication following surgery for removal of a pituitary tumour.

SOMATROPIN (HUMAN GROWTH HORMONE)

Following Minkowski's observation, in 1887, that pituitary tumours were associated with the characteristic overgrowth of the hands and feet in patients with acromegaly, evidence accrued to support the belief that a growth hormone was produced by the pituitary. Schäfer reported in 1908 that feeding rats on pituitary tissue accelerated their growth,[24] while Robertson at Berkeley found that feeding young mice on an alcoholic extract of anterior pituitary lobes not only accelerated growth but even extended it beyond normal limits.[25] Herbert Evans and Joseph Long, at the University of California, accelerated the growth of rats by injecting a saline extract of anterior pituitary lobes.[26] These early experiments have been criticised because of the lack of controlled conditions and failure to take into account the influence of pituitary feedback.[27] However, in 1922 at Berkeley, Philip Smith fed pituitary to hypophysectomised tadpoles (i.e. in which the pituitary had been surgically removed) and thereby restored normal growth.[28] This approach became the basis of modern procedures for determining the activity of pituitary extracts. An additional difficulty was that the early extracts of pituitaries were contaminated with other hormones. Choh Hao Li and Evans eventually isolated pure growth hormone from ox pituitaries in 1944. Later, the term 'somatotropin' was introduced to describe the hormone, but early hopes that it would be of value in the treatment of human pituitary dwarfs were soon dashed.

In 1954, Grace Pickford at Yale demonstrated that fish growth hormone was active in fish, but not rats, suggesting that growth hormone might be species-specific.[30] Choh Hao Li then purified both human and monkey growth hormones and found that their chemical structures differed from that of the other growth hormones. Monkey hormone was then shown to be active in monkeys, whereas bovine and porcine varieties were not. Finally, human hormone (prepared independently by Li and by Maurice Raben of Tufts University in Boston) was shown to be effective in patients.[31] Li determined the chemical structure of somatotropin in 1966; this was later revised.[32] It was a globular protein containing 191 amino acid residues.

Growth hormone accounts for about 15% of dried cadaver pituitary. Yet, for many years, supplies of somatotropin from cadavers could not satisfy the demand. Tragically, some batches were belatedly found to be contaminated with a virus that caused several deaths from Creutzfeldt–Jakob encephalopathy.[33] This led to the withdrawal of somatotropin obtained from pooled cadaver pituitaries. Subsequently, genetically modified bacteria were used to produce human growth hormone.[34] This was given the approved name of 'somatropin' and replaced somatotropin from cadavers. It is used successfully in the treatment of growth failure in children.

GONADOTROPHIN

When Evans and Long injected an extract of the anterior lobe of the pituitary into rats in 1921, they had been surprised to detect stimulation not only of growth but also of the gonads.[26] Seven years later, Evans reported that an acidic extract had maximal effects on sexual maturation.

Gonadotrophin extracted from cadaver pituitary was used to induce ovulation in infertile women. Until 1985, the gonadotrophins were extracted from cadaver pituitary glands. It was then discovered that these were also contaminated with the virus that causes Creutzfeldt–Jakob disease. Since then, gonadotrophin has been obtained by extraction from urine, being known as chorionic gonadotrophin.

MENOTROPHIN

This follicle-stimulating hormone is a glycoprotein of molecular weight around 67 000 that was obtained pure from sheep anterior pituitaries by Li[35] in 1949. It promoted the development of the ovarian follicle and hence was used in women who, as a consequence of hypopituitarism, were unable to ovulate. It received the approved name of 'menotrophin'. The hormone extracted from the urine of post-menopausal women differs slightly from the human pituitary hormone and is known as 'urofollitrophin'.[36] It has similar uses to follicle-stimulating hormone.

CORTICOTROPHIN

In 1926, Philip Smith removed the pituitary gland of rats without causing brain damage. Atrophy of the thyroid and adrenal glands ensued, but this was overcome by daily injections prepared from the anterior pituitary of rats.[37] This encouraged James Collip to introduce an extract for clinical use in 1933 for the treatment of adrenal insufficiency.[38] Further refinement of the extraction procedures led to the isolation in 1943 from sheep and pig pituitaries of an impure peptide that was named corticotrophin.[39,40]

In 1943, James Murphy and Ernest Sturm at the Rockefeller Institute in New York reported that removal of the adrenals greatly increased the susceptibility of rats to transplanted lymphatic leukaemia.[41] Their findings suggested that adrenal hormones inhibit the development of leukaemia. The following year, they showed that the development of transplanted leukaemia cells in mice was inhibited by both deoxycortone acetate and corticotrophin. F.R. Heilman and Edward Kendall at the Mayo Clinic injected large doses of cortisone into mice bearing transplanted tumours of lymphatic origin and also found that the growth of these were inhibited.[42]

Thomas Dougherty and Abraham White at the Department of Anatomy and Physiological Chemistry of Yale University demonstrated that adrenal stimulation with corticotrophin caused shrinkage of lymph nodes.[43] A second important observation was that adrenal stimulation induced a rapid, but short-lived, diminution in the number of white blood cells. Lloyd Law of the Roscoe B. Jackson Memorial Laboratory at Bar Harbor in Maine and Robert Spiers at the University of Wisconsin-Madison demonstrated in 1947 that naturally occurring leukaemic white cells in mice were also sensitive to adrenal hormones released by injection of corticotrophin.[44]

A team of doctors at the Memorial Hospital in New York headed by Cornelius Rhoads carried out a major clinical trial of corticotrophin.[45] The results were dramatic; patients with chronic leukaemias, advanced Hodgkin's disease (malignancy of lymphatic cells) or sarcomas underwent remission and regained good health as the number of malignant cells declined. Unfortunately, these effects were of temporary duration. Sidney Farber, a pathologist at the Harvard Medical School, reported similar results with acute leukaemia patients at the Children's Medical Center in Boston.[46] Tragically, the remissions again proved to be only temporary. Nonetheless, this was a major step towards the present situation where most children with acute lymphocytic leukaemia, and a majority of all patients with advanced Hodgkin's disease, can be cured by combining chemotherapy with radiotherapy. Cortisone replaced corticotrophin, but was in turn superseded by prednisolone, which had fewer side effects. This was combined with the administration of cytotoxic drugs.

The value of corticotrophin in the treatment of malignant diseases and also its use by Philip Hench as an alternative to the scarce cortisone in the treatment of rheumatoid arthritis stimulated further work on the purification of corticotrophin (then known as ACTH, i.e. adrenocorticotrophic hormone).

In 1955, Li and his colleagues at Berkeley established the structure of corticotrophin isolated from sheep as a peptide containing 39 amino acids.[47] Human corticotrophin was isolated in 1963.[48] Eight years later, its structure was determined.[49]

Corticotrophin is no longer used in therapy, the direct application of corticosteroids being preferred when steroid therapy is indicated since the outcome is more predictable. However, it is administered in the diagnosis of adrenocortical insufficiency to determine its effect on hydrocortisone levels in the plasma.

The first 24 amino acids of corticotrophin are identical in all species and retain the full biological activity of the entire molecule. Since the occurrence of hypersensitivity (allergic) reactions after administration of corticotrophin derived from animal pituitaries is attributable to the region of the molecule containing amino acids 25 to 39, a molecule containing the first 24 amino acids was synthesised in 1961.[50] Given the name 'tetracosactrin', it had identical properties to corticotrophin, but was less immunogenic. Its main use is in the diagnosis of adrenocortical insufficiency.

Lutropin is a peptide hormone that was isolated in 1940 from sheep pituitaries by Li[51], and from pig pituitaries by Shedlovsky.[52] It is a glycoprotein of molecular weight around 30 000. The amino acid sequence of bovine lutropin hormone (LH) was determined in 1971[53] and that of human LH in 1978.[54] In the female it stimulates the production of progesterone by the ovaries and, in conjunction with follicle-stimulating hormone (FSH), releases oestrogen from ovarian follicles. It induces ovulation and then facilitates the conversion of the follicular cells into the corpus luteum. In the male, LH stimulates the interstitial cells of the testes to synthesise testosterone. Lutropin is not used in therapy since injection of gonadorelin produces the required rise in LH levels.

REFERENCES

1. J. von Mering, O. Minkowski, Diabetes mellitus nach Pankreasextirpation. *Centralblatt für klinische Medicin, Leipzig*, 1889; **10**: 393–4.
2. L.W. Ssobolew, Zur normalen und pathologishen Morphologie der inneren Secretion der Bauchspeicheldruse. *Arch. Path. Anat. Physiol. Klin. Med.*, 1902; **168**: 91–128.
3. G. Zuelzer, Ueber Versuche einer specifischen Fermenttherapie de Diabetes. *Z. Exp. Path. Ther.*, 1908; **5**: 307–18.
4. E.L. Scott, On the influence of intravenous injections of an extract of the pancreas on experimental pancreatic diabetes. *Am. J. Physiol.*, 1912; **29**: 306–10.
5. N.C. Paulesco, Action de l'extrait pancréatique injecte dans le sang, chez un animal diabétique. *C. R. Soc. Biol.*, 1921; **85**: 555–9.
6. I. Murray, Paulesco and the isolation of insulin. *J. Hist. Med. Allied Sci.*, 1971; **26**: 150–7.
7. F.G. Banting, C.H. Best, Pancreatic extracts. *J. Lab. Clin. Med.*, 1922; **7**: 464–72.
8. H. Brown, F. Sanger, R. Kitai, The structure of pig and sheep insulins. *Biochem. J.*, 1955; **60**: 556.
9. J.J. Abel, Crystalline insulin. *Proc. Natl. Acad. Sci.*, 1926; **12**, 132–6.
10. H.C. Hagedorn, B.N. Jensen, N.B. Krarup, I. Wodstrup, Protamine insulinate. *J. Am. Med. Assoc.*, 1936; **106**: 177–80.
11. J.I. Harris, F. Sanger, M.A. McNaughton, Species differences in insulin. *Arch. Biochem. Biophys.*, 1956; **65**: 427.
12. C.P. Kimball, J.R. Murlin, Aqueous extracts of pancreas. *J. Biol. Chem.*, 1923; **58**: 337–46.
13. W.W. Bromer, L.G. Sinn, A. Staub, O.K. Behrens, The amino acid sequence of glucagon. *J. Am. Chem. Soc.*, 1956; **78**: 3858–9.
14. G. Oliver, E.A. Schäfer, On the physiological actions of extracts of pituitary body and certain other glandular organs. *J. Physiol.*, 1895; **18**: 277.
15. W.H. Howell, The physiological effects of the hypophysis cerebri and infundibular body. *J. Exp. Med.*, 1898, 3: 245–58.
16. H.H. Dale, On some physiological actions of ergot. *J. Physiol.*, 1906; **34**: 163–206.
17. R. von den Velden, Nierenwirkung von hypophysenextrakten beim menshen. *Klin. Wochenschr.*, 1913; **50**: 2083–6.
18. F. Farini, Diabete insipido ed opoterapia. *Gazz. Osp. Clin.*, 1913; **34**: 1135–9.
19. O. Kamm, The active principles of the posterior lobe of the pituitary gland. I. The demonstration of the presence of two active principles. II. The separation of the two principles and their concentration in the form of potent solid preparations. *J. Am. Chem. Soc.*, 1928; **50**: 573–601.
20. V. du Vigneaud, Trail of sulfur research: from insulin to oxytocin. *Science*, 1956; **123**: 967–74.
21. V. du Vigneaud, C. Ressler, J.M. Swan, *et al.* The synthesis of an octapeptide amide with the hormonal activity of oxytocin. *J. Am. Chem. Soc.*, 1953, **75**: 4879–80.
22. V. du Vigneaud, H.C. Lawler, E.A. Popenoe, Enzymatic cleavage of glycinamide from vasopressin and a proposed structure for this pressor-antidiuretic hormone of the posterior pituitary. *J. Am. Chem. Soc.*, 1953, **75**: 4880–1.

23. V. du Vigneaud, D.T. Gish, P.G. Katsoyannis, G.P. Hess, Synthesis of the pressor-antidiuretic hormone, arginine-vasopressin. *J. Am. Chem. Soc.*, 1958; **80**: 3355–8.
24. E.A. Schäfer, *Proc. Roy. Soc. Lond., Ser. B*, 1908; **81**: 442.
25. T.B. Robertson, Experimental studies on growth. *J. Biol. Chem.*, 1916; **24**: 397.
26. H.M. Evans, J.A. Long, The effect of the anterior lobe administered intraperitoneally upon growth, maturity and oestrus cycles of the rat. *Anat. Record*, 1921; **21**: 62.
27. P.S. Brown, Anterior pituitary hormones: definition, measurement and use, in *Discoveries in Pharmacology*, Vol. 2, eds M.J. Parnham, J. Bruinvels. Oxford: Elsevier; 1984, pp. 369–89.
28. P.E. Smith, I.P. Smith, *Anat. Record*, 1922; **23**: 38.
29. C.H. Li, H.M. Evans, The isolation of pituitary growth hormone. *Science*, 1944; **99**: 183–4.
30. G.E. Pickford, The response of hypophysectomized male killifish to purified fish growth hormone, as compared with the response to purified beef growth hormone. *Endocrinology*, 1954; **55**: 274–87.
31. M.S. Raben, Treatment of a pituitary dwarf with human growth hormone. *J. Clin. Endocrinol. Metab.*, 1958; **18**: 901–3.
32. H.D. Niall, Revised primary structure for human growth hormone. *Nature New Biol.*, 1971; **230**: 90–1.
33. P. Brown, D.C. Gajdusek, C.J. Gibbs Jr, D.M. Asher, Potential epidemic of Creutzfeldt–Jakob disease from human growth hormone therapy. *New Engl. J. Med.*, 1985; **313**: 728–31.
34. D.V. Goeddel, H.L. Heyneker, T. Hozumi, *et al.*, Direct expression in *Escherichia coli* of a DNA sequence coding for human growth hormone. *Nature*, 1979; **281**: 544–8.
35. C.H. Li, H.M. Evans, M.E. Simpson, Isolation of pituitary follicle-stimulating hormone (FSH). *Science*, 1949; **109**: 445–6.
36. M.L. Taymor, T. Tamada, M. Soper, W.F. Blatt, Immunologic relationships between urinary and pituitary follicle-stimulating hormone. *J. Clin. Endocrinol. Metab.*, 1967; **27**: 709–14.
37. P.E. Smith, *Anat. Rec.*, 1926; **32**: 221.
38. J.B. Collip, E.M. Anderson, D.L. Thompson, The adrenotropic hormone of the anterior pituitary lobe. *Lancet*, 1933; **2**: 347.
39. C.H. Li, H.M. Evans, M.E. Simpson, Adrenocorticotrophic hormone. *J. Biol. Chem.*, 1943; **149**: 413.
40. G. Sayers, A. White, C.N.H. Long, Preparation and properties of pituitary adrenotropic hormone. *J. Biol. Chem.*, 1943; **149**: 425–36.
41. J.B. Murphy, E. Sturm, Adrenals and susceptibility to transplanted leukaemia of rats. *Science*, 1943; **98**: 568.
42. F.R. Heilman, E.C. Kendall, The influence of II-dehydro-17-hydroxycorticosterone (compound E) on the growth of a malignant tumor in the mouse. *Endocrinology*, 1944; **34**: 416–20.
43. T.F. Dougherty, A. White, Effect of pituitary adrenotropic hormone on lymphoid tissue. *Proc. Soc. Exp. Biol. Med.*, 1943; **53**: 132–3.
44. L.W. Law, R. Spiers, Responses of spontaneous lymphoid leukemias in mice to injection of adrenal cortical extracts. *Proc. Soc. Exp. Biol. Med.*, 1947; **66**: 226–30.
45. K. Siguira, C.C. Stock, K. Dobriner, C.P. Rhoads, *Cancer Res.*, 1950; **10**: 244.
46. S. Farber, *Proceedings of the First Clinical ACTH Conference*, ed. J.R. Mote. London: J. & A. Churchill; 1950, p. 328.
47. C.H. Li, I.I. Geschwind, R.D. Cole, *et al.*, Amino-acid sequence of alpha-corticotropin. *Nature*, 1955; **176**: 687–9.
48. B.T. Pickering, R.N. Andersen, P. Lohmar, *et al.*, Adrenocorticotropin. XXVII. On the presence of pig-type adrenocorticotropin in sheep. *Biochem. Biophys. Acta*, 1963; **74**: 763–73.
49. L. Gráf, S. Bajusz, A. Patthy, *et al.*, Revised amide location for porcine and human adrenocorticotropic hormone. *Acta Biochim. Biophys. Acad. Sci. Hung.*, 1971; **6**: 415–18.
50. H. Kappeler, R. Schwyzer, Die Synthese eines Tetracosapeptides mit der Aminosäuresequenz eines Lochaktiven Abbauproduktes des β-Corticotropins (ACTH) aus Schweinhypophysen. *Helv. Chim. Acta*, 1961; **44**: 1136.
51. C.H. Li, M.E. Simpson, H.M. Evans, Interstitial cell inhibiting hormone. II *Endocrinology*, 1940; **27**: 803–8.
52. T. Shedlovsky, A. Rothen, R.O. Greep, *et al.*, *Science*, 1940; **92**: 178.
53. H. Papkoff, M.R. Sairam, C.H. Li, Amino acid sequence of the subunits of ovine pituitary interstitial cell-stimulating hormone. *J. Am. Chem. Soc.*, 1971; **93**: 1531–2.
54. H.T. Keutmann, B. Dawson, W.H. Bishop, R.J. Ryan, *Endocrin. Res. Commun.*, 1978; **5**: 57.

15

Sex Hormones

During the period of enthusiasm for organotherapy at the end of the nineteenth century, the Viennese gynaecologist Rudolf Chrobak gave tablets prepared from cows' ovaries to his patients in an attempt to overcome the undesirable effects that ensued after surgical removal of their ovaries.[1] Chrobak then asked his assistant, Emil Knauer, to investigate the effects of transplantation of ovarian tissue in animals. He began by extirpating the ovaries of rabbits and showing that this led to involution of the uterus and cessation of the oestrus cycles. However, when he reimplanted the ovaries elsewhere in the animals, normal ovarian function was restored.[2] His work inspired others to administer desiccated ovaries or extracts of these to patients in the hope of relieving menopausal disorders. The most potent of the extracts available before the First World War were prepared by Henri Iscovesco in Paris[3] and by Otfried Fellner in Vienna.[4] Both used fat solvents (alcohol, ether and acetone) to obtain extracts that produced sexual changes in castrated animals. The Swiss manufacturer Ciba marketed an ovarian extract in 1913. The techniques used in the preparation of these extracts would have failed to separate the compounds produced by the two distinct components of the ovary, viz. the oestrogens produced in the follicles and progesterone produced in the corpus luteum.

In the recently reprinted 1910 edition of his perennial best seller *Old Age Deferred*, the Czech physician Arnold Lorand cited the use in the preceding two decades of ovarian extracts to relieve menopausal symptoms. Such extracts were introduced when organotherapy was at the height of its popularity and would have been of questionable value. The availability of oestrogens in the 1930s permitted more reliable treatment to be given for the alleviation of menopausal symptoms such as hot flushes and decrease of libido.[5] The case for such therapy was advanced further in 1940 when the American endocrinologist Fuller Albright[6] revealed that menopausal osteoporosis was caused by lack of oestrogens. However, it was not until long after oral contraceptives had become popular that hormone replacement therapy (HRT) began to be routinely used in menopausal women.[7] Recently there have been confusing reports concerning the balance between the risks and benefits of HRT.

OESTROGENS

Early attempts to isolate sex hormones therapeutically failed through lack of a cheap, quick assay that could determine the amount of hormone present as the extraction procedures were being refined. Edgar Allen and Edward Doisy of Washington University in St Louis[8] overcame this problem in 1923 by taking advantage of an important discovery made six years before by Charles Stockard and George Papanicolaou at Cornell Medical College.[9] The Cornell researchers had found that changes in the appearance of the cells lining the vaginal wall in rodents closely paralleled the phases of the oestrus cycle. By microscopic examination of cells scraped from the vagina in immature mice or rodents, Allen and Doisy were able to monitor the effects of ovarian extracts. Such was the sensitivity of this simple technique that it afforded an accurate assay for oestrogenic activity. It enabled Doisy to make advances in isolating the pure ovarian hormone, although he faced major difficulties due to the small amount present in the follicular fluid of the porcine ovaries that he extracted.

In 1927, the Berlin gynaecologists Selman Aschheim and Bernhard Zondek used the Allen and Doisy method of detecting ovarian hormone in an attempt to devise a pregnancy test based on changes in the urinary excretion pattern of female hormone.[10] They found that the amount of hormone in urine increased markedly with the onset of pregnancy. With the publication of the Aschheim–Zondek findings, biochemists realised that an alternative source of female hormone had been found and a race to isolate the hormone ensued.

Oestrone

In August 1929, Doisy isolated a crystalline oestrogenic hormone from pregnant mares' urine by first subjecting the urine to acid hydrolysis (which was later shown to hydrolyse both sulfate esters and glucuronides of the hormone) and then taking advantage of its acidic nature to successively partition the hormone between alkaline solution and ether in order to leave behind impurities in the decanted liquors.[11] His work had been supported by Parke, Davis & Company. Two months later, the isolation of 20 mg of this same hormone was reported by Adolf Butenandt from the laboratory of Adolf Windaus at the University of Göttingen.[12] He had worked on a syrupy extract that the Schering–Kahlbaum Company of Berlin had prepared from nearly 2000 litres of pregnancy urine. Early in the following year, Ernst Laquer and his colleagues at the University of Amsterdam also isolated the hormone.[13] The outcome of these investigations was that Parke, Davis & Company in the United States and Schering–Kahlbaum in Germany were able to provide endocrinologists with oestrone and related hormones. In 1935, it was agreed by a committee of the League of Nations that the hormone should be known as 'oestrone'.

After J.D. Bernal had proved by means of X-ray crystallography that the structure proposed by Windaus and Heinrich Wieland for the sterol known as 'ergosterol' was wrong, Otto Rosenheim and Harold King at University College in London deduced the correct one and then went on to propose that the oestrogens may be related to it.[14] The proof was provided after Girard,[15] in Paris, had developed a special reagent that combined with the carbonyl group in oestrone to form a water-soluble derivative. This enabled large amounts of oestrone to be isolated. At the Research Institute of the Cancer Hospital, London (later the Chester Beatty Research Institute), James Cook and his colleagues used material supplied by Girard to confirm the proposed structures for oestrone and oestriol in 1938.

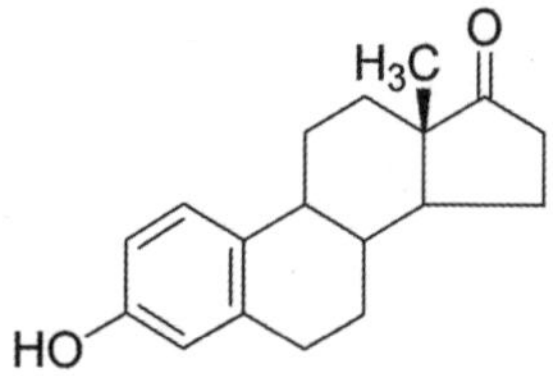

oestrone

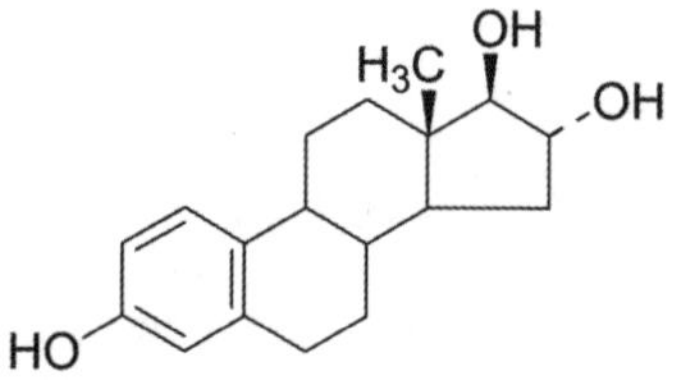

oestriol

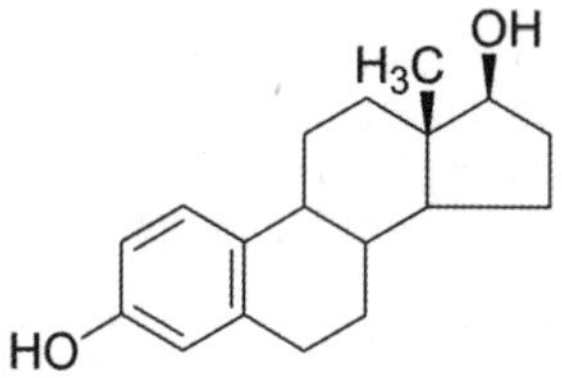

oestradiol

Oestrone was originally used to treat menopausal symptoms, primary amenorrhoea and hormone deficiency states. It was administered either in an oily injection or by mouth, but was superseded by other oestrogens in the 1950s. Oestrone is not very potent by mouth since it undergoes first-pass metabolism in the liver after absorption from the gut to form oestriol, which is then glucuronidated or sulfated to produce an inactive ester that is excreted in the urine.

Oestriol

The isolation of a second oestrogenic hormone from human pregnancy urine was reported in the summer of 1930 by Guy Marrian, a research student in the Department of Physiology at University College London.[16] This was later named 'oestriol', on account of the three hydroxyl groups in its structure. The structure of oestriol was confirmed by Cook at the same time as oestrone. When administered by injection, it seemed less potent than oestrone, but this is now known to be due to its short duration of action. Oestriol therefore has to be formulated as slow-release implants and transdermal patches.

Oestradiol

Erwin Schwenk and Fritz Hildebrandt, colleagues of Butenandt, found that hydrogenation of oestrone formed oestradiol, a new substance that was 8–10 times as potent as oestrone.[17] Three years later, in 1935, Doisy showed that it was a natural hormone by extracting 12 mg of it from 4 tons of sows' ovaries.[18]

Oestradiol undergoes first-pass metabolism in the liver to form oestriol, which is then either glucuronidated or sulfated to form an inactive polar ester that is excreted in the urine. This means that an oral dose had to be 5–10 times greater than that given by intramuscular injection.

PROGESTERONE

Auguste Prenant, a histologist at the University of Nancy, presented histological evidence to the Société de Biologie in 1898 which indicated that the corpus luteum in the ovary was a gland of internal secretion. Ludwig Fraenkel at the University of Breslau subsequently demonstrated that the corpus luteum played a role in ensuring that the developing embryo attached to the wall of the uterus. The French investigators Paul Bouin and Albert Ansel established that this was achieved by inducing changes in the mucosal lining of the uterus in readiness for pregnancy.[19] Twenty years later George Corner and his doctoral student Willard Allen, at the University of Rochester, exploited this observation. By monitoring the changes in the uterine lining, they showed that these were dependent on hormone levels and could form the basis of a reliable bioassay for the presence of hormone in extracts prepared from corpora lutea from sows. The following year they prepared a potent extract that prevented abortion in rabbits after surgical removal of the corpus luteum.[20]

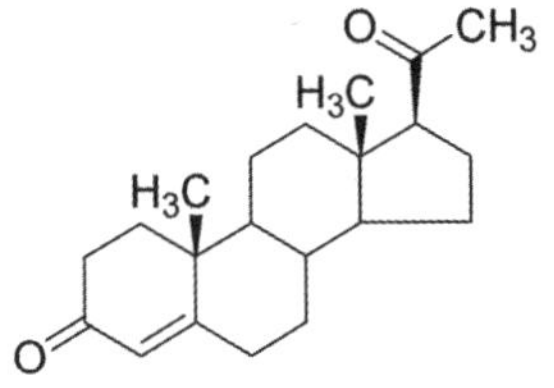

progesterone

In 1934, Adolf Butenandt at the University of Gottingen[21], Willard Allen and Oskar Wintersteiner at Columbia University[22], Karl Slotta, Heinrich Ruschig and Erich Fels at the University of Vienna[23] and Max Hartmann and Albert Wettstein at the Ciba laboratories in Basle[24] almost simultaneously announced the isolation of the active principle from the corpus luteum. Its chemical structure was established by Butenandt.[25] Because the hormone was capable of maintaining gestation, it was given the name 'progesterone'.

MALE HORMONES

The work of Brown-Séquard on testicular extracts may have been controversial, yet it stimulated other French researchers to investigate the internal secretions of the testicles. Among these was Albert Pézard, whose studies began at the Faculté de Médecine in Paris in 1911. He found that the development of the cock comb was dependent upon testicular function, receding on castration but growing after implantation of testicular tissue grafts. Pézard reported his findings to the International Congress of Sex Research held in Berlin in 1926.[26] The following year, Fred Koch and his research student Lemuel McGee, at the Department of Physiology in the University of Chicago, developed a simple assay for male hormone based on the work of Pézard.[27] They demonstrated that bovine testicle extracts prepared using fat solvents could be assayed by their ability to induce growth of the capon's comb. The availability of this assay was to enable the testicular hormone to be isolated.

After Aschheim and Zondek had shown large amounts of female hormone to be present in pregnancy urine, S. Loewe[28] at Dorpat University confirmed that male hormone was to be found in male urine. Butenandt realised that this would be a suitable source from which to extract the hormone, using the cock comb assay to measure the strength of extracts. In 1930, he isolated 15 mg of pure hormone from 25 000 litres of urine after having removed the phenolic oestrogenic fraction.[29] He named the hormone androsterone and elucidated its structure four years later.[30] It was then synthesised from cholesterol at the Ciba Laboratories in Basle by the Yugoslavian chemist Leopold Ruzicka in 1934.[31] Until evidence to the contrary appeared, it was generally assumed that androsterone was the sole hormone produced in the testicles.

Funded by the Organon Company, Ernst Laqueur led a group of researchers at the University of Amsterdam, which included a Hungarian pharmacologist, John Freud, who had previously worked on the cock comb assay with Pézard in Paris. Freud and Laqueur discovered that extracts of male hormone from urine and from testicular tissue that had been standardised so as to exhibit identical potency in the cock comb assay did not have the identical activity when their effects on castrated rats were compared. While both types of extract induced growth in the rat seminal vesicles, the effect of the urinary extract was markedly less than that of the testicular extract. The possibility that a different male hormone might be present in testicular tissue was then seriously considered and led to isolation of 5 mg of testosterone from nearly 1 ton of bovine testicles in 1935. The paper reporting this was submitted on 27 May 1935. It revealed that the new hormone was 10 times as potent as androsterone in the cock comb test and about 70 times more potent in the castrated rat seminal vesicle test.[32] It concluded that androsterone was a urinary metabolite of testosterone, which was the true male hormone.

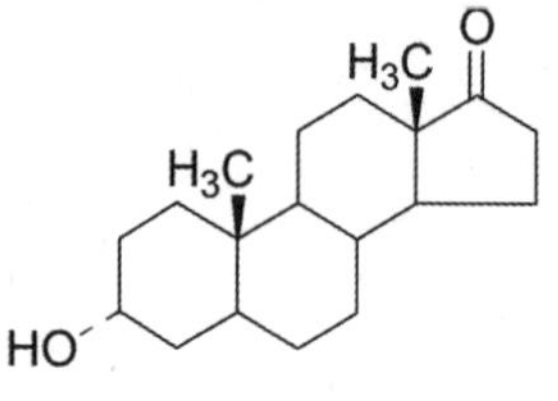

androsterone

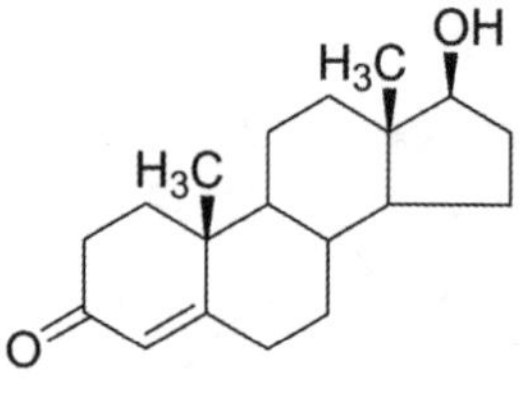

testosterone

The chemical structure of testosterone was elucidated by Ruzicka and Wettstein through its synthesis from cholesterol. They applied for a patent for Ciba on the synthetic process before submitting a paper to *Helvetica Chimica Acta* on 31 August 1935.[33] Butenandt and Hanisch had submitted a paper to *Chemische Berichtes* one week earlier.[34] Before these papers appeared in print, the structure of testosterone was revealed by Kåroly Gyula David, who rushed to publish on behalf of the Amsterdam group.[35] Inadvertently, he omitted the name of his colleague who had actually determined the structure before departing on a holiday, namely Bernard Josephy (later changed to B. J. Brent). No correction of this omission was ever published and the matter may not have come to light were it not for the account presented by Marius Tausk.[36]

Testosterone is administered in hypogonadism and other conditions where there is a deficiency of the hormone, such as that arising from testicular or pituitary disease. As it is rapidly metabolised by the liver after oral dosing, it was for many years formulated as sublingual tablets, but nowadays it is usually administered in the form of patches, intramuscular implants or by depot injection of its esters formulated in oil.

REFERENCES

1. R. Chrobak, Einverleibung von Eierstockgewebe. *Zentralbl. Gynaekol.*, 1896; **20**: 521–4.
2. E. Knauer, Die ovarien Transplanatation. *Arch. Gynaek.*, 1900; **60**: 322.
3. H. Iscovesco, Le lipoide utero-stimulant de l'ovaire. *C. R. Soc. Biol. Paris*, 1912; **73**: 104.
4. O.O. Fellner, Experimentelle Untersuchungen über die Wirkung von Gewebsextrakten aus der Plazenta und den weiblichen Sexualorganen auf das Genitale. *Arch. Gynakol.*, 1913; **100**: 641.
5. S.H. Geist, F. Spillman, The therapeutic use of amniotin in the menopause. *Am. J. Obstet. Gynecol.*, 1932; **23**: 697–707.
6. G.S. Gordon, Fuller Albright and postmenopausal osteoporosis: a personal appreciation. *Perspect. Biol. Med.*, 1981; **24**: 547–60.
7. WHO Scientific Group on Research on the Menopause in the 1990s. WHO Technical Report Series 866, Geneva; 1994.
8. E. Allen, E. Doisy, An ovarian hormone: preliminary reports on its localization, extraction and partial purification and action in test animals. *J. Am. Med. Ass.*, 1923, **81**: 819–21.
9. G.N. Papanicolaou, C.R. Stockard, The existence of a typical oestrous cycle in the guinea pig; with a study of its histological changes. *Am. J. Anat.*, 1917; **22**: 225–83.
10. S. Aschheim, B. Zondek, Hypophysenvorderlappen Hormon und Ovarialhormon im Harn von Schwangeren. *Klin. Wochenschr.*, 1927; **6**: 1322.
11. E.A. Doisy, C.D. Veler, S.A. Thayer, The preparation of the crystalline ovarian hormone from the urine of pregnant women. *J. Biol. Chem.*, 1930; **86**: 499–509.
12. A. Butenandt, Über Progynon ein krystallisertes weibliches Sexualhormon, *Naturwiss.*, 1929; **17**: 878.
13. E. Dingemanse, S.E. de Jongh, S. Kober, E. Laqueur, *Deutsch. Med. Wochenschr.*, 1930; **56**: 301.
14. O. Rosenheim, H. King, The ring-system of sterols and bile acids. *J. Soc. Chem. Ind.*, 1932; **51**: 464–6.
15. A. Girard, G. Sandulesco, Une nouvelle série de réactifs du groupe carbonyl, leur utilisation a l'extraction des substances cétoniques et la characterisation microchimique des aldéhydes et cétones. *Helv. Chim. Acta*, 1936; **19**: 1095.
16. G.F. Marrian, Early work on the chemistry of pregnanediol and the oestrogenic hormones. *J. Endocrinol.*, 1966; **35**: 6–16.
17. E. Schwenk, F. Hildebrandt, Ein neues isomers Follikelhormon aus Stutenharn. *Naturwiss.*, 1932; **20**: 658.
18. D.W. MacCorquodale, S.A. Thayer, E.A. Doisy. The isolation of the principal estrogenic substance of liquor folliculi. *J. Biol. Chem.*, 1936; **115**: 435–48.
19. P. Bouin, A. Ansel, Recherches sur les fonctions du corps jaune gestatif. I. Sur le déterminisme de la préparation de l'utérus a la fixation de l'oeuf. *J. Physiol. Path. Gen.*, 1910; **12**: 1.
20. W.M. Allen, G.W. Corner, Maintenance of pregnancy in rabbits after very early castration, by corpus luteum extracts. *Proc. Soc. Exp. Biol. Med.*, 1930; **27**: 403.
21. A. Butenandt, U. Westphal, Zur Isolierung und Charakterisierung des Corpus-luteum-Hormons. *Ber.*, 1934; **67**: 1440.
22. W.M. Allen, O. Wintersteiner, Crystalline progestin. *Science*, 1934; **80**: 190.
23. K. Slotta, H. Ruschig, E. Fels, Reindarstellung der Hormone aus dem Corpus-luteum. *Ber.*, 1934; **67**: 1270.

24. M. Hartmann, A. Wettstein, Ein krystallisiertes Hormon aus Corpus-luteum. *Helv. Chim. Acta*, 1934; **17**: 878.

25. A. Butenandt, U. Westphal, H. Cobler, Über einen Abbau des Stigmasterins zu corpus-luteum-wirksamen Stoffen; ein Beitrag zur Konstitution des Corpus-luteum-Hormons. *Ber.*, 1934; **67**: 1611–16.

26. A. Pézard, in *Verhandlungen des I. Internationalen Kongresses für Sexualforschung*, Band I, ed. M. Marcuse. Berlin: Marcus und Weber; 1927, p. 178.

27. L.C. McGee, The effect of the injection of a lipoid fraction of bull testicle in capons. *Proc. Inst. Med. Chicago*, 1927; **6**: 242.

28. S. Loewe, H.E. Voss, F. Lange, A. Wähner, *Klin. Wochenschr.*, 1928; **7**: 1376.

29. A. Butenandt, Ueber die chemische Untersuchung der Sexuallhormone. *Angew. Chem.*, 1931; **44**: 905.

30. A. Butenandt, K. Tscherning, Über Androsteron, Krystallisiertes mannliches Sexualhormon. I. Bolierung und Reindarstellung aus Munnerharn. *Z. Physiol. Chem.*, 1934; **229**: 167–84.

31. L. Ruzicka, M.W. Goldberg, J. Meyer, *et al.*, Ueber die Synthesis des Testikelhormon (Androsteron) und stereoisomerer desselben durch Abbau hydrierter Sterine. *Helv. Chim. Acta*, 1943; **17**: 1395.

32. E. Laqueur, K. David, E. Dingemanse, *et al.*, Ueber mannliches Hormon. Unterschied von Androsteron aus Harn und Testosteron aus Testis. *Acta Brev. Neerland.*, 1935; **4**: 5.

33. L. Ruzicka, A. Wettstein, Ueber die kunstliche Herstellung des Testikelhormons Testosteron (Androsten-3-on-17-ol). *Helv. Chim. Acta*, 1935; **18**: 1264.

34. A. Butenandt, G. Hanisch, Über Testosterone. Umwand-lung des Dehydro-Androsterons in Androstendiol und Testosterone; ein Weg zur Darstellung des Testosterons aus Cholestrin. *Hoppe Seylers Z. Physiol. Chem.*, 1935; **237**: 89.

35. K. David, Ueber des Testosteron, des Kristallisierte Manniche Hormon aus Steerentestes. *Acta Brev. Neerland Physiol. Pharmacol. Microbiol.*, 1935: **5**: 85–6.

36. M. Tausk, Androgens and anabolic steroids, in *Discoveries in Pharmacology*, Vol. 2, eds. M.J. Parnham, J. Bruinvels. Amsterdam: Elsevier; 1984, pp. 307–19.

Adrenal Cortex Hormones

Julius Rogoff and George Stewart of Western Reserve University in Cleveland reported in 1927 that dogs from which the outer cortex of the adrenal gland had been removed could be kept alive by repeated intravenous injections of saline extracts of canine adrenal cortex.[1] Frank Hartman at the University of Buffalo, who similarly prolonged the life of adrenalectomised cats by administering aqueous extracts of adrenal glands from cattle, proposed the name 'cortin' for the extracts.[2] These early preparations had only slight and variable activity and hence were unsuitable for clinical application.

In 1930, Hartman and Katherine Brownell developed a potent preparation which was the first successfully used to maintain the life of patients with Addison's disease.[3] Although comatose when treatment began, one of the first patients to be given their extract was revived and lived eight months until contracting pneumonia.[4] Wilbur Swingle and Joseph Pfiffner at Princeton also prepared potent preparations of cortin by initially extracting with ethanol.[5] The residue obtained on evaporation of the ethanol under reduced pressure was then extracted with benzene, thereby greatly reducing contamination with adrenaline. This procedure permitted complete adrenal glands to be utilised, rather than cortices separated from medullae. The extract was given to patients at the Mayo Clinic and, following a successful outcome, Parke, Davis & Company marketed a product based on this preparation.[6] Further improvements in the extraction procedure were introduced, with Marvin Kuizenga of the Upjohn Company developing what became the standard method of purification. In this, a dried acetone extract was dissolved in water and then fatty contaminants were removed with petroleum ether before final extraction of the aqueous solution with ethylene dichloride. This left behind the adrenaline and phospholipids in the aqueous solution, giving the first cortin preparations totally free of adrenaline.[7]

The improved extraction techniques were based upon developments in assay procedures that became more refined, enabling faster progress to be made. In 1927, Carl and Gerty Cori at the State Institute for the Study of Malignant Disease in Buffalo had demonstrated that adrenalectomy depleted carbohydrate stores.[8] Sidney Britton and Herbert Silvette at the University of Virginia then found that the resulting hypoglycaemia could be reversed by injection of adrenal extract, which came to be described as glucocorticoid activity.[9] This led to the development of assays for the hormone based on its ability to restore blood glucose levels to normal. Another assay procedure originated from the recognition by Emil Baumann and Sarah Kurland at the Montefiore Hospital in New York of serious loss of sodium through renal excretion in patients with Addison's disease.[10] They had temporarily prolonged the lives of adrenalectomised cats by administering sodium chloride. Robert Loeb at Columbia University subsequently found that restoration of sodium levels to normal in animals could serve as an assay for adrenocortical extracts and indicate their mineralocorticoid activity.[11] Assays introduced by Everse and De Fremery, who tested Reichstein's extracts,[12] and by Dwight Ingle at the Mayo Clinic in Rochester[13] were based on the profound loss of muscle strength observed in patients with Addison's disease and in adrenalectomised animals.

The breakthrough came in 1936 when a crystalline compound from the adrenal cortex was isolated by Tadeus Reichstein at the Pharmaceutical Institute in the University of Basel.[14]

Drug Discovery. A History. W. Sneader.
©2005 John Wiley & Sons Ltd

Although this lacked cortin-like activity, it did exhibit androgenic activity. When it was discovered to be a steroid, it opened up an entirely new perspective on the field of adrenocortical hormones by raising the possibility that other steroids might be present in the adrenal cortex. Reichstein named this compound 'adrenosterone' in recognition of its steroid nature.

The first pure, crystalline compound to be extracted from adrenal tissue and possessing adrenocortical hormonal activity was corticosterone. It was isolated in 1937 by Reichstein and his colleagues, who found that 1 mg of it had the activity of 50–100 g of the whole gland.[15] As it has only weak anti-inflammatory combined with strong mineralocorticoid activity, corticosterone found no therapeutic application.

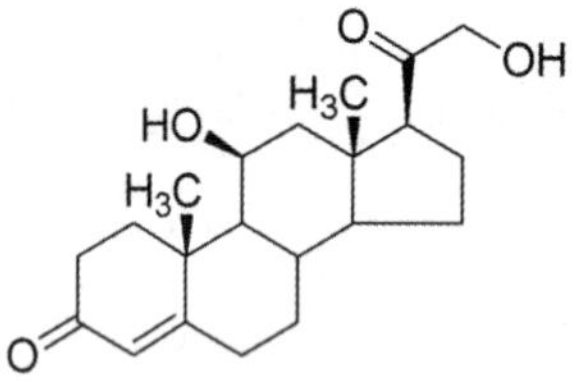

corticosterone

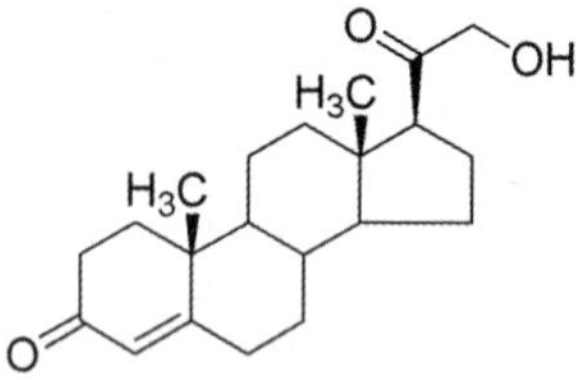

deoxycorticosterone

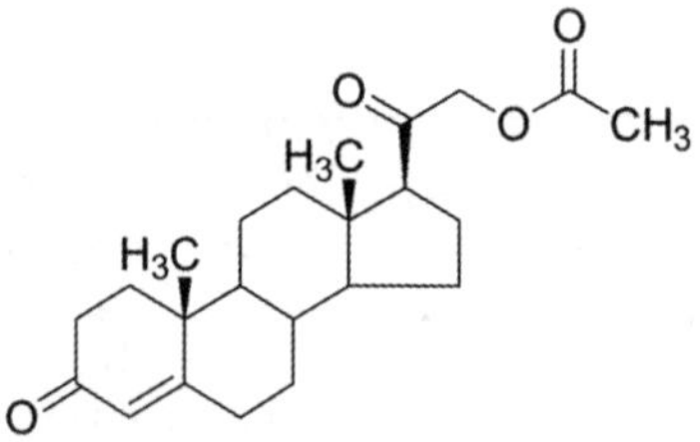

deoxycorticosterone acetate

In 1938, Reichstein isolated deoxycorticosterone from an adrenal extract. Although this was the fifth active compound extracted from adrenals, Reichstein had already synthesised it the previous year in an attempt to establish whether a substituent at the 11-position of the steroid nucleus was required for adrenocortical activity. He prepared it from 3-hydroxy-5-etiocholenic acid, which was readily obtainable by oxidative degradation of cholesterol. The penultimate stage of the synthesis had resulted in the formation of the acetate ester of deoxycorticosterone. Removal of the acetate function in the final synthetic stage resulted in the formation of only half of the amount of deoxycorticosterone that was expected. Although a plentiful supply of deoxycorticosterone was obtained by further improvements in synthetic procedures, much higher yields of its acetate were consistently obtained. Most conveniently, the acetate was to prove ideal for formulation in an oily injection, being broken down in the plasma (through esterase hydrolysis) to release free deoxycorticosterone.

The synthesis of deoxycorticosterone acetate produced sufficient quantities for it to become the first adrenocortical steroid available for clinical use – despite being inactive by mouth because of hepatic metabolism after absorption from the gut. Ciba formulated it as an oily injection in 1939, for use as a mineralocorticoid in patients with Addison's disease. It was administered in conjunction with adrenocortical extract, this combination being considered superior to the extract on its own.

THE INTRODUCTION OF CORTISONE

In 1941, American intelligence agents picked up a rumour that Germany was purchasing vast quantities of adrenal glands from slaughterhouses in Argentina, purportedly for the production of extracts that would enable Luftwaffe pilots to withstand the stress of flight at high altitudes! Despite remarkable progress in the isolation and structural elucidation of corticosteroids, insufficient quantities had been obtained for any of them to be used therapeutically, other than deoxycorticosterone acetate. For this reason, Kendall and other leading steroid chemists in the United States were invited by the Office of Scientific Research and Development to collaborate with the pharmaceutical industry in the synthesis of an active corticosteroid for the US Air Force to investigate. The one selected was 11-dehydrocorticosterone, which had been isolated in 1936 by Kendall (as his 'Compound A') and found to be active.[17] Kendall contacted Merck & Company in Rahway, New Jersey, and, even though military interest had dried up when it was revealed that the rumours from Argentina were without foundation, the collaboration between Merck and Kendall continued even after the war was over.

The synthesis of an 11-oxygenated corticosteroid presented a formidable challenge since no synthetic precursor containing an oxygen at the 11-position was available. The American researchers had been making slow progress until, in 1943, Reichstein reported his synthesis of 11-dehydrocorticosterone from deoxycholic acid. This required 40 stages to remove both the acidic side chain and the 12-oxygen prior to introducing an 11-oxygen and then the desired side chain at the 17-position, followed by introducing unsaturation in ring A.[18] This synthesis was quite unsuitable for commercial production and its complexity discouraged most researchers from proceeding further. Kendall, to his great credit, was undeterred. His perseverance paid off when, collaborating with Lewis Sarett from Merck, he dramatically reduced the complexity of the Reichstein synthesis, allowing Sarett to prepare 5 g of 11-dehydrocorticosterone for clinical trials.[19] The results of these were disappointing; the compound had no clinical advantage over the much less expensive deoxycorticosterone acetate.

Despite this setback, Merck called a meeting of its clinical consultants. It was agreed that an attempt should be made to synthesise a few grams of Kendall's 'Compound E', 17-hydroxy-11-dehydrocorticosterone (now known as 'cortisone'), of which Sarett had already obtained 1–2 mg from a 37-step synthesis. Compound E had been isolated in 1936 and found to be active in Ingle's test on muscular strength.[17]

The insertion of a hydroxyl group in the 17-position proved a difficult task, but Sarett eventually managed to prepare 5 g of Compound E in 1948.[20] Merck & Company distributed 300 samples of Compound E for clinical trial that same year. They were subsequently approached by a colleague of Kendall, Philip Hench, who was head of the Department of Arthritis at the Mayo Clinic. Seven years before, he had told Kendall that when patients with rheumatoid arthritis contracted jaundice, their painful symptoms were markedly diminished for a time. A similar state of affairs often occurred when women with arthritis became pregnant, their pains returning after childbirth. Believing these and other similar observations to be instances of stress inducing the release of a protective hormone, Hench had suggested to Kendall that once adequate supplies of a corticosteroid hormone became available, it should be tried in rheumatoid arthritis.

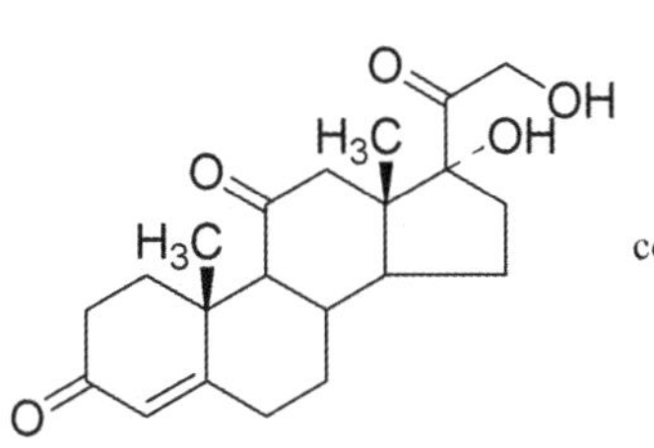

cortisone (Kendall's Compound E)

Hench was given 1 g of Compound E by Sarett, all that had remained of the 5 g that had been synthesised. On the 21 September 1948, he injected 100 mg into a young woman who was desperately ill with arthritis and whose condition was giving rise to great concern. She showed no signs of improvement until after her third daily injection, when the response was truly dramatic. She awoke the following morning to find herself totally free of pain on moving, something she had not experienced for over five years! After a further week of treatment she was able to walk out of the hospital unaided. An urgent request for more Compound E was then sent to Merck. The company responded magnificently. Improvements to the synthesis allowed the production of 1000 g within only a few weeks. As a result, Hench was able to proceed with a limited trial of the hormone in 13 arthritic patients for six months. Once again, remarkable results were achieved, but only so long as the injections continued.[21] It was evident that Compound E did not cure rheumatoid arthritis. It could only suppress the distressing symptoms, but for patients who did not respond to any other therapy this in itself seemed miraculous. Compound E was now renamed as 'cortisone'. In 1950, the Nobel Prize in Physiology or Medicine was jointly awarded to Kendall, Reichstein and Hench for their work on it.

The press seized on this dramatic new treatment for arthritis with zeal. Some newspapers hastily described the drug as 'vitamin E', thereby creating an undeserved (and persistent) reputation for the vitamin as a miracle drug! To avoid further confusion, Compound E was given the approved name of cortisone. When the success of cortisone therapy became known, there was a worldwide cortisone famine, which the pharmaceutical industry was hard pressed to overcome. To meet the demand, Merck scaled up Sarett's process and by the end of 1949 were able to offer limited amounts of cortisone acetate to physicians at a price of $200/g. A year later, they had produced 1000 kg and reduced the price to $35/g. Within 5 years, the range of clinical applications for cortisone was to embrace over 50 different conditions, including asthma, allergies and skin diseases. The initial overenthusiasm for the drug had, by then, been tempered somewhat as clinicians experienced at first hand the problems resulting from its mineralocorticoid activity, which seriously upset mineral and fluid balance. Nevertheless, it took many years before it was generally accepted that, because of its side effects, systemic therapy with cortisone and related compounds should be reserved for severely ill patients and those who required hormone replacement therapy in Addison's disease or after pituitary surgery.

Hydrocortisone

Reichstein isolated 17β-hydroxycorticosterone (as 'Substance M') in 1937,[22] with Kendall isolating it (as 'Compound F') soon afterwards.[23] It was found to possess both glucocorticoid and mineralocorticoid activity, later being shown to be the active form of cortisone in the body and interconvertible with cortisone.

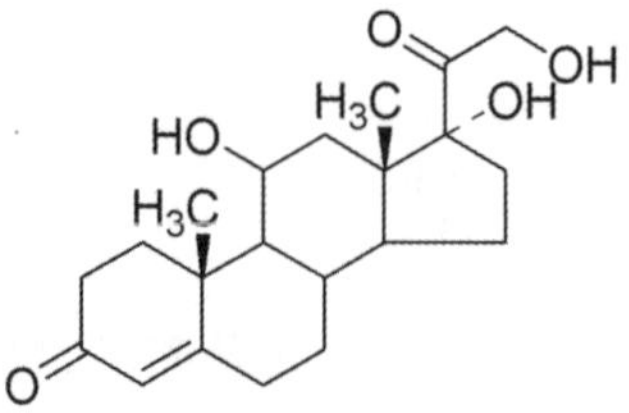

hydrocortisone

Hydrocortisone became available in amounts adequate for clinical study in 1949 after N.L. Wendler and Max Tishler developed a practical synthesis for Merck & Company.[24] Before

Merck & Company was able to produce it on a commercial scale, G.D. Searle & Company set up a novel process that had been developed by Gregory Pincus and Oscar Hechter at the Worcester Foundation for Experimental Biology. This consisted of hundreds of perfusion cells, each holding cattle adrenal glands immersed in a serum-like solution into which was pumped cheap biochemical precursors, from which the glands synthesised 17β-hydroxycorticosterone. Searle ignored the costs involved and distributed clinical supplies without charge. The corticosteroid then received the approved name of hydrocortisone. Meanwhile, Upjohn put more than 150 chemists to work on its production. They succeeded by using a mould known as *Rhizopus nigricans* to convert cheaply available progesterone to 11-hydroxyprogesterone which, in turn, was readily made into hydrocortisone.[25]

ALDOSTERONE

Swiss and American chemists realised that biologically active material remained in the solvent mother liquors after the 30 or so known corticosteroids had been recovered from adrenal extracts. Yet it was James Tait, a newly appointed lecturer in medical physics at the Middlesex Hospital in London, who made the next advance. In 1948, Tait and Sylvia Simpson tested the effect of Allen and Hanbury's Eucortone®, a brand of adrenal extract, on mineral and water balance. It was known that this was disturbed in patients receiving therapy with this extract. Although Tait and Simpson had no grant to support their study, they did have access to radioisotopes of sodium and potassium. This meant that they could study mineral balance after administering Eucortone® to animals from which the adrenals had been removed. By applying the recently developed technique of paper chromatography, they were able to separate fractions from Eucortone® that were biologically active, but it still took them three years to finally obtain a new hormone that had an unprecedented ability to promote salt retention.[26] They gave this the name 'electrocortone', but were persuaded to rename it as 'aldosterone' after establishing its full chemical structure in collaboration with Reichstein and Wettstein.[27]

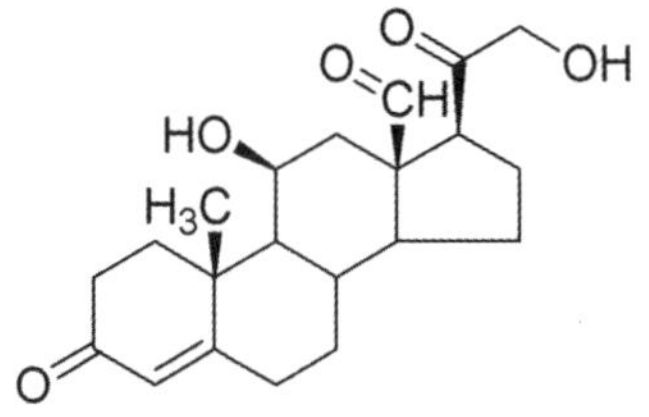

aldosterone

Aldosterone is secreted by the adrenal cortex in response to raised circulating levels of angiotensin II. It acts on the distal tubules of the kidney, stimulating reabsorption of sodium and water. There is an associated loss of potassium. Aldosterone is a major causative factor in certain types of oedema.

REFERENCES

1. G.N. Stewart, J.M. Rogoff, The influence of extracts of adrenal cortex on the survival period of adrenalectomized dogs and cats. *Am. J. Physiol.*, 1929; **91**: 254–64.
2. F.A. Hartman, C.J. MacArthur, W.E. Hartman, Substance which prolongs life of adrenalectomized cats. *Proc. Soc. Exp. Biol. Med.*, 1927; **25**: 69–70.
3. F.A. Hartman and K. A. Brownell, The hormone of the adrenal cortex. *Science*, 1930; **72**: 76.
4. F.A. Hartman, A.H. Aaron, J.E. Culp, The use of cortin in Addison's disease. *Endocrinology*, 1930; **14**: 438–42.

5. W.W. Swingle, J.J. Pfiffner, The revival of comatose adrenalectomized cats with an extract of the suprarenal cortex. *Science*, 1930; **72**: 75–6.

6. L.G. Rowntree, P.H. Greene, W.W. Swingle, J.J. Pfiffner, Addison's disease. *J. Am. Med. Ass.*, 1931; **96**: 231–5.

7. G.F. Cartland, M.H. Kuizenga, The preparation of extracts containing the adrenal cortex hormone. *J. Biol. Chem.*, 1936; **116**: 57–64.

8. C.F. Cori, G.T. Cori, The fate of sugar in the animal body. VII. The carbohydrate metabolism of adrenalectomized rats and mice. *J. Biol. Chem.*, 1927; **74**: 473–94.

9. S.W. Britton, H. Silvette, The effects of cortico-adrenal extract on carbohydrate metabolism in normal animals. *Am. J. Physiol.*, 1932; **100**: 693–700.

10. E. Baumann, S. Kurland, Changes in the inorganic constituents of blood in suprarenalectomised cats and rabbits. *J. Biol. Chem.*, 1927; **71**: 281–302.

11. R.F. Loeb, D.W. Atchley, E.M. Benedict, J. Leland, Electrolyte balance studies in adrenalectomized dogs with particular reference to the excretion of sodium. *J. Exp. Med.*, 1933; **57**: 775–92.

12. J.W.R. Everse, P. De Fremery, On a method of measuring fatigue in rats and its application for the testing of the suprarenal cortical hormone (cortin). *Acta Brev. Neerland. Physiol. Pharmacol. Microbiol.*, 1932; **2**: 152–4.

13. D.J. Ingle, Work capacity of adrenalectomized rat treated with cortin. *Am. J. Physiol.*, 1936; **116**: 622.

14. T. Reichstein, Über Cortin, das Hormon der nebbennieren Rinde (X). I. Mitteilung. *Helv. Chim. Acta*, 1936; **19**: 29–63.

15. P. De Fremery, E. Laqueur, T. Reichstein, R.W. Spanhoff, I.E. Uyldert, Corticosterone, a crystallized compound with the biological activity of the adrenocortical hormone. *Nature*, 1937; **139**: 26.

16. T. Reichstein, J. von Euw, Ueber Bestandtiele der nebennierinde Isolierung der Substanzen Q (Desoxycorticosterone) and R sowie weitere Stoffe. *Helv. Chim. Acta*, 1938; **21**: 1197.

17. H.L. Mason, C.S. Myers, E.C. Kendall, The chemistry of crystalline substances isolated from the suprarenal gland. *J. Biol. Chem.*, 1936; **114**: 613–31.

18. A. Lardon, T. Reichstein, Über Bestandteile der Nebennierenrinde und ver-wandte Stoffe. Teilsynthese des II-Dehydro-corticosterons. *Helv. Chim. Acta*, 1943; **26**: 747–55.

19. L.H. Sarett, Partial synthesis of pregnene-4-triol-17(β),20(β),21-dione-3,11 and pregnene-4-diol-17(β),21-trione-3,11,20. *J. Biol. Chem.*, 1946; **162**: 601-31.

20. L. Sarett, A new method for the preparation of 17(α)-hydroxy-20 ketopregnanes. *J. Am. Chem. Soc.*, 1948; **70**: 1454–8.

21. P.S. Hench, E.C. Kendall, C.H. Slocumb, H.F. Polley, The effect of a hormone of the adrenal cortex (17-hydroxy-11-dehydrocorticosterone: compound E) and of pituitary adrenocorticotrophic hormone on rheumatoid arthritis. *Ann. Rheum. Dis.*, 1949; **8**: 97–104.

22. T. Reichstein, Über Bestandtelle der Nebennieren-Rinde (X). Zur Kenntnis des Cortico-sterons. *Helv. Chim. Acta*, 1937; **20**: 953–69.

23. L. Mason, W.M. Hoehn, E.C. Kendall, Chemical studies of the suprarenal cortex. IV. Structures of compounds C, D, E, F, and G. *J. Biol. Chem.*, 1938; **124**: 459–74.

24. N.L. Wendler, R.P. Graber, R.E. Jones, M. Tishler, Synthesis of 11-hydroxylated cortical steroids; 17α-hydroxycorticosterone. *J. Am. Chem. Soc.*, 1950; **72**: 5793–4.

25. D.H. Peterson, H.C. Murray, Microbiological oxygenation of steroids at carbon ll. *J. Am. Chem. Soc.*, 1952; **74**: 1871–2.

26. S.A. Simpson, J.F. Tait, A. Wettstein, *et al.*, Isolierung eines neuen kristallisierten Hormons aus Nebennerien mit besonders hoher Wirksamkeit auf den Mineralsoffwechsel. *Experientia* 1953; **9**: 333–5.

27. S.A. Simpson, J.F. Tait, A. Wettstein, *et al.*, Aldosteron. Isolierung und Eigenschaften. Über Bestandteile der Nebennierenrinde und verwandte Stoffe. *Helv. Chim. Acta*, 1954; **37**: 1163.

Prostaglandins

In 1930, Raphael Kurzrok and Charles Lieb of the Department of Obstetrics and Gynecology at Columbia University in New York observed that the uteri of women undergoing artificial insemination sometimes contracted violently and on other occasions relaxed.[1] Five years later, Ulf von Euler at the Karolinska Institute in Stockholm detected an acid in extracts of monkey, sheep and goat seminal vesicles that lowered blood pressure and caused smooth muscle to contract.[2] He named it 'prostaglandin', but no further work was carried out until one of his students, Sune Bergström, began purifying the extract using a Craig countercurrent extraction apparatus. He was then able to establish that unsaturated fatty acids free of nitrogen were present. After Bergström had obtained a $100 000 grant from the Upjohn Company in the mid-1950s, he and his colleagues began using combined gas chromatographic mass spectrum analysis to isolate small amounts of several different highly potent prostaglandins and then to elucidate their structures. The prostaglandins were subsequently classified as types A to F according to the pattern of functions in the cyclopentane ring, with each type being given a subscript that indicated the number of unsaturated centres in the side chains. A further refinement was to add the suffix α or β to indicate the stereochemical configuration of ring substituents.

In 1957, Bergström isolated crystals of alprostadil (prostaglandin E_1) from sheep prostate glands.[3] Within five years, the structures of alprostadil, dinoprost (prostaglandin $F_{2\alpha}$) and dinoprostone (prostaglandin E_2) had been determined.[4] Dinoprostone is one of the most commonly occurring and most potent of the mammalian prostaglandins, but it and the other prostaglandins of the E series proved to be chemically unstable because of the presence in the cyclopentane ring of a β-hydroxy carbonyl, which readily underwent an acid or base catalysed elimination reaction.

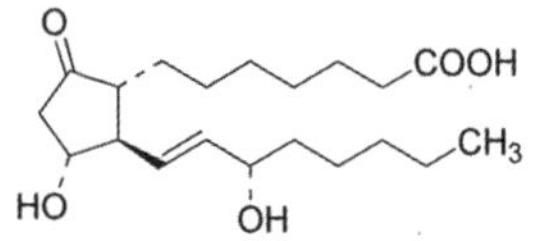

alprostadil (prostaglandin E_1)

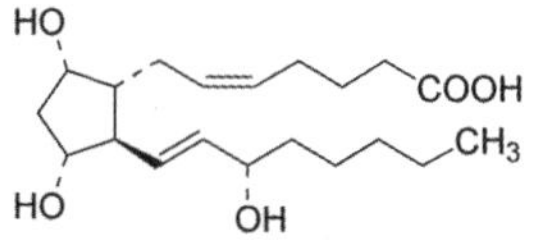

dinoprost (prostaglandin $F_{2\alpha}$)

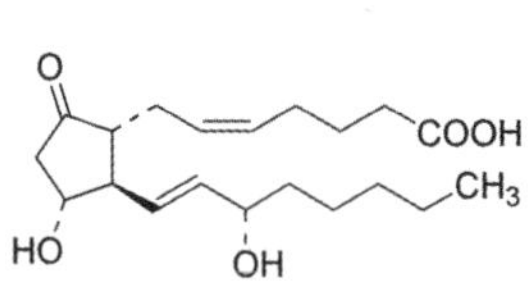

dinoprostone (prostaglandin E_2)

Phillip Beal and his colleagues at the Upjohn laboratories achieved the first synthesis of a natural prostaglandin in 1965.[5] They synthesised alprostadil four years later.[6] When Elias Corey synthesised dinoprostone at Harvard in 1970, he also succeeded in reducing its cyclic carbonyl group to obtain dinoprost.[7] The availability of synthetic prostaglandins allowed clinical investigations to be considered for the first time and the Upjohn Company then supplied prostaglandins free of charge to researchers with the expectation of significant therapeutic advances. However, only three of the sixteen naturally occurring prostaglandins have proved to be of any clinical value.

Hope that prostaglandins could be used in the treatment of peptic ulcer was aroused by a report in 1967 that some of the E series inhibited gastric acid secretion.[8] This was dashed when it was found that they were inactive by mouth and had only a very brief duration of action when injected. This was due primarily to metabolic oxidation of the 15-hydroxyl group in the side chain. Furthermore, like many other pharmacologically active molecules occurring in the body, the prostaglandins of the E series lacked specificity of action and so produced unwanted effects such as vasodilation, which caused facial flushing, headache and hypotension. The vasodilating properties of alprostadil have been exploited to dilate the ductus arteriosus in neonates with congenital heart disease.

Because of its ability to induce contraction of the uterus, intravenous infusion of dinoprostone was found to be an effective way of inducing labour.[9] It is now given by the vaginal route, either as pessaries, tablets or gels.[10] Dinoprostone was also licensed in 1972 for the induction of abortion after the first trimester of pregnancy.[11]

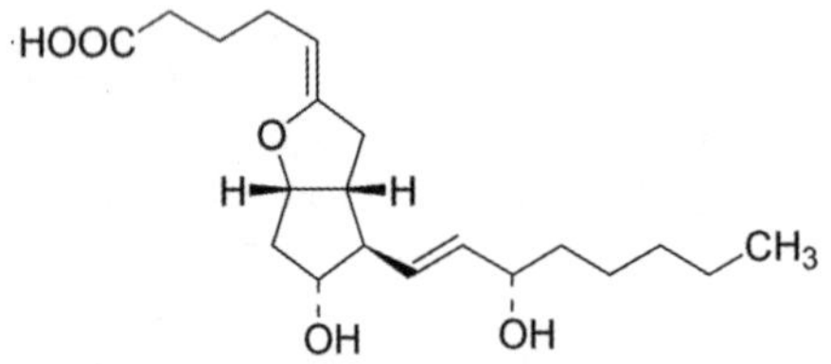

epoprostenol (prostacyclin)

Epoprostenol (prostacyclin) is a prostanoid (i.e. prostaglandin-like substance) that was isolated from microsomes of pig and rabbit aorta in 1976 by John Vane[12] of Burroughs Wellcome. It occurs in the walls of blood vessels, where it produces vasodilation and prevents clotting. Like the prostaglandins it is rapidly metabolised, half of it being destroyed within three minutes of injection.

In 1944, the artificial kidney was invented in Holland, the use of which only became possible as a result of the introduction of pure heparin to stop clotting of the blood. The later development of heart–lung bypass techniques in surgery also depended on the availability of heparin. However, in patients exposed to a high risk of bleeding, the use of heparin in such procedures may be hazardous. In those situations, epoprostenol may serve as an alternative since its effects wear off quickly when it is withdrawn.

REFERENCES

1. R. Kurzrok, C.C. Lieb, Biochemical studies of human semen. II. The action of semen on the human uterus. *Proc. Soc. Exp. Biol. Med.*, 1930; **28**: 268–72.
2. U.S. von Euler, Über die spezifische blutdrucksendende Substanz des menschlichen Prostata- und Samenblasensekretes. *Klin. Wochenschr.*, 1935; **14**, 1182–3.
3. S. Bergström, J. Sjövall, The isolation of prostaglandin. *Acta Chem. Scand.*, 1957; **11**: 1086.
4. S. Bergström, R. Rhyage, B. Samuelsson, J. Sjövall, The structure of prostaglandin E, F₁ and F₂. *Acta Chem. Scand.*, 1962; **16**: 501–2.
5. P.F. Beal III, J.C. Babcock, F.H. Lincoln, A total synthesis of a natural prostaglandin. *J. Am. Chem. Soc.*, 1966; **88**: 3131–3.
6. W.P. Schneider, U. Axen, F.H. Licoln, J.E. Pike, J. Thompson, Synthesis of prostaglandin E1 and related substances. *J. Am. Chem. Soc.*, 1969; **91**: 5372–8.

7. E.J. Corey, T.K. Schaaf, W. Huber, *et al.*, Total synthesis of prostaglandins F2α and E2 as the naturally occurring forms. *J. Am. Chem. Soc.*, 1970; **92**: 397–8.
8. A. Robert, J.E. Nezamis, J.P. Phillips, Inhibition of gastric secretion by prostaglandins. *Am. J. Digest. Dis.*, 1967; **12**: 1073–6.
9. M.P. Embrey, The effect of prostaglandins on the human pregnant uterus. *J. Obstet. Gynaecol.*, 1969; **76**: 783–9.
10. G.C. Liggins, Controlled trial of induction of labor by vaginal suppositories containing prostaglandin E2. *Prostaglandins*, 1979; **18**: 167–72.
11. I. Craft, Intra-amniotic urea and prostaglandin E2 for abortion. A clinical study to determine the efficacy of using a variable prostaglandin dosage. *Prostaglandins*, 1973; **4**: 755–63.
12. S. Moncada, R. Gryglewski, S. Bunting, J.R. Vane, An enzyme isolated from arteries transforms prostaglandin endoperoxides to an unstable substance that inhibits platelet aggregation. *Nature*, 1976; **263**: 663–5.

Hormone Analogues

The introduction of structurally related analogues of the alkaloids and other plant products cleared the way for similar developments in the field of hormone analogues. However, it is mainly during the last half century that important therapeutic advances have been made. Among these are several drugs that have saved millions of lives and have made their manufacturers undreamt of profits. The same cannot be said of plant product analogues. In part, this is due to the need for patients to take some hormone-derived drugs on a continuing basis. It is also a reflection of the poisonous nature of the plant prototypes from which the analogues were derived.

ADRENALINE ANALOGUES

The chemical and physiological similarities between tyramine and adrenaline persuaded George Barger and Henry Dale to investigate a variety of amines of similar structure.[1] Many were found to produce physiological effects similar to those observed on stimulation of the sympathetic nervous system. For this reason, Dale described them as 'sympathomimetic amines'. None of these compounds appeared to have any significant clinical advantage over adrenaline.

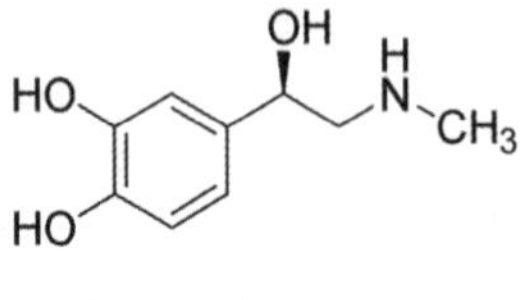

adrenaline

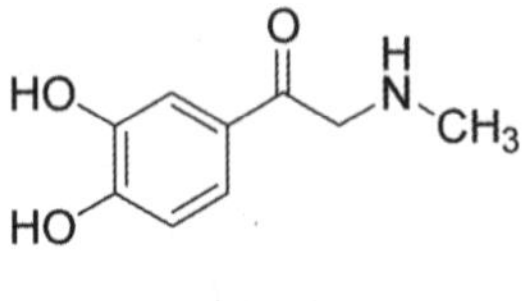

adrenalone

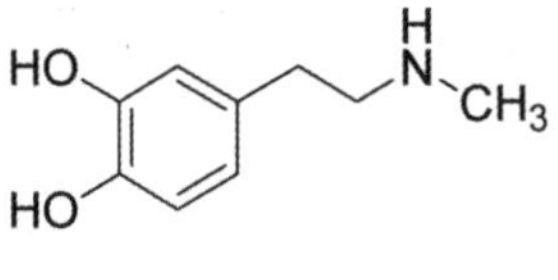

deoxyepinephrine

The only adrenaline analogues used before the First World War were inconsequential, namely adrenalone, deoxyepinephrine and adrenalone. The last of these was prepared by Friedrich Stolz as an intermediate in the synthesis of adrenaline and marketed by Hoechst as a topical haemostatic agent and nasal decongestant.[2] Deoxyepinephrine was synthesised by Frank Pyman.[3] Now obsolete, it was marketed for many years by Burroughs Wellcome as a vasoconstrictor and pressor agent. Although having only about one-tenth the potency of adrenaline, it had a longer duration of action.

Drug Discovery. A History. W. Sneader.
©2005 John Wiley & Sons Ltd

Isoprenaline (Isoproterenol)

The introduction of ephedrine, which had a structural resemblance to adrenaline, stimulated interest in the clinical potential of adrenaline analogues, particularly in Germany. In 1927, the Berlin chemist Helmut Legerlotz synthesised both oxedrine (synephrine) and phenylephrine as analogues of adrenaline in which one hydroxyl had been removed from the catechol ring. The former was licensed to C.H. Boehringer of Ingelheim, Germany, who introduced it in 1930 as a vasoconstrictor,[4] while the latter was marketed by Frederick Stearns & Company of Detroit as a decongestant.[5] It was injected to maintain blood pressure during anaesthesia as it was less likely than adrenaline to reduce the perfusion of blood into vital organs.

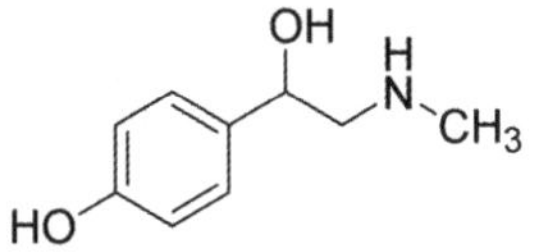

oxedrine

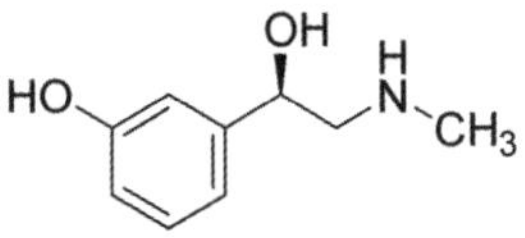

phenylephrine

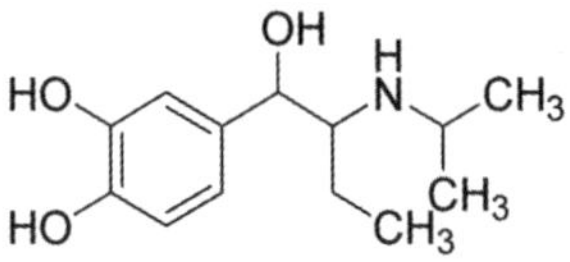

isoetharine

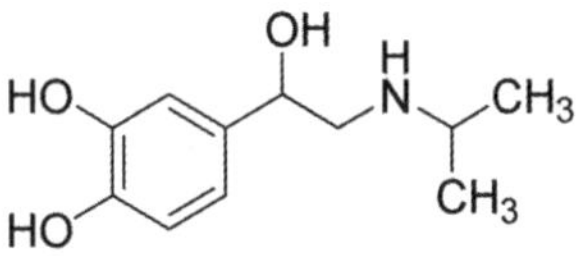

isoprenaline

Isoetharine was developed by Max Bockmühl, Gustav Erhart and Leonhard Stein at the Höchst laboratories of I.G. Farbenindustrie in 1934.[6] The patent claimed it was a bronchodilator with reduced effects on blood pressure and decreased toxicity compared to adrenaline. Many years later, once the selective action of isoprenaline on beta-adrenoceptors had been recognised, it was realised that isoetharine had similar selectivity.[7] Next, Heribert Konzett at the University of Vienna examined analogues of adrenaline in which the methyl group attached to its nitrogen was replaced with either an ethyl, n-propyl, isopropyl or butyl substituent.[8] As was the case with isoetharine, the isopropyl analogue was the most promising compound, being ten times as potent a bronchodilator as epinephrine when injected but free of its hypertensive effects. Boehringer marketed it in 1941 under the name Aludrin®, but little attention was given to it due to the wartime situation. It was not until after the US State Department investigated the research conducted during the war by German companies that the clinical merit of the drug was fully appreciated.[9] It was given the approved name of isoprenaline in 1951 (isoproterenol in the United States). Being a catecholamine, it did not cause insomnia as did the more lipid-soluble ephedrine, which penetrated into the brain. For the next 20 years it was the drug of choice for the relief of asthma attacks. Tragically, its introduction in convenient aerosol form during the 1960s may have led to an estimated 3000 deaths among asthmatic teenagers in the United Kingdom alone. If so, it was probably due to the effects on the heart of accidental overdosing.

In the course of examining drugs on smooth muscle to see if they could relieve menstrual cramps, Raymond Ahlquist of the Medical College of Georgia noticed that the effects of isoprenaline and noradrenaline differed. The latter made smooth muscle contract while isoprenaline caused it to relax. Both activities were shown by adrenaline, but varied with the site of action. Ahlquist recognised a parallel with cholinergic receptors, which had been classified in 1914 by Dale as being either nicotinic or muscarinic on the basis of whether nicotine or muscarine stimulated a response. Ahlquist explained the different actions of adrenaline analogues by introducing the concept of two types of adrenoceptors, namely alpha-receptors and beta-receptors.[10] Noradrenaline was an alpha-receptor agonist whereas isoprenaline was a beta-receptor agonist, while adrenaline was a mixed alpha- and beta-receptor agonist. Ahlquist's contribution ultimately transformed the understanding of the action of sympathomimetic amines and cleared the way for major therapeutic developments in this field, yet for many years his claims were universally ignored.

Orciprenaline

The success of isoprenaline served to highlight one of its disadvantages, even before its cardiac toxicity became a matter of concern. Because isoprenaline was a catecholamine like adrenaline, it was susceptible to enzymes in the gut wall and throughout the body that rapidly deactivated catecholamines. As a result, isoprenaline was inactive by mouth and too short-acting to be employed in the same manner as ephedrine for the prophylaxis of asthma.

Emil Eidebenz and Maxmilian Depner of Chemische Werke Albert in Wiesbaden-Biebrich applied for a patent in 1943 for adrenaline analogues in which one of the two hydroxyl groups had been moved to an adjacent position on the benzene ring. The patent was finally granted ten years later.[11] The compounds, including the orcinol isomer of adrenaline, had similar activity to adrenaline but were more stable since they were no longer catecholamines. This approach was not exploited clinically until the next development came in 1961 from the laboratories of the Boehringer subsidiary Dr Karl Thomae GmbH, where Otto Thomae and Karl Zelle introduced orciprenaline (metaproterenol), which incorporated the ring system previously used by Eidebenz and Depner.[12] Although this produced only a slight change in molecular topography, it conferred resistance to enzymic destruction since the catechol system had been replaced by one which, while similar enough to act on the catecholamine receptors in the bronchi, was now unable to fit on to the active site of the enzymes that deactivated isoprenaline and all other catecholamines, viz. sulfokinase in the gut wall and catechol O-methyltransferase elsewhere.[13] Furthermore, as the hydroxyl substituents still remained on the ring, the drug was now too polar to enter the brain, hence patients did not experience the problem of insomnia found with ephedrine. Orciprenaline was marketed by Boehringer, as was fenoterol, an analogue prepared the following year.[14] The selective action of fenoterol on the beta$_2$-adrenoceptors of the lung was not recognised until after the revelation of similar selectivity in salbutamol. Its degree of selectivity was not as great as with salbutamol or terbutaline.

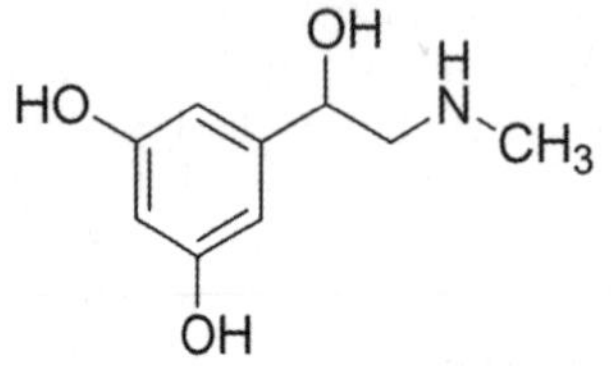

orcinol isomer of adrenaline

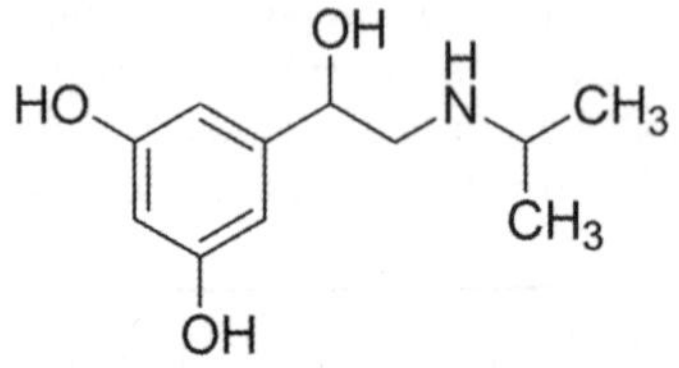

orciprenaline

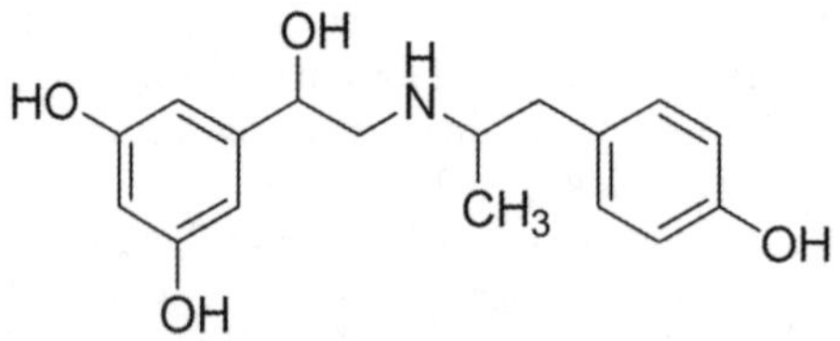

fenoterol

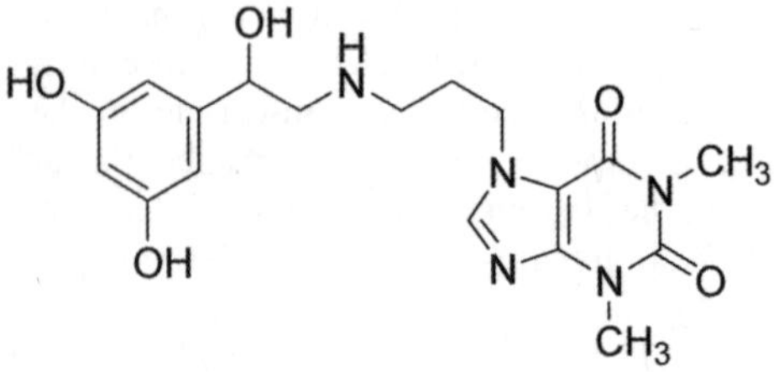

reproterol

Reproterol[15] was designed to combine the actions of orciprenaline and theophylline on the lung. Again, a selective action on the beta$_2$-adrenoceptors in the lung was not recognised until after the discovery of such selectivity in salbutamol. Reproterol turned out to have similar pharmacological properties to salbutamol and terbutaline.

Terbutaline

Terbutaline is an analogue of orciprenaline developed in 1966 by Kjell Wetterlin and Leif Svensson of the Draco division of Astra in Lund, Sweden. They replaced the isopropyl group in orciprenaline with a tertiary butyl group.[16] This both enhanced potency and introduced a high degree of selectivity for the beta$_2$-adrenoceptors in the lung, which was comparable with that of salbutamol (see below).[17] This meant that the heart rate was no longer increased as it had been with earlier bronchodilators. It has been the principal commercial rival to salbutamol, with little to distinguish between the two drugs in the clinic.

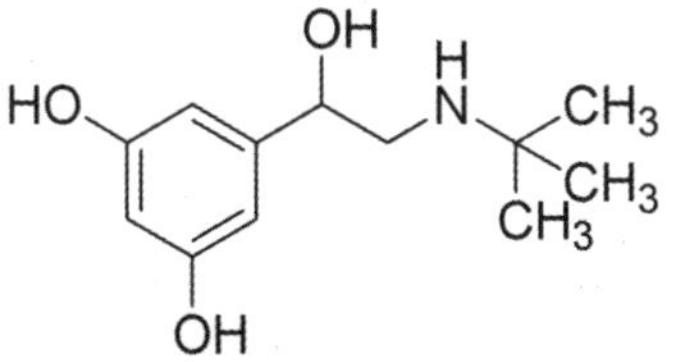

terbutaline

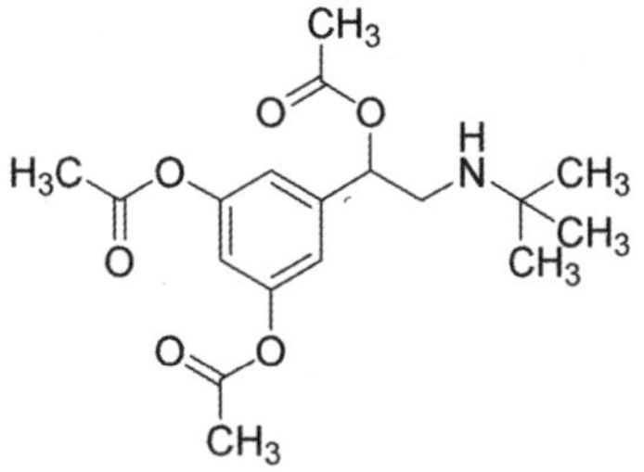

bambuterol

Bambuterol[18] was developed as a long-acting prodrug of terbutaline. Plasma cholinesterase and oxidative enzymes promote the slow hydrolysis of the pharmacologically inert bambuterol into terbutaline, making it suitable for once-daily, oral dosage in the prophylaxis of asthma attacks.

Salbutamol (Albuterol)

An alternative approach to changing the position of one of the catechol hydroxy groups (as in orciprenaline) was taken in 1964 when Larsen and Lish of the Mead Johnson Research Center in Evansville, Indiana, replaced the hydroxyl group in the 3-position of the benzene ring by the chemically similar methanesulfonamide group to form soterenol. This was an effective bronchodilator.[19]

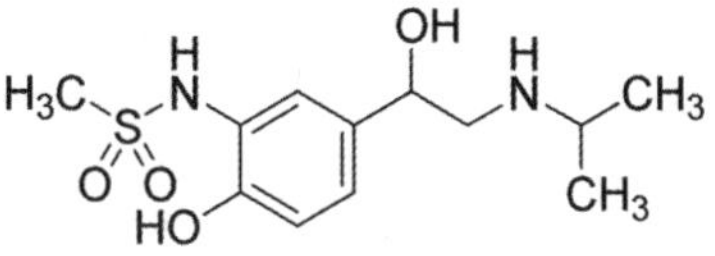

soterenol

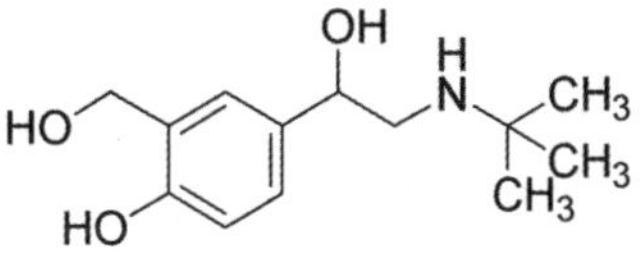

salbutamol

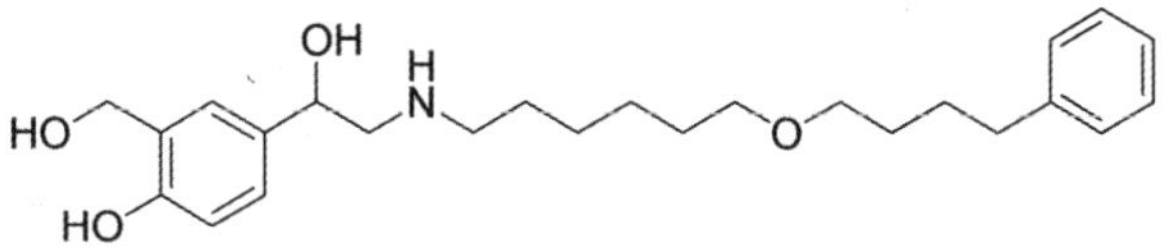

salmeterol

David Hartley, David Jack, Lawrence Luntz and Alexander Ritchie of Allen & Hanbury's in Ware (a division of Glaxo) developed salbutamol in 1967.[20] The hydroxyl group of isoprenaline was deliberately replaced with a hydroxymethyl group to obtain resistance to destruction by catechol O-methyltransferase. As anticipated, the duration of action was lengthened by this manoeuvre, while potency remained high – unlike with orciprenaline.

However, a second modification consisted of the introduction of a tertiary butyl group into the molecule in place of the isopropyl group attached to the side chain nitrogen atom. This unexpectedly increased affinity for the beta$_2$-adrenoceptors in the lung and reduced that for beta$_1$-adrenoceptors in the heart. So marked was this selectivity in salbutamol that within only a few years it had displaced isoprenaline as the drug of choice for the control of asthmatic attacks.

Glaxo researchers later sought a longer-acting analogue of salbutamol that could confer protection against obstruction of the airways throughout the night so that the sleep of patients was not disturbed. They achieved this by attaching a non-polar chain on to the amine nitrogen, which caused the drug to partition mainly into the lipid bilayer of the bronchial cell membrane, with only a small amount remaining in the aqueous biophase. The lipid bilayer acted as a depot from which the drug was able to slowly diffuse out to produce continuing stimulation of the beta$_2$-adrenoceptor. Salmeterol turned out to be not only longer acting but also 5–10 times as potent as salbutamol, permitting twice-daily inhalation to prevent bronchospasm.[21] It was introduced clinically in 1988.

ADRENALINE ANTAGONISTS

The first drug to antagonise the effects of adrenaline on the heart was serendipitously discovered by Irwin Slater of Eli Lilly in Indianapolis while testing analogues of isoprenaline as potential long-acting bronchodilators. These were screened for their ability to relax tracheal strips that had been contracted by exposure to pilocarpine in order to simulate the bronchoconstriction of asthma. Between test runs, the strips were treated with adrenaline to ensure they were still responsive. However, those that had been exposed to dichloroisoprenaline did not relax when the adrenaline was added. This antagonism of adrenaline by dichloroisoprenaline was reported by Slater at a scientific meeting in 1957.[22] Neil Moran, at Emory University in Atlanta, then requested samples of the new drug to investigate its effects on the heart. He found that while dichloroisoprenaline antagonised the changes in heart rate and muscle tension produced by adrenaline, it still mimicked the activity of the hormone to some extent.[23]

Moran's report on the cardiac effects of dichloroisoprenaline interested James Black, who had been invited to join ICI Pharmaceuticals Division (now incorporated in AstraZeneca) at Alderley Park in Cheshire after seeking a grant from the company to further his investigations into coronary artery disease. Black believed that an alternative way of treating angina would be to find a drug that reduced the need of the heart for oxygen rather than merely increasing its supply through the use of vasodilators such as the nitrates. He realised that the heart rate was a major factor influencing the demand of the heart for oxygen, so had been seeking a drug to protect the hearts of patients with coronary disease against the effects of adrenaline and noradrenaline. On reading Moran's paper, he realised it should be possible to find an analogue of dichloroisoprenaline devoid of intrinsic action when bound to the receptors in the heart (which belonged to the type described by Raymond Ahlquist as beta-adrenoceptors).[24]

John Stephenson synthesised the first effective beta-adrenoceptor blocker for Black in February 1960, by replacing the bulky chlorine atoms of dichloroisoprenaline with a second benzene ring to form pronethalol.[25] The drug was active when given by mouth. A clinical trial on 30 patients confirmed pronethalol to be effective in angina, although mild side effects were noted. It was also somewhat short-acting. The therapeutic value, however, was not limited to the treatment of angina. Black's belief, that it might prevent atrial fibrillation and atrial or ventricular tachycardias through diminution of the response to emotional or exercise-induced sympathomimetic activity, was confirmed in the clinic.[26] What had not been anticipated, however, was the discovery that the drug had a marked hypotensive effect in anginal patients who had been taking it for several months. A subsequent report confirmed the value of pronethalol in reducing blood pressure in hypertensive patients.[27]

Shortly after, long-term toxicity testing in mice revealed that pronethalol could cause cancer of the thymus gland.[28] Consequently, when the drug was marketed late in 1963, its use was limited to patients whose lives were seriously at risk. As a result, pronethalol was quickly superseded by another ICI product, propranolol. This, as well as all the other beta-blockers developed since then, retained the anti-anginal, anti-arrhythmic and antihypertensive properties of pronethalol.

Propranolol was synthesised in order to determine the effect of increasing the distance between the alcoholic hydroxyl group on the side chain and the aromatic ring of pronethalol. When it was found to be non-carcinogenic as well as ten times as potent as pronethalol it satisfied all the desired clinical criteria for a beta-adrenoceptor antagonist and so, in 1964, became the first one on the market licensed for general use.[29] It remains in use as an anti-anginal, anti-arrhythmic and antihypertensive drug.

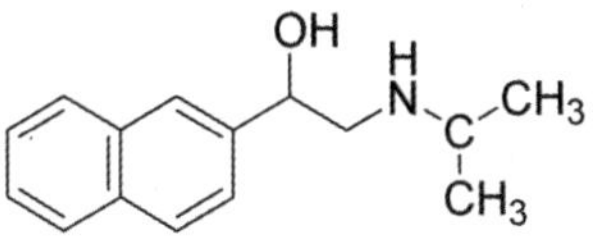

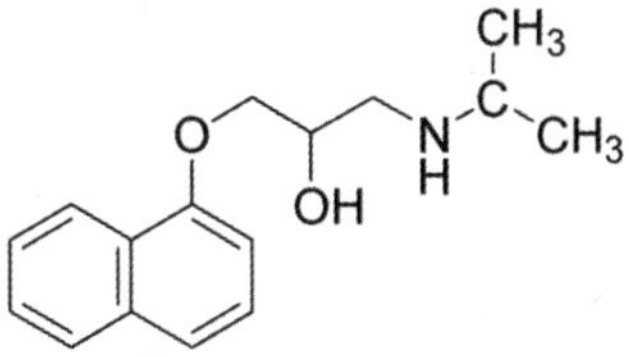

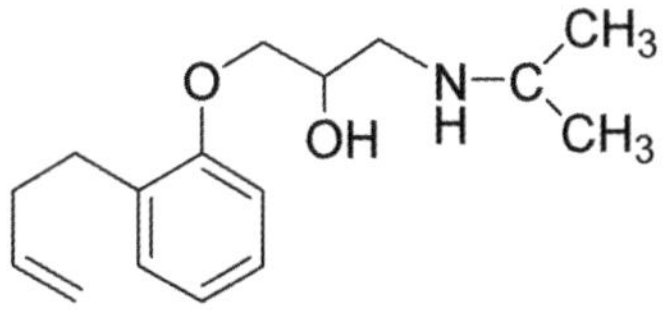

At AB Hässle in Göteborg, Sweden, analogues of dichloroisoprenaline were investigated as potential anti-arrhythmic agents. Alprenolol emerged as an effective agent that could also be used in the management of hypertension.[30] Oxprenolol, a successful beta-blocker developed by Ciba, retained some partial agonist activity when it interacted with $beta_1$ and $beta_2$-adrenoceptors.[31] The stimulation of the $beta_1$-adrenoceptors in the heart meant that it tended to cause less slowing of the heart rate than those beta-blockers that did not exhibit partial agonist activity (i.e. instrinsic sympathomimetic activity). This feature was useful in some patients with mild peripheral vascular problems. Because of its stimulating action on $beta_1$-adrenoceptors in peripheral blood vessels, oxprenolol was less likely to induce a feeling of coldness in the extremities, an undesirable feature of some beta-blockers. However, the stimulation of $beta_2$-adrenoceptors in the lungs did not render oxprenolol sufficiently safe for use in asthmatic patients.

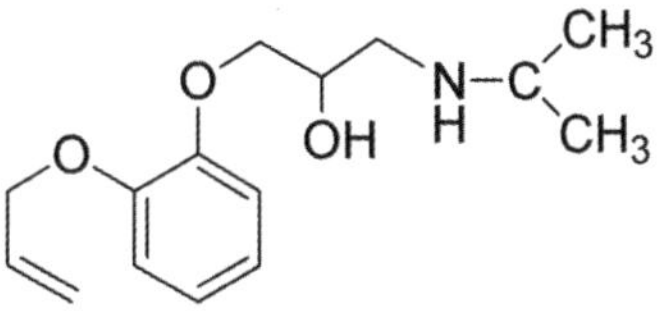

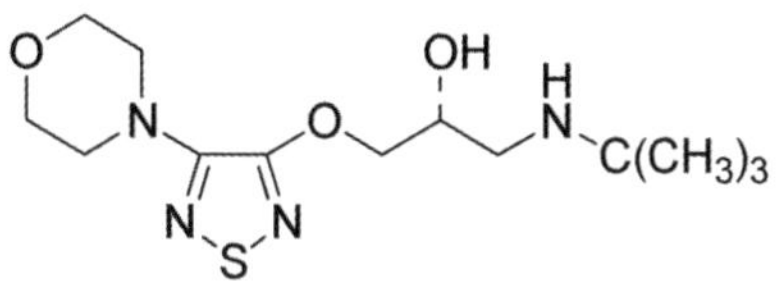

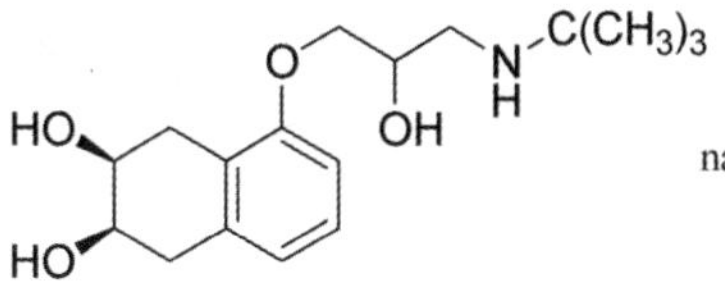

In timolol, not only was the aromatic system changed, but the N-isopropyl substituent was replaced by a tert-butyl group (cf. salbutamol).[32] Timolol has been used as a beta-blocker both in the management of hypertension and by direct application to the eye for the reduction of intraocular pressure in chronic simple glaucoma. Other similar drugs are used in glaucoma.

A series of oxygenated compounds prepared from the dihydro analogue of propranolol were evaluated at the Squibb Institute for Medical Research in Princeton, resulting in the introduction of nadolol.[33] The low lipophilicity of nadolol prevented it from entering the central nervous system, hence patients did not experience the sleep disturbance and nightmares associated with more lipophilic beta-blockers. Another advantage of the low lipophilicity was reduced entry into liver cells, ensuring a longer duration of action than most beta-blockers.

Cardioselective Beta-blockers

Sotalol was synthesised as a potential beta-adrenoceptor agonist by Larsen of Mead Johnson in 1960.[34] As Larsen had previously worked with sulfonamides, he synthesised sotalol in the belief that it would be worth replacing an acidic phenolic group of isoprenaline with an acidic sulfonamide group. Pharmacological evaluation indicated that sotalol not only relaxed tracheal, uterine and intestinal muscle but also blocked the tachycardia produced by the action of adrenaline on the heart in a similar manner to dichloroisoprenaline. Due to its polar nature, sotalol did not enter the brain to cause vivid dreams, nor undergo hepatic metabolism.[35]

On learning of the enhanced safety margin associated with the use of sotalol, ICI chemists prepared a series of its analogues, which included practolol.[36] Unexpectedly, it did not antagonise the peripheral vasodilation caused by isoprenaline in anaesthetised dogs, despite having the usual effects on the heart. It was the first time a beta-blocker was found that was relatively selective for heart receptors. This appeared to avoid a major clinical problem with beta-blockers, namely the risk of inducing bronchospasm in patients with asthma or obstructive airways disease through blockade of beta$_2$-adrenoceptors in the lungs. Consequently, practolol was marketed in 1970 for use in asthmatic patients. Tragically, it caused a serious oculomucocutaneous reaction in a small number of patients on long-term oral therapy, some of whom were blinded. In 1975 its use was restricted to specialised hospital units. Fortunately, alternative drugs soon appeared and practolol was abandoned. Like practolol, these are relatively selective with regard to their action on beta$_2$-adrenoceptors in the lung and on beta$_1$-adrenoceptors in the heart.

Atenolol was found to be a relatively cardioselective analogue of practolol.[37] For a long time it was the most frequently prescribed beta-blocker and the third-best-selling drug in the world (after ranitidine and cimetidine). Part of its success can be attributed to its low lipophilicity, which prevents it from entering the central nervous system.

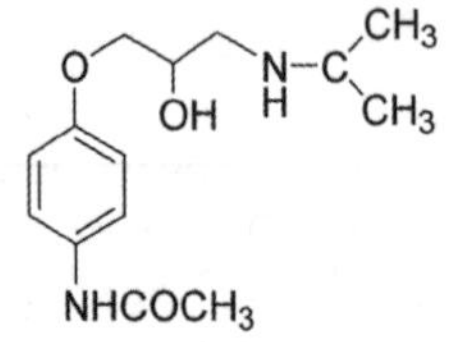

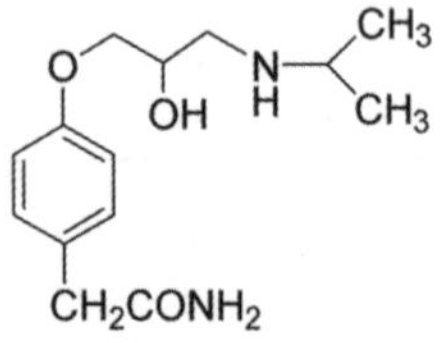

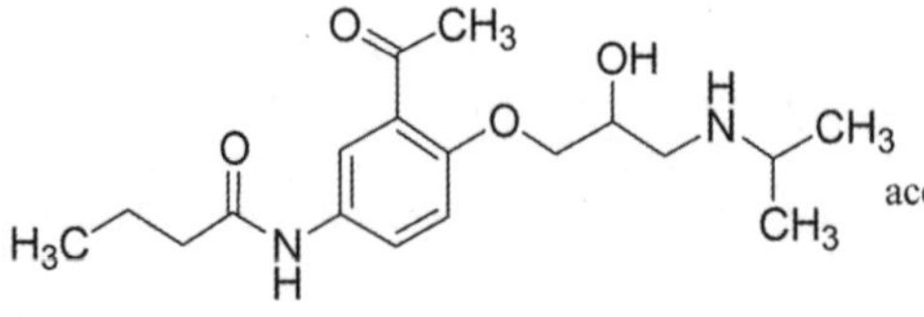

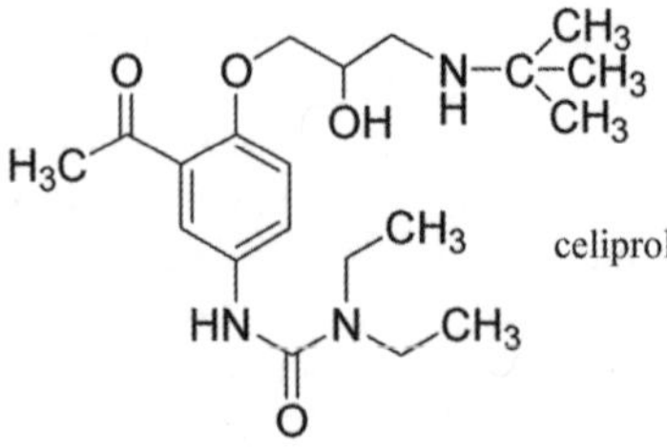

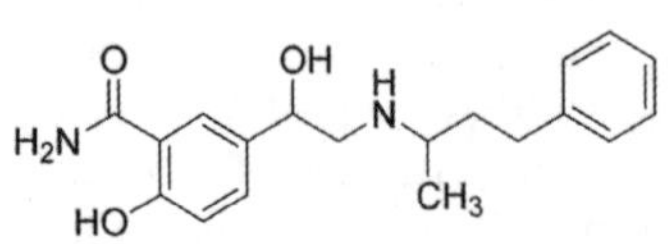

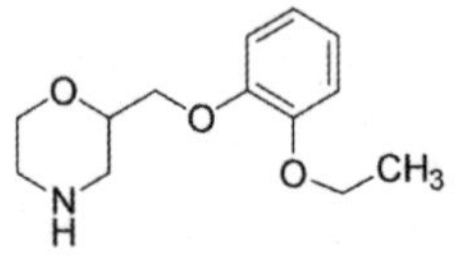

Several alternatives to atenolol have been developed. May & Baker's acebutolol did not have as much selectivity for cardiac receptors such as found with more recently introduced analogues of practolol.[38] However, it retained partial agonist activity. Chemie Linz introduced celiprolol as the first relatively cardioselective beta-blocker in which the partial agonist activity was limited to the beta$_2$-adrenoceptors.[39] Labetalol, introduced by Allen & Hanbury, blocked the action of sympathomimetic amines both at beta-adrenoceptors and also at alpha$_1$-adrenoceptors.[40] Drugs exerting either of these actions are generally antihypertensive agents; hence it was believed that a compound such as labetalol would exert its effects on blood pressure at a lower level of adrenoceptor blockade.

The tendency of propanolol and other lipophilic beta-blockers to produce vivid dreams or even hallucinations in patients receiving large doses interested ICI chemists. To explore the nature of this phenomenon, they prepared analogues in which lipophilicity was further increased by linking the hydroxyl group to the nitrogen atom through a two-carbon chain to form a morpholine ring. The resulting compounds were psychoactive, and in 1969 viloxazine was patented as an antidepressant drug.[41] It turned out to be a relatively non-sedating antidepressant that inhibited both noradrenaline and 5-HT reuptake in the brain. It was particularly of value in patients troubled by the anticholinergic and cardiac side effects of antidepressant drugs such as imipramine and amitriptyline.

OESTROGEN ANALOGUES

A variety of lipophilic ester prodrugs of oestradiol have been developed for formulation in oil to provide a long-lasting effect when administered as a depot injection. The first was oestradiol benzoate, which was patented by Schering–Kahlbaum of Berlin in 1936. The depot effect lasted for up to five days.[42] After the ester had been slowly released from the site of injection, it entered the circulation and underwent hydrolysis in the plasma to release oestradiol. Other similar esters of oestradiol include the dipropionate, valerate, enanthate and undecanoate. However, the administration of oestradiol by long-acting injections of its esters in oil is now considered to be an inappropriate way of managing menstrual disorders or for hormone replacement therapy in the menopausal patient, since the slow decline in plasma oestrogen levels is unlike that of the natural menstrual cycle.

oestradiol

oestradiol benzoate

Schering researchers Hans Inhoffen and Walter Hohlweg found that an ethinyl group introduced into the 17-position of progesterone conferred oral activity, so they decided to do likewise with oestrone to form ethinyloestradiol.[43] When injected into rats this showed similar potency to oestradiol, but given by mouth it turned out to be at least 20 times as potent as oral doses of the natural oestrogens. This was due to the acetylenic function at the 17-position of the steroid nucleus sterically blocking deactivation by liver enzymes. Ethinyloestradiol remains in use for the treatment of menopausal symptoms, despite more than half a century having passed since its introduction into the clinic. It is also found as the oestrogenic component of oral contraceptives.

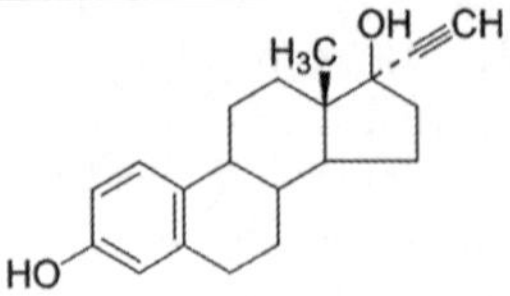

ethinyloestradiol

Synthetic Oestrogens

In 1932, Otto Rosenheim and Harold King had proposed that oestrogens had a similar structure to ergosterol, the structure of which had been established. Charles Dodds and Colin Leslie Hewett at the Middlesex Hospital in London joined forces with James Cook, an organic chemist working at the Cancer Hospital in London, to synthesise a tricyclic phenanthrene compound which they assumed would be an analogue of oestrone. This was found to have some oestrogenic activity. Although this tricyclic compound was the first substance of known constitution to exhibit oestrogenic activity, clinical application was out of the question.[44] Nevertheless, its synthesis represented an important landmark.

Cook and Dodds next found oestrogenic activity in several other phenanthrenes. Then they discovered that the phenanthrene ring was not essential for activity.[45] From this point onwards, they were able to demonstrate activity in a variety of compounds containing two benzene rings linked together via a short carbon chain, including diphenylethane and diphenylstilbene. One of the most potent was dihydroxystilbene. To establish whether both of its benzene rings were required for oestrogenic action, one was replaced by a methyl group. No complicated synthetic procedure was necessary to obtain the desired analogue as it was commercially available under the name 'anol', being cheaply prepared from the essential oil anethole. At first anol seemed to be as potent as any of the naturally occurring hormones. A letter was sent to the editor of *Nature*, in April 1937, to inform the scientific community of this remarkable development.

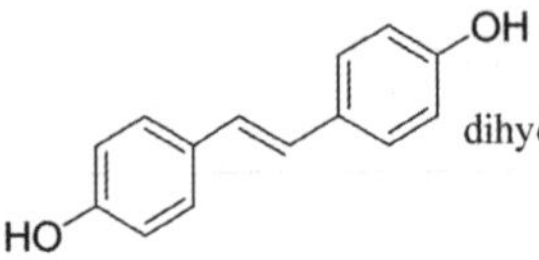

dihydroxystilbene

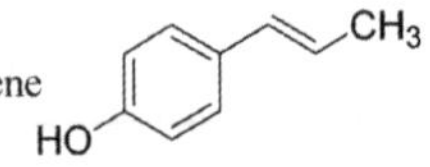

anol

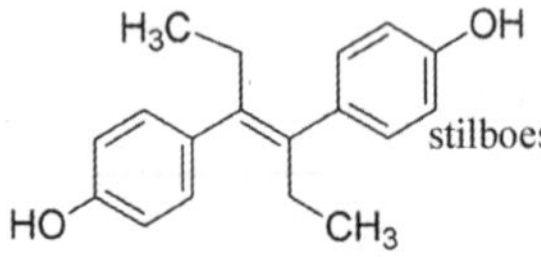

stilboestrol

Within weeks, several workers had written to Dodds confirming his findings, while others wrote to say they had been unable to demonstrate any activity at all, even with high doses of anol. Walter Schoeller of the Schering–Kahlbaum laboratories then sent details of experiments that showed that some batches of anol were contaminated with an impurity formed during its formation from anethole. On crystallisation of crude anol from chloroform, this impurity remained in the mother liquor, proving to be highly oestrogenic. Dodds now collaborated with a group at the Dyson Perrins Laboratory in Oxford University, led by Robert Robinson, in an effort to synthesise this as-yet-unidentified impurity. They decided it could be a stilbene derivative formed by the condensation of two molecules of anol. Several such compounds were then prepared. Early in January 1938, the British researchers reported in *Nature* that one of these, subsequently known as 'stilboestrol' (diethylstilbestrol), was two or three times as potent as oestrone.[46] Later, it was discovered to be almost as potent as oestradiol when injected, but around five times as potent when given by mouth. This was due to its enhanced resistance towards metabolic deactivation. As it was cheap to synthesise and well tolerated in patients, stilboestrol provided gynaecologists, for the first time, with a substance that was in plentiful supply and could be used to deal with oestrogen deficiency, especially in the menopausal patient. It was widely prescribed.

In his first publication about stilboestrol, Dodds drew attention to a structural resemblance between it and the natural oestrogens. Three years later, using X-ray crystallography, Dorothy Crowfoot established that the molecular dimensions were almost identical to those of oestradiol, especially with regard to the distance between the hydroxyl groups at either end of both molecules.[47]

In 1971, two reports were published describing an increased incidence of vaginal and cervical adenocarcinomas in young women whose mothers had previously been given stilboestrol or similar synthetic oestrogens during the first trimester of their pregnancies in order to prevent unwanted abortion. Fear of a horrendous human tragedy that would have eclipsed even the thalidomide disaster led the US government to investigate the situation. It was found that the incidence of the adenocarcinomas was between 0.01 and 0.1% of those exposed. Up to five million American women are estimated to have taken the drug before its withdrawal, so up to as many as 5000 of their daughters could have developed adeno-carcinomas.[48] The situation is being monitored.

Paradoxically, stilboestrol is also used in the treatment of cancer. Following his isolation of testosterone, Ernst Laqueur and others administered it in unsuccessful attempts to treat enlarged prostate glands in elderly men. The belief that overgrowth of this gland must be due to diminished production of male hormone was challenged by Wugmeister, a Polish endocrinologist working in Milan. He had read of the detection of female hormones in male urine and also of the ability of such hormones to inhibit release of gonad-stimulating hormone from the pituitary gland. He suggested that a deficiency of female hormone might permit excessive release of gonadotrophic hormone from the pituitary, causing prostatic enlargement. In 1937, he put his hypothesis to the test by treating 23 patients with large doses of oestrone. His patients showed marked improvement, with clear reduction in prostate size. Wugmeister published his findings in *Paris Medical*,[49] where they were seen by Pierre Kahle and Emile Maltry of the School of Medicine at Louisiana State University. They then carried out a similar study using the newly introduced stilboestrol. Their results, published in 1940, confirmed Wugmeister's claims and introduced this new form of therapy for benign prostatic overgrowth into the United States.[50] They then initiated a trial of stilboestrol in prostatic cancer that was not completed until 1942, by which time others had already published evidence to show this was highly effective.

While Kahle and Maltry were carrying out their studies, Charles Huggins at the University of Chicago made a far-reaching chance observation. He had been collecting prostatic fluid from dogs for several years as part of an enquiry into the biochemistry of seminal fluid. Some of his experiments involved castrating the dogs prior to measuring the effect of testosterone injections on prostatic fluid production. It was in the course of one such study that Huggins noticed that castration caused regression of tumours that had spontaneously arisen in elderly dogs. After confirming that castration consistently caused shrinkage of prostatic tumours in dogs, Huggins realised that this indicated these tumours were hormone-dependent. He then administered stilboestrol in order to produce chemical castration. Again, regression of the tumours was observed.[51] A year later, Huggins published the results of a trial of stilboestrol in patients with prostatic cancer that had spread into their bones. These results were most encouraging, the patients exhibiting marked regression of tumours and alleviation of pain.[52] This was the first demonstration of a synthetic drug having induced an unquestionable improvement in a malignant disease. For this and his subsequent studies, Huggins was belatedly awarded the Nobel Prize for Medicine and Physiology in 1966.

Huggins' pioneering investigations were soon followed by studies of the effects of hormones on various types of tumours, notably that conducted by Alexander Haddow at the Chester Beatty Research Institute of the Royal Cancer Hospital in London. From there, in 1944, he reported that stilboestrol and some of its analogues had beneficial effects in patients with breast cancer.[53] This type of therapy was ultimately shown to be effective in up to 40% of postmenopausal women.

Tamoxifen

Following Dodds' disclosure of the oestrogenic activity of diphenylethane and diphenyl-stilbene, John Robson and Alexander Schonberg of the Department of Pharmacology at the University of Edinburgh found that triphenylethylene had oestrogenic activity, although only about one-ten-thousandth that of oestrone. The interesting feature, however, was that when it was given by mouth it was just as active as by injection, and the effects of a single dose lasted about one week. This was reported in 1937, and the following year the Edinburgh researchers showed that replacement of the sole ethylenic hydrogen atom with a chlorine atom increased potency two-thousandfold. Robson and Schonberg prepared more analogues, and early in 1942 reported that DBE, a bromine-substituted triphenylethylene with ethoxy groups attached to two of the benzene rings, was a possible alternative to stilboestrol.[54] However, John Davies and his associates at ICI, who had followed the earlier reports by the Edinburgh group, were now working along similar lines. The outcome was that a British patent was awarded to ICI for the chloro-substituted triphenylethylene with methoxy groups on all of the benzene rings, namely chlorotrianisene.[55] It was longer-acting than the natural oestrogens because its high lipophilicity resulted in deposition in fat tissue, from which it was slowly released.

Although chlorotrianisene was a weak oestrogen, it antagonised the ability of oestradiol to induce pituitary enlargement in rats. It was established that it was an anti-oestrogen which blocked oestrogen receptors in the hypothalamus. This interfered with feedback of oestrogen levels via the hypothalamic–pituitary axis.[56] Chlorotrianisene was only a partial antagonist, retaining some intrinsic oestrogenic activity.

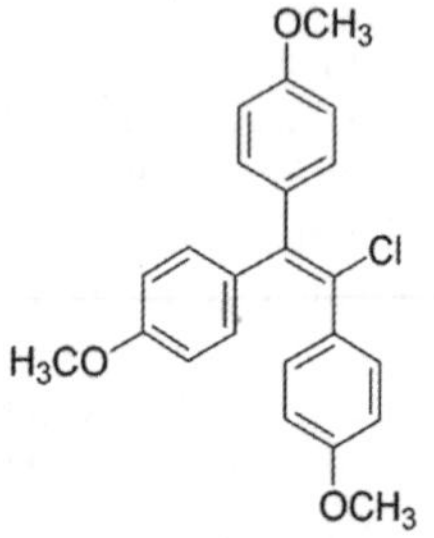

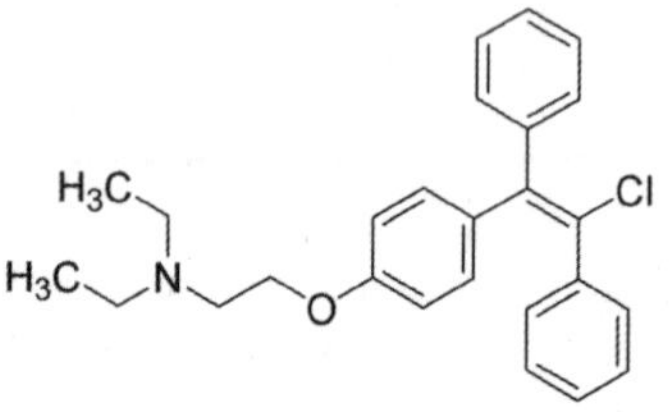

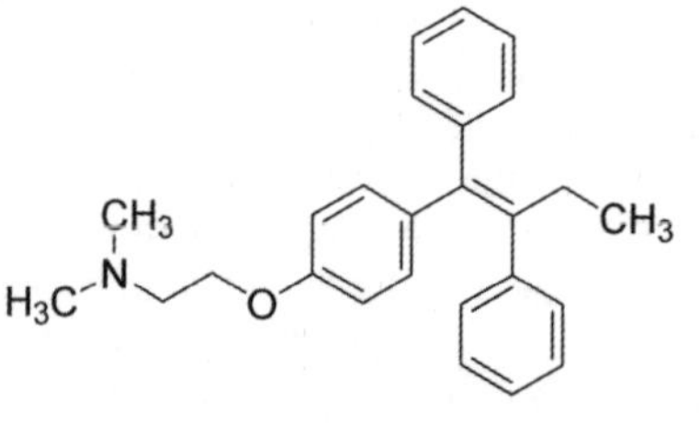

Clomiphene, now known to be a mixture of two geometric isomers, had only weak oestrogenic activity.[57] It was a more potent anti-oestrogen than chlorotrianisene and was capable of stimulating gonadotrophin release by inhibiting the negative feedback of gonadal oestrogens on the hypothalmus. As it had been found to cause enlargement of the ovaries in women, it was used to induce ovulation in women whose infertility was due to ovulatory failure. Experience has shown that considerable caution must be exercised when using clomiphene, otherwise multiple births may ensue.

Tamoxifen was synthesised by Dora Richardson of ICI Pharmaceuticals Division in 1962.[58] When its two geometric isomers were evaluated by Michael Harper and Arthur Walpole, they found that while the *cis* isomer had oestrogenic activity, the *trans* isomer had only weak oestrogenic properties. Further investigation revealed the latter to act as an anti-oestrogen. As an analogue of stilboestrol, tamoxifen was active when given by mouth. It was therefore evaluated in rats as a postcoital oral contraceptive.[59] However, when tested in women, it induced ovulation instead.[60]

The anti-oestrogenic activity of tamoxifen was exploited in the treatment of oestrogen-dependent breast cancer, convincingly demonstrated in 1971 by a clinical trial at the Christie Hospital and Holt Radium Institute in Manchester.[61] Two years later, tamoxifen was licensed in the United Kingdom for the treatment of breast cancer. Since then it has become established as the drug of choice for treating breast cancer in women who respond to hormone therapy. A large amount of evidence from randomised trials with tamoxifen has confirmed that it reduces the risk of reoccurrence of the disease and improves overall survival in both pre- and postmenopausal women, irrespective of whether or not they have received cytotoxic chemotherapy.[62] However, it can cause serious side effects and there is a risk of contracting endometrial cancer.

Formestane

Formestane is an analogue of androstenedione that competes with it at the active site of aromatase enzyme to inhibit the final stage of oestrogen biosynthesis.[63]

formestane

Formestane was introduced in 1984, being given intramuscularly as a depot injection every two weeks in patients in whom tamoxifen therapy had failed to control breast cancer. The introduction of orally active aromatase inhibitors has led to a decline in its use.

PROGESTERONE ANALOGUES

Progesterone was inactive by mouth because of rapid metabolism in the liver, but it was injected intramuscularly in the hope of preventing miscarriages during the early months of pregnancy. It is now generally believed that low levels of progesterone are not normally a cause of miscarriage. Progesterone later became available in the form of implants, pessaries and suppositories but it was expensive to produce and relatively large amounts had to be administered.

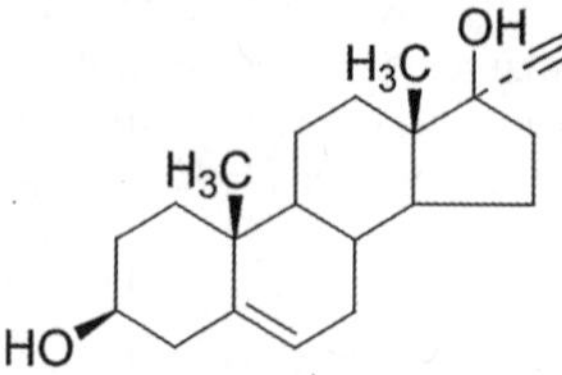

ethinylandrostenediol

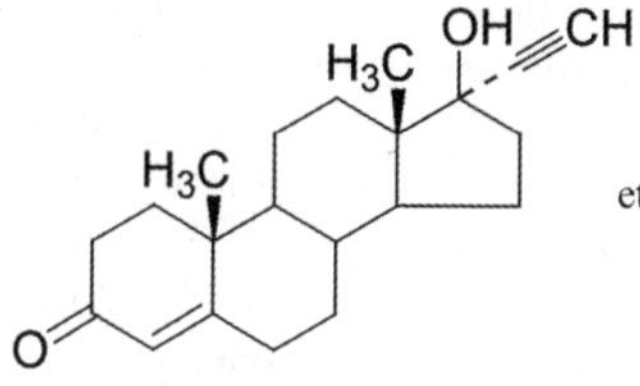

ethisterone

A potent, orally active analogue of progesterone was obtained by Leopold Ruzicka and Klaus Hofmann at the Zurich Eidgenossiche Technische Hochschule,[64] and also by Willy Logemann and his colleagues at the Schering laboratories in Berlin.[65] These groups independently added potassium acetylide to dehydroepiandrosterone, a cheaply available, cholesterol-derived steroid intermediate. Their objective had been to introduce at the 17-position of the steroid nucleus a two-carbon acetylenic side chain which could subsequently be converted to the progesterone side chain. In the event, the stereochemistry at the 17-position of the resulting ethinylandrostenediol turned out to be the opposite of that in progesterone. The Schering group converted ethinylandrostenediol into ethisterone.[66] This bore enough resemblance to progesterone to become established as a most successful orally active progestogen since disubstitution at the 17-position blocked metabolism by liver enzymes. Unfortunately, an inherent chemical resemblance to testosterone was responsible for virilising side effects of ethisterone and its analogues, a contraindication to use in pregnant women bearing a female foetus.

Norsteroids as Oral Progestogens

The first norprogestogen, in which the methyl group normally found in the 19-position was absent, was prepared in very low yield (0.7%) from the naturally occurring heart stimulant strophanthidin by Maximilian Ehrenstein at the University of Pennsylvania in 1944.[67] His small supply of crude 19-norprogesterone was tested on two rabbits by Willard Allen (who with Oskar Wintersteiner had isolated progesterone ten years earlier).[68] It was as active as the natural hormone. Further progress was not possible until Arthur Birch developed a practical synthesis of norsteroids at the University of Oxford in 1950.[69] Using this procedure, Carl Djerassi and Luis Miramontes synthesised norethisterone (norethindrone) at the Mexico City laboratories of Syntex SA in 1951.[70] It was marketed as a progestogen for the treatment of amenorrhoea, dysfunctional uterine bleeding and endometriosis.

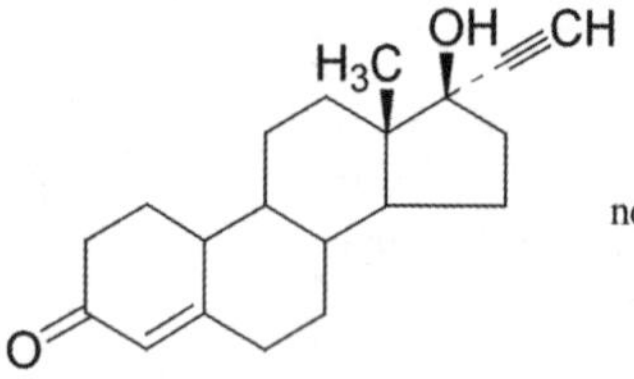

norethisterone

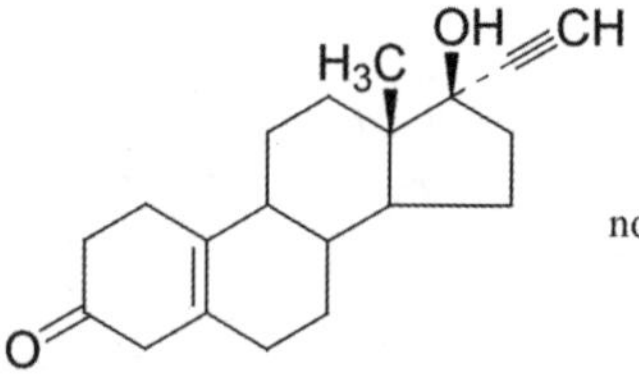

norethynodrel

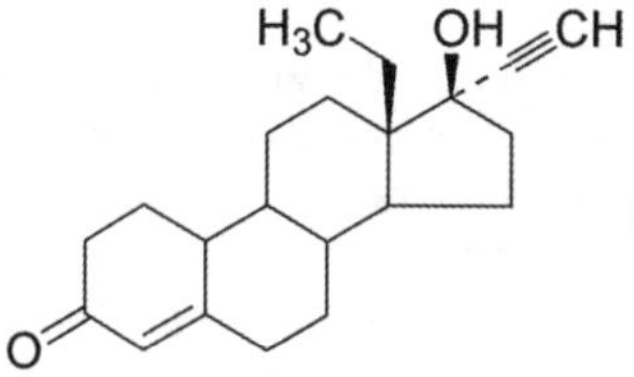

levonorgestrel

Frank Colton of G.D. Searle & Company in Chicago synthesised a series of norsteroids that included norethynodrel, for which a patent was granted in 1954.[71] Both norethisterone and norethynodrel were developed for the treatment of gynaecological conditions and it was only after Gregory Pincus found norethynodrel to be the ideal progestogen for incorporation in an oral contraceptive formulation that a lucrative market was created for norsteroids. Neither norethisterone nor norethynodrel had been synthesised with the aim of developing an oral contraceptive.

At the Wyeth Pharmaceutical Company in Pennsylvania in 1960, Herchel Smith, a former student of Birch, synthesised an analogue of norethisterone in which an ethyl group replaced the 18-methyl group.[72] This new steroid, norgestrel, was more potent than either norethisterone or norethynodrel, its activity residing in the isomer known as levonorgestrel.

Oral Contraceptives

Leo Loeb revealed in 1911 that one of the functions of the mammalian corpus luteum was to inhibit ovulation. A quarter of a century later, A.W. Makepeace, George Weinstein and Maurice Friedman at the University of Pennsylvania successfully prevented ovulation in rabbits in oestrus by injecting progesterone for five days before mating. They attributed this outcome to pituitary inhibition and cited 15 previous reports of similar use of corpus luteum extracts of widely varying quality.[73] This proof of the capability of progesterone to prevent conception was not exploited at that time for two reasons. Firstly, public opinion would have been against such a development and, secondly, the need to inject the progesterone would have severely limited its value. Yet, the following year Schering chemists developed the first orally active progestogen, namely ethisterone.

The investigation of potent progestogens as oral contraceptives did not take place until 13 years later when Gregory Pincus (see below), the author of another article that appeared in the same issue of the *American Journal of Physiology* as the University of Pennsylvania paper, became involved after being approached not by a pharmaceutical company but by Margaret Sanger, founder of the Planned Parenthood Federation of America.

The Federation was established by Sanger in 1916 under the name of the American Birth Control League. She dedicated her life to campaigning for the right of women to have access to effective means of contraception. She faced hostility on many fronts and was even imprisoned for her activities at a time when contraception was illegal in the United States. Sanger met Pincus over dinner in New York early in 1951, shortly after 75-year-old philanthropist Katharine Dexter McCormick, heiress to the International Harvester Company fortune, had written to Sanger asking her views on where financial support was most needed and what the outlook was for birth control research. Pincus, an authority on mammalian fertilisation, had established a small private laboratory in Worcester, Massachusetts, known as the Worcester Foundation for Experimental Biology. When asked by Sanger what the prospects were of science developing a perfect contraceptive, his reply so encouraged Sanger that she and McCormick decided to visit his laboratory. The result of that visit was that the Worcester Foundation received an initial $40 000 funding for research into oral contraception from McCormick. Over the next decade, that sum rose to 2 million dollars. This investment resulted in the development of what some consider may be the single most significant contribution of science to society in the twentieth century.[74]

The initial approach taken by Pincus and his assistant Min Chueh Chang was similar to that taken in 1937 by Makepeace, Weinstein and Friedman except that, instead of injecting female rabbits with progesterone, they gave very large oral doses (more than 5 mg) in order to compensate for its extensive metabolism in the liver. Again, when the rabbits were subsequently placed in mating cages, none of them ovulated and no pregnancies occurred.

Shortly after completing his first experiments, Pincus happened to meet John Rock, a gynaecologist from Harvard University. Rock had been trying to stimulate the growth of underdeveloped ovarian tubes or uteri in infertile women by administering female hormones to mimic the stimulation in the growth of these organs produced by hormonal changes at the onset of pregnancy. In order to achieve the response he required, his patients were sequentially taking large oral doses of stilboestrol followed by progesterone, for three months. This regimen led to a total suppression of menstruation during treatment. Even though many of the women later conceived, this amenorrhoea was emotionally distressing. Pincus suggested to Rock that an alternative approach would be to give progesterone solely for three weeks and then withdraw medication so as to induce menstruation. Treatment could restart on the fifth day of the subsequent cycle.

When Rock tried this method, it proved acceptable to most of his patients. Several became pregnant after the three month course was completed. However, one in five of the women in the trial experienced 'breakthrough bleeding'. This was due to the inadequacy of orally administered progesterone. Further, the hormone failed to maintain total suppression of ovulation for more than around 85% of the time that it was being taken, despite the use of 300 mg daily doses. It was evident that a more potent oral progestogen would be required, so in September 1953 Pincus invited leading pharmaceutical manufacturers to send him supplies of orally active progestogens that they had investigated. By the end of the year nearly 200 compounds had been examined at Worcester in both rabbits and rats. Fifteen were confirmed as potent ovulation inhibitors suitable for Rock to test in the clinic. These compounds were supplied to 50 infertile women. All were prevented from ovulating with doses in the order of 10–50 mg, a marked improvement over progesterone. In six of the patients, sufficient stimulation of their underdeveloped ovarian tubes or uteri occurred, permitting conception after the course of treatment was completed, the so-called 'Rock rebound'. By now it was clear to Rock and Pincus that they had developed not only a remedy for one common type of infertility but also an effective oral anti-ovulatory therapy that could potentially satisfy Margaret Sanger's expectations.[75]

The first large-scale clinical trial of a progestogen as an oral contraceptive began in Puerto Rico in April 1956. Norethynodrel was selected as it was the most potent of the anti-ovulatory compounds that had been examined. It was also free of androgenic activity. There were few reports of breakthrough bleeding during the early phase of the trial, but later numerous reports were received of this occurring towards the end of the treatment cycles. Thorough investigation revealed that these reports coincided with the introduction of purer batches of norethynodrel from the Searle laboratories. It was then discovered that the original batches had been contaminated with around 1–2% of the 3-methyl ether of ethinyloestradiol, from which the norethynodrel had been synthesised. This was a potent oestrogen, and when small amounts of it were deliberately incorporated in tablets containing pure norethynodrel, the problem of breakthrough bleeding was overcome. Later, this oestrogen was given the approved name of 'mestranol'.

After two years, the preliminary results of the Puerto Rico trial were analysed. None of the 221 married women of proven fertility who had taken the tablets became pregnant, but there were several pregnancies among the small number of women who had dropped out of the trial, confirming the return of normal fertility. Finally, after examination of the clinical records of 1600 women who had received the tablets, the US Food and Drug Administration gave G.D. Searle & Company permission in May 1960 to market their combined progestogen–oestrogen oral contraceptive containing 9.58 mg of norethynodrel and 150 μg of mestranol. The following year, Schering introduced a similar product in Sweden containing 4 mg of norethisterone and 50 μg of ethinyloestradiol. The dose chosen for both these products was much higher than necessary. Today, for example, there is a combined oral contraceptive on the market containing 1 mg of norethisterone and 20 μg of ethinyloestradiol. The diminution of the dose over the last 40 years has generally resulted in a reduction of side effects.

There was considerable concern during the early 1970s that chronic administration of oral contraceptives might cause some deaths as a result of cardiovascular changes. One theory was that residual androgenic activity in synthetic progestogens could be responsible for changes in the blood lipids (fats), leading to arterial disease, the chemical structures of progestogens having structural features in common with testosterone. In attempting to avoid this possibility, Organon introduced desogestrel, a compound synthesised in 1975 following reports that introduction of certain substituents in the 11-position of progestogens markedly enhanced potency.[76] It was the first new progestogen to be launched in the United Kingdom for over a decade when it was marketed in combination with ethinyloestradiol in 1982. Within a year of its introduction, however, concern was also being expressed about the possible hazard of an increased risk of breast cancer in women who had been taking highly potent progestogens.

Mifepristone

At the laboratories of Roussel–UCLAF in Romainville, France, Georges Teutsch developed a synthetic route for 11β-substituted steroids in 1975. He used this to synthesise a series of novel compounds that were examined as potential glucocorticoids. The two compounds that were found to bind most tightly to the rat hepatic glucocorticoid receptor during *in vitro* studies, RU 25055 and RU 25593, were unexpectedly found to be wholly devoid of activity in another experimental screening model in which glucocorticoid activity was correlated with induced tyrosine amino transferase in rat hepatoma tissue cells. Further investigation revealed that both test compounds antagonised the action of dexamethasone in this model. The possibility that this could be due to antiglucocorticoid activity was considered, but other possible explanations for the behaviour of the compounds also existed. A variety of tests for antiglucocorticoid activity proved inconclusive.

In 1980, specific screens for antiglucocorticoid activity were established by the company following a renewal of interest in the activity of the two compounds synthesised by Teutsch. Several more were synthesised, the sixth of which totally blocked the activity of dexamethasone while still being devoid of glucocorticoid activity. When this compound, RU 38486, was subjected to evaluation in a battery of pharmacological models over the next few months, it consistently behaved as an antiglucocorticoid.

Teutsch knew that in the initial receptor binding studies of his 11β-substituted steroids, the two antiglucocorticoids had also tightly bound to the progesterone receptor. He therefore requested that they be screened to determine their effect at this receptor. When absolutely no activity on the rabbit endometrium was demonstrated, the possible parallel with the results in the glucocorticoid models immediately struck him. At the beginning of 1981, his suspicion that RU 38486 was an antiprogestin was confirmed.[77] Further evaluation in rats revealed this to be a potent antagonist of progesterone since it deformed the receptor and thus prevented the activation that normally ensued after progesterone was attached. RU 38486 was given the name 'mifepristone' and was patented in 1982.[78]

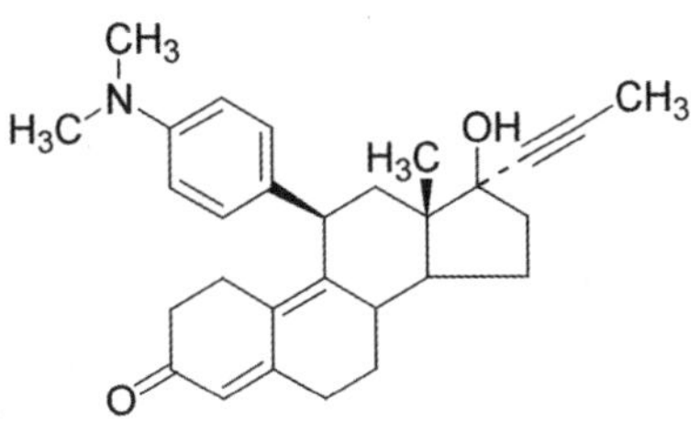

mifepristone

The potential of an antagonist of a hormone that was essential for the maintenance of pregnancy was obvious. Tests on small animals and primates confirmed that administration of mifepristone daily for four days interrupted early gestation. In a trial at the University Hospital in Geneva in 1982, 11 volunteers who were between six and eight weeks pregnant were given 200 mg of mifepristone by mouth daily for four days. Abortion was complete in 3–5 days in eight cases and in eight days in another. In two cases, the procedure failed. These two failures were addressed by taking into account one of the important effects of progesterone, which is to render the uterus less responsive to excitatory agents such as prostaglandins. Since mifepristone made the uterus more susceptible to the action of prostaglandins by antagonising progesterone, the supplementation of it with a prostaglandin analogue to stimulate the uterus to contract was investigated. Several studies found that when a single dose of 600 mg of mifepristone was followed two days later by such an analogue, misoprostol, more than 95% of early pregnancies aborted. This combination of drugs was licensed for use in France in 1987 and in the United Kingdom four years later. Its introduction, not unexpectedly, was surrounded by controversy and it was not licensed for use in the United States until 2000.

Mifepristone proved to be of no clinical value as an antiglucocorticoid since on continued administration its effects were overcome by an increase in corticotrophin and hydrocortisone levels arising from biofeedback via the hypothalamic–pituitary axis.

Medroxyprogesterone Acetate

In 1954, Karl Junkmann of Schering AG reported that the acetylation of the 17-hydroxyl group of ethisterone provided a derivative suitable for formulating in oil for injection intramuscularly as a depot medication.[79] There resulted widespread interest in preparing the acetates (and other esters) of various hydroxy-steroids. One such ester, Upjohn's 17-acetoxyprogesterone, proved to be a promising progestogen even though its hydroxy precursor was inactive. Unfortunately, it turned out that no significant prolongation of action was obtained by formulating it in oil. The Upjohn researchers, however, made the unexpected discovery that their acetoxy derivative was orally active, an observation that had been missed by the Schering group, who were primarily interested in the oil solubility of such esters. Several manufacturers then competed to prepare the 6-methyl derivative of 17-acetoxyprogesterone. Priority of discovery for this derivative, subsequently known as medroxyprogesterone acetate, went to Upjohn in the United States for a patent application submitted on 23 November 1956.[80] Syntex, however, submitted its application in France on 8 September 1956.[81]

Medroxyprogesterone acetate proved to be around 25 times as potent as ethisterone. When formulated as an oily injection, it formed a deposit of steroid in the body near the site of intramuscular injection. Small quantities of drug were then slowly released over a period of several months to give prolonged contraceptive cover.

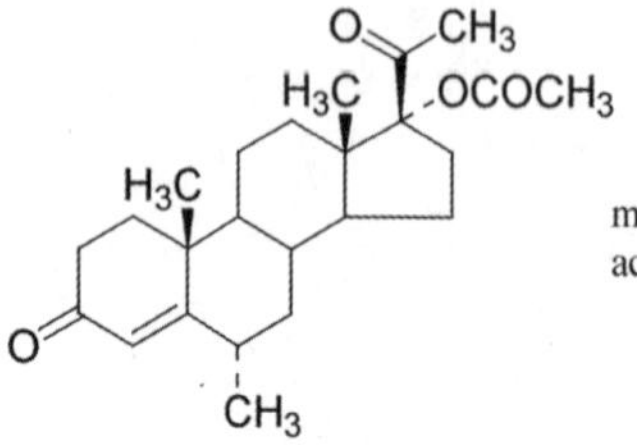

medroxyprogesterone acetate

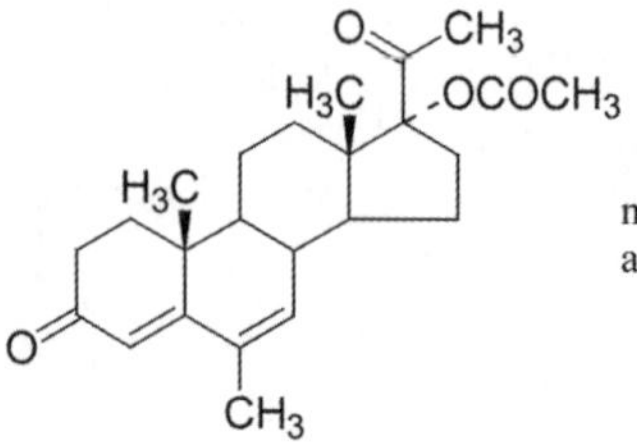

megestrol acetate

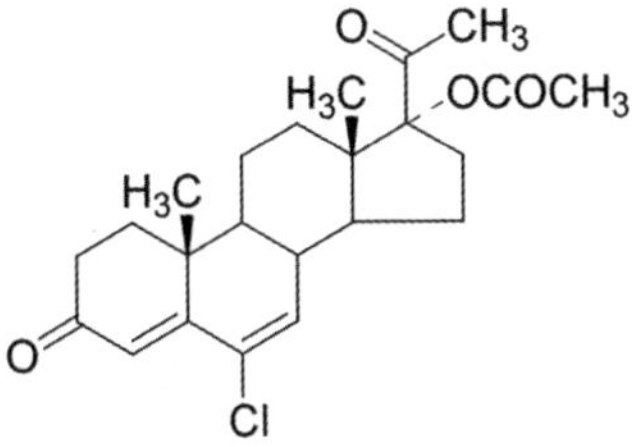

chlormadinone acetate

Vladimir Petrow of the British Drug Houses oxidised medroxyprogesterone acetate, a drug he had developed as an oral contraceptive, to introduce an extra double bond at the 6-position. He did not expect this to have useful progestational activity, since the analogous derivative of progesterone was less potent than progesterone itself. Surprisingly, routine screening in rabbits revealed this new compound, megestrol acetate, to be the most potent ovulation inhibitor discovered up to that time (1958).[82] Syntex and E. Merck AG both found an alternative way of blocking metabolism at the 6-position, namely by inserting a chlorine atom. They independently prepared chlormadinone acetate.[83,84]

TESTOSTERONE ANALOGUES

Following the introduction of oestradiol benzoate in an oily depot formulation, testosterone phenylacetate was likewise prepared for intramuscular injection.[85] It has been replaced by a variety of other testosterone esters used for the same purpose, either on their own or as mixtures, including testosterone propionate, which is a shorter-acting ester formulated as an oily injection for administration two or three times a week. It is also formulated as a component of mixtures containing several esters with varying duration of action, such as the phenylpropionate, enanthate and undecanoate.

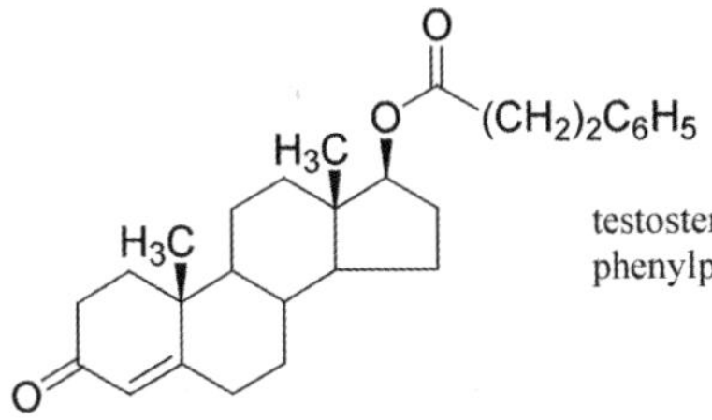

testosterone phenylpropionate

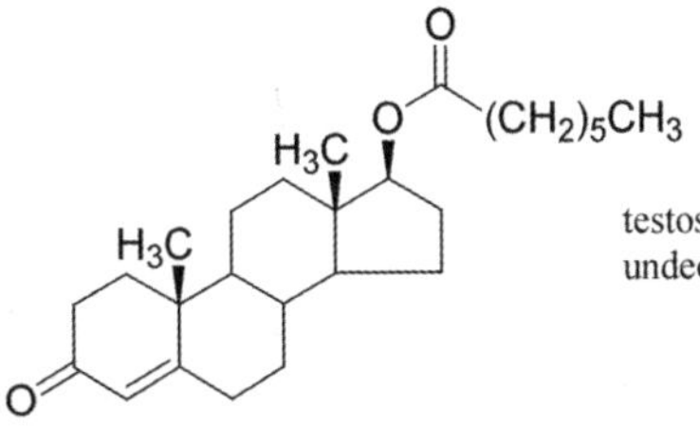

testosterone undecanoate

Testosterone undecanoate is also formulated in capsules given by mouth. Because of its short half-life, thrice-daily dosing is required.

Anabolic Steroids

At the University of Rochester in 1931, John Murlin tried to determine whether the higher metabolic rate in the male than the female was related to hormonal activity. The administration of androgenic male urine extract to castrated dogs appeared to increase the basal metabolic rate, but this could not be confirmed by Murlin. Charles Kochakian and Murlin then refined the study by examining the effect of a potent androgenic male urine extract on the abnormal levels of nitrogen (derived from protein metabolism) to be found in the urine of castrated dogs.[86] This extract had no significant effect on the metabolic rate, but it did reduce urinary nitrogen levels, indicating that protein metabolism had been decreased by the androgen. This was followed by an increase in the body weight of the dogs. When testosterone became available, Kochakian found that it had a similar influence on nitrogen balance.[87] A similar effect with testosterone in patients with endocrine deficiencies was subsequently reported by Kenyon.[88] Due to its hormonal effects, however, testosterone could only be administered as a muscle-building drug to mature male patients.

In 1948, G.D. Searle & Company initiated a long-term project to exploit the muscle-building properties of testosterone, hoping to obtain an anabolic agent free of hormonal activity. The screening procedure compared any increase in the weight of the seminal vesicles and the prostate glands of castrated rats with that of the levator ani muscle. In an ideal anabolic agent, only the weight of the latter would increase. This screen was applied to more than 1000 steroids over a 7 year period. Eventually, when Colton's norsteroids were examined in 1955, norethandrolone emerged as having similar anabolic activity to testosterone, with only one-sixteenth the potency as an androgen.[89] Furthermore, it was orally active since the alkyl substituent at the 17β-position stabilised the 17β-hydroxyl substituent against oxidation in the liver. After a satisfactory clinical trial, norethandrolone was released for use in debilitated patients in the expectation that enhanced utilisation of protein would build up wasting muscles. This expectation has been realised only to a very limited extent, even with newer drugs. Growing concern about the risk of cholestatic jaundice resulted in the withdrawal of norethandrolone from the market in the 1980s.

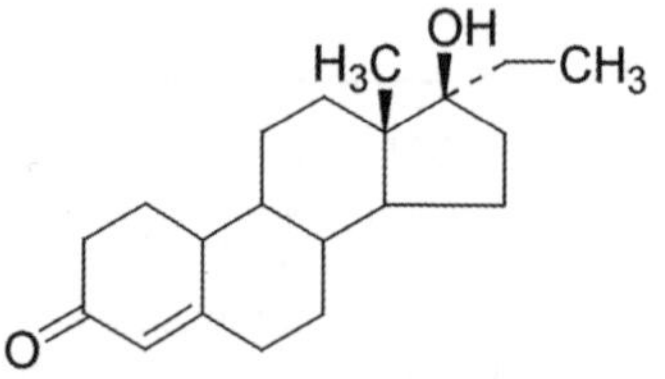

norethandrolone

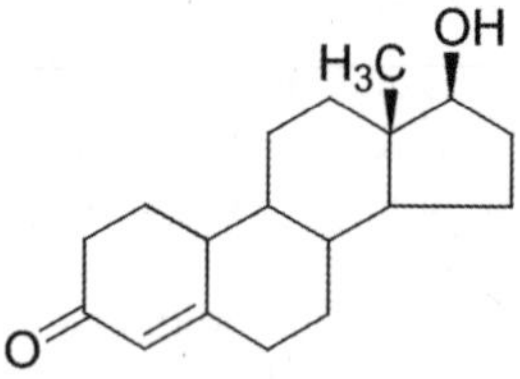

nandrolone

Nandrolone was one of the norsteroids originally synthesised by Birch.[90] As it was an analogue of norethandrolone, it had anabolic activity. Unlike norethandrolone, it does not possess a 17α-ethyl substituent to stabilise the 17β-hydroxyl substituent against oxidation in the liver. Consequently, nandrolone is inactive by mouth. It is administered as an injection in oil. Nandrolone decanoate is used in an oily depot injection.

Finasteride

An inherited condition in which apparently hermaphrodite children develop into males at puberty was found, in 1974, to be caused by a deficiency of 5α-reductase, the enzyme that converts testosterone to the more potent dihydrotestosterone.[91] Subsequent studies revealed that while most organs were responsive to testosterone and thus developed normally as testosterone levels rose at puberty, some were responsive only to dihydrotestosterone. The lack of development of the prostate gland throughout life in those with the 5α-reductase deficiency syndrome was attributed to the lack of responsiveness of the prostate to testosterone.

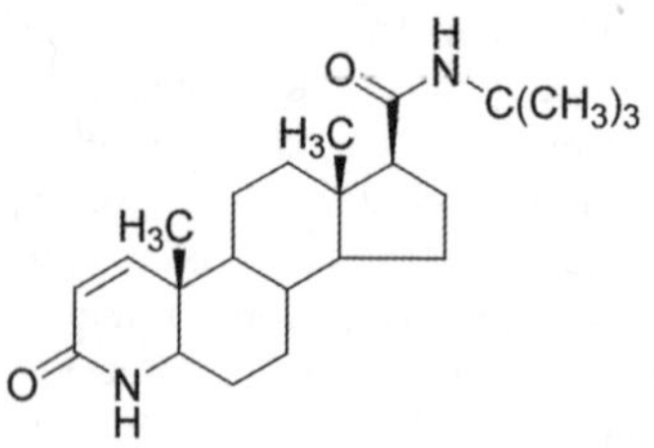

finasteride

Chemists at the Merck Research Laboratories postulated that 5α-reductase inhibitors could be of value in the treatment of benign prostatic hyperplasia, an overgrowth of prostatic tissue commonly found in males over the age of 50. The resulting enlargement of the gland causes obstruction of urinary flow through the urethra, frequently requiring surgical intervention. In 1985, they found finasteride, an azasteroid (i.e. incorporating a nitrogen atom), to be a potent, orally active inhibitor of 5α-reductase.[92] It has proved to be of value in the treatment of benign prostatic enlargement, where it can stop progression of the disease in mild to moderate cases, even to the extent of reducing the volume of the prostate. It thereby avoids the need for surgery so long as treatment is continued.

CORTISONE ANALOGUES

When cortisone and hydrocortisone were introduced, there was no demand for any new analogues to be prepared. Both hormones seemed to offer all that could be expected of corticosteroid therapy, bearing in mind that the activity of a number of existing corticosteroids had already been carefully evaluated in several laboratories. Sarett and colleagues pointed out that the decade preceding the introduction of cortisone witnessed the introduction of a wide range of vitamins. Their success had strengthened the long-held view that natural products were inherently superior to synthetic drugs, thereby discouraging the quest for synthetic analogues of cortisone.[93] A complementary viewpoint to that of Sarett can, with hindsight, be taken today: the successful development of therapeutically significant synthetic corticosteroids was a key stimulus for medicinal chemists to synthesise analogues of other natural products. Nevertheless, the first two analogues of cortisone that were introduced were, in fact, the unexpected products of attempts to find novel means of synthesising the natural hormone.

The first useful analogue of hydrocortisone was discovered when the Squibb chemist Joseph Fried attempted to synthesise hydrocortisone from its isomer epicortisol. These two compounds differed only in the stereochemical orientation of the hydroxyl group at the 11-position of the molecule. In order to achieve the conversion, Fried first had to introduce a bromine atom at the 9-position. Routine screening revealed that 9β-bromoepicortisol had one-third of the activity of cortisone. As epicortisol was feebly active, it appeared that bromination had increased potency. Fried next prepared 9α-halogenated analogues of hydrocortisone, the potency of which turned out to be inversely proportional to the size of the halogen substituent. This indicated that fluorine would be an ideal halogen to use in preference to the bulky bromine. When 9α-fluorohydrocortisone was tested, it was found to be about 10 times as potent as cortisone in relieving rheumatoid arthritis.[94] Unfortunately, the salt- and water-retaining (mineralocorticoid) activity was increased by a factor of 300–800. This showed – for the first time – that molecular manipulation might alter the balance of mineralocorticoid and glucocorticoid activity in corticosteroids. It also meant that patients became moon-faced and oedematous, hence its clinical role was limited to the management of adrenal insufficiency in Addison's disease or hypopituitarism. The new steroid received the approved name of fludrocortisone.

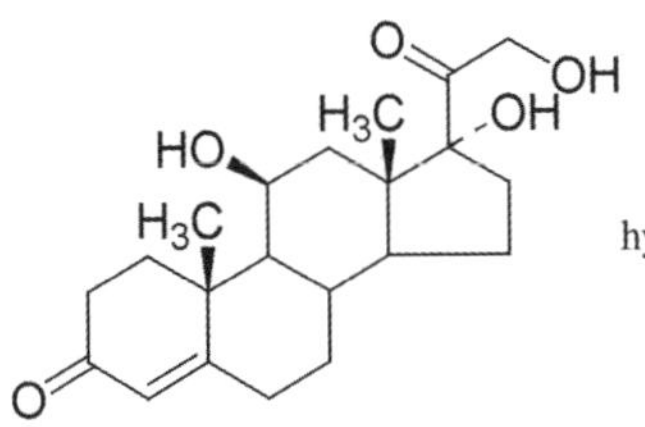

hydrocortisone

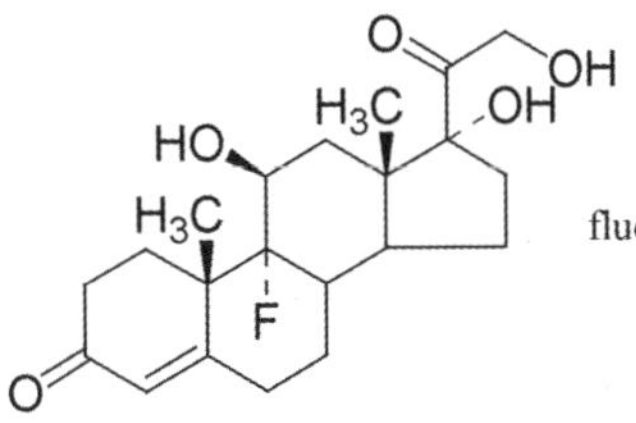

fludrocortisone

During an attempted synthesis of hydrocortisone, chemists of the Schering Corporation in Broomfield, New Jersey, protected the sensitive 11-hydroxyl group of the steroid ring by acetylation. At the penultimate stage in the synthesis, what was expected to be routine removal of the 11-acetate by chemical hydrolysis instead resulted in decomposition of the molecule. The only way round this was to attempt bacterial hydrolysis. Unexpectedly, in addition to ester hydrolysis, oxidation of the steroid ring A also resulted in the introduction of a second double bond. Two novel steroids were isolated, namely prednisone and prednisolone.[95] A preliminary clinical trial of them at the National Institutes of Health indicated that both compounds were about five times as potent as hydrocortisone in the treatment of arthritis and inflammatory conditions, yet no more likely to cause salt retention. The duration of action of both was about three times as long as that of hydrocortisone. Encouraged by this outcome, the company took the decision to risk pouring $100 000 worth of cortisone into a large fermentation tank.[96] The gamble paid off handsomely when prednisone generated more than $20 million in sales for Squibb within a year of its introduction in 1955. For many years, prednisone was as widely prescribed as prednisolone, but ultimately the latter compound was preferred once it was generally realised that it was also the active metabolite of prednisone.

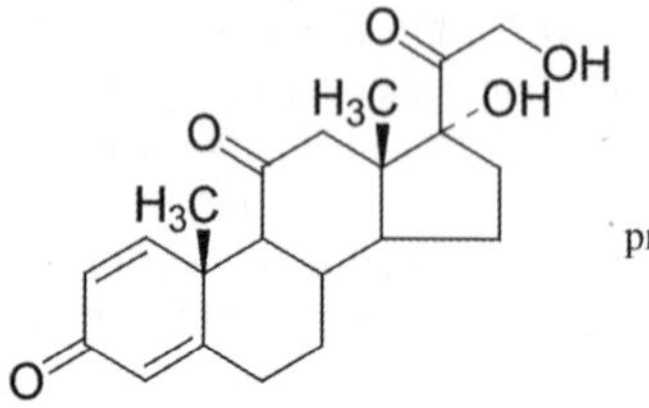

prednisone

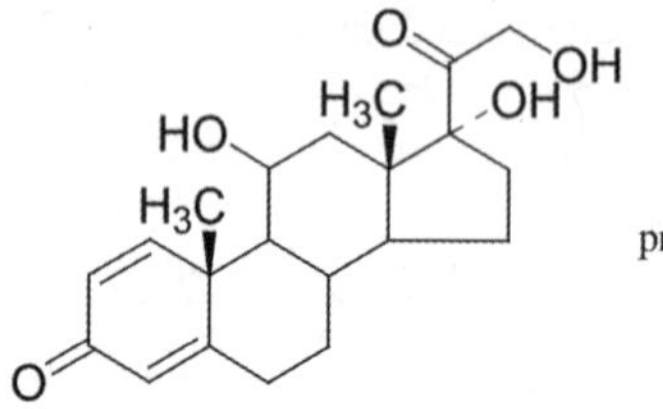

prednisolone

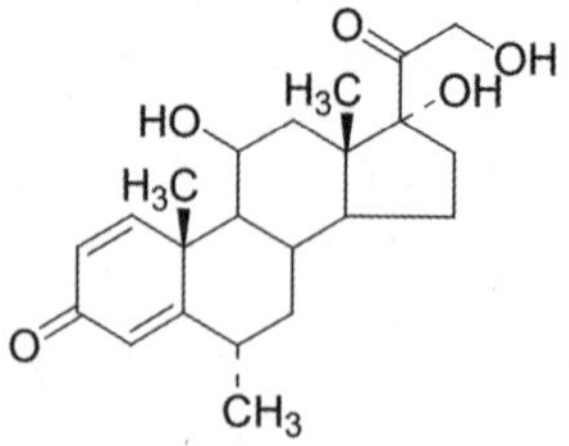

methylprednisolone

Because of its diminished effect on mineral balance, prednisolone has remained the most commonly prescribed steroid for oral anti-inflammatory therapy. Its instant success in the market-place undermined much of the effort of several leading pharmaceutical companies to develop their own syntheses of cortisone or hydrocortisone. However, now that it was known that the potency can be increased by the introduction of a 9α-fluoro substituent and also that glucocorticoid activity can be enhanced through the placing of an extra double bond in ring A, the pharmaceutical industry rapidly set about preparing more corticosteroid analogues. One of the first, methylprednisolone, was designed in the wake of a report that hydrocortisone was metabolically deactivated by oxidation in the liver to its 6-hydroxy derivative. This gave researchers at Upjohn the idea of blocking the metabolism of prednisolone by placing a methyl substituent on the vulnerable 6-position to hinder the fit of the steroid at the surface of the oxidative enzymes. Methylprednisolone proved to be slightly more potent than prednisolone.[97]

Following their isolation from the urine of a boy with an adrenal tumour of corticosteroids substituted with a hydroxyl group at the 16α-position, researchers at Lederle Laboratories, led by Seymour Bernstein, synthesised several steroids of this type so that their biological activity could be examined. They found that the 16α-hydroxyl group decreased mineralocorticoid activity, an observation that was to have far-reaching implications. It was immediately exploited to overcome the major shortcoming of fludrocortisone by also placing a second double bond in ring A. This resulted in the introduction of triamcinolone at the beginning of 1958.[98] Triamcinolone was also synthesised at the same time by Squibb, so a cross-licensing agreement was made between the two companies in order to avoid patent difficulties.

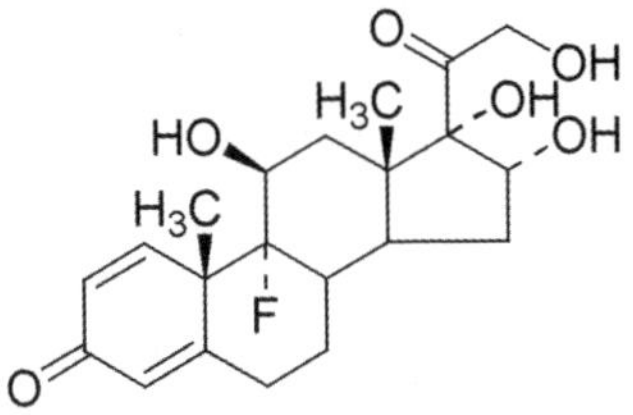

triamcinolone

Clinicians found triamcinolone to be largely free from mineralocorticoid activity, while still as potent an anti-inflammatory agent as prednisolone. Yet it did not live up to its early promise. There was a distinct tendency to cause nausea, dizziness and other unwelcome effects.

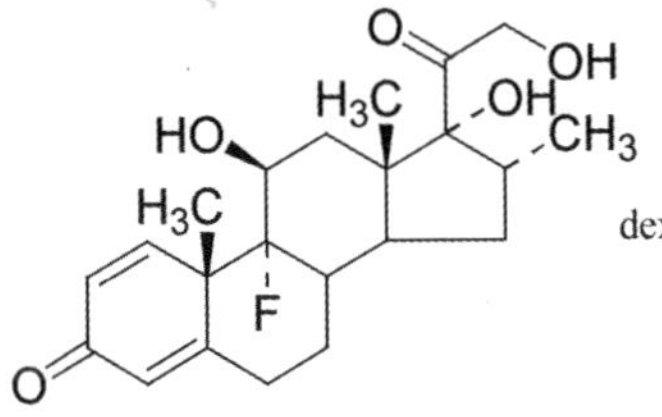

dexamethasone

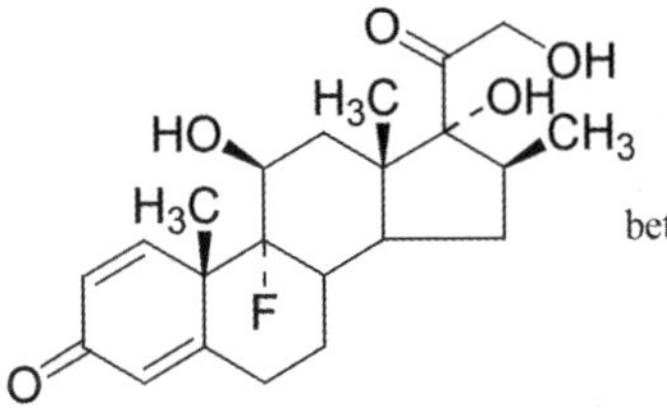

betamethasone

In an attempt to hinder metabolic reduction of the carbonyl at C20 in the side chain attached to corticosteroids, Merck chemists employed a technique previously developed during an investigation when it had been found that introduction of a 16α-methyl group hindered the rapid metabolic reduction of the carbonyl function at C20 on the side chain. They achieved their objective of lengthening the duration of action, but of even greater significance, however, was the discovery that the 16α-methyl group not only acted similarly to a hydroxyl group in that position by depressing mineralocorticoid activity but it also enhanced anti-inflammatory activity. The outcome was the introduction of dexamethasone in 1958.[99] It was six times as potent as prednisolone and 25 times as potent as hydrocortisone, with a duration of action of 36–54 hours (hydrocortisone being about 8–12 hours). Dexamethasone is used when long-term medication with a potent systemic corticosteroid is required, as well as in shock and cerebral oedema. Merck also introduced betamethasone, the isomer of dexamethsasone in which there was a 16β-methyl group.[100] This alteration made little difference to the biological activity, both drugs being six or seven times as potent as prednisolone, with practically no salt-retaining activity at all. Neither of these drugs had the unwelcome side effects previously seen with triamcinolone.

Topical Corticosteroids

The first reports of corticosteroids being applied to the skin appeared in 1952. Since then, this has become a popular means of treating inflammatory conditions such as psoriasis. Although hydrocortisone and its acetate were initially used, the range of products available for treating severe conditions was widened by the development of highly lipophilic, potent compounds. The first of these was triamcinolone acetonide, introduced by Lederle.[101]

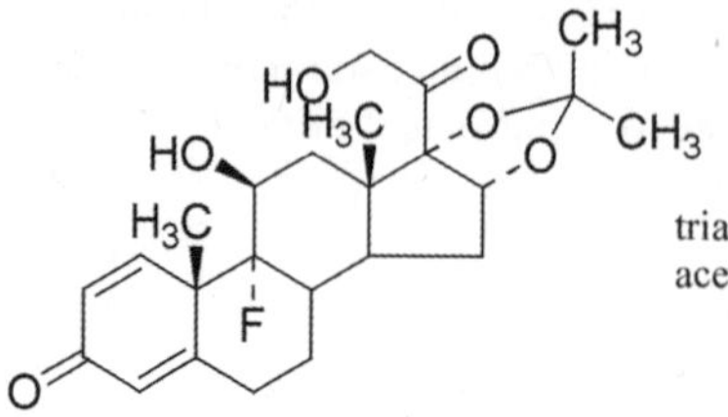

triamcinolone
acetonide

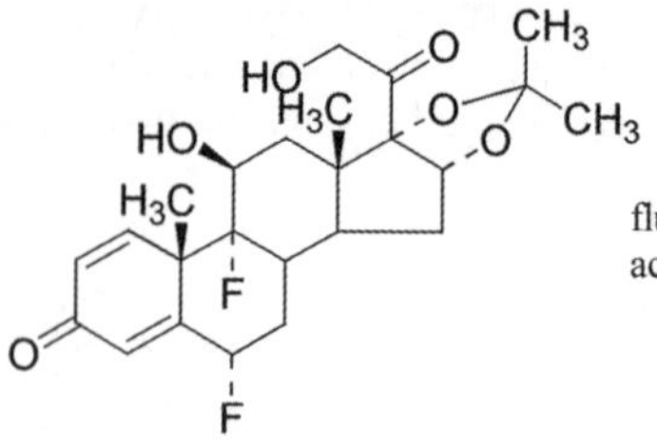

fluocinolone
acetonide

The fact that the acetonide was easily prepared by stirring a suspension of triamcinolone in acetone containing a trace of perchloric acid encouraged other companies to develop similar compounds for topical treatment of severe dermatological conditions. Syntex, for example, placed a fluorine atom at the 6-position of triamcinolone to prévent metabolic oxidation and thereby increase potency in fluocinolone acetonide.[102] Squibb introduced halcinonide, a chlorine-containing analogue of triamcinolone acetonide.[103] As this was a highly potent topical anti-inflammatory agent, it was reserved for the treatment of severe skin conditions that did not respond to milder agents. A British patent on fluocinonide, which was the acetate of fluocinolone acetonide, was granted to the Olin Mathieson Company in 1963, while Syntex obtained a US patent for the same compound the following year. Several other potent corticosteroids have been marketed.

Inhaled Corticosteroids

In 1958, Merck applied for a patent on beclomethasone, a glucocorticoid with similar potency to dexamethasone.[104] Although this was granted four years later, the drug does not appear to have been marketed. However, in the early 1970s Glaxo introduced beclomethasone dipropionate in an aerosol for the treatment of asthma.

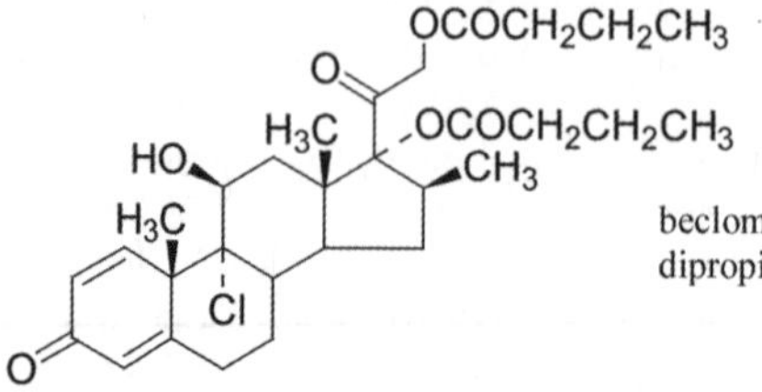

beclomethasone
dipropionate

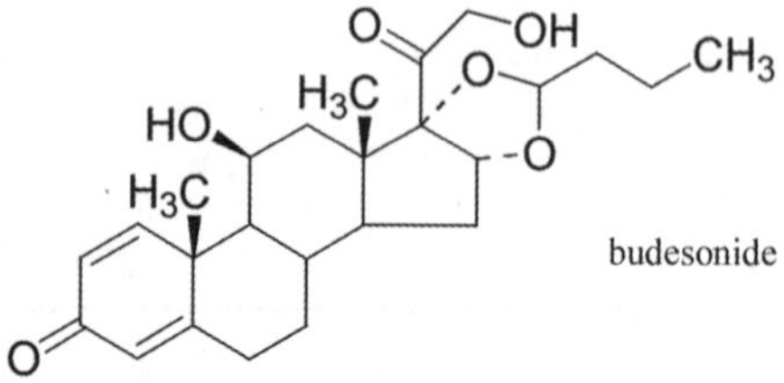

budesonide

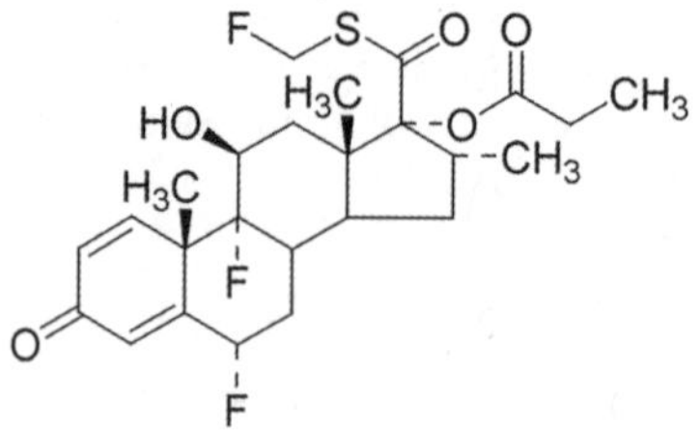

fluticasone propionate

Beclomethasone dipropionate represented a major advance over oral administration of drugs such as prednisolone since the effects were largely confined to the lungs.[105] Since then, other topical steroids have been similarly formulated, including budesonide[106] and fluticasone propionate.[107] With the newer agents, very little of the drug was absorbed, which was of advantage since up to 80% of any inhaled dose of any drug may be swallowed.

ALDOSTERONE ANALOGUES

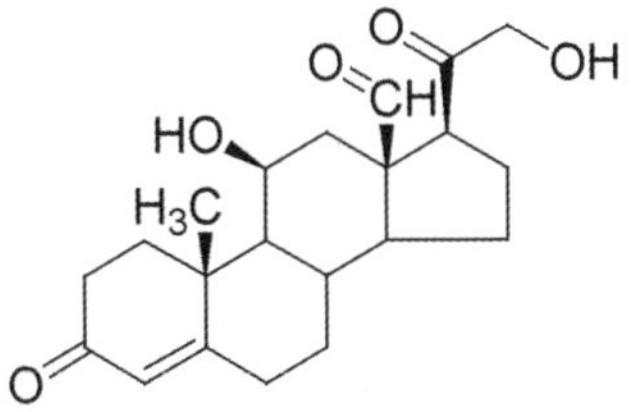

aldosterone

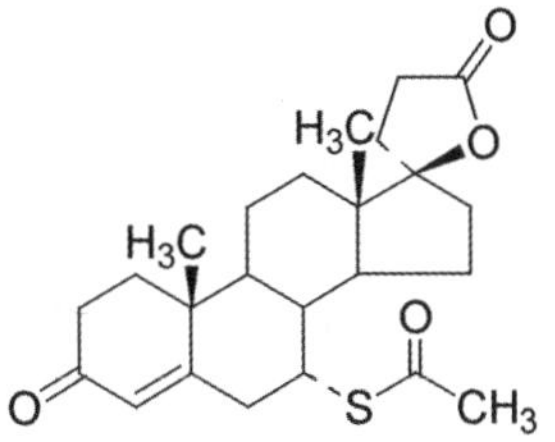

spironolactone

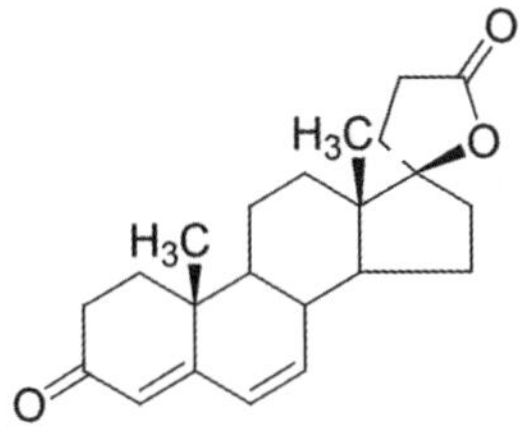

canrenone

Spironolactone was prepared in 1959 by researchers at the laboratories of G.D. Searle & Company in Chicago as an antagonist of aldosterone, which blocks the aldosterone receptor.[108] As it increased the excretion of sodium ions, chloride and water while retaining potassium and magnesium ions, spironolactone was used as a potassium-conserving diuretic, either alone or in combination with other diuretics. In recent years a growing awareness of the role of aldosterone in cardiovascular disease has led to the use of spironoloctone in heart failure.[109]

Canrenone is an active metabolite of spironolactone, which was synthesised at the same time as spironolactone. When parenteral spironolactone treatment is required, canrenone is used in the form of the potassium salt of the open-chain free acid formed from the lactone, viz. potassium canrenoate. This is rapidly converted *in vivo* to canrenone. Canrenone is better tolerated than spironolactone itself, presumably because of the absence of the thiol ester function.

POSTERIOR PITUITARY HORMONE ANALOGUES

Sandoz researchers Stephan Guttmann and Roger Boissonnas prepared analogues of oxytocin and vasopressin in which one or more amino acids were replaced. This initially led to the introduction in 1960 of felypressin as a vasoconstrictor. It is mainly used in combination with prilocaine to prolong local anaesthetic action.[110]

Removal of the free amino group in position 1 of vasopressin was shown to increase antidiuretic activity 2–3 times. Boissonnas then discovered that when D-arginine was also put in place of the L-arginine in position 8, pressor activity was reduced by a factor of more than 1500. In the resulting compound, desmopressin, the antidiuretic activity was almost 4000 times greater than the pressor activity and hence it had no vasoconstrictor action in clinical use.[111] It was also longer-acting than vasopressin. When first marketed for the treatment of diabetes insipidus, desmopressin was administered either by injection or by nasal spray. Tablets are now available, but because of enzymatic destruction they have only one-hundredth of the bioavailability of the hormone administered parenterally. Nonetheless, the introduction of a tablet containing a peptide is significant since such molecules are usually completely digested.

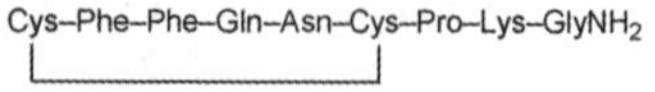

Cys–Phe–Phe–Gln–Asn–Cys–Pro–Lys–GlyNH₂ felypressin

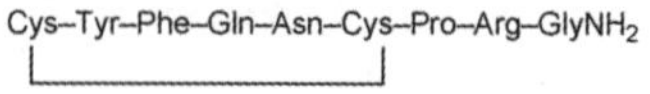

Cys–Tyr–Phe–Gln–Asn–Cys–Pro–Arg–GlyNH₂ vasopressin

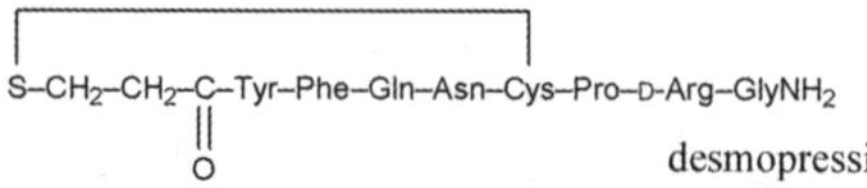

S–CH₂–CH₂–C–Tyr–Phe–Gln–Asn–Cys–Pro–ᴅ-Arg–GlyNH₂
‖
O
desmopressin

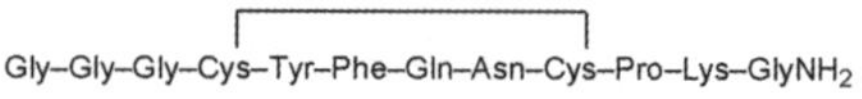

Gly–Gly–Gly–Cys–Tyr–Phe–Gln–Asn–Cys–Pro–Lys–GlyNH₂
terlipressin

Terlipressin was developed as a prodrug of porcine lysine–vasopressin for intravenous administration as an emergency measure to stop bleeding from oesophageal varices. These varicose veins in the lower part of the oesophagus occur as a consequence of portal hypertension in patients with cirrhosis of the liver. In this situation, it is the vasoconstrictor action of vasopressin that is exploited as the glycyl groups are enzymatically cleaved from terlipressin to give a sustained release of the active hormone over a period of 4–6 hours after injection.

HISTAMINE H₂ RECEPTOR ANTAGONISTS

The early antihistamines (see Chapter 28, section on Classical Antihistamines) were unable to antagonise the release of gastric acid caused by the action of histamine, leading to the suggestion by Bjorn Folkow and Georg Kahlson in Gothenburg that there might be more than one type of histamine receptor.[112] This view was reinforced by experiments with alkyl-substituted histamine analogues that did not have equal activity on histamine receptors in different tissues.[113,114] James Black thought there might be a parallel between the situation regarding histamine and that which pertained for adrenaline at the outset of his previous work on the beta-blockers. Having recently moved to the Smith Kline and French Research Institute in Welwyn, Hertfordshire, he speculated as to whether the failure of antihistamines to prevent gastric acid release might be due to their inability to block a different histamine receptor. This raised the possibility that it could be possible to treat gastric ulcers with a new type of antihistamine. The receptors that were blocked by the classical antihistamines were later given the name histamine H₁ receptors,[115] while those in the stomach that were not affected by classical antihistamines were to be named histamine H₂ receptors.[116]

In 1964, Black and Michael Parsons set up an assay procedure to detect the proposed new type of antihistamine, measuring the effect of experimental compounds in preventing the production of acid in the stomach of anaesthetised rats. Robin Ganellin was in charge of the project to find the desired antagonist, assisted by Graham Durant and John Emmett. For four years they synthesised and tested over 200 histamine analogues, but none met their objective of fitting the histamine receptor in the stomach without triggering acid release. The first breakthrough came with a compound that had been synthesised at the start of the project, namely guanylhistamine. Its activity as a blocker of histamine-induced acid release had been overlooked because it mimicked the action of histamine by producing some acid release. This was an example of 'partial agonism' as previously found with dichloroisoprenaline. Within a few days of this being recognised, Emmett found that isosteric replacement of a nitrogen by a sulfur atom to form the isothiourea analogue provided enhanced potency as a blocking agent, although the acid-releasing activity still remained.

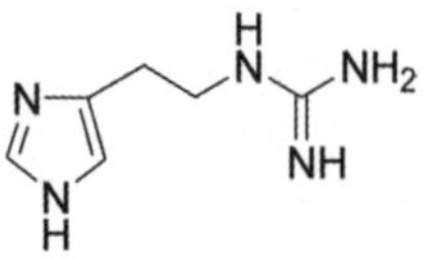

guanylhistamine

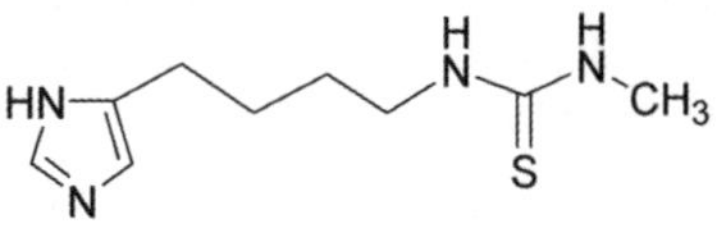

burimamide

Eventually it was realised that so long as the side chain could ionise to acquire an electronic charge, as histamine did, then the acid-releasing activity would remain. This was avoided by replacing the ionisable moiety with a thiourea function. The polarity of the thiourea group resembled that of the amino group in histamine, facilitating the fit of the drug to the gastric receptor.[117] This compound, known as burimamide, was tested in 1970 and proved to be solely a histamine blocker which, on injection, prevented the release of gastric acid. Burimamide inhibited the release of gastric acid in animals and man, but it lacked sufficient potency to be administered by mouth.

By careful consideration of electronic effects influencing the disposition of the hydrogen atom attached to one or other of the nitrogens in the imidazole ring, a derivative was obtained in which an electron-donating methyl group induced protonation on the adjacent nitrogen atom. This was metiamide, which was ten times as potent as burimamide and so could be administered by mouth. Unfortunately, during early clinical trials involving 700 patients, a few cases of a blood disorder, granulocytopenia, were reported and the drug was abandoned even though the condition was reversible.

metiamide

cimetidine

Suspicion fell on the thiourea group, so other polar groups were examined, resulting in the development of cimetidine, which was synthesised in 1972.[118] The Smith Kline and French researchers found that either a nitrile or a nitrovinyl group attached to the guanidino nitrogen atom withdrew the electrons that would otherwise confer basicity and resulted in the molecule being ionised in the body. Cimetidine was shown to be effective in reducing acid release in animals and was introduced into clinical practice in 1976, 12 years after Black's initial experiments. It was a remarkable achievement which even today still represents medicinal chemistry at its best. Ganellin has written a detailed account of the discovery of cimetidine.[119] In the clinic, cimetidine fulfilled its early promise of healing peptic ulcers by reducing gastric acid release. It soon became the world's best-selling drug with sales exceeding $1000 million by 1983. It was eventually overtaken by its analogue ranitidine.

Cimetidine occasionally caused problems through its ability to enter liver cells and bind via its nitrile function to the haem moiety in microsomal cytochrome P450. This interfered with the oxidative metabolism of a variety of drugs.[120] For those drugs with a wide margin of safety it was not a problem, e.g. benzodiazepines and beta-blockers. However, when drugs with a narrow therapeutic window were being taken, then serious problems could arise, e.g. phenytoin, warfarin or theophylline. Also, some elderly patients experienced confusion after taking cimetidine due to its ability to enter the central nervous system. Some male patients developed gynaecomastia or the even rarer impotence from binding of cimetidine to androgen receptors, but this was dose-related and rarely occurred at normal dose levels.

During the early stages of the development of cimetidine, Black and his colleagues investigated a variety of alternatives to the imidazole ring but none proved effective. Smith, Kline and French had been remarkably non-secretive by publishing their ongoing work which was to lead to the development of cimetidine. This enabled a Glaxo research team led by Roy Brittain and David Jack to tackle the problem posed by the lack of oral activity of burimamide. They began by preparing tetrazole isosteres of the imidazole ring in burimamide since the company had considerable experience of tetrazole syntheses. This resulted in the preparation of AH 15475, a tetrazole that had similar potency to burimamide, but in addition exhibited evidence of oral activity in the dog.

Although many tetrazoles were synthesised and tested, no compound emerged as superior to AH 15475. By this time, early 1976, Smith Kline and French had published their first reports on metiamide and its successor cimetidine. The project was continued by Glaxo because it was recognised that the aminotetrazole ring of AH 15475 (pK_a of 1, approximately) lacked the overt basicity of the imidazole ring in burimamide (pK_a of 6.8), yet the compound was just as potent. This indicated that non-basic heterocycles might be considered as alternatives to imidazole.

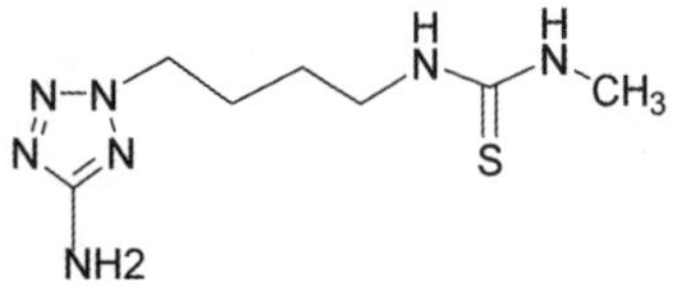

AH 15475

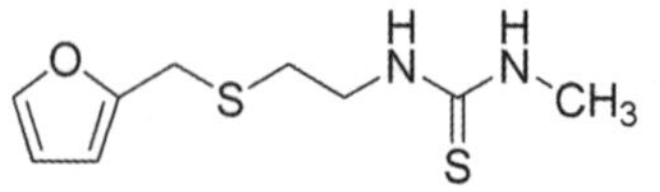

AH 18166

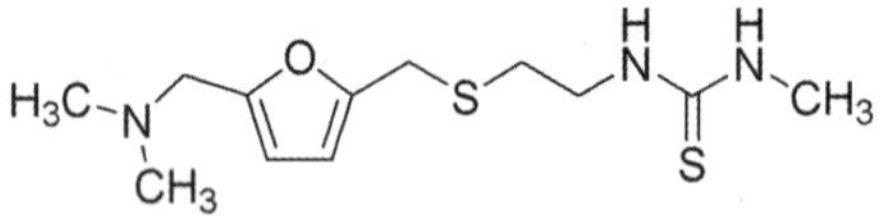

AH 18665

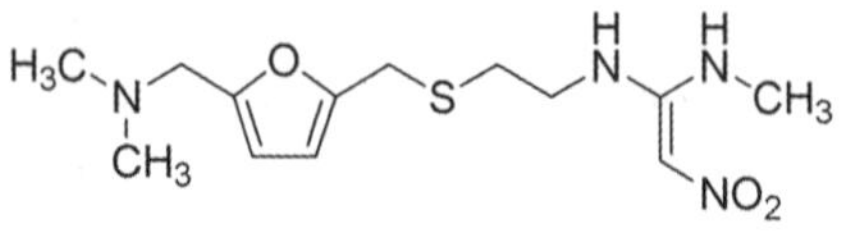

ranitidine

The first heterocycle that was utilised was 2-furfuryl mercaptan since supplies were commercially available. Incorporation of the side chain of metiamide gave AH 18166. This was insoluble and thus gave erratic results during biological evaluation. The problem was overcome by taking advantage of the ability of furans to undergo the Mannich reaction, resulting in the formation of an analogue with a basic side chain that would confer water solubility by permitting salt formation, namely AH 18665. This proved to be as potent as metiamide in various *in vitro* tests. However, concern that AH 18665 would produce agranulocytosis, as had metiamide, led the Glaxo researchers to test its cyanoguanidine analogue, AH 18801. This proved to be very similar to cimetidine in its pharmacological profile. When this turned out to have a low melting point and lacked crystallinity, the Glaxo chemists recalled their previous experience with one of the tetrazole analogues in which replacement of the cyano substituent with a nitrovinyl group overcame this very problem. To their dismay, this did not produce the crystalline compound they had hoped for when a similar alteration was made to the structure of AH 18801. Equally unexpected was the discovery that this new nitrovinyl compound was much more potent than cimetidine. With the approved name of ranitidine, it was destined to become the best-selling drug in the world for many years to come.[121] A detailed account of its discovery has been published.[122]

Ranitidine did not bind to cytochrome P450 since it lacked the nitrile function as it had been replaced in the attempt to obtain a crystalline compound. Furthermore, insertion of the basic dimethylamino substituent to permit water solubility had made ranitidine more polar than cimetidine. This rendered it less likely to enter the brain and cause confusion. Nor did ranitidine produce breast enlargement or impotence from binding of cimetidine to androgen receptors. As many physicians preferred to prescribe it rather than cimetidine because of the reduced likelihood of side effects, for many years it was the best-selling drug in the world, with cimetidine taking second place.

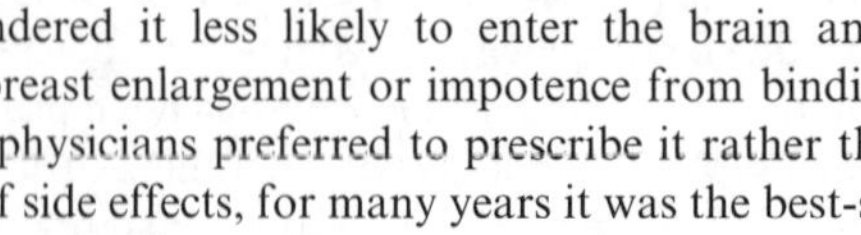

famotidine

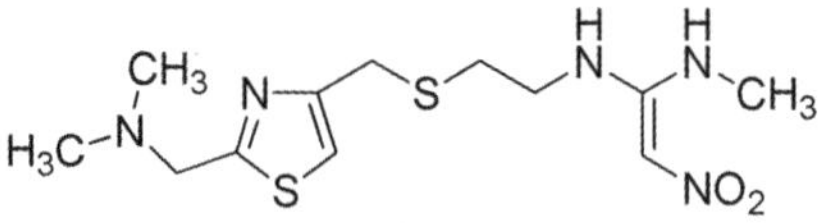

nizatidine

In famotidine, the inherent basicity of the terminal guanidine function is removed by the incorporation of an electron-withdrawing sulfonamido group rather than a nitrile or nitrovinyl group.[123] Famotidine has about 20 times the potency of cimetidine as an H_2 antagonist and is longer-acting, but for most patients this is of little significance since, at equipotent doses, H_2-blockers are all equally effective in the treatment of peptic ulcers. Nizatidine[124] has nine times the potency of cimetidine as an H_2 receptor antagonist,[125] but again this is of little clinical significance.

5-HT ANALOGUES

Once 5-hydroxytryptamine had been implicated in both inflammatory processes and as a chemotransmitter in the central nervous system, pharmaceutical companies showed interest in its potential for therapeutic exploitation. In more recent times, interest has centred on the development of agonists and antagonists displaying receptor selectivity.

Indometacin

Researchers at the Merck Institute for Therapeutic Research, West Point, Pennsylvania, found indometacin to be the most efficacious among 350 indole compounds screened in animals because of the possibility that serotonin might be involved in the inflammatory process.[126] It proved to be both the most potent and the most toxic of the non-steroidal anti-inflammatory drugs. As side effects such as dizziness, headache or gastrointestinal disturbances were common, Merck subsequently introduced sulindac, a prodrug that was metabolised in the liver to form an active metabolite by conversion of the sulfoxide function to a sulfide.[127] It lacked the high potency of indometacin but was well tolerated, apart from gastrointestinal disturbance occurring in some patients.

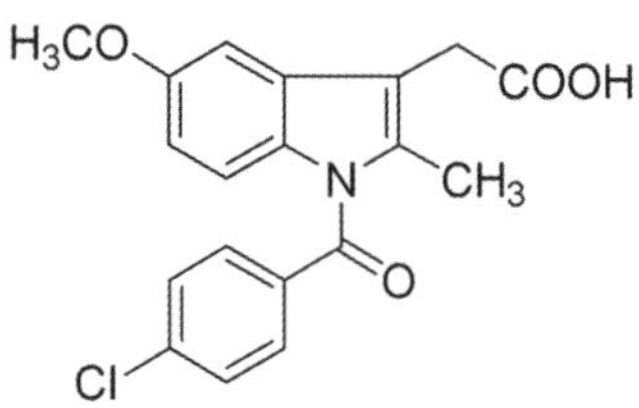

indometacin

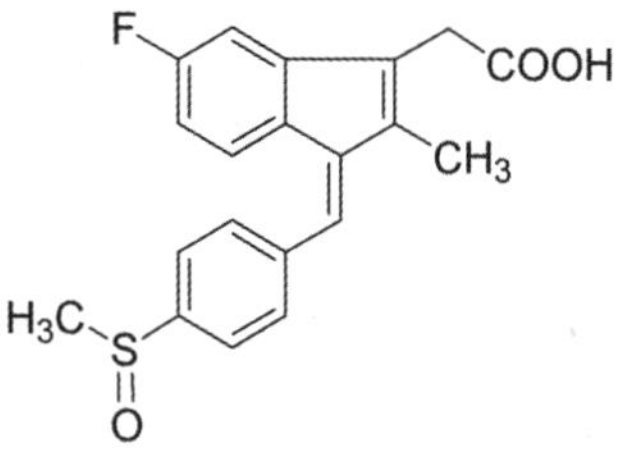

sulindac

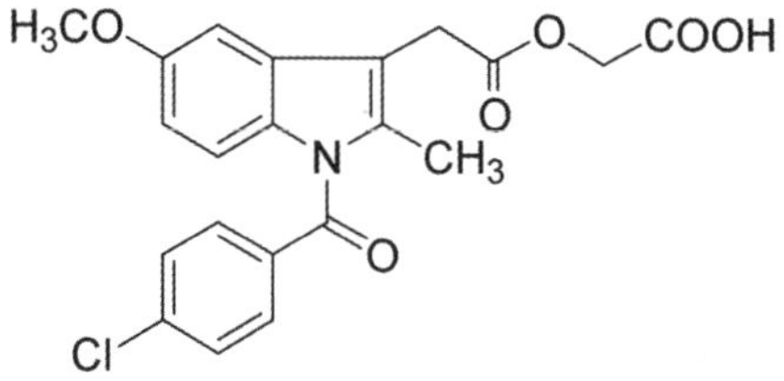

acemetacin

E. Merck and Company in Germany developed acemetacin, the glycolic acid ester of indometacin, in an attempt to provide a safer drug.[128] However, as acemetacin was metabolised to form indometacin, it retained its disadvantages, although gastrointestinal side effects may possibly have been reduced.

Benzydamine

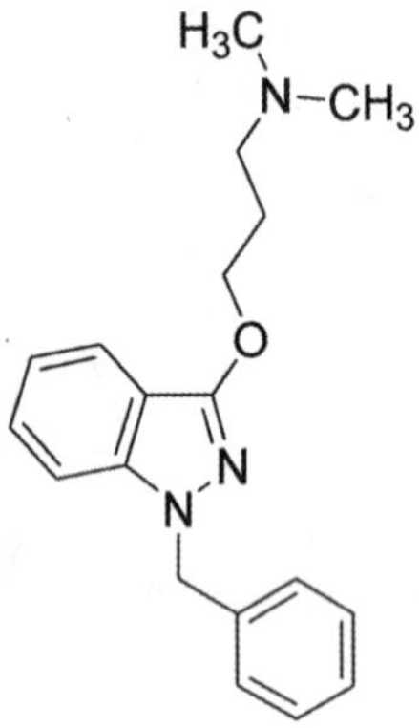

benzydamine

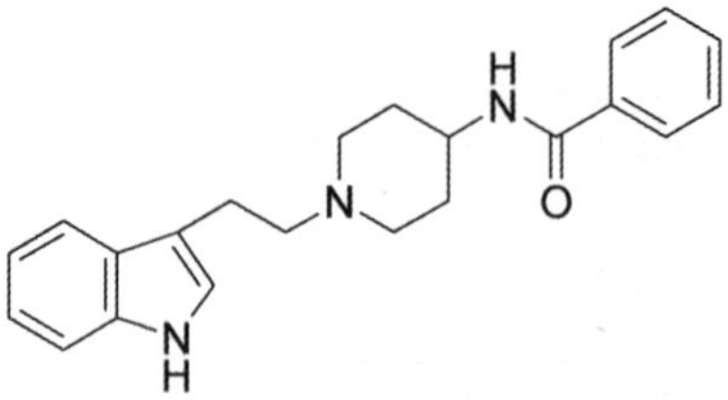

indoramin

·At the research laboratory of Angelini Francesco in Rome, 34 indazoles were synthesised as isosteric analogues of serotonin in which one of its ring nitrogen atoms had been moved.[129] Some of these incorporated molecular features previously found in antihistamines. As little was known about the activity of indazoles, the compounds were screened for anticonvulsive, antispasmodic, analgesic, anti-inflammatory and anaesthetic activity. Several were active and had low toxicity, with benzydamine emerging as an analgesic, anti-inflammatory agent. It was incorporated in topical preparations such as creams and mouthwashes.

Indoramin was synthesised by John Archibald, John Cavalla and their colleagues at the John Wyeth British laboratories near Maidenhead, following the company's discovery that compounds featuring the indolylethylpiperidine moiety reduced blood pressure.[130] It was marketed in the United Kingdom in 1981 and is an alpha D_1-adrenoceptor antagonist with potent antihypertensive activity.

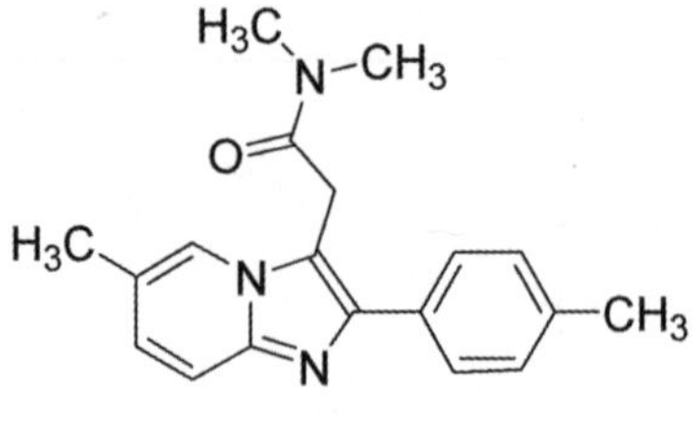

zolpidem

Zolpidem[131] is an indole isostere developed by Jean-Pierre Kaplan and Pascal George of Synthélabo at Vitry-sur-Seine, France. It is a hypnotic that acts primarily at benzodiazepine omega-receptors. Its hypnotic properties are similar to those of the benzodiazepines, but with a less impressive safety profile.

Selective 5-HT Agonists and Antagonists

John Gaddum at the University of Oxford discovered that 5-HT acted on the isolated guinea pig ileum by two distinct mechanisms.[132] The first was directly on smooth muscle, while the second was mediated through the release of acetylcholine from cholinergic nerve endings and was blocked by morphine. The first mechanism was described as involving the serotonin D-receptor, whereas the second was said to involve the serotonin M-receptor. The M-receptor was later renamed as the 5-HT$_3$ receptor. A variety of indole analogues that were 5-HT antagonists had been developed by the late 1970s, yet none of them had selectively blocked the 5-HT$_3$ receptor.

Glaxo researchers recognised that existing methods of testing, such as on the guinea pig ileum, were indirect insofar as they depended upon the detection of the release of one or more neurohormones. Considering this to be unreliable, they instead measured the depolarisation of the rat vagus nerve, a direct indicator of 5-HT$_3$ receptor stimulation. This enabled them to identify 5-HT$_3$ receptor blockers in amines derived from indolylpropanone. After an initially promising series of imidazoles proved to be inactive by mouth, the decision was taken to limit the flexibility of the indolylpropanone side chain by linking it to an adjacent position in the indole ring, thus forming a third ring to give a tetrahydrocarbazolone. When this gave encouraging results, the corresponding analogue of the most promising of the earlier imidazoles was synthesised. It proved active by mouth and in 1990 became the first 5-HT$_3$ antagonist to be introduced into the clinic, with the approved name of 'ondansetron'.[133,134] It was prescribed as an anti-emetic agent to control vomiting induced when chemotherapy releases 5-HT from enterochromaffin cells in the upper gastrointestinal tract. The released 5-HT then acts on both local 5-HT$_3$ receptors as well as on those in the chemoreceptor-trigger zone of the brain to stimulate the vagus nerve and so cause vomiting.

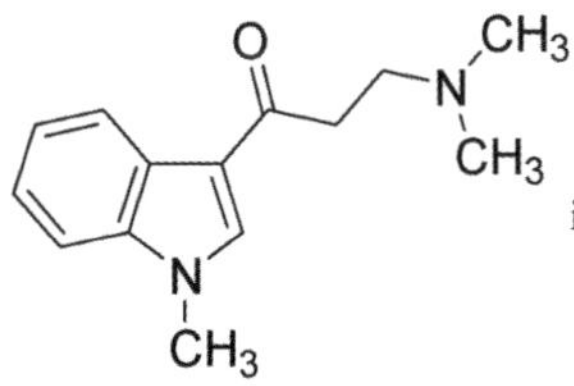

indolylpropanone 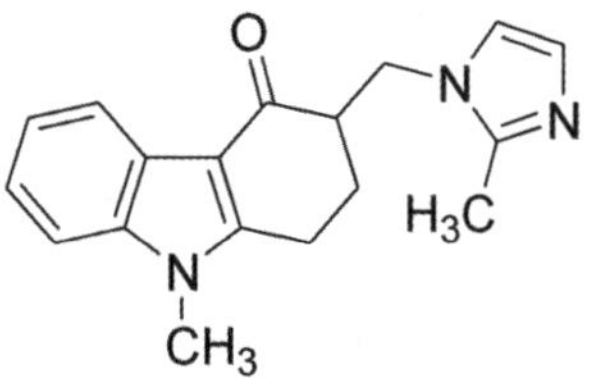   ondansetron

Investigations by John Fozard and his colleagues at the Merrell Dow Research Institute in Strasbourg in the late 1970s revealed that cocaine (and metoclopramide) blocked 5-HT$_3$ receptors.[135] Sandoz researchers then tried to find potent and selective blockers based on the structure of 5-HT itself. They began by extending the ethylamine side chain and then made further analogues in which the side chain nitrogen was provided by incorporation of the tropine ring system (which was similar to that of cocaine). Among these was tropisetron.[136] It proved to be a highly selective, potent 5-HT$_3$ receptor blocker, which was of particular value in dealing with nausea and vomiting caused by anticancer drugs such as cisplatin. This action was due to its ability to block presynaptic 5-HT$_3$ receptors in peripheral neurones which mediate the emetic reflex. Tropisetron was marketed in 1992. Granisetron is a similar drug developed at the Beecham Research Laboratories and introduced into the clinic in 1991.[137]

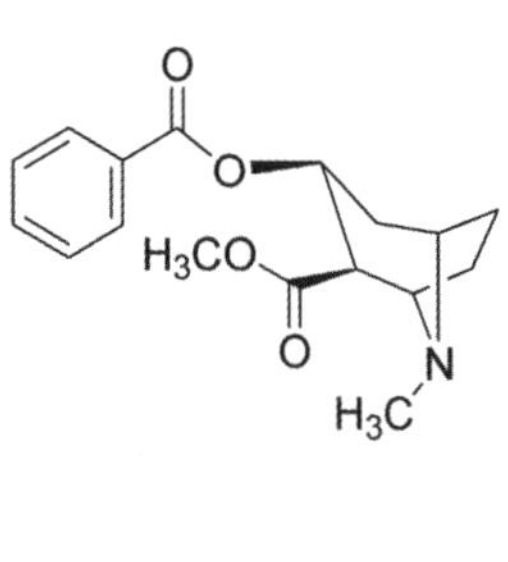

cocaine 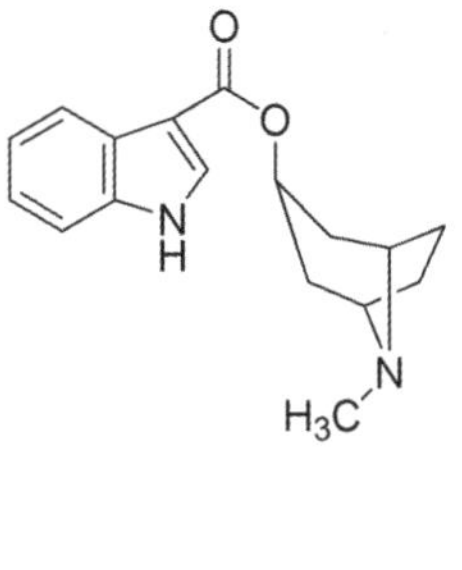   tropisetron

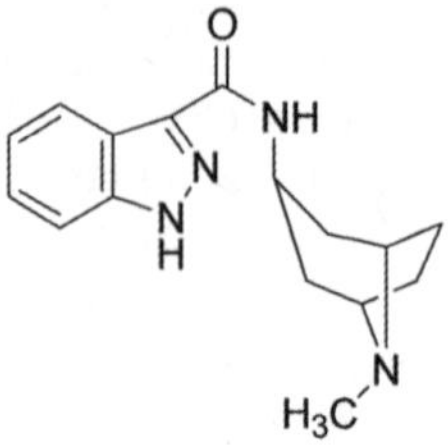

granisetron

Following the suggestion that the antimigraine action of the semi-synthetic alkaloid methysergide was due to its acting as a 5-HT agonist rather as an antagonist, Glaxo researchers decided to develop a non-ergot 5-HT$_1$ agonist. They began by making analogues of 5-HT in which the hydroxyl group was replaced by other substituents. This resulted in compounds that, with one exception, were not only less potent but were also equiactive at 5-HT$_2$ receptors However, activity at 5-HT$_3$ receptors was reduced. The sole compound that was twice as potent at 5-HT$_1$ receptors was also only one-twenty-fifth as potent at 5-HT$_2$ receptors. This was 5-carboxamidotryptamine (5-CT). When tested in a dog to confirm that it constricted the carotid bed, it was found to do the opposite in causing vasodilation and a drop in blood pressure. The conclusion was drawn that it was now also acting on a new type of receptor that was then designated as the 5-HT$_7$ receptor. Alkylating 5-CT resulted in analogues that were more active at the 5-HT$_1$ receptor than at the 5-HT$_7$ one. Further modification led to AH 25086, a compound that was highly selective for the HT$_1$ receptor and which produced constriction of the canine carotid bed. Following extensive testing, AH 25086 was found to relieve migraine in the clinic. Unfortunately, it was inactivated in the liver after being absorbed from the gut after oral administration. This was avoided by making the N-dimethyl derivative at the same time as replacing the amide side chain with a sulfonamide, as this had already been found to be just as effective. The resulting compound, sumatriptan, was one-fifth as active as 5-HT, but it was highly selective.[138] It acts on the carotid arteries, which are stretched by an increased blood flow during a migraine attack, reducing dilation of these vessels.[139] Although it and other triptans should be avoided in patients with cardiac disease, sumatriptan represented a major advance in the treatment of migraine.

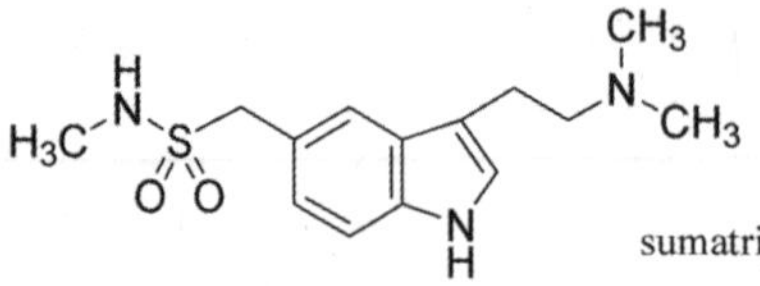

sumatriptan

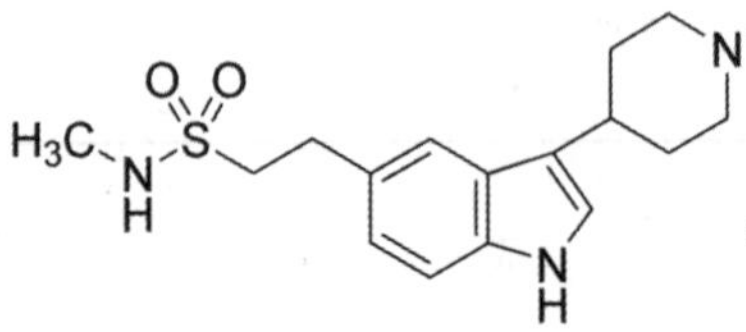

naratriptan

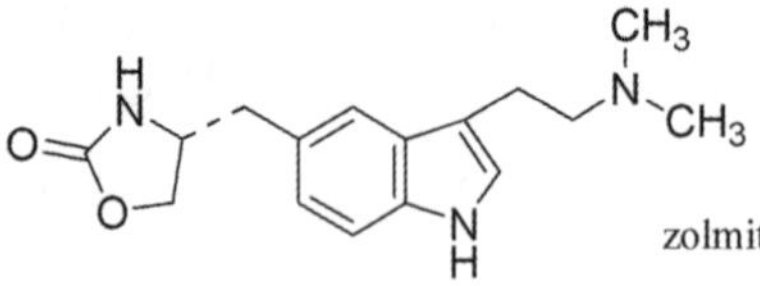

zolmitriptan

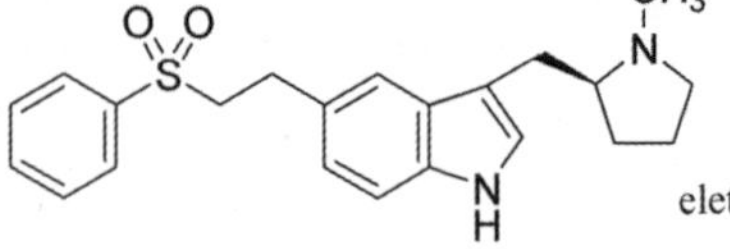

eletriptan

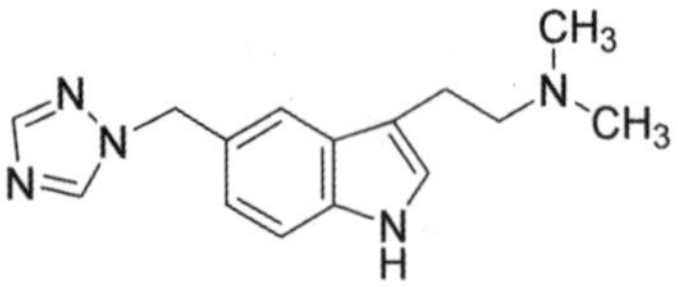

rizatriptan

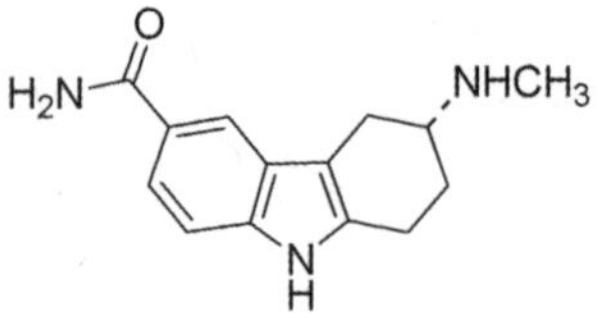

frovatriptan

A number of other orally active triptans have been introduced, including naratriptan,[140] which has a longer lasting action, zolmitriptan,[141] which has a faster onset of action, and also eletriptan,[142] rizatriptan[143] and frovatriptan.[144]

AMINO ACID ANALOGUES

Gustav Stein and Karl Pfister at the Merck Sharpe and Dohme laboratories in Rahway, New Jersey, synthesised methyldopa as an analogue of the amino acid phenylalanine in 1953.[145] It was found by Ted Sourkes at McGill University to be an *in vitro* inhibitor of the enzyme dopa decarboxylase, which plays a major role in catecholamine biosynthesis.[146] In 1960, it was discovered that methyldopa was capable of reducing blood pressure and that its pharmacological profile justified a clinical trial.[147] Since then, its main use has been in controlling raised blood pressure during pregnancy. Inititally, it was assumed that its action was attributable to its effects on the peripheral synthesis of catecholamines, but subsequent studies revealed that it was due to a direct effect on the brain.

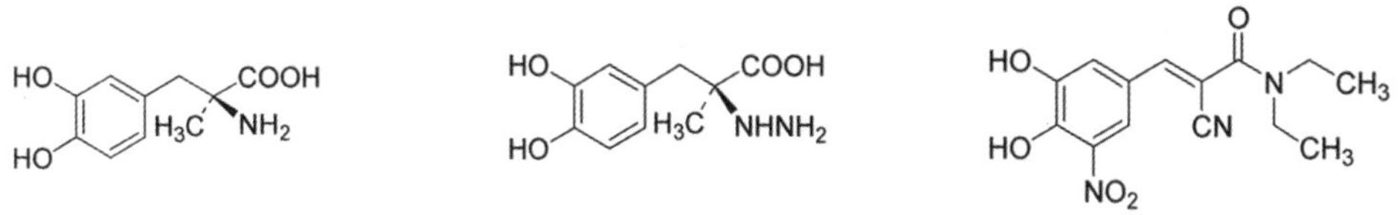

methyldopa   carbidopa   entacapone

Once its ability to reduce blood pressure was recognised, Merck chemists synthesised analogues of methyldopa as potential hypotensive agents.[148] Carbidopa was found to inhibit peripheral conversion of levodopa to dopamine, but it had no action within the central nervous system since it could not pass the blood–brain barrier. By good fortune the first studies on levodopa were then being conducted. It was realised that carbidopa opened up the possibility of using much lower doses of levodopa and reducing the side effects caused by the actions of dopamine outside the central nervous system. Carbidopa was subsequently marketed in a combined formulation with levodopa. Entacapone is a similar drug introduced in the 1990s.[149]

GABA analogues

Baclofen is an analogue of GABA that was synthesised by Ciba chemists soon after GABA was shown to have a role in the central nervous system.[150] It is a specific agonist that binds to spinal $GABA_B$ receptors (a subtype of the GABA receptors) to relax skeletal muscles, making it a useful antispastic agent in multiple sclerosis or spinal disease.[151] As it is a GABA analogue, higher doses exert a sedating effect.

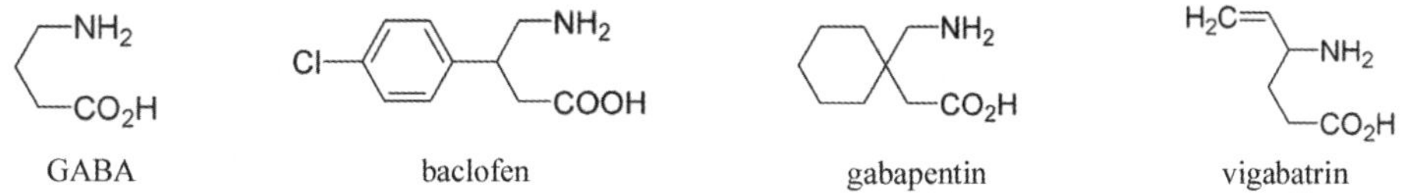

GABA   baclofen   gabapentin   vigabatrin

Gabapentin was designed by Parke, Davis and Company researchers as a GABA analogue that would readily enter the brain.[152] Examination of its three-dimensional structure, however, shows that the spatial positioning of its hydroxyl and amino groups does not correspond with those in GABA and it is now thought to act by binding to a brain protein.[153] It was introduced in the United Kingdom in 1993, licensed for use in conjunction with other anticonvulsants for the management of partial epilepsy.

Vigabatrin[154] is an analogue of GABA that chemically reacts with GABA transaminase to irreversibly inhibit this enzyme, which normally degrades GABA. It is given to patients with partial epilepsy who do not respond to other drugs.[155]

PROSTAGLANDIN ANALOGUES

Being aware that removal of the metabolically sensitive 15-hydroxyl group of alprostadil resulted in loss of pharmacological activity, Paul Collins and his co-workers at G.D. Searle examined the effect of placing this group on the 16-position. The resulting compound had similar antisecretory activity to alprostadil, but was just as short-acting and feebly active by mouth since it was still a substrate for the dehydrogenase enzyme that deactivated natural prostaglandins. After it was discovered that enzymatic oxidation of the 15-hydroxyl group in the side chain of alprostadil could be sterically hindered by placing either one methyl group on the C-15 atom or two methyl groups on the adjacent C-16 atom, Collins placed a methyl group in his active analogue to determine whether it would block enzymatic deactivation. The resulting compound, prepared as the methyl ester, was misoprostol.[156] When administered intravenously, it inhibited gastric acid secretion and was thirty-five times as potent as the analogue without the 16-methyl group. Misoprostol was also orally active and had a greatly increased duration of action. Subsequent studies confirmed that the active compound was the free carboxylic acid.

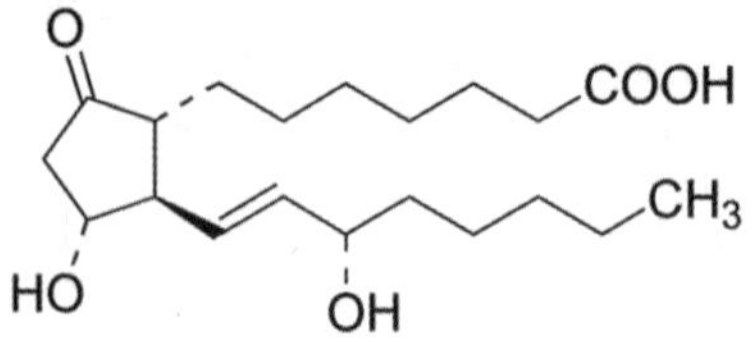

alprostadil

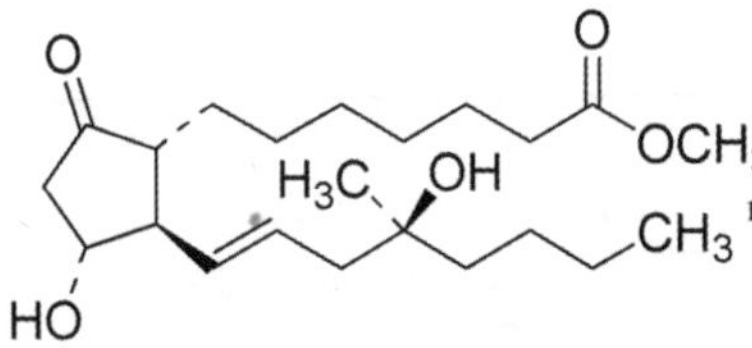

misoprostol

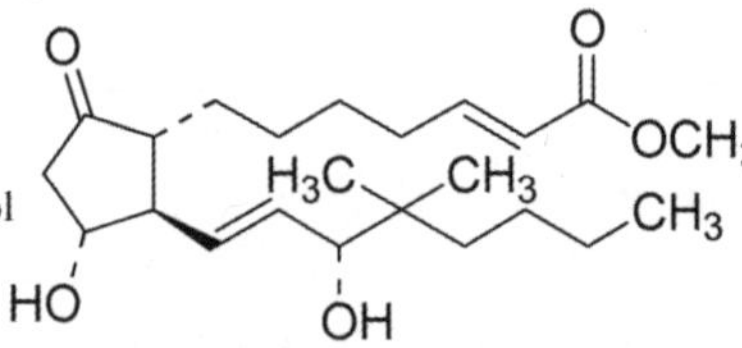

gemeprost

In 1975, André Robert reported that prostaglandins administered to rats concomitantly with indometacin could prevent the intestinal damage caused by indometacin on its own.[157] He suggested that the administered prostaglandins provide mucosal protection and overcome the effects of the indometacin on internal prostaglandin biosynthesis. Consequently, misoprostol was introduced to prevent gastrointestinal ulcers in patients receiving prolonged treatment with indometacin and other non-steroidal anti-inflammatory drugs.

The placing of a second methyl group on the C-16 atom of alprostadil by researchers at the Ono Pharmaceutical Company in Osaka, Japan, provided a longer-acting prostaglandin, namely gemeprost.[158] By administering it in a pessary to ripen and soften the cervix before abortion is induced, the problems associated with lack of specificity of this prostaglandin analogue are deftly avoided.

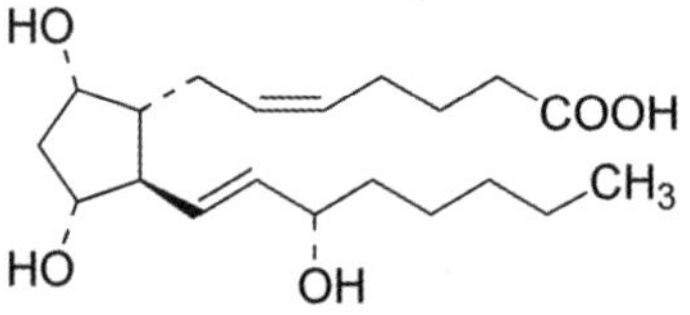

dinoprost

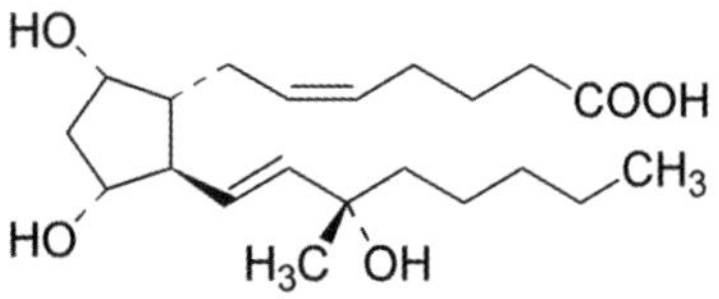

carboprost

The brief duration of action of dinoprost was overcome by Upjohn chemists in 1971 by placing a methyl group on the 15-position to prevent enzymatic oxidation of the hydroxyl group.[159] The resulting compound, carboprost, is administered to stop post-partum bleeding that has not ceased after ergometrine and oxytocin administration.

REFERENCES

1. G. Barger, H.H. Dale, Chemical structure and sympathomimetic action of amines. *J. Physiol. (London)*, 1910; **41**: 19–59.
2. F. Stolz, *Ber.*, 1904; **37**: 4152.
3. F.L. Pyman, CXLII. Isoquinoline derivatives. Part I. Oxidation of laudanosine. *J. Chem. Soc.*, 1909; **95**: 1266–75.
4. Ger. Pat., 1931: 566578 (to Boehringer).
5. US Pat., 1934: 1932347 (to Frederick Stearns & Co.).
6. Ger. Pat. 1936; 638650 (to I.G. Farbenindustrie).
7. A.M. Lands, F.P. Lildueña, J.I. Grant, E. Ananenko, The pharmacological action of some analogs of 1-(3,4-dihydroxyphenyl)-2-amino-1-butanol). *J. Pharmacol. Exp. Ther.*, 1950; **99**: 45.
8. H. Konzett, New highly active broncholytic compounds of the adrenaline series. *Arch. Exp. Path. Pharmakol.*, 1940; **197**: 27–40.
9. O.H. Siegemund, H.R. Granger, A.M. Lands, *J. Pharmacol. Exp. Ther.*, 1947; **90**: 254.
10. R.P. Ahlquist, A study of the adrenotropic receptors. *Am. J. Physiol.*, 1948; **153**: 586–600.
11. Ger. Pat. 1953: 865315 (to Chemische Werke Albert).
12. US Pat. 1967: 3341594 (to Boehringer).
13. H.H. Pelz, Metaproterenol, a new bronchodilator: comparison with isoproterenol. *Am. J. Med. Sci.*, 1967; **253**: 321–34.
14. US Pat. 1867: 3341593 (to Boehringer).
15. K.L. Rominger, W. Pollmann. Vergleichende pharmakokinetik von Fenoterol Hydrobromid bei Ratte, Hund und Mensch. *Arzneim.-Forsch.*, 1972; **22**: 1190–6.
16. Belg. Pat. 1968: 704932 (to Draco).
17. J. Bergman, H. Persson, K. Wetterlin, 2 new groups of selective stimulants of adrenergic beta-receptors. *Experientia.*, 1969; **25**: 899–901.
18. B.K. Pedersen, L.C. Laursen, Y. Gnosspelius, *et al.*, Bambuterol: effects of a new anti-asthmatic drug. *Eur. J. Clin. Pharmacol.*, 1985; **29**: 425–7.
19. A.A. Larsen, W.A. Gould, H.R. Roth, *et al.*, Sulfonanilides. II. Analogs of catecholamines. *J. Med. Chem.*, 1967; **10**: 462–72.
20. D. Hartley, D. Jack, L. Luntz, A.C.H. Ritchie, A new class of selective stimulants of beta-adrenergic receptors. *Nature*, 1968; **219**: 861–2.
21. M. Johnson, The pharmacology of salmeterol. *Lung*, 1990; **168**: 115–16.
22. C.E. Powell, I.H. Slater, Blocking of inhibitory adrenergic receptors by a dichloro analogue of isoproterenol. *J. Pharm. Exp. Ther.*, 1958; **122**: 480–8.
23. N.C. Moran, M.E. Perkins, Adrenergic blockade of the mammalian heart by a dichloro analogue of isoproterenol. *J. Pharm. Exp. Ther.*, 1958; **12**: 223–7.
24. R.G. Shanks, in *Discoveries in Pharmacology*, Vol. 2: *Haemodynamics, Hormones and Inflammation*, eds M.J. Parnham, J. Bruinvels. Oxford: Elsevier; 1984, p. 37.
25. J.W. Black, J.S. Stephenson, Pharmacology of a new adrenergic beta-receptor antagonist. *Lancet*, 1962; **2**: 311–14.
26. J.P. Stock, N. Dale, Beta-adrenergic receptor blockade in cardiac arrhythmias. *Br. Med. J.*, 1963; **2**: 1230–3.
27. B.N.C. Prichard, P.M.S. Gillam, Use of propranolol (inderal) in treatment of hypertension. *Br. Med. J.*, 1964; **2**: 725–7.

28. R. Howe, Carcinogenicity of 'Alderlin' (pronethalol) in mice. *Nature*, 1965; **207**: 594–5.
29. Belg. Pat. 1964: 640312 (to ICI).
30. A. Brandstrom, H. Corrodi, U. Junggren, T.E. Jonsson, Synthesis of some beta-adrenergic blocking agents. *Acta Pharm. Suec.*, 1966; **3**: 303–10.
31. Belg. Pat. 1966: 669402 (to Ciba).
32. B.K. Wasson, W.K. Gibson, R.S. Stuart, *et al.*, β-Adrenergic blocking agents. 3-(3-Substituted-amino-2-hydroxypropoxy)-4-substituted-1,2,5-thiadiazoles. *J. Med. Chem.*; 1972; **15**: 651.
33. M.E. Condon, C.M. Cimarusti, R. Fox, *et al.*, Nondepressant beta-adrenergic blocking agents. 1. Substituted 3-amino-1-(5,6,7,8-tetrahydro-1-naphthoxy)-2-propanols. *J. Med. Chem.*, 1978; **21**: 913–22.
34. A.A. Larsen, P.M. Lish, A new bio-isostere: alkylsulphonamidophenethanolamines. *Nature*, 1964; **203**: 1283–4.
35. P.M. Lish, J.H. Weikel, K.W. Dungan, Pharmacological and toxicological properties of two new-adrenergic receptor antagonists. *J. Pharmacol. Exp. Ther.*, 1965; **149**: 161–73.
36. L.H. Smith, *J. Appl. Chem. Biotechnol.*, 1978; **28**: 201.
37. J.F. Giudicelli, J.R. Boissier, Y. Dumas, C. Advenier, [Comparative effects of several beta-adrenergic inhibitors on pulmonary resistance in guinea pigs]. *C. R. Soc. Biol. Paris*, 1973; **167**: 232–7.
38. M.F. Cuthbert, K. Owusu-Ankomah, Effect of M & B 17803A, a new-adrenoceptor blocking agent, on the cardiovascular responses to tilting and to isoprenaline in man. *Br. J. Pharmacol.*, 1971; **43**: 639–48.
39. J. Bonelli, D. Magometchnigg, G. Hitzenberger, G. Kaik, Haemodynamic characterisation of a new beta-receptor blocking agent, celiprolol, at rest and during ergometer exercise, compared with propranolol. *Wien. Klin. Wochenschr.*, 1978; **90**: 350–4.
40. J.E. Clifton, I. Collins, P. Hallett, *et al.*, Arylethanolamines derived from salicylamide with alpha- and beta-adrenoceptor blocking activities. Preparation of labetalol, its enantiomers and related salicylamides. *J. Med. Chem.*, 1982; **25**: 670–9.
41. K.B. Mallion, A.H. Todd, R.W. Turner, *et al.*, 2-(2-Ethoxyphenoxymethyl)tetrahydro-1,4-oxazine hydrochloride, a potent psychotropic agent. *Nature*, 1972; **238**: 157.
42. U.S. Pat. 1936: 2054271 (to Schering–Kahlbaum).
43. H.H. Inhoffen, W. Hohlweg, New female glandular derivatives active per os. *Naturwiss.*, 1938; **26**: 96.
44. J.W. Cook, E.C. Dodds, C.L. Hewett, A synthetic oestrus-exciting compound. *Nature*, 1933; **131**: 56–7.
45. E.C. Dodds, W. Lawson, Synthetic oestrogenic agents without the phenanthrene nucleus. *Nature*, 1936; **137**: 996.
46. E.C. Dodds, L. Goldberg, W. Lawson, R. Robinson, Oestrogenic activity of certain synthetic compounds. *Nature*, 1938; **141**: 247–8.
47. C.H. Carlisle, D. Crowfoot, A determination of molecular symmetry in the α-β-diethyl-dibenzyl series. *J. Chem. Soc.*, 1941: 6–9.
48. R.M. Giusti, K. Iwamoto, E.E. Hatch, Diethylstilbestrol revisited. A review of the long-term health effects. *Ann. Intern. Med.* 1995; **122**: 778–88.
49. I. Wugmeister, Le traitement de l'hypertrophie de la prostate par doses massives de Folliculine. *Paris Med.*, 1937; **1**: 535–6.
50. P. Kahle, E. Maltry, Treatment of hyperplasia of prostate with diethylstilbestrol and diethylstibestrol dipropionate: preliminary result. *New Orleans Med. Surg. J.*, 1940, **93**: 121.
51. C. Huggins, P.J. Clark, Quantitative studies of prostatic secretion. II. The effect of castration and of estrogen injection on the normal and on the hyperplastic prostate glands of dogs. *J. Exp. Med.*, 1940; **72**: 747–61.
52. C. Huggins, C.V. Hodges, Studies on prostatic cancer. I. The effect of castration, of estrogen and of androgen injection on serum phosphatases in metastatic carcinoma of the prostate. *Cancer Res.*, 1941; **1**: 293–7.
53. A. Haddow, J.M. Watkinson, E. Paterson, Influence of synthetic oestrogens upon advanced malignant disease. *Br. Med. J.*, 1944; **2**: 393–8.
54. J.M. Robson, A. Schonberg, A new synthetic oestrogen with prolonged action when given orally. *Nature*, 1942; **150**: 22–3.
55. Br. Pat. 1944: 561508 (to ICI).
56. J. Kato, T. Kobayashi, C.A. Villee, Effect of clomiphene on the uptake of estradiol by the anterior hypothalamus and hypophysis. *Endocrinol.*, 1968; **82**: 1049–52.
57. US Pat. 1959: 2914563 (to Merrell).
58. G.R. Bedford, D.N. Richardson, Preparation and identification of *cis* and *trans* isomers of a substituted triarylethylene. *Nature*, 1966; **212**: 733–4.
59. M.J.K. Harper, A.L. Walpole, Mode of action of ICI 46,474 in preventing implantation in rats. *J. Endocrinol.*, 1967; **37**: 83–92.

60. A. Klopper, M. Hall, New synthetic agent for the induction of ovulation: preliminary trials in women. *Br. Med. J.*, 1971; **1**: 152–4.
61. M.P. Cole, C.T.A. Jones, I.D. Todd, A new antiestrogenic agent in late breast cancer. An early appraisal of ICI 46,474. *Br. J. Cancer*, 1971; **25**: 270–5.
62. Early Breast Cancer Trialists' Collaborative Group, Tamoxifen for early breast cancer: an overview of the randomised trials. *Lancet*, 1998; **351**: 1451–67.
63. D.A. Marsh, H.J. Brodie, W. Garrett, *et al.*, Aromatase inhibitors. Synthesis and biological activity of androstenedione derivatives. *J. Med. Chem.*, 1985; **28**: 788–95.
64. L. Ruzicka, K. Hofmann, Uber die Anlagerung von Acetylen an die 17-stangige Ketogruppe bei trans-androsteron und D5-trans-dehydro-androsterone. *Helv. Chim. Acta*, 1937; **20**: 1280–2.
65. J. Kathol, W. Logemann, A. Serini, Ein Übergang aus der Androstan-Reihe in die Pregnan-Reihe. *Naturwissen.*, 1937; **25**: 682.
66. H.H. Inhoffen, W. Logemann, W. Hohlweg, A. Serini, Sex hormone series. *Ber.*, 1938; **71**: 1024–32.
67. M. Ehrenstein, Investigations on steroids. VIII. Lower homologs of hormones of the pregnane series: 10-nor-11-desoxy-corticosterone acetate and 10-norprogesterone. *J. Org. Chem.*, 1944; **9**: 435–6.
68. W.M. Allen. M. Ehrenstein, 10-Nor-progesterone, a physiologically active lower homolog of progesterone. *Science*, 1944; **100**: 251.
69. A.J. Birch, Hydroaromatic steroid hormone. I. 10-Nortestosterone. *J. Chem. Soc.*, **1950**: 367.
70. C. Djerassi, L.E. Miramontes, G. Rosenkranz, G. Sondheimer, Steroids. LIV. Synthesis of 19-nor 17-alpha ethynyl testosterone and 19-nor 17-alpha methyl testosterone. *J. Am. Chem. Soc.*, 1954; **76**: 4092–6.
71. US Pat. 1954: 2691028 (to Searle).
72. H. Smith, G.A. Hughes, G.H. Douglas, *et al.*, Totally synthetic (+ −) 13-alkyl-3-hydroxy and methoxy-gona-1,3,5 (10)-trien-17-ones and related compounds. *Experientia*, 1963; **9**: 394–6.
73. A.W. Makepeace, G.L. Weinstein, M. H. Friedman, The effect of progestin and progesterone on ovulation in the rabbit. *Am. J. Physiol.*, 1937; **119**: 512–16.
74. B. Asbell, *The Pill. A Biography of the Drug That Changed the World.* London: Random House; 1995.
75. G. Pincus, Progestational agents and the control of fertility. *Vitam. Horm.*, 1959; **17**: 307.
76. A.J. van den Broek, C. van Bokhoven, P.M.J. Hobbelen, J. Leemhuis, 11-Alkykidene steroids in the 19-nor series. *Rec. Trav. Chim.*, 1975; **94**: 36.
77. G. Teutsch, R. Deraedt, D. Philibert, Mifepristone, in *Chronicles of Drug Discovery*, Vol. 3, ed. D. Lednicer. Washington: American Chemical Society; 1993, p. 1.
78. W. Herrmann, R. Wyss, A. Riondel, *et al.*, Effer d'un steroide anti-progestérone chez la femme: interuption du cycle menstruel et de la grossesse au début. *C. R. Acad. Sci.*, 1982; **294**: 933–8.
79. K. Junkmann, Über protrahiert wirksame Gestagene. *Arch. Exp. Pathol. Pharmakol.*, 1954; **223**: 244–53.
80. US Pat. 1968: 3377364 (to Upjohn).
81. Fr. Pat. 1962: 1 295307 (to Syntex)
82. B. Ellis, D.N. Kirk, V. Petrow, B. Waterhouse, D.M. William, Modified steroid hormones. XVII. Some 6-methyl-4,6-dienes. *J. Chem. Soc.*, 1960: 2828–33.
83. US Pat. 1969: 3485852 (to Syntex).
84. Ger. Pat. 1960: 1075114 (to E. Merck).
85. K. Miescher, H. Kagi, C. Scholz, *et al.*, Weitere untersuchungen über die Wirkungverstärkung mänlicher Sexualhormon. *Biochem. Z.*, 1937; **294**: 39.
86. C.D. Kochakian, J.R. Murlin, The effect of male hormone on the protein and energy metabolism of castrate dogs. *J. Nutrition*, 1935; **10**: 437–59.
87. C.D. Kochakian,Testosterone and testosterone acetate and the protein and energy metabolism of castrate dogs. *Endocrinology*, 1937; **21**: 750–5.
88. A.T. Kenyon, K. Knowlton, I. Sandiford, *et al.*, A comparative study of the metabolic effects of testosterone propionate in normal men and women and in eunuchoidism. *Endocrinology*, 1940; **26**: 26–45.
89. F.B. Colton, L.N. Nysted, B. Riegel, A.L. Raymond, 17-Alkyl-19-nortestosterones. *J. Am. Chem. Soc.*, 1957; **79**: 1123–7.
90. A.J. Birch, The reduction of organic compounds by metal–ammonia solutions. *Quart. Rev. Chem. Soc.*, 1950; **4**: 69–93.
91. P.C. Walsh, J.D. Madden, M.J. Harrod, *et al.*, Familial incomplete male pseudohermaphroditism, type 2: decreased dihydrotestosterone formation in pseudovaginal perineoscrotal hypospadias. *New Engl. J. Med.*, 1974; **291**: 944–9.
92. G.H. Rasmusson, G.F. Reynolds, N.G. Steinberg, *et al.*, Azasteroids: structure–activity relationships for inhibition of 5-α-reductase and of androgen receptor binding. *J. Med. Chem.*, 1986, **29**: 2298–315.

93. L.H. Sarett, A.A. Patchett, S. Steelman, The effects of structural alteration on the antiinflammatory properties of hydrocortisone. *Prog. Drug Res.*, 1963; **5**: 11–154.
94. J. Fried, E.F. Sabo. 9α-Fluoro derivatives of cortisone and hydrocortisone. *J. Am. Chem. Soc.*, 1954; **76**: 1455–6.
95. H.L. Herzog, A. Nobile, S. Tolksdorf, *et al.*, New anti-arthritic steroids. *Science*, 1955; **121**: 176.
96. T. Mahoney, *The Merchants of Life*. New York: Harper; 1959; p. 260.
97. G.B. Spero, J.J. Thompson, B.J. Magerlein, *et al.*, Adrenal hormones and related compounds. IV. 6-Methyl steroids. *J. Am. Chem. Soc.*, 1956; **78**: 6213.
98. S. Bernstein, R.H. Lenhard, W.S. Allen, *et al.*, 16-Hydroxylated steroids. IV. The synthesis of the 16α-hydroxy derivatives of 9α-halo-steroids. *J. Am. Chem. Soc.*, 1956; **78**: 5693–4.
99. G.E. Arth, J. Friend, D.B.R. Johnston, *et al.*, 16-Methylated steroids. II. 16α-Methyl analogs of cortisone, a new group of anti-inflammatory steroids. 9α-halo derivatives. *J. Am. Chem. Soc.*, 1958; **80**: 3161–3.
100. D. Taub, R.D. Hoffsommer, H.L. Slates, N.L. Wendler. 16β-Methyl cortical steroids. *J. Am. Chem. Soc.*, 1958; **80**: 4435.
101. J. Fried, A. Borman, W.B. Kessler, *et al.*, Cyclic 16α,17α-ketals and acetals of 9α-fluoro-16α-hydroxy-cortisol and -prednisolone. *J. Am. Chem. Soc.*, 1958; **80**: 2338–9.
102. J.S. Mills, A. Bower, C. Djerassi, H.J. Ringold. Steroids. CXXXVII. Synthesis of a new class of potent cortical hormones. 6,9-difluoro-16-hydroxyprednisolone and its acetonide. *J. Am. Chem. Soc.*, 1960; **82**: 3399–404.
103. S. Bernstein, R.B. Brownfield, R.H. Lenhard, *et al.*, Hydroxylated steroids. XXIII. 21-Chloro-16α-hydroxycorticoids and their 16α,17α-acetonides. *J. Org. Chem.*, 1962; **27**: 690–2.
104. Br. Pat. 1962: 912378 (to Merck).
105. H.M. Brown, G. Storey, W.H. George, Beclomethasone dipropionate: a new steroid aerosol for the treatment of allergic asthma. *Br. Med. J.*, 1972; **1**: 585–90.
106. A. Thalén, R. L. Brattsand, Synthesis and anti-inflammatory properties of budesonide, a new non-halogenated glucocorticoid with high local activity. *Azneim.-Forscrh.*, 1979; **29**: 1687–90.
107. G.H. Phillips, Structure–activity relationships of topically active steroids: the selection of fluticasone propionate. *Respir. Med.*, 1990; **84**(Suppl. A): 19–23.
108. J.A. Cella, R.C. Tweit, Steroidal aldosterone blockers. II. *J. Org. Chem.*, 1959; **24**: 1109–19.
109. J.A. Delyani, Mineralocorticoid receptor antagonists: the evolution of utility and pharmacology. *Kidney Int.*, 2000; **57**: 1408–11.
110. S. Guttmann, R.A. Boissonnas, Synthèse de dix analogues de l'oxytocine et de la lysine-vasopressine contenant de la sérine, de l'histidine ou du tryptophane en position 2 ou 3. *Helv. Chim. Acta*, 1960; **43**: 200–16.
111. R.L. Huguenin, R.A. Boissonnas, Synthèses de la desamino1-Arg8-vasopressine et de la desamino1-Phe2-Arg8-vasopressin, deux analogues possedant une active antidiuretique plus élévee et plus selective que celle des vasopressines naturelles. *Helv. Chim. Acta*, 1966; **49**: 695–705.
112. B. Folkow, K. Haeger, G. Kahlson, Observations on reactive hyperaemia as related to histamine, on drugs antagonizing vasodilatation induced by histamine. *Acta Physiol. Scand.*, 1948; **15**: 264.
113. G.A. Alles, B.B. Wisegarnier, M.A. Shull, Comparative physiological actions of some β-9-imidazolyl-4)alkylamines. *J. Pharm. Exp. Ther.*, 1943; **77**: 54.
114. M.I. Grossmann, C. Robertson, C.E. Rosiere, The effect of some compounds related to histamine on gastric acid secretion. *J. Pharm. Exp. Ther.*, 1952; **104**: 277–83.
115. A.S.F. Ash, H.O. Schild, Receptors mediating some actions of histamine. *Br. J. Pharmacol.*, 1966; **27**: 427–9.
116. J.W. Black, W.A.M. Duncan, G.J. Durant, *et al.*, Definition and antagonism of histamine H2-receptor. *Nature*, 1972; **236**: 385–90.
117. C.R. Ganellin, Medicinal chemistry and dynamic structure–activity analysis in the discovery of drugs acting at histamine H2 receptors. *J. Med. Chem.*, 1981; **24**: 913–20.
118. R.W. Brimblecombe, W.A. Duncan, G.J. Durant, *et al.*, The pharmacology of cimetidine, a new histamine H2-receptor antagonist. *Br. J. Pharmacol.*, 1975; **53**: 435P–436P.
119. C.R. Ganellin, in *Medicinal Chemistry*, 2nd edn, eds C.R. Ganellin, S.M. Roberts. London: Academic Press; 1993, p. 228.
120. S. Rendic, V. Sunjic, R. Toso, F. Kajfez, Interaction of cimetidine with liver microsomes. *Xenobiot.*, 1979; **9**: 555–64.
121. J. Bradshaw, R.T. Brittain, J.W. Clitherow, *et al.*, Ranitidine (AH19065): a new potent, selective histamine H2-receptor antagonist. *Br. J. Pharmacol.*, 1979; **66**: 464P.
122. J. Bradshaw, in *Chronicles of Drug Discovery*, Vol 3, ed. D. Lednicer. Washington: American Chemical Society; 1993, p. 45.
123. M. Takeda, T. Takagi, Y. Yashima, H. Maeno, Effect of a new potent H2-blocker. *Artzneim.-Forschr.*, 1982; **32**: 734–7.

124. Eur. Pat. Applic. 1982, 49618 (to Eli Lilly).
125. K. Bemis, A. Bendele, J. Clemens, et al., General pharmacology of nizatidine in animals. Arzneim.-Forschr., 1989; **39**: 240–50.
126. T.Y. Shen, T.B. Windholz, A. Rosegay, et al., Non-steroid anti-inflammatory agents. J. Am. Chem. Soc., 1963; **85**: 488–9.
127. R.F. Shuman, S.H. Pines, W.E. Shearin, et al., A sterically efficient synthesis of (Z)-5-fluoro-2-methyl-1-(p-methylthiobenzylidene)-3-indenylacetic acid and its S-oxide, sulindac. J. Org. Chem., 1977; **42**: 1914–19.
128. Ger. Pat. 1972: 2234651 (to Troponwerke).
129. G. Palazzo, G. Corsi, L. Baiocchi, B. Silvestrini, Synthesis and pharmacological properties of 1-substituted 3-dimethylaminoalkoxy-1H-indazoles. J. Med. Chem., 1966; **9**: 38 41.
130. J.L. Archibald, B.J. Alps, J.F. Cavalla, J.L. Jackson, Synthesis and hypotensive activity of benzamidopiperidylethylindoles, J. Med. Chem., 1971; **14**: 1054-59.
131. S. Arbilla, H. Depoortere, P. George, S.Z. Langer, Pharmacological profile of the imidazopyridine zolpidem at benzodiazepine receptors and electrocorticogram in rats. Arch. Pharmacol., 1985; **330**: 248–51.
132. J.H. Gaddum, Z.P. Picarelli, Two kinds of tryptamine receptor. Br. J. Pharmacol. Chemother., 1957; **12**: 323–8.
133. US Pat. 1987: 4695578 (to Glaxo).
134. A. Butler, J.M. Hill, S.J. Ireland, et al., Pharmacological properties of GR38032F, a novel antagonist at 5-HT3 receptors. Br. J. Pharmacol., 1988; **94**: 397–412.
135. J.R. Fozard, A.T.M. Mobarok Ali, G. Newgrosh, Blockade of serotonin receptors on autonomic nerves by (-) cocaine and some related compounds. Eur. J. Pharmacol., 1979; **59**: 195–210.
136. B.P. Richardson, G. Engel, P. Donatsch, P.A. Stadler, Identification of serotonin M-receptor subtypes and their specific blockade by a new class of drugs. Nature, 1985; **316**: 126–31.
137. G.J. Kilpatrick, B.J. Jones, M.B.Tyers, Identification and distribution of 5-HT$_3$ receptors in rat brain using radioligand binding. Nature, 1987; **330**: 746.
138. P.P.A. Humphrey, W. Feniuk, M.J. Perren, et al., GR43175, a selective agonist for the 5-HT$_1$-like receptor in dog isolated saphenous vein. Br. J. Pharmacol., 1988; **94**: 1123–32.
139. A. Doenicke, J. Brand, V.I. Perrin, Possible benefit of GR43175, a novel 5-HT$_1$-like receptor agonist, for the acute treatment of severe migraine. Lancet, 1988; **1**: 1309–11.
140. US Pat 1989: 4997841 (to Glaxo).
141. WO Pat.1991: 9118897 (to Wellcome Foundation).
142. US Pat. 1992: 5545644 (to Pfizer).
143. US Pat. 1992: 5298520 (to Merck Sharp & Dohme).
144. US Pat. 1993: 5464864 (SmithKline Beecham).
145. US Pat. 1959: 2868818 (to Merck & Company).
146. T.L. Sourkes, Inhibition of dihydroxy-phenylalanine decarboxylase by derivatives of phenylalanine. Arch. Biochem. Biophys., 1954; **51**: 444–56.
147. J.A. Oates, L. Gillespie, S. Udenfriend, A. Sjoerdsma, Decarboxylase inhibition and blood pressure reduction by alpha methyl-3,4-dihydroxy-DL-phenylalanine. Science, 1960; **131**: 1890–1.
148. M. Sletzinger, J.M. Chemerda, F.W. Bollinger, Potent decarboxylase inhibitors. Analogs of methyldopa. J. Med. Chem., 1963; **6**: 101–3.
149. E. Nissinen, I.B. Lindén, E. Schultz, P. Pohto, Biochemical and pharmacological properties of a peripherally acting catechol-O-methyltransferase inhibitor entacapone. Arch. Pharmacol., 1992; **346**: 262–6.
150. Swiss Pat. 1968: 449046 (to Ciba).
151. R.N. Brogden, T.M. Speight, G.S. Avery, Baclofen: a preliminary report of its pharmacological properties and therapeutic efficacy in spasticity. Drugs, 1974; **8**: 1–14.
152. G.D. Bartoszyk, N. Meyerson, W. Reimann, et al., Gabapentin, in New Anticonvulsant Drugs, eds B.S. Meldrum, R.J. Porter. London: Libbey; 1986, pp. 147–64.
153. C.P. Taylor, Mechanisms of action of gabapentin. Rev. Neurol., 1997; **153**: 39–45.
154. US Pat. 1976: 3960927 (to Richardson-Merrell).
155. L. Gram, O.M. Larsson, A. Johnsen, A. Schousboe, Experimental studies of the influence of vigabatrin on the GABA system. Br. J. Clin. Pharmacol., 1989; **27** (Suppl. 1): 13S–17S.
156. P.W. Collins, E.Z. Dajani, D.R. Driskill, et al., Synthesis and gastric antisecretory properties of 15-deoxy-16-hydroxyprostaglandin E analogs. J. Med. Chem., 1977; **20**: 1152–9.
157. A. Robert, An intestinal disease produced experimentally by a prostaglandin deficiency. Gastroenterology, 1975; **69**: 1045–7.
158. Ger. Pat.1977: 2700021 (to Ono).
159. G. Bundy, F. Lincoln, N. Nelson, et al., Novel prostaglandin syntheses. Ann. N.Y. Acad. Sci., 1971; **180**: 76–90.

19

Vitamins

During the siege of Paris in 1871 the eminent French chemist Jean Baptiste Dumas published his observations on the disastrous consequences of feeding starving infants on artificial milk prepared by emulsifying fat in a sweetened, albuminous solution. This led to concern about dietary requirements, particularly with regard to mineral content. One of the leading investigators at that time was Gustav von Bunge at the University of Dorpat. Being particularly interested in the role of alkaline salts in the diet, he set his student Nicholas Lunin the task of establishing the effect of a sodium carbonate supplement on mice maintained on an artificial diet consisting of pure protein from milk (casein) and cane sugar. The mice receiving sodium carbonate survived from 12 to 30 days, in contrast to the 11 to 21 days of those denied the supplement. In an attempt to keep his mice alive longer, Lunin fed the mice on casein, milk fat, milk sugar (lactose) and salts, each of the ingredients being purified and then incorporated in the same proportion as in natural milk. This artificial milk did not improve survival rates, yet the mice thrived on powdered natural milk. This led Lunin to conclude that milk must contain unknown substances essential for growth and the maintenance of good health.[1] Soon after publishing his results, Lunin left Dorpat to take up a clinical appointment at St Petersburg. Although his findings had presented a direct challenge to generally accepted views, he did not pursue them any further. Consequently, they were overlooked for many years. A similar fate befell a study by Cornelis Pekelharing of the University of Utrecht in 1905 as it was published in Dutch.[2] This showed that mice could be kept healthy on an artificial diet to which had been added a small amount of whey (obtained by removal of fat and casein from milk).

Scant attention was paid to the results of feeding animals on artificial diets until 1906 when Gowland Hopkins of the University of Cambridge initiated his classical experiments on feeding rats with purified diets, with and without supplements of natural foods. These studies not only confirmed the findings of the earlier workers but also revealed that minute amounts of unknown substances present in normal food were essential for healthy nutrition. Hopkins kept the rats in good health on a diet consisting of casein, fat, starch, sugar and inorganic salts with a little milk added. Carefully controlled experiments showed that enough milk to increase the amount of solids in the diet by as little as 1% was quite sufficient. When Hopkins published his results in 1912, after a delay of five years caused by his ill health, his eminence in the scientific community ensured that the significance of the unknown accessory factors in food would never again be overlooked.[3] By then, other work casting light on diseases caused through dietary deficiencies had given added significance to his conclusions.

WATER-SOLUBLE VITAMINS

Concerned at the high mortality from beriberi among soldiers living in barracks in Sumatra and prisoners in Java, the Dutch government in 1886 sent the bacteriologist Cornelis Pekelharing and a neurologist, Cornelis Winkler, to Batavia (now Djakarta) in the Dutch East Indies to investigate the cause of the malady. Beriberi was a progressive disease that manifested itself by peripheral neuritis, emaciation, paralysis and then cardiac failure. Its

Drug Discovery. A History. W. Sneader.
©2005 John Wiley & Sons Ltd

incidence had greatly increased since 1870, especially among those living in barracks and prisons for more than three months. Two views prevailed as to its aetiology, namely that it was either of bacterial origin or else caused by a toxin in the rice that formed the staple diet of all those who were afflicted. That rice diets were the key to understanding the cause of the disease should have been evident from the study conducted in 1873 by van Leent, a Dutch naval surgeon who traced the high mortality from beriberi among Indian crews to their rice diet.[4] Simply by putting them on the same diet as their European shipmates, he achieved a dramatic reduction in the morbidity of the disease.

Pekelharing and Winkler tentatively concluded that beriberi was of bacterial origin, and they recommended the disinfection of barracks and prisons. On returning to The Netherlands they left behind their assistant, Christiaan Eijkman, to isolate the causative organism. He was given a small laboratory in the Dutch military hospital in Batavia where he spent several months studying inoculation of fowls with body fluids from beriberi victims. Although his early studies were inconclusive, after a few months some of the fowls began to stagger around as if intoxicated. Eijkman recognised this as being due to polyneuritis (nerve degeneration) comparable to that seen in humans suffering from beriberi. Surprisingly, it was not confined to fowls that he had inoculated. Six months later, the fowl disease disappeared as mysteriously as it had begun. At this point, Eijkman realised that the diet of the fowls had changed shortly before they became diseased. When he had begun his investigations, the birds were fed on cheap rice. Later, Eijkman had been able to acquire supplies of the more expensive milled, polished rice fed to patients in the military hospital. When a new director of the hospital was appointed, Eijkman was told to cease this practice. It was the change back to unmilled rice that led to the disappearance of the disease in the fowls.

Eijkman now carried out an experiment which involved feeding milled and unmilled rice to two groups of fowls. The results proved conclusively that milling removed some unknown ingredient that prevented polyneuritis, present in the germ or pericarp. Nevertheless, Eijkman was not yet convinced that this simple explanation could account for the occurrence of beriberi in humans. Further investigations were required before this conclusion could be drawn, but he was unwilling to experiment upon humans. Meanwhile, he published his findings in the same obscure Dutch East Indies journal as had van Leent.[5] It was another seven years before he presented a short account in a prominent German journal in 1897.[6]

Eijkman asked his friend Adolphe Vorderman, the inspector of health in Java, to ascertain the frequency of beriberi in the 101 prisons on the island, as well as the type of rice being consumed. Vorderman discovered that in the years 1895–1896, in those prisons where the inmates prepared their own crude rice, the disease was rarely seen. However, in prisons where machine-milled, polished rice was supplied there was a high incidence of beriberi. It was even found that the degree of milling influenced the incidence of disease. Eijkman then showed that the outer layers of rice, the silver-skin, could prevent polyneuritis in fowls. In 1897, Vorderman demonstrated in a prison in the village of Tulung Agung that beriberi could be eliminated by replacing polished rice with unmilled rice.

When Eijkman returned to The Netherlands in 1896 to take up a post at the University of Utrecht, he was convinced that beriberi was caused by gut microbes converting some constituent of the carbohydrate-rich rice into a toxic substance that was then absorbed into the circulation. An antidote to this was presumed to be present in the outer layers of the rice that were removed during milling.

Gerrit Grijns, who was Eijkman's successor at Batavia, showed that polyneuritis could be produced in fowls by feeding them on unmilled rice that had been autoclaved for two hours at 120 °C. Unmilled rice that had not been exposed to this destructive treatment consistently protected the birds against the disease. Grijns also found that feeding fowls on vegetables known to prevent beriberi protected them against the occurrence of polyneuritis. By means of these and other experiments, he finally came to the conclusion that in order to maintain the functional integrity of their nervous system it was essential for the diets of animals and humans to contain a protective substance that was present in unmilled rice, meat and vegetables.[7] If the diet was deficient in this protective substance, beriberi resulted.

Thiamine

In 1906, Eijkman established that the protective substance was soluble in water and had a low molecular weight. Several unsuccessful attempts were made to isolate it in crystalline form. In one by Casimir Funk, a Polish scientist working in the Lister Institute in London, a crystalline compound extracted from rice polishings cured experimentally induced polyneuritis in pigeons.[8] Subsequently, it was established that these crystals had probably consisted of nicotinic acid contaminated with the true antineuritic substance. Believing he had isolated the genuine protective factor and that it was an amine, Funk proposed that it be called 'beriberi vitamine'. His new term was altered to 'vitamin' by his colleague Jack Drummond in 1920, after it was realised that none of the protective substances then isolated were amines. Drummond proposed that different vitamins be distinguished by the use of letters of the alphabet, the antineuritic vitamin being described as vitamin B.[9]

A reliable method of detecting the antineuritic substance was devised by Barend Jansen, who was based in a new laboratory in Batavia, built for the Dutch East Indies Medical Service. With considerable persistence, he established that small birds known as 'bondols' (*Munia maja*) were much more susceptible to polyneuritis than were fowls. When fed on polished rice for only ten days, they developed polyneuritis, instantly detectable by their flying in characteristic circles. By 1920, Jansen had perfected a reliable assay which involved measuring the ability of his rice extracts to prevent, rather than cure, polyneuritis induced by feeding the bondols on polished rice.[10] Due to its instability, years of patient work were required before Jansen and Donath finally isolated crystals of the pure vitamin in 1926.[11] Samples of the crystalline vitamin were sent to Eijkman at Utrecht, where he demonstrated that the addition of 2–4 mg of these to every kilogram of polished rice restored its full antineuritic value. Three years later, Eijkman and Hopkins were jointly awarded the Nobel Prize for Physiology and Medicine for their work on vitamins.

The Jansen–Donath extraction process was too expensive for commercial exploitation. In 1933, Robert Williams developed a new isolation process for the antineuritic vitamin, by then known as vitamin B$_1$. His work was carried out privately in his spare time while he was the chemical director of the Bell Telephone Laboratories in the United States, but had spent more than 20 years researching the vitamin while at the Bureau of Chemistry set up in the Philippines by the US Army Medical Commission for the study of tropical disease. Williams contacted Ralph Major, director of research at the Merck Laboratories in Rahway, with the outcome that collaboration with the company began at once. Early in 1936, Williams presented his proposal for the chemical structure of the vitamin. Shortly after, he and Joseph Cline of Merck published a preliminary account of their synthesis which confirmed the validity of the proposed structure.[12] Immediately after this appeared in print, Williams learned that Hans Andersag and Kurt Westphal at the Elberfeld laboratories of I.G. Farben had synthesised the vitamin some months earlier.[13] It was also synthesised around this time by Alexander Todd and Franz Bergel[14] at the Lister Institute in London. In 1937, Williams licensed Merck to produce the vitamin commercially by his synthetic process. Much of the profit from the Williams patents was devoted to a fund that supported nutritional research.

thiamine (vitamin B$_1$)

In 1937, several European countries accepted Jansen's proposal that the anti-berberi vitamin should be known henceforth as 'aneurine'. The US Council on Pharmacy and Chemistry rejected this since their policy was to avoid drug names with any therapeutic connotation. Williams suggested that it be called thiamine, a term that highlighted the presence of the sulfur atom. This was not universally accepted until the International Union of Pure and Applied Chemistry's Commission on the Nomenclature of Biological Chemistry approved the name in 1951.

Riboflavine

Until 1919, it was generally believed that there were only two water-soluble vitamins, namely vitamins B and C. Elmer McCollum at the University of Wisconsin had discovered the existence of a fat-soluble vitamin in 1913 while feeding rats on artificial diets.[15] Continuing his nutritional studies, he then found that if the commercially supplied lactose (milk sugar) that he had been using was purified by recrystallisation, the rats exhibited manifest evidence of a growth disorder due to a dietary deficiency that he could correct merely by supplementing the diet with the aqueous mother liquor from which the lactose had been crystallised. This led him to conclude that a water-soluble vitamin existed, whereupon he then learned of the earlier studies by Eijkman and Grijns. After repeating their work with polished rice, he came to the conclusion that the antineuritic vitamin and his rat growth-promoting vitamin were identical.

In a review article that appeared in the *Journal of Biological Chemistry* in 1919, H.H. Mitchell of the University of Illinois questioned the assertion that the antineuritic vitamin and the rat growth vitamin were identical.[16] He claimed that all the evidence was circumstantial. During the next few years, experimental results tended to support Mitchell's contention and researchers began to speak of the vitamin B complex.

In 1927, the British Committee on Accessory Food Factors distinguished between vitamin B$_1$, the antineuritic vitamin isolated the previous year, and the more heat-stable vitamin B$_2$. The following year, Harriet Chick and M.H. Roscoe of the Lister Institute developed an assay for vitamin B$_2$ activity. This involved measuring the effect of test material on young rats fed on a diet deficient in the vitamin B complex, but to which vitamin B$_1$ had been added. In the absence of any vitamin B$_2$ supplementation, the rats exhibited loss of hair from the eyelids, sealing of the eyelids by a sticky exudate, dermatitis, blood-stained urine and stunted growth. Each assay took three or four weeks to complete. Tedious as this must have been, it was sufficient to encourage Richard Kuhn of the Institute of Chemistry in the University of Heidelberg and Theodore Wagner-Jauregg of the Kaiser Wilhelm Institute for Medical Research to join with the paediatrician Paul György, an émigré Hungarian, in his attempt to isolate pure vitamin B$_2$. Wagner-Jauregg noticed that all extracts that proved active by this assay procedure exhibited an intense yellow–green fluorescence, the intensity of which was proportional to potency. When attempts were first made to isolate the fluorescent material, the growth-promoting activity of the extracts deteriorated. It was then realised that other growth-promoting vitamins must have been present prior to refinement of the crude vitamin B$_2$, thus pointing the way to the discovery of further members of the vitamin B complex. The biological assay procedure had to be modified to allow for this, with the outcome that the yellow–green fluorescent material was isolated from spinach, kidney and liver, proving identical in each case. The vitamin was crystallised at Heidelberg in 1933 and named 'riboflavine'[18] (American workers at one time referred to it as vitamin G; confusingly, the term vitamin B_2 has been retained despite the fact that this was originally applied to crude preparations containing several members of the vitamin B complex).

Two years later, Richard Kuhn at Heidelberg[19] and Paul Karrer at the University of Zurich[20] almost simultaneously synthesised the vitamin. The latter's process was adapted by Hoffmann–La Roche for commercial production. In 1937, Karrer was awarded a Nobel Prize for Chemistry, shared with Norman Haworth for their work on vitamins. The following year, Kuhn was similarly honoured, but the Nazi government in Germany made him decline the award.

riboflavine

Nicotinamide

When the distinction between vitamin B_1 and the crude vitamin B_2 had first been made, it was generally assumed that the latter was the pellagra-preventing vitamin (P-P factor). The Spaniard Gaspar Casal described pellagra in the early eighteenth century and attributed it to consumption of diets rich in maize.[21] The disease was given its name in 1771 by the Italian physician Francesco Frapolli on account of the characteristic skin changes (*pelle* = skin, *agra* = dry) that it caused.[22] These were in addition to gastrointestinal disturbances and degenerative changes in the central nervous system that ultimately lead to insanity.

It was not until an epidemic swept through the southern United States in the early years of the twentieth century that the cause of pellagra was subjected to experimental scrutiny. An extensive series of clinical and epidemiological studies was initiated in 1914 by a team from the US Public Health Service, led by Joseph Goldberger. The initial conclusion was that diets rich in maize were to blame, this being consistent with the earlier demonstration by Edith Willcock and Hopkins that young mice failed to grow on diets in which zein from maize was the sole source of protein, zein being deficient in tryptophan.[23] However, in 1920, Carl Voegtlin and his colleagues from the pharmacology division of the Public Health Service discovered that pellagra could be cured by administration of dried yeast or aqueous extracts of yeast, these preparations being known to be a rich source of vitamin B.[24] Further experiments indicated that the active material in the yeast was not destroyed by heating the yeast at 52 °C. Since Voegtlin had previously shown that vitamin B_1 had no beneficial value in pellagra, Goldberger and his colleagues concluded that vitamin B_2 must be the P-P factor.[25]

After vitamin B_2 was found to be a complex mixture and riboflavine to have no P-P activity, the search for the true P-P factor was intensified. It was greatly facilitated through the earlier recognition by T.N. Spencer, a veterinarian from Concord in North Carolina, that a disease in dogs known as 'black tongue' was the canine counterpart of human pellagra. Goldberger then developed an assay for the P-P factor based on prevention of 'black tongue' in dogs. He was able to demonstrate that liver was one of the richest sources of the P-P factor. Conrad Elvehjem and his associates in the Department of Agricultural Chemistry at the University of Wisconsin-Madison finally isolated the P-P factor from fresh liver in 1937.[26] The vitamin was immediately recognised to be nicotinamide (niacinamide), a substance already being studied by biochemists. With Wayne Woolley, Elvehjem demonstrated that both nicotinamide and nicotinic acid (niacin) were capable of preventing and curing 'black tongue' in dogs.[27] Human trials followed at once, which proved highly successful. Since nicotinic acid had been synthesised by Albert Ladenburg 40 years earlier, there was no problem in producing large amounts for the treatment of pellagra. The amide was readily prepared from the acid.

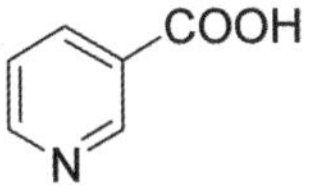

Rudolf Altschul and Abram Hoffer at the University of Saskatchewan discovered in 1955 that high doses of nicotinic acid lowered serum cholesterol levels in humans.[28] It was the first drug ever used for this purpose, but vasodilation was a dose-limiting side effect.[29] The vasodilatory activity, however, has been exploited in remedies for chilblains and in the formulation of counter-irritant creams.

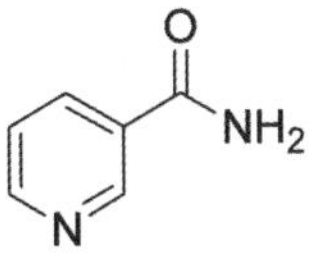

Pyridoxine

The isolation of further members of the vitamin B complex rapidly followed that of nicotinamide. The first of these was discovered as a consequence of studies on young animals deliberately deprived of the B group of vitamins other than those already known. The main difficulty facing the researchers was that of unravelling the complex pathological changes arising from deficiency of unidentified vitamins. It was while working at Cambridge University in 1934 that Paul György suggested that one such unidentified vitamin could protect rats from a specific type of skin lesion. He proposed the name vitamin B_6 for this rat antidermatitis factor, which he and T.W. Birch did much to characterise chemically. Early in 1938, Samuel Lepkovsky[30] of the College of Agriculture at the University of California, Berkeley, informed György that he and John Keresztesy[31] of Merck and Company were each independently about to submit papers describing the crystallisation of the vitamin. This magnanimous gesture enabled György[32], now at Western Reserve University in Cleveland, to publish his own account of the crystallisation shortly after.

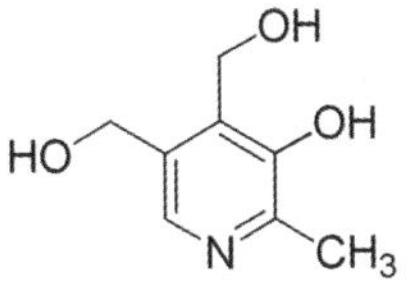

Within a year of its isolation, both Karl Folkers[33] of Merck and Company and also Kuhn[34] at Heidelberg had established the chemical structure of vitamin B_6 and had then synthesised it. György proposed that it henceforth be known as 'pyridoxine'. It should, in passing, be mentioned that following American government indications that it favoured the supplementation of foods and cereals with vitamins, Merck and Company had invested heavily in equipment to separate the B vitamins from natural sources such as yeast. The success of their own and rival chemists rapidly rendered this obsolete.

Pantothenic Acid

In 1939, Thomas Jukes,[35] a colleague of Lepkovsky at the University of California, and also Woolley, Waisman and Elvehjem[36] at the University of Wisconsin, simultaneously discovered that pantothenic acid was the hitherto unidentified vitamin whose deficiency had been shown to cause dermatitis in chickens. This had been isolated the previous year by Roger Williams at the University of Oregon during investigations into nutrients essential for the growth and replication of cultured yeast cells.[37] It had taken him four years to isolate and purify this new yeast growth factor after differentiating traces of it from other essential nutrients present in food extracts. It was synthesised by Merck chemists in 1940.[39]

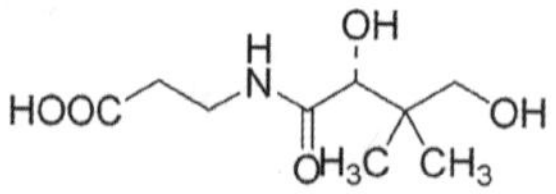

pantothenic acid

Biotin

Roger Williams' pioneering studies on yeast growth factors had begun in 1919 when he tried unsuccessfully to develop a new type of assay for vitamin B_1 activity. Nevertheless, his work stimulated microbiologists to isolate growth factors for other organisms, including bacteria. In 1931, a Canadian scientist, W.L. Miller, detected the presence of two yeast growth factors in malt, namely Bios I and II. He identified the former as inositol, a sugar long known to be present in muscle. Five years later, Kögl and Tönnis isolated crystals believed to be Bios II from boiled duck egg yolks. They named these 'biotin'.[40]

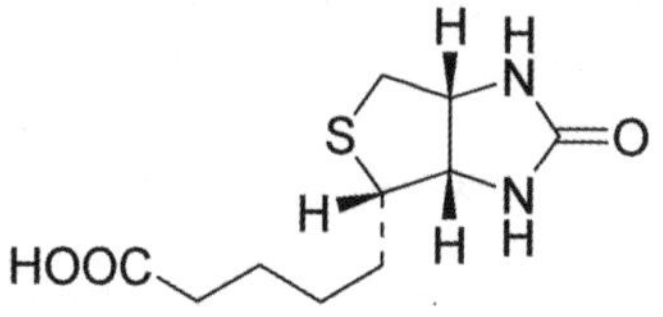

biotin

Vincent Du Vigneaud and Donald Melville[41] determined the chemical structure of biotin in 1942, which was synthesised a year later by Karl Folkers and his colleagues at Merck.[42]

Cyanocobolamin

The discovery of cyanocobolamin (vitamin B_{12}) came about as a consequence of fundamental studies into bile pigment metabolism and its relation to liver disease by George Whipple that began in 1914 at the University of California Medical School in San Francisco.[43] He found it necessary to extend his investigations to cover the rate of formation of haemoglobin, the pigment in red blood cells from which the bile pigments are derived. He did this simply by draining blood from dogs and waiting to see how long it took for haemoglobin levels to return to their original level. When this unexpectedly revealed that diet influenced the rate of haemoglobin regeneration, Whipple's interest in liver disease led him to examine the effect of feeding liver to his anaemic dogs. It proved to have a more powerful effect than any other food. On taking up a new appointment at the University of Rochester, New York, Whipple refined his techniques to confirm the earlier studies.[44] His results came to the attention of

George Minot at Harvard, a clinician who had been investigating the influence of diet on patients with pernicious anaemia, since some of its symptoms resembled those of beriberi and pellagra. Pernicious anaemia was at that time an incurable disease characterised by failure of normal red cell formation, with death within a few years.

When Minot fed liver to a few patients with pernicious anaemia their condition improved, but the results were hardly conclusive. A detailed investigation followed, with 45 patients receiving enormous daily doses of liver by mouth.[45] When the trial was completed in May 1926 the results were startling. Many patients showed obvious signs of improvement within a week, their red blood cell count being restored to satisfactory levels within two months. For this outstanding contribution, Whipple, Minot and Murphy received the Nobel Prize for Medicine and Physiology in 1934.

Eating as much as half a kilogram of liver each day was a daunting prospect for anyone, let alone a sick patient. To overcome this, Edwin Cohn at the Harvard Medical School prepared an extract that was marketed by Eli Lilly and Company in 1928. Two years later, Lederle Laboratories introduced a more refined extract for intramuscular injection. This was far more satisfactory since the real cause of pernicious anaemia was a defect in gut absorption processes. A single injection once every one to three weeks proved adequate.

During the early 1940s it was realised that the loss of activity when liver extracts were decolorised with charcoal was due to adsorption of the active material. This was ultimately turned to advantage when it was shown that under certain conditions the active material could be eluted from charcoal to give a much purer product. This paved the way for eventual isolation of crystals of the active principle in 1948 by Lester Smith of Glaxo Laboratories in the United Kingdom[46] and also by Karl Folkers and his colleagues at Merck and Company in the United States.[47] Later in the year, the Merck researchers isolated the vitamin from a strain of *Streptomyces griseus* used in streptomycin production.[48] This meant that a cheap source had been found and commercial production began in 1949.

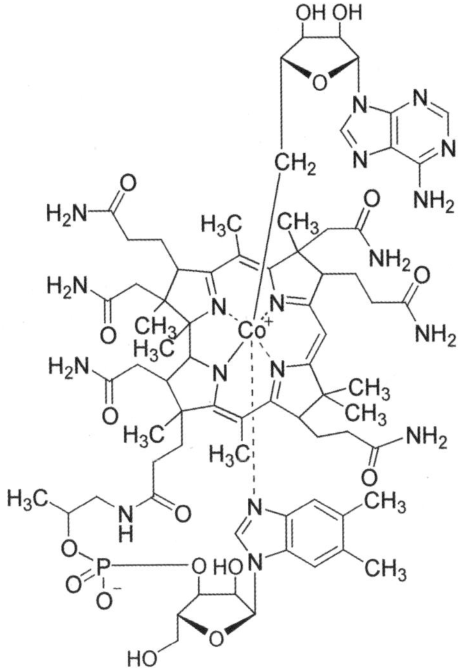

cyanocobolamin

As the vitamin turned out to be a cobalt-containing molecule it received the approved name of cyanocobalamin, although it became widely known as vitamin B_{12}. Its chemical structure was elucidated in 1955 through the collaboration of chemists from the University of Cambridge, led by Alexander Todd, with X-ray crystallographers from the University of Oxford, led by Dorothy Hodgkin, and a team from Glaxo, led by Lester Smith.[49,50,] Todd was awarded the Nobel Prize for Chemistry in 1957 and Hodgkin in 1964.

Folic Acid

In 1930, Lucy Wills and S.N. Talpade of the Haffkine Institute in Bombay found that undernourished mothers of premature babies were consuming diets deficient in the vitamin B complex. They thought this might account for the manifestation of pernicious anaemia-like symptoms during pregnancy. Wills went on to study textile workers in Bombay who had developed a form of anaemia resembling pernicious anaemia except for the absence of neurological complications. She described this as 'tropical macrocytic anaemia' because of the presence of many large, immature blood cells.[51] In contrast to pernicious anaemia, this macrocytic anaemia responded positively to treatment with a proprietary brand of yeast extract (Marmite®) that was rich in the vitamin B complex. When monkeys were fed on diets similar to those eaten by the anaemic patients, they also developed the disease. Administration of yeast extract or liver also cured the monkeys, but injections of liver extract normally used in treating pernicious anaemia proved worthless both in monkeys and humans suffering from the macrocytic anaemia.[52] Evidently, the purification of the liver extract had removed a protective factor that was different from vitamin B_{12}.

The significance of Wills' results was not appreciated. However, in 1935 similar observations on monkeys were noted by Paul Day of Little Rock University, Arkansas, in the course of feeding experiments designed to produce cataracts from riboflavine deficiency.[53] He blamed a dietary deficiency when his monkeys developed anaemia and died from complications. After Day had managed to correct the purported deficiency with either yeast supplements or whole livers, he proposed that the protective factor be called vitamin M. In the absence of a convenient assay system using small animals with a short lifespan, Day was unable to consider the isolation of the vitamin.

In 1939, Albert Hogan and Ernest Parrott at the University of Missouri found that chickens fed on a simple diet sometimes became anaemic and failed to grow. Abnormalities in their red blood cells were traced to variations in the quality of the commercial liver extract incorporated in the feedstuff. The evidence pointed to deficiency of an unidentified B complex vitamin that they described as vitamin B_c.[55] Unlike Wills and Day, Hogan was able to conduct assays for vitamin activity and so proceed with its isolation from liver. In the autumn of 1940 he approached Parke, Davis and Company, who put a team of scientists on to the project. It took two and a half years before crystals of the anti-anaemic factor were isolated. It turned out to be an acid.[56] In the interim, events had moved rapidly.

In attempting to devise artificial media that would permit determination of the exact nutritional requirements of bacteria such as *Lactobacillus casei*, Esmond Snell and William Peterson at the University of Wisconsin found little bacterial growth with a hydrolysed casein-based culture medium, unless plant or animal extracts were incorporated.[57] Further investigation revealed that yeast extract was the richest source of growth-promoting material, and in 1939 an active fraction was separated from this source by means of elution chromatography. Peterson, assisted by Brian Hutchings and Nestor Bohonos, went on to obtain a *Streptococcus lactis* growth-stimulating fraction from liver.[58] Snell transferred from Wisconsin to work with Roger Williams, now at the University of Texas, where, with the assistance of Herschel Mitchell, they isolated from spinach a concentrate of a *Lactobacillus casei* growth factor that they named folic acid (Latin: *folium* = leaf).[59] Elvehjem and Hart at

Wisconsin then found it to be capable of preventing anaemia in chickens.[60] It seemed that it must be the same as Hogan's anti-anaemic factor, vitamin B_c, as the physical properties of the two substances were similar. Hogan confirmed that the substances had identical biological properties.

In 1938, Robert Stockstad and P.D.V. Manning of the Californian-based Western Condensing Company were involved in formulating a diet that would be suitable for assaying riboflavine on chickens when they came to the conclusion that an unknown dietary growth factor existed.[61] Tentatively, they described it as the U factor and stated that it was present in certain yeast extracts.[62]

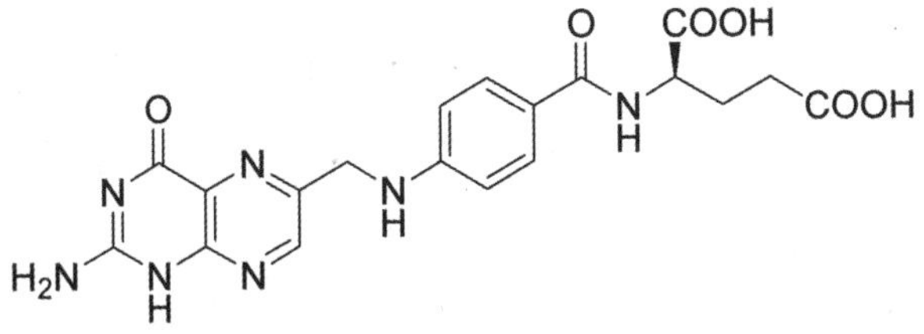

folic acid

In 1941, Stockstad was recruited by Lederle Laboratories to work on liver extracts at their Pearl River research centre. Two years later, he isolated crystals of the *L. casei* growth factor from 1.5 tons of liver.[63] These proved to be identical to the vitamin B_c that had just been described by Hogan and the Parke, Davis and Company researchers. The structure was determined by Lederle researchers,[64] who went on to achieve the total synthesis of folic acid in August 1945.[65]

Folinic Acid

A growth factor required by *Leuconostoc citrovorum*, known as either 'citrovorum factor', 'leucovorin' or 'folinic acid', was isolated[66] and subsequently synthesised by Lederle chemists.[67] It acts as an antagonist of antifolate drugs and has been used clinically to treat methotrexate toxicity. Such treatment is described as folinic acid rescue.

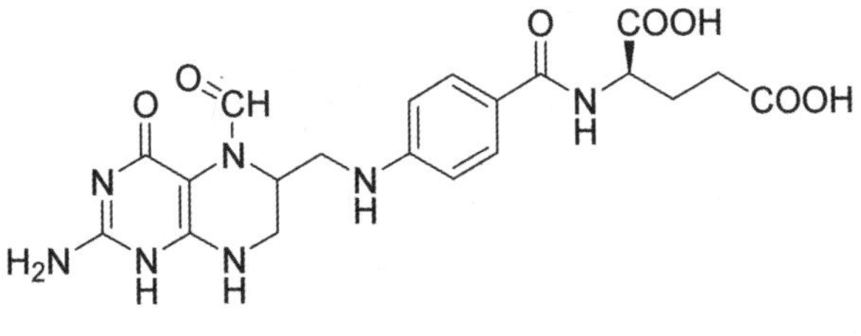

folinic acid

Folinic acid enters human cells by the folate uptake pathway. Once inside the cells it is rapidly metabolised to 5-methylenetetrahydrofolate, an active form of folic acid that can participate in a one-carbon transfer system to convert uracil to thymine. Folinic acid is thus able to bypass the blocking of tetrahydrofolate formation caused by antifolates.

Ascorbic Acid (Vitamin C)

The investigations into the nature of the antineuritic vitamin were directly responsible for the discovery that absence from the diet of another water-soluble vitamin was the cause of scurvy. In 1536, the French explorer Jaques Cartier had vividly described the nature of this disease that afflicted all but ten of the 110 men aboard his three ships wintering in the frozen St Lawrence River. The victims' weakened limbs became swollen and discoloured, while their putrid gums bled profusely. The captain of his ship learned from an Indian how to cure the sailors with a decoction prepared from the leaves of an evergreen tree. Miraculously, so it seemed, the remedy proved successful. Nearly 30 years later, the Dutch physician Ronsseus advised that sailors consume oranges to prevent scurvy and in 1639 John Woodall, one of England's leading physicians, recommended lemon juice as an anti-scorbutic.[68]

Notwithstanding these earlier developments, it is the Scottish naval surgeon James Lind who is remembered as the first person to conduct a controlled clinical trial, through which he proved that scurvy could be cured by drinking lemon juice.[69] In May 1747, he tested a variety of reputed remedies on 12 scorbutic sailors quartered in the sick bay of the fourth-class ship called the *Salisbury*. Two were restricted to a control diet, but each of the others were additionally given one of the substances under trial. The two seamen who were provided with two oranges and a lemon each day made a speedy recovery, one of them being fit for shipboard duties in only six days. The only others to show any signs of recovery were those who had been given cider. Lind observed no improvement in the condition of those who had been given either oil of vitriol (dilute sulfuric acid), vinegar, sea-water or only the control diet. He drew the obvious conclusions, which were acted upon by Captain James Cook on his second voyage round the world. Although Cook was at sea for three years, none of his crew died from scurvy thanks to adequate provision of lemon juice, as well as fresh fruit and vegetables. Surprisingly, it was not until 1795 that the Admiralty finally agreed to Lind's demands for a regular issue of lemon juice on British ships. The effect of this action was dramatic; in 1780, there had been 1457 cases of scurvy admitted to Haslar Naval Hospital, but only two admissions took place between 1806 and 1810. The situation then deteriorated for over a century until it was discovered that cheaper lime juice that had been introduced had only about a quarter of the antiscorbutic activity of the lemon juice that it had widely displaced.

In 1899, Stian Erichsen wrote in *Tidsskrift for den Norske Laegeforening* (the Journal of the Norwegian Medical Association) that a mysterious illness afflicting sailors on very long voyages was caused by lack of fresh food. Concerned at the growing incidence of this disease, which had similarities to both beriberi and scurvy, the Norwegian Navy asked Axel Holst and Theodor Frolich of Christiana University in Oslo to investigate the matter. Holst visited Gerrit Grijns in Batavia before returning to carry out experiments on guinea pigs. Fortunately for him, guinea pigs are exceptionally sensitive to ascorbic acid deficiency, and so he and Frolich readily induced a condition analogous to human scurvy by feeding their animals on polished rice. This was not alleviated by giving the guinea pigs rice polishings, but fresh fruit or vegetables known to cure scurvy restored them to good health. On the basis of these findings, Holst argued, in 1907, that in addition to the antineuritic dietary protective substance postulated by Grijns there must also exist an antiscorbutic one, and the disease among Norwegian sailors could be prevented by appropriate dietary measures.[70]

Holst and Frolich went on to demonstrate that the antiscorbutic factor was soluble in water. They showed that, like the antineuritic substance, it was of low molecular weight. They also found that when foods were subjected to drying, the antiscorbutic principle was destroyed. Their pioneering studies were confirmed by work on monkeys carried out at the Lister Institute during the First World War in the wake of outbreaks of scurvy among British troops serving in the Middle East; this was despite the provision of lime juice. Only then was it recognised that the juice of West Indian limes had poor antiscorbutic activity in comparison to that of lemons. In 1920, Jack Drummond proposed that the antiscorbutic protective substance be called vitamin C until its chemical structure was established.

Working at the Lister Institute, Sylvester Zilva began to prepare concentrated extracts of the vitamin in 1918. Five years later, he introduced a highly potent concentrate.[71] At the

University of Pittsburgh, Charles King obtained a more stable form of this by removing traces of heavy metals that catalysed oxidation. In the autumn of 1931, after four years of intensive investigations, King finally isolated pure crystals of the vitamin from lemon juice.[72] Tests with these showed that a daily dose of 0.5 mg could prevent a guinea pig becoming scorbutic on a diet deficient in the vitamin. The crystals turned out to be very similar to an acidic carbohydrate isolated in Gowland Hopkins' laboratory at Cambridge in 1928 from adrenal glands, cabbages and oranges by Albert Szent-Györgyi, a Hungarian biochemist who had been awarded a Rockefeller Fellowship to investigate oxidation–reduction processes in the adrenals. The possibility that his new compound, then thought to be a hexuronic acid, might be the antiscorbutic vitamin had apparently been ruled out by results obtained by Zilva, but King's successful isolation of the vitamin reopened the issue.

Assisted by Joseph Svirbely of King's department, Szent-Györgyi found that 1 mg daily of his hexuronic acid protected guinea pigs against scurvy.[73] King gained further confirmation of the identity of his vitamin with Szent-Györgyi's acid. In order to establish the nature of the vitamin, Szent-Györgyi initiated a collaborative programme with the University of Birmingham, a leading centre in the field of carbohydrate chemistry. It soon became evident that the vitamin could not be a hexuronic acid, and Szent-Györgyi and Norman Haworth then proposed that it should be known as ascorbic acid.

$$CH_2OH$$
$$HC-OH$$

ascorbic acid

In a letter appearing in the _Journal of the Society of Chemistry and Industry_ on 10 March 1933, Edmund Hirst published the correct structure for ascorbic acid as determined by him and his colleagues at Birmingham. A race to synthesise the vitamin began, and on 11 July 1933 Tadeus Reichstein at the Eidgenossische Technische Hochschule (ETH) in Zurich submitted a letter to the editor of _Nature_ giving a preliminary account of his successful synthesis of ascorbic acid. It did not appear in print until over five weeks later, by which time a preliminary account of a synthesis by Haworth and Hirst had already appeared in the _Journal of the Society of Chemistry and Industry_ on 4 August 1933, having been submitted only three days earlier! Reichstein, however, amassed a fortune from patent royalties after Hoffmann–La Roche began commercial production of synthetic ascorbic acid in 1934. Szent-Györgyi received the Nobel Prize for Medicine and Physiology in 1937 for his work on the biochemical role of ascorbic acid, while Haworth shared the Nobel Prize for Chemistry.

FAT-SOLUBLE VITAMINS

In the course of examining the effect on rat growth of varying the mineral content of artificial diets, Elmer McCollum and his assistant Marguerite Davis at the University of Wisconsin noted that normal growth patterns could be maintained for only 70–120 days. However, when natural diets were reintroduced, normal growth was restored. Many experiments had to be carried out before suspicion fell on the nature of the fat content of the artificial diet. To confirm that a fat-soluble accessory factor was present in only certain foods, McCollum and Davis supplemented the deficient diet with ether extracts of foods containing fat. This proved that the factor was to be found in butterfat and egg yolk, but not in lard. When they reported their findings much surprise was engendered in nutritional circles as it had universally been believed that the role of fats in the diet was solely to produce energy, the qualitative differences between them being of no consequence.[75] Later, when McCollum detected a water-soluble accessory food factor in milk, he named the two factors he had discovered as 'fat-soluble A' and 'water-soluble B'. These terms were changed in 1920 by Jack Drummond to vitamin A and B respectively, the latter being identical to the antineuritic vitamin first discovered by Eijkman.

Retinol (Vitamin A)

McCollum's findings were immediately pursued at Yale University by two of the leading American nutritionalists, Thomas Osborne and Lafayette Mendel. They noticed that in animals fed on a diet deficient in the fat-soluble factor, a characteristic eye disease occurred. This had been observed in malnourished animals before, but had not been considered of any particular significance. Once the association with vitamin deficiency became evident, attitudes quickly changed as researchers realised that many clinical reports had associated eye disorders with nutritional factors. Of particular relevance was one published in 1904 by M. Mori, a Japanese ophthalmologist. This described an eye disease characterised by dryness of the conjunctiva (xerophthalmia) frequently seen among infants fed on cereals and beans, but never found among the children of fishermen.[76] The author of the report stated that the disease was due to lack of fat in the diet and could rapidly be cured by administering cod liver oil. Osborne and Mendel were able to demonstrate that both butter fat and cod liver oil could alleviate the ophthalmic disorder in their experimental animals.[77] Not long after, an outbreak of serious eye disease, sometimes blinding, occurred among Dutch children fed on fat-free skimmed milk because of wartime measures to ensure increased export of butter. These infants were cured with cod liver oil supplements and full cream milk. At Wisconsin, S. Mori subsequently carried out extensive microscopic studies on the eyes of rats prepared for him by McCollum, thereby elucidating the pathology of xerophthalmia.[78] He found that the dryness of the eyes was due to changes (keratinisation) in the cells lining the tear glands. As a result, the tears were unable to exercise their protective role against bacteria, and infection of the inner surfaces of the eyelids ensued. In severe cases, the infection spread into the eye, causing ulceration of the cornea. The interest aroused by the discovery of the relationship between eye disease and vitamin A deficiency drew attention to old reports of night-blindness being cured by the eating of liver. Biochemists eventually established that the vitamin was converted to the pigment in the retina known as visual purple (rhodopsin).

In 1924, Jack Drummond at University College London developed a steam distillation process to separate vitamin A from other unchanged fats remaining in cod liver oil after boiling in alcoholic potassium hydroxide (to saponify biologically inactive fats). The following year, he and Otto Rosenheim exploited their discovery that isolation of the vitamin could be greatly facilitated by measuring the intensity of the purplish colour it produced on reacting with arsenic trichloride. In collaboration with Isidor Heilbron at the University of Liverpool, Drummond made further improvements by developing a high-vacuum distillation technique that ultimately yielded almost pure vitamin.[79,80] In 1929, after it was discovered that livers of other types of fish were often richer sources of vitamin A than cod liver, Abbott Laboratories and Parke, Davis and Company jointly began to process halibut liver oil for its vitamin content. The resulting product, though of high potency, was not particularly palatable on account of its strong fishy smell.

retinol

In 1931, Paul Karrer at Zurich University introduced adsorption chromatography to isolate a viscous yellow oil consisting of almost pure vitamin A.[81] With this, he determined the chemical structure, reporting it two years later. However, it was not until 1937 that pure vitamin A, retinol, was crystallised by Harry Holmes and Ruth Corbet of Oberlin College, Pennsylvania, using fractional freezing and cold filtration. In 1947, Otto Isler of Hoffmann–La Roche introduced a commercial synthesis of the vitamin, as a consequence of which fish liver oil extraction processes are no longer in use.[82]

As long ago as 1925 it was observed that rats fed on a vitamin A-deficient diet developed dyskeratotic skin conditions.[83] This finding was not exploited until 1959 when the Berlin dermatologist Gunter Stüttgen showed that retinol palmitate inhibited the growth of benzpyrene-induced tumours in mice when administered systemically, but not if applied topically – even though it penetrated the stratum corneum of the skin.[84] When the same thing happened on treating various dyskeratoses, Stüttgen came to the conclusion that retinol was only effective after it had undergone metabolic activation. He then collaborated with Hoffmann–La Roche to arrange a study with the major metabolite of retinol, which is now known as 'tretinoin'. This showed it to be effective when applied topically in a variety of skin conditions.[85] Unfortunately, healing was preceded by local irritation, which militated against clinical acceptance of the drug. However, it was reported in 1969 that tretinoin did not cause irritation when used to treat acne vulgaris.[86] Consequently, this and treatment of photodamaged skin became its principal clinical application. It is now also given by mouth in acute promyelocytic leukaemia, resulting in three out every four patients remaining disease free after five years.[87] The mechanism of action is unknown, but its administration restores the ability of defective granulocyte precursors in the bone marrow to develop normally.

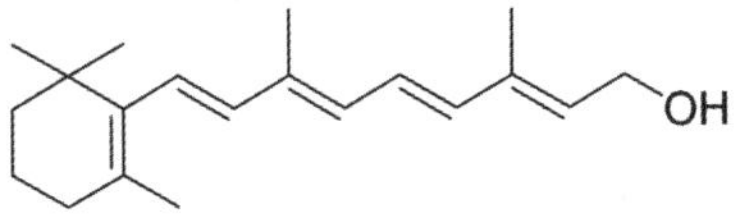

retinol

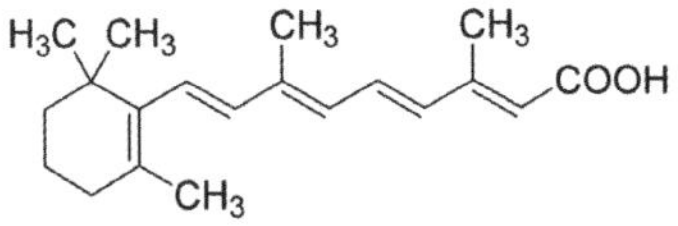

tretinoin (*trans*-retinoic acid)

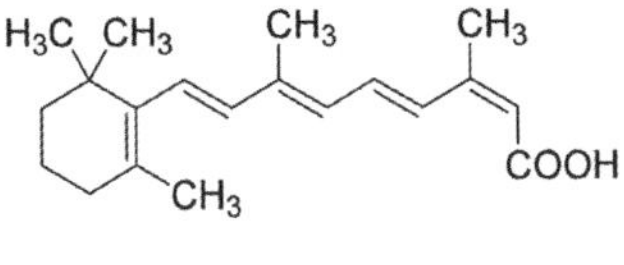

isotretinoin (*cis*-retinoic acid)

Isotretinoin, the synthetic geometric isomer of tretinoin, was found to be just as effective in the treatment of acne, but when given by mouth it had a greater safety margin.[88] Unfortunately, it is teratogenic and so cannot be prescribed for women of child-bearing age unless effective contraceptive cover is provided.

Vitamin D$_2$ (Ergocalciferol, Calciferol)

In 1912, Gowland Hopkins suggested that rickets might be yet another of the diseases caused by deficiency of an accessory food factor.[89] Outwardly, rickets (rachitis) was characterised by deformity of the limbs of infants arising from failure of calcium phosphate to be deposited at the growing ends of their bones. Unchecked, the disease not infrequently involved the central nervous system, which could be fatal. Although known for centuries, rickets reached epidemic proportions early in the twentieth century in the industrial cities of Northern Europe and America. This spurred Hopkins to recommend to the newly formed Medical Research Committee that it should designate rickets as a subject for special study. He recommended research should be undertaken by one of his former students, Edward Mellanby. The Committee agreed and Mellanby began work in 1914. Travelling between London and Cambridge, where he had access to a colony of puppies, he painstakingly conducted hundreds of feeding experiments in an attempt to identify the type of diet that induced rickets. In 1918, he was able to inform the Physiological Society that he could produce rickets in puppies by feeding them for three or four months on either a diet of milk, rice, oatmeal and salt, or on milk and bread. By adding a variety of foods to these rachitic diets, Mellanby was able to confirm that animal fats such as butter, suet and cod liver oil had antirachitic activity.[90] The latter was a Northern European folk remedy that became esteemed as a tonic in the late eighteenth century, since when it had been widely employed in the palliation of debilitating diseases such as tuberculosis and rheumatism. The Parisian physician Armand Trousseau referred to its use for treating rickets in his *Clinique Médicale de l'Hôtel-Dieu de Paris*, published in 1861. However, it was not until after Mellanby had offered experimental proof of the value of cod liver oil that any significant reduction in the incidence of the disease was recorded. By the early 1930s the disease was no longer seen in London.

Mellanby believed that the antirachitic vitamin and vitamin A were identical, although he recognised that the evidence was not altogether conclusive. In an attempt to settle the issue, he took advantage of Hopkins' new observation that vitamin A activity of hot butterfat was destroyed by bubbling oxygen through it. When Mellanby treated both butterfat and cod liver oil in this manner, he found the latter retained antirachitic activity. He was undecided as to whether this proved the existence of a second fat-soluble vitamin or merely reflected the presence either of a larger initial amount of vitamin A in the cod liver oil or else of an antioxidant. McCollum, who had moved from Wisconsin to Johns Hopkins University, set out to settle the matter by experimenting on rats he had already made rachitic by feeding them on artificial diets containing an unfavourable balance of calcium and phosphorus. He heated cod liver oil in a current of air for a prolonged period to ensure oxidation of all the vitamin A present and then demonstrated that the oil still retained its protective antirachitic action. His results were published in 1922.[91] They conclusively proved the non-identity of vitamin A and the antirachitic factor, which was named vitamin D in 1925 as it was the fourth one to have been discovered.

McCollum's demonstration that vitamin D deficiency was the cause of rickets did not settle one outstanding matter. In 1919, a Berlin physician, Kurt Huldschinsky, had cured rickets in children by exposing them to ultraviolet light emitted from a mercury vapour lamp.[92] His results were corroborated the following year in Vienna by Chick's group of lady doctors and scientists who, at the end of the war, had been sent from the Lister Institute to assist during a severe epidemic of rickets that affected four out of every five infants in the city. They found the disease did not develop in children exposed to adequate sunlight.[93] Further confirmation came from New York, where Hess at the College of Physicians and Surgeons at Columbia University cured rachitic infants by exposing them to sunlight or ultraviolet radiation. Hess suggested that the antirachitic principle might be formed by the action of ultraviolet light on a putative provitamin. He went on to make the surprising discovery, announced in June 1924, that irradiation of certain foods could confer antirachitic properties on them.[94] Before his report appeared in print in October of that year, Harry Steenbock of the University of Wisconsin published similar findings.[95] He took out patents to cover the processing of foods by ultraviolet light, assigning these to a body established in 1925 to enable the vast sums earned from license fees to be used in support of research in Wisconsin. This was the Wisconsin Alumni Research Foundation, which earned more than $14 million from Steenbock's patents during the next 20 years.

Hess and workers in several other laboratories soon established that the substance converted into vitamin D when vegetable oils were irradiated was to be found among the plant sterol fraction. He then went to Göttingen to work on the isolation of provitamin D under the guidance of Adolf Windaus. In 1927, Hess and Windaus, with the assistance of the Göttingen physicist Robert Pohl, established that the provitamin D was a known substance, namely ergosterol.[96] The following year, Windaus was awarded the Nobel Prize for Chemistry in recognition of this and his earlier work on sterols. In collaboration with the Elberfeld laboratories of I.G. Farbenindustrie, in 1932 Windaus isolated the product formed by irradiation of ergosterol.[97] He named it vitamin D_2, to distinguish it from what he had previously thought was the pure vitamin, namely its complex with lumisterol (a precursor also formed by irradiation of ergosterol). Windaus renamed that complex, calling it vitamin D_1 and then elucidated the chemical structure of vitamin D_2 in 1936.[98]

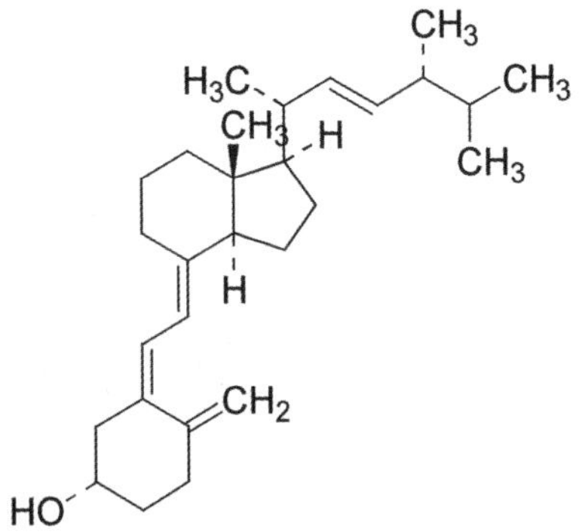

vitamin D_2

A vitamin D_2 complex was also isolated in 1932 by Askew and his colleagues[99] at the National Institute for Medical Research, London. At that time, this was believed to be homogeneous, and was mistakenly assumed to be identical to Windaus' vitamin D_2. It was given the name 'ergocalciferol'. The term 'vitamin D' is now used as a generic term to describe any substance that can be converted in the body into the active antirachitic metabolite 1,25-dihydroxy-cholecalciferol.

Vitamin D_3 (Cholecalciferol, Calciol)

Irradiation of other sterols was found to generate antirachitic products such as vitamin D_3, or cholecalciferol, which was formed from 7-dehydrocholesterol.[100] Windaus found this sterol present in skin, thereby solving the mystery of how exposure to sunlight could prevent or cure rickets.[101] Vitamin D_3 is the product formed on irradiation of foods from animal sources. It was synthesised in 1966 by Hector DeLuca and his colleagues at the University of Wisconsin-Madison.[102]

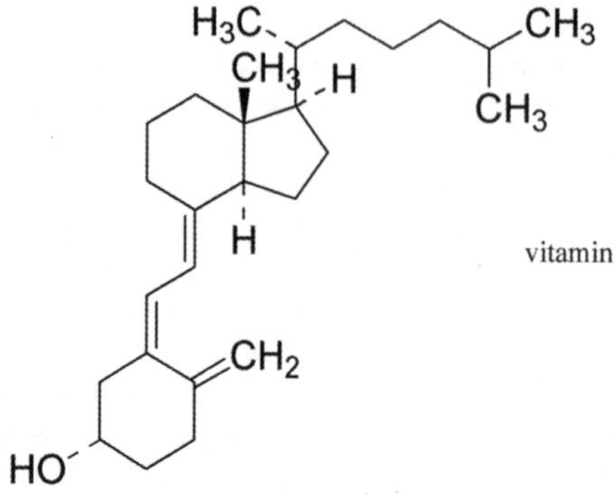

vitamin D₃

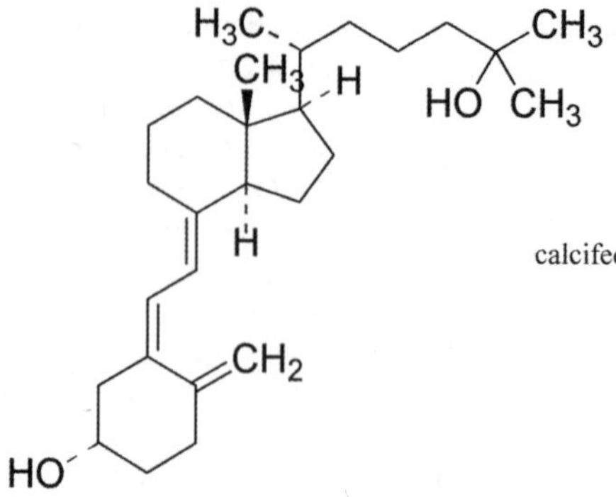

calcifediol

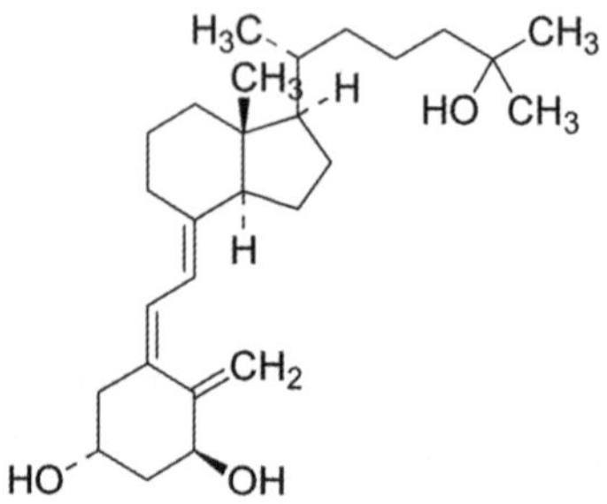

calcitriol

DeLuca identified calcifediol as the major circulating metabolite of cholecalciferol.[103] Since it was more potent than cholecalciferol and had a faster onset of action, it was introduced in the early 1970s for treating hypocalcaemia in hypoparathyroid patients and in those on renal dialysis. DeLuca also isolated calcitriol, another active metabolite of cholecalciferol.[104] As it is more polar than other vitamin D analogues, its duration of action is shorter.

α-Tocopherol (Vitamin E)

In 1922, Herbert Evans and Katharine Bishop of the University of California, San Francisco, announced that normal pregnancies did not occur in rats kept for long periods on an artificial diet supplemented with all known vitamins.[105] Few offspring were produced as most foetuses were resorbed a few days after conception. Evans suggested that this arose from deficiency of a substance that was eventually to become known as vitamin E. Much interest was aroused six years later when Evans and George Burr discovered that paralysis occurred in young rats whose mothers had been maintained on low levels of the vitamin during pregnancy.[106] Wheat germ oil, a rich source of the vitamin, could cure the paralysis if administered to the rats shortly after their birth. Other workers later suggested this paralysis was a form of muscular dystrophy, leading to much controversy over the possible role of the vitamin in that disease. Matters were complicated by the instability of the vitamin preparations, which were sensitive to oxidation.

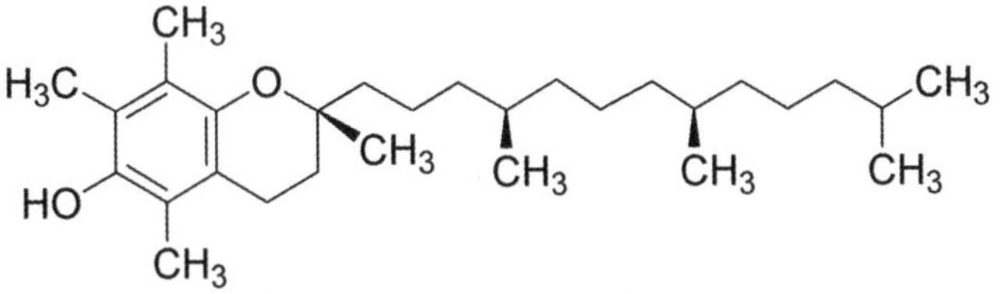

α-tocopherol

The pure vitamin was isolated by Evans and his colleagues in 1936 from a wheat germ oil concentrate.[107] It was given the name α-tocopherol (Greek: *tokos* = childbirth, *pherein* = to bear). The chemical structure was established two years later by Fernholz of Merck and Company[108] and the vitamin was synthesised shortly after by Paul Karrer.[109] Availability of the pure vitamin from natural or synthetic sources enabled researchers to establish whether it had any role in human nutrition or therapeutics. None of the many claims made for its therapeutic value in human diseases has ever been substantiated. The main value of α-tocopherol appears to be its safety as an antioxidant for use by the pharmaceutical and food processing industries.

Phytomenadione (Vitamin K₁)

Henrik Dam carried out a series of experiments at the University of Copenhagen in 1929 to establish whether chickens could synthesise cholesterol, doubts previously having existed about this.[110] He was able to confirm that cholesterol was indeed synthesised, but in the course of proving it he found that his chickens began to haemorrhage after two or three weeks on a fat-free diet supplemented with the known fat-soluble vitamins. Samples of their blood showed delayed coagulation. Dam doubted that this could be a form of scurvy since chickens were already known not to require vitamin C. Nevertheless, he addded lemon juice to their diet, but to no avail. Only large amounts of cereals and seeds in the diet afforded protection. In 1934, he reported the existence of a new accessory food factor and then went on to show, in the following year, that this was fat-soluble, but different from vitamins A, D or E. He described it as vitamin K since it was required for blood coagulation.[111] A substance with similar activity was discovered shortly after by H.J. Almquist and Robert Stockstad at Berkeley. They had discovered that alfalfa meal contained a factor that protected chickens against a scurvy-like haemorrhagic disease induced by being fed on diets in which the source of protein was sardine meal. Almquist was able to demonstrate that meat meal was satisfactory because its slower processing allowed bacterial production of an antihaemorrhagic factor. After this had finally been isolated, a report submitted to *Science* was rejected.[35] It was belatedly sent to *Nature* where it appeared a few weeks after Dam's paper had been published.[112]

Dam sought the assistance of Paul Karrer at the University of Zurich in purifying vitamin K. They isolated it in 1939 as an impure oil.[113] At the same time, a team led by Edward Doisy at the St Louis University School of Medicine separated two forms of the vitamin from vegetable and animal sources, naming them vitamin K₁ and vitamin K₂ respectively. They obtained pure vitamin K₁ as crystals from alfafa and promptly determined its structure[114] and then synthesised it.[115] The synthesis was also accomplished by Louis Fieser at Harvard[116] and by Almquist.[117] Vitamin K₁ later received the approved name of 'phytomenadione', but it is also known as 'phylloquinone'. Henrik Dam and Edward Doisy were awarded the Nobel Prize for Physiology and Medicine in 1943 for the discovery of vitamin K.

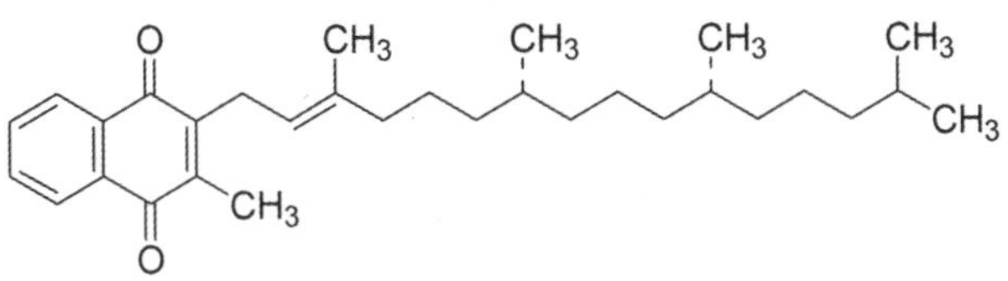

phytomenadione
(vitamin K₁)

menaquinone 4

Vitamin K_2 was extracted from fishmeal by Doisy and his colleagues.[118] It consisted of closely related compounds known as 'menaquinones', which are synthesised in the intestines by bacteria. They have up to 15 isoprene units in their side chain. The structure of menaquinone 4, which has four isoprene units, is shown.

REFERENCES

1. N. Lunin, Ueber die Bedeutung der anorganischen Salze für die Ernährung des Thieres. *Z. Physiol. Chem.*, 1881; **5**: 31–9.
2. C.A. Pekelharing, Over onze kennis van de Waarde der Voedings middelen nit chemische fabrieken. *Ned. Tijdschr. Geneesk.*, 1905; **41**: 111–24.
3. F.G. Hopkins, Feeding experiments illustrating the importance of accessory factors in normal diets. *J. Physiol.*, 1912; **44**: 425–60.
4. F.J. van Leent, Mededeelingen over Beri-Beri. *Geneesk. Tijdschr. v. Nederland-Indie*, 1880; **9**: 272–310.
5. C. Eijkman, Polineuritis bij Hoenderen. *Geneesk. Tijdschr. v. Nederland-Indie*, 1890; **30**: 295–334.
6. C. Eijkman. Eine beri-beri-aehnliche Krankheit der Huehner. *Arch. Pathol. Anat.*, 1897; **148**: 523–32.
7. G. Grijns, Over polineuritis gallinarum. *Geneesk. Tijdschr. v. Nederland.-Indie*, 1901; **41**: 3–110.
8. C. Funk, The etiology of the deficiency diseases: beri-beri, polyneurites in birds, epidemic dropsy, scurvy, experimental scruvy in animals, infantile scurvy, sheep beri-beri, pellagra. *J. State Med.*, 1912; **20**: 341–68.
9. J.C. Drummond, The nomenclature of the so-called accessory food factors (vitamins). *Biochem. J.*, 1920; **14**: 660.
10. B.C.P. Jansen, Early nutritional researches on beriberi leading to the discovery of vitamin B1. *Nutrition Abstracts and Rev.*, 1956; **26**: 1–14.
11. B.C.P. Jansen, W.F. Donath, Over de isoleering van het anti-beriberi-vitamine. *Geneesk. Tidskr. v. Nederland.-Indie*, 1927; **66**: 810–27.
12. J.K. Cline, R.R. Williams, J. Finkelstein, Synthesis of vitamin B1. *J. Am. Chem. Soc.*, 1937; **59**: 1052–4.
13. H. Andersag, K. Westphal, Über die synthsis des antineuritische Vitamins. *Ber.*, 1937; **70**: 2035–54..
14. A.R. Todd, F. Bergel, Aneurin. Part VII. A synthesis of aneurin. *J. Chem. Soc.*, 1937: 364–7.
15. E.V.McCollum, M. Davis, Necessity of certain lipins in the diet during growth. *J. Biol. Chem.*, 1913; **15**: 167–75.
16. H.H. Mitchell, On the identity of the water-soluble growth-promoting vitamine and the antineuritic vitamine. *J. Biol. Chem.*, 1919; **40**: 399–413.
17. H. Chick, M. Roscoe, The dual nature of water-soluble vitamin B. II. *Biochem. J.*, 1928; **22**: 790–9.
18. R. Kuhn, P. György, T.Wagner-Jauregg, Über eine neue Klasse von Naturfarbstoffen. *Chem. Ber.*, 1933; **66**: 317–20.
19. R. Kuhn, *et al.*, *Naturwissen.*, 1935; **23**: 260.
20. P. Karrer, K. Schöpp, F. Benz, Synthesen von Flavinen. IV. *Helv. Chim. Acta*, 1935; **18**: 426–9.
21. G. Casal, *Historia Natural y Medica de el Principado Asturias*. Madrid: M. Martin; 1762.
22. F. Frapolli, *Animadversiones in Morbum, vulgo Pellagram*. Milan: J. Galeatium; 1771.
23. E.G. Willcock, F.G. Hopkins, The importance of individual amino acids in metabolism: Observations on the effect of adding tryptophane to a diet in which zein is the sole nitrogenous constituent. *J. Physiol.*, 1906; **35**: 88–102.
24. C. Voegtlin, M.H. Neil, A. Hunter, The influence of vitamins on the course of pellagra. *U.S. Public Health Serv. Bull.*, 1920; **116**: 7–35.
25. J. Goldberger, G.A. Wheeler, Experimental black tongue of dogs and its relation to pellagra. *U.S. Public Health Rep.*, 1928; **43**: 172.
26. C.A. Elvehjem, R.J. Madden, F.M. Strong, D.W. Woolley, The isolation and identification of the anti-black tongue factor. *J. Biol. Chem.*, 1938; **123**: 137.

27. D.W. Woolley, H.A. Waisman, C.A.Elvehjem, Nature and partial synthesis of the chick antidermatitis factor. *J. Am. Chem. Soc.*, 1939; **61**: 977–7.

28. R. Altschul, A. Hoffer, J. D. Stephen, Influence of nicotinic acid on serum cholesterol in man. *Arch. Biochem. Biophys.*, 1955; **54**: 558–9.

29. W.B. Parsons, Jr, R.W.P. Achor, K.G. Berge, B.F. McKenzie, N.W. Barker, Changes in concentration of blood lipids following prolonged administration of large doses of nicotinic acid to persons with hypercholesterolemia: preliminary observations. *Proc. Staff. Meet. Mayo Clinic*, 1956; **31**: 377–90.

30. S. Lepkovsky, The isolation of factor one in crystalline form. *J. Biol. Chem.*, 1938; **124**: 125–8.

31. J.C. Keresztesy, J.R. Stevens, Vitamin B-6. *J. Am. Chem. Soc.*, 1938; **60**: 1267–8.

32. P. György, Crystalline vitamin B6. *J. Am. Chem. Soc.*, 1938; **60**: 983–4.

33. S.A Harris, E.T. Stiller, K. Folkers, Synthesis of vitamin B6. *J. Am. Chem. Soc.*, 1939; **61**: 1245–7.

34. R. Kuhn, K. Westphal, G. Wendt, O.Westphal, *Naturwissen.*, 1939; **27**: 469.

35. T.H. Jukes, Vitamin K – a reminiscence. *Trends Biol. Sci.*, 1980; **5**: 140–1.

36. D.W. Woolley, H.A. Waisman, C.A. Elvehjem, Nature and partial synthesis of the chick antidermatitis factor. *J. Am. Chem. Soc.*, 1939; **61**: 977–8.

37. R.J. Williams, J.H. Truesoail, H.H. Weinstock, *et al.*, Pantothenic acid. II. Its concentration and purification from liver. *J. Am. Chem. Soc.*, 1938; **60**: 2719–23.

38. R.J. Williams, C.M. Lyman, G.H. Goodyear, J.H. Triesdall, 'Pantothenic acid', a growth determinant of universal biological occurrence. *J. Am. Chem. Soc.*, 1933; **55**: 2912–27.

39. E.T. Stiller, S.A. Harris, J. Finkelstein, J.C. Keresztesy, K. Folkers, Pantothenic acid. VIII. The total synthesis of pure pantothenic acid. *J. Am. Chem. Soc.*, 1940; **62**: 1785–90.

40. F.Kögl, B.Tönnis, *Z. Physiol. Chem.*, 1936; **242**: 43.

41. V. Du Vigneaud, K. Hofmann, D.B. Melville, On the structure of biotin. *J. Am. Chem. Soc.*, 1942; **64**: 188–9.

42. S.A. Harris, D.E. Wolf, R. Mozingo, *et al.*, Biotin. V. Synthesis of *dl*-biotin, *dl*-allobiotin and *dl*-epi-allobiotin. *J. Am. Chem. Soc.*, 1945; **67**: 2096–2100.

43. G.W. Corner, *George Hoyt Whipple and his Friends: the Life Story of a Nobel Prize Winner.* Philadelphia: Lippincott; 1963.

44. G.H. Whipple, F.S. Robscheit-Robbins, Blood regeneration in severe anemia: I. Standard basal ration bread and experimental methods. *Am. J. Physiol.*, 1925; **72**: 395–407.

45. G.R. Minot, W.P. Murphy, Treatment of pernicious anemia by a special diet. *J. Am. Med. Ass.*, 1927; **89**: 759.

46. L. Smith, Purification of anti-pernicious anemia factors from liver. *Nature*, 1948; **161**: 638–9.

47. E.L. Rickes, N.G. Brink, F.R. Koninszy, *et al.*, Crystalline vitamin B$_{12}$. *Science*, 1948; **107**: 396–7.

48. E.L. Rickes, N.G. Brink, F.R. Koninszy, *et al.*, Comparative data on vitamin B$_{12}$ from liver and from a new source *Streptomyces griseus. Science*, 1948; **108**: 634.

49. D.C. Hodgkin, J. Pickworth, J.H. Robertson, *et al.*, The crystal structure of the hexacarboxylic acid derived from B-12 and the molecular structure of the vitamin. *Nature*, 1955; **176**: 325.

50. R. Bonnett, J.R. Cannon, A.W. Johnson, *et al.*, The structure of vitamin B12 and its hexacarboxylic acid degradation product. *Nature*, 1955; **176**: 328–30.

51. L. Wills, Treatment of pernicious anemia of pregnancy and 'tropic anemia' with special reference to yeast extract as curative agent. *Br. Med. J.*, 1931; **1**: 1059–64.

52. L. Wills, A. Stewart, Experimental anaemia in monkeys, with special reference to macrocytic nutritional anaemia. *Br. J. Exp. Pathol.*, 1935; **16**: 444.

53. P.L. Day, W.C. Langston, C.F. Shukers, Leukopenia and anemia in the monkey resulting from vitamin deficiency. *J. Nutr.*, 1935; **9**: 637–44.

54. W.C. Langston, W.J. Darby, C.F. Shukers, P.L. Day. Nutritional cytopenia (vitamin M deficiency) in the monkey. *J. Exp. Med.*, 1938; **68**: 923–40.

55. A.G. Hogan, E.M. Parrott, Anemia in chicks caused by a vitamin deficiency. *J. Biol. Chem.*, 1940; **132**: 507–17.

56. J.J. Pfiffner, S.B. Binkley, E.S. Bloom, *et al.*, Isolation of the antianemia factor (Bc) incrystalline form from liver. *Science*, 1943; **97**: 404–5.

57. E.E. Snell, W.H. Peterson, Growth factors for bacteria X. Additional factors required by certain lactic acid bacteria. *J. Bacteriol.*, 1940; **39**: 273–85.

58. B.L. Hutchings, N. Bohonos, W.H. Peterson, Growth factors for bacteria. XIII. Purification and properties of an eluate factor required by certain lactic acid bacteria. *J. Biol. Chem.*, 1941; **141**: 521–8.

59. H.K. Mitchell, E.E. Snell, R.J. Williams, The concentration of 'folic acid'. *J. Am. Chem. Soc.*, 1941; **63**: 2284.

60. R.C. Mills, G.M. Briggs, C.A. Elvehjem, E.B. Hart, *Proc. Soc. Exp. Biol. Med.*, 1942; **49**: 186.

61. E.L.R. Stockstad, P.D.V. Manning, Evidence of a new growth factor required by chicks. *J. Biol. Chem.*, 1938; **125**: 687–96.

62. E.L.R. Stockstad, P.D.V. Manning, R.E. Rogers, The relation between factor U and vitamin B6. *J. Biol. Chem.*, 1940; **132**: 463.
63. E.L.R. Stockstad, Some properties of a growth factor for *Lactobacillus casei*. *J. Biol. Chem.*, 1943; **149**: 573–4.
64. J.H. Mowat, J.H. Boothe, B.L. Hutchings, *et al.*, The structure of the liver *L. casei* factor. *J. Am. Chem. Soc.*, 1948; **70**: 14–18.
65. R.B. Angier, J.H. Boothe, B.L. Hutchings, *et al.*, The structure and synthesis of the liver *L. casei* factor. *Science*, 1946; **103**: 667–9.
66. H.E. Sauberlich, C.A. Baumann, A factor required for the growth of *Leuconostoc citrovorum*. *J. Biol. Chem.*, 1948; **176**: 165–73.
67. D.B. Cosulich, B. Roth, J.M. Smith, *et al.*, Chemistry of leucovorin. *J. Am. Chem. Soc.*, 1952; **74**: 3252–63.
68. A.J. Lorenz, The conquest of scurvy. *J. Am. Dietetic Ass.*, 1954; **30**: 665–70.
69. J. Lind, *Treatise on Scurvy* [containing a reprint of the 1st edn of *A Treatise of the Scurvy*, with additional notes], eds C.P. Stewart, D. Guthrie. Edinburgh: University Press; 1953.
70. A. Holst, T. Frolich, Experimental studies relating to ship beriberi and scurvy. *J. Hygiene*, 1907; **7**: 634.
71. S.S. Zilva, The antiscorbutic factor in lemon juice. *Biochem. J.*, 1924; **18**: 632–7.
72. W.A. Waugh, C.G. King, Isolation and identification of vitamin C. *J. Biol. Chem.*, 1932; **97**: 325–31.
73. J.L. Svirbely, A. Szent-Györgyi, The chemical nature of vitamin C. *Biochem. J.*, 1932; **26**: 865–80.
74. E.V. McCollum, The paths to the discovery of vitamins A and D. *J. Nutrition*, 1967; **91** (Suppl. 1), 11–16.
75. E.V. McCollum, M. Davis, The nature of the dietary deficiencies of rice. *J. Biol. Chem.*, 1915; **23**: 181–230.
76. M. Mori, Ueber den sogenanten Hikan (Xerosis conjunctivae infantum eventuell Keratomalacie). *Jb. Kinderheilkd.*, 1904; **59**: 175–95.
77. T.B. Osborne, L.B. Mendel, Amino acids in nutrition and growth. *J. Biol. Chem.*, 1914; **17**: 325.
78. S. Mori. Primary changes in eye of rats which result from deficiency of fat-soluble A in diet. *J. Am. Med. Ass.*, 1922; **79**: 197–200.
79. R.A. Morton, I. Heilbron, Absorption spectrum of vitamin A. *Biochem. J.*, 1928; **22**: 987.
80. J.C. Drummond, *Biochem. J.*, 1932; **26**: 1178.
81. P. Karrer, R. Mörf, K. Schöpp, Zur Kenntnis des Vitamins-A aus Fischtranen. *Helv. Chim. Acta.*, 1931; **14**: 1036–40.
82. O. Isler, W. Huber, A. Ronco, M. Kofler, Synthésis des vitamin A. *Helv. Chim. Acta.*, 1947; **30**: 1911.
83. S.B. Wolbach, P.R. Howe, Tissue changes following deprivation of fat-soluble A vitamin. *J. Exp. Med.*, 1925; **42**: 753–77.
84. G. Stüttgen, H. Krause, Der nachweis von trikiummarkiertem vitamin A in den Schichten der Haut nach lokaler Applikation. *Hautarzt*, 1959; **10**: 504.
85. G. Stüttgen, Zur Lokalbehandlung von Keratosen mit Vitamin A Säure. *Dermatologica*, 1962; **124**: 65–80.
86. A.M. Kligman, J.E. Fulton Jr, G. Plewig, Topical vitamin A acid in acne vulgaris. *Arch. Dermatol.*, 1969; **99**: 469–76.
87. M.S. Tallman, MJ.W. Anderson, C.A. Schiffer, *et al.*, All-trans retinoic acid in acute promyelocytic leukaemia: long-term outcome and prognostic factor analysis for the North American Intergroup protocol. *Blood*, 2000; **100**: 4298–302.
88. A.B. Barua, M.C. Ghosh, Preparation and properties of 4-oxo-retinoic acid and its methyl ester. *Tetrahedron Lett.*, 1972; 1823–5.
89. F.G. Hopkins, Feeding experiments illustrating the importance of accessory factors in normal diets. *J. Physiol.*, 1912; **44**: 425–60.
90. E. Mellanby, The part played by an 'accessory factor' in the production of experimental rickets. A further demonstration of the part played by accessory food factors in the aetiology of rickets. *J. Physiol.*, 1918; **52**: 11–53.
91. E.V. McCollum, N. Simmonds, J.E.Becker, Studies on experimental rickets. XXI. An experimental demonstration of the existence of a vitamin which promotes calcium deposition. *J. Biol. Chem.*, 1922; **53**: 293–312.
92. K. Huldschinsky, Heilung von Rachitis durch Künstliche Höhensonne. *Deut. Med. Wochenschr.*, 1919; **45**: 712–13.
93. H. Chick, The discovery of vitamins. *Prog. Food Nutrition*, 1975; **1**: 1–20.
94. A.F. Hess, M. Weinstock, Antirachitic properties imparted to lettuce and to growing wheat by ultraviolet irradiation. *Proc. Soc. Exp. Biol. Med.*, 1924; **22**: 5.
95. H. Steenbock, The induction of growth promoting and calcifying properties in a ration by exposure to light. *Science* 1924; **60**: 224–5.

96. A. Windaus, A. Hess, O. Rosenheim, R. Pohl, T.A. Webster, *Chem. Ztg.*, 1927; **51**: 113.
97. A. Windaus, O. Linsert, A. Lüttringhaus, G. Weidlich, Über das kristalliserte Vitamin D_2. *Annalen*, 1932; **492**: 226–41.
98. A. Windaus, W. Thiel, Über die Konstitution des Vitamins D_2. *Annalen*, 1936; **521**: 160–75.
99. F.A. Askew, R.B. Bourdillon, H.M. Bruce, *et al.*, The distillation of vitamin D. *Proc. Roy. Soc. London*, 1931; **B107**: 76–90.
100. A. Windaus, *et al.*, *Z. Physiol. Chem.*, 1936; **241**: 100.
101. A. Windaus, A.F. Bock. Über das Provitamin aus dem Sterin der Schweineschwarte. *Zeitschrift für Physiologische Chemie*, 1937; **245**: 168–70.
102. P.F. Neville, H.F. DeLuca, The synthesis of [1,2-3H]vitamin D_3 and tissue localization of a 0.25 mg (10 IU) dose per rat. *Biochemistry*, 1966; **5**: 2201–7.
103. J.W. Blunt, H.F. DeLuca, H.K. Schnoes, 25-Hydroxycholecalciferol. A biologically active metabolic of vitamin D_3. *Biochemistry*, 1968; **7**: 3317–22.
104. M.F. Holick, H.K. Schnoes, H.F. DeLuca, *et al.*, Isolation and identification of 1,25-dihydroxycholecalciferol. *Biochemistry*, 1971; **10**: 2799–804.
105. H. Evans, K.S. Bishop, On the existence of a hitherto unrecognized dietary factor essential for reproduction. *Science*, 1922; **56**: 650–1.
106. H.M. Evans, G.O. Burr, On the amount of vitamin B required during lactation. *J. Biol. Chem.*, 1928; **76**: 263–72.
107. H. Evans, O.H. Emerson, G.A. Emerson, The isolation from wheat germ oil of an alcohol, α-tocopherol, having the properties of vitamin E. *J. Biol. Chem.*, 1936; **113**: 319–32.
108. E. Fernholz, The thermal decomposition of α-tocopherol. *J. Am. Chem. Soc.*, 1937, **59**: 1154–5.
109. P. Karrer, H. Fritzsche, B.H. Ringier, H. Salomon, α-Tocopherol. *Helv. Chim. Acta*, 1938; **21**: 520–3.
110. H. Dam, Cholesterinstoffwechsel in Huhnereiern und Huchnhen. *Biochem. Z.*, 1929; **215**: 475.
111. H. Dam, The antihaemorrhagic vitamin of the chick. *Nature*, 1935; **135**: 652–3.
112. H.J.Almquist, R. Stockstad, Dietary haemorrhagic disease in chicks. *Nature*, 1935: **136**: 31.
113. H. Dam, A. Geiger, J. Glavind, *et al.*, Isolierung des Vitamins K in hochgereinigter Form. *Helv. Chim. Acta*, 1939; **22**: 310–13.
114. S.B. Binkley, D.W. MacCorquodale, S.A. Thayer, E.A. Doisy, The isolation of vitamin K_1. *J. Biol. Chem.*, 1939; **130**: 219–34.
115. D.W. MacCorquodale, L.C. Cheney, S.B. Binkley, *et al.*, The constitution and synthesis of vitamin K_1. *J. Biol. Chem.*, 1939; **131**: 357.
116. L.F. Fieser, Synthesis of 2-methyl-3-phytyl-1,4-naphthoquinone. *J. Am. Chem. Soc.*, 1939; **61**: 2559–61.
117. H.J. Almquist, A.A. Klose, Synthetic and natural antihemorrhagic compounds. *J. Am. Chem. Soc.*, 1939; **61**: 2557–8.
118. R.W. McKee, S.B. Binkley, S.A. Thayer, *et al.*, The isolation of vitamin K_2. *J. Biol. Chem.*, 1939; **131**: 327–44.

20

Antimetabolites

Following the introduction of the sulfonamides and antihistamines in the late 1930s and the 1940s, the idea that drugs may compete with natural metabolites at receptors gained wide recognition. The 'lock and key' analogy, popularised by Paul Ehrlich at the beginning of the twentieth century to explain how chemotherapeutic drugs interacted with their receptors, now came back into vogue. An essential feature of that hypothesis was the necessity for a potential drug to have a structural similarity to the substrate, or metabolite, that normally attached to the receptor in question. The drug could then block access to the receptor by the natural substrate and so act as an antimetabolite.

Only one vitamin has provided analogues that are used therapeutically as antimetabolites, namely folic acid. When antifolates were found to induce remissions in children with leukaemia, it encouraged chemists to synthesise a diverse range of antimetabolites that interfered with cell division. This remains a significant driving force behind the design of novel chemotherapeutic drugs.

FOLIC ACID ANALOGUES

Richard Lewisohn, a surgeon at the Mount Sinai Hospital in New York, initiated a series of experiments in 1937, hoping to establish why primary tumours of the spleen were rarely encountered.[1] He began by injecting a concentrated beef spleen extract into mice bearing transplanted tumours (sarcoma-180) and observed complete regression of these in 60% of the animals. The extract had to be given subcutaneously as it was highly irritant and caused thrombosis if injected into a vein. It had no effect on spontaneous (i.e. naturally occurring) tumours in mice. Noting that the spleens of the healed mice were often enlarged, Lewisohn prepared extracts of these as well. Since the mouse spleen extract was dilute and non-irritant, it could be safely injected intravenously. Unlike the original beef extract, this induced complete regression of spontaneous breast tumours in 30% of mice.[2] Although tumours reappeared in a quarter of the mice, never before had a non-toxic substance produced such a result.

Although encouraged by his findings, Lewisohn recognised that quantity production of the 'healed mouse extract' was impracticable. He began an extensive screening programme in 1939 to find other agents that could cause regression of tumours. An obvious step was to examine the recently discovered vitamins of the B group that were present in liver. Using yeast extract as a cheap, convenient source of these, he and his colleagues obtained positive results. Intensive investigations began in January 1941, requiring the use over the next three or four years of some 12 000 mice bearing sarcoma-180, and half that number bearing spontaneous tumours. When the yeast extract was injected intravenously once daily for up to ten weeks, approximately one-third of spontaneous breast adenocarcinomas regressed completely. Measurable changes were also detected when autopsies were conducted on mice that received four intravenous injections over a 48 hour period beginning a week or so after transplantation with sarcoma-180. With the entry of the United States into the war, it became impossible for Lewisohn to continue to obtain from Germany the brewer's yeast used in the preparation of

Drug Discovery. A History. W. Sneader.
©2005 John Wiley & Sons Ltd

his extracts. Seeking alternative sources of active material, he found that barley extract was as efficacious as yeast extract. By this time such was the nature of his findings that the International Cancer Research Foundation, which had sponsored Lewisohn's investigations, asked Cornelius Rhoads, Director of the Memorial Hospital in New York, to conduct an independent enquiry.[3] This was duly entrusted to one of the senior scientists at the hospital, in 1943. His attempts to confirm Lewisohn's findings were unsuccessful, no significant difference being detected either in the growth of sarcoma-180 in treated and untreated mice, or in the regression of spontaneous breast cancer.

Early in 1944, it occurred to Lewisohn that the active substance in yeast and barley extracts might be the newly isolated folic acid. His subsequent finding of antitumour activity in a folic acid concentrate did not settle the matter, since it contained many impurities. However, Lewisohn was able to obtain the support of Lederle Laboratories, who supplied small amounts of the scarce crystalline fermentation *Lactobacillus casei* factor they had just isolated. In an initial test on seven mice, each receiving 0.25 µg by intravenous injection, Lewisohn observed the greatest inhibition of tumour growth he had ever seen. Extensive studies on mice with spontaneous breast cancer confirmed the initial results, these being published in January 1945.[4] Subsequently, Lederle researchers discovered that the material supplied to Lewisohn was not identical to the liver *L. casei* factor. It was pteroyltriglutamic acid, whereas the liver factor was pteroylglutamic acid (now always described as folic acid).[5] When the Mount Sinai Hospital researchers tested the liver *L. casei* factor (i.e. folic acid), they found it to be ineffective against spontaneous breast cancer in mice.[6] This persuaded Lederle to synthesise both the di- and triglutamates for clinical trials. These were appropriately named Diopterin® and Teropterin®, respectively, and were sent to Farber for clinical evaluation. He and colleagues in three hospitals associated with the Harvard Medical School began by administering the drugs to 90 patients with a variety of malignancies that offered no hope of cure. The purpose of this phase I trial was to determine the toxicology, appropriate dosage and suitable routes of administration of the folic acid conjugates in humans.

After cautious initial studies, Teropterin® was given daily in doses up to 150 mg intramuscularly, or 500 mg intravenously, the average length of treatment being 35 days. Neither local nor systemic adverse effects were encountered. Although reluctant to draw conclusions about the efficacy of the treatment after so short a period of observation, Farber and his colleagues,[7] in a brief preliminary report published in December 1947, suggested that the temporary improvements observed in some patients warranted further investigations.

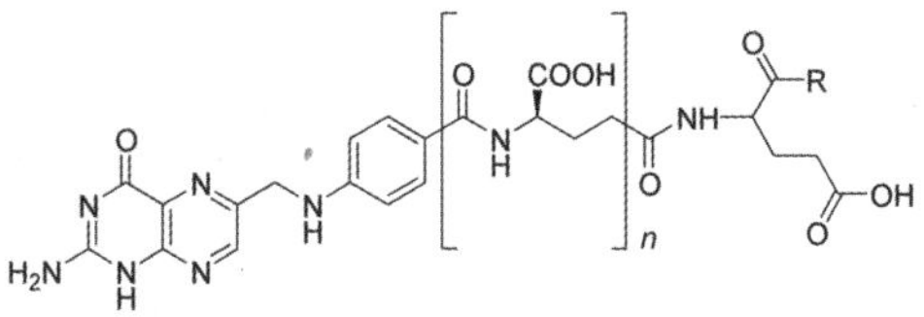

Folic acid	$n=0$	R= —OH
Diopterin	$n=0$	R= —NH—CH(CO$_2$H)(CH$_2$CH$_2$CO$_2$H)
Teropterin	$n=2$	R= —OH

During that phase I trial, biopsies of bone marrow had been routinely taken from several patients with acute leukaemia. Unexpectedly, these showed that the folates had accelerated the progress of the disease towards its fatal termination. On reviewing the situation, Farber speculated whether this 'acceleration phenomenon' could be put to advantage in either of two distinct ways.[8] The first would be to use the folic acid conjugates to stimulate leukaemic cells to grow and divide, thereby rendering them more susceptible to nitrogen mustard chemotherapy. The second possibility would be to administer one of the antagonists of folic acid that had been newly developed by Yellepragada Subba Row and his colleagues at Lederle Laboratories.

A variety of antagonists of vitamins, hormones and cell metabolites had been synthesised after Donald Woods of Oxford University discovered in 1940 that sulfonamides exerted their antibacterial action by antagonising the role of 4-aminobenzoic acid, a growth factor for bacteria.[9] Such antagonists were described as antimetabolites. This phenomenon has first been reported in 1931 by Judah Quastel at Cambridge when he introduced the term 'competitive inhibition' to describe how malonic acid inhibited the oxidation of succinic acid by bacteria, and also in brain and muscle tissue.[10]

In 1947, Lederle researchers reported the synthesis and biological properties of a crude methylated derivative of folic acid ('x-methyl pteroylglutamic acid', a mixture of methylated products) that was a weak antimetabolite.[11] It had a depressant action on the blood-forming elements of the bone marrow in rats and mice, as was anticipated from prior knowledge of the effects of folic acid deficiency. The extent of this, however, was greater than previously seen, the animals becoming not only anaemic but also exhibiting a reduction in the number of white blood cells. When tested on a patient with chronic leukaemia, the antimetabolite was not potent enough to produce any worthwhile result. The same chemists who had synthesised folic acid at Lederle Laboratories now prepared more antimetabolites, the first of which, pteroylaspartic acid, was sent to Farber.[12]

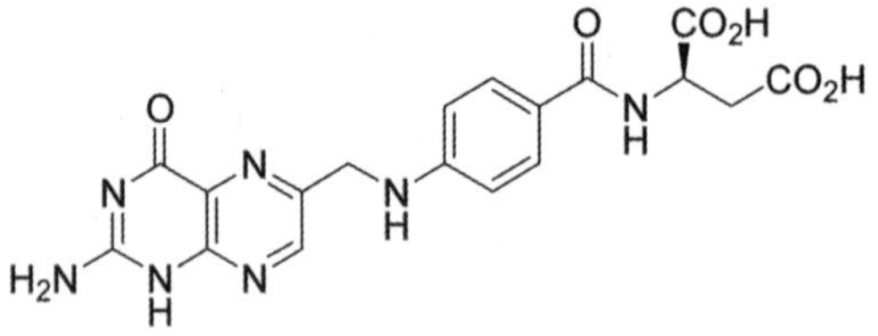

pteroylaspartic acid

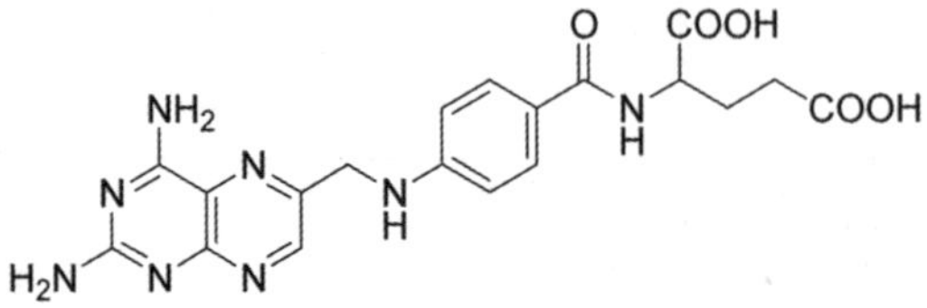

aminopterin

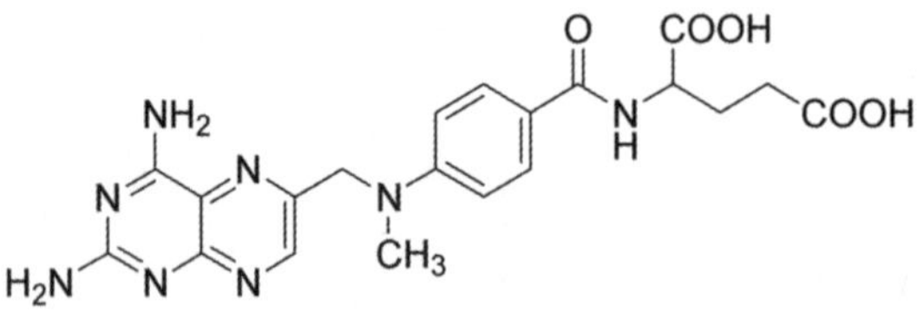

methotrexate

At the Children's Medical Center in Boston on 28 March 1947, Farber administered pteroylaspartic acid to a four year old girl dying of acute myeloid leukaemia. She received 40 mg daily intramuscular injections until her death a week later. Post-mortem examination revealed that although the treatment had not been started in time to save her life, the number of leukaemic cells in her bone marrow had been drastically reduced. Commenting on this, in their report in the *New England Journal of Medicine* a year later, Farber and his colleagues stated, 'A change of this magnitude in such a short time has not been encountered in the marrow of leukaemic children in our experience.' The deaths throughout the world, despite treatment with the best available therapy, of that four year old girl and thousands after her were inevitable until the lessons painfully learned from them finally enabled Farber's successors to conquer the most dreaded of all diseases of childhood.

Following the discovery that replacing the oxygen substituent on the 4-position of the pteridine ring by an amino group enhanced the potency of folic acid antagonists,[13] aminopterin was synthesised at the Bound Brook laboratories of the American Cyanamid Company in New Jersey (Lederle Laboratories was the other main division).[14]

Farber had treated a further 14 children with pteroylaspartic acid before receiving aminopterin in November 1947. It was the first potent folic acid antagonist he had worked with. During the next six months, 16 children with acute leukaemia were treated with it. Many of them had been moribund at the onset of therapy, yet complete remissions were obtained in ten cases. These were the first sustained remissions ever obtained in leukaemic patients. The white cell counts had apparently returned to normal levels, with either a marked reduction or complete disappearance of the malignant blast cells (lymphoblasts). The red cell count had also approached normal values. Toxic effects were certainly present, the most frequent being a troublesome stomatitis affecting the rapidly dividing lining cells of the mouth, leading to painful ulceration.[15] Notwithstanding this, aminopterin quickly became established as a major advance in the treatment of acute leukaemia.

In the summer of 1947, the Bound Brook researchers developed methotrexate.[16] Once there was evidence that it was safer than aminopterin, it rendered all other antifolates obsolete. It also proved effective in treating choriocarcinoma, lymphomas and some solid tumours.

Two schools of thought now developed over the management of the leukaemic children in whom remission had been achieved. Some haematologists stopped administration of aminopterin as soon as the blood and bone marrow appeared normal, restarting treatment only when leukaemic cells reappeared, as they inevitably did. This process was repeated until no further improvement could be achieved. Farber had spurned this type of intermittent therapy, and opted for maintenance therapy with repetitive administration of aminopterin at regular intervals until resistance to the drug occurred. In this way, he kept his patients alive for an average of eight or nine months, and one in every hundred or so appeared to be cured. In the succeeding years, it became evident that Farber's approach had been correct since remission did not represent elimination of the disease, but was merely a reduction in the number of leukaemic cells in the body to a level at which they were no longer detectable. It has been estimated that something in the order of 1000 million leukaemic cells were present at this stage; without further treatment these repeatedly divided until symptoms were experienced when the number had increased a thousandfold. Only by further exposure to regular cycles of therapy could the number of these remaining cells be kept under control. The problem facing Farber was to establish the maximum dosage of methotrexate that could be administered for several days without destruction of too many normal blood-forming elements in the bone marrow. When the white cell count fell below the critical level, the patient was exposed to the risk of death from overwhelming infection. To strike the right balance required constant monitoring of the bone marrow and blood.

Following the development of a variety of other antileukaemic drugs, including cortisone or prednisolone, an aggressive approach aimed at an outright cure of leukaemia in children was pioneered by Donald Pinkel, one of Farber's former associates.[17] Appointed medical director of the newly opened St Jude's Children's Research Hospital in Memphis, Tennessee, in 1962, he administered different combinations of several cytotoxic drugs to randomly allocated groups of patients who had received no prior treatment. The progress of these groups revealed the optimum treatment schedules. Among the advances made by Pinkel was his recognition that the frequent deaths from meningeal leukaemia among long-term survivors who had received chemotherapy was due to the persistence of small numbers of leukaemic cells present in the brain at the onset of the disease. These had not been destroyed because of the poor penetration of the central nervous system by the cytotoxic drugs. Simply by irradiating the craniums of children after they first entered remission, he virtually eliminated this complication. By the early 1970s, more than half his patients were still alive five years after diagnosis of the disease. Today, the majority of children with acute lymphocytic leukaemia reach this stage, and most of them are cured. Prior to the introduction of the antifolate drugs, no child ever survived beyond three months after diagnosis.

In the 1970s, there was growing recognition of the success of combination chemotherapy in treating leukaemia. It strongly influenced the approaches being taken towards cancer chemotherapy. At the same time, medicinal chemists acquired an appreciation of the potential of antimetabolites as chemotherapeutic agents. By the end of the twentieth century there were many more of them in the pharmacopoeia.

MODIFIED PURINES

Soon after the introduction of sulfanilamide into clinical practice it was discovered that its antibacterial activity was antagonised by pus, as well as by tissue or yeast extracts. In 1940, Donald Woods at Oxford University had the idea that the substance responsible for this antagonism might be structurally similar to sulfanilamide, thereby acting in much the same manner as physostigmine and tubocurarine did in antagonising the action of acetylcholine, i.e. by competing with it for an unidentified receptor site. Woods showed that 4-aminobenzoic acid was a very effective antagonist of sulfanilamide, which led him to propose that sulfanilamide acted as an antimetabolite of 4-aminobenzoic acid.[9] He emphasised that the antagonism depended on the close structural relationship between the two compounds.

George Hitchings of the Wellcome Research Laboratories in Tuckahoe, New York, appreciated the implications of the antimetabolite hypothesis. He had gained his PhD from Harvard in 1933 for work in the then-unfashionable field of nucleic acid metabolism. He joined Wellcome in 1942 and started to prepare and test potential antimetabolites of the pyrimidines required for the synthesis of nucleic acids. This was long before Watson and Crick had elucidated the central role of DNA in cellular reproduction. Hitchings' motivation had simply been his recognition that as all cells required nucleic acids to divide, it might be possible to block reproduction of rapidly dividing bacteria, protozoa or tumours by interfering with their synthesis of nucleic acids.

Hitchings and his colleagues initially examined more than 100 analogues of thymine for their ability to inhibit the growth of *L. casei*, an organism used in screening for drugs with antifolate or antipyrimidine activity.[18] Although 5-bromouracil was found to have potent activity, it was never introduced into the clinic. The first success for the Wellcome researchers was not with an analogue derived from a pyrimidine, but with one from a purine.

Mercaptopurine

The work at Tuckahoe on nucleic acid antimetabolites continued with the examination of analogues of adenine, one of the two purines in DNA. Diaminopurine was synthesised in 1948 by Gertude Elion and found to inhibit the growth of *L. casei*.[19] Following encouraging animal tests, it was evaluated at the Sloane–Kettering Institute, then the leading centre for screening of potential anticancer agents against transplanted tumours.[20] Subsequently, it underwent a full clinical trial at the Memorial Hospital in New York. Although an occasional remission was seen in children with acute leukaemia, diaminopurine was clearly inferior to aminopterin and methotrexate.

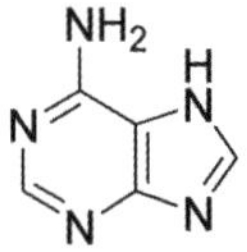

adenine 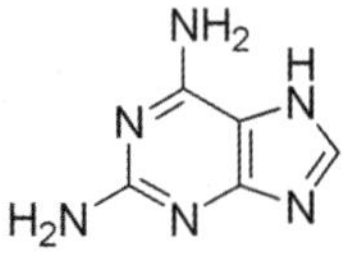   diaminopurine 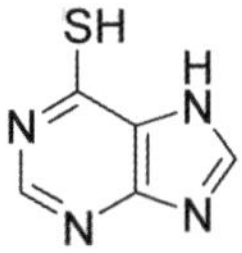   mercaptopurine

Elion synthesised mercaptopurine in 1950 as a chemical intermediate for the subsequent preparation of more aminopurines.[21] When routine screening unexpectedly revealed it to possess outstanding activity as an inhibitor of the growth of *L. casei*,[22] it was sent to the Sloane–Kettering Institute for further evaluation. This led on to a clinical trial at the Memorial Hospital, where it was established that mercaptopurine was the safest and most effective antileukaemic agent yet to have been discovered. Used on its own, the average remission lasted about one year.[23] Of particular significance was the fact that a remission was even induced in one patient who had relapsed after failing to respond to further treatment with a folic acid antagonist. This indicated that there was no cross-resistance between the two classes of compounds and so provided the basis for the introduction of the combination chemotherapy that was to have such dramatic results in children with acute leukaemia.

By 1956, it was evident that a combination of cortisone with either methotrexate or mercaptopurine produced more remissions than any single drug, while also extending the duration of the remissions. By the end of the decade, the mean survival rate for children with acute leukaemia who had received intensive chemotherapy in specialised centres had risen beyond one and a half years. Advances since then have raised the cure rate to between 70 and 80% for those children with the commonest type of acute leukaemia who receive properly supervised combination chemotherapy.

Azathioprine

Much of the mercaptopurine administered to patients was rapidly metabolised before it could exert any therapeutic effect. This occurred through the action of the enzyme xanthine oxidase. Analogues containing substituents on the sulfur atom were synthesised by Elion and Hitchings in the hope of obtaining a prodrug that would provide a sustained release of mercaptopurine. The most active of these was azathioprine, synthesised in 1957.[24] However, it gave disappointing results in a clinical trial. Nevertheless, it later became an important drug in its own right after the discovery of a new application for mercaptopurine.

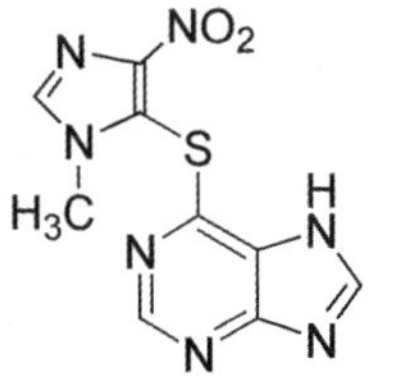

azathioprine

Peter Medawar at University College London stimulated interest in immunological tolerance with his Nobel Prize winning work in the 1950s.[25] At Tufts University School of Medicine in Boston, William Dameshek then had the idea that if a drug could be found that would be more effective than cortisone in depressing the immune response, it might be possible to carry out bone marrow transplantation in patients with aplastic anaemia, leukaemia or radiation damage.[26] Assisted by Robert Schwartz, he examined a variety of drugs to assess their effect on the ability of rabbits to produce antibodies against injected human serum albumin. Mercaptopurine turned out to be highly effective, but when Roy Calne at Harvard Medical School tried it in dogs receiving kidney transplants, it was unable to induce the same degree of immunological tolerance.[27] Fortunately, Calne noted that the transplanted kidneys functioned for much longer than usual, no other drug having a comparable effect. In the light of this development, Hitchings set up a screening programme in which a variety of drugs was tested for their capacity to inhibit the haemagglutinin reaction in mice challenged with foreign red blood cells.[28] From this screen, azathioprine emerged, in 1961, as the most effective drug. Calne went on to pioneer its use in human transplant surgery, and it became the most commonly used cytotoxic immunosuppressant until the introduction of ciclosporin.

Allopurinol

Having failed to find any analogues of mercaptopurine that were superior in the treatment of leukaemia, Elion and Hitchings turned to trying to enhance the activity of mercaptopurine by finding a compound that would inhibit xanthine oxidase, the enzyme responsible for its rapid destruction after administration. Allopurinol, in which one of the purine nitrogens had been moved to form a pyrazolopyrimidine,[29] emerged from *in vitro* testing and was shown to diminish the conversion of mercaptopurine to inactive thiouric acid. Furthermore, it was devoid of cytotoxicity.

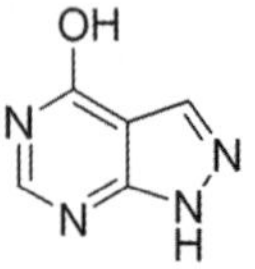

allopurinol

A clinical trial organised by Wayne Rundles of Duke University School of Medicine in North Carolina then found that, despite an enhancement of the antitumour activity of mercaptopurine in the presence of allopurinol, no therapeutic advantage was detected. However, Rundles recognised the potential of a xanthine oxidase inhibitor in interfering with the biosynthesis of uric acid. He tested it in the treatment of gout, where it prevented the formation of uric acid crystals in joints and thus brought relief to the victims of this painful condition.[30] Allopurinol was then marketed for this purpose in 1966.

Thiabendazole

One of the leading exponents of antimetabolite therapy was Wayne Woolley, who worked at the Rockefeller Institute in New York. In 1944, he noted the resemblance of benzimidazole to adenine and speculated that it might act as an adenine antimetabolite. He even demonstrated that it could inhibit the growth of bacteria and fungi and that this could be reversed by either adenine or guanine.[31] After this, several papers were published describing the *in vitro* activity of benzimidazole, but in 1953 researchers at the University of Michigan reported that subcutaneous injections of benzimidazole reduced mortality in mice experimentally infected with poliomyelitis virus.[32] This encouraged researchers at Merck to synthesise benzimidazoles as potential antiviral drugs. Routine screening of these using a variety of assays for chemotherapeutic activity revealed that 2-phenylbenzimidazole had anthelminitic activity. Hundreds of analogues were then examined, from which thiabendazole emerged as one of the most potent chemotherapeutic agents ever discovered. A concentration as low as one part in a million was capable of preventing the development of *Ascaris* eggs *in vitro*.[33] It became the drug of choice for the treatment of strongyloidiasis, a gut infection caused by the roundworm *Stronglyoides stercoralis*, which occurs widely in tropical regions. Unless effective eradication is instituted, the larvae penetrate the gut wall to invade the tissues and a cycle of autoinfection ensues.

 benzimidazole 2-phenylbenzimidazole thiabendazole

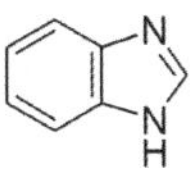

Merck researchers went on to seek an analogue of thiabendazole that would be more resistant to metabolic inactivation, hoping that this would permit higher plasma levels and thus better tissue penetration. This was to be achieved by preventing enzymatic hydroxylation at the 5-position of the ring system. However, straightforward substitution with alkyl, aryl and halo substituents was to no avail. When the 5-amino analogue of thiabendazole was found to lack activity, it was presumed that this was due to metabolic conversion of the amino group. To avoid this, carbamates were synthesised, resulting in the preparation of cambendazole.[34] It proved to be a potent, broad-spectrum anthelmintic, which is valuable for the control of gastrointestinal nematode infection in animals.

 cambendazole mebendazole

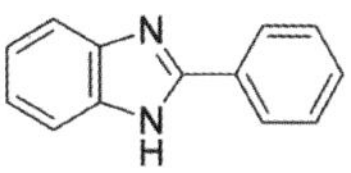

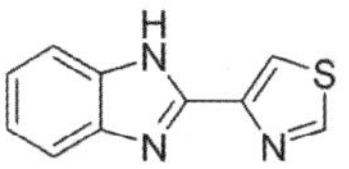

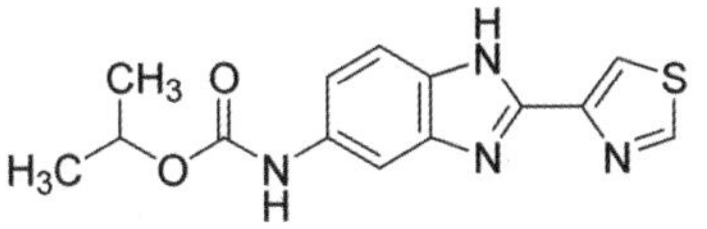

A similar approach enabled Janssen researchers to introduce mebendazole, a broad-spectrum anthelmintic drug of which a single oral dose can eliminate threadworm, roundworm, whipworm or hookworm infections.[35]

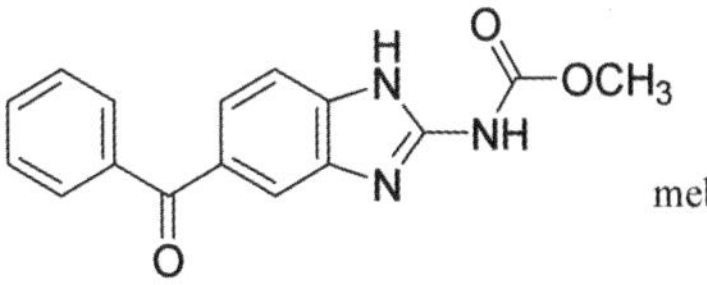

MODIFIED PYRIMIDINES

When Abraham Cantarow and Karl Paschkis at the Jefferson Medical College in Philadelphia reported in 1954 that radioactive uracil was more rapidly absorbed into experimental rat liver tumours than into normal liver cells,[36] it caught the attention of Charles Heidelberger at the McArdle Laboratory for Cancer Research in the University of Wisconsin, Madison. He had previously shown that the fluorine atom in the metabolic poison fluoroacetic acid was responsible for inhibition of a vital enzyme, so he now decided to test the effects of incorporating a fluorine atom into uracil. He asked Robert Duschinsky and Robert Schnitzer of Hoffmann–LaRoche in Nutley, New Jersey, to synthesise fluorouracil and other similar pyrimidines. In 1957, they reported that fluorouracil had potent activity against transplanted tumours in rats and mice.[37] Subsequent clinical studies have shown that its principal therapeutic role is in the treatment of gastrointestinal tumours and in combination chemotherapy for breast cancer.

 fluorouracil , floxuridine (FdUMP)

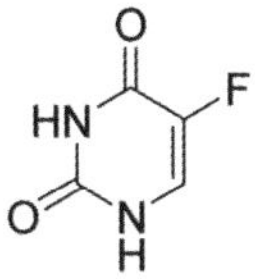

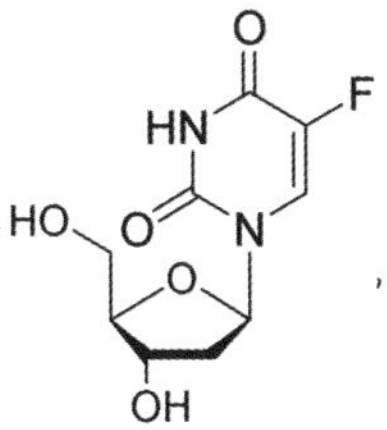

Heidelberger was able to demonstrate that fluorouracil was metabolised to 5-fluorodeoxyuridine monophosphate (FdUMP). This then inhibited thymidylate synthetase, the enzyme that otherwise methylates the uracil in deoxyuridine monophosphate to form the thymine that is incorporated into DNA. This led Heidelberger to test FdUMP, which was subsequently given the approved name of 'floxuridine'.[38] Although active against some tumours and viruses, it has been much less successful than fluorouracil in the clinic.

A major problem with fluorouracil was its variable oral bioavailability, leading to unpredictable results in the clinic. This is due mainly to it being metabolised in the gut wall and in the liver after absorption from the gut. The enzyme responsible for this is dihydropyrimidine dehydrogenase. Several prodrugs have been introduced that avoid metabolism by this particular enzyme, the first being tegafur (ftorafur). This was synthesised at the Taiho Pharmaceutical Company, Tokuahima, in Japan. It is converted to fluorouracil in the liver after absorption from the gut, without involvement of dihydropyrimidine dehydrogenase.[39] Tegafur is used in colorectal cancer.

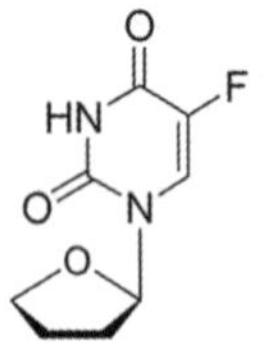

tegafur 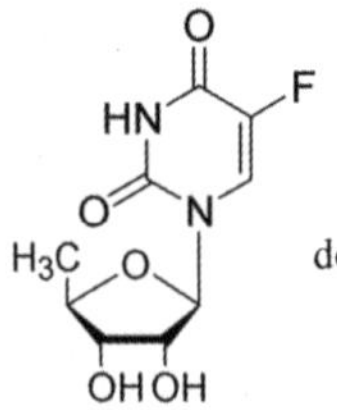   doxifluridine 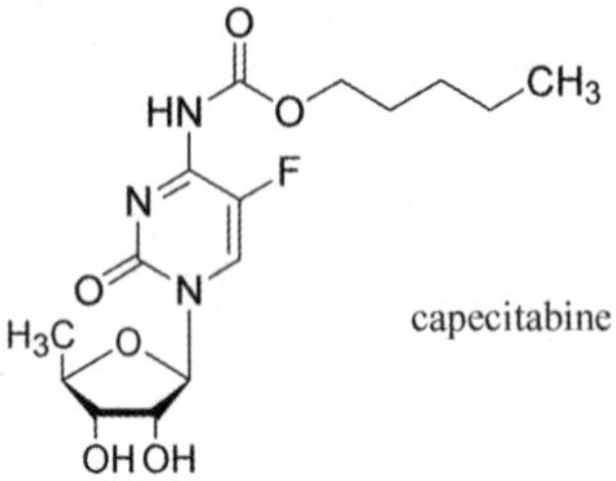   capecitabine

Doxifluridine was developed by Hoffmann–La Roche at Nutley.[40] It releases fluorouracil in the presence of uridine phosphorylase, of which relatively high amounts were found in sarcoma-180 and thus accounted for the high activity against this solid tumour.[41] Capecitabine is a prodrug designed by Hoffmann–La Roche chemists to release doxifluridine only after it had been absorbed from the gut. This reduces the risk of intestinal damage caused when fluorouracil or doxifluridine are given by mouth.[42] The principal use of capecitabine at present is in metastatic colorectal cancer.

Flucytosine, which was also prepared by Hoffmann–La Roche chemists, is selectively converted to fluorouracil in fungi by cytosine deaminase, an enzyme absent from human cells.[43] This has made it a valuable agent in the treatment of systemic yeast infections such as candidiasis, cryptococcosis and torulopsosis. These can be particularly hazardous in cancer patients whose immune system has been compromised as a consequence of exposure to intensive chemotherapy.

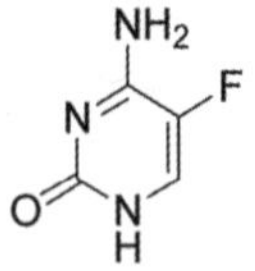

flucytosine 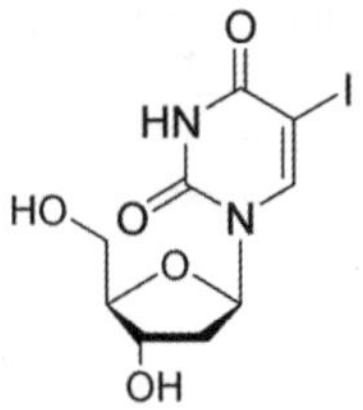   idoxuridine

William Prusoff of the Department of Pharmacology at Yale prepared idoxuridine.[44] Although this was too toxic for systemic use, it could be topically applied to treat skin infections caused by herpes simplex. In infected cells, viral thymidine kinase converts idoxuridine into its monophosphate, which then interferes with cell division. As mammalian thymidine kinase is much more selective, it does not accept idoxuridine as a substrate. However, the exact mechanism of action of idoxuridine is not yet fully understood.

Purine synthesis involves conversion of 5-aminoimidazole ribotide to its 4-carboxamide, which in turn is converted to the purine known as inosinic acid.

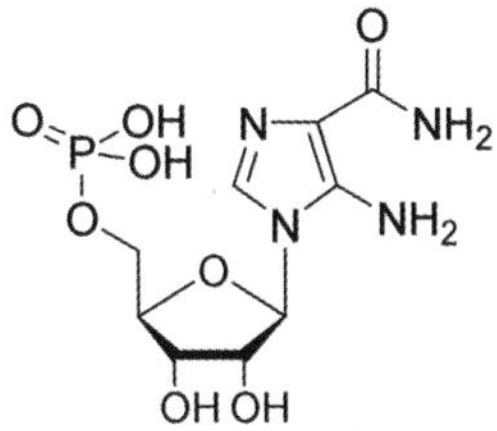

5-aminoimidazole-4-carboxamide ribotide

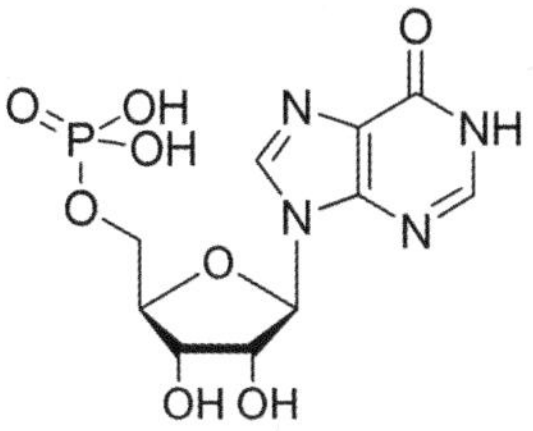

inosinic acid

Dacarbazine was synthesised by Fulmer Shealy at the Southern Research Institute, Birmingham, Alabama. It was intended to be an antimetabolite by virtue of its structural similarity to the 5-aminoimidazole-4-carboxamide moiety in the precursor of inosinic acid.[45] However, it was found that dacarbazine underwent metabolic N-demethylation in the liver and the resulting monomethyl compound spontaneously decomposed, forming 5-aminoimidazole-4-carboxamide and diazomethane. The latter is a well-known alkylating agent and *in vivo* it reacts with the 5-position of guanine in DNA. Since the alkylating activity is not limited to attack on guanine, dacarbazine proved to be highly toxic. Its use has been restricted to treatment of melanoma and in combination chemotherapy of Hodgkin's disease and soft tissue sarcomas.

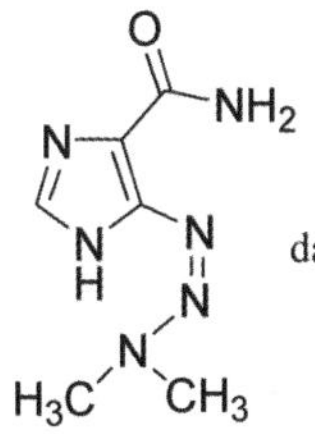

dacarbazine

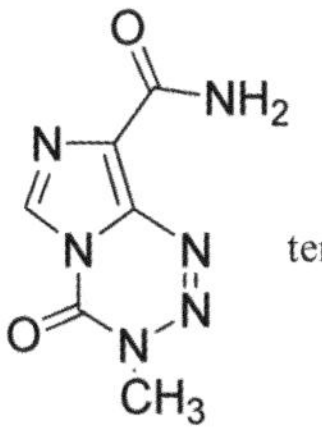

temozolomide

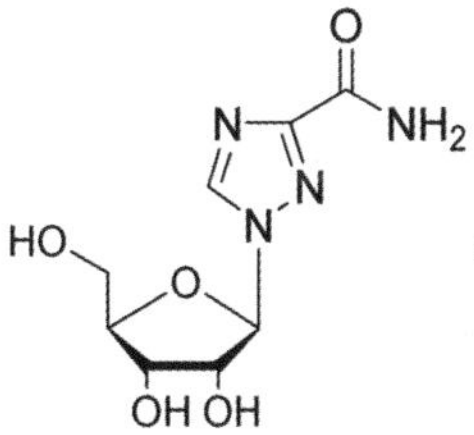

ribavirin

Dacarbazine is chemically unstable, especially in the presence of light. This necessitates it being administered by injection. To avoid this, temozolomide was developed by Malcolm Stevens and his colleagues in the Department of Pharmacy at the University of Aston, in association with Cancer Research UK and May & Baker Limited.[46] It is currently given by mouth to patients with malignant glioma, the commonest and most lethal form of brain tumour, who have not benefited from other therapy. It remains to be seen whether other types of tumour will also respond to temozolomide.

Ribavirin is an analogue of the precursor of inosinic acid that was prepared by Roland Robins and his colleagues at ICN Pharmaceutical Corporation in Irvine, California.[47] It was the first broad-spectrum antiviral drug, active against a variety of DNA and RNA viruses. It has been given by mouth in hepatitis C (in conjunction with interferon) and Lassa fever, as well as by inhalation in children with severe bronchiolitis caused by the respiratory syncytial virus.

MODIFIED NUCLEOSIDES

In 1951, Werner Bergman and Robert Feeney isolated spongothymidine, a novel type of thymine derivative, from a sponge (*Tethya crypta*) collected in the shallow waters off Elliot Key in Florida. Four years later, Bergman and David Burke isolated spongouridine and established the structure of both compounds, describing them as 'spongonucleosides'.[48]

Because the spongonucleosides resembled the nucleosides involved in DNA synthesis, there was interest in the possible anticancer activity of these and related synthetic compounds. The only one that turned out to have useful clinical activity was cytarabine, which was synthesised in 1959 by Richard Walwick, Walden Roberts and Charles Dekker in the Biochemistry Department of the University of California at Berkeley.[49] It was shown to possess activity against a transplanted sarcoma-180 by John Evans and his colleagues of the Upjohn Company.[50] It has since found extensive use in the treatment of acute leukaemia.

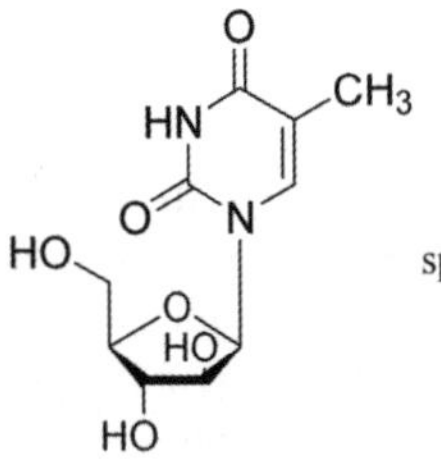

spongothymidine 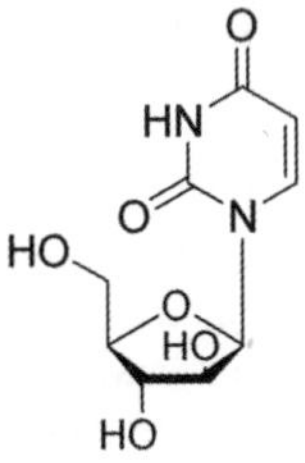   spongouridine 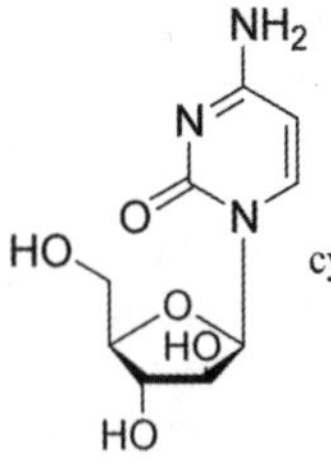   cytarabine

In 1960, Bernard Randall Baker and his colleagues at the Stanford Research Institute in California synthesised adenine arabinoside, which became known as 'vidarabine'.[51] Although intended to be an anticancer drug, it found some value as an antiviral agent. By this time, it was apparent that the biological activity of drugs designed to interfere with DNA synthesis did not necessarily follow expectations. Screening against cancer cells, viruses, fungi, bacteria and even protozoal parasites was essential if potentially valuable compounds were not to be overlooked. That lesson had already been learned by Paul Ehrlich at the beginning of the twentieth century when his arsphenamine turned out to be an effective remedy for syphilis despite having been rejected as an antitrypanosomal agent.

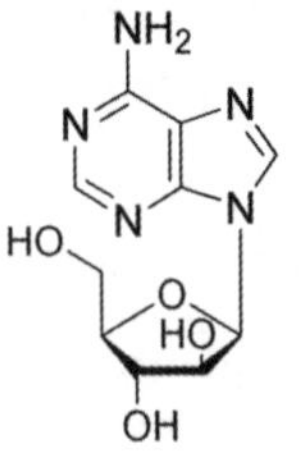

vidarabine 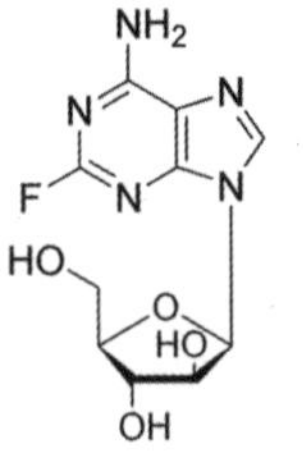   fludarabine 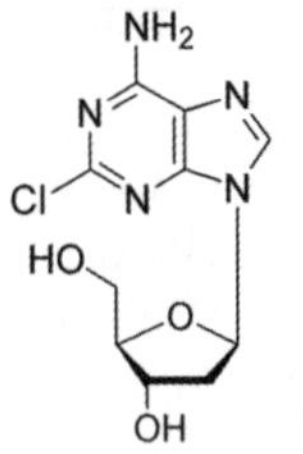   cladribine 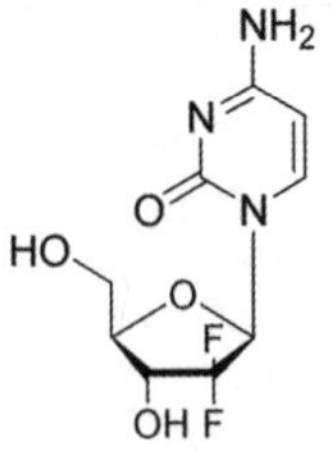   gemcitabine

Following the discovery of the activity of fluorouracil in 1957, John Montgomery and Kathleen Hewson at the Southern Research Institute synthesised 2-fluoroadenosine. It was so highly cytotoxic that it could not be used medicinally. Eleven years later, after the introduction of vidarabine, they prepared the arabinose analogue of 2-fluoroadenosine, which is now known as 'fludarabine'.[52] It is of value in B-cell chronic lymphocytic leukaemia. Cladribine is a deoxyadenosine analogue that was synthesised at Brigham Young University, Boston.[53] It is given by intravenous infusion in hairy cell leukaemia or in some cases of chronic lymphocytic leukaemia.

In the early 1980s, Larry Hertel at the Eli Lilly laboratories in Indianapolis synthesised a number of deoxyribosides in which both hydrogen atoms in the 2'-position of the sugar rings were replaced by fluorine atoms. These were prepared as potential antiviral agents, but in 1990 it was reported that 2',2'-difluorodeoxycytidine was a very potent inhibitor of human leukaemia cells grown in culture.[54] It is now known as gemcitabine and is administered intravenously in metastatic non-small lung cell carcinoma and pancreatic cancer.

Acyclic Nucleosides

In 1968, Burroughs Wellcome decided to extend their research to include antiviral agents after Frank Schabel at the Southern Research Institute in Birmingham, Alabama, had reported that adenine arabinoside was active against both DNA and RNA viruses.[55] The arabinosides of 2,6-diaminopurine and of guanine were then prepared and found to be active against DNA and RNA viruses. As one of the company chemists, Howard Schaeffer, had previously demonstrated that the intact sugar ring of such nucleosides was not essential for binding to enzymes, acyclic analogues were synthesised and tested.[56] In 1977, it was reported that one of them, aciclovir, had outstanding activity against the herpes virus.[57]

Aciclovir is activated only in cells infected with herpes virus since these contain the key enzyme, viral thymidine kinase, that selectively converts the drug into a monophosphate. This, in turn, is converted by normal intranuclear kinases to aciclovir triphosphate. The triphosphate is incorporated into the DNA chain where it blocks further DNA chain extension as it lacks the key 2' and 3' carbon atoms needed for this to occur. The remarkable selective toxicity of aciclovir is due to the incapability of normal cellular thymidine kinase to accept it as a substrate.

Aciclovir is administered to treat superficial herpes simplex infections such as cold sores (by topical application) and genital herpes (by oral administration), or for treating life-threatening herpes varicella-zoster (chickenpox) infection in immunocompromised patients. Its safety having been demonstrated in over 30 million patients, aciclovir became available from United Kingdom pharmacies without prescription for the treatment of cold sores in 1993.

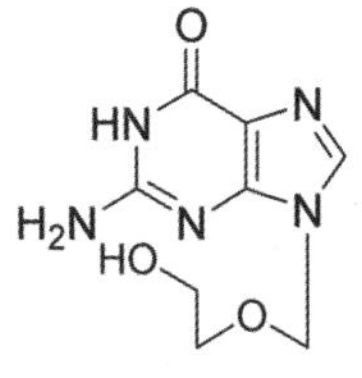

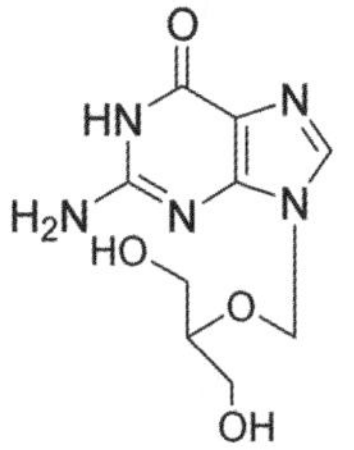

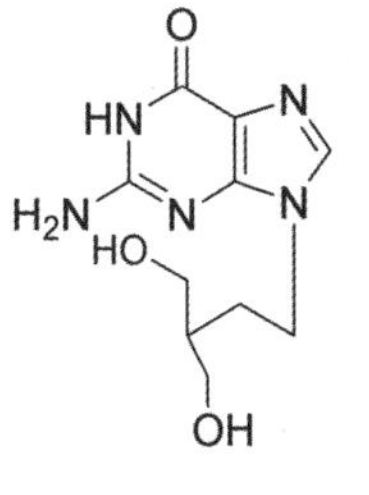

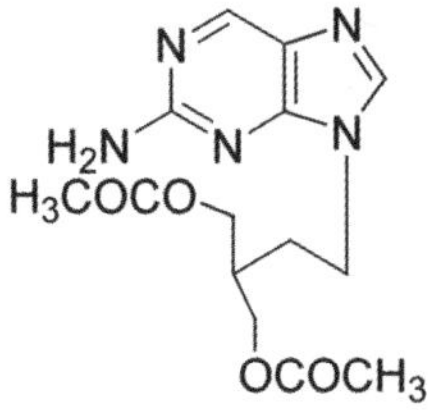

Ganciclovir was synthesised at the Syntex laboratories in 1980 as an analogue of aciclovir with a closer structural resemblance to natural nucleosides than had aciclovir.[58] As a consequence, it turned out to be a better substrate for both viral and normal kinases involved in the formation of its triphosphate. This led both to a wider spectrum of antiviral activity and a greater degree of toxicity towards uninfected cells. This toxicity has limited its clinical application to life-threatening infections that do not respond to aciclovir or other agents. In particular, it has been administered in attempts to prevent blindness from peripheral retinitis in AIDS patients infected with the cytomegalovirus.

Penciclovir is an analogue of ganciclovir with similar activity to aciclovir against herpes simplex.[59] It appears to have a safety profile not unlike that of aciclovir because of its lack of effect on DNA synthesis in uninfected cells.[60] The closer structural similarity to natural nucleosides makes it a superior substrate for viral thymidine kinase, resulting in an enhancement of phosphorylation. Any advantage gained from this is counterbalanced by the reduced activity of penciclovir triphosphate as an inhibitor of DNA synthesis.

Famciclovir is a prodrug of penciclovir that was introduced in 1994.[61] It was prepared at Beecham Research Laboratories in order to overcome the poor oral bioavailability of penciclovir. After absorption, it is metabolised in the liver, the ester functions being hydrolysed and the purine ring oxidised, to form penciclovir. It only needs to be administered three times a day by mouth, whereas aciclovir requires five daily doses.

Zidovudine

The clinical success of cytarabine, and to a lesser extent of vidarabine, drew attention to the potential of nucleoside analogues with a modified ribose ring. The importance of one such compound, azidothymidine, was to be overlooked for 20 years.

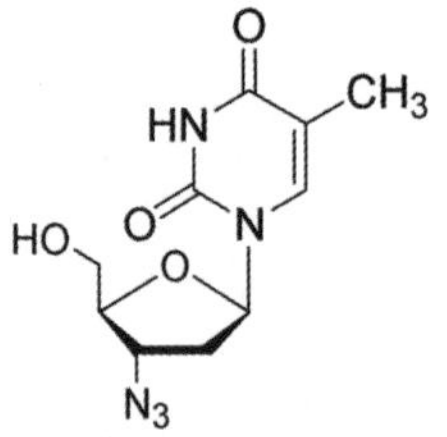

zidovudine (azidothymidine)

Azidothymidine was synthesised at the Michigan Cancer Foundation in 1964 by Jerome Horwitz as a potential antileukaemic drug.[62] Its azido group was considered to be an appropriate substitute for the hydroxyl group in thymidine. After it gave negative results and was toxic when tested against L1210 leukaemia in mice, there seemed to be no point in further investigation. Ten years later, however, Wolfram Ostertag at the Max Planck Institute in Göttingen found azidothymidine shared the ability of another thymidine analogue, bromodeoxyuridine, to inhibit replication of the Friend leukaemia virus in cultured mouse cells.[63] Even though this was a retrovirus, a type that transmits its genetic information through RNA rather than DNA, there was little interest in his report since retroviruses were then unknown in humans. The situation dramatically changed in 1983 when scientists at the Institut Pasteur in Paris isolated the retrovirus that is now known as the human immunodeficiency virus (HIV). By April of the following year, it had become apparent that HIV was responsible for the rapidly developing epidemic of the acquired immunodeficiency syndrome (AIDS). In those who had contracted AIDS, the HIV infected their T4 white blood cells that were crucial for development of an immune response to infections. The death rate among victims soared throughout the world.

The Burroughs Wellcome researchers were able to respond swiftly to the crisis because of their previous experience in screening for antiviral activity. In June 1984, virologist Marty St Clair set up a programme to identify drugs that had the potential to attack HIV by screening them against immortalised mouse cells infected with either the Friend leukaemia virus or the

Harvey sarcoma virus. These retroviruses were considered to be sufficiently comparable to HIV for the task in hand. The responsibility for selecting appropriate compounds for screening was given to nucleoside chemist Janet Rideout. Azidothymidine was one of the 14 compounds she chose. Her decision was to set in motion a sequence of events that successfully delivered a life-prolonging drug to AIDS patients 18 months later.

When the first laboratory results for azothymidine were obtained on 16 November 1984, they were so exceptional that St Clair thought she had omitted to add the Friend leukaemia virus to the cultures! When the results of testing against the Harvey sarcoma virus with a lower concentration of the drug were received two weeks later, they showed that there had been almost complete inhibition of virus replication. This aroused considerable interest and resulted in the speedy collation of all the previous findings when the company had examined azidothymidine as a potential antibacterial agent. These revealed that it had low toxicity when given to rats for two weeks. There was also evidence that it acted on DNA synthesis in bacteria but not in mammals. The decision was now taken to send samples of azidothymidine to the US National Cancer Institute (NCI) in Bethesda, Maryland, for further evaluation by Samuel Broder and Hiroaki Mitsuya. They were the first researchers to develop a method of testing compounds for activity against HIV, growing the virus in immortalised human T4 cells. Within two weeks of receiving azidothymidine from Burroughs Wellcome, they had concluded that it was highly effective.[64]

The results from the NCI were confirmed at Duke University and the findings of the three laboratories involved were submitted for publication on 28 June 1985.[65] The following month the Food and Drug Administration permitted a phase I clinical trial to proceed. The purpose was to establish safe dose levels with different routes of administration. The results indicated that the drug was safe enough for further investigation.[66] In January 1986, a randomised, double-blind clinical trial on 282 patients began. It had been planned to last for 24 weeks, but after only 16 weeks the independent monitoring board stopped the trial since it had become apparent that among those receiving the drug only one of 145 had died, whereas 16 of those receiving placebo (dummy capsules) had died.[67] As a result of these findings, supplies of azidothymidine were released in October 1986 for treatment on a named patient basis, with the Food and Drug Administration granting a product licence the following spring. Azidothymidine now received the approved name of 'zidovudine'. Two major trials, the US Veterans Affairs Cooperative Study[68] and the Anglo French Concorde Study,[69] confirmed its ability to prolong the lives of those with AIDS or delay its development in those infected with HIV. Since these trials were completed, the practice has been to administer zidovudine as one of a cocktail of anti-HIV drugs. The outcome of this is to prolong the lives of those infected indefinitely, so long as the medication is continued.

As HIV is a retrovirus it contains RNA and must form viral DNA if it is to take over control and replicate inside the host cell. The enzyme that transcribes RNA into a DNA copy inside the host cell is known as reverse transcriptase. It is inhibited by zidovudine triphosphate which, once incorporated in the viral DNA chain, stops more nucleotides from being added since the azido group prevents the creation of the phosphodiester linkages required for the completion of the DNA chain. However, zidovudine triphosphate also affects mammalian DNA polymerase, for which it has one-hundredth the affinity of that for viral reverse transcriptase. This causes considerable toxicity, notably dose-dependent suppression of bone marrow, resulting in anaemia and leucopenia, which causes treatment to be abandoned in many patients.

Dideoxynucleosides

Jerome Horwitz and his colleagues in Chicago prepared the dideoxynucleoside now known as 'zalcitabine' three years after they synthesised zidovudine.[70] This was also shown by Samuel

Broder and Hiroaki Mitsuya to prevent replication of HIV.[71] The active form of zalcitabine is its triphosphate. In the nucleus this terminates DNA chain extension since there is no 3′-hydroxy group on its sugar ring to allow formation of phosphodiester linkages. It is now used in combination with other antretroviral drugs in the treatment of AIDS, as is its sulfur analogue lamivudine.[72]

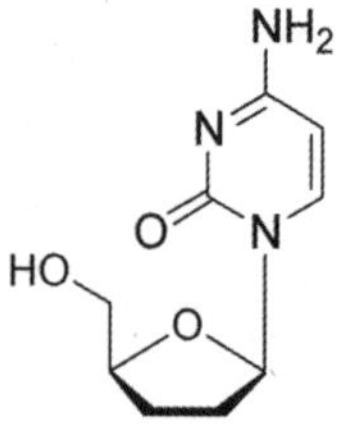

zalcitabine

lamivudine

Dideoxyadenosine, the corresponding adenosine analogue of zalcitabine, was synthesised by Roland Robbins in 1964.[73] It was shown to have anti-HIV activity at the same time as zalcitabine, but was less potent and caused kidney damage. Didanosine was prepared from it by enzymatic oxidation and was also found to be active against HIV, but without causing kidney damage.[74] It is also used in combination with other antiretroviral drugs for AIDS.

dideoxyadenosine

didanosine

abacavir

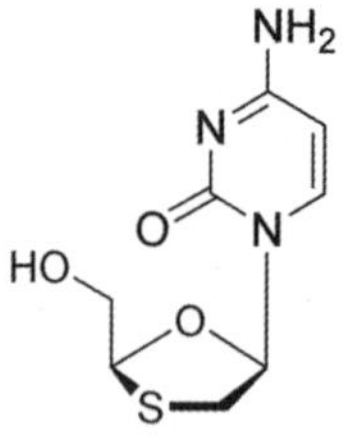

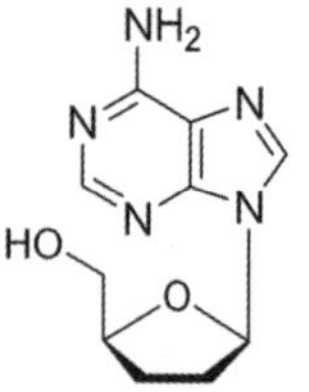

Glaxo Wellcome researchers at Stevenage in the United Kingdom and in North Carolina (to where Wellcome had moved from Tuckahoe in 1970) collaborated in the investigation of potential anti-HIV drugs in which the oxygen atom in the sugar ring had been replaced by a carbon in order to obtain compounds that were more stable *in vivo* by virtue of not possessing a glycosidic linkage. Minimal activity was first observed in the carbocyclic analogues of dideoxyadenosine and four other nucleosides. Many nucleoside analogues were prepared and examined, but the only one that had significant activity and satisfied the requirements for use in the clinic was the 2′,3′-didehydro analogue of dideoxyadenosine. Insertion of a cyclopropyl group on its 6-amino nitrogen of the adenine ring increased lipophilicity and thereby enhanced brain penetration. The resulting compound, known as abacavir, is used in combination with other antiretroviral drugs in the treatment of AIDS.

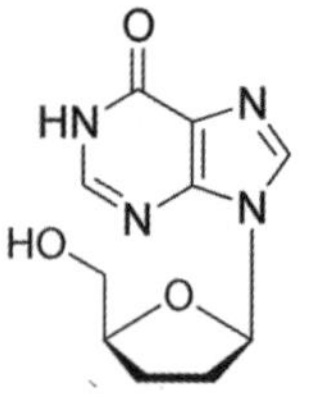

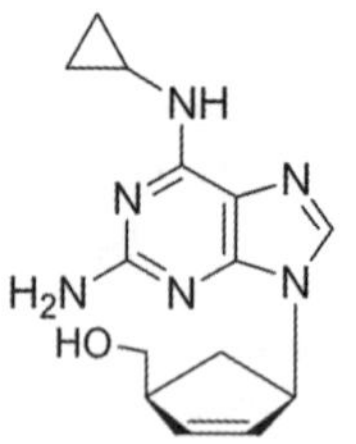

HIV PROTEASE INHIBITORS

Soon after the discovery of HIV, it was confirmed that it functioned like other retroviruses, so that once the double-stranded DNA is formed by the virus it is inserted into the host genome, a process facilitated by the action of viral integrase. The altered DNA then produces a polyprotein that is broken down to form smaller proteins which function as enzymes essential for HIV multiplication. Viral proteases that promoted this process were characterised in the mid-1980s.[78]

In 1987, investigators at the Smith Kline and French Laboratories in Philadelphia prepared _Escherichia coli_ recombinant HIV protease and confirmed not only that the cleavage of the polyprotein in HIV was catalysed by HIV protease but also that among the enzymes formed were reverse transcriptase, viral integrase and HIV protease itself.[79] In February 1989, Merck Sharp and Dohme Research Laboratories in Rahway, New Jersey, published the three-dimensional structure of HIV-1 protease.[80] This major breakthrough was achieved so quickly because of the extensive work that had been taking place in commercial, academic and government laboratories around the world.[81]

Roche Products in the United Kingdom was one of several groups that now sought non-toxic inhibitors of HIV protease. The approach they followed was similar in principle to that taken previously in the design of the ACE inhibitor captopril (see page 280), except that this time non-hydrolysable analogues of the dipeptides Phe–Pro and Tyr–Pro were sought since these occurred at the points of cleavage of the viral polyprotein that was formed in cells infected by HIV. This in itself was auspicious, since no mammalian protease was able to cleave such moieties; hence the expectation was that any effective inhibitor would be selective for the viral protease. Disappointing results were obtained with the first analogues, in which the amide linkage R–CO–NH–R' (where R and R' are peptide chains) had simply been reduced to form R–CHOH–NH–R'. However, promising inhibitors were found when hydroxyethyla-mines with an interposed carbon atom, R–CHOH–CH$_2$–NH–R', were prepared. The stereochemical configuration of the carbon atom bearing the hydroxyl substituent proved critical. Ultimately, the most active inhibitor, saquinavir, was obtained when the proline residue was replaced by decahydroisoquinoline-3-carbonyl.[82] Saquinavir was the first HIV protease inhibitor to be marketed in the United States, in December 1995. An oral dose of 600 mg was prescribed three times a day in combination with a reverse transcriptase inhibitor and lamivudine. By this time, the importance of combining drugs with different modes of action in order to minimise the risk of viral resistance had been widely recognised.

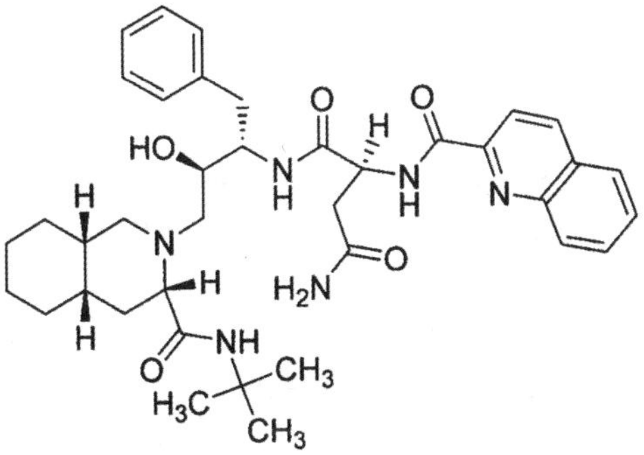

saquinavir

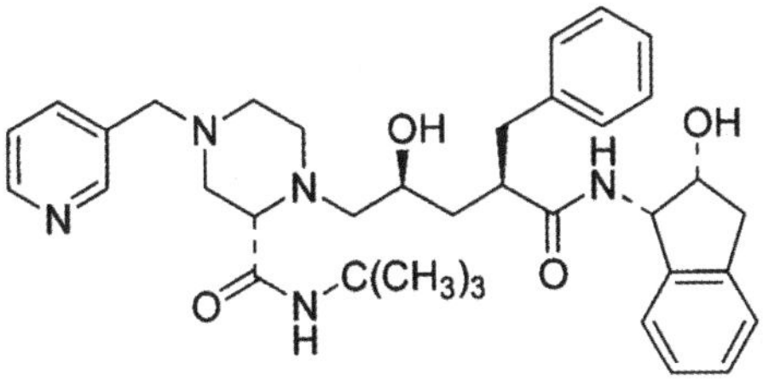

indinavir

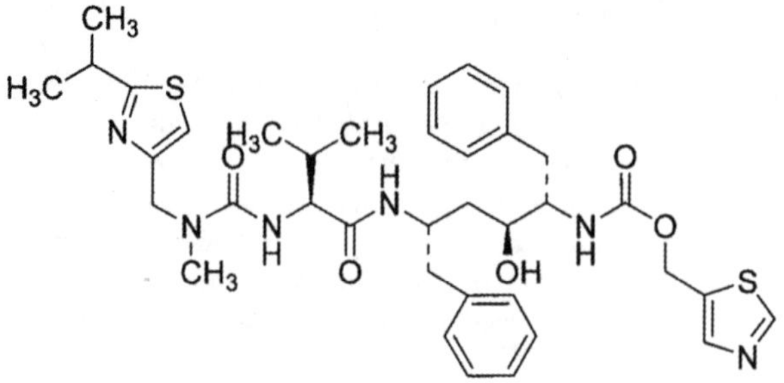

ritonavir

Merck's indinavir incorporated a hydroxyethylene moiety (R–CHOH–CH$_2$CH$_2$–R′) and included a basic piperazine ring to enhance oral bioavailability.[83] It was also administered thrice daily by mouth. Abbott Laboratories introduced ritonavir, the first protease inhibitor to be marketed in the United Kingdom.[84] It was administered twice daily.

Clinical results have confirmed that treatment combining protease inhibitors with reverse transcriptase inhibitors dramatically reduces the HIV viral load to innocuous levels. Since an estimated 10^{12} viruses are estimated to be formed daily before treatment, this is remarkable. As a consequence, AIDS can now be held at bay for many years – so long as chemotherapy is continued. The cost of treatment has been very high and in many countries it has been unaffordable and so patients continue to die of AIDS. Even when public pressure on companies persuaded them to waive their patent rights, lack of a robust local infrastructure for providing health care often negated the benefits of therapy.

NEURAMINIDASE INHIBITORS

The influenza virus consists of DNA coated with two proteins, neuraminidase and haemagglutinin. After the DNA replicates inside an infected cell, the haemagglutinin of the daughter viruses binds to glycolipid on the outside surface of the infected cell. Infection of other cells can only take place after neuraminidase has promoted the breakdown of the bonds holding the haemagglutinin to the sugar moiety of the glycolipid on the cell surface. The importance of inhibiting neuraminidase was recognised in the 1960s by Meindl and Tuppy, who prepared 2-deoxy-2,3-dehydro derivatives of sialic acid (also known as N-acetylneuraminic acid), the natural substrate for the enzyme.[75] These were inactive in live animals.

Once neuraminidase had been crystallised by Graeme Laeve at the John Curtin School of Medical Research in the Australian National University, molecular models revealed a slot in the enzyme that was common to all strains of the virus. Attempts to develop inhibitors that would bind within this slot were initiated in 1986 by Mark van Itzstein at the Victorian College of Pharmacy in Melbourne and Peter Coleman of the ANC. Their approach was to use molecular modelling on a computer to establish how sialic acids bound to neuraminidase. This revealed that binding of the 2-deoxy-2,3-dehydro sialic acids made by Meindl and Tuppy could be enhanced by changing a hydroxyl group to an amino group. Further enhancement of activity was achieved by increasing base strength by replacing this amino group with a guanidine group to form zanamavir.[76]

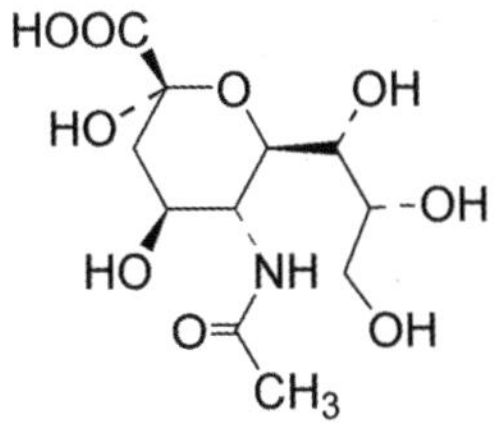

sialic acid

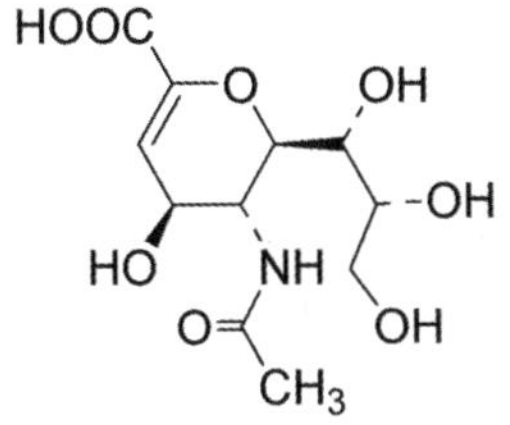

2-deoxy-2,3-dehydro sialic acid

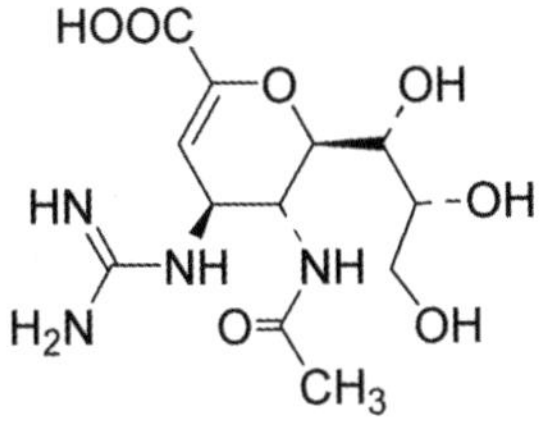

zanamavir

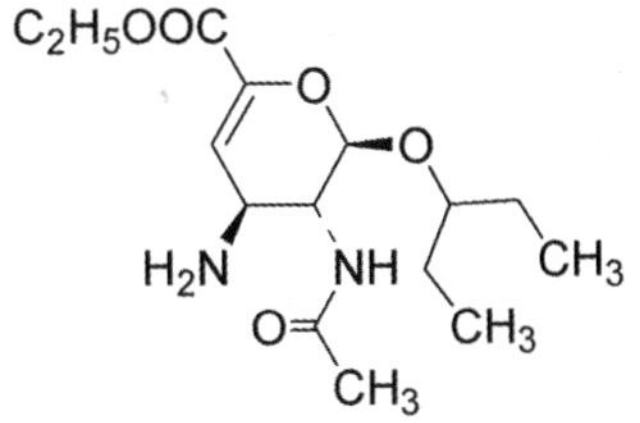

oseltamavir

Zanamavir was evaluated at the laboratories of Biota in Melbourne and shown to be of value in the treatment of influenza.[77] The development and commercialisation of it was subsequently licensed to Glaxo Wellcome, who marketed it in 1999. As a direct consequence of its origin from a carbohydrate, zanamir was too polar to be taken by mouth. Instead, it had to be administered from a dry powder inhaler. When inhaled twice daily for five days, it reduced influenza symptoms by 1–2.5 days.

Compound GS 4071 was developed as a neuraminidase inhibitor with enhanced lipophilicity. Even though the strongly basic guanidine group and the polar hydroxyls of zanamavir were absent it also had poor oral bioavailability. Unlike zanamavir, its ethyl ester was now sufficiently lipophilic to serve as an orally active prodrug. It received the name 'oseltamavir'. The activity is similar to that of zanamavir, but it can be given by mouth to relieve the symptoms of influenza.

THE SAFETY OF DRUGS

The development of the HIV protease inhibitors generated considerable enthusiasm in some quarters, where it was naively believed that a host of inhibitors of proteases involved in other systems would soon become available. The failure of this to happen reflects the basic problem with all drug research – not merely that involving antimetabolites or substrate analogues – namely that while it is relatively easy to design active drugs, it remains exceedingly difficult to develop safe, effective medicinal compounds. No matter how much specificity for a single target a drug molecule may appear to have, there can never be a guarantee that it will not unexpectedly interact with a receptor site or enzyme somewhere else in the body.

REFERENCES

1. R. Lewisohn, Review of the work of the laboratory, 1937–1945, in _Approaches to Tumour Chemotherapy_, ed. F.R. Moulton. Washington: American Association for the Advancement of Science; 1947, pp. 139–47.

2. R. Lewisohn, C. Leuchtenberger, R. Leuchtenberger, D. Laszlo, Effect of intravenous injection of yeast extract on spontaneous breast adenocarcinoma in mice. *Proc. Soc. Exp. Biol. Med.*, 1940; **43**: 558–61.

3. R. Lewisohn, C. Leuchtenberger, R. Leuchtenberger, D. Laszlo, 'Folic acid', a tumour growth inhibitor. *Proc. Soc. Exp. Biol. Med.* 1944; **55**: 204–5.

4. R. Leuchtenberger, C. Leuchtenberger, D. Laszlo, R. Lewisohn, Folic acid, a tumour growth inhibitor. *Science*, 1945; **101**: 45.

5. J.H. Boothe, J.H. Mowat, B.L. Hutchings, *et al.*, Pteroic acid derivatives. II. Pteroyl-γ-glutamylglutamic acid and pteroyl-γ-glutamyl-γ-glutamylglutamic acid. *J. Am. Chem. Soc.*, 1948; **70**: 1099–102.

6. R. Lewisohn, C. Leuchtenberger, R. Leuchtenberger, J.C. Keresztesy, The influence of liver *L. casei* factor on spontaneous breast cancer in mice. *Science*, 1946; **104**: 436–7.

7. S. Farber, E.C. Cutler, J.W. Hawkins, *et al.*, The action of pteroylglutamic acid conjugates on man. *Science*, 1947; **106**: 619–21.

8. S. Farber, L.K. Diamond, R.D. Mercer, *et al.*, Temporary remission in acute leukaemia in children produced by folic acid antagonists. *N. Engl. J. Med.*, 1948; **238**: 787–93.

9. D.D. Woods, The relation of *p*-aminobenzoic acid to the mechanism of the action of sulphanilamide. *Br. J. Exp. Path.*, 1940; **2**: 74–90.

10. J.H. Quastel, A.H.M. Wheatley, XVI. Biological oxidations in the succinic acid series. *Biochem. J.*, 1931; **25**: 1171–228.

11. A.L. Franklin, E.L.R. Stockstad, M. Belt, T.H. Jukes, Biochemical experiments with a synthetic preparation having an action antagonistic to that of pteroylglutamic acid. *J. Biol. Chem.*, 1947; **169**: 427–35.

12. B.L. Hutchings, J.H. Mowat, J.J. Oleson, *et al.*, Pteroylaspartic acid, an antagonist for pteroylglutamic acid. *J. Biol. Chem.*, 1947; **170**: 323–8.

13. L.J. Daniel, L.C. Norris, Growth inhibition of bacteria by synthetic pterins. I. Studies with *Streptococcus faecalis, Lactobacillus casei*, and *Lactobacillus arabinosus. J. Biol. Chem.*, 1947; **169**: 689–97.

14. D. Seeger, J. Smith, M. Hultquist, Antagonist for pteroylglutamic acid. *J. Am. Chem. Soc.*, 1947; **69**: 2567.

15. S. Farber, L.K. Diamond, R.D. Mercer, *et al.*, Temporary remissions in acute leukaemia in children produced by folic acid antagonist 4-aminopteroyl-glutamic acid. *N. Engl. J. Med.*, 1948; **238**: 787–93.

16. D. Seeger, D.B. Cosulich, J.M. Smith Jr, M.E. Hultquist, Analogs of pteroylglutamic acid. III. 4-Amino derivatives. *J. Am. Chem. Soc.*, 1949; **71**: 1753–8.

17. S. Farber, R. Toch, F.M. Sears, D. Pinkel, Advances in chemotherapy of cancer in man. *Adv. Cancer Res.*, 1956: **4**: 1–71.

18. G.H. Hitchings, E.A. Falco, M.B. Sherwood, The effect of pyrimidines on the growth of *L. casei. Science*, 1945; **102**: 251–2.

19. G.H. Hitchings, G.B. Elion, H. Vander Werff, A.E. Falco, Pyrimidine derivatives as antagonists of pteroylglutamic acid. *J. Biol. Chem.*, 1948; **174**: 765–6.

20. J.H. Burchenal, A. Bendich, G.B. Brown, G.B. Elion, G.H. Hitchings, C.P. Rhoads, C.C. Stock, Preliminary studies on the effect of 2,6-diaminopurine on transplanted mouse leukemia. *Cancer*, 1949; **2**: 119–20.

21. G.B. Elion, E. Burgi, G.B. Hitchings, Studies on condensed pyrimidine systems. IX. The synthesis of some 6-substituted purines. *J. Am. Chem. Soc.*, 1952; **74**: 411–14.

22. G.B. Elion, G.H. Hitchings, H. Vander Werff, Antagonists of nucleic acid derivatives. VI. Purines. *J. Biol. Chem.*, 1951; **192**: 505–18.

23. G.H. Hitchings, C.P. Rhoads, 6-Mercaptopurine. *Ann. N.Y. Acad. Sci.*, 1954; **60**: 183.

24. US Pat. 1962: 3056785 (to Burroughs Wellcome).

25. R.E. Billingham, L. Brent, P.B. Medawar, Activity acquired tolerance of foreign cells. *Nature*, 1953; **172**: 603–6.

26. R. Schwartz, W. Dameshek, Drug-induced immunological tolerance. *Nature*, 1959; **183**: 1682–3.

27. R.Y. Calne, G.P.J. Alexander, J.E. Murray, Study of the effect of drugs in prolonging survival of homologous renal transplants in dogs. *Ann. N.Y. Acad. Sci.*, 1962; **99**: 743.

28. H.C. Nathan, S. Bieber, G.B. Elion, G.H. Hitchings, Detection of agents which interfere with the immune response. *Proc. Soc. Exp. Biol. Med.*, 1961; **107**: 796–9.

29. R.K. Robins, Potential purine antagonists. I. Synthesis of some 4,6-substituted pyrazolo [3,4-*d*] pyrimidines. *J. Am. Chem. Soc.*, 1956; **78**: 784–90.

30. R.W. Rundles, Allopurinol in the treatment of gout. *Ann. Int. Med.*, 1966; **64**: 229–58.

31 D.W. Woolley, Some biological effects produced by benzimidazole and their reversal by purines. *J. Biol. Chem.*, 1944; **152**: 225–32.

32. G.C. Brown, D.E. Craig, A. Kandel, Effect of benzimidazole on experimental poliomyelitis in mice and monkeys. *Proc. Soc. Exp. Biol. Med.*, 1953; **83**: 408–11.

33. H.D. Brown, A.R. Matzuk, I.R. Ilves, *et al.*, Antiparasitic drugs. IV. 2-(4′-Thiazolyl)-benzimidazole, a new anthelmintic. *J. Am. Chem. Soc.*, 1961; **83**: 1764–5.

34. D.R. Hoff, M.H. Fisher, R.J. Bochis, *et al.*, A new broad-spectrum anthelmintic: 2-(4-thiazolyl)-5-isopropoxycarbonylamino-benzimidazole. *Experientia*, 1970; **26**: 550–1.

35. J.P. Brugans, D.C. Thienpont, I. Van Wijngaarden, O.F. Vanparijs, V.L. Schuermans, H.L. Lauwers, Mebendazole in enterobiasis. Radiochemical and pilot clinical study in 1,278 subjects. *J. Am. Med. Ass.*, 1971; **217**: 313–16.

36. R.J. Rutman, A. Cantarow, K. Paschkis, Studies in 2-acetylaminofluorene carcinogenesis. III. The utilization of uracil-2-C14 by preneoplastic rat liver and rat hepatoma. *Cancer Res.*, 1954; **14**: 119.

37. C. Heidelberger, N.K. Chaudhuri, P. Danneberg, *et al.*, Fluorinated pyrimidines, a new class of tumor-inhibitory compounds. *Nature*, 1957; **179**: 663.

38. C. Heidelberger, L. Griesbach, O. Cruz, *et al.*, Fluorinated pyrimidines. VI. Effects of 5-fluorouridine and 5-fluoro-2′-deoxyuridine on transplanted tumors. *Proc. Soc. Exp. Biol. Med.*, 1958; **97**: 470–5.

39. M. Yasumoto, I. Yamawaki, T. Marunaka, S. Hashimoto, Studies on antitumor agents. 2. Syntheses and antitumor activities of 1-(tetrahydro-2-furanyl)-5-fluorouracil and 1,3 bis(tetrahydro-2-furanyl)-5-fluorouracil. *J. Med. Chem.*, 1978; **21**: 738–41.

40. A.F. Cook, M.J. Holman, M.J. Kramer, P.W. Trown, Fluorinated pyrimidine nucleosides. 3. Synthesis and antitumor activity of a series of 5′-deoxy-5-fluoropyrimidine nucleosides. *J. Med. Chem.*, 1979; **22**: 1330–5.

41. H. Ishitsuka, M. Miwa, K. Takemoto, *et al.*, Role of uridine phosphorylase for antitumor activity of 5′-deoxy-5-fluorouridine. *Gann.*, 1980; **71**: 112–23.

42. US Pat. 1995: 5472949 (to Hoffmann–La Roche).

43. R. Duschinsky, E. Pleven, The synthesis of 5-fluoropyrimidines. *J. Am. Chem. Soc.*, 1957; **79**: 4559.

44. W.H. Prusoff, Synthesis and biological activity of iododeoxyuridine, an analog of thymidine. *Biochem. Biophys. Acta*, 1959; **32**: 295–6.

45. Y.F. Shealy, C.A. Krauth, J.A. Montgomery, Imidazoles. I. Coupling reactions of 5-diazoimidazole-4-carboxamide. *J. Org. Chem.*, 1962; **27**: 2150–4.

46. M.F.G. Stevens, J.A. Hickman, R. Stone, *et al.*, Antitumour imidazotetrazines. 1. Synthesis and chemistry of 8-carbamoyl-3-(2-chloroethyl)imidazo[5,1-*d*]-1,2,3,5-tetrazin-4(3H)-one, a novel broad-spectrum. *J. Med. Chem.*, 1984; **27**: 196–201.

47. J.T. Witkowski, R.K. Robins, R.W. Sidwell, L.N. Simon, Design, synthesis, and broad spectrum antiviral activity of 1-beta-D-ribofuranosyl-1,2,4-triazole-3-carboxamide and related nucleosides. *J. Med. Chem.*, 1972; **15**: 1150–4.

48. W. Bergmann, D.C. Burke, Contributions to the study of marine products. XXXIX. The nucleosides of sponges. III. Spongothymidine and spongouridine. *J. Org. Chem.*, 1955; **20**: 1501–7.

49. R. Walwick, W. Roberts, C. Dekker, Cyclization during phosphorylation of uridine and cytidine. *Proc. Chem. Soc.*, 1959: 84.

50. US Pat. 1963: 3116282 (to Upjohn).

51. W.W. Lee, A. Benitez, L. Goodman, B.R. Baker, Potential anticancer agents. XL. Synthesis of the β-anomer of 9-(*d*-arabinofuranosyl)-adenine. *J. Am. Chem. Soc.*, 1960; **82**: 2648–9.

52. J.A. Montgomery, K. Hewson, Nucleosides of 2-fluoroadenine. *J. Med. Chem.*, 1969; **12**: 498–504.

53. L.F. Christensen, A.D. Brown, M.J. Robins, A. Bloch, Synthesis and biological activity of selected 2,6 disubstituted (2-deoxy-alpha and beta-D-erythro-pentofuranosyl) purines. *J. Med. Chem.*, 1972; **15**: 735–9.

54. L.W. Hertel, G.B. Boder, J.S. Kroin, *et al.*, Evaluation of the antitumor activity of gemcitabine (2′,2′-difluoro-2′-deoxycytidine). *Cancer Res.*, 1990; **50**: 4417–22.

55. F.M. Schabel Jr, The antiviral activity of 9-*b*-D-arabinofuranosyladenine (ara-A). *Chemotherapy*, 1968; **13**: 321–38.

56. H.J. Schaeffer, L. Beauchamp, P. De Miranda, *et al.*, 9-(2-Hydroxymethylocymethyl) guanine activity against viruses of the herpes group. *Nature*, 1978; **272**: 583–5.

57. G.B. Elion, P.A. Furman, J.A. Fyfe, *et al.*, An orally active acyclic nucleoside with inhibitory activity towards several herpes viruses. *Proc. Natl Acad. Sci. USA*, 1977; **74**: 5716–20.

58. J.C. Martin, C.A. Dvorak, D.F. Smee, *et al.*, 9-[(1,3-Dihydroxy-2-propoxy)methyl]guanine: a new potent and selective antiherpes agent. *J. Med. Chem.*, 1983; **26**: 759–61.

59. A. Weinberg, B.J. Bate, H.B. Masters, *et al.*, *In vitro* activities of penciclovir and acyclovir against herpes simplex virus types 1 and 2. *Antimicrob. Agents Chemother.*, 1992; **36**: 2037–8.

60. D. Sutton, M.R. Boyd, Comparative activity of penciclovir and acyclovir in mice infected intraperitoneally with herpes simplex virus type 1 SC16. *Antimicrob. Agents Chemother.*, 1993; **37**: 642–5.

61. R.A. Vere Hodge, D. Sutton, M.R. Boyd, M.R. Harnden, R.L. Jarvest, Selection of an oral prodrug (BRL 42810; famciclovir) for the antiherpes virus agent BRL 39123 [9-(4-hydroxy-3-hydroxy-methylbut-l-yl)guanine; penciclovir]. *Antimicrob. Agents Chemother.*, 1989; **33**: 1765–73.

62. J.P. Horwitz, Nucleosides. The monomesylates of thymidine. *J. Org. Chem.*, 1964; **29**: 2076.

63. W. Ostertag, G. Roesler, C.J. Krieg, *et al.*, Induction of endogenous virus and of thymidine kinase by bromodeoxyuridine in cell cultures transformed by Friend virus. *Proc. Natl Acad. Sci. USA*, 1974; **71**: 4980–5.

64. Some of the information appearing here was obtained from the 1998 trial decision of the Federal Court of Canada in Apotex Inc. v. Wellcome Foundation Ltd (1998), 145 F.T.R. 161. An electronic copy is available at http://decisions.fct-cf.gc.ca/fct/1998/t-3197-90.html.

65. H. Mitsuya, K.J. Weinhold, P.A. Furman, *et al.*, 3′-Azido-3′deoxythymidine (BWA509U). An agent that inhibits the infectivity and cytopathic effect of human T lymphotropic virus type III/lymphadenopathy-associated virus *in vitro*. *Proc. Natl Acad. Sci. USA*, 1985; **82**: 7096–100.

66. R. Yarchoan, K.J. Weinhold, H.K. Lyerly, *et al.*, Administration of 3′-azido-3′-deoxythymidine, an inhibitor of HTLV-III/LAV replication, to patients with AIDS or AIDS-related complex. *Lancet*, 1986; **1**: 575–80.

67. M.A. Fischl, D.D. Richman, M.H. Grieco, *et al.*, The efficacy of azidothymidine (AZT) in the treatment of patients with AIDS-related complex. A double-blind, placebo controlled trial. *N. Engl. J. Med.*, 1987; **317**: 185–91.

68. J.D. Hamilton, P. Hartigan, M. Simberkoff, A controlled trial of early versus late treatment with zidovudine in symptomatic human immunodeficiency virus infection. Results of the Veterans Affairs Cooperative Study. *N. Engl. J. Med.*, 1992; **326**: 437–43.

69. Concorde Coordinating Committee, Concorde: MRC/ANRS randomised double-blind controlled trial of immediate and deferred zidovudine in symptom-free HIV infection. *Lancet*, 1994; **343**: 871–81.

70. J.P. Horwitz, J. Chua, M. Noel, J.T. Donati, Nucleosides. XI. 2′,3′-Dideoxycytidine. *J. Org. Chem.*, 1967; **32**: 817–18.

71. H. Mitsuya, S. Broder, Inhibition of the *in vitro* infectivity and cytopathic effect of human T-lymphotrophic virus type III/lymphadenopathy associated virus (HTLV-III/LAV) by 2′,3′-dideoxynucleosides. *Proc. Natl Acad. Sci. USA*, 1986; **83**: 1911–15.

72. J.A.V. Coates, N. Cammack, H.J. Jenkinson, *et al.*, The separated enantiomers of 2′-deoxy-3′-thiacytidine (BCH 89) both inhibit human immunodeficiency virus. *Antimicrob. Agents Chemother.*, 1992; **36**: 202–5.

73. M.J. Robins, R.K. Robins, The synthesis of 2′-dideoxyadenosine from 2′-deoxyadenosine. *J. Am. Chem. Soc.*, 1964; **86**: 3585–6.

74. W. Plunkett, S.S. Cohen, Two approaches that increase the activity of analogs of adenine nucleosides in animal cells. *Cancer Res.*, 1975; **35**: 1547–54.

75. P. Meindl, H. Tuppy, [2-Deoxy-2,3-dehydrosialic acids. II. Competitive inhibition of *Vibrio cholerae* neuraminidase by 2-deoxy-2,3-dehydro-*N*-acylneuraminic acids]. *Z. Physiol. Chem.*, 1969; **350**: 1088–92.

76. M. von Itzstein, W.Y. Wu, G.B. Kok, *et al.*, Rational design of potent sialidase-based inhibitors of influenza virus replication. *Nature*, 1993; **363**: 418–23.

77. F.G. Hayden, J.J. Treanor, R.F. Betts, *et al.*, Safety and efficacy of the neuraminidase inhibitor GG167 in experimental human influenza. *J. Am. Med. Ass.*, 1996; **275**: 295–9.

78. Y. Yoshinaka, I. Katoh, T.D. Copeland, S. Oroszlan, Murine leukemia virus protease is encoded by the gag-pol gene and is synthesized through suppression of an amber termination codon. *Proc. Natl Acad. Sci. USA*, 1985; **82**: 1618–22.

79. C. Debouck, J.G. Gorniak, J.E. Strickler, *et al.*, Human immunodeficiency virus protease expressed in *Escherichia coli* exhibits autoprocessing and specific maturation of the gag precursor. *Proc. Natl Acad. Sci. USA*, 1987; **84**: 8903–6.

80. M.A. Navia, P.M. Fitzgerald, B.M. McKeever, *et al.*, Three-dimensional structure of aspartyl protease from human immunodeficiency virus HIV-1. *Nature*, 1989; **337**: 615–20.

81. A. Wlodawer, J. Vondrasek, Inhibitors of HIV-1 protease: a major success of structure-assisted drug design. *Ann. Rev. Biophys. Biomolec. Struct.*, 1998; **27**: 249–84.

82. N.A. Roberts, J.A. Martin, D. Kinchington, *et al.*, Rational design of peptide-based HIV proteinase inhibitors. *Science*, 1990; **248**: 358–61.

83. B.D. Dorsey, R.B. Levin, S.L. McDaniel, *et al.*, L-735,524: the design of a potent and orally bioavailable HIV protease inhibitor. *J. Med. Chem.*, 1994; **37**: 3443–51.

84. D.J. Kempf, H.L. Sham, K.C. Marsh, Discovery of ritonavir, a potent inhibitor of HIV protease with high oral bioavailability and clinical efficacy. *J. Med. Chem.*, 1998; **41**: 602–17.

Blood and Biological Products*

Towards the end of the nineteenth century, several European investigators found that the clotting of blood was accelerated by material extracted from body tissues with fat solvents. In 1911, however, Doyon in France discovered that aqueous extraction of de-fatted dog liver yielded an anticoagulant that he called antithrombine.[1] He studied its effects over the next 15 or so years, but his findings were overlooked when William Howell, the Professor of Physiology at Johns Hopkins University in Baltimore, isolated a similar material in 1922.[2] Howell called his product 'heparin', a name he had previously given to an inferior anticoagulant extracted from liver by fat rather than aqueous solvents.[3] This earlier product had been discovered in Howell's laboratory in 1916 after Jay McLean, a pre-medical student, had let his extracts of fat-soluble coagulating substance deteriorate and so unmask the presence of an anticoagulant.[4] Howell spent some time trying to isolate and purify McLean's phospholipid anticoagulant before switching to the use of aqueous solvents for liver extraction.

HEPARIN

During the 1920s, Howell studied the chemistry and physiological activity of his water-soluble heparin, characterising it as a sulfur-containing polysaccharide.[5] He even persuaded the Baltimore firm Hynson, Westcott, and Dunning to sell it for laboratory investigations. This product is now known to have contained less than 1–2% of active material. Attempts were made to use it as an anticoagulant while transfusing blood into patients, but side effects were caused by impurities. Nevertheless, this brought it to the attention of researchers in Canada and Sweden.

In 1928, David Scott and Arthur Charles at the Connaught Laboratories in the University of Toronto joined in the effort to prepare samples of heparin that would be sufficiently pure for clinical investigations. By 1933, they had completed an investigation into the amounts of heparin in different animal tissues, which pointed to beef lung as the best source of the anticoagulant.[6] However, the product extracted from this was different to that originally introduced by Howell. During the next three years, Scott and Charles worked with this new heparin and made a series of advances that finally provided material pure enough for clinical studies. This received recognition when an international standard for the sodium salt of heparin was published in 1935. Heparin prepared by Albert Fischer at the University of Copenhagen[7] and by Erik Jorpes at the Caroline Institute in Stockholm[8] also met the new standard.

---

*In view of the limited number of compounds in this chapter, drug prototypes and their analogues are considered together.

Drug Discovery. A History. W. Sneader.
©2005 John Wiley & Sons Ltd

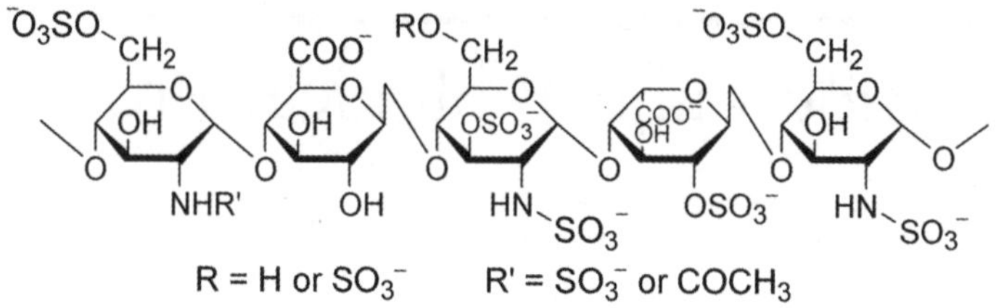

heparin

R = H or SO_3^- R' = SO_3^- or $COCH_3$

Jorpes found heparin was an acidic sulfated polysaccharide that interfered with thrombin formation.[9] The number of sugar units varied according to the source of the heparin, but was usually in the range of from 12 to 20.[10]

Gordon Murray[11] initiated clinical trials in Toronto with a view to establishing the value of the purified heparin in preventing thrombosis after injuries. Around the same time, Clarence Crafoord injected patients in Stockholm with it to prevent postoperative thrombosis.[12] Charles Best went on to use heparin in exchange transfusion of blood between patients and donors in Toronto, just two years before the outbreak of the Second World War, when this technique was to save many lives. Purified heparin also facilitated the introduction of the artificial kidney in 1944, as well as the development of heart–lung bypass techniques in surgery.

The recognition that only part of the heparin molecule was required to inhibit blood clotting factor Xa led to the manufacture of low molecular weight heparin fractions by gel filtration.[10] These retained the activity of heparin while overcoming its short duration of action. Later, enoxaparin was prepared by depolymerisation of porcine heparin.[13] Apart from only requiring once daily injections, it is less likely to cause haemorrhage since it does not possess the longer-chain polysaccharide residues that inhibit other clotting factors.[14]

ALTEPLASE (TISSUE-TYPE PLASMINOGEN ACTIVATOR)

Plasmin is present in blood to prevent unwanted clotting by catalysing the breakdown of the fibrin polymer that provides the framework of a blood clot. Plasmin is formed from plasminogen, a process that occurs after plasminogen has been activated by forming a complex with fibrin. After certain tissue fragments had been shown to possess plasminogen-activating ability, a soluble fraction possessing this activity was extracted and purified.[15] Cloning and expression of complementary DNA (cDNA) for human tissue-type plasminogen activator permitted commercial production of alteplase.[16]

Alteplase was shown to be a potent thrombolytic agent by virtue of its ability to activate plasminogen, thus breaking down fibrin in the thrombus.[17] An early clinical study in patients with coronary occlusion following myocardial infarction showed that it was an effective drug for dissolving clots in coronary arteries.[18] It was marketed for this purpose in 1988.

APROTININ

Aprotinin is a non-specific serine protease inhibitor originally found in bovine lung tissue and pancreas at the University of Munich in 1930.[19] Six years later a trypsin inhibitor was also isolated from bovine pancreas.[20] In 1959 Bayer patented aprotinin obtained from parotid glands where it occurred in greater concentrations.[21] After purification, aprotinin was found to be a single-chain polypeptide containing 58 amino acids.[22] In addition to its antifibrinolytic activity arising from its ability to inhibit plasmin, aprotinin inhibits various other proteolytic enzymes, including trypsin and kallikreins. It is licensed for treatment of life-threatening haemorrhage arising from hyperplasminaemia and also for the prevention of blood loss during surgery.

EPOETIN

Following the introduction of thyroid extract and adrenaline, there was considerable interest in finding humoral factors that controlled other functions in the body. The first evidence of the presence of a factor that regulated red cell formation was obtained at the University of Paris in 1906 by Carnot and Deflandre, who reported the presence in the serum of recently bled animals of what they called 'hemopoietine', a substance that provoked marked hyperglobulism in normal animals.[23] Other workers were unable to replicate the work of Carnot and Deflandre until 1953, when Erslev modified their method by infusing large volumes of plasma from extremely anaemic rabbits into normal rabbits.[24] This study renewed interest in isolating what was now called 'erythropoietin'. In 1960, erythropoietin was obtained from the plasma of anaemic sheep[25] and a decade later from the urine of anaemic humans.[26] Human erythropoietin, which differed slightly from that from sheep, was purified in 1977 by researchers at the University of Chicago using chromatographic and immunoadsorption techniques, and a partial amino acid sequence was established.[27] From this development, cDNA probes were constructed in order to identify the gene; it was then cloned and expressed in bacteria (*Escherichia coli*).[28] Once recombinant human erythropoietin became available, therapeutic application was possible.[29] The recombinant product was given the approved name of 'epoietin'.

The full structure of epoietin was determined in 1985 after it was shown to be an acidic glycoprotein featuring hexoses and hexosamines with sialic acid.[30] The protein consisted of 166 amino acids, the sequence of which was deduced from the structure of the nucleotide sequence of cloned foetal hepatic erythropoietin cDNA. Two types are commercially available, alpha-epoietin and beta-epoietin. They are interchangeable for clinical purposes.

The first results from trials of epoietin to compensate for the deficiency of erythropoietin in patients with chronic renal failure as a result of damage to the sites of its formation in the kidneys were reported in 1986.[31] Since then, the availability of epoietin has greatly reduced the need for blood transfusions in patients with chronic renal failure.

COLONY-STIMULATING FACTORS

Bone marrow cells divide rapidly to generate the billions of red and white cells that replace blood cells as they degenerate. However, bone marrow harvested from human donors could not be made to develop *in vitro* until the 1960s.

Following the discovery that colonies of malignant sarcomas, but not normal fibroblasts, grew on an agar medium, Ray Bradley at the Peter MacCallum Hospital in Melbourne tried to grow thymic lymphomas. This approach failed. He attempted to stimulate growth of colonies by inserting a variety of tissue fragments or cell suspensions in the agar layer. When a suspension of bone marrow cells was incorporated in the agar, colonies of cells at last grew. However, these were not found in the lymphoma cell layer as expected, but instead were developing within the bone marrow layer in the agar. This striking observation led Bradley and Don Metcalf to conduct a detailed examination which revealed that the new colonies consisted of macrophages whose growth was stimulated by substances diffusing from a variety of different tissues incorporated in the agar feeder layer. From these observations, it was realised that haemopoietic cell proliferation required stimulation by unknown factors and that a technique for investigating these was at hand.[32] Metcalf and Bradley went on to show that blood serum contained growth factors that stimulated marrow cells to divide. Further investigation in several laboratories confirmed that serum contained a variety of these colony-stimulating factors. Each was originally named after the specific type of cell whose growth it was found to stimulate, but subsequent work has shown that each colony factor can stimulate growth of more than one cell type.

The first stage of progress towards isolation of haemopoietic growth factors was to devise methods for culturing the specific types of cell lines involved. This was followed by the isolation of glycoproteins that proved to be the factors that stimulated growth. The breakthrough which finally opened the way for therapeutic application of these glycoproteins was the cloning of the genes regulating their production.[33]

The first of the stimulating factors to be marketed was granulocyte colony-stimulating factor. This was a relatively selective factor that mainly stimulated the proliferation and differentiation of bone marrow stem cells to form neutrophils. Metcalf and his associates purified it from mouse lung conditioned media[34] and subsequently obtained the human factor.[35] Human factor cDNA was cloned[36] and then expressed in bacteria (*E. coli*), thus permitting commercial production.[37] The recombinant granulocyte colony-stimulating factor was given the approved name of 'filgrastim'. In contrast to the natural product, it has an amino-terminal methionine residue and was not *O*-glycosylated.

Filgrastim was developed by the American biotechnology company Amgen, working in collaboration with Hoffmann–La Roche. The first report of its use in the clinic appeared in 1987, when it was shown to reduce the number of episodes of infection during chemotherapy for advanced lung cancer. It was licensed in 1991 for shortening the duration of neutropenia following cancer chemotherapy or bone marrow transplantation.

Granulocyte macrophage colony-stimulating factor was purified in 1977 from murine lung conditioned media.[38] Seven years later, highly purified material was isolated from a leukaemic human T-cell line in sufficient amounts to permit partial amino acid sequencing to be carried out.[39] From this development, it became possible to construct cDNA probes to identify the gene from a DNA library. Human factor cDNA was then cloned and expressed in bacteria (*E. coli*), permitting commercial production.[40] The recombinant granulocyte macrophage colony-stimulating factor, which was developed by scientists at Schering Plough's Palo Alto Research Institute, received the approved name of 'molgramostin' and was marketed in 1992. It stimulates the production of monocytes, from which macrophages develop, and also that of granulocytes. Like filgrastim, it has been used in the clinic to shorten the duration of neutropenia following cancer chemotherapy or bone marrow transplantation.

INTERFERONS

The interferons are a family of proteins found by Alick Isaacs and Jean Lindenmann at the National Institute for Medical Research in London to confer resistance to a variety of viral infections.[41] The discovery was made during an investigation into viral interference, being produced when inactivated influenza virus interacted with chick chorioallantoic membrane. Further investigation revealed that live viruses could stimulate cells to produce interferons that are species specific. These were found to exert complex effects on immunity and cell function. Three antigenically distinct interferons with different physical properties, respectively known as interferon alpha, beta and gamma, have been introduced into medicine as they have been found to exhibit antitumour activity.

When cultures of human leucocytes were exposed to a virus, interferons were produced. The major one, interferon alpha, was purified in 1974.[42] An amino acid analysis was carried out in 1979, permitting cDNA to be cloned and expressed in bacteria. Commercial production was then feasible, subtypes also being produced.[43,44] Two interferon alpha products have been evaluated in the treatment of viral infections and malignancies, including hairy cell leukaemia and Kaposi's sarcoma in AIDS patients. They have received product licenses for such uses and several other purposes.

When cultures of human fibroblasts were exposed to a virus, interferon was produced. Interferon beta was isolated and then purified.[45] Following the establishment of the amino acid sequence, cloning of cDNA and expression in bacteria led to the production in 1980 of recombinant interferon beta.[46–48] It has been used in the treatment of colorectal cancer. Recombinant interferon gamma has been similarly obtained.[49] It is given together with antibiotics to patients with chronic granulomatous disease.

UROKINASE

In 1861, the physiologist Ernst Wilhelm von Brucke in Vienna discovered that human urine had proteolytic activity.[50] This was confirmed by Sahli, who noted the ability of urine to decompose fibrin.[51] However, it was not until 1947 that the enzyme responsible for these observations was isolated from urine by Gwyn MacFarlane in Oxford.[52] The name 'urokinase' was given to the enzyme and it was shown to activate plasminogen to form plasmin rather than simply to digest fibrin. A crystalline preparation was obtained in 1965.[53] It is used to dissolve pulmonary embolisms.

BILE ACIDS

Chenodeoxycholic acid (chenodiol) was isolated in 1924 from goose gall by Adolf Windaus[54] and human gall by Heinrich Wieland.[55] Its complete structural configuation was elucidated by Hans Lettré at the University of Göttingen.[56]

In 1968, William Admirand and Donald Small at Boston University Medical School established that in patients with gallstones their bile was saturated with cholesterol, sometimes even exhibiting microcrystals, whereas this was not the case in normal people.[57] It was then found that biliary levels of cholic acid and chenodeoxycholic acid were lower in patients with cholesterol gallstones than in normal people. Leslie Thistle and John Schoenfield at the Mayo Clinic in Rochester, Minnesota, then administered individual bile salts by mouth for four months and found that chenodeoxycholic acid reduced the amount of cholesterol in the bile.[58] This led to a national collaborative study in the United States, which confirmed the effectiveness of chenodeoxycholic acid in bringing about dissolution of gallstones in selected patients. However, recent developments such as laparoscopic cholecystectomy and endoscopic biliary techniques have curtailed the role of chenodeoxycholic acid and ursodeoxycholic acid in the treatment of cholelithiasis.

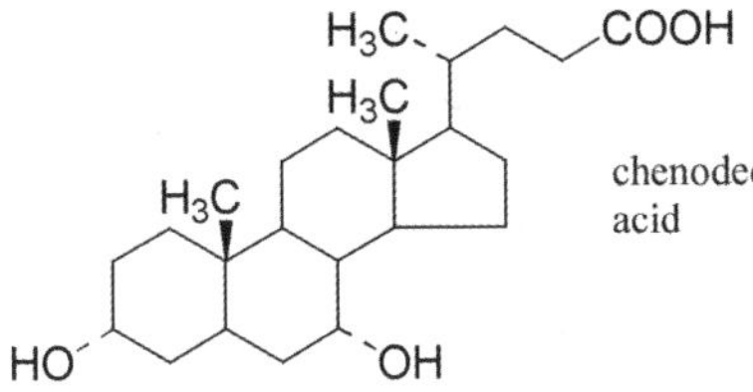

chenodeoxycholic acid

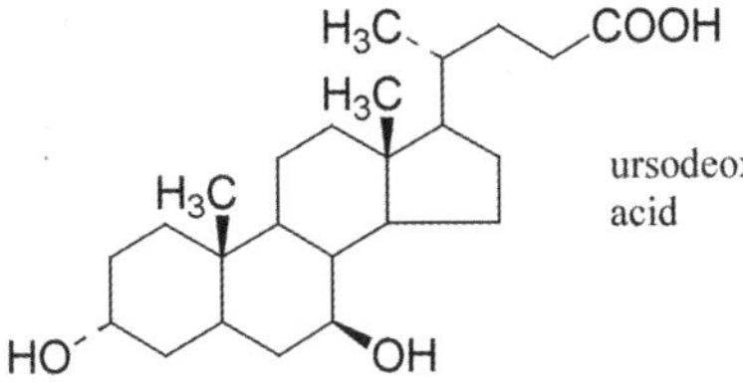

ursodeoxycholic acid

Researchers in Japan investigated ursodeoxycholic acid (ursodiol) since it was a metabolite of chenodeoxycholic acid. Ursodeoxycholic acid had been isolated from bear bile in 1927 at Okayama University by Shoda, a product with a long history of use as a cholagogue in Japanese folk medicine.[59] The structure was elucidated by Iwasaki.[60] It proved to be four times as potent as chenodeoxycholic acid and of similar efficacy in selected patients with gallstones.[61]

Hypocholesterolaemic Drugs

The French biochemist Joseph Redel synthesised 80 compounds that supposedly retained features of rings C and D of dehydrocholic acid. However, as these compounds contained an aromatic system, any resemblance to deoxycholic acid must be considered fanciful. Notwithstanding, the clinician Jean Cottet found that phenylethylacetic acid and several other disubstituted acetic acids lowered cholesterol levels in rats and humans.[62] In 1953, Cottet and his colleagues published three papers on the clinical trials of these acids. Four years later, ICI researchers screened a variety of similar compounds that the company had previously prepared as plant hormone analogues and discovered high hypocholesterolaemic activity in clofibrate.[63]

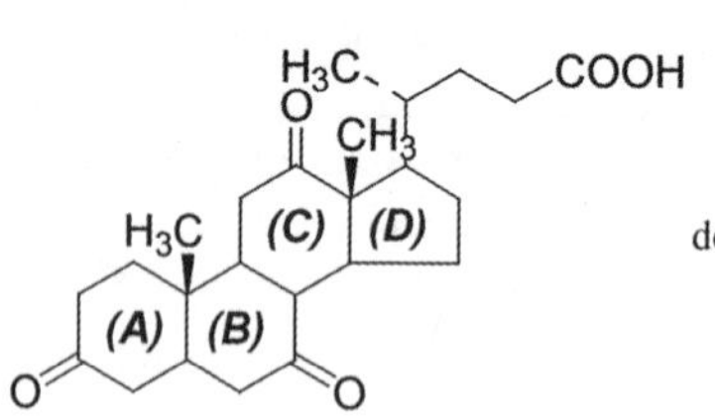

dehydrocholic acid

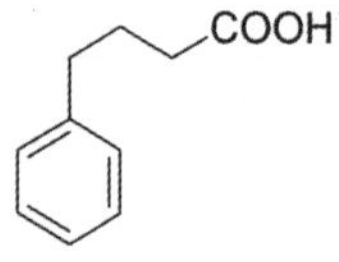

phenylethylacetic acid

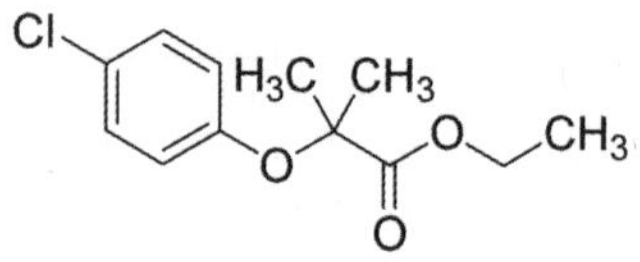

clofibrate

With increasing concern about the relationship between cholesterol levels and cardiac disease, clofibrate soon became the most frequently prescribed lipid-lowering drug. It was particularly recommended for use in the treatment of Type III hyperlipoproteinaemia. However, a major World Health Organization sponsored clinical trial found an association between chronic use of clofibrate and excess non-cardiovascular and total mortality. Consequently, it was replaced by safer analogues.

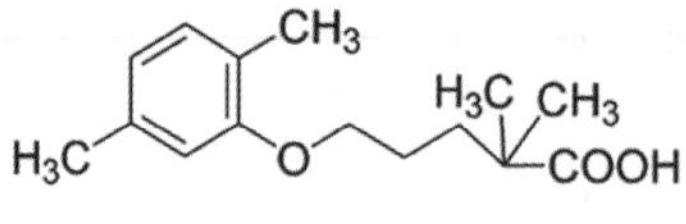

gemfibrozil

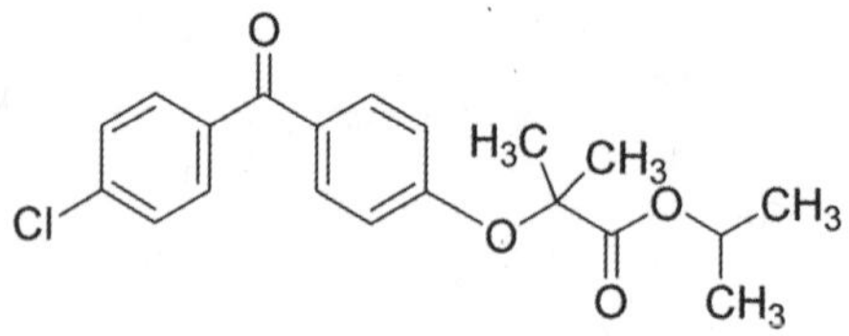

fenofibrate

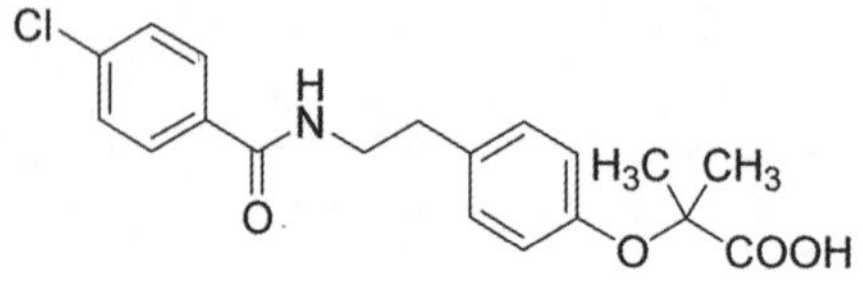

bezafibrate

Gemfibrozil was introduced by Parke, Davis and Company.[64] It was of value in the prevention of coronary heart disease in patients with very high cholesterol levels.[65] Fenofibrate was more potent than gemfibrozil, but this is of little clinical significance.[66] Bezafibrate had similar properties to fenofibrate.[67]

Antidiabetic Drugs

In 1975, researchers at the Takeda Laboratories in Osaka synthesised 71 clofibrate analogues containing a biphenyl ether moiety as they were aware that both this and alkanoic acids were often present in experimental hypolipidaemic drugs. Several of the compounds they made had not only hypolipidaemic properties but also gave positive results when screened for hypoglycaemic activity. A chlorine atom at the alpha-position and an aryloxy or aralkyloxy substituent on the beta-aryl moiety enhanced hypoglycaemic activity in mice. Further testing in obese and diabetic mice revealed that AL-321 increased the insulin sensitivity of adipose tissue.[68] Analogues of this were prepared, of which the most promising was a thiazolidine derivative, AL-294. Once again, more analogues were prepared, resulting in the development of ciglitazone.[69] Although intended for clinical evaluation, it was finally considered to lack potency. Yet again, more analogues were prepared, finally resulting in the marketing of pioglitazone almost a quarter of a century after work had begun on clofibrate analogues.[70]

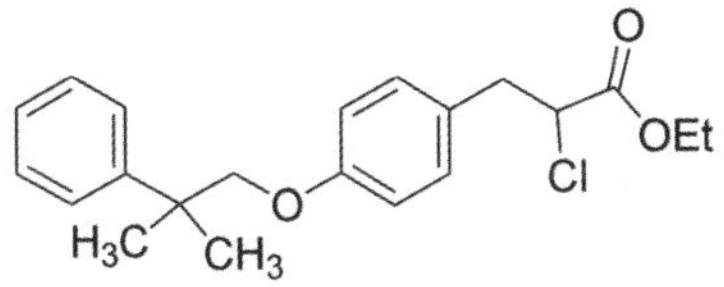

AL-321

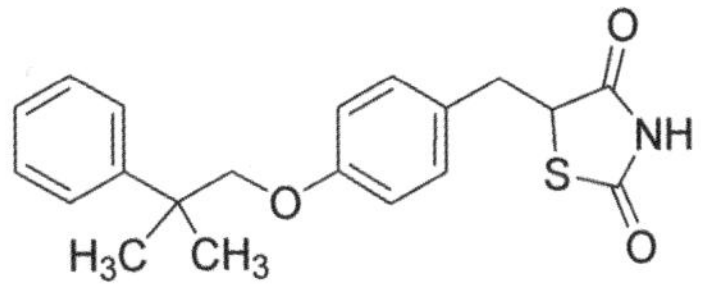

AL-294

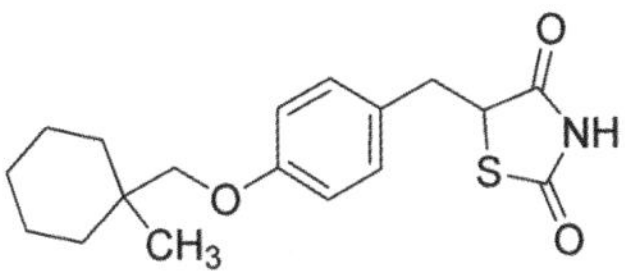

ciglitazone

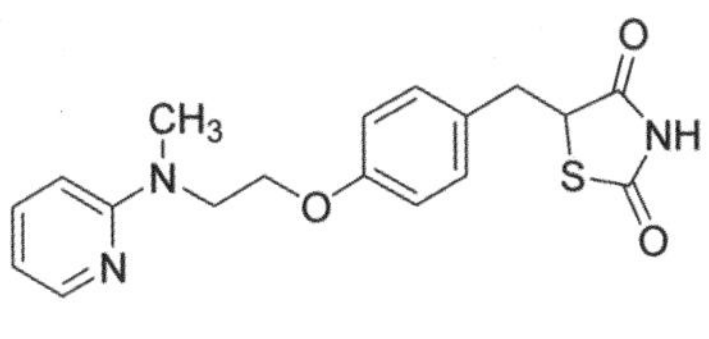

rosiglitazone

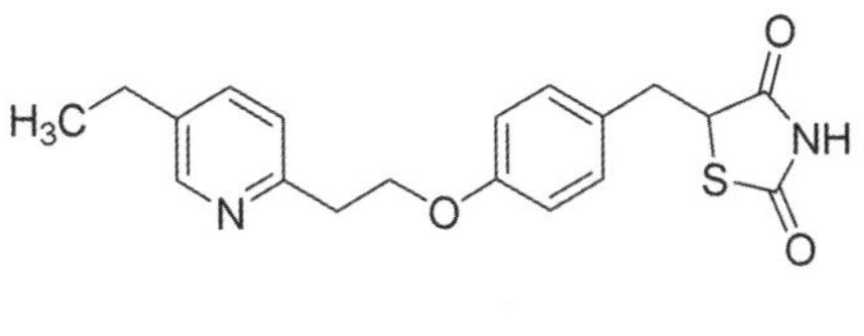

pioglitazone

SmithKline Beecham researchers enhanced the potency of ciglitazone by increasing its lipophilicity. When a urea or thiourea moiety was introduced into the ether side chain, a tenfold increase in potency was observed. Further modification involved the incorporation of the urea moiety into a heterocyclic ring. Rosiglitazone emerged from this approach.[71] Both it and pioglitazone lower blood sugar levels by reducing peripheral resistance to insulin in patients with type 2 diabetes. They are prescribed in conjunction with another oral hypoglycaemic drug such as metformin or a sulfonylurea.

GUANIDINE AND ITS ANALOGUES

C.K. Watanabe at Yale found that removal of the parathyroid gland resulted in the appearance of abnormal amounts of guanidine in the blood and claimed this was responsible for the drop in blood sugar levels following parathyroidectomy. He also demonstrated in rabbits that blood sugar levels fell after injection of guanidine.[72] When guanidine was tested as a potential antidiabetic agent it proved to be toxic.[73]

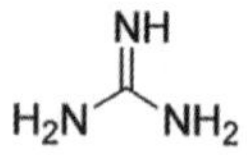

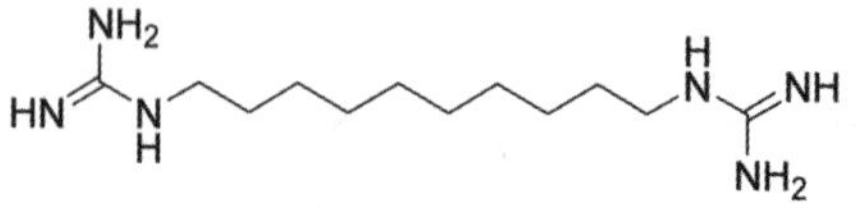

Karl Slotta at the Chemistry Institute at the University of Vienna synthesised a series of compounds with guanidine groups at each end of a long polymethylene chain.[74] They were more potent than the earlier monoguanidine compounds and less toxic. Following a clinical trial in 1926, one of these was marketed as Synthalin® by Schering AG of Berlin for use as an oral hypoglycaemic agent in mild cases of diabetes. As a result of adverse reports about Synthalin®, the related Synthalin B® was introduced with the claim that it was safer. However, the high incidence of side effects discouraged diabetics from taking either of these and they were finally withdrawn in the early 1940s because of evidence of liver toxicity.

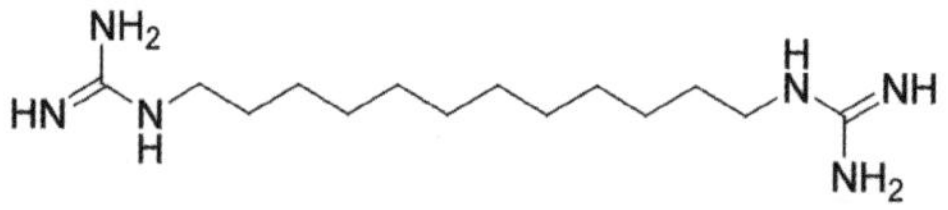

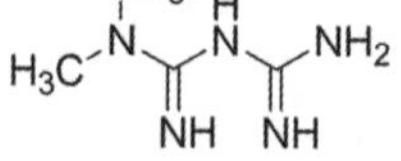

The introduction of the sulfonylureas as oral antidiabetic agents renewed interest in the guanidines. In 1957, Seymour Shapiro, Vincent Parrino and Louis Freedman of the US Vitamin Corporation in Yonkers, New York, synthesised the biguanide known as 'metformin'. It was one of more than 200 biguanides that had been examined.[75] Metformin does not act by stimulating the islet cells of the pancreas to produce more insulin, as do the sulfonylureas. Rather, it decreases gluconeogenesis and increases peripheral glucose utilisation.

Evidence obtained in 1928 indicated that trypanosomes required relatively large amounts of glucose in order to reproduce.[76] Seven years later, Hildrus Poindexter at the Howard University School of Medicine in Washington DC demonstrated that survival of animals infected with trypanosomes was prolonged if their blood glucose levels were kept depressed by insulin injections.[77] At the University of Szeged, Nikolaus von Jancso then exposed mice infected with *Trypanosoma brucei* to Synthalin® and several of its analogues.[78] These turned out to have a trypanocidal action. At the Liverpool School of Tropical Medicine, Warrington Yorke and E.M. Lourie were quick to note that Synthalin® killed the trypanosomes at dose levels that did not significantly lower blood sugar in mice. They demonstrated that Synthalin® was even active against trypanosomes growing on culture media rich in glucose. This indicated that the trypanocidal action of the amidines was a direct one, and had nothing to do with lowering blood sugar in the host animal.[79] The Liverpool findings were then reported to Harold King at the National Institute for Medical Research, which led to his preparing analogues of Synthalin® as potential trypanocidal agents. Among these were several diamidines that turned out to be powerful trypanocides that cured mice and rabbits infected with *T. rhodesiense*. The analogue with one extra methylene group in the chain was found to be particularly active in mice.[80]

In 1937, Arthur Ewins of May and Baker was invited to participate in the research programme. He arranged for the preparation of diamidines in which the polar amidine groups were separated by an intermediate chain consisting of two benzene rings rather than polymethylene groups, as had previously been the case. The first of his compounds proved to be active, so the series was extended, with a large number of analogues being examined. Many of these were trypanocidal, especially stilbamidine and pentamidine.[81] As pentamidine had greater water solubility than stilbamidine, it was preferred for intramuscular injections.

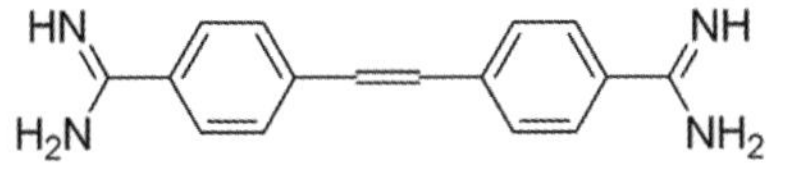

stilbamidine

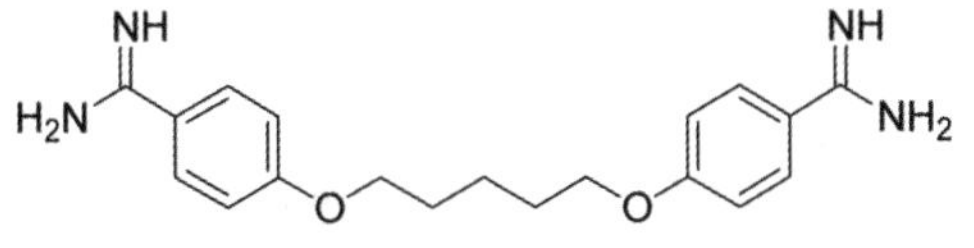

pentamidine

By 1940, the diamidines had been tested on over 400 patients suffering from sleeping sickness or the related tropical disease, leishmaniasis. Because of their polar nature, they were unsuitable for treating advanced forms of sleeping sickness in which there was central nervous system involvement. Nevertheless, these diamidines, prepared as a result of close collaboration between academia and industry, are still used against trypanosomiasis. They are also active against a range of protozoa, including *Pneumocystis carinii*.[82] Pentamidine isethionate has proved to be a life-saving drug when administered by aerosol into the lungs of AIDS patients infected with this organism.

It has been known since 1938 that amidoximes (i.e. *N*-hydroxyamidines) also have trypanocidal activity.[83] After reading a report on the activity of some of these compounds, Mull synthesised Su-4029 at the Ciba laboratories in Basle. This proved inactive against trypanosomes, but was found to possess antihypertensive activity in dogs after being routinely submitted to Ciba's general screening programme.

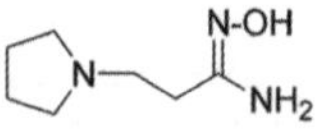

A clinical trial confirmed the antihypertensive activity of Su-4029, but also revealed that it induced fever. Mull then prepared several analogues.[84] One of these was guanethidine, which is still sometimes administered along with a diuretic or beta-blocker in cases of resistant hypertension. It acts both by blocking the release of noradrenaline from postganglionic adrenergic neurones and (unlike its analogues) also depletes the nerve endings of noradrenaline. Adrenergic blockers such as guanethidine and its analogues are no longer popular because they cause postural hypotension. Several analogues of guanethidine were marketed in the mid-1960s. One of these was debrisoquine, which is still occasionally used in combination with a beta-blocker or diuretic in the treatment of hypertension.[85]

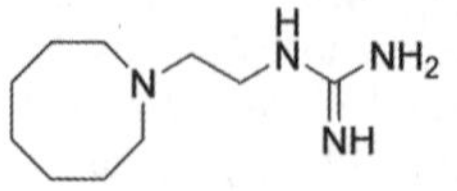

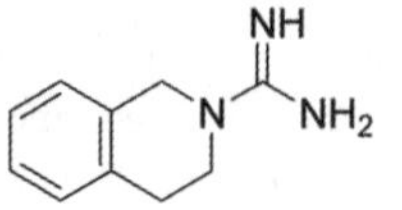

PYROPHOSPHORIC ACID ANALOGUES

At the University of Berne, Herbert Fleisch discovered that plasma, urine and saliva inhibited calcium phosphate precipitation because they contained inorganic pyrophosphate, a substance not previously known to be present in these fluids.[86] Pyrophosphate was then shown to reduce both the dissolution of calcium phosphate crystals and also their aggregation into larger clusters.[87] When injected into rats, pyrophosphate inhibited experimentally induced deposition of calcium in soft tissue, but had no effect on bone resorption (loss of calcium phosphate).[88] This lack of effect on bone resorption was assumed to be due to rapid hydrolysis of pyrophosphate by alkaline phosphatase.

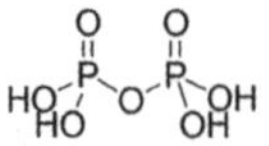

Fleisch prepared biphosphonate analogues of pyrophosphate in which the metabolically sensitive central oxygen atom was isosterically replaced by a carbon atom. Low concentrations of many of these were shown to behave like pyrophosphate by inhibiting precipitation of calcium phosphate and the aggregation of its crystals.[89] It was found that this was due to a common affinity for the solid phase of calcium salts, to which they strongly bound.[90] The clinical implication was evident since pyrophosphate is present in plasma, teeth and bone. Topical application of the new biphosphonates in rats reduced the formation of dental tartar, which in part consisted of calcium phosphate and carbonate.[91] This explains why Procter and Gamble, a company with a major interest in toothpaste formulation, was collaborating with Fleisch. A dental hygiene product was later marketed.

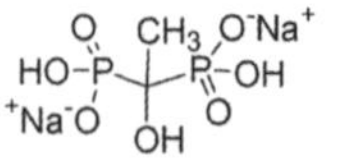

Safety was a paramount factor in selecting a biphosphonate for clinical testing. Among the biphosphonates that had been prepared, the tendency to interfere with normal calcification in bone, cartilage and dentine varied. Although this was reversible, it had to be taken into consideration. After careful evaluation, disodium etidronate was selected for evaluation in the clinic after it had been shown to inhibit hydroxyapatite dissolution *in vitro* and bone resorption *in vivo*.[92] The prime interest lay in the role of disodium etidonate in Paget's disease, a metabolic disorder involving excessive bone resorption with abnormal calcification and bone formation. When given by mouth, the drug effectively improved or even abolished bone pain and stopped bone resorption and restored serum calcium levels to near-normal values. These early clinical studies used doses of disodium etidronate that produced abnormal bone mineralisation, which caused deposition of a layer of osteoid tissue on the bones. This led to pain and fractures. Reduction of the dose resulted in some bone resorption, but a balance was achieved by reducing the dosage and administering the drug in six-monthly cycles. Disodium etidronate was then licensed in the United Kingdom for use in Paget's disease. The license was later extended to include treatment of vertebral osteoporosis and the bone pain and hypercalcaemia caused by malignancies such as breast cancer or multiple myeloma.

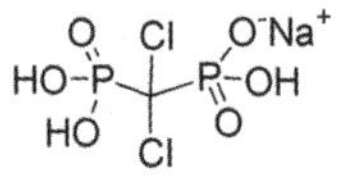

sodium clodronate 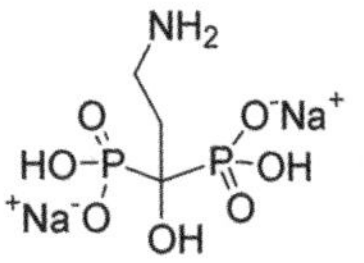   disodium pamidronate

Sodium clodronate was prepared at the same time as disodium etidronate. It is somewhat more effective when used in patients who have hypercalcaemia associated with malignancy. However, Fleisch found that lengthening of the carbon skeleton of disodium etidronate and also incorporating a terminal amino group on the carbon chain markedly increased potency.[93] Disodium pamidronate was then shown to be more effective than sodium clodronate in the treatment of hypercalcaemia of malignancy, but it could only be administered parenterally as it was poorly absorbed from the gut.

UNDECENOIC ACID (UNDECYLENIC ACID)

Knowing that human sweat had a reputation for inhibiting the growth of fungi, Peck and Rosenfeld tested the antifungal activity of fatty acids present in sweat and claimed that propionic, butyric, lactic and ascorbic acids were of value in the treatment of fungal infections.[94]

undecenoic acid

A systematic examination of a wider range of fatty acids by researchers at Johns Hopkins Hospital showed that antifungal activity increased with chain length and was optimal in formulations at pH 5.[95] Foley and Lee later demonstrated that the most potent of the fatty acids was undecenoic acid.[96] It is used as a topical antifungal agent.

ANGIOTENSIN-CONVERTING ENZYME

In 1898 at the Karolinska Institute, Robert Tigerstedt and his student Per Bergman discovered that a hypertensive agent was present in a saline extract of kidney tissue.[97] Named 'renin', the physiological significance of this substance became clearer in 1934 after Cleveland physician Harry Goldblatt demonstrated that it was involved in renal hypertension.[98] By 1940, it had been established that renin was an enzyme released from the kidney into the blood, where it promoted the formation of a pressor peptide called angiotensin.[99] This peptide was shown to exist in two forms, which were then isolated.[100,101] Angiotensin I was a decapeptide without pressor activity, whereas angiotensin II was an octapeptide that turned out to be the most potent blood pressure raising substance in the human body. The hydrolysis of angiotensin I into angiotensin II in the plasma was catalysed by a zinc-containing enzyme named angiotensin-converting enzyme (ACE), which was purified in 1956.[102]

Further investigation into the clinical role of angiotensin II led to the discovery that, in addition to its potent pressor activity, it also acted directly on the adrenals to stimulate release of aldosterone. This produced sodium retention, thereby enhancing the ability of angiotensin II to raise blood pressure.[103]

ACE Inhibitors

While working in the Institute of Basic Medical Sciences at the Royal College of Surgeons of England in the 1960s, John Vane was investigating the causes of hypertension. When he tested an extract of the venom of the Brazilian arrowhead viper, *Bothrops jararaca*, which had been brought to London by Sergio Ferreira, one of his post-doctoral students, he discovered that it could block the formation of angiotensin II from angiotensin I by inhibiting ACE.[104] Despite leading clinicians not sharing his opinion, Vane was able to convince researchers at the Squibb Institute for Medical Research in New Jersey that an ACE inhibitor offered a novel approach to the control of high blood pressure. Miguel Ondetti and his colleagues then fractionated the viper venom and elucidated the structure of several peptides.[105] One of the most active at inhibiting ACE was teprotide, which had already been isolated in Vane's laboratory.[106] It was now synthesised and investigated thoroughly in animals by the Squibb researchers. The first clinical study showed that when teprotide was administered intravenously in patients with elevated plasma renin levels, it was an effective hypotensive agent.[107] Further studies revealed that it has similar activity in hypertensive patients with normal renin levels. As was to be expected with any peptide, teprotide was inactive by mouth. This ruled out any likelihood of its routine use in the clinic as a hypotensive drug, but pointed the way forward for an attempt to synthesise an orally active analogue.

The Squibb research team next screened about 2000 non-peptides in a vain attempt to find an orally active ACE inhibitor suitable for clinical evaluation. A new approach was then taken after the publication of a paper by Byers and Wolfenden about the inhibition by benzylsuccinic acid of another zinc-containing metallopeptidase, the digestive enzyme carboxypeptidase A.[108] Byers and Wolfenden had found that natural substrates were bound to the active site of carboxypeptidase A by their C-terminal phenylalanine residue. As benzylsuccinic acid had a similar enough structure to phenylalanine it bound to the enzyme active site, but since it had no amide function that could be hydrolysed it remained bound to the active site, thereby blocking access to it by the natural substrate.

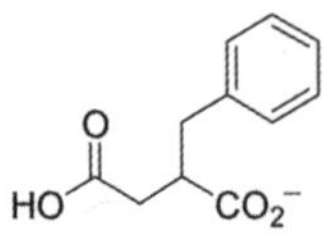

benzylsuccinic acid

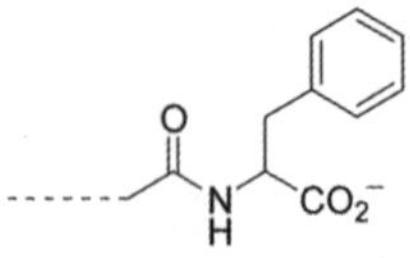

phenylalanine residue

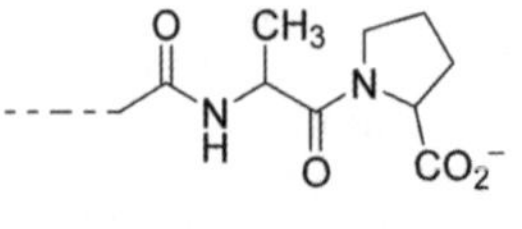

alanylproline residue

The Squibb researchers knew that hydrolysis of angiotensin I involved the splitting off of a dipeptide rather than a single C-terminal amino acid as in the phenylalanine bound to carboxypeptidase A. This meant that in an inhibitor of ACE, binding to the active site would only occur if the distance between the amino acid moiety and the second carboxylic acid group was extended. Since it was already known that peptide inhibitors with an alanylproline residue at the C-terminal end had the greatest affinity for the active site of ACE, the Squibb team synthesised succinoyl-L-proline and tested its binding affinity for ACE. This confirmed that succinoyl-L-proline was a specific inhibitor, albeit with low potency. This was enhanced by introducing a methyl substituent adjacent to the amide function, with the appropriate stereochemistry to maintain similarity with the peptide inhibitors. The resulting compound, D-2-methylsuccinoyl-L-proline, was still not potent enough to be considered as a candidate compound for clinical investigations and it required considerable effort and ingenuity to enhance its potency.

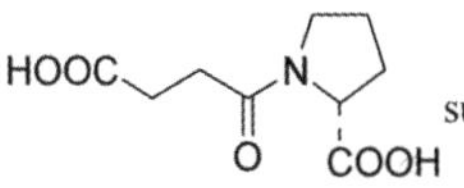

HOOC ... succinoyl-L-proline

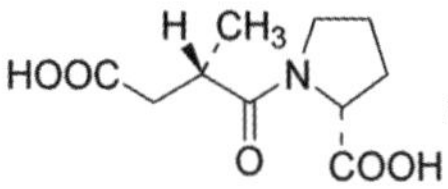

HOOC ... H CH₃ ... D-2-methylsuccinoyl-L-proline

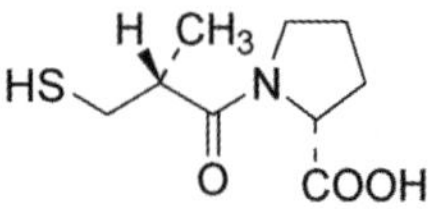

HS ... H CH₃ ... captopril

The Squibb researchers believed that interaction between the carboxyl group in the carboxyalkanoyl moiety and the zinc atom in the enzyme active site could be increased, but this proved difficult to achieve until its carboxyl was replaced with a thiol group. This resulted in a one-thousand-fold increase in inhibitory activity for captopril. This was the first non-peptide ACE inhibitor suitable for introduction into the clinic.[109] When it was marketed in the United Kingdom in 1981, the Committee on Safety of Medicines limited the product licence to use only in patients with severe hypertension who had not responded to standard therapy. This limitation was imposed because early clinical trials had shown captopril to have the potential to cause renal damage or granulocytopenia. After intensive postmarketing surveillance revealed that side effects were mainly of a minor nature and of relatively low incidence, the licence was extended in 1985, to cover its use in the treatment of mild to moderate hypertension. Captopril and other ACE inhibitors are now firmly established in the treatment of heart failure and in hypertension when thiazide diuretics or beta-blockers are either ineffective or not tolerated.

Investigators at Merck reconsidered the use of succinoyl-L-proline as a non-hydrolysable substitute for the alanylproline residue and examined derivatives of glutamyl-L-proline. This led to the discovery of high activity in compound I.

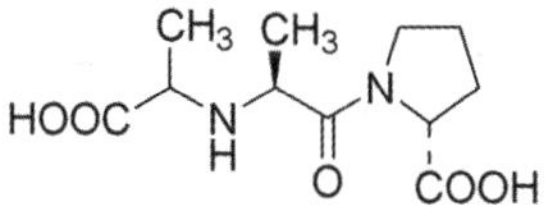

HOOC ... CH₃ CH₃ ... compound I

HOOC ... CH₃ ... enalaprilat

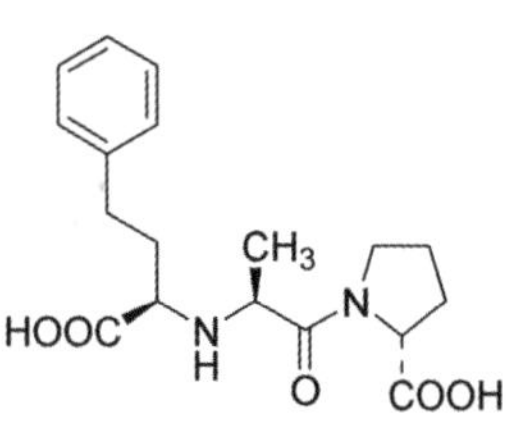

C₂H₅O₂C ... CH₃ ... enalapril

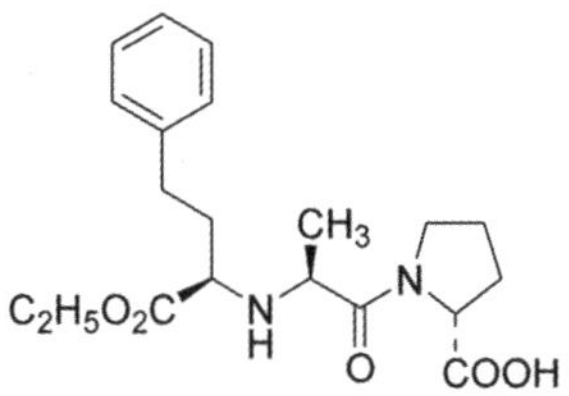

Taking into account what was known about the specificity of the active site of ACE, the Merck researchers introduced hydrophobic amino acid side chains in place of one of the methyl groups in compound I. Of these, phenylalkyl substituents were found to increase inhibitory activity. Maximal activity occurred in enalaprilat, which was 2000 times more potent than compound I. When administered intravenously, it was more potent than captopril. However, oral bioavailability was disappointing. This was overcome by masking the carboxyl group through conversion to its ethyl ester to form enalapril.[110] This was well absorbed from the gut before undergoing hydrolysis in the liver to release enalaprilat.

Merck scientists also decided to determine whether or not the best substrates for ACE were those based on the alanylproline residue at the C-terminal end of peptide ACE inhibitors. Accordingly they examined a variety of dipeptides containing either alanine or proline connected to a variant amino acid. After their results had confirmed the importance of retaining proline at the C-terminal, it was established that lysine could replace the alanine in enalaprilat. The resulting compound, lisinopril, was unexpectedly found to be well absorbed from the gut, unlike enalaprilat. It seems likely that the presence of the lysine residue in lisinopril permits peptide-carrier-mediated transport from the gut into the portal circulation.[111]

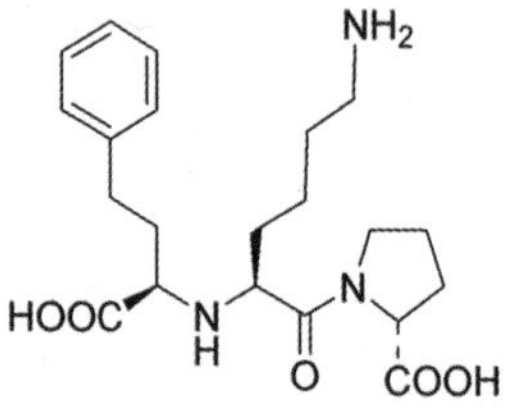

lisinopril

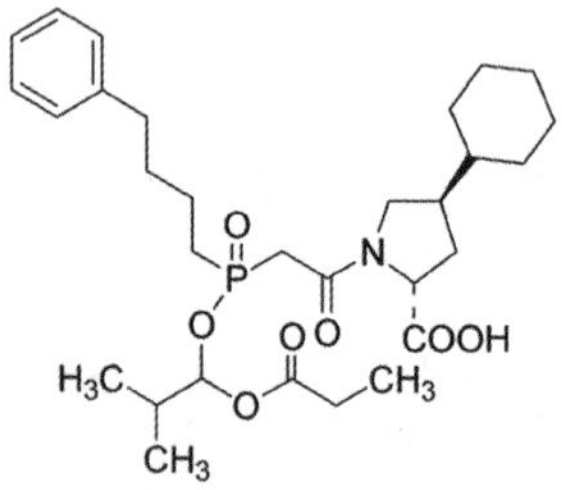

fosinopril

Squibb investigators established that the potent zinc-binding ligand phosphonic acid could be incorporated into ACE inhibitors. In addition, they were aware that the lipophilicity of the proline ring could be increased with retention of activity. They therefore synthesised fosinopril in 1982.[112] It was found to be different to other ACE inhibitors insofar as, after·being hydrolysed to the active fosinoprilat, it was eliminated roughly equally by both the liver and the kidneys. This meant that it was suitable for administration to patients with reduced renal function.

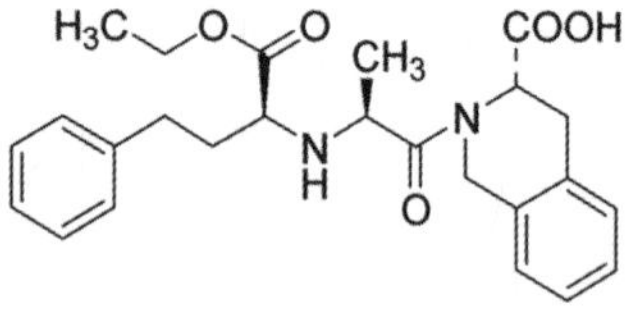

quinapril

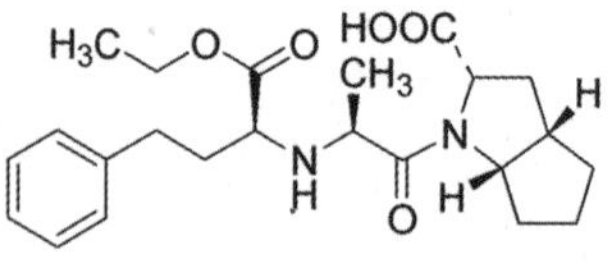

ramipril

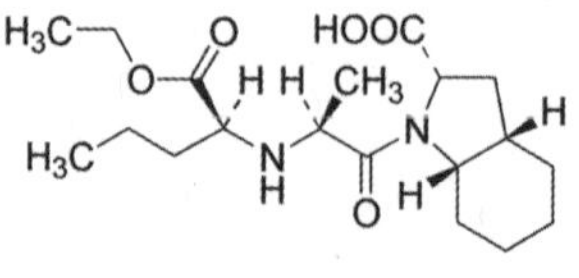

perindopril

Several companies independently discovered that the proline ring in ACE inhibitors could be enlarged without any loss of activity. This led to the introduction of quinapril,[113] ramipril[114] and perindopril.[115]

REFERENCES

1. M. Doyon, A. Morel, A. Policard, Estraition directe de l'antithrombine du foie: influence de la congelation. *C. R. Soc. Biol.*, 1911; **70**: 341–4.
2. W.H. Howell, Heparin, an anticoagulant. *Am. J. Physiol.*, 1922; **63**: 434–5.
3. W.H. Howell, E. Holt, Two new factors in blood coagulation – heparin and pro-antithrombin. *Am. J. Physiol.*, 1918; **47**: 328–41.
4. J. McLean, The thromboplastic action of cephalin. *Am. J. Physiol.*, 1916; **41**: 250–57.
5. W.H. Howell, E. Holt, The purification of heparin and its chemical and physiological reactions. *Bull. Johns Hopkins Hosp.*, 1928; **42**: 199–206.
6. D. Scott, A. Charles, Studies on heparin. III The purification of heparin. *J. Biol. Chem.*, 1933; **102**: 437–48.
7. A. Schmitz, A. Fischer, Uber die chemische Natur des Heparins. III. Einige Untersuchungen zur Konstitution des Heparins. *Z. Physiol. Chem.*, 1933; **216**: 274–80.
8. E. Jorpes, On heparin, its chemical nature and properties. *Acta Med. Scand.*, 1936; **88**: 427–33.
9. J.E. Jorpes, S. Bergstrom, Heparin: a mucoitin polysulphonic acid. *J. Biol. Chem.*, 1937, **118**: 447–57.
10. E.A. Johnson, B. Mulloy, The molecular-weight range of mucosal-heparin preparations. *Carbohydrate Res.*, 1976; **51**: 119–217.
11. D.W.G. Murray, L.B. Jaques, T.S. Perrett. Heparin. *Surgery*, 1937; **2**: 163–87.
12. C. Crafoord, Preliminary report on post-operative treatment with heparin as a preventive of thrombosis. *Acta Chir. Scand.*, 1937; **79**: 407–26.
13. Eur. Pat. 1981: 40144 (to Pharmindustrie).
14. M. Aiach, A. Michaud, J.I. Balian, *et al.*, A low molecular weight heparin derivative. *Thromb. Res.*, 1983; **31**: 611–621.
15. F. Bachmann, A.P. Fletcher, N. Alkjaersig, S. Sherry, Partial purification and properties of the plasminogen activator from pig heart. *Biochemistry*, 1964; **3**: 1578–85.
16. D. Pennica, W.E. Holmes, W.J. Kohr, *et al.*, Cloning and expression of human tissue-type plasminogen activator cDNA in *E. coli. Nature*, 1983; **301**: 214.
17. W. Weimar, J. Stibbe, A.J. Vanseyen, *et al.*, Specific lysis of an ileofemoral thrombus by administration of extrinsic (tissue-type) plasminogen-activator. *Lancet*, 1981; **2**: 1018–20.
18. M. Verstraete, M. Bory, D. Collen, *et al.*, Randomized trial of intravenous recombinant tissue-type plasminogen activator versus intravenous streptokinase in acute myocardial infarction. *Lancet*, 1985; **1**: 842–7.
19. H. Kraut, E.K. Frey, E. Werle, Über die Inaktivierung des Kallikreins. *Z. Physiol. Chem.*, 1930; **192**: 1–21.
20. M. Kunitz, J.H. Northrup, Isolation from beef pancreas of crystalline trypsinogen, trypsin, trypsin inhibitor, and inhibitor–trypsin compound. *J. Gen Physiol.*, 1936; **19**: 991–1007.
21. U.S. Pat. 1959: 2890986 (to Bayer).
22. B. Kassell, M. Laskowski Sr, The basic trypsin inhibitor of bovine pancreas. V. The disulfide linkages. *Biochem. Biophys. Res. Commun.*, 1966; **20**: 463–8.
23. P. Carnot, C. Deflandre, Sur l'activité hématopoietique des serum au cours de la régénération du sang. *C. R. Acad. Sci.*, 1906; **143**: 384–6.
24. A. Erslev, Humoral regulation of red cell production. *Blood*, 1953; **8**: 349–57.
25. W.F. White, C.W. Gurney, E. Goldwasser, L.O. Jacobson, Studies on erythropoietin. *Recent Progr. Horm. Res.*, 1960; **16**: 219–62.
26. J. Espada, A. Gutnisky, A new method for concentration of erythropoietin from human urine. *Biochem. Med.*, 1970; **3**: 475–84.
27. T. Miyake, C.K.-H. Kung, E. Goldwasser, Purification of human erythropoietin. *J. Biol. Chem.*, 1977; **252**: 5558–64.
28. F.K. Lin, S. Suggs, C.H. Lin, *et al.*, Cloning and expression of the human erythropoietin gene. *Proc. Natl Acad. Sci. USA*, 1985; **82**: 7580–4.
29. J.C. Egrie, T.W. Strickland, J. Lane, *et al.* Characterization and biological effects of recombinant human erythropoietin. *Immunobiology*, 1986; **172**: 213–24.
30. M.S. Dordal, F.F. Wang, E. Goldwasser, The role of carbohydrate in erythropoietin action. *Endocrinology*, 1985; **116**: 2293–9.
31. C.G. Winearls, D.O. Oliver, M.J. Pippard, Effect of human erythropoietin derived from recombinant DNA on the anaemia of patients maintained by chronic haemodialysis. *Lancet*, 1986; **2**: 1175–8.
32. T.R. Bradley, D. Metcalf, The growth of mouse bone marrow cells *in vitro. Aust. J. Exp. Biol. Med. Sci.*, 1966; **44**: 287–99.
33. D. Metcalf, The hemopoietic regulators an embarrassment of riches. *Bioessays*, 1992; **14**: 799–805.

34. N.A. Nicola, D. Metcalf, M. Matsumoto, G.R. Johnson, Purification of a factor inducing differentiation in murine myelomonocytic leukemia cells. Identification as granulocyte colony-stimulating factor. *J. Biol. Chem.*, 1983; **258**: 9017–23.
35. N.A. Nicola, C.G. Begley, D. Metcalf, Identification of the human analogue of a regulator that induces differentiation in murine leukaemic cells. *Nature*, 1985; **314**: 625–8.
36. S. Nagata, M. Tsuchiya, S. Asano, *et al.*, Molecular cloning and expression of cDNA for human granulocyte colony-stimulating factor. *Nature*, 1986; **319**: 415–18.
37. Y. Komatsu, T. Matsumoto, T. Kuga, *et al.*, Cloning of granulocyte colony-stimulating factor cDNA from human macrophages and its expression in *Escherichia coli*. *Jap. J. Cancer Res.*, 1987; **78**: 1179–81.
38. A.W. Burgess, J. Camakaris, D. Metcalf, *et al.*, Purification and properties of colony-stimulating factor from mouse lung-conditioned medium. *J. Biol. Chem.*, 1977; **252**: 1998–2003.
39. L.G. Sparrow, D. Metcalf, M.W. Hunkapiller, *et al.*, Purification and partial amino acid sequence of asialo murine granulocyte-macrophage colony stimulating factor. *Proc. Natl Acad. Sci. USA*, 1985; **82**: 292–6.
40. K. Kaushansky, P.J. O'Hara, K. Berkner, *et al.*, Genomic cloning, characterization, and multilineage growth-promoting activity of human granulocyte-macrophage colony-stimulating factor. *Proc. Natl Acad. Sci. USA*, 1986; **83**: 3101–5.
41. A. Isaacs, J. Lindenmann, Virus interference. I. The interferon. *Proc. Roy. Soc.*, 1957; **B147**: 258–67.
42. C.B. Anfinsen, S. Bose, L. Corley, D. Gurari-Rotman, Partial purification of human interferon by affinity chromatography. *Proc. Natl Acad. Sci. USA*, 1974; **71**: 3139–42.
43. M. Streuli, S. Nagata, C. Weismann, At least three human type alpha interferons: structure of alpha 2. *Science*, 1980; **209**: 1343–7.
44. D.V. Goeddel, D.W. Leung, T.J. Dull, *et al.*, The structure of eight distinct cloned human leukocyte interferon cDNAs. *Nature*, 1981; **290**: 20–26.
45. E. Knight Jr, Interferon: purification and initial characterization from human diploid cells. *Proc. Natl Acad. Sci. USA*, 1976; **73**: 520–3.
46. T. Taniguchi, L. Guarente, T.T. Roberts, *et al.*, Expression of the human fibroblast interferon gene in *Escherichia coli*. *Proc. Natl Acad. Sci. USA*, 1980; **77**: 5230–3.
47. R. Derynck, J. Content, E. DeClercq, *et al.*, Isolation and structure of a human fibroblast interferon gene. *Nature*, 1980; **285**: 542–7.
48. D.V. Goeddel, H.M. Shepard, E. Yelverton, *et al.*, Synthesis of human fibroblast interferon by *E. coli*. *Nucleic Acid Res.*, 1980; **8**: 4057–74.
49. P.W. Gray, D.W. Leung, D. Pennica, *et al.*, Expression of human immune interferon cDNA in *E. coli* and monkey cells. *Nature*, 1982; **295**: 503–8.
50. E. von Brucke, S.B. Akad, *Wiss. Wien*, 1861; **43**: 601.
51. W. Sahli, *Arch. Physiol.*, 1886; **38**: 35.
52. R.G. MacFarlane, J. Pilling, Fibrinolytic activity of normal urine. *Nature*, 1947; **159**: 779.
53. A. Lesuk, L. Terminiello, J.H. Traver, Crystalline human urokinase: some properties. *Science*, 1965; **147**: 880–2.
54. A. Windaus, A. Bohne, E. Schwarzkopf, *Z. Physiol. Chem.*, 1924; **140**: 177.
55. H. Wieland, G. Reverey, *Z. Physiol. Chem.*, 1924; **140**: 186.
56. H. Lettré, *Ber.*, 1935; **68**: 766.
57. W.H. Admirand, D.M. Small, The physicochemical basis of cholesterol gallstone formation in man. *J. Clin. Invest.*, 1968; **47**: 1043–52.
58. J.L. Thistle, L.J. Schoenfield, Induced alterations in composition of bile of persons having cholelithiasis. *Gastroenterol.*, 1971; **61**: 488–96.
59. M. Shoda, Über die Ursodeoxycholsäure aus Barengallen und ihre physiologische Wirkung. *J. Biochem. (Japan)*, 1927; **7**: 505–17.
60. T. Iwasaki, Über die Konstitution der Ursodeoxycholsäure. *Z. Physiol. Chem.*, 1936; **244**: 181–93.
61. Tokyo Cooperative Gallstone Study Group, Efficacy and indications of ursodeoxycholic acid treatment for dissolving gallstones. A multicenter double-blind trial. *Gastroenterol.*, 1980; **78**: 542–8.
62. J. Redel, J. Cottet, Action hypocholestérolemiante de quelques acides acétiques disubstutués. *C. R. Acad. Sci.*, 1953; **236**: 2553–5.
63. J.M. Thorp, W.S. Waring, Modification of metabolism and distribution of lipids by ethyl chlorophenoxyisobutyrate. *Nature*, 1962; **194**: 948–9.
64. Ger. Pat. 1969: 1925423 (to Parke, Davis).
65. A.H. Kissebah, S. Alfarsi, P.W. Adams, *et al.*, Transport kinetic of plasma free fatty acid, very low density lipoprotein triglycerides and apoprotein in patients with endogenous hypertriglyceridaemia. *Artherosclerosis*, 1976; **24**: 199.
66. Ger. Pat. 1973: 2250327 (to Orchimed).
67. US Pat. 1973, 3781327 (to Boehringer).

68. Y. Kawamatsu, H. Asakawa, T. Saraie, et al., II. Synthesis and biological activities of 2-chloro-3-arylpropionic acids. Arzneimittelforschung., 1980; **30**: 585–9.
69. T. Sohda, K. Mizuno, E. Imamiya, et al., Studies on antidiabetic agents. II. Synthesis of 5-[4-(1-methylcyclohexylmethoxy)-benzyl]thiazolidine-2,4-dione (ADD-3878) and its derivatives. Chem. Pharm. Bull., 1982; **30**: 3580–600.
70. T. Sohda, Y. Kawamatsu, T. Fujita, et al., [Discovery and development of a new insulin sensitizing agent, pioglitazone]. Yakugaku Zasshi., 2002; **122**: 909–18.
71. C.C. Barrie, M.A. Cantello, G.P. Cawthorne, et al., [[ω-(Heterocyclylamino)alkoxy]benzyl]-2,4-thiazolidinediones as potent antihyperglycemic agents. J. Med. Chem., 1994; **37**: 3977–85.
72. C.K. Watanabe, Studies in the metabolism changes induced by administration of guanidine bases. I. Influence of injected guanidine hydrochloride upon blood sugar content. J. Biol. Chem., 1918; **33**: 253–65.
73. E. Frank, M. Northmann, A. Wagner, Über synthetisch dargestellte Korper mit insulinartiger Wirkung auf den normalen und den diabetischen Organismus. Klin. Wochenschr., 1926; **5**: 2100–7.
74. K.H. Slotta, R., Tschesche, Die blutzucker-senkende Wirkung der Biguanides. Ber., 1929; **62**: 1398–405.
75. S.L. Shapiro, V. Parrino, L. Freedman, Hypoglycemic agents. I. Chemical properties of β-phenethylbiguanide. A new hypoglycemic agent. J. Am. Chem. Soc., 1959; **81**: 2220–5.
76. K. Schern, Biochem. Ztschr., 1928; **193**: 264.
77. H.A. Poindexter, Further observations on the relation of certain carbohydrates to Trypanosoma equiperdum metabolism. Parasitol., 1935; **21**: 292–301.
78. N. von Jancso, H. von Jancso, Chemotherapeutische Schnellfestigung von Trypanosomen durch Ausschaltung der näturlich Abwehrkräfts. Z. Immun. Forschr., 1935; **85**: 81.
79. E.M. Lourie, W. Yorke, Studies in chemotherapy. XXI. The trypanocidal action of certain aromatic diamidines. Ann. Trop. Med., 1939; **33**: 289–304.
80. H. King, E.M. Lourie, W. Yorke, Studies in chemotherapy, XIX: Further report on new trypanocidal substances. Ann. Trop. Med. Parasitol., 1938; **32**: 177–92.
81. J.N. Ashley, H.J Barber, A.J. Ewins, et al., A chemotherapeutic comparison of the trypanocidal action of some aromatic diamidines. J. Chem. Soc., 1942: 103–16.
82. V.T. DeVita, M. Emmer, A. Levine, B. Jacobs, C. Berard, Pneumocystis carinii pneumonia: successful diagnosis and treatment of two patients with associated malignant processes. N. Engl. J. Med., 1969; **280**: 287–91.
83. I.D. Lamb, A.C. White, Some amidines and amidoximes with trypanocidal activity. J. Chem. Soc., **1939**: 1253–7.
84. R.P. Mull, R.A. Maxwell, A.J. Plummer, Antihypertensive activity of hexahydro-1-azepinepropionamidoxime. Nature, 1957; **180**: 1200–1.
85. W. Wenner, 1,2,3,4-Tetrahydroisoquinoline derivatives with antihypertensive properties. J. Med. Chem., 1965; **8**: 125–6.
86. H. Fleisch, S. Bisaz, Isolation from urine of pyrophosphate, a calcification inhibitor. Am. J. Physiol., 1962; **203**: 671–5.
87. H. Fleisch, J. Maerki, R.G.G. Russell, Effect of pyrophosphate on dissolution of hydroxyapatite and its possible importance in calcium homeostasis. Proc. Soc. Exp. Biol. Med., 1966; **122**: 317–20.
88. D. Schibler, R.G.G. Russell, H. Fleisch, Inhibition by pyrophosphate and polyphosphate of aortic calcification induced by vitamin D_3 in rats. Clin. Sci., 1968; **35**: 363–72.
89. H. Fleisch, R.G.G. Russell, S. Bisaz, et al., The inhibitory effect of phosphonates on the formation of calcium phosphate crystals in vitro and on aortic and kidney calcification in vivo. Eur. J. Clin. Invest., 1970; **1**: 12–18.
90. A. Jung, S. Bisaz, H. Fleisch, The binding of pyrophosphate and two diphosphonates by hydroxyapatite crystals. Calcif. Tissue Res., 1973; **11**: 269–80.
91. W.W. Briner, M.D. Francis, J.S. Widder, The control of dental calculus in experimental animals. Int. Dent. J., 1971; **21**: 61–72.
92. M.D. Francis, R.G.G. Russell, H. Fleisch, Diphosphonates inhibit formation of calcium phosphate crystals in vitro and pathological calcification in vivo. Science, 1969; **165**: 1264–6.
93. R. Schenk, P. Eggli, R. Felix, et al., Quantitative morphometric evaluation of the inhibitory activity of new aminobisphosphonates on bone resorption in the rat. Calcif. Tissue Int., 1986; **38**: 342–9.
94. S.M. Peck, H. Rosenfeld, The effects of hydrogen ion concentration, fatty acids and vitamin C on the growth of fungi. J. Invest. Dermatol., 1938; **1**: 237–65.
95. E.L. Keeney, E. Lankford, L. Ajello, Bacteriostatic and bactericidal effects of fatty acid salts in broth cultures. Bull. Johns Hopkins Hosp., 1945; **77**: 437–9.
96. E.J. Foley, F. Hermann, S.W. Lee, The effects of the pH on the antifungal activity of fatty acids and other agents. Preliminary report. J. Invest. Dermatol., 1947; **8**: 1.
97. R. Tigerstedt, P.G. Bergman, Niere und Kreislauf. Scand. Arch. Pharmacol., 1898; **8**: 223–71.

98.   H. Goldblatt, J. Lynch, R.F. Hanzal, W.W. Summerville, Studies on experimental hypertension. *J. Exp. Med.*, 1934; **59**: 347–9.

99.   I.H. Page, The discovery of angiotensin. *Persp. Biol. Med.*, 1975; **18**: 456–62.

100. L.T. Skeggs, W.H. Marsh, J.R. Kahn, N.P. Shumway, The purification of hypertension I. *J. Exp. Med.*, 1954; **100**: 363–70.

101. W.S. Peart, The isolation of a hypertension. *Biochem. J.*, 1956; **62**: 520–7.

102. L.T. Skeggs, W.H. Marsh, J.R. Kahn, N.P. Shumway, The preparation and function of the hypertensin-converting enzyme. *J. Exp. Med.*, 1956; **103**: 295–9.

103. W.S. Peart, The renin–angiotensin system. *Pharmacol. Rev.*, 1965; **17**: 143–82.

104. K.K.F. Ng, J.R. Vane, Fate of angiotensin I in the circulation. *Nature*, 1968; **218**: 144–50.

105. M.A. Ondetti, N.J. Williams, E.F. Sabo, *et al.*, Angiotensin-converting enzyme inhibitors from the venom of *Bothrops jararaca*. Isolation, elucidation of structure, and synthesis. *Biochemistry*, 1971; **10**: 4033–9.

106. S.H. Ferreira, L.J. Greene, V.A. Alabaster, *et al.*, Activity of various fractions of bradykinin potentiating factor against angiotensin I converting enzyme. *Nature*, 1970; **225**: 379–80.

107. H. Gavras, H.R. Brunner, J.H. Laragh, *et al.*, An angiotensin-converting enzyme inhibitor to identify and treat vasoconstrictor and volume factors in hypertensive patients. *N. Engl. J. Med.*, 1974; **291**: 817–21.

108. L.D. Byers, R. Wolfenden, Binding of the by-product analog benzylsuccinic acid by carboxypeptidase A. *Biochemistry*, 1973; **12**: 2070–8.

109. M.A. Ondetti, B. Rubin, D.W. Cushman, Design of specific inhibitors of angiotensin-converting enzyme: new class of orally active antihypertensive agents. *Science*, 1977; **196**: 441–4.

110. A.A. Patchett, E. Harris, E.W. Tristram, *et al.*, A new class of angiotensin-converting enzyme inhibitors. *Nature*, 1980; **288**: 280–3.

111. D.I. Friedman, G.L. Amidon, Intestinal absorption mechanism of dipeptide angiotensin converting enzyme inhibitors of the lysyl-proline type: lisinopril and SQ 29,852. *J. Pharm. Sci.*, 1989; **78**: 995–8.

112. J. Krapcho, C.Turk, D.W. Cushman, *et al.*, Angiotensin-converting enzyme inhibitors. Mercaptan, carboxyalkyl dipeptide, and phosphinic acid inhibitors incorporating 4-substituted pralines. *J. Med. Chem.*, 1988; **31**: 1148–60.

113. S. Klutchko, C.J. Blankley, R.W. Fleming, *et al.*, Synthesis of novel angiotensin converting enzyme inhibitor quinapril and related compounds. A divergence of structure–activity relationships for non-sulfhydryl and sulfhydryl types. *J. Med. Chem.*, 1986; **29**: 1953–61.

114. V. Teetz, R.Geiger, R.Henning, H.Urbach, Synthesis of a highly active angiotensin converting enzyme inhibitor: 2-[N-[(S)-1-ethoxycarbonyl-3-phenylpropyl]-L-alanyl]-(1S,3S,5S)-2-azabicy-clo[3.3.0]octane-3-carboxylic acid (Hoe 498). *Arzneimittel.-Forsch.*, 1984; **34**: 1399–401.

115. M. Vincent, G. Rémond, B. Portevin, *et al.*, Stereoselective synthesis of a new perhydroindole derivative of chiral iminodiacid, a potent inhibitor of angiotensin converting enzyme. *Tetrahedron Lett.*, 1982; **23**: 1677–80.

Antibiotics

In 1877, Louis Pasteur and Jules Joubert showed that animals inoculated with a mixture of anthrax bacilli and common bacteria did not contract anthrax.[1] Pasteur went on to suggest that microbes liberated materials that might be exploited therapeutically and soon after the Viennese pathologist Victor Babès detected the secretion of substances from one bacterial species that could kill those of another.[2,3] Ten years later, Garré demonstrated an antibiotic effect arising from diffusion when two different species of bacteria were streaked closely in parallel across an agar plate.[4] That same year, Rudolf Emmerich was giving a demonstration to his class at the University of Munich when he found that a guinea pig he had innoculated with *Vibrio cholera* did not develop cholera. It transpired that the animal had previously been injected with *Streptococcus erysipelatis*. He followed this up by showing that anthrax could be experimentally prevented through prior inoculation with a culture of *Bacillus anthracis*.[5]

PYOCYANASE

Eduard von Freudenreich reported from the Pasteur Institute in 1888 that typhoid bacilli often failed to grow in filtered broths in which bacteria had previously been cultured.[6] He went on to demonstrate that *Pseudomonas aeruginosa* (then known as *Bacillus pyocyaneus*) was particularly effective in antagonising the growth of typhoid bacilli and some other bacteria. The following year, Bouchard showed that inoculation of rabbits with *P. aeruginosa* protected them against anthrax.[7] Ten years later, Rudolf Emmerich and Oscar Löw began treating hundreds of patients with pyocyanase, a powdered material that they mistakenly believed to be an enzyme.[8] It was probably obtained by allowing cultures of *P. aeruginosa* to incubate for six weeks, during which time the bacteria decomposed and liberated the antibiotic into solution. Emmerich never disclosed how he prepared his pyocyanase, so most subsequent workers had to purchase commercially prepared material. It was claimed that this destroyed a variety of pathogenic bacteria, including the causative organisms of diphtheria, anthrax, plague and typhoid. Although the therapeutic use of pyocyanase became a controversial matter, its use continued until 1913 when the commercial product suddenly ceased to have any activity. Investigations continued and it was found that the active material was highly lipophilic in nature. Edward Doisy and his colleagues at St Louis University School of Medicine in 1945 isolated five lipophilic antibacterial substances from cultures of *P. aeruginosa* active mainly against Gram-negative bacteria.[9] Three of them were subsequently identified by Ibert Wells at Syracuse University, New York, as 2-heptyl, 2-nonyl and 2-(Δ^1-nonenyl-)-4-hydroxyquinoline.[10] The arrival of penicillin meant that there was no further interest in them.

TYROTHRICIN

At the Rockefeller Institute in New York in 1939, Rene Dubos isolated a bactericidal, protein-free extract from *Bacillus brevis*.[11] This material, tyrothricin, was soon shown to consist mainly

Drug Discovery. A History. W. Sneader.
©2005 John Wiley & Sons Ltd

of a cyclic decapeptide called tyrocidin, together with a similar compound called gramicidin S, which was 50 times as potent.[12] Their structures were elucidated by Richard Synge at the Lister Institute in London using paper chromatography in one of its earliest applications.[13,14] Although both components were able to protect mice against pneumococci, they were too toxic for general use.

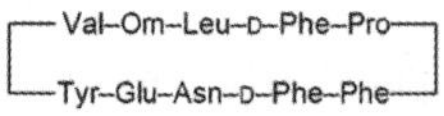

```
┌─ Val–Orn–Leu–D–Phe–Pro ─┐
│                         │         tyrocidin
└── Tyr–Glu–Asn–D–Phe–Phe ─┘
```

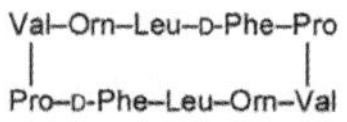

```
Val–Orn–Leu–D–Phe–Pro
 │                   │            gramicidin S
Pro–D–Phe–Leu–Orn–Val
```

Tyrothricin was suitable only for topical application. It was marketed in the USA by Sharp & Dohme in 1942 for treatment of Gram-positive infections. Since then, it has been universally used in throat lozenges as a non-prescription antibiotic, but the commercial success of this type of product has been largely due to the incorporation of benzocaine, a topical anaesthetic that soothes sore throats.

FUNGAL ANTIBIOTICS

The first scientific report of the antimicrobial activity of the familiar green *Penicillium* mould found on oranges or jam appeared in 1870. John Burdon-Sanderson of St Mary's Hospital in London, who was one of the first British physicians to accept Pasteur's ideas on the germ theory of disease, noted that bacteria did not grow and produce turbidity in sterilised culture solutions that had become contaminated by an air-borne *Penicillium* mould.[15] A series of experiments confirmed this, although Burdon-Sanderson had misinterpreted his observations by concluding that only fungi, and not bacteria, could cause aerial contamination of culture solutions.[16]

In January 1895, Vicenzo Tiberio of the Sanitorio Militare Marittimo Hospital in Naples published an account of work on the antibacterial properties of moulds.[17] An extract of a mould that he identified as *P. glaucum* had inhibited the growth of staphylococci and other pathogenic bacteria. He gave a detailed description of its action on infected rabbits, guinea pigs and rats. The following year, his countryman Bartolomeo Gosio isolated a crystalline product from a mould that he thought was *P. glaucum*.[18] There was no connection between these two investigations in Italy, for Gosio had been examing fungal growth on spoiled maize in an attempt to identify the cause of pellagra. Nevertheless, since the chemical properties of the crystalline compound indicated it was a phenol, he tested its antiseptic properties on cultures of the anthrax organism. This showed it to be an inhibitor of their growth. Lack of material prevented him from carrying out animal experiments, but the phenolic crystalline antibiotic was named 'mycophenolic acid' in 1913. It seems probable that the early claims of antibacterial activity among varieties of *Penicillium* moulds relate to this compound, the species described as *P. glaucum* most likely really being *P. brevi-compactum*.

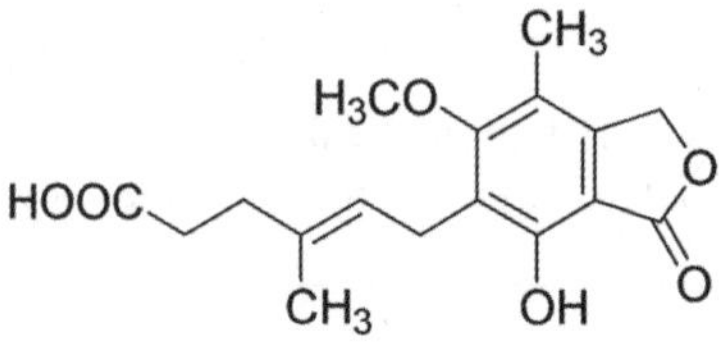

mycophenolic acid

mycophenolate mometil

Several more reports of antibacterial activity associated with *P. glaucum* were published before mycophenolic acid was investigated by Harold Raistrick at the Department of Biochemistry at the London School of Hygiene and Tropical Medicine in the early 1930s. Its structure was elucidated in 1952, and found to be chemically unrelated to benzylpenicillin.[19] Although mycophenolic acid was too toxic for clinical application as an antibiotic, in 1969 it was shown to be a non-competitive inhibitor of inosine monophosphate dehydrogenase (IMPDH), causing cessation of purine synthesis.[20] This led to its clinical evaluation as an immunosuppressant in transplant surgery[21] and psoriasis.[22] When taken by mouth, the morpholinoethyl ester, known as 'mycophenolate mometil', had enhanced bioavailability.[23] It is a prodrug that releases mycophenolic acid after absorption from the gut. It has become accepted as a useful immunosuppressant.

Penicillins

Alexander Fleming, director of the Inoculation Department at St Mary's Hospital in Paddington, London, was extremely fortunate when a remarkable combination of events occurred in the summer of 1928 and permitted him to discover the antibacterial activity of penicillin. The probable sequence of these events has been reconstructed in an account written by his former assistant, Ronald Hare.[24] Firstly, on the floor beneath Fleming's laboratory a colleague worked with moulds required for the production of vaccines to treat allergies, and it seems likely that one of these was wafted through the air into Fleming's laboratory to settle on a petri dish covered with a layer of agar impregnated with staphylococci. Secondly, this mould was a rare strain of *Penicillium notatum* that produced significant amounts of penicillin. Thirdly, Fleming left his culture plate on his work bench instead of placing it in an incubator at body temperature to ensure bacterial growth. Fourthly, an exceptionally cool spell followed when Fleming went on holiday at the end of July, which favoured growth of the mould in preference to that of the staphylococci. Fifthly, the climatic conditions changed later in the month, by which time the mould had produced sufficient penicillin to kill bacteria in its vicinity. This rise in temperature allowed colonies of staphylococci to grow elsewhere on the culture plate, thus enabling Fleming to observe a zone of inhibition of staphylococcal growth when he returned to the laboratory on 3 September. The original plate is kept in the British Museum, Fleming having treated it with formaldehyde vapour to preserve it.

As a leading authority on antiseptics, Fleming was ideally qualified to make a realistic assessment of the significance of the effect of the mould on bacterial growth. He prepared subcultures of the mould and gave the name 'penicillin' to the filtrate of the broth in which these had been grown for one or two weeks at room temperature. Two assistants, Frederick Ridley and Stuart Craddock, were given the task of preparing sufficient quantities of penicillin for bacteriological studies. They were also asked to obtain information about the chemical nature of the active substance, this initially being assumed to be an enzyme. They devised an efficient process for growing the mould in large, flat-sided bottles from which the juice below the surface growth was drained and filtered. As boiling destroyed the activity of the mould juice, acidification and evaporation at 4 °C under reduced pressure was used. The resulting concentrate was taken up in alcohol, in the process of which much of it was precipitated. The activity was found in the alcohol, ruling out the likelihood of penicillin being an enzyme. The alcohol extract was quite potent, but its activity gradually disappeared over a period of weeks. This later discouraged Fleming from pursuing the therapeutic possibilities of penicillin.

Fleming carried out tests to find which types of bacteria were sensitive to the penicillin extract. He established that it could lyse a variety of major pathogens, but the marked insensitivity of *Haemophilus influenzae* to it particularly interested Fleming. This had proved to be a difficult organism to isolate from infected patients since cultures of it were readily overgrown by other bacteria, preventing the preparation of vaccines. At that time, many believed (wrongly) that this organism was the cause of influenza and other respiratory conditions. In an attempt to kill off interfering bacteria, Fleming incorporated penicillin in cultures of *H. influenzae* prepared from throat swabs. His ploy worked admirably, and for the next ten years or so was the principal purpose for which penicillin was used. Indeed, the title of Fleming's first paper on penicillin was 'Cultures of a *Penicillium*, with special reference to their use in the isolation of *B. influenzae*'.[25]

In December 1928, Fleming examined the effect of penicillin on slides of bacteria growing in the presence of blood or serum, in the presence of which practically every known antiseptic had been rendered worthless, other than for topical application where sufficiently high concentrations could be achieved. When this test indicated that the activity of penicillin was diminished, he no longer expected it to be active against generalised infections or those in deep wounds. He subsequently learned from Craddock that penicillin injected into a rabbit disappeared from its blood within 30 minutes, which was most discouraging as it had been established that penicillin required around four hours to act. Fleming went on to investigate the potential of penicillin as an antiseptic for topical application. He began by treating Craddock's infected nasal antrum with penicillin, but to no avail. A second disappointment followed when it failed to cure an infected amputation stump and the patient died from septicaemia. Fleming did have one success with penicillin when he applied it to the eye of another of his assistants who had contracted a pneumococcal conjunctivitis. This time the infection rapidly cleared.

At the beginning of April 1929, Fleming and Craddock incubated the organs of a newly killed rabbit in a liquid culture of staphylococci for 24 hours and then transferred them to a penicillin solution for a similar period. Examination of slices of the organs revealed that staphylococci which penetrated deep within the organs had survived. It was concluded that penicillin could not penetrate beyond the surface of organs. This was a misleading interpretation since it failed to consider the possibility of circulating blood being able to carry penicillin into the tissues. It cast further doubt on the clinical value of penicillin. Unfortunately, Fleming never administered penicillin to infected mice and it was left to others to discover its outstanding therapeutic efficacy. When he did inject penicillin into a rabbit and a mouse, it was solely to confirm its lack of toxicity.

After submitting his paper to the *British Journal of Experimental Pathology*, Fleming carried out little further research on the clinical potential of penicillin other than occasionally applying it locally to treat infections. His subsequent researches with it were mainly as a reagent in bacteriological investigations. As a biographer of Fleming has pointed out, this ensured that cultures of the original strain of *P. notatum* were available when requested by laboratories, including that of the School of Pathology at Oxford.[26]

Harold Raistrick at the London School of Hygiene and Tropical Medicine assigned Percival Clutterbuck to the isolation of penicillin. The bacteriological assays were carried out by Reginald Lovell. Even though Raistrick spoke to Fleming on the telephone several times, he was never told about Ridley and Craddock's isolation procedure. Eventually, an ether extract of penicillin was obtained, but evaporation of the solvent resulted in the loss of its activity.[27] Baffled by this, Raistrick abandoned the project since there was no good reason to deploy the extensive resources that would be required to pursue the investigation further.

The next attempt to purify penicillin came from Fleming's laboratory in 1934. By now Ridley and Craddock had both left the Inoculation Department, but Holt, a chemist, had joined the staff. He made quick progress, thanks to Clutterbuck's discovery that penicillin could be extracted into organic solvents from slightly acid solutions. This enabled him to recover penicillin from an amyl acetate extract by partitioning it with a very slightly alkaline solution, a procedure that had not been tried by anyone else. The instability of penicillin in this solution led him to abandon his work on it after only a few weeks, without publishing his findings.

In the summer of 1938 at the Sir William Dunn School of Pathology at Oxford, Ernst Chain became interested in naturally occurring antibacterial substances. Penicillin was one of three substances that particularly interested him as it seemed likely to be an enzyme. Through a stroke of good fortune he then encountered a colleague carrying a mould dish along a corridor and learnt that it contained penicillin supplied by Fleming. He now believed a detailed study of how it lysed the bacterial cell wall would afford useful information on the structure of this wall, but he had no reason to believe that it would be of any particular therapeutic value. He was encouraged to pursue this idea by Howard Florey, an Australian who held the Chair of Pathology at Oxford.[28] Florey had encountered penicillin in 1932, when a colleague employed it in three cases of skin infections. Florey, Chain and their colleagues published an account of the subsequent work on penicillin at Oxford in 1949.[29] Since then, books by Hare,[24] Macfarlane,[26] Clarke[30] and others have enabled a detailed picture to be assembled about the momentous events that were to be of so much benefit to mankind, yet which took place during the darkest days of Western civilisation as war again swept across Europe.

Chain used a culture of Fleming's original mould provided by a colleague who had propagated it at Oxford for several years after it had been obtained for use as a bacteriological reagent. Unaware of the previous work by Holt, Chain rediscovered the solvent extraction procedure. While doing this, he realised that the physical properties of penicillin indicated that it could not be an enzyme as he had anticipated. It seemed to be a small molecule.

Work reported in 1939 from the Rockefeller Institute in New York on tyrothricin gave an added dimension to the research on penicillin. With Chain's research grant about to expire, Florey now sought financial support for the penicillin project on the grounds that penicillin was active against pathogenic bacteria. He argued that since previous workers had used only crude penicillin, every effort should be made to obtain it in pure form so that the effects of injecting it intravenously could be assessed. Thus the ultimate discovery of the value of penicillin actually depended upon that of an earlier antibiotic of questionable therapeutic benefit – and an inspired Australian!

In October 1939, the Medical Research Council awarded Chain an annual grant of £300 plus £100 for materials, this to be for each of three years. Shortly after, the Rockefeller Foundation awarded Florey a magnificent annual grant of $5000 for five years, plus an initial sum to purchase equipment. This large sum guaranteed the viability of the project and must rank as the finest investment ever made by a charitable foundation. Of course, it had also been the work on tyrothricin at the Rockefeller Institute that led to the potential clinical role of penicillin being evaluated. This transatlantic co-operation was soon to progress much further.

The Rockefeller Foundation also awarded a fellowship to Norman Heatley, whose ability to improvise apparatus for the large-scale production of penicillin was to prove invaluable. With support for the project secured, work on penicillin proceeded at a fast pace under Florey's skilled administration, with the entire staff of the School of Pathology becoming involved. Heatley put considerable effort into finding the best means of producing mould juice and recovering active material from it. It was not, however, until 1941 that Chain established the superiority of amyl acetate as an extraction solvent; this had also been previously discovered by Holt. This solvent then replaced ether in Heatley's mechanised countercurrent extraction system, which involved transference of penicillin in downward-flowing streams of filtered, acidified mould juice into upward-flowing streams of the immiscible solvent in the same glass tubes. A reversal of the procedure was utilised for the back-transfer of the penicillin to slightly alkaline, aqueous solution. Recovery of penicillin from this had thwarted Holt in 1934, but the following year the process of freeze-drying had been developed in Sweden and was now introduced at Oxford. It enabled Heatley to recover a brown, dry powder containing around 5 Oxford units per mg, which, although far from pure, was suitable for biological studies. By mid-March 1940, Chain had in his possession a supply of about 100 mg of this powder. A substantial portion was then injected intraperitoneally into two mice, with no ill effects being observed. From here on, Florey took over the complex biological, toxicological and clinical investigations, gambling the entire resources of his department on the production and purification of penicillin, while having no proof that it would be effective when injected.[31] It had only been proven to be a local antiseptic and practically all antiseptics were ineffective when injected. Neither Fleming, Raistrick nor Holt had ventured further, but a major development had occurred since they worked on penicillin. In 1935, I.G. Farben introduced Prontosil®, the synthetic antibacterial sulfonamide that was the first drug ever to cure systemic infections. Its success had revolutionised medical thinking, else Florey would never have seriously considered proceeding as he did.

The crucial experiment with penicillin took place on Saturday 25 May 1940. At 11 am, Florey injected each of eight mice with a lethal amount of virulent streptococci, a technique used by Gerhardt Domagk during the development of Prontosil®. At noon, two mice were injected subcutaneously with 10 mg of penicillin and two others received half this amount. These latter two received four more similar injections during the next ten hours. Heatley stayed in the laboratory that night and watched the untreated mice die. One that received a single dose of penicillin died two days later, while all the others survived. This greatly encouraged all those engaged on the project, and they proceeded to conduct extensive biological studies with the crude powder. In July, it was confirmed that penicillin was also effective in mice infected with either staphylococci or *Clostridium septicum*, the causative organism of gas gangrene. The results of the investigations were published in the *Lancet* on 24 August 1940.[32] Only brief details were given, but the fact that it appeared at all is evidence of the lack of importance attached to penicillin in comparison with projects considered to be of military importance, such as a major one concerning antimalarials. Ten days after this paper appeared Fleming visited Oxford to make his first contact with Florey's department. Florey must have been disappointed that this was the sole response to his paper, for he had hoped that publication of his preliminary findings might impress the pharmaceutical industry. Earlier that summer, he had approached Burroughs Wellcome to ask if they would produce penicillin on a large enough scale for him to begin clinical trials. This offer was declined because their facilities were stretched to the limit in trying to prepare blood plasma and also meet the requirements of the armed forces for vaccines. It was felt that penicillin would not be as important for the war effort.

By January 1941, the School of Pathology had become a factory for the production of penicillin. Chain and Abraham now tried the new technique of adsorption chromatography, which enabled penicillin to be adsorbed on to a column of powdered alumina after a solution of it was poured in. The impurities passed down through the column with the solvent, but penicillin was retained on the alumina. It was recovered by solvent extraction of the alumina, the material obtained in this manner having an activity of 50 units per mg. As a result of this development, a single intravenous injection of 100 mg of penicillin was administered to a woman dying of cancer. The sole adverse reaction was a bout of shivering followed by a fever, which was due to impurities that were then removed by chromatography. Further studies on volunteers provided valuable information about the best way of administering penicillin. For example, it was learned that penicillin was destroyed by the acidity of gastric juice, so could not be given by mouth. Since the kidneys were found to excrete penicillin rapidly from the body, it had to be given by slow intravenous drip in order to maintain adequate bactericidal levels in the blood and tissues.

The first attempt to treat a patient with a life-threatening infection took place at the Radcliffe Infirmary, Oxford, on 12 February 1941. The patient was a 43 year old policeman with a mixed staphylococcal and streptococcal infection that had spread throughout his body and had already necessitated removal of his eye. He had not responded to sulfapyridine, yet only 24 hours after receiving 200 mg of penicillin intravenously, followed by 100 mg every three hours, his condition improved rapidly. On the third day of treatment, the supply of penicillin ran out. More was obtained from the School of Pathology, where all the urine voided by the patient had been extracted to recover the penicillin. By this expedient, it proved possible to continue the injections for a further three days. The policeman's health remained good for the next ten days, but a residual lung infection then flared up and he died on 15 March. Before his death occurred, two other patients responded to penicillin treatment. The fourth was a child with a severe infection behind the eye. This disappeared after large doses of penicillin were administered, but the child died when an artery in the brain ruptured as a result of damage caused prior to treatment with penicillin. Two other very ill patients recovered.

The second paper on penicillin appeared in the *Lancet* on 16 August 1941.[33] It gave details of penicillin production, animal results and clinical reports. By the time it was published, Florey and Heatley were in the United States seeking to arrange large-scale production of penicillin for extensive clinical trials.[34] Before departing, Florey had approached the Boots Company and ICI, but nothing had come of this. When the Rockefeller Foundation offered to pay the expenses for Florey and a colleague to visit the United States to discuss penicillin production, Florey referred the matter to the secretary of the Medical Research Council. He was advised to proceed with the visit as no British manufacturer was in a position to produce penicillin because of the wartime situation.

Soon after their arrival in the United States, Florey and Heatley met Charles Thom at the US Department of Agriculture, who told them that the only way of producing large quantities of penicillin was in deep fermentation tanks similar to those used by brewers. He put Florey in touch with Robert Coghill, head of the fermentation division at the Department of Agriculture's Northern Regional Research Laboratory in Peoria, Illinois. Coghill agreed to attempt to grow the *Penicillium* in deep culture tanks if Heatley would work on the project for several months. Production of penicillin began the following day, using a culture Florey had brought from Oxford. The yield of penicillin was increased twelvefold within six weeks by including corn steep liquor in the culture medium. This was a syrupy waste product from the manufacture of corn starch. A further improvement was strict control of the acidity of the culture to minimise penicillin degradation.

A few weeks after arriving in the United States, Florey met the research directors of Merck and Company, E.R. Squibb and Sons, Charles Pfizer and Company and Lederle Laboratories. His mission was greatly assisted by the presentation of a paper on penicillin a few months earlier by Martin Dawson, Karl Meyer and Gladys Hobby of Columbia University College of Physicians and Surgeons. This had been given at the annual conference of the American Society for Clinical Investigation, an abstract being published in their widely read journal.[35] The paper described how the authors had grown cultures of penicillin obtained from the Oxford team and briefly outlined clinical results on testing a crude preparation on patients. Several American companies had also conducted exploratory work on penicillin before the arrival of Florey and Heatley.[36]

Merck agreed to proceed with penicillin production at once and also to exchange information with other interested parties, but the other companies reserved their position. After further meetings, Squibb and Pfizer joined in the collaborative effort. Another consortium calling itself the Midwest Group was formed when Abbott Laboratories, Eli Lilly and Company, Parke, Davis and Company and the Upjohn Company also agreed to exchange information on penicillin. Wyeth Laboratories took up penicillin production near Philadelphia by growing the mould in cellars where mushrooms had previously been cultivated for the gourmet market. They became the largest producer of penicillin until deep fermentation processes were introduced.

The first large-scale clinical use of penicillin in America was unrehearsed. On the night of 28 November 1942, over 500 people perished in a disastrous fire at the Coconut Grove night club in Boston. As soon as it was known that there were 220 badly burned casualties, the Committee on Medical Research authorised the release of supplies of penicillin in an attempt to reduce the anticipated mortality among the survivors. The drug exceeded all expectations, but the public were not told since penicillin was classified as a US military secret.

At Peoria, intensive effort was put into finding a strain of the *Penicillium* mould that would grow in deep tanks and also deliver a higher yield of penicillin. The US Army Transport Command delivered thousands of samples from all over the world, either in the form of soil or as cultures of soil organisms. Despite this effort, one of the best improvements in penicillin yield was obtained with a strain of *P. chrysogenum* growing on a cantaloup melon from the fruit market in Peoria! As part of a collaborative effort involving Minnesota, Stanford and Wisconsin Universities, as well as the Carnegie Institution at Cold Spring Harbor, this strain was irradiated with X-rays, producing mutants that provided even higher yields of penicillin. At the University of Wisconsin, one of the tens of thousands that were to be examined was found to produce 500 units of penicillin per millilitre. This became the standard strain for wartime production of penicillin in America.

It took two years of intense effort before large amounts of penicillin could be produced in deep fermentation tanks containing 12 000 gallons of mould juice. When this was eventually achieved, the impact was staggering. During the first five months of 1943, American manufacturers had delivered 400 million units of penicillin, but during the next seven months 20 000 million units were produced by deep culture. In January 1944, Charles Pfizer and Company alone prepared 4000 million units, yet by the end of that year they had become the world's largest producer of penicillin, turning out 100 000 million units a month! When the invasion of occupied France began in June 1944, enough penicillin was available to satisfy all military requirements.

Following approaches from Fleming, the British Ministry of Supply established a Penicillin Chemical Committee late in 1942 to co-ordinate industrial manufacture. Boots, Burroughs Wellcome, Glaxo, ICI, Kemball Bishop and May and Baker eventually produced penicillin using the surface culture method. Nevertheless, when in March 1943 Florey and his wife published the results in 187 cases, most of the penicillin had been made at Oxford. Only enough had been available to treat 17 of these patients by intravenous injection; the others received the drug by direct local application. Several pharmaceutical companies in both Britain and the United States had been reluctant to commit themselves to the production of penicillin by a fermentation process. As recently as 1938, the Lederle division of the American Cyanamid Company had lost millions of dollars when a plant for the production of pneumonia vaccines was rendered obsolete in only eight months by the introduction of sulfapyridine. In the 1940s there was a general expectation that penicillin would shortly be synthesised and produced on as large a scale as the sulfonamides. In the event this was not to be so and Merck, the company that made the largest commitment to penicillin synthesis, lost out to rivals who stuck to fermentation processes.

Preliminary work towards determination of the chemical structure of penicillin was initiated when Robert Robinson and his colleagues at the Dyson Perrins chemistry laboratory at Oxford joined forces with Florey's group early in 1942. Their first investigations had to be carried out with penicillin that was only about 50% pure. Further purification by means of crystallisation was essential if the structure was to be determined. Oskar Wintersteiner and his group at the Squibb Laboratories achieved this in the summer of 1943. After the news of this reached Oxford, Florey's group managed to crystallise their penicillin, only to find it was different to that isolated in the United States. This led to the realisation that there were variant forms of the antibiotic, with different side chains. The Oxford material was named penicillin F (it was shown to be 2-pentenylpenicillin). The American product obtained from deep fermentation was designated as penicillin G. This was the one that came into routine clinical use, later receiving the approved name of 'benzylpenicillin'. Five more variant penicillins were identified the following year.

The Committee on Medical Research set up a project to deal with penicillin synthesis, sponsoring research in American universities, independent foundations and industrial laboratories. Roger Adams of the University of Illinois was in overall charge. In the United Kingdom, the Committee for Penicillin Synthesis was established, permitting a formal exchange of information between British and American scientists while maintaining strict secrecy about the chemical nature of penicillin.[37]

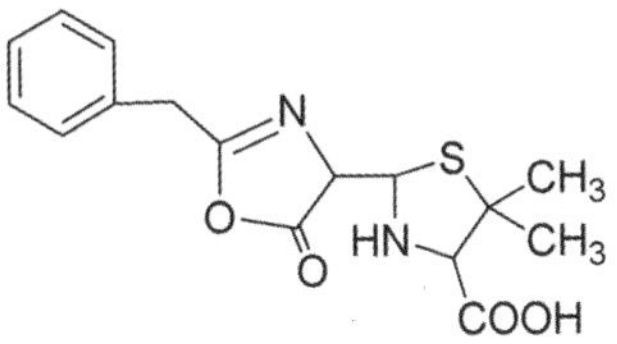

oxazolone–thiazolidine structure

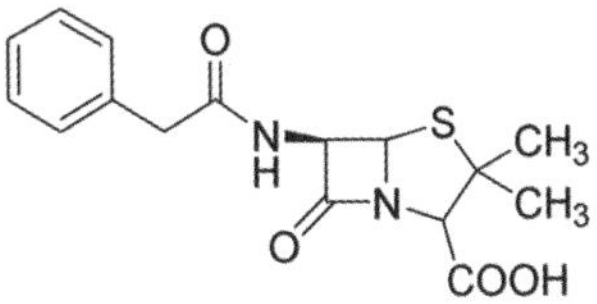

benzyl-penicillin

By late 1943 chemists at Oxford and the Merck laboratories had concluded that penicillin had one of two possible chemical structures. One was a five-membered oxazolone ring joined to a thiazolidine ring, while the other had a rare four-membered beta-lactam ring fused at two points to the thiazolidine ring. Attempts were made to synthesise each of these compounds, with more than a thousand chemists in 39 university and industrial laboratories being involved. Most of their efforts were in vain, but the Merck and Oxford groups did obtain trace amounts of synthetic penicillin. This did not settle the issue of which chemical structure was correct, as the possibility of interconversion existed. This was resolved in 1945 when Dorothy Hodgkin at Oxford employed X-ray crystallographic studies to confirm that penicillin contained the beta-lactam ring system.[38] However, with the ending of the war and the winding up of the collaborative programmes, all the laboratories decided to abandon the synthetic work.[39]

In 1948, benzylpenicillin was formulated as its sparingly soluble procaine salt for use as a depot intramuscular injection.[40] It provided effective levels in the tissues for up to 24 hours and became the favoured form of penicillin for the treatment of syphilis, rendering the arsenicals obsolete.

Cephalosporins

Following the clinical introduction of penicillin, Guiseppe Brotzu, the director of the Instituto d'Igiene in Cagliari, Sardinia, began to search for antibiotic-producing organisms. Believing that the self-purification of sea water near a local sewer might be due in some measure to microbial antagonism, Brotzu sampled the water and isolated a mould, *Cephalosporium acremonium*, which inhibited the growth of typhoid bacilli and other pathogens growing on agar plates. He then prepared hundreds of subcultures of the mould until he had isolated a strain that conferred high antibacterial activity on filtrates of mould juice. Adding alcohol to an extract from the juice precipitated inactive material, leaving a solution that had a wider spectrum of antibacterial action than penicillin and active against Gram-negative as well as Gram-positive bacteria. When preliminary clinical studies were carried out by direct application of filtered mould juice to boils and abscesses caused by staphylococci and streptococci, the results were most encouraging. The concentrated extract was then injected into patients with typhoid, paratyphoid and brucellosis, again with good results. Brotzu published his findings in a pamphlet describing the work being conducted at the Instituto d'Igiene.[41] This did not receive a wide circulation, but a copy was passed to Florey, resulting in a culture of *Cephalosporium* being sent to Oxford in September 1948.

Florey made arrangements for the mould to be grown in deep culture tanks at the Medical Research Council's Antibiotic Research Station at Clevedon in Somerset. At Oxford, Heatley then extracted an acidic antibiotic from the *Cephalosporium* mould juice. By July 1949, Edward Abraham and H.S. Burton[42] had isolated an antibiotic from Heatley's solvent extract, but were disappointed to find that it was active only against Gram-positive bacteria. Accordingly, they designated it 'cephalosporin P'. Its chemical structure was not established until 1966, after it had been shown to consist of at least five components, of which the major one was designated cephalosporin P$_1$.[43]

cephalosporin P$_1$

penicillin N

A second antibiotic produced by *Cephalosporium* spp. growing in sewage was detected in mould juice that had already been extracted with organic solvent.[44] Since this had the activity of Brotzu's original material, it was named cephalosporin N, reflecting its activity against Gram-negative as well as Gram-positive organisms. It proved difficult to isolate, but Abraham and Guy Newton showed it to be an unstable penicillin and renamed it penicillin N.[45] Shortly after, Newton assisted the staff of the Antibiotic Research Station to increase the yield of the antibiotic, thus finally permitting its isolation in fairly pure form by Abraham in 1955.[46] It turned out to have only a hundredth of the activity of benzylpenicillin against Gram-positive bacteria, but was much more active against Gram-negative organisms. It now became evident, for the first time, that the nature of the side chain in a penicillin could have a marked effect on the spectrum of antibacterial activity. In the late 1950s, Abbott Laboratories prepared sufficient supplies of penicillin N for its clinical value in typhoid to be established. The results were satisfactory, but the introduction of the semi-synthetic penicillins rendered it obsolete.

While completing degradation studies to confirm their proposed chemical structure for penicillin N, Newton and Abraham separated three contaminants from a crude sample of the antibiotic. The third of these was isolated as its crystalline sodium salt. Surprisingly, it exhibited weak antibiotic activity, and so was given the name 'cephalosporin C'. Abraham and soon Newton discovered that it was chemically related to benzylpenicillin, but had a much greater ability to withstand the destructive action of beta-lactamases produced by bacteria. It was also less toxic, holding out the promise of being able to inject large doses to destroy staphylococci that had become resistant to penicillin by producing beta-lactamases.

Major problems had to be overcome before cephalosporin C could be produced on a scale large enough to meet the anticipated demand. These were overcome at the Antibiotic Research Station. Patents on cephalosporin C production were taken out by the National Research Development Corporation, which had been set up in 1949 to exploit discoveries from British universities and government laboratories.[47] Licences were issued to Glaxo Laboratories and several foreign manufacturers. These provided a rich dividend for the British taxpayer when cephalosporin C later became the principal starting material for the production of semi-synthetic cephalosporins.[48]

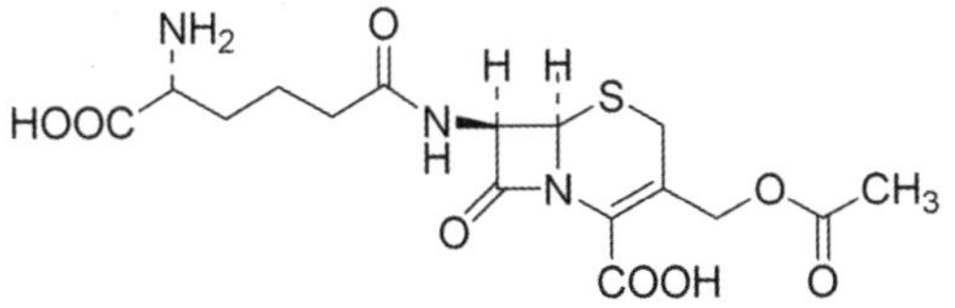

cephalosporin C

Plans to market cephalosporin C had to be scrapped in 1960 when Beecham Research Laboratories introduced methicillin, a penicillin that was resistant to beta-lactamases. The following year, Abraham and Newton elucidated its chemical structure.[49] Like benzylpenicillin, it possessed a sensitive beta-lactam ring, but fused to a dihydrothiazine rather than a thiazolidine ring. The future of the cephalosporins was assured when it was discovered that superior analogues could be prepared in much the same way as was to happen with the penicillins.

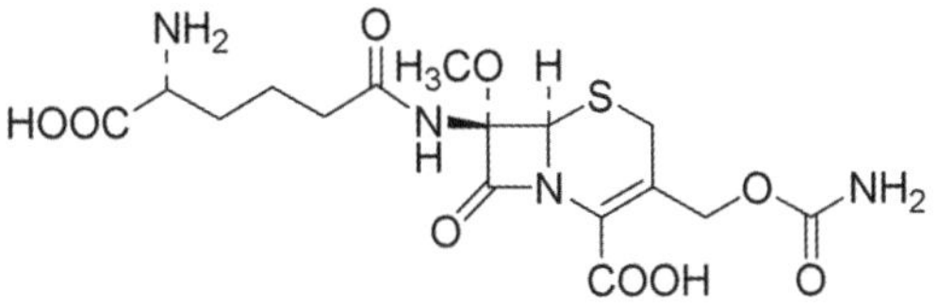

cephamycin C

Cephamycin C was first isolated from *Streptomyces clavuligerus* at the Merck Sharpe & Dohme Research Laboratories in 1971.[50] It became the most widely studied of the natural cephamycins since it was resistant to beta-lactamases. This enhanced its activity against certain Gram-negative organisms (other than pseudomonal strains) that were not susceptible to cephalosporins, though it lacked activity against Gram-positive bacteria.

Griseofulvin

The failure of newly planted conifers to grow on Wareham Heath in Dorset during the 1930s was attributed to the presence in the soil of a substance toxic to the fungi whose presence was essential for normal tree development. During the Second World War, Brian, Hemming and McGowan investigated this at ICI's Jealott Hill Research Station in Bracknell.[51] They discovered that the common soil microbes were largely absent from Wareham Heath soil and in their place was an abundance of *Penicillia*. When cultures of these were screened against a fungus known as *Botrytis allii*, a strain of *Penicillium janczewskii* produced a peculiar distortion of the germ tubes of the fungus, even at high dilutions. An active substance was extracted with chloroform and crystallised from alcohol, this being named 'curling factor' until identified as griseofulvin,[52] a mould metabolite previously isolated from *Penicillium griseofulvum* by Harold Raistrick, though not screened for antibiotic activity.[53]

The chemical structure of griseofulvin was elucidated in 1952 by John Grove and his colleagues at the Akers Research Laboratories of ICI.[54] Its stereochemistry was established later[55] and the first synthesis was reported in 1960.[56]

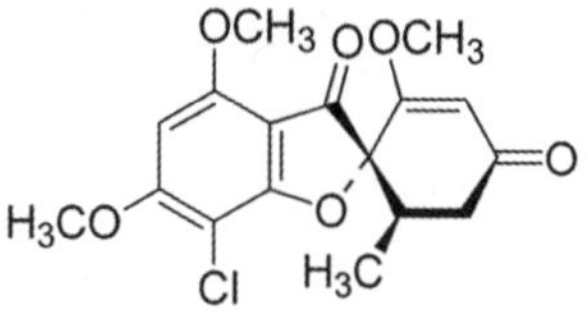

griseofulvin

ICI had intended to introduce griseofulvin for eradication of fungal diseases in plants, but its early promise was never fulfilled. Matters might have rested there had not a group of Polish scientists at the Municipal Hospital in Poznan published a report in November 1957 describing their spectacular success in treating fungal infections with salicylhydroxamic acid.[57] Their clinical studies broke new ground, for they took the unprecedented step of administering the compound by mouth. This was noted by James Gentles, a mycologist in the Department of Bacteriology at Anderson's College in the University of Glasgow, who had been seeking a remedy for fungal infection of the feet of coal miners. He administered griseofulvin by mouth to guinea pigs that had developed severe lesions after having been infected with *Microsporum canis*. The beneficial effects of the antibiotic were evident within four days.[58] Further studies revealed that griseofulvin could prevent and cure ringworm in cattle. Furthermore, when griseofulvin was administered by mouth over a prolonged period of time, it selectively concentrated in keratin and so was of value in the treatment of dermatophyte infections of skin or nails. Griseofulvin then became the first orally administered antifungal drug to be marketed when ICI and Glaxo introduced it in 1959, the latter having developed a fermentation plant that permitted economical production.

Ciclosporin

The Sandoz Company in Basle began testing fungal metabolites for cytostatic activity in 1957, employing two screening tests introduced by Hartmann Stähelin. These involved detection of the inhibition of growth of either murine P-815 mastocytoma cells or chick embryo fibroblasts. Thousands of filtered fungal cultures were examined, leading to several active compounds being isolated, including the cytochalasins, brefeldin A, verrucarin A, anguidine and chlamydocin.[59] Like many other fungal metabolites obtained by pharmaceutical companies, none of these ever found a place in medicine. However, in 1965 the sesquiterpene ovalicin was isolated from the culture fluid of *Pseudeurotium ovalis* and shown to have potent cytostatic activity. Shortly after this, Sandoz widened its interests to include immunology. Sandor Lazary then set up a screening programme to find an immunosuppressant drug with less toxicity than azathioprine. As ovalicin had reduced the weight of the spleen in mice, Stähelin asked for it to be tested in the mouse haemagglutinin test, which reflected the degree of antibody production. Ovalicin proved to be a potent immunosuppressant in this and a variety of other tests in animals. Significantly, it was not toxic to the cells of the bone marrow, which distinguished it from existing immunosuppressants. Unfortunately, when tested in humans it turned out to be too toxic and had to be abandoned.

Sandoz introduced a general screening programme in 1970, through which products prepared in its various laboratories could be submitted to a battery of screens for a broad range of activity. Screening for immunosuppressive activity was included, involving a novel procedure developed by Stähelin in which both activity against leukaemia L1210 cells and also a haemagglutination assay could be conducted in the same mouse. The latter involved injecting a mouse with red blood cells from sheep to produce an immune response. Compounds being screened for immunosuppressant activity were then injected intraperitoneally for four days. Nine days later, blood was taken from the mouse and the level of antibodies in the serum was measured.

Among 20 compounds submitted to Stähelin for screening in December 1971 was an extract from *Tolypocladium inflatum*, a new strain of fungi imperfecti that had been active when tested for antifungal activity. It was then produced in greater amounts and an extract was partially purified by Arthur Rüegger and his colleagues. This product was not active *in vivo* against pathogenic fungi, but when it was screened for immunosuppressive activity in the mouse the results were quite different. The antibody count in the haemagglutinin test was reduced by a factor of 1024, indicating a strong immunosuppressant effect, while there was no activity in the L1210 test. Further tests on P-815 mastocytoma cells were also negative. This placed the product in a unique category. All known immunosuppressant drugs had achieved their effects by a non-specific suppression of mitosis, not just in lymphocytes but in all rapidly dividing somatic cells. The indication now was that this product acted selectively on lymphocytes involved in the immune response.

The fermentation process was now improved and the individual components of the product were separated by column chromatography, resulting in the isolation of the pure peptides cyclosporin A and B[60] and the determination of their structures.[61] Since most of the immunosuppressant activity was associated with cyclosporin A, only it was used in further researches. It now has the approved name of 'ciclosporin'.

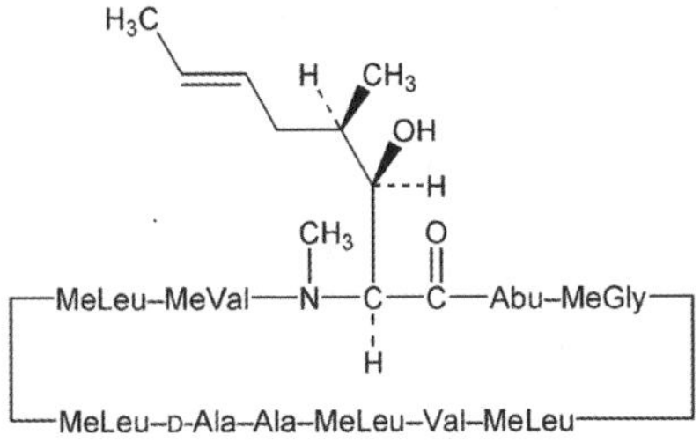

ciclosporin

The first report revealing the work that had been carried out at the Sandoz laboratories appeared in 1976.[62] Samples of the drug were then supplied to Roy Calne (who had been closely involved in the development of azathioprine) for clinical trial in patients receiving kidney transplants at Addenbrooks Hospital in Cambridge, and also to Ray Powles at the Leukaemia Unit in the Royal Marsden Hospital, London, where its role in bone marrow transplantation was to be assessed. Their findings appeared in 1978.[63,64] Since then, the value of ciclosporin in dealing with the commonest cause of death after transplantation, namely rejection, has been confirmed by its successful use in hundreds of thousands of patients. It has to be given for at least six months after transplantation, the main side effects involving the kidneys.

ACTINOMYCETAL ANTIBIOTICS

The actinomycetes are ubiquitous soil organisms that have features in common with both bacteria and fungi. In 1917, Greig-Smith[65] detected antibiotic substances diffusing from actinomycetes soil samples gathered in New South Wales. Rudolf Lieske then reported that while some actinomycetes lysed the cells of *Staphylococcus aureus* and *S. pyogenes*, they had no effect on other bacteria.[66]

At first unaware of the prior work of Lieske, André Gratia and his group at the Pasteur Institute in Brussels began publishing a series of papers in which substances diffusing from various species of *Streptothrix* and *Actinomyces* were employed to lyse bacteria in culture in order to liberate antigens.[67] The resulting 'mucolysates' were administered to patients in attempts to induce formation of antibodies that would produce immunity against the specific bacterium from which they had been prepared.[68] These experiments were not particularly successful. Maurice Welsch at the Rockefeller Institue later gave the name 'actinomycetin' to the soluble protein-like mucolysate.[69]

Streptomycin and Other Aminoglycoside Antibiotics

During his work leading to the isolation of tyrothricin, Rene Dubos had kept in contact with Selman Waksman, his former teacher in the Department of Microbiology at the New Jersey Agricultural Experiment Station in Rutgers University. As a result, Waksman became convinced that his own wide experience in dealing with soil microbes could be used to good effect in seeking further antibiotic-producing organisms. He chose to begin with actinomycetes, he and his students starting with a preliminary survey of these organisms. This confirmed findings of earlier investigators by establishing that, out of 244 freshly isolated cultures from soil, over 100 had antimicrobial activity, of which 49 were highly active. Waksman then used the pyocyanase isolation procedure to obtain from *Actinomyces antibioticus* a crystalline antibiotic known as actinomycin A, the first antibiotic ever isolated from an actinomycete.[70] After routine bacteriological tests, it was sent to the nearby laboratories of Merck and Company in Rahway for further investigation. Though considerable effort was spent in attempting to elucidate its chemical structure, and many animals were used to assess its toxicity and potential clinical scope, it proved too toxic for human application.

Waksman was now awarded an annual grant of $9600 from the Commonwealth Fund to support his search for antibiotics. Further funds were forthcoming from Mrs Albert Lasker. Waksman, like Florey and Ehrlich before him, thus received his main financial support for his chemotherapeutic research from charitable foundations. The similarity goes further, for all three subsequently witnessed their discoveries being rapidly brought to the clinic as a consequence of the readiness of major pharmaceutical firms to commit large sums of money in the exploitation of these discoveries.

In 1941, Waksman isolated two more antibiotics, clavacin (patulin) and fumigatin.[71] They were less toxic than actinomycin A, but still unsuitable for clinical application. When it became evident that penicillin was active only against Gram-positive bacteria, Waksman concentrated on the search for an antibiotic to treat Gram-negative infections. He appeared to have achieved a breakthrough when he isolated streptothricin, as it was lethal to bacteria unaffected by penicillin and appeared to be safe enough for human trials. Several industrial concerns were given cultures of the actinomycete that produced it, and pilot plant production was begun. Unfortunately, chronic toxicity testing revealed that several days after streptothricin was injected into animals a variety of toxic effects appeared as a result of kidney damage. Work on it had to be abandoned.

Early in 1943, Waksman decided to focus his activities on finding an antibiotic that could be used to treat tuberculosis, a major scourge that caused millions of deaths each year. The problem he faced was that the causative organism, *Mycobacterium tuberculosis*, grew very slowly, making the screening of cultures impracticable on the scale Waksman was operating on. Acting on a suggestion from his son, he instead screened his cultures against the faster growing *Mycobacterium phlei*, a non-pathogenic organism. Any cultures that were active were then examined further in animals infected with tuberculosis. Waksman also enriched his numerous soil samples with *Mycobacterium tuberculosis* in order to encourage elaboration of actinomycetes that produced an antituberculosis antibiotic.

After having cultured thousands of strains of actinomycetes since the start of the project, Waksman found what he wanted in September 1943. By a twist of fate, this was an actinomycete that he and his former professor had been the first to isolate 28 years earlier at Rutgers, namely *Streptomyces griseus*. Within four months, a new antibiotic had been isolated in a concentrated form, and its activity against *Mycobacterium tuberculosis* confirmed. It also proved effective against microbes causing plague, brucellosis and various forms of bacterial dysentery. Although its chemical properties turned out to be similar to those of streptothricin, the new antibiotic was free of renal toxicity. It was given the name 'streptomycin'.[72] The chemical structure was elucidated in 1947[73] and confirmed its aminoglycoside nature. It was later established that streptomycin and other aminoglycoside antibiotics disrupt translation from messenger RNA (mRNA) to protein in bacterial ribosomes, thereby blocking protein biosynthesis.

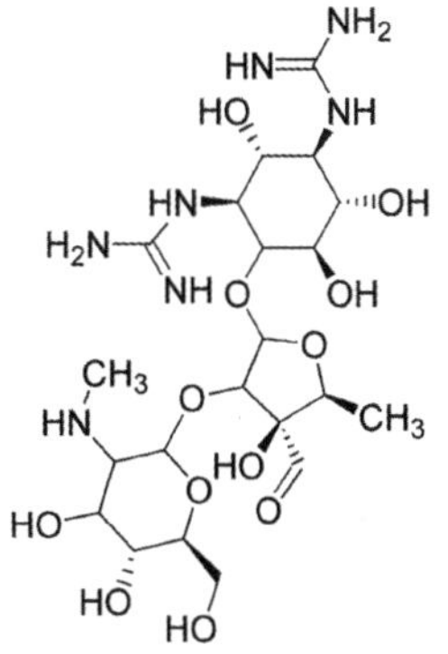

streptomycin

After initial tests on guinea pigs infected with tuberculosis, the first clinical trials with streptomycin on patients began. When the early results proved encouraging, it was agreed that the US National Research Council should co-ordinate large-scale trials in order to hasten progress, such was the pressing need for a cure for tuberculosis. The cost of the co-ordinated programme was met entirely by the pharmaceutical industry. By the time the trials were completed and large-scale production of streptomycin had begun, the US industry alone had spent 20 times as much again on production plant.[74]

For many years streptomycin was used as a first-line drug for the treatment of tuberculosis, but large doses were found to damage the aural nerve, causing deafness. Fortunately, the discovery of synthetic antituberculosis drugs eased this problem by permitting them to be used in combination with smaller doses of streptomycin.

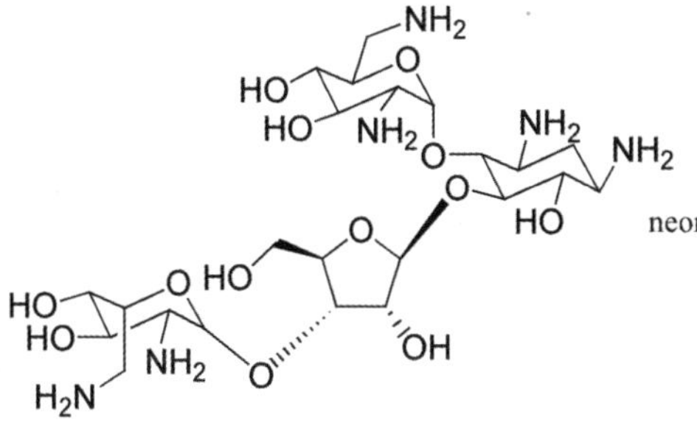

neomycin B (framycetin)

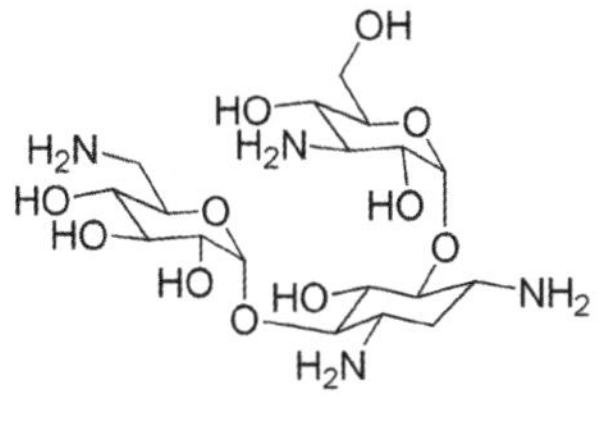

kanamycin A

Waksman isolated neomycin from *Streptomyces fradiae* in 1949.[75] It was an antibiotic complex, the component used therapeutically being neomycin B. The structure of neomycin B was fully elucidated in 1962[76] and it was synthesised in 1987.[77] Framycetin, an antibiotic isolated by French researchers in 1954, was shown to be identical to neomycin B.[78] Because of its neurotoxicity, neomycin is restricted to topical use for skin, ear and eye infections.

Kanamycin is an antibiotic complex isolated in 1957 by Hamao Umezawa at the Institute of Microbial Chemistry in Tokyo, where he had been running an antitumour screening programme utilising the transplanted Yoshida sarcoma in mice.[79] The structure of the major component, kanamycin A, was established in 1958[80] and it was synthesised ten years later.[81] It is another toxic aminoglycoside antibiotic which is reserved for the treatment of penicillin-resistant staphylococcal and serious Gram-negative infections resistant to gentamicin.

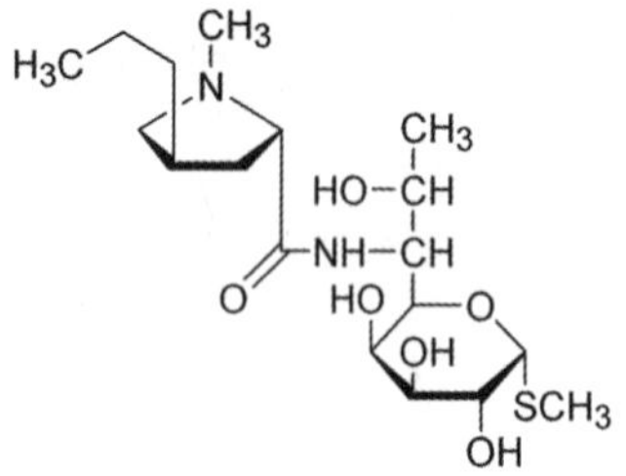

lincomycin

Lincomycin was isolated in 1962 from *Steptomyces lincolnensis* in a soil sample gathered in Lincoln, Nebraska.[82] Its structure was determined two years later and it was synthesised in 1970.[83] It was a highly toxic drug and has been superseded by its derivative clindamycin, which has similar activity but is absorbed from the gut more efficiently.

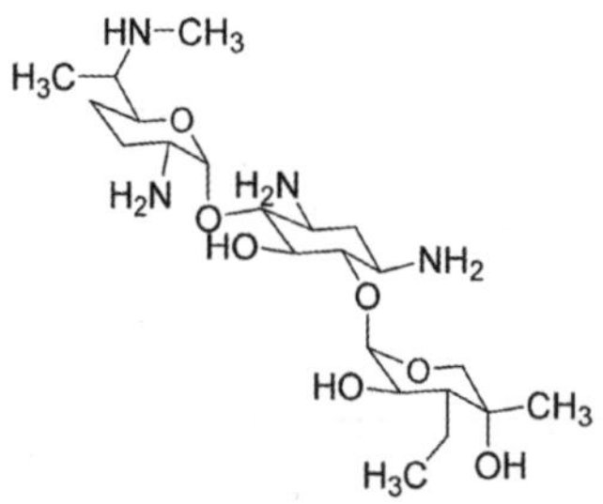

gentamicin C$_1$

Gentamicin C$_1$ was isolated from a complex of antibiotics produced by *Micromonospora purpurea* and *M. echinospora* by researchers at the Schering Corporation in Bloomfield, New Jersey, in 1963.[84] Its structure was determined in 1967.[85] Gentamicin became the aminoglycoside of choice for a variety of purposes, being active against pseudomonal infections, unlike kanamycin. However, resistance gradually developed and has limited its value.

Several other aminoglycosides have been introduced into medicine, including amikacin, netilmicin and tobramycin.

Chloramphenicol

Yale botanist Paul Burckholder received a grant of $5000 from Parke, Davis and Company in 1943 to screen soil samples for antibiotic activity against six selected types of bacteria. He examined more than 7000 samples from all over the world, including one collected by his friend Derald Langham, an agricultural geneticist who was carrying out research near Caracas in Venezuela. An unknown antibiotic-producing actinomycete was isolated from this sample and given the name *Streptomyces venezuelae*. A culture was sent to Parke, Davis and Company in Detroit, where John Ehrlich and Quentin Bartz[86,87] isolated chloramphenicol in 1947. It turned out to be an orally active, broad-spectrum antibiotic.

The first patients to receive chloramphenicol were victims of an epidemic of typhus sweeping through Bolivia. In December 1947, 22 patients in the General Hospital at La Paz, of whom at least five were close to death, were given injections of the new antibiotic by Eugene Payne, a clinical researcher from Parke, Davis and Company. All were cured. Similar results were obtained in Kuala Lumpur, but this time there was some confusion that resulted in patients with typhoid also receiving injections of chloramphenicol. Their rapid recovery was unprecedented.

chloramphenicol

Parke, Davis and Company researchers rapidly elucidated the chemical structure of chloramphenicol and then synthesised it.[88] By 1949, large amounts of chloramphenicol were being manufactured, with sales that year exceeding $9 million. They increased fivefold over the next two years, turning Parke, Davis and Company into the largest pharmaceutical company in the world. Some eight million patients were treated with this apparently safe antibiotic before reports that patients had died from aplastic anaemia caused by it appeared in leading medical journals. The incidence has been estimated as between 1 in 20 000 and 1 in 100 000, with 80% of the victims dying. Had the drug not been rushed on to the market, the number of cases would hardly have caught the public eye. However, the *Journal of the American Medical Association* published a warning against the promiscuous use of chloramphenicol and then issued a press release that was widely reported. The available evidence was studied by the Food and Drug Administration, with the result that Parke, Davis and Company were permitted to continue selling the drug, but with a warning on the package. This stated that prolonged or intermittent use could cause blood disorders. Although sales dropped markedly for a year or two, the drug regained some of its popularity when the problem of resistant strains of staphylococci became more common, but the arrival of the broad-spectrum penicillins largely replaced it for many purposes. Today, chloramphenicol is reserved for the treatment of typhoid, salmonella, meningitis and rickettsial infections. It is also used topically in treating eye and ear infections. No superior analogue ever emerged, despite the synthesis of many compounds. It acts by binding exclusively to the 50S subunit of bacterial ribosomes, inhibiting the enzyme peptidyl transferase. This prevents peptide bond formation.

The Tetracycline Antibiotics

Benjamin Duggar, a 71 year old retired botany professor from the University of Wisconsin, became a consultant to Lederle Laboratories at Pearl River, New York, in 1943 to investigate a plant alleged to have antimalarial activity. A year later he was asked to supervise the screening of hundreds of soil samples in order to find a safer antibiotic than streptomycin for treatment of tuberculosis. In the summer of 1945, a sample was received from William Albrecht of the University of Missouri, where Duggar had been a member of faculty 40 years earlier. This contained a golden actinomycete with antibiotic activity. Duggar named it *Streptomyces aureofaciens*.[89] The antibiotic was isolated and later given the approved name of 'chlortetracycline' after its chemical structure was elucidated. Tests showed it to be an orally active, broad-spectrum antibiotic with a therapeutic profile similar to that of chloramphenicol, but with no value in tuberculosis. By December 1948, large-scale production in fermentation tanks had begun. The process was patented and the antibiotic was put on the market shortly before chloramphenicol.[90]

Robert Woodward at Harvard and chemists from the Pfizer laboratories in Brooklyn established the chemical structure of chlortetracylcine in 1952 at the same time as that of oxytetracycline, in what is generally considered to be a classic example of structural elucidation.[91] The first total chemical synthesis of the tetracyclines was reported in 1959 by Lederle chemists.[92]

Like other tetracyclines, chlortetracycline inhibits protein synthesis by binding to the 30S subunit of bacterial ribosomes and interfering with the binding of aminoacyl transfer RNA (tRNA). The selectivity for bacteria arises from a transport process that occurs in both Gram-positive and negative bacteria, but tetracyclines also diffuse by passive transport into bacterial cells and, if the concentration is sufficiently high, into mammalian cells, where protein synthesis can also be impaired. However, chlortetracycline has poor oral bioavailability and a shorter half-life than other tetracyclines, hence it is less appropriate than those for the treatment of systemic infections.

tetracycline, R = H
chlortetracycline, R = Cl

oxytetracycline

The introduction of chlortetracycline and chloramphenicol threatened the market position of Pfizer as the leading producer of penicillin, the price of which was already plummeting. In response, the company set up a team of 11 researchers and 45 assistants who went on within the next 18 months to examine almost 100 thousand soil samples from around the world. The antibiotic that emerged was cultured from a soil sample containing *Streptomyces rimosus*, which was collected near Pfizer's Terre Haute factory! This was followed by the rapid isolation of oxytetracycline.[93] It was quickly purified and found to have similar properties to chlortetracycline.[94] A patent was applied for at the end of November 1949.[95]

Pfizer decided to market oxytetracycline themselves, rather than supply it in bulk to others as it had done with all its previous products. Lacking a sales force of medical representatives who could call on physicians, they indulged in what proved to be a much criticised campaign of direct advertising to the general public. The cost of developing oxytetracycline had been about $4 million, but in two years the company spent almost double this amount on advertising. This resulted in their obtaining a quarter of the American market for broad-spectrum antibiotics in 1951, roughly the same share as chloramphenicol.

In 1952, Lederle researchers isolated tetracycline from the actinomycete they were using to produce chlortetracycline,[96] and Bristol Laboratories obtained it from *Streptomyces viridifaciens.*[97] Around the same time, Lloyd Conover of Pfizer obtained tetracycline when the chlorine atom of chlortetracycline was removed by catalytic hydrogenation.[98] These three American companies all applied for patents in 1952–1953, with Lederle and Pfizer agreeing to cross-license each other. All the applications were rejected, but Pfizer and Bristol fought the decision and succeeded in obtaining patents in 1955. A subsequent agreement made between the tetracycline producers and their licensees was criticised by the US Federal Trade Commission in 1958, on the grounds that it was an attempt to eliminate competition and fix prices. This was denied by the companies.

The three original tetracylines were sought after eagerly when first introduced. They acquired a popular reputation as miracle drugs, vying with penicillin. Their commercial success was due to their oral activity, the first penicillin suitable for oral administration not being introduced until the mid-1950s. However, the proportion of these early tetracyclines absorbed from the intestine was only a fraction of the total dose, leading to disturbance of the normal bacterial flora of the gut. In hospitalised patients, this gradually led to replacement of the flora by strains of resistant bacteria, leading to a decline in their use in hospital practice. Nevertheless, some tetracyclines remain the drugs of choice in trachoma, psitacosis and similar chlamydial infections, as well as in those caused by rickettsia, mycoplasma and brucella. They are also effective against *Haemophilus influenzae* and so may be of value in chronic bronchitis.

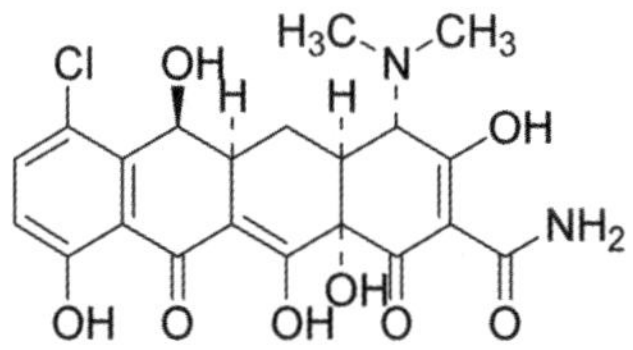

demeclocycline

In 1957, demeclocycline was isolated from mutant strains of *Streptomyces aureofaciens* by Lederle researchers.[99] It had improved chemical stability due to the absence of an unstable tertiary alcohol function in the tetracycline ring system, the alcohol function now being secondary. This resulted in enhanced blood levels and encouraged chemists to develop semi-synthetic tetracyclines in which there was no tertiary alcohol function.

Macrolide Antibiotics

Robert Bunch and James McGuire of Eli Lilly and Company isolated erythromycin in 1952 from a strain of *Streptomyces erythreus* cultured from a soil sample collected in Iloilo in the Philippines.[100] The chemical structure was determined in 1956.[101] It was the first of the macrolide antibiotics (so named because of the presence of a large ring in their chemical structure) to be introduced into the clinic. It binds to the 50S subunit of bacterial ribosomes, inhibiting the translocation stage of protein synthesis. This type of action makes it a bacteriostatic rather than a bacteriocidal antibiotic.

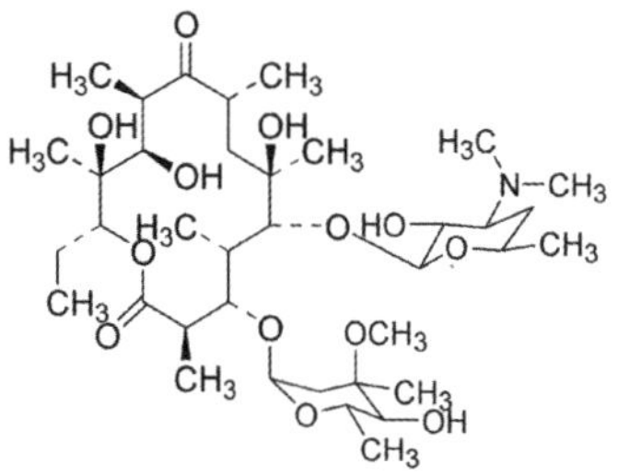

erythromycin

Although the chemical structure of erythromycin bears no relation to penicillin, its spectrum of action is similar. This renders it of value in patients allergic to penicillin and in the treatment of penicillin-resistant staphylococcal infections. The principal use of erythromycin is in the treatment of Gram-positive infections of the skin, soft tissues and respiratory tract.

Erythromycin is unstable in gastric juice, so oral administration leads to erratic plasma levels. Furthermore, the decomposition products are responsible for a high incidence of nausea and epigastric pain. Several esters and salts of erythromycin that are insoluble in gastric juice have been introduced to avoid gastric acid decomposition. Also, by formulating erythromycin base in enteric-coated tablets, degradation is largely prevented.

A novel macrolide was discovered by Surendra Sehgal and his colleagues at the Ayerst laboratories in Montreal when they isolated an antifungal antibiotic from *Streptomyces hygroscopicus* in 1972. They named it 'rapamycin' because the streptomycete had been found in a soil sample collected seven years earlier by the Canadian Medical Expedition to Easter Island, the local name of which was Rapa Nui.[102] The structure was determined in 1978 and it was synthesised some years later.[103]

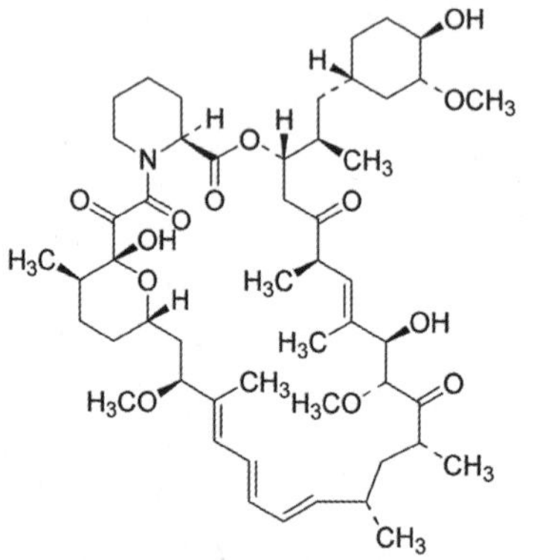

sirolimus (rapamycin)

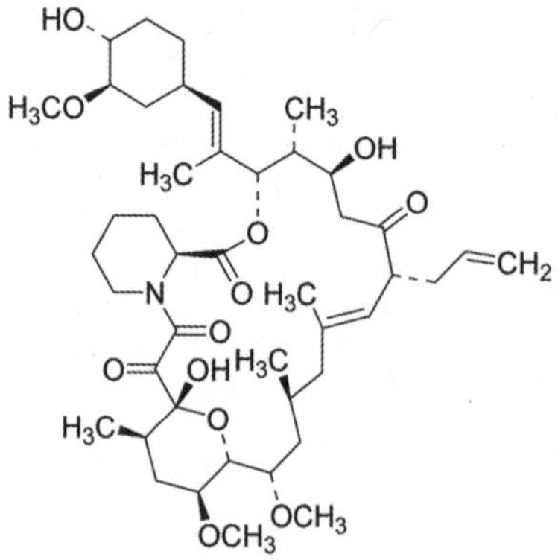

tacrolimus

Rapamycin was found to be active against *Candida albicans* and some other fungi, but when detailed studies were conducted it was found that it had a powerful suppressant effect on the immune system. This resulted in rapamycin being dropped as a potential antifungal, but then it was investigated as an immunosuppressant drug. In addition, tests conducted by the National Cancer Institute revealed that it possessed promising activity as a cytostatic drug when used in combination with cytotoxic agents. With such exciting prospects for rapamycin, the sudden closure of the Montreal research facility and termination of all projects must have been a terrible blow for the staff concerned. Fortunately, Sehgal had the foresight to harvest a final batch of the antibiotic from the fermentation tanks and store it safely when he was transferred to the Princeton laboratories of the company. However, it was not until Ayerst was merged with Wyeth in 1987 and new management was in place that he was able to persuade anyone to reinstate the project. Fortunately, steady progress was then made and rapamycin was licensed by the Federal Drug Authority (FDA) in 1999 for use in kidney transplantation. It now has the international non-proprietary name (INN) of sirolimus and is being clinically investigated for its potential in various types of cancer.

Researchers at the Fujisawa Pharmaceutical Company in Osaka, Japan, isolated an antibiotic from *Streptomyces tsukubaensis* in a soil sample taken from the foot of Mt Tsukuba, near Tokyo, in 1984.[104] The structure was elucidated in 1987,[105] with a total synthesis being achieved within two years by Ichiro Shinkai and his colleagues at Merck.[106]

After it was found to inhibit T-cell activation, tacrolimus was investigated as an immunosuppressant.[107] Positive results were followed by clinical trials at the University of Pittsburg in patients whose liver transplants were being rejected after suppressant therapy. Within ten years of its isolation, tacrolimus was licensed for use in liver and kidney transplant patients. It is more toxic than ciclosporin.

Another macrolide antibiotic was discovered in 1977 as the major component in a mixture of avermectins isolated from *Streptomyces avermitilis* by Merck researchers. The components were separated and then their structures were elucidated.[108] Abamectin was selected as it had anthelmintic activity, but it is only used in animals.[109] Its chemical reduction product, ivermectin, will be discussed in the next chapter.

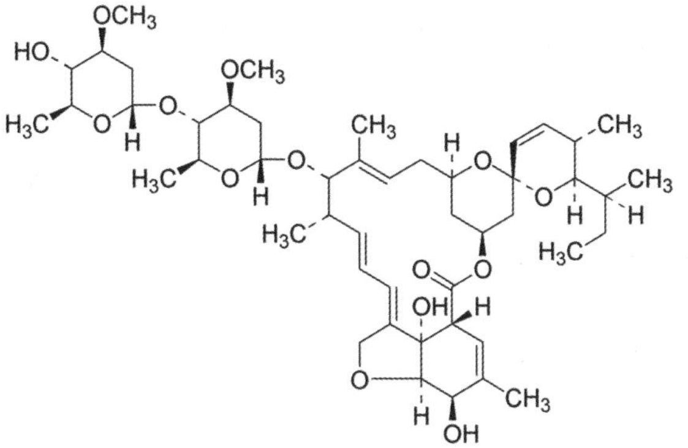

abamectin

Rifamycin Antibiotics

Following their entry into the field of penicillin production, Lepetit Research Laboratories of Milan established a screening programme to discover new antibiotics. This led to the isolation from a soil sample collected on the Cote d'Azur in the summer of 1957 of a strain of *Streptomyces* that, when cultured, exhibited high activity against Gram-positive bacteria and *Mycobacterium tuberculosis*.[110] The micro-organism was initially classified as *Streptomyces mediterranei*, but later evidence led to its reclassification as *Nocardia mediterranei*.

The available chemical evidence from investigation of a brown powder obtained from the fermentation broth indicated that its activity was due to a group of at least five novel antibiotics. These were given the generic name of 'rifamycins', this term being chosen from the title of a Jules Dassin gangster film, namely 'Rififi' (French argot: *rififi* = a struggle).[111] Due to stability problems, the only crystalline antibiotic isolated from the original mixture was rifamycin B.[110] It constituted less than 10% of the total antibiotic titre. However, another strain of *N. mediterranei* was found to produce high yields of this.

Rifamycin B had moderate antibiotic activity when administered to animals, but this was not sufficient to justify it being considered for potential therapeutic application. However, aqueous solutions of the antibiotic acquired stronger antibacterial activity on standing. Investigation of this phenomenon led to the clinical introduction of rifamycin SV as it was active against a wide range of organisms, including *Mycobacterium tuberculosis*, *M. leprae*, Gram-positive bacteria and some Gram-negative bacteria.

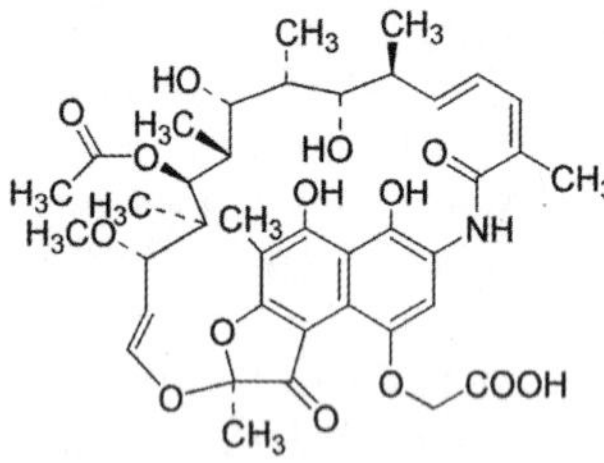

rifamycin B

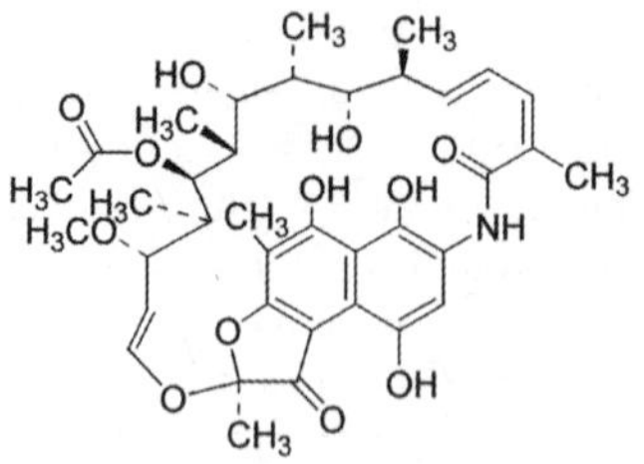

rifamycin SV

Victor Prelog and Wolfgang Oppolzer of the Swiss Federal Institute of Technology (ETH) in Zurich determined the chemical structure of rifamycin B in 1964.[112]

Polyene Antibiotics

Nystatin, an antifungal polyene antibiotic produced by various species of *Streptomyces* including *S. noursei* and *S. aureus*, was introduced by Squibb after its discovery in 1950 by Elizabeth Hazen and Rachel Brown during an extensive soil survey conducted by the New York State Department of Health.[113] The correct structure was established in 1976.[114] As nystatin is too toxic for systemic treatment, its use has been limited to either topical application for *Candida albicans* infection of the skin or, as it is not absorbed from the gut, oral administration to treat intestinal candidiasis.

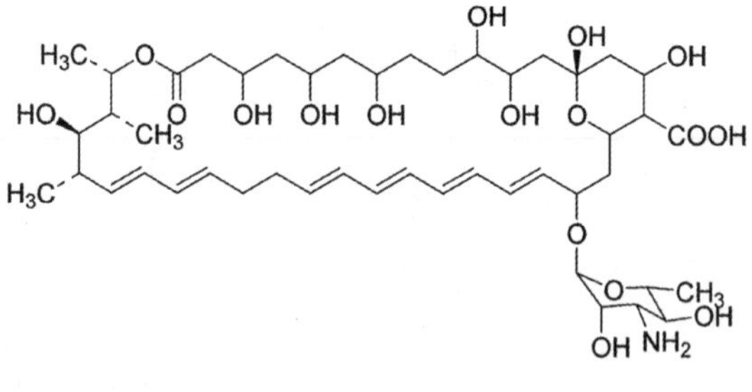

nystatin

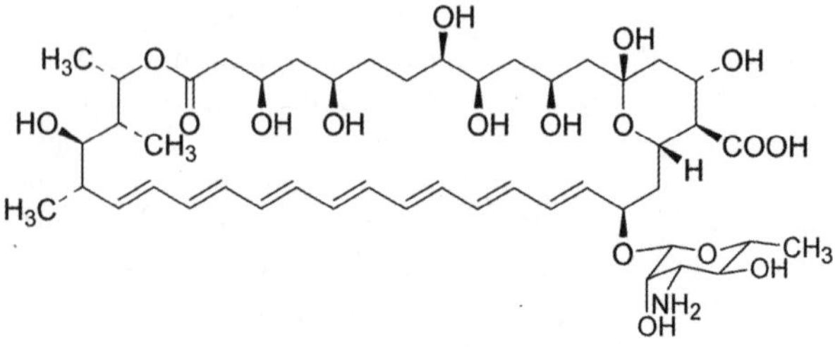

amphotericin

Amphotericin was isolated in 1953 from a strain of *Streptomyces nodosus* growing in a sample of Orinoco river soil that had been sent from Venezuela to the Squibb Institute for Medical Research.[115] The complete structure was confirmed in 1976.[114]

Amphotericin is the only polyene antibiotic safe enough to be administered parenterally, being used to combat life-threatening systemic fungal infections in patients whose immune system is compromised as a consequence of intensive cancer chemotherapy or AIDS. It is structurally related to nystatin.[116] It enters the fungal cell membrane, where it disrupts membrane integrity by binding to ergosterol. This causes loss of ions which are critical for the normal functioning of the fungal cell. Unfortunately, amphotericin can also damage human cell membranes, making it one of the most toxic drugs in current clinical use when administered parenterally.

Azomycin

Nakamura and Umezawa isolated azomycin in 1953 from an unidentified species of *Streptomyces*.[117] Its structure was determined by Nakamura[118] and it was synthesised in 1965.[119]

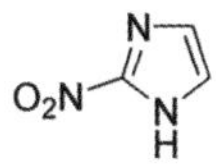

azomycin

As azomycin was structurally similar to the trichomonicidal agents aminitrozole and 2-amino-5-nitrothiazole, Rhône Poulenc researchers tested it and found it to have similar activity, but it turned out to be too toxic for clinical exploitation. Analogues were then prepared and metronidazole was found to be not only an antiprotozoal drug but also an antibacterial one. This is discussed in the next chapter.

Vancomycin and Teicoplanin

Vancomycin was isolated from *Streptomyces orientalis* at the Lilly laboratories in 1955,[120] but its structure was not correctly assigned until 1981.[121] Vancomycin was effective against both aerobic and non-aerobic Gram-positive bacteria, especially resistant staphylococci, acting by disrupting bacterial cell wall synthesis. Early preparations were contaminated with impurities that caused a high incidence of side effects. As a result of its purification, vancomycin became the drug of choice for the oral treatment of pseudomembranous colitis since it was not absorbed from the gut.

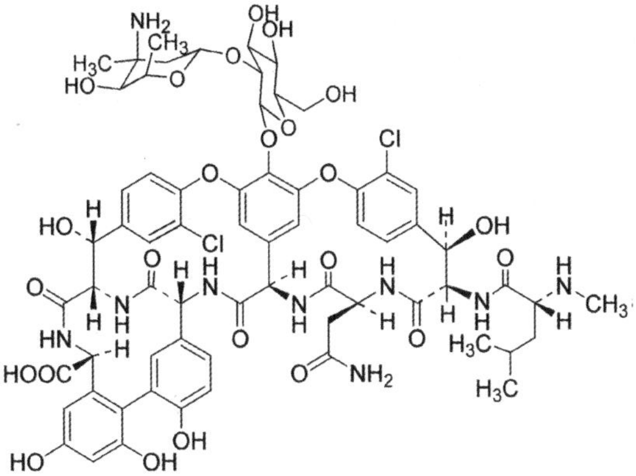

vancomycin

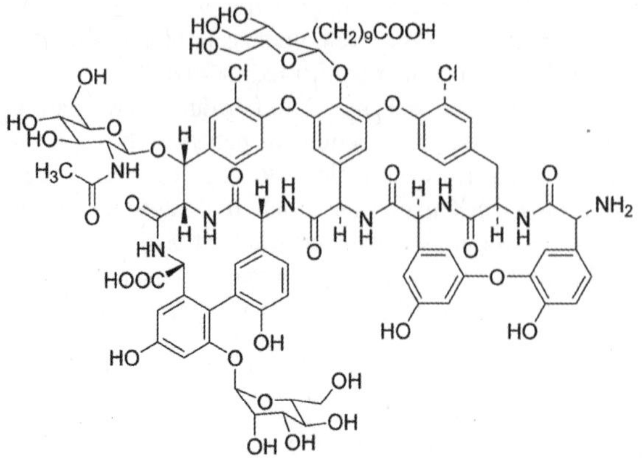

teicoplanin A$_{2-3}$

Teicoplanin is a complex of five antibiotics isolated from *Actinoplanes teichomycetius* by Lepetit researchers in 1976.[122,123] The components were separated in 1983 and then identified.[124] The long side chain attached to one of the sugar rings varied in each of the five components, teicoplanin A$_{2-3}$ being shown here. Teicoplanin has similar activity to vancomycin, but has a longer half-life and hence can be administered once daily. As it is less irritant than vancomycin, it can be injected intramuscularly or intravenously.

Thienamycin

The carbapenems are a group of exceedingly potent antibiotics isolated from streptomycetes, more than 40 having been reported by 1990.[125] Their potency may be due to superior penetration into bacterial cells. The most effective of them is thienamycin, which was isolated in 1976 from *Streptomyces cattleya* by Merck investigators in the course of screening for inhibitors of peptidoglycan synthesis.[126] Its structure was determined two years later.[127] Early indications were that thienamycin was resistant to beta-lactamases and was as active as gentamicin against Gram-negative organisms. However, it had poor activity when administered by mouth and was chemically unstable. Consequently, it is only used in the form of a stable derivative known as imipenem (discussed in the next chapter).

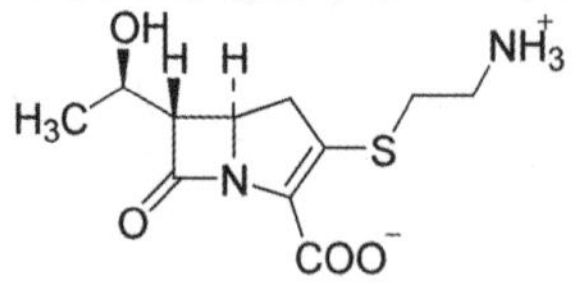

thienamycin

Clavulanic Acid

Concern about the growing problem of resistance to penicillins and cephalosporins led Rolinson of Beecham Pharmaceuticals to consider the possibility that β-lactamase inhibitors might be produced by micro-organisms. A screen was set up to spot fungal or actinomycete extracts that could restore the sensitivity towards benzylpenicillin of a β-lactamase-producing strain of *Klebsiella aerogenes*. If a β-lactamase inhibitor was present in an extract, no growth of *K. aerogenes* would occur. When isolates of *Streptomyces clavuligerus* exhibited high β-lactamase inhibitory activity, clavulanic acid was extracted.[128] Although it had no significant antibacterial activity, it was a potent inhibitor of β-lactamases from a variety of bacterial species. When combined with amoxycillin, it overcame resistance from organisms that produced β-lactamases.[129] This combination proved satisfactory in clinical trials and is still prescribed for treatment of respiratory or urinary tract infections. Combinations of clavulanic acid and other antibiotics have been marketed.

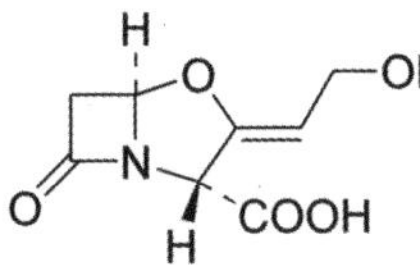

clavulanic acid

CYTOTOXIC ANTIBIOTICS FROM STREPTOMYCETES

After Selman Waksman had isolated actinomycin A, Hans Brockmann at Göttingen University obtained a second actinomycin antibiotic from *Streptomyces chrysomallus* in 1949, which he named actinomycin C.[130] Later, this was shown to be a complex consisting of three components, named actinomycin CD_1, CD_2 and CD_3. Following a report from the Sloane–Kettering Institute that Waksman's actinomycin A had detectable activity against sarcoma-180, Brockmann sent his crude actinomycin C to the Bayer Institute for Experimental Pathology at Elberfeld, where it was found to inhibit tumour growth.[131] It was subsequently found to be effective in some patients with lymphatic tumours at dose levels that did not produce toxic effects, though this was disputed. Nevertheless, actinomycin C was marketed for some years, principally for use in Hodgkin's disease, until rendered obsolete by more effective agents.

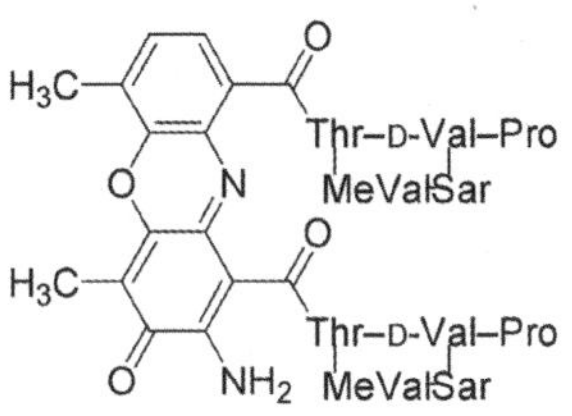

dactinomycin

In 1953, Waksman isolated actinomycin D from *Streptomyces parvullus*, the first organism to yield a single actinomycin rather than a complex mixture.[132] It turned out that this was identical to actinomycin CD_1, and both of these terms were later replaced by the approved name of 'dactinomycin'. Pure dactinomycin, which constituted about 10% of the actinomycin C complex, was superior to actinomycins CD_2 and CD_3 as an antitumour agent. Its chemical structure was established in 1957[133] and it was synthesised by Brockmann in 1964.[134] The other actinomycins have similar structures, differing only in two of their amino acids.

Dactinomycin showed marked antitumour activity against many transplanted mouse tumours, occasionally causing total regression of some of them. As soon as toxicological studies were completed, a clinical trial was initiated in children with acute leukaemia. Disappointingly, dactinomycin was ineffective. Nevertheless, encouraging results were achieved in children with advanced Wilm's tumour, rhabdomyosarcomas (muscle tumours), Ewing's tumour of the bone and Hodgkin's disease.[135] The prospects of survival for children with Wilm's tumour of the kidney were completely transformed when a combination of radiotherapy and sustained chemotherapy with dactinomycin on its own or with vincristine was introduced. When maintained for up to two years, 90% of children are cured, which is three times as many as was the case prior to the introduction of dactinomycin The main value of it in adults appears to be for the treatment of soft tissue sarcomas and testicular teratomas.

Bleomycin

In 1962, Umezawa detected antitumour activity while screening culture filtrates of *Streptomyces verticillus*. He managed to isolate the bleomycins from this, a complex group of glycopeptides of which bleomycin A_2 is the main component of the product used clinically.[136,137]

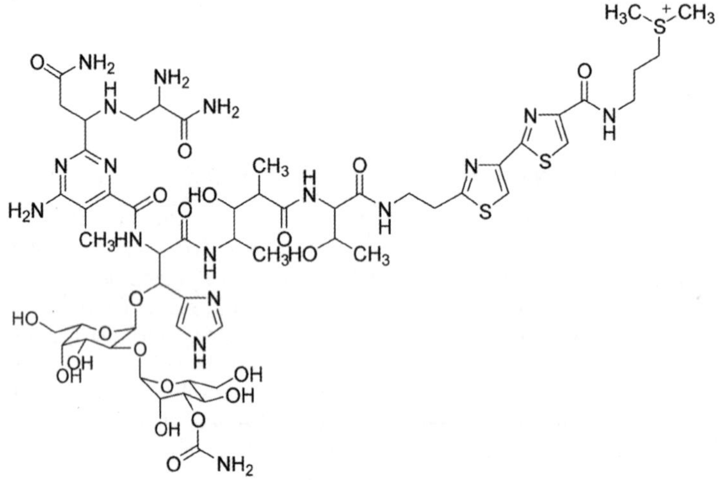

bleomycin

The most striking feature of bleomycin is the fact that it is one of the few anticancer drugs that does not cause bone marrow depression. However, it may cumulatively damage the lung, especially in elderly patients. Bleomycin is administered parenterally in the treatment of squamous cell carcinoma (cancer in flattened lining cells), head and neck tumours, non-Hodgkin's lymphoma and testicular teratomas.

Anthracycline Antibiotics

Daunorubicin (daunomycin, rubidomycin) was isolated in 1962 from *Streptomyces peucetius* by Di Marco and his colleagues of the Farmitalia Company in Milan.[138] The chemical structure was established two years later.[139] This particular streptomycete had been investigated after extracts of a soil sample from Apulia (where the Peucetii and Daunii tribes had lived in ancient times) yielded material with activity against cultures of murine Ehrlich carcinoma. At the Instituto Nazionaleper lo Studio e le Cura dei Tumori in Milan, where Di Marco had previously worked, daunorubicin was shown to exhibit actinomycin-like activity against tumours, intercalating with DNA to prevent it from serving as a template for replication.

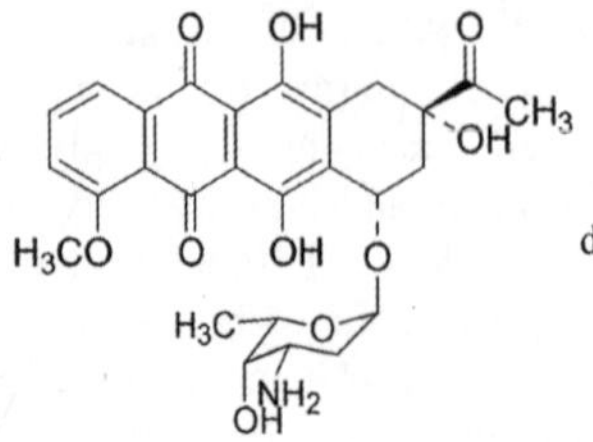

daunorubicin

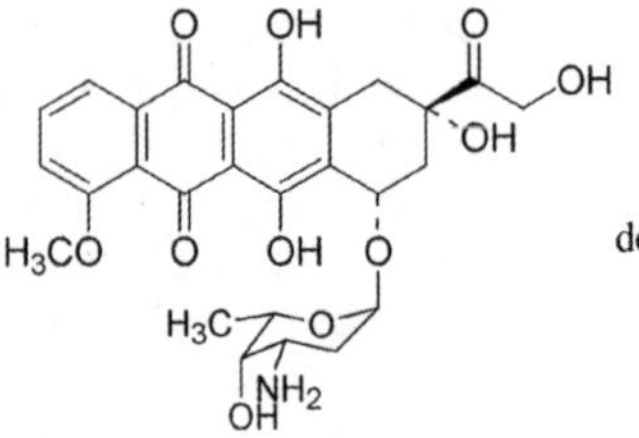

doxorubicin

When administered to animals with tumours, daunorubicin was more active than actinomycin C. Its clinical value was limited by severe toxicity to the heart, but it found some use in the combination chemotherapy of leukaemia, particularly in helping to induce remissions in cases where the patient had failed to respond to safer drugs.

Another antibiotic isolated by Farmitalia scientists from *S. peucetius* in 1967 was named 'doxorubicin', which turned out to be the 14-hydroxy derivative of daunorubicin.[140] This apparently minor chemical difference transformed it into one of the most successful antitumour drugs ever discovered, but the problem of cardiotoxicity remained. Patients who received a cumulative dose in excess of $550 \, mg/m^2$ were at risk of having a heart attack induced by the affinity of the drug for cardiac muscle. Nevertheless, when doxorubicin is administered with care it is of considerable value in the treatment of acute leukaemias, lymphomas (non-Hodgkin's) and many solid tumours. Aclarubicin, which has similar uses, was one of 20 aclacinomycins isolated from *Streptomyces galilaeus* by Umezawa in 1974.[141]

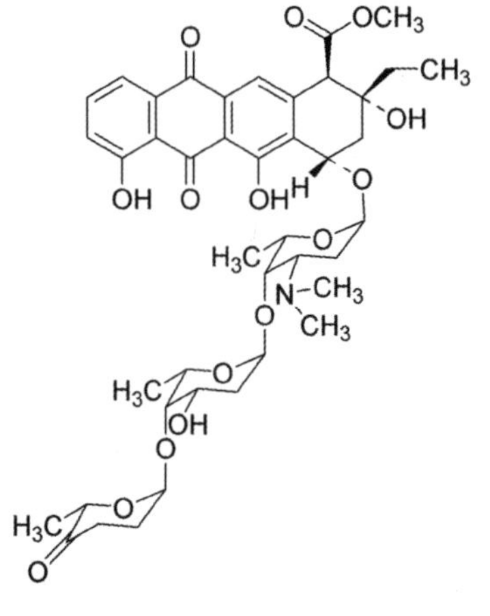

aclarubicin

MONOBACTAM ANTIBIOTICS

Selective screening techniques applied to soil micro-organisms resulted in the isolation of sulfazecin and other novel monobactams from the bacteria *Pseudomonas acidophila* and *P. mesoacidophila* by Akira Imada at the Takeda laboratories in Osaka[142] and by Richard Sykes and his colleagues at the Squibb Institute for Medical Research.[143] Sykes also isolated SQ 26180.

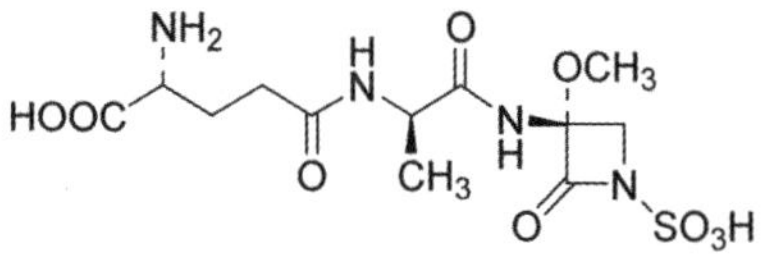

sulfazecin

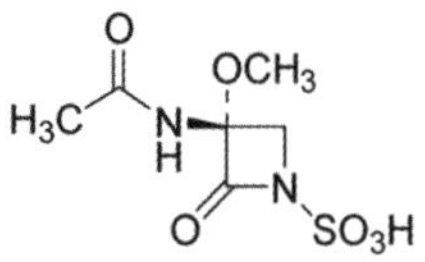

SQ 26180

Although these natural antibiotics were active against Gram-negative bacteria, they were not considered to be effective enough for clinical application.

REFERENCES

1. L. Pasteur, J.F. Joubert, Charbon et septicemia. *C. R. Acad. Sci.*, 1877; **85**: 101–15.
2. L. Pasteur, *C. R. Acad. Sci.*, 1880; **90**: 952.

3. V. Babès, J. Connais. *Med. Prat. (Paris)*, 1885; **7**: 321.
4. C. Garré, Ueber Antagonisten unter Bacterien. *Correspondenzblatt Schweiz Aertze*, 1887; **17**: 385.
5. R. Emmerich, Die Heilung des Milzbrandes. *Arch. Hyg.*, 1887; **6**: 442–501.
6. E. von Freudenreiche, De l'antagonisme des bactéries et de l'immunité qu'il confère aux milieux de culture. *Ann. Inst. Pasteur*, 1888; **2**: 200–6.
7. C. Bouchard, L'influence qu'exerce sur la maladie charbonneuse l'inoculation du bacille pyocyanique. *C. R. Acad. Sci.*, 1889; **108**: 713–14.
8. R. Emmerich, O. Löw, Bakteriolytische Enzyme als Ursache der erworbenen Immunität und die Heilung von Infectionskrankheiten durch Dieselben. *Z. Hyg. u. Infektionskr.*, 1899; **31**: 1.
9. E.E. Hays, I.C. Wells, P.A. Katzman, *et al.*, Antibiotic substances produced by *Pseudomonas aeruginosa*. *J. Biol. Chem.*, 1945; **159**: 725–50.
10. I.C. Wells, Antibiotic substances produced by *Pseudomonas aeruginosa*. Syntheses of Pyo Ib, Pyo Ic, and Pyo III. *J. Biol. Chem.*, 1952; **196**: 331–40.
11. R.J. Dubos, Bactericidal effect of an extract of a soil bacillus on Gram-positive bacteria. *Proc. Soc. Exp. Biol. Med.*, 1939; **40**: 311–12.
12. R.D. Hotchkiss, R.J. Dubos, The isolation of bactericidal substances from cultures of *Bacillus brevis*. *J. Biol. Chem.*, 1941; **141**: 155–62.
13. A.H. Gordon, A.J.P. Martin, R.L.M. Synge, Amino-acid composition of tyrocide. *Biochem. J.*, 1943; **37**: 313–18.
14. R.L.M. Synge, 'Gramicidin S' overall chemical characteristics and amino-acid composition. *Biochem. J.*, 1945; **39**: 363–7.
15. J. Burdon-Sanderson, 13th Report of the Medical Officer of the Privy Council, Append 5. H.M.S.O.; 1871, p. 48.
16. S.J. Selwyn, Pioneer work on the 'penicillin phenomenon', 1870–1876. *Antimicrob. Chemother.*, 1979; **5**: 249–55.
17. V. Tiberio, Sugli Estratti di Alcune Muffe. *Annali di Igiene Sperimentale*, 1895; **5**: 91–103.
18. B. Gosio, Ricerdre batteriologiche chirmide sulle alterazioni del mais. *Rivista d'Ingienne e Sanita Pubblica Ann.*, 1896; **7**: 825–68.
19. J.H. Birkinshaw, H. Raistrick, D.J. Ross, Studies in the biochemistry of micro-organisms. 86. The molecular constitution of mycophenolic acid, a metabolic product of *Penicillium brevi-compactum* Dierckx. Part III. Further observations on the structural formula for mycophenolic acid. *Biochem. J.*, 1952; **50**: 630–4.
20. T.J. Franklin, J.M. Cook, The inhibition of nucleic acid synthesis by mycophenolic acid. *Biochem. J.*, 1969; **113**: 515.
21. P. Halloran, T. Matthew, S. Tomlanovich, *et al.*, Mycophenolate mofetil in renal allograft recipients: a pooled efficacy analysis of three randomized, double-bind, clinical studies in prevention of rejection. The International Mycophenolate Mofetil Transplant Study Groups. *Transplantation*, 1977; **63**: 39–47.
22. E.L. Jones, W.W. Epinette, V.C. Hackney, *et al.*, Treatment of psoriasis with oral mycophenolic acid. *J. Invest. Dermatol.*, 1975; **65**: 537.
23. W.A. Lee, L. Gu, A.R. Mikszal, *et al.*, Bioavailability improvement of mycophenolic acid through amino ester derivitization. *Pharm. Res.*, 1990; **7**: 161–6.
24. R. Hare, *The Birth of Penicillin*. London: Allen and Unwin; 1970.
25. A. Fleming, Cultures of a penicillium, with special reference to their use in the isolation of *B. influenzae*. *Br. J. Exp. Pathol.*, 1929; **10**: 226–36.
26. R. G. Macfarlane, *Alexander Fleming: The Man and the Myth*. Chatto and Windus, London, 1984.
27. P.W. Clutterbuck, R. Lovell, H. Raistrick, The formation from glucose by members of the *Penicillium chrysogenum* series of a pigment, an alkali-soluble protein and penicillin – the antibacterial substance of Fleming. *Biochem. J.*, 1932; **26**: 1907–18.
28. R.G. Macfarlane, *Howard Florey*. London: Oxford University Press; 1979.
29. E.P. Abraham, E. Chain, H.W. Florey, *et al.*, in *Antibiotics*, Vol. 2, eds H.W. Florey, E. Chain, N.G. Heatley. London: Oxford University Press; 1949.
30. R.W. Clark, *The Life of Ernst Chain: Penicillin and Beyond*. London: Weidenfeld and Nicolson; 1985.
31. H.W. Florey, E.P. Abraham, The work on penicillin at Oxford. *J. Hist. Med.*, 1951; **6**: 302–17.
32. E.B. Chain, H.W. Florey, A.D. Gardner, *et al.*, Penicillin as a chemotherapeutic agent. *Lancet*, 1940; **2**: 226–8.
33. E.P. Abraham, E. Chain, C.M. Fletcher, *et al.*, Further observations on penicillin. *Lancet*, 1941; **2**: 177–89.
34. A.E. Elder, *The History of Penicillin Production*. New York: American Institute of Chemical Engineers; 1970.
35. M.H. Dawson, G.L. Hobby, K. Meyer, E. Chaffee, Penicillin as a chemotherapeutic agent. *J. Clin. Invest.*, 1941; **20**: 434.

36. G.L. Hobby, *Penicillin: Meeting the Challenge.* New Haven, Connecticut: Yale University Press, 1985.

37. J.P. Swann, The search for synthetic penicillin during World War II. *Br. J. Hist. Sci.*, 1983; **16**: 154–90.

38. D. Crowfoot, C. Bunn, B. Rogers-Low, A. Turner-Jones, X-ray crystallographic investigation of the structure of penicillin, in *Chemistry of Penicillin*, eds H.T. Clarke, J.R. Johnson, R. Robinson. Princeton, New Jersey: Princeton University Press; 1949.

39. H.T.J. Clarke, R. Robinson, J.R.R. Johnson, *The Chemistry of Penicillin: Report on a Collaborative Investigation by American and British Chemists.* Princeton, New Jersey: National Academy of Sciences; 1949.

40. N.P. Sullivan, A.T. Symmes, H.C. Miller, H.W. Rhodehamel Jr, New penicillin for prolonged blood levels. *Science*, 1948; **107**: 169–70.

41. G. Brotzu, Richerche sudi un nuovo antibiotico. *Lavori Dell'Instituto D'Igiene du Cagliari*, 1948: 1–11.

42. H.S. Burton, E.P. Abraham, Isolation of antibiotics from a species of Cephalosporium; cephalosporins P1, P2, P3, P4, and P5. *Biochem. J.*, 1951; **50**: 168–74.

43. T.G. Halsall, E.R.H. Jones, G. Lowe, C.E. Newall, Cephalosporin P1. *Chem. Commun.*, 1966: 685–7.

44. R.Y. Gottshall, J.M. Roberts, L.M. Portwood, J.C. Jennings, Synnematin, and antibiotic produced by Tilachlidium. *Proc. Soc. Exp. Biol. Med.*, 1951; **76**: 307–11.

45. E.P. Abraham, G.G.F. Newton, K. Crawford, *et al.*, Cephalosporin N: a new type of penicillin. *Nature*, 1953; **171**: 343.

46. E.P. Abraham, G.G.F. Newton, B.H. Olson, *et al.*, Identity of cephalosporin N and synnematin B. *Nature*, 1955; **176**: 551.

47. Br. Pat.1959: 810189 (to National Research Development Corporation).

48. E.P. Abraham, P.B. Loder, *Cephalosporins and Penicillins*, ed. E.H. Flynn, New York: Academic Press; 1972.

49. E.P. Abraham, G.G.F. Newton, The structure of cephalosporin C. *Biochem. J.*, 1961; **79**: 377–93.

50. R. Nagarajan, L.D. Boeck, M. Gorman, R.L. Hamill, *et al.*, β-Lactam antibiotics from Streptomyces. *J. Am. Chem. Soc.*, 1971; **93**: 2308–10.

51. P.W. Brian, H.G. Hemming, J.C. McGowan, Origin of a toxicity to Mycorrhiza in Wareham Heath. *Nature*, 1945; **155**: 637–8.

52. J.F. Grove, J.C. McGowan, Identification of griseofulvin and 'curling factor'. *Nature*, 1947; **160**: 574.

53. A.E. Oxford, H. Raistrick, P. Simonart, Griseofulvin, $C_{17}H_{17}O_6Cl$, a metabolic product of *Penicillium griseofulvum* Dierks. *Biochem. J.*, 1939; **33**: 240–8.

54. J.F. Grove, J. MacMillan, T.P.C. Mulholland, J. Zeally, Griseofulvin. Part III. The structures of the oxidation products $C_9H_9O_5Cl$ and $C_{14}H_{15}O_7Cl$. *J. Chem. Soc.*, 1952: 3967–77.

55. J. MacMillan, Griseofulvin. Part XIV. Some alcoholytic reactions and the absolute configuration of griseofulvin. *J. Chem. Soc.*, 1959: 1823–30.

56. A. Brossi, M. Baumann, M. Gerecke, E. Kyburz, Syntheseversuche in der Griseofulvin-Reihe. *Helv. Chim. Acta*, 1960; **43**: 1444–7.

57. J. Alkiewicz, Z. Eckstein, H. Halweg, *et al.*, Fungistatic activity of some hydroxamic acids. *Nature*, 1957; **180**: 1204–5.

58. J.C. Gentles, Experimental ringworm in guinea pigs: oral treatment with griseofulvin. *Nature*, 1958; **182**: 476–7.

59. H.F. Stähelin, The history of cyclosporin A (Sandimmune) revisited: another point of view. *Experientia*, 1996; **52**: 5–13.

60. A. Rüegger, M. Kuhn, H. Lichti, *et al.*, Cyclosporin A. Ein immunosuppresiv wirkamer Peptidmetabolit aus Tricoderma polysporium (Link ex Pers). Rafai. *Helv. Chim. Acta*, 1976; **59**: 1075–92.

61. T.J. Petcher, H.P. Weber, A. Rüegger, Crystal and molecular structure of an iodo-derivative of the cyclic undecapeptide cyclosporin A. *Helv. Chim. Acta*, 1976; **59**: 1480–9.

62. J.F. Borel, C. Feurer, H.U. Gubler, H. Stähelin, Biological effects of cyclosporin A: a new antilymphocytic agent. *Agents Actions*, 1976; **6**: 468–75.

63. R.Y. Calne, D.J.G. White, S. Thiru, D.B. Evans, P. McMaster, D.C. Dunn, G.N. Craddock, B.D. Pentlow, K. Rolles, Cyclosporin A in patients receiving renal allografts from cadaver donors. *Lancet*, 1978; **1**: 1323–7.

64. R.L. Powles, A.J. Barrett, H. Clink, H.E.M. Kay, J. Sloane, T.J. McElwain, Cyclosporin A for the treatment of graft-versus-host disease in man. *Lancet*, 1978; **2**: 1327–31.

65. R. Greig-Smith, Contributions to our knowledge of soil fertility, XV. The action of certain micro-organisms upon the numbers of bacteria in the soil. *Proc. Linn. Soc. NSW*, 1917; **42**: 162–6.

66. R. Lieske, *Morphologie und Biologie der Strahlenpilze.* Leipzig: Borntraeger; 1921; pp.138–43.

67. A. Gratia, S. Dath, Propriétés bacteriolytiques de certaines moisissures. *C. R. Soc. Biol.*, 1925; **91**: 1442–3; 1925; **92**: 461–2.
68. A. Gratia, La dissolution des bacteries et ses applications therapeuthiques. *Bull. Acad. Roy. Med. Belg.*, 1934; **14**: 285–300.
69. M. Welsch, Bactericidal substances from sterile culture-media and bacterial cultures. *J. Bacteriol.*, 1941; **42**: 801–14.
70. S.A. Waksman, H.B. Woodruff, Bacteriostatic and bactericidal substances produced by soil Actinomyces. *Proc. Soc. Exp. Biol. Med.*, 1940; **45**: 609–14.
71. S.A. Waksman, W.B. Geiger, The nature of the antibiotic substances produced by *Aspergillus fumigatus*. *J. Bacteriol.*, 1944; **47**: 391–7.
72. A. Schatz, E. Bugie, S.A. Waksman, Streptomycin, a substance exhibiting antibiotic activity against Gram-positive and Gram-negative bacteria. *Proc. Soc. Exp. Biol. Med.*, 1944; **55**: 66–9.
73. N.G. Brink, K. Folkers, in *Streptomycin*, ed. S.A. Waksman. Baltimore, Maryland: Williams & Wilkins; 1949, p. 55.
74. S.A. Waksman (ed.), *Streptomycin, Nature and Practical Applications*. Baltimore, Maryland: Williams & Wilkins; 1949.
75. S.A. Waksman, H.A. Lechevalier, Neomycin, a new antibiotic active against streptomycin-resistant bacteria, including tuberculosis organisms. *Science*, 1949; **109**: 305–7.
76. K.L. Rinehart Jr, M. Hitchins, A.D. Argoudelis, Chemistry of the neomycins. Neomycins B and C. *J. Am. Chem. Soc.*, 1962; **84**: 3218–20.
77. T. Usui, H. Umezawa, Total synthesis of neomycin B. *J. Antibiot.*, 1987; **40**: 1464–7.
78. K.L. Rinehart Jr, Chemistry of the neomycins. V. Differentiation of the neomycin complex. Identity of framycetin and neomycin B. Compounds obtained from methyl neobiosaminide B. *J. Am. Chem. Soc.*, 1960; **82**: 3938–46.
79. H. Umezawa, M. Ueda, M. Maeda, *et al.*, Production and isolation of a new antibiotic: kanamycin. *J. Antibiot.*, 1957; **10A**: 181–8.
80. M.J. Cron, D.L. Evans, F.M. Palermiti, *et al.*, Kanamycin. V. The structure of kanosamine. *J. Am. Chem. Soc.*, 1958; **80**: 4741–2.
81. S. Umezawa, K. Tatsuta, S. Koto, The total synthesis of kanamycin A. *J. Antibiot.*, 1968; **21**: 367–8.
82. D.J. Mason, A. Dietz, C. Deboer, Lincomycin, a new antibiotic. I. Discovery and biological properties. *Antimicrob. Agents Chemother.*, 1962: 554–9.
83. B.J. Magerlein, Lincomycin. X. The chemical synthesis of lincomycin. *Tetrahedron Lett.*, 1970; **1**: 33–6.
84. M.J. Weinstein, G.M. Luedemann, E.M. Oden, *et al.*, Gentamicin, a new antibiotic complex from Micromonospora. *J. Med. Chem.*, 1963; **6**: 463–4.
85. D.J. Cooper, M.D. Yudis, R.D. Guthrie, A.M. Prior, The gentamicin antibiotics. Part I. Structure and absolute stereochemistry of methyl garosaminide. *J. Chem. Soc. (C)*, 1971: 960–3.
86. J. Ehrlich, Q.R. Bartz, R.M. Smith, *et al.*, Chloramphenicol, a new antibiotic from soil actinomycetes. *Science*, 1947; **106**: 417.
87. Q.R. Bartz, Isolation and characterization of chloromycetin. *J. Biol. Chem.*, 1948; **172**: 445–50.
88. J. Controulis, M.C. Rebstock, H.M. Crooks Jr, Chloramphenicol (Chloromycetin). IV. Synthesis. *J. Am. Chem. Soc.*, 1949; **71**: 2463–8.
89. B.M. Duggar, Aureomycin: a product of the continuing search for new antibiotics. *Ann. N.Y. Acad. Sci.*, 1948; **51**: 177–81.
90. US Pat. 1949: 2482055 (to American Cyanamid).
91. C.R. Stephens, L.H. Conover, F.A. Hochstein, *et al.*, Terramycin. VIII. Structure of aureomycin and terramycin. *J. Am. Chem. Soc.*, 1952; **74**: 4976–7.
92. J.H. Boothe, A.S. Kende, T.L. Fields, R.G. Wilkinson, Total synthesis of tetracyclines. I. (±)-Dedimethylamino-12a-deoxy-6-demethylanhydrochlorotetracycline. *J. Am. Chem. Soc.*, 1959; **81**: 1006–7.
93. A.C. Finlay, G.L. Hobby, S.Y. P'an, *et al.*, Terramycin, a new antibiotic. *Science*, 1950; **111**: 85.
94. P.P. Regna, I.A. Solomons, The chemical and physical properties of terramycin. *Ann. N.Y. Acad. Sci.*, 1950; **53**: 229–37.
95. US Pat. 1950: 2516080 (to Pfizer).
96. J.H. Boothe, J. Morton II, J.P. Petisi, *et al.*, Tetracycline. *J. Am. Chem. Soc.*, 1953; **75**: 4621.
97. US Pat. 1955: 2712517 (to Bristol Laboratories).
98. L.H. Conover, Terramycin. XI. Tetracycline. *J. Am. Chem. Soc.*, 1953; **75**: 4622–3.
99. J.R.D. McCormick, N.O. Sjolander, U. Hirsch, *et al.*, A new family of antibiotics: the demethyltetracyclines. *J. Am. Chem. Soc.*, 1957; **79**: 4561–3.
100. US Pat. 1953: 2653889 (to Lilly).

101. K. Gerzon, E.H. Flynn, M.V. Sigal, *et al.*, Erythromycin. VIII. Structure of dihydroerythronolide. *J. Am. Chem. Soc.*, 1956; **78**: 6396–408.
102. C. Vezina, A. Kudelski, S.N. Sehgal, Rapamycin (AY-22,989), a new antifungal antibiotic. I. Taxonomy of the producing streptomycete and isolation of the active principle. *J. Antibiot.*, 1975; **28**: 721–6.
103. K.C. Nicolaou, T.K. Chakraborty, A.D. Piscopio, *et al.*, Total synthesis of rapamycin. *J. Am. Chem. Soc.*, 1993; **115**: 4419–20.
104. T. Kino, H. Hatanaka, M. Hashimoto, *et al.*, FK-506, a novel immunosuppressant isolated from a Streptomyces. I. Fermentation, isolation, and physico-chemical and biological characteristics. *J. Antibiot.*, 1987; **40**: 1249–55.
105. H. Tanaka, A. Kuroda, H. Marusawa, *et al.*, Structure of FK506, a novel immunosuppressant isolated from Streptomyces. *J. Am. Chem. Soc.*, 1987; **109**: 5031–3.
106. T.K. Jones, S.G. Mills, R.A. Reamer, *et al.*, Total synthesis of immunosuppressant (−)-FK-506. *J. Am. Chem. Soc.*, 1989; **111**: 1157–9.
107. T. Kino, H. Hatanaka, S. Miyata, *et al.*, FK-506, a novel immunosuppressant isolated from a Streptomyces. II. Immunosuppressive effect of FK-506 *in vitro*. *J. Antibiot. (Tokyo)*, 1987; **40**: 1256–65.
108. G. Albers-Schönberg, B.H. Arison, O.D. Hensens, *et al.*, Avermectins. Structure determination. *J. Am. Chem. Soc.*, 1981; **103**: 4216–21.
109. J.R. Egerton, D.A. Oslind, L.S. Blair, *et al.*, Avermectins, new family of potent anthelmintic agents. Efficacy of the B1 component. *Antimicrob. Agents Chemother.*, 1979; **15**: 372–8.
110. P. Sensi, P. Margalith, M.T. Timbal, Rifamycin, a new antibiotic. Preliminary report. *Farmaco, Ed. Sci.*, 1959; **14**: 146–7.
111. P. Sensi, Rifampicin, in *Chronicles of Drug Discovery*, ed. J.S. Bindra, D. Lednicer, Chichester: John Wiley; 1982, p. 201.
112. W. Oppolzer, V. Prelog, P. Sensi, [The composition of rifamycin B and related rifamycins]. *Experientia*, 1964; **20**: 336–9.
113. E.L. Hazen, R. Brown, Fungicidin, an antibiotic produced by a soil actinomycete. *Proc. Soc. Exp. Biol. Med.*, 1951; **76**: 93–7.
114. R. C. Pandey, K. L. Rinehart, Polyene antibiotics. VII. Carbon-13 nuclear magnetic resonance evidence for cyclic hemiketals in the polyene antibiotics amphotericin B, nystatin A1, tetrin A, tetrin B, lucensomycin, and pimaricin1,2. *J. Antibiot.*, 1976; **29**: 1035–42.
115. M. Oura, T.H. Sternberg, E.T. Wright, A new antifungal antibiotic, amphotericin B. *Antibiot. Annual*, 1955–1956; **3**: 566–73.
116. J. Dutcher, The discovery and development of amphotericin B. *Dis. Chest.*, 1968; **54** (Suppl. 1): 296–8.
117. K. Maeda, T. Osato, H. Umezawa, A new antibiotic, azomycin. *J. Antibiot.*, 1953; **6**: 182.
118. S. Nakamura, Structure of azomycin, a new antibiotic. *Pharm. Bull.*, 1955; **3**: 379–83.
119. G.C. Lancini, E. Lazzari, The synthesis of azomycin (2-nitroimidazole). *Experientia*, 1965; **21**: 83.
120. M.H. McCormick, W.M. Stark, G.E. Pittenger, *et al.*, Vancomycin, a new antibiotic. I. Chemical and biologic properties. *Antibiot. Annual*, **1955–56**: 606–11.
121. C.M. Harris, H. Kopecka, T.M. Harris, Vancomycin: structure and transformation to CDP-I. *J. Am. Chem. Soc.*, 1983; **105**: 6915–22.
122. M.R. Bardone, M. Paternoster, C. Coronelli, Teichomycins, new antibiotics from *Actinoplanes teichomyceticus* nov. sp. II. Extraction and chemical characterization. *J. Antibiot.*, 1978; **31**: 170–7.
123. F. Parenti, G. Beretta, M. Berti, V. Arioli, Teichomycins, new antibiotics from *Actinoplanes teichomyceticus* Nov. Sp. I. Description of the producer strain, fermentation studies and biological properties. *J. Antibiot.*, 1978; **31**: 276–83.
124. J.C.J. Barna, D.H. Williams, D.J.M. Stone, *et al.*, Structure elucidation of the teicoplanin antibiotics. *J. Am. Chem. Soc.*, 1984; **106**: 4895–902.
125. A.G. Brown, M.J. Pearson, R. Southgate, in *Comprehensive Medicinal Chemistry*, Vol. 2, eds C. Hansch, P.G. Sammes, J.B. Taylor. Oxford: Pergamon Press; 1990, p. 665.
126. J.S. Kahan, F.M. Kahan, R. Goegelman, *et al.*, Thienamycin, a new beta-lactam antibiotic. I. Discovery, taxonomy, isolation and physical properties. *J. Antibiot.*, 1979; **32**: 1–12.
127. G. Albers-Schönberg, B.H. Arison, O.D. Hensens, *et al.*, Structure and absolute configuration of thienamycin. *J. Am. Chem. Soc.*, 1978; **100**: 6491–9.
128. C. Reading, M. Cole, Clavulanic acid: a beta-lactamase-inhiting beta-lactam from *Streptomyces clavuligerus*. *Antimicrob. Agents Chemother.*, 1977; **11**: 852–7.
129. P.A. Hunter, K. Coleman, J. Fisher, D. Taylor, *In vitro* synergistic properties of clavulanic acid, with ampicillin, amoxycillin and ticarcillin. *J. Antimicrob. Chemother.*, 1980; **6**: 455–70.
130. H. Brockmann, N. Grubhofer, Actinomycin C. *Naturwissen.*, 1949; **36**: 376–7.

131. C. Hackmann, Experimentelle Untersuchugen über die Wirkung von actinomycin C (HBF 386) bei bösartigen Geschwülsten. *Z. Krebsforsch.*, 1952; **58**: 607–13.
132. R.A. Manaker, F.J. Gregory, L.C. Vining, S.A. Waksman, *Antibiot. Ann.*, 1954–1955: 853.
133. E. Bullock, A.W. Johnson, Actinomycin. Part V. The structure of actinomycin D. *J. Chem. Soc.*, 1957; 3280–5.
134. H. Brockmann, C. Manegold, *Naturwissen.*, 1964; **51**: 383.
135. S. Farber, G. D'Angio, A. Evans, A. Mitus, Clinical studies on actinomycin D with special reference to Wilm's tumor in children. *Ann. N.Y. Acad. Sci.*, 1960; **89**: 421–5.
136. H. Umezawa, Bleomycin and other antitumor antibiotics of high molecular weight. *Antimicrob. Agents Chemother.*, 1965; **19**: 1079–85.
137. H. Umezawa, K. Maeda, T. Takuchi, Y. Okami, New antibiotics, bleomycin A and B. *J. Antibiot.*, 1966; **19**: 200–9.
138. A. Di Marco, M. Gaetani, P. Orezzi, *et al.*, Daunomycin: a new antibiotic of the rhodomycin group. *Nature*, 1964; **201**: 706–7.
139. F. Arcamone, G. Franceschi, P. Orezzi, *et al.*, Daunomycin. I. The Structure of Daunomycinone. *J. Am. Chem. Soc.*, 1964; **86**: 5334–5.
140. F. Arcamone, G. Franceschi, S. Penco, A. Selva, Adriamycin (14-hydroxy daunorubicin), a novel antitumor antibiotic. *Tetrahedron Lett.*, 1969; **13**: 1007–10.
141. T. Oki, Y. Matsuzawa, A. Yoshimoto, *et al.*, New antitumor antibiotics, aclacinomycins A and B. *J. Antibiot.*, 1975; **28**: 830–4.
142. A. Imada, K. Kitano, K. Kintaka, *et al.*, Sulfazecin and isosulfazecin, novel beta-lactam antibiotics of bacterial origin. *Nature*, 1981; **289**: 590–1.
143. R.B. Sykes, C.M. Cimarusti, D.P. Bonner, *et al.*, Monocyclic beta-lactam antibiotics produced by bacteria. *Nature*, 1981; **291**: 489–91.

Antibiotic Analogues

Just as with plant products and biochemical substances from mammalian sources, analogues of naturally occurring antibiotics were prepared in order to overcome disadvantages that had become manifest. The earliest of these were analogues of benzylpenicillin.

6-AMINOPENICILLANIC ACID

John Sheehan, one of the leading Merck chemists working on the synthesis of penicillin, continued his work in this area after he left the company to become Professor of Chemistry at Massachusetts Institute of Technology. With his research financially supported by Bristol Laboratories, Sheehan devised novel techniques that would make it possible to synthesise the unstable beta-lactam ring, the stumbling block that had thwarted all previous attempts. In 1957, he finally synthesised phenoxymethylpenicillin.[1] The overall yield was around 1%, but within two years Sheehan had increased this to more than 60%. His synthesis enabled him to prepare 6-aminopenicillanic acid, the key to making analogues with novel side chains by reacting it with acid chlorides. This was a major improvement on the only alternative method of obtaining new penicillins, namely through addition of a chemical precursor to the liquor in which the *Penicillium* mould grew.

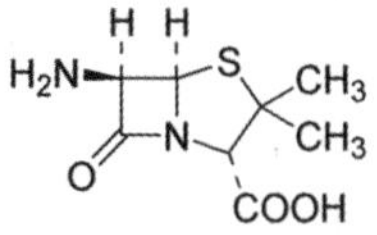

6-aminopenicillanic acid

Sheehan's synthesis of 6-aminopenicillanic acid would have been difficult to scale-up and develop for commercial application, but the problem was avoided by a separate development the following year. Ralph Batchelor, Peter Doyle, John Nayler and Rolinson of the recently established Beecham Research Laboratories at Betchworth, Surrey, made a remarkable discovery.[2] They were newcomers to the field of penicillin research and had been advised by Ernst Chain to prepare 4'-aminobenzylpenicillin as it might be possible to make novel derivatives from this. In the course of extracting the new penicillin, its acetyl derivative crystallised out of solution as a pure compound. Since the crystals could easily be converted to the desired 4'-aminobenzylpenicillin, it was obvious that addition of an acetylating agent to the mould juice would facilitate the isolation process by converting all of the 4'-amino compound to the less soluble acetylamino derivative. When this was done, a discrepancy appeared in the microbiological assay of the mould juice, which now indicated enhanced antibacterial activity. In an inspired interpretation of this, the Beecham researchers concluded that 6-aminopenicillanic acid must have been present in the mould juice. Only after conversion to its acetyl derivative did it exhibit sufficient activity to affect the assay result. Subsequent tests confirmed that 6-aminopenicillanic acid was always present in mould juice, and was a stable substance, contrary to expectation. The Beecham team quickly exploited their discovery

Drug Discovery. A History. W. Sneader.
©2005 John Wiley & Sons Ltd

by developing methods of obtaining large quantities of this key intermediate by fermentation and their first patent was applied for in August 1958.

The availability of large amounts of 6-aminopenicillanic acid was a turning point in antibiotic chemotherapy, for it was now easier to make novel semi-synthetic penicillins rather than indulging in expensive screening of soil samples in the hope of finding that rare product – a non-toxic antibiotic. Most of the new antibiotics introduced since the early 1960s have been intended for the treatment of conditions that would not be expected to respond to penicillin therapy, such as non-bacterial infections and cancer.

Since Beecham Research Laboratories employed Sheehan's method of converting 6-aminopenicillanic acid into therapeutically useful penicillins, meetings were held with him and representatives of Bristol Laboratories. Early agreement was reached whereby both companies would collaborate in the development of semi-synthetic penicillins. After this, harmony between the three parties rapidly dissipated. Bristol and Beecham went their own ways, and for the next 20 years there were interminable legal wrangles between Sheehan and Beecham over patent rights to the new semi-synthetic penicillins. The matter is too complex to be dealt with here, but suffice it to say that the US Board of Patent Interferences ruled in favour of Sheehan in 1979. He has written his own account of what transpired.[3]

Phenoxymethylpenicillin

The ability to synthesise new penicillins at will permitted the shortcomings of benzylpenicillin (penicillin G) to be tackled. It had been the only penicillin in general use until the mid-1950s, when phenoxymethylpenicillin (penicillin V) was introduced. This was one of a number of penicillins originally obtained in 1948 by scientists at the Lilly Research Laboratories after they had pioneered the technique of adding different chemical precursors to the culture medium in which the *Penicillium* mould was fermented. Its true value was not recognised until four years later when Ernst Brandl, a chemist in an Austrian penicillin production plant who was attempting to extract phenoxymethylpenicillin, noticed that it had the unique property of being resistant to degradation by dilute acid.[4] This meant that it should withstand exposure to gastric acid, thus avoiding the extensive degradation when benzylpenicillin was taken by mouth. Eli Lilly and Company exploited this by introducing the potassium salt of phenoxymethylpenicillin as the first reliable orally active penicillin. It is not used to treat serious infections as it is less active than benzylpenicillin and plasma levels vary after oral administration. It is prescribed mainly for respiratory infections in children.

The acid stability of phenoxymethylpenicillin and similar penicillins containing a hetero-atom in the side chain α-carbon arises from reduced interaction between the amide carbonyl and the beta-lactam group, which otherwise triggers decomposition.

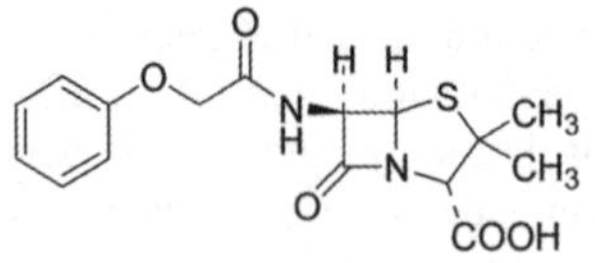

phenoxymethylpenicillin

Beta-Lactamase Resistant Penicillins

Beecham Research Laboratories introduced methicillin in 1961.[5] It was the first penicillin not inactivated by penicillinases, enzymes produced by strains of staphylococci that were resistant to penicillins. These enzymes split the beta-lactam ring open, and were of several types. The Beecham team established that stability towards attack by them was achieved by placing bulky substituents in close proximity to the labile part of the beta-lactam ring. This is described by chemists as steric hindrance. In the particular case of methicillin, *ortho*-disubstitution on the benzene ring gave optimal stability in the presence of beta-lactamases but antibacterial potency was reduced.

methicillin

oxacillin, R, R' = H
cloxacillin, R = H; R' = Cl
dicloxacillin, R, R' = Cl
flucloxacillin, R = F; R' = Cl

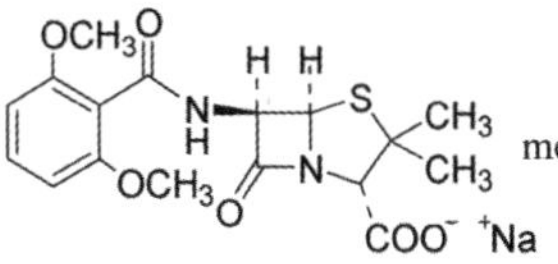

Methicillin was acid-sensitive and had to be injected. It was replaced by oxacillin in which Beecham chemists obtained steric hindrance and acid resistance by incorporating an isoxazole ring in the side chain.[6] Four analogues were synthesised by the Beecham team, namely oxacillin, cloxacillin, dicloxacillin and flucloxacillin.

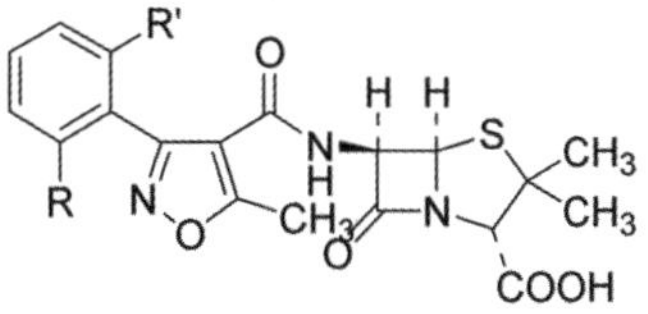

Extended Spectrum Semi-synthetic Penicillins

Unlike benzylpenicillin itself, the early semi-synthetic penicillins were not active against Gram-negative bacteria. Analysis of analogues of phenoxymethylpenicillin by Beecham chemists indicated that acid stability was enhanced by side chain substituents that attracted electrons. The most suitable substituent proved to be the amino group, the compound with this being called 'ampicillin'.[7] It had a wider spectrum of activity against Gram-negative bacteria than had its parent compound, benzylpenicillin, though activity against Gram-positive bacteria was somewhat reduced. These differences were due to variation in the ability of the penicillins to penetrate into bacteria. All penicillins, however, act in the same manner, namely by disrupting bacterial cell wall synthesis.

Ampicillin was an outstanding commercial success, but it had two drawbacks. The more serious was an inability to avoid destruction by bacteria capable of producing beta-lactamases, notably *Staphylococcus aureus* and *Escherichia coli*. Many microorganisms that were once sensitive to ampicillin became resistant by producing beta-lactamases. The other drawback was the polar nature of ampicillin, which reduced its ability to be absorbed from the gut. Less than half the dose was absorbed, with the result that some patients experienced diarrhoea through upset of the normal bacterial flora in the gut by unabsorbed ampicillin. This problem was largely overcome by the introduction by Beecham of the phenolic analogue amoxicillin, which had superior absorption characteristics.

ampicillin

amoxicillin

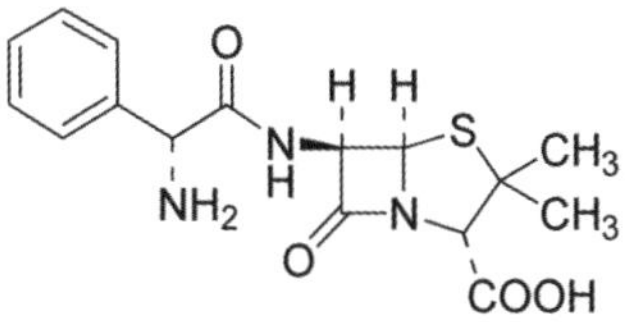

A different approach to overcoming the poor absorption of ampicillin was taken by chemists at Leo Pharmaceutical Products in Denmark when they masked the polar carboxylic acid function in ampicillin to form pivampicillin.[8] The resulting ester decomposed after absorption from the gut, liberating ampicillin. Other similar prodrugs of ampicillin were then introduced, including talampicillin and bacampicillin.

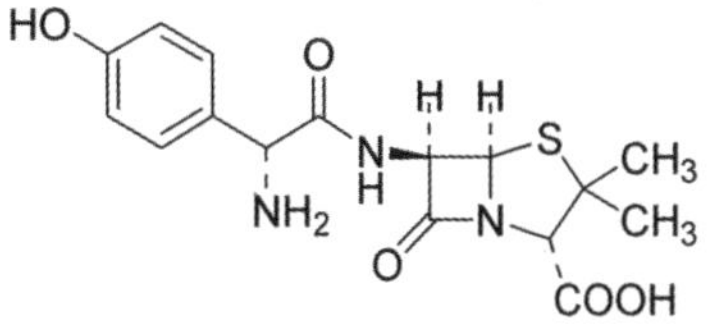

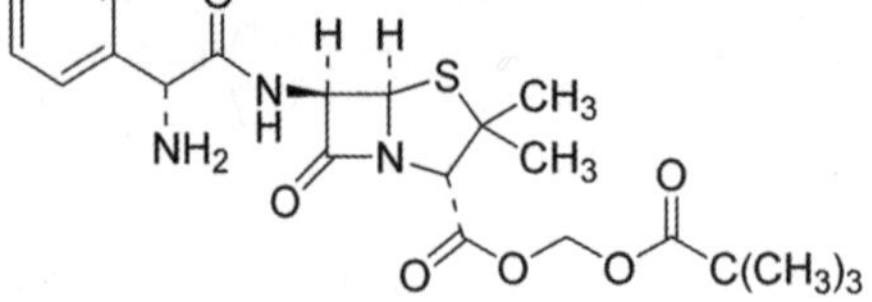

pivampicillin

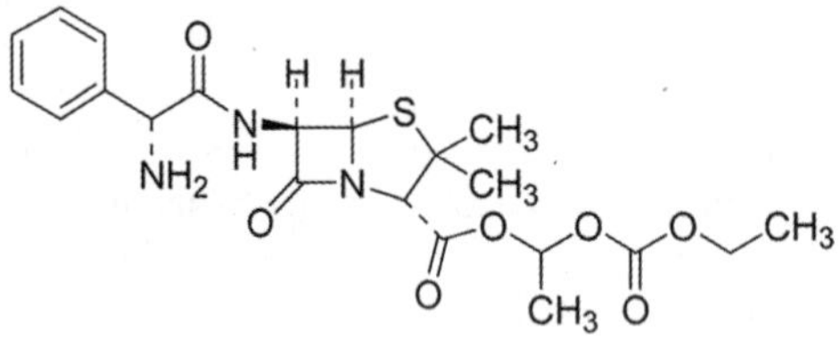

bacampicillin

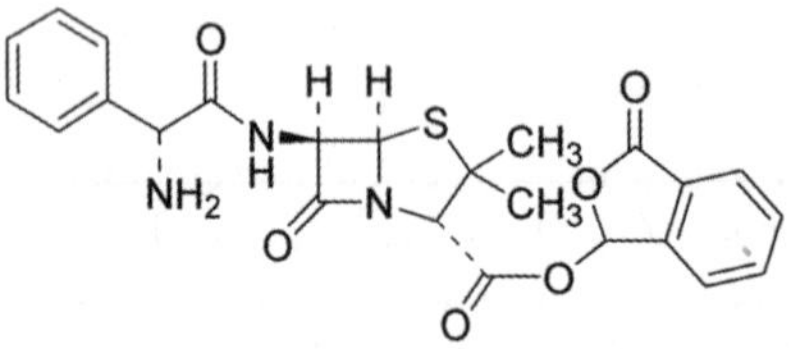

talampicillin

Both Pfizer and Beecham chemists reasoned that as the introduction of the basic amino group into the side chain of benzylpenicillin afforded a derivative with a wider spectrum of activity, namely ampicillin, it would be worth while to examine the effect of introducing an acidic carboxyl function. The resulting compound, carbenicillin, was obtained in 1964.[9] It turned out to be one of the first penicillins to have high activity against *Pseudomonas aeruginosa*, a particularly troublesome pathogen that could be life-threatening in severely burned patients and in those whose immune system was compromised.

Because of the polar carboxyl group on its side chain, carbenicillin was poorly absorbed from the gut. Beecham chemists then prepared ticarcillin, which incorporated the thiophene ring previously introduced into semi-synthetic cephalosporins. Ticarcillin had strong activity against *P. aeruginosa* and so superseded carbenicillin. It is sensitive to beta-lactamases, but a combination product combining it with the beta-lactamase inhibitor clavulanic acid has been marketed.

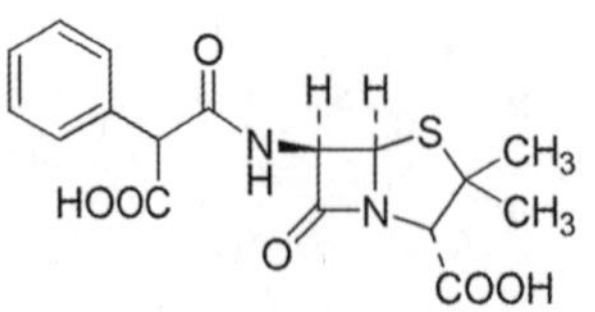

carbenicillin

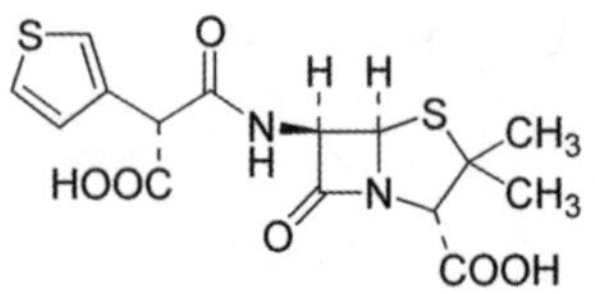

ticarcillin

Attempts to enhance the activity of ampicillin against Gram-negative bacteria by preparing simple *N*-acylated analogues resulted in a reduction in potency, though bulkier acyl groups were more promising.[10] When the ureidopenicillin known as 'azlocillin' was synthesised by Bayer chemists at Elberfeld in 1971, it proved to be a broad-spectrum antibiotic.[11] It was one of the most effective analogues of ampicillin for use against *Pseudomonas* infections, being even more active than ticarcillin, but it and other ureidopenicillins lack resistance to beta-lactamases.[12]

azlocillin

piperacillin

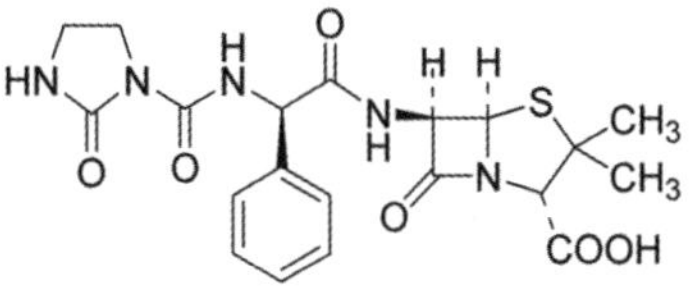

Piperacillin is a broad-spectrum ureidopenicillin which is injected for the treatment of severe infections, especially pseudomonal septicaemia, or for peri-operative prophylaxis.[13] It is synergistic with aminoglycoside antibiotics and hence is often administered concurrently with one of them.

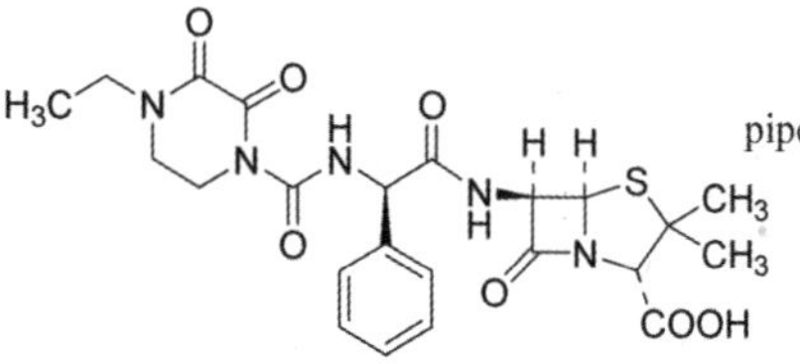

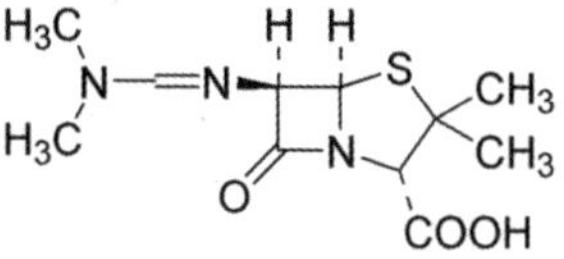

6β-dimethylformamidinopenicillanic acid

mecillinam

pivmecillinam

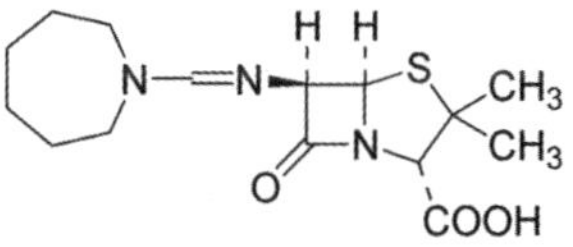

During an attempted synthesis of penicillin analogues with variant ring systems, 6β-dimethylformamidinopenicillanic acid was prepared at Leo Laboratories as a chemical intermediate. Although the synthesis of novel ring systems was unsuccessful, the novel intermediate was routinely screened. Unexpectedly, it was found to be an active antibacterial agent. A large number of analogues of it were then synthesised and tested, resulting in the introduction of mecillinam (amdinocillin).[14] It was highly effective by the parenteral route against severe infections caused by Gram-negative enteric bacteria, including salmonellae, though ineffective against pseudomonal infections. The prodrug pivmecillinam was introduced at the same time as mecillinam.[15] Masking of the polar carboxyl group enhanced lipophilicity, with a consequent enhancement of gut absorption. It is principally used to treat urinary tract infections.

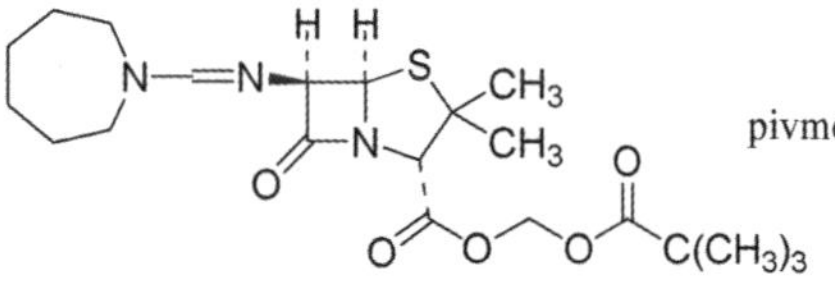

β-Lactamase Inhibitors

Pfizer chemists devised a method of removing the amino group from 6-aminopenicillanic acid and then used the product to form the β-lactamase inhibitor sulbactam.[16] It had only weak intrinsic antibacterial activity and inhibited a narrower range of β-lactamases than did clavulanic acid.[17]

sulbactam

tazobactam

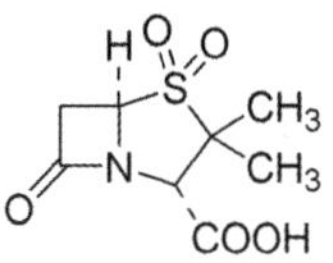

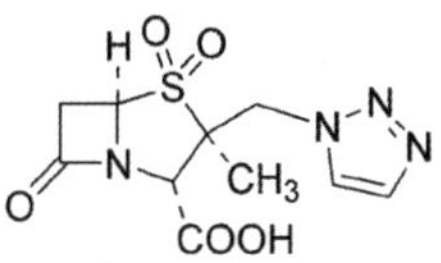

Sulbactam was formulated in combination with ampicillin in both oral and parenteral products for use against organisms that had become resistant to ampicillin.[18] These products were withdrawn from the United Kingdom market.

Tazobactam is an analogue of sulbactam that was developed by Taiho Pharmaceuticals of Tokyo.[19] It is effective against similar β-lactamases to those inhibited by clavulanic acid, but it also inhibits those from the enterobacteriaceae. It is combined with piperacillin in a parenteral formulation that is active against organisms otherwise resistant to piperacillin.[20]

CEPHALOSPORIN ANALOGUES

Edward Abraham and Guy Newton at Oxford were able to prepare 7-aminocephalosporanic acid from cephalosporin C in 1961, but only in low yield.[21] By analogy with 6-aminopenicillanic acid, it was then feasible to prepare a range of novel semi-synthetic cephalosporins. This became a reality the following year when Robert Morin and his colleagues at the Lilly Research Laboratories developed a method for obtaining good yields of 7-aminocephalosporanic acid, which enabled them to synthesise cephalothin.[22] This was the first semi-synthetic cephalosporin to be marketed. Like other so-called 'first-generation' cephalosporins subsequently developed, it had potent activity against Gram-positive bacteria, but was only mildly effective against Gram-negative bacteria. As cephalothin was poorly absorbed from the gut, it had to be injected. Frequent injections or continuous infusion was necessary because the acetoxy ester rapidly hydrolysed in the presence of plasma enzymes. This liberated a hydroxyl group that spontaneously reacted with the nearby carboxyl group to form an inactive lactone.

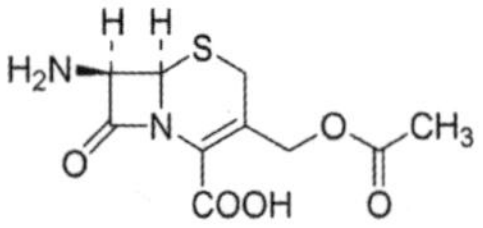

7-aminocephalosporanic acid

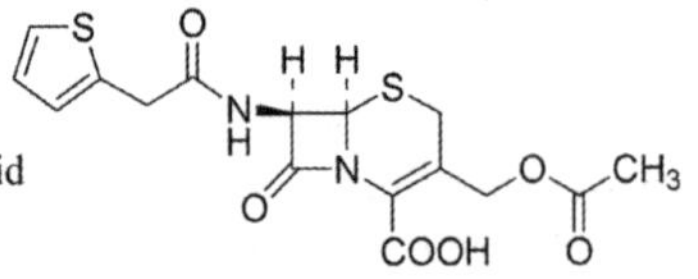

cephalothin

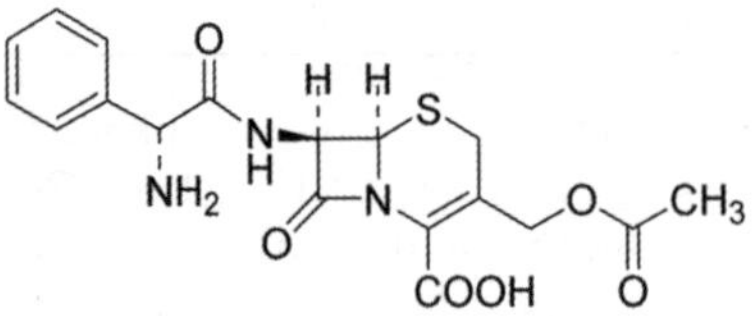

cephaloglycin

Lilly Research Laboratories next incorporated the phenylglycine side chain that Beecham Research Laboratories had previously found to confer acid stability on the beta-lactam ring of ampicillin. For cephalosporins, this also conferred some resistance to beta-lactamases as well as enhancing the activity against Gram-negative bacteria at the expense of Gram-positive activity.[23] These changes were first seen in cephaloglycin, an orally active drug that required frequent dosing because of the presence of the metabolically labile acetoxy ester.[24] This particular problem was overcome simply by removing the acetoxy ester, as first seen when Lilly introduced cefalexin in 1967.[25] Although less potent than cephaloglycin, it was the first cephalosporin to be completely absorbed after oral administration. Takeda Laboratories subsequently developed cefadroxil, the phenolic analogue of cephalexin, akin to the conversion of ampicillin to amoxicillin.[26]

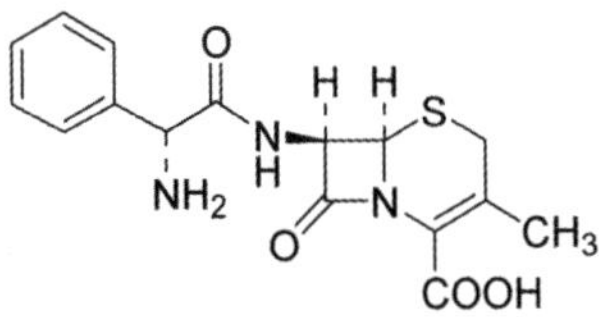

cefalexin

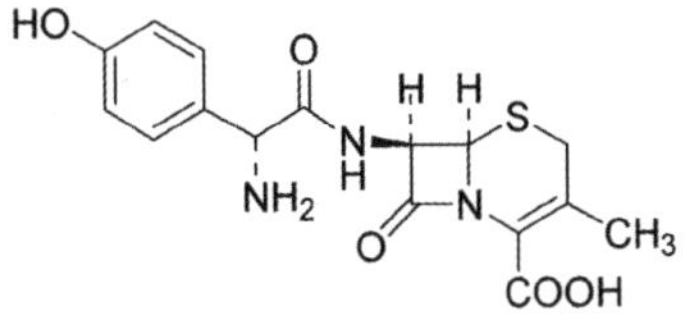

cefadroxil

At Oxford in 1961, Abraham and Newton had not only discovered that in aqueous solution the acetoxyl group of cephalosporin C could be displaced by pyridine but also that this modification altered the antibacterial spectrum.[27] Glaxo took advantage of this to prepare cephaloridine, which had a longer duration of action than cephalothin.[28] Unfortunately, this change introduced a degree of nephrotoxicity.

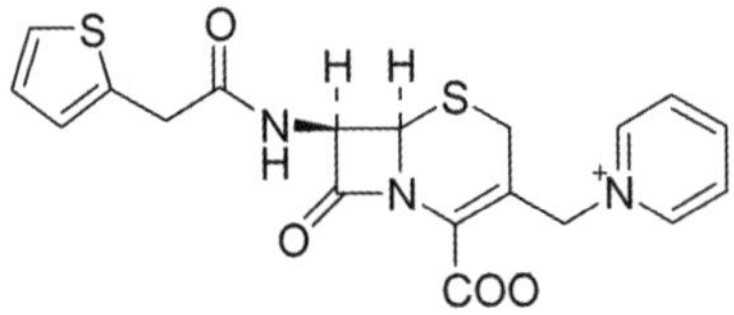

cephaloridine

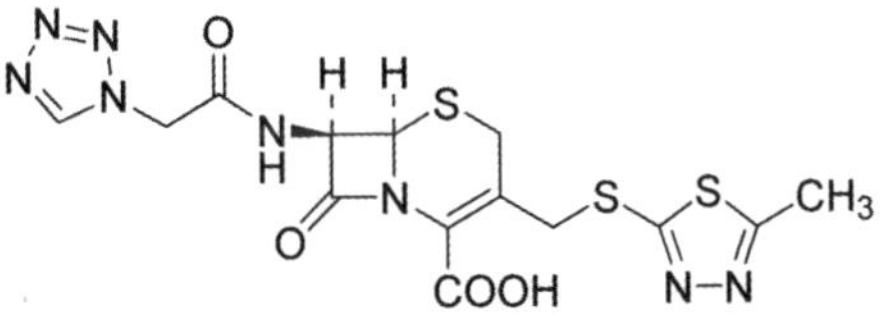

cefazolin

Kariyone and his colleagues at the Fujisawa Pharmaceutica laboratories in Osaka subsequently found that heterocyclic thiol-containing compounds could also displace the acetoxyl group. This led to their development of cefazolin, which incorporated a metabolically stable heterocyclic thiol that prolonged the plasma half-life and hence blood levels after parenteral administration.[29] Despite being more potent than cephalothin when injected, cefazolin was poorly absorbed after oral administration and had disappointing activity against Gram-negative infections.

Second-generation Cephalosporins

The 'second-generation' cephalosporins were characterised by a broadened spectrum of activity and a reduced tendency to be deactivated by beta-lactamases when compared with earlier cephalosporins. The methyltetrazole–thiomethyl moiety of cefamandole proved superior to the thiadiazole–thiomethyl of cefazolin, not only with regard to enhancement of plasma half-life but also for antibacterial potency.[30] However, as cefamandole lacked a phenylglycine side chain, it was inactive by mouth.

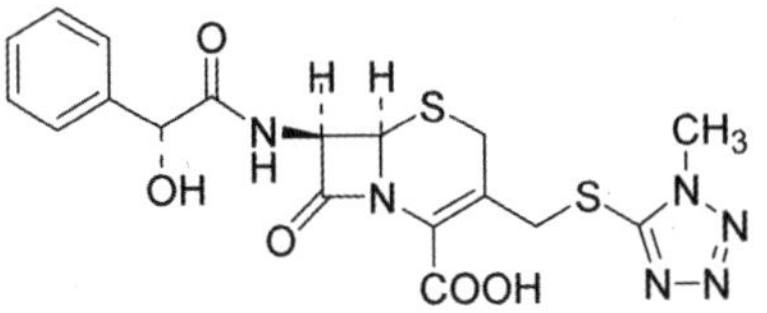

cefamandole

The 3'-carbamate ester in Glaxo's cefuroxime ensured its metabolic stability, with plasma half-life being enhanced and the frequency of dosing reduced.[31] However, of much more significance was the incorporation of an α-oximino side chain to form a *syn*-oxime, which made it the first cephalosporin to exhibit significantly enhanced resistance towards beta-lactamases. This rendered it particularly of value in treating infections caused by Gram-negative organisms resistant to penicillins; consequently it was widely prescribed in hospitals.[32] The only drawback preventing its wider use was that it could not be given by mouth as it lacked a phenylglycine side chain. This was addressed when Lilly marketed cefaclor in 1979.[33] It became one of the best selling drugs in the world until its patent expired in 1992. In response, Glaxo introduced cefuroxime axetil, an acetoxyethyl ester prodrug of cefuroxime. The enhanced lipophilicity conferred by the esterification favoured absorption from the gut.[34] The intact axetil ester then underwent enzymatic hydrolysis to liberate free cefuroxime both during passage across the intestinal wall and transport to the liver.

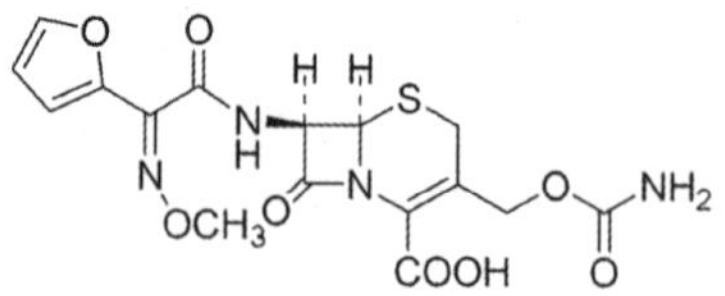

cefuroxime

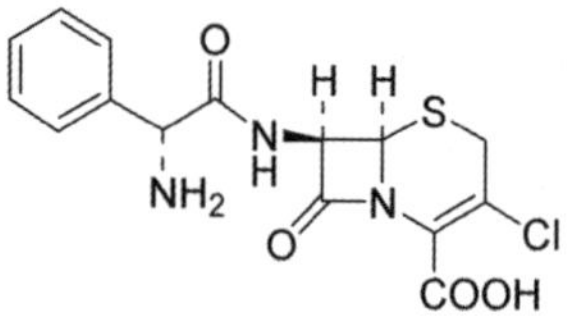

cefaclor

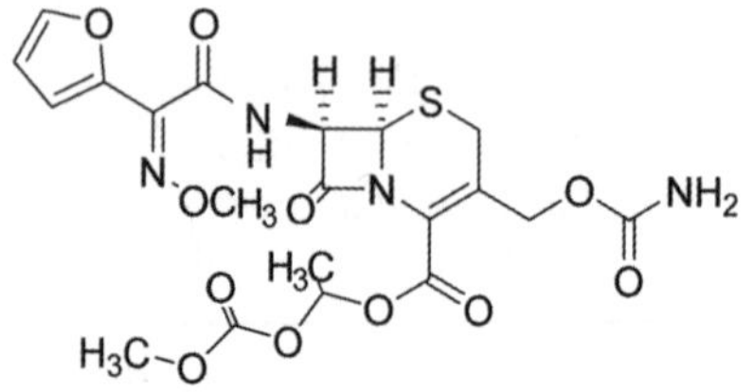

cefuroxime axetil

Third-generation Cephalosporins

The so-called 'third-generation' cephalosporins featured a wider spectrum of activity than earlier ones. Their importance lay in their outstanding activity against specific Gram-negative pathogens, in some cases even including *Pseudomonas aeruginosa*. This was partly due to their resistance to beta-lactamases as they all contained the α-oximino side chain first seen in cefuroxime. Their activity against Gram-positive bacteria, however, was inferior to that of some 'second-generation' drugs. Furthermore, as they all lacked a phenylglycine side chain some had to be injected. This was the case for cefotaxime,[35] developed in Japan by Takeda, and also Glaxo's ceftazidime.[36] The latter was of value in treating septicaemia and also had enhanced activity against *P. aeruginosa*. Ceftriaxone, developed by Hoffmann–La Roche, had a longer half-life, which permitted once-daily parenteral administration for septicaemia, meningitis or pneumonia.[37]

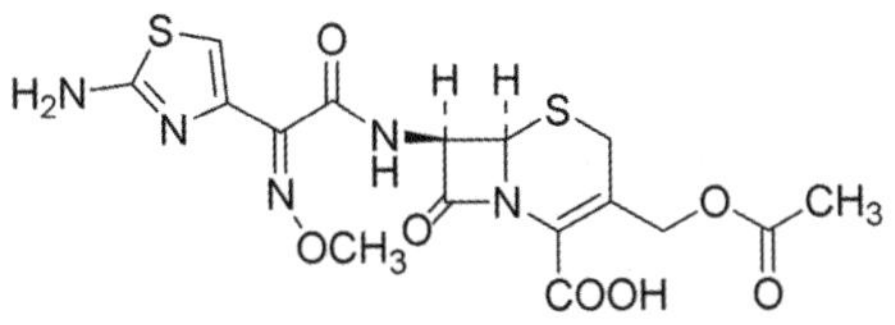

cefotaxime

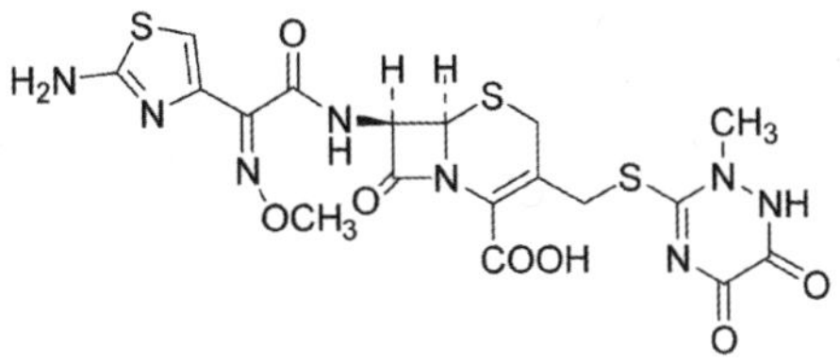

ceftriaxone

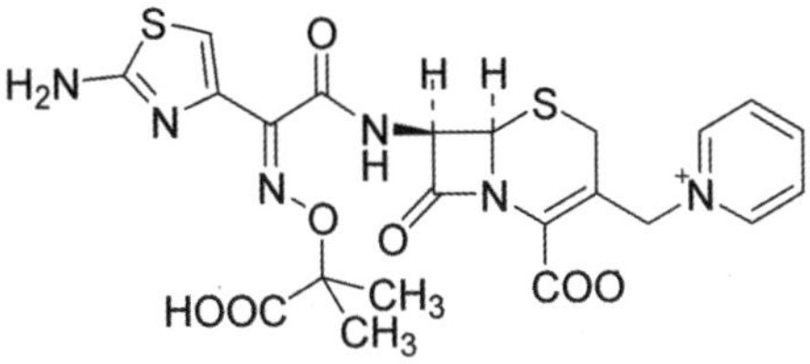

ceftazidime

Researchers from Toho University School of Medicine in Tokyo and the Fujisawa Pharmaceutical Company developed cefixime, the first of the third-generation cephalosporins that could be administered by mouth.[38] It had a long plasma half-life, permitting once- or twice-daily dosing. Unfortunately, there was a higher incidence of gastrointestinal disturbances than with other cephalosporins or penicillins. Cefpodoxime proxetil was developed at the Episome Institute in Gunma, Japan, and found to be well absorbed after oral administration.[39]

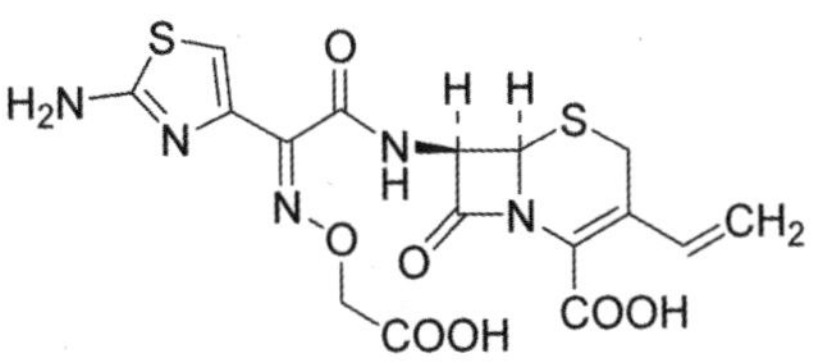

cefixime

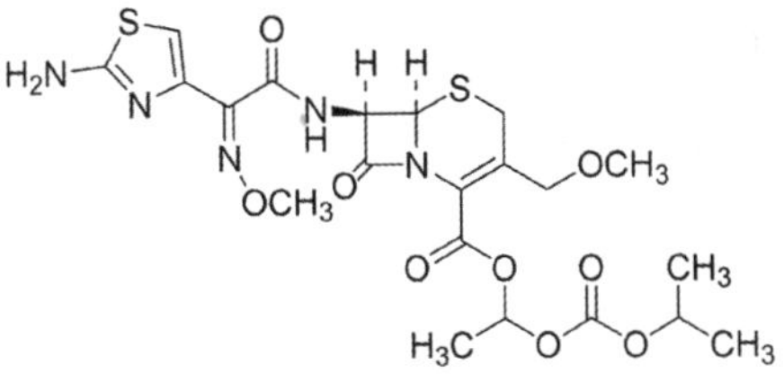

cefpodoxime proxetil

Fourth-generation Cephalosporins

The description 'fourth-generation' cephalosporins has been applied to those that are resistant to destruction by beta-lactamases and have high potency against not only Gram-positive as well as Gram-negative bacteria but also *P. aeruginosa*. The first of these was cefepime, which was developed at the Bristol–Myers Research Institute in Tokyo.[40] Cefpirome is another, which was introduced by Hoechst.[41] Both must be given by injection or infusion.

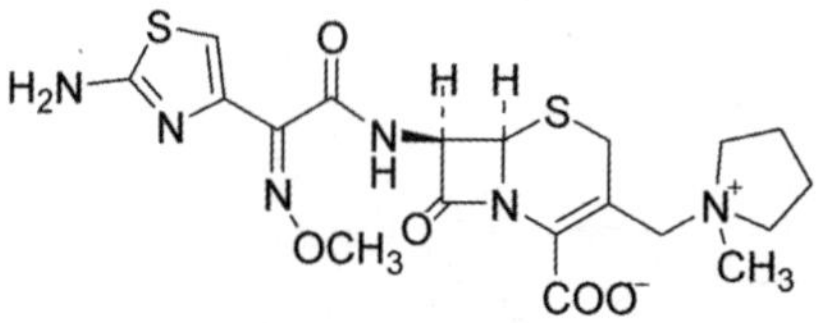

cefepime

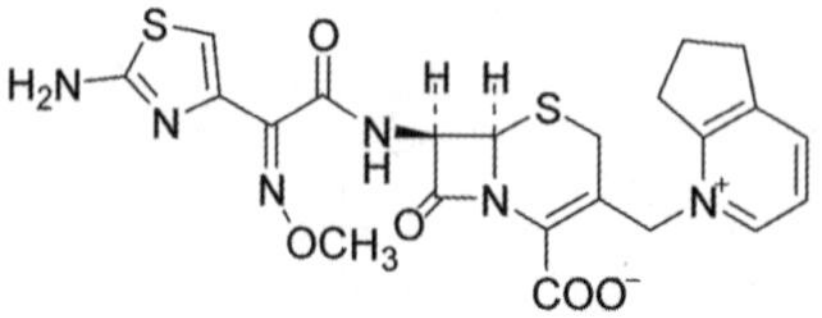

cefpirome

Cefoxitin

The premise that the lack of activity of cephamycin C against Gram-positive organisms might parallel that of the closely related cephalosporin C led Merck chemists to replace the D-aminoadipoyl side chain with others that had previously widened the spectrum of activity when incorporated into penicillins or cephalosporins.

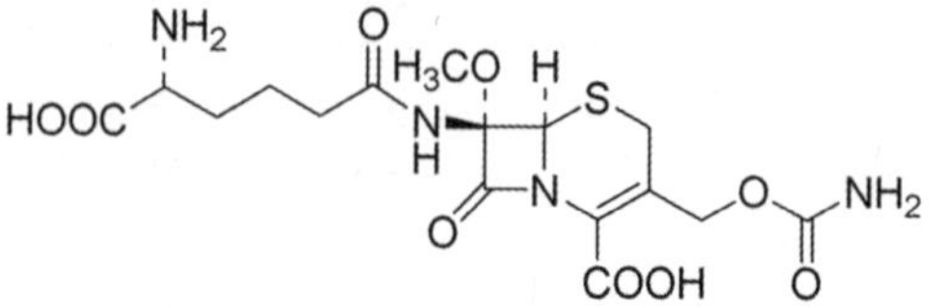

cephamycin C

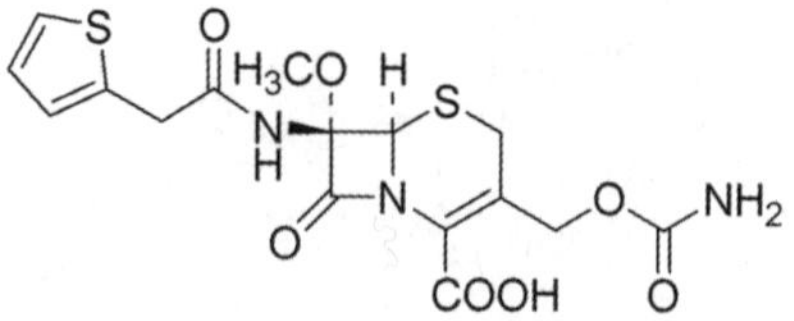

cefoxitin

From over three hundred 7α-methoxycephalosporins that were prepared, cefoxitin emerged with dramatically increased activity against Gram-positive organisms, while retaining activity against Gram-negative bacteria and resistance to destruction by beta-lactamases. Its activity against bacteria found in the bowel has led to its use in peritonitis.

AMINOGLYCOSIDE ANALOGUES

In marked contrast to the situation with penicillins and cephalosporins, there are very few semi-synthetic aminoglycoside antibiotics. The first one of any significance was amikacin, an analogue of kanamycin developed by the Bristol–Banyu Research Institute in Tokyo.[43] It was designed to have greater resistance to bacterial enzymes that inactivated kanamycin by phosphorylating and adenylating its hydroxyl groups or acylating its amino groups. The activity is otherwise similar to that of kanamycin, and it is prescribed in the treatment of serious infections caused by organisms resistant to gentamicin.

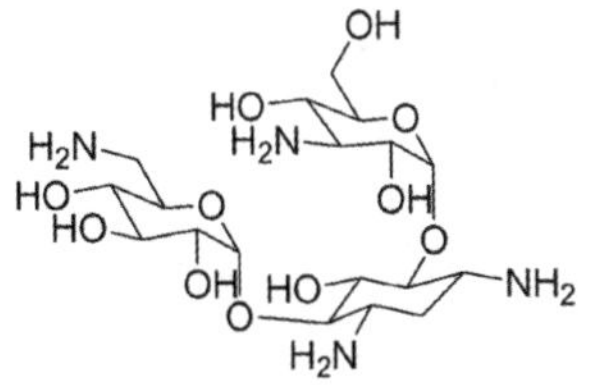

kanamycin A

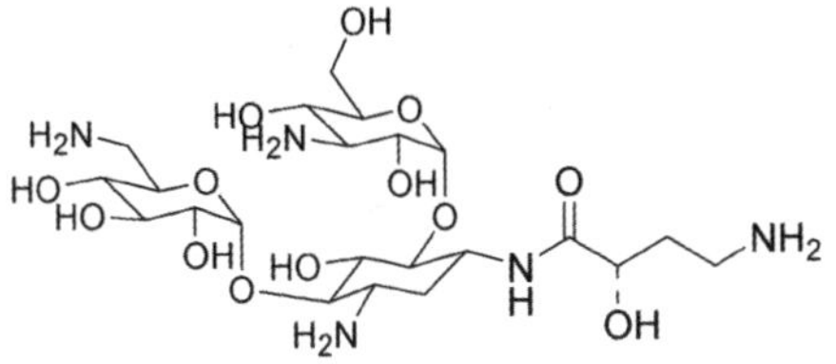

amikacin

Clindamycin is the 7-chloro analogue of lincomycin, from which it was synthesised in 1966 by Upjohn chemists Robert Birkenmeyer and Fred Kagan.[44] It is active against many Gram-positive organisms, including resistant staphylococci, as well as some Gram-negative bacteria and anaerobic organisms such as *Bacteroides fragilis*. As it has good tissue penetration, clindamycin has been prescribed for the treatment of staphylococcal bone and joint infections. Since it can cause pseudomembranous colitis, its use is restricted to serious infections where the organism is known to be susceptible. Pseudomembranous colitis can be fatal, especially in elderly patients.

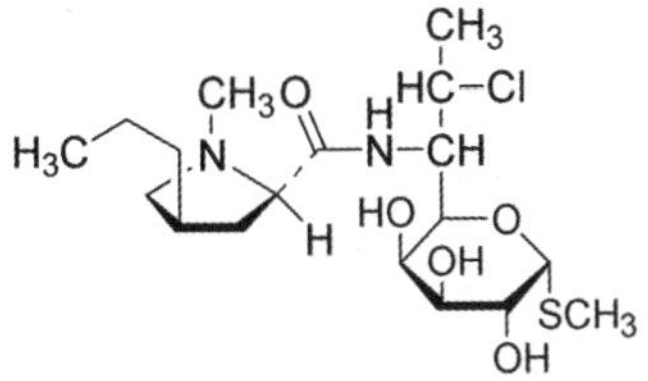

clindamycin

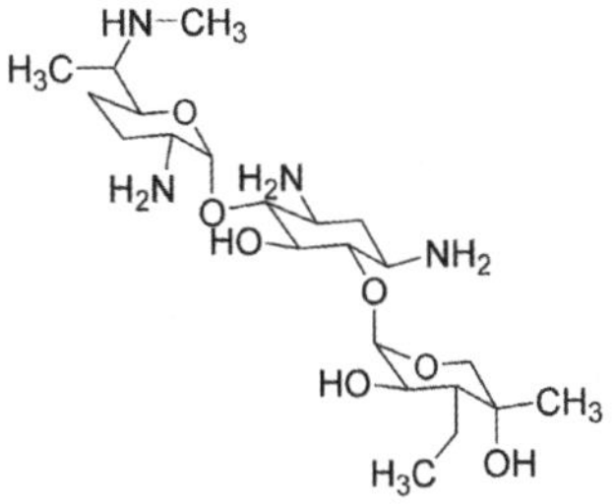

gentamicin C₁

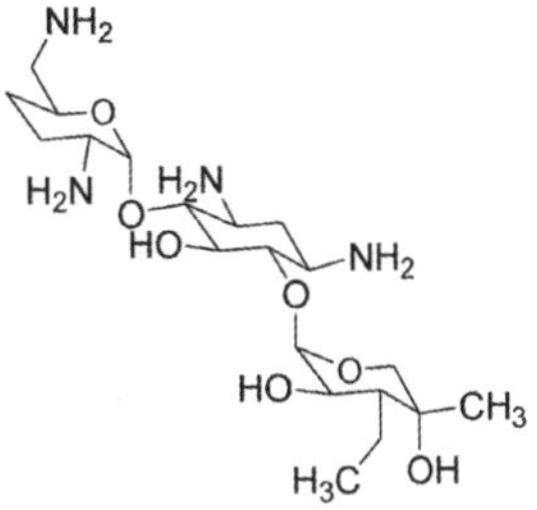

netilmicin

Netilmicin is a demethylated analogue of gentamicin C₁ that was synthesised by Schering Corporation chemists in Bloomfield, New Jersey, in 1975.[45] It had similar activity to gentamicin, but with a significantly lower level of renal toxicity, which made it the aminoglycoside of choice in elderly patients or those with renal failure.

TETRACYCLINE ANALOGUES

Liquid formulations of the tetracyclines presented a variety of problems, particularly as they had low water solubility. Dissolving them in mildly acid solution caused epimerisation at position 4, resulting in reversal of the stereochemistry of the dimethylamin substituent, with greatly diminished activity. Acid-catalysed elimination of the tertiary hydroxyl group was also a problem. For these reasons, injections had to be used as soon as possible after reconstitution of the hydrochloride salt in water. These difficulties were overcome by reacting the carboxamido group with formaldehyde and amines to form water-soluble tetracyclines such as lymecycline.[46] This prodrug, developed in Italy in 1959 by Carlo Erba chemists Willy Logemann and Francesco Lauria, was formulated in capsules for oral administration. It liberated tetracycline *in vivo*.

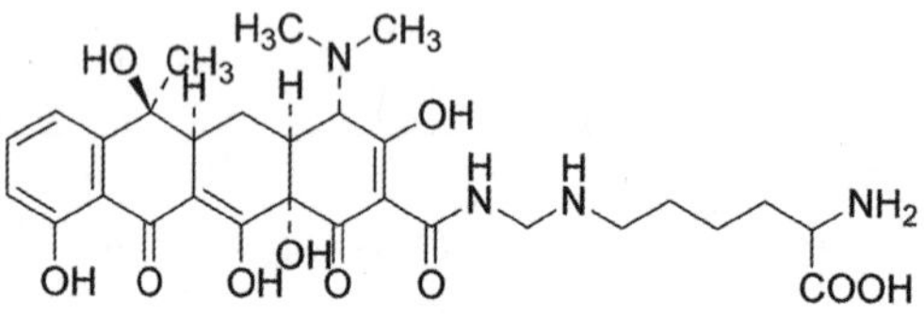

lymecycline

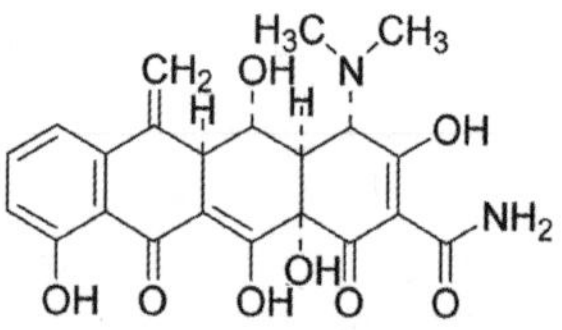

methacycline

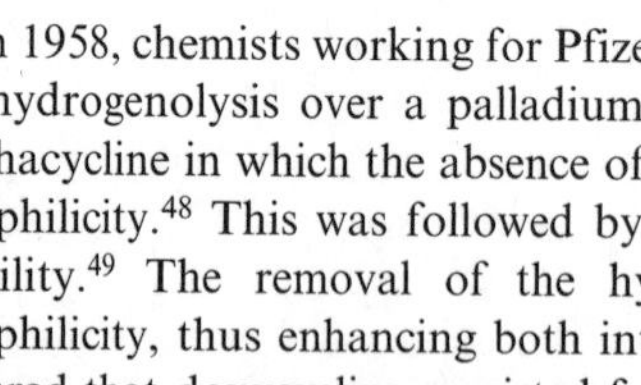

doxycycline

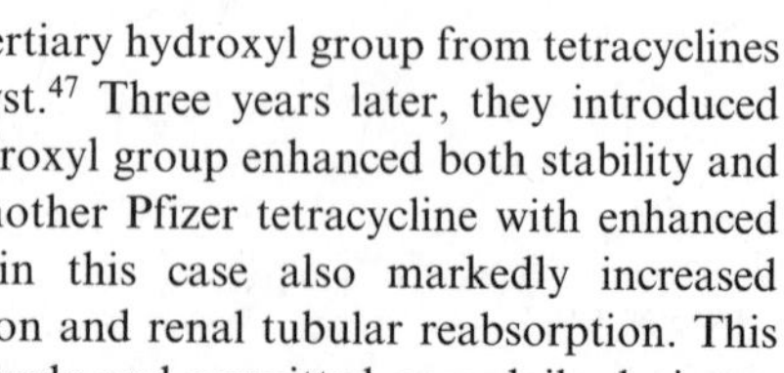

minocycline

In 1958, chemists working for Pfizer removed the tertiary hydroxyl group from tetracyclines by hydrogenolysis over a palladium charcoal catalyst.[47] Three years later, they introduced methacycline in which the absence of the tertiary hydroxyl group enhanced both stability and lipophilicity.[48] This was followed by doxycycline, another Pfizer tetracycline with enhanced stability.[49] The removal of the hydroxyl group in this case also markedly increased lipophilicity, thus enhancing both intestinal absorption and renal tubular reabsorption. This ensured that doxycycline persisted for longer in the body and permitted once-daily dosing.

The stability of these 6-deoxytetracyclines towards acid permitted the use of synthetic reagents that would have destroyed other tetracyclines. This allowed novel tetracyclines to be made, but few of them had significant clinical superiority. The exception was minocycline, which was synthesised by the Lederle Division of American Cyanamid in 1965.[50] Its increased lipophilicity enhanced its ability to penetrate into various tissues, but also permitted its entry into the central nervous system to cause nausea, ataxia, dizziness and vertigo. On the credit side, however, minocycline penetrated into the cerebrospinal fluid for the prophylaxis of meningococcal meningitis. Furthermore, it achieved a high enough concentration in tears and saliva to eliminate the meningococcal carrier state. Other tetracyclines could not do this.

ANALOGUES OF MACROLIDE ANTIBIOTICS

Clarithromycin, the 6-O-methyl derivative of erythromycin, is a semi-synthetic aminoglyco-side antibiotic that was introduced by the Taisho Pharmaceutical Company of Japan.[51] It was designed to have enhanced acid stability through conversion of the tertiary alcohol at the 6-position to an ether, thus preventing intramolecular reaction with the ketone at the 9-position. Clarithromycin had a similar antimicrobial spectrum to that of erythromycin, though more effective against *Haemophilus influenzae*. It was better absorbed from the gut, with a lower incidence of gastrointestinal side effects. It was also more potent, with a longer half-life that allowed twice-daily dosing rather than the four times a day required with erythromycin.

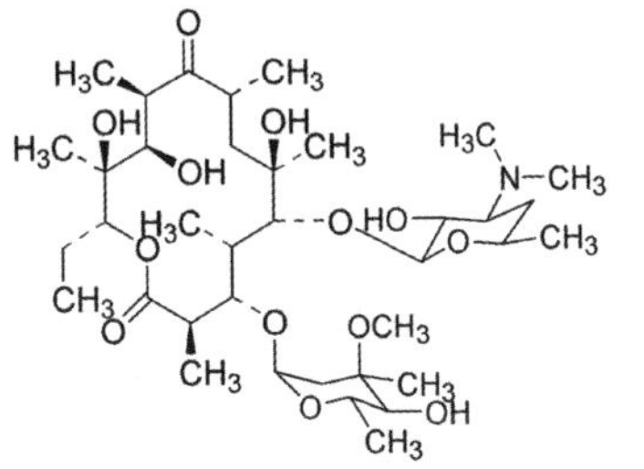

erythromycin

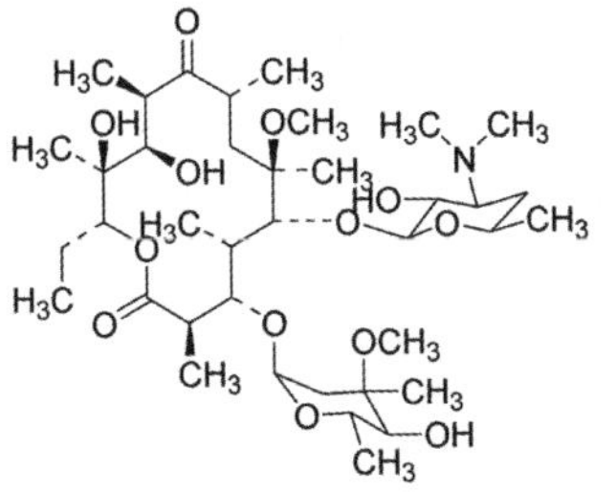

clarithromycin

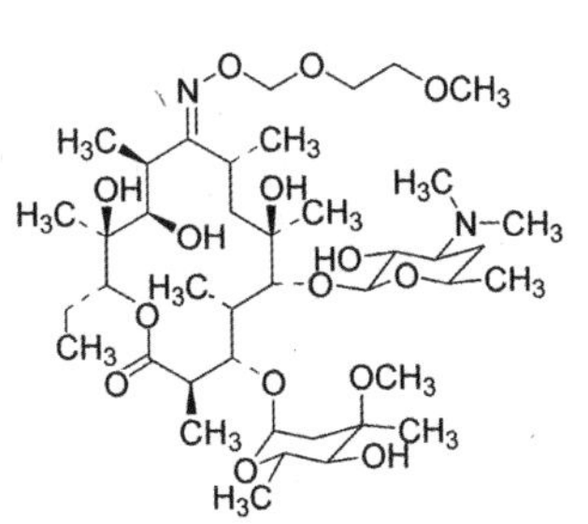

roxithromycin

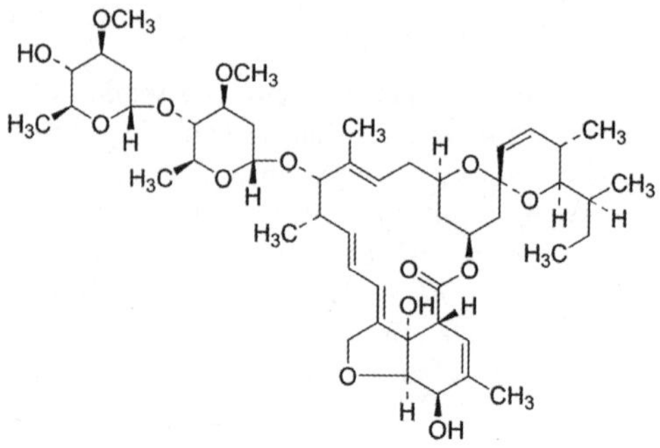

azithromycin

Roxithromycin, an ether oxime of erythromycin, was a similar acid-stable analogue with excellent bioavailability after oral administration.[52] Compared with erythromycin, it had reduced *in vitro* antimicrobial activity, but this was compensated by its markedly higher plasma concentrations. Roxithromycin is used in the treatment of both soft tissue and respiratory tract infections.

Gabrijela Kobrehel and Slobodan Djokic of Pliva in Zagreb patented azithromycin, a semi-synthetic analogue of erythromycin, in 1982.[53,54] It contained an expanded ring system that was acid-stable. The drug was then developed in collaboration with Pfizer. Its better absorption from the gut than erythromycin, coupled with a plasma half-life of around 70 hours, permitted once-daily dosing.

Although azithromycin was possibly inferior to erythromycin against Gram-positive bacteria, it was more effective against a number of Gram-negative organisms, including *H. influenzae*. Side-effects were also less troublesome than those of erythromycin, being mainly gastrointestinal.

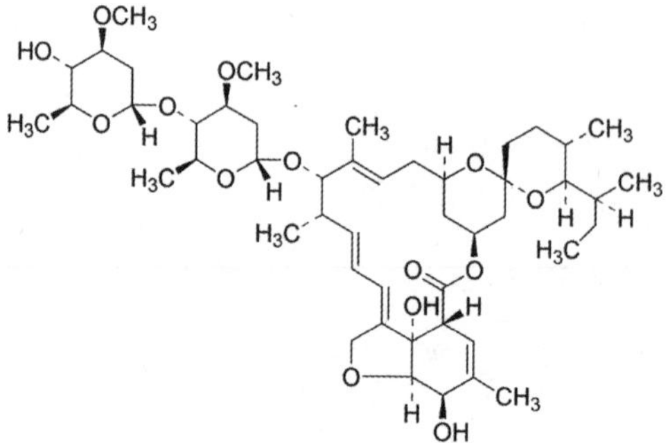

ivermectin

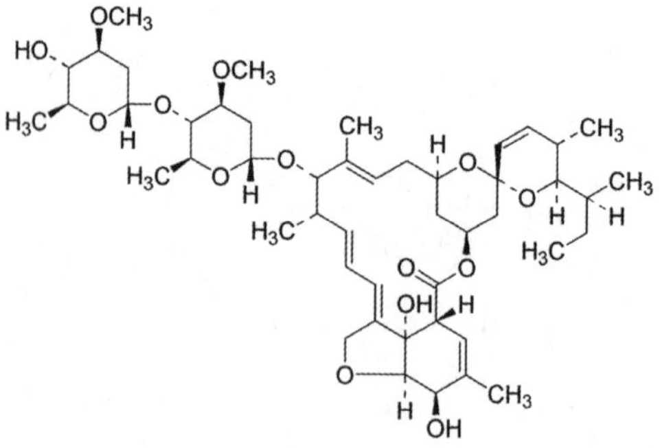

abamectin

The semi-synthetic macrolide antibiotic ivermectin was prepared in the Merck laboratories by chemical reduction of avermectin.[55] It was a mixture consisting mainly of the product shown here, differing only from abamectin in the absence of the double bond at the 22-position.

In 1974, the World Health Organization (WHO) launched a campaign to eliminate river blindness, a disease caused by infection with *Onchocerca volvulus*, a nematode worm carried by a black fly that bred in fast-flowing rivers. When the WHO campaign began, at least 17 million people were infected in Africa and Central America, of whom hundreds of thousands had lost their sight. By use of insecticide sprays and administration of ivermectin the disease was eradicated by the end of the century. The ivermectin was supplied free of charge to the WHO by Merck. By the mid-1990s it was being taken by 65 million people to prevent reproduction of the worm. It acted by augmenting the action of the inhibitory neurotransmitter gamma-aminobutyric acid on the reproductive tract of the female nematode.

RIFAMYCIN ANALOGUES

The elucidation of the structure of rifamycin B in 1963 opened the door to a joint programme of research between Lepetit and Ciba, in which several hundred of its derivatives were prepared. A starting point was the hypothesis that the low activity of rifamycin B could be attributed to the presence of an ionisable glycolic acid function at the 4-position, rendering it highly polar and so hindering penetration through the bacterial cell wall. It was argued that the absence of the glycolic acid side chain accounted for the enhanced potency of rifamycin SV. This hypothesis was proved to be correct when Lepetit chemists in Milan synthesised several esters, amides and hydrazides that exhibited enhanced activity.[56]

Another key strategy adopted in the synthetic programme addressed the rapid elimination of rifamycin B through biliary excretion. This was dealt with by preparing analogues with extended side chains, several of which were well absorbed after oral administration. This led to the marketing of rifamide in some countries.[57]

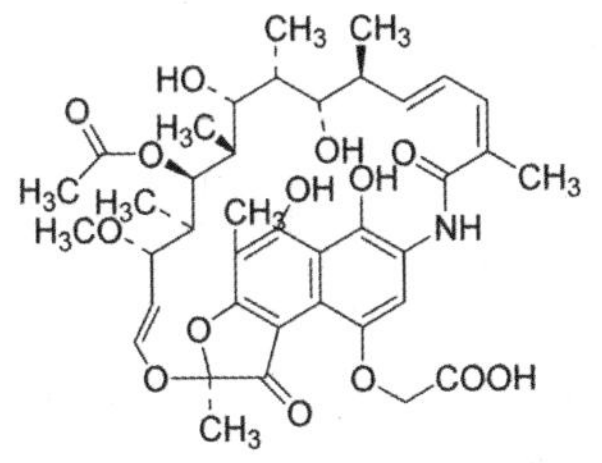

rifamycin B

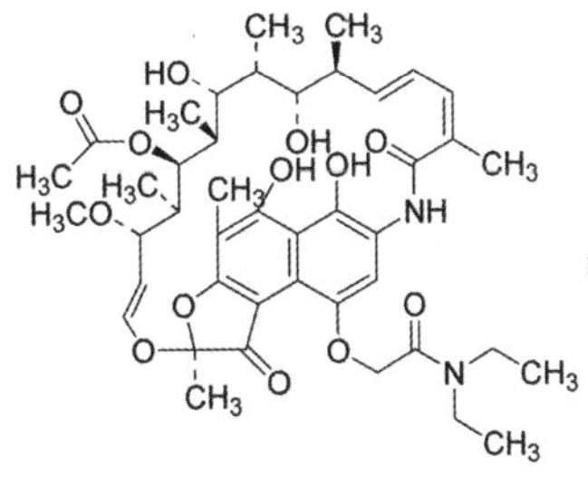

rifamide

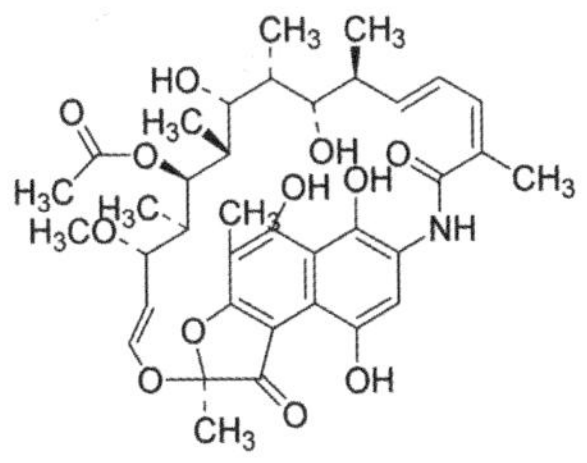

rifamycin SV

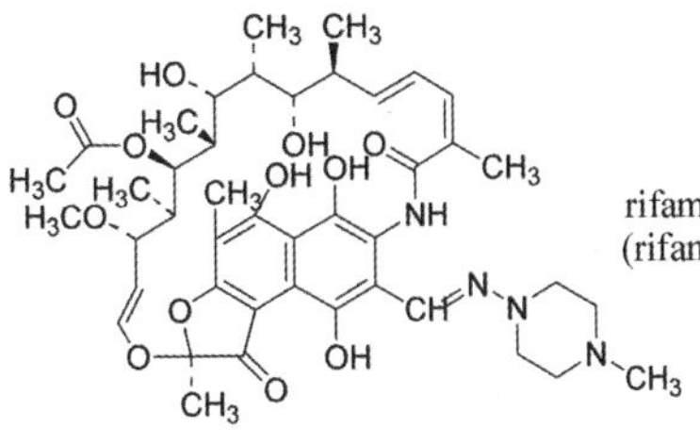

rifampicin (rifampin)

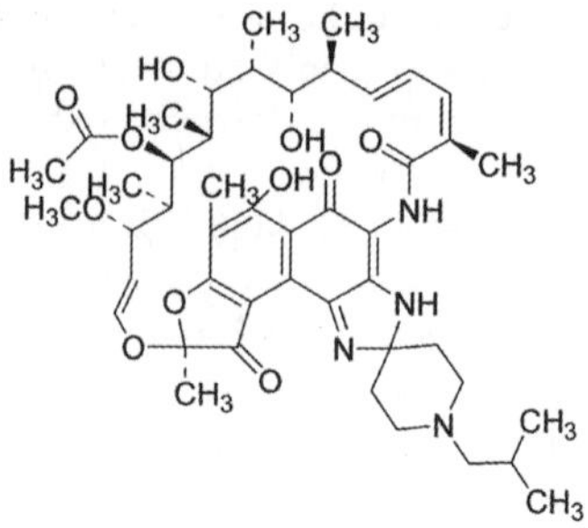

rifabutin

Derivatives at the 3-position of rifamycin SV were also investigated, including the 3-dimethylaminomethyl compound.[58] This was superior to rifamycin SV when administered by mouth, but the plasma levels varied considerably. It was also highly susceptible to oxidation, with 3-formyl rifamycin being its principal oxidation product. When this was subjected to biological testing, high levels of antibacterial activity were recorded. A series of highly potent derivatives were then prepared from 3-formyl rifamycin, including imines, hydrazones, oximes and hydrazide-hydrazones. Of these, rifampicin (also known as 'rifampin') was found to be the most promising.[59] It was introduced in 1966 and became established as a valuable drug in the treatment of both tuberculosis and leprosy. It is now an essential component of various combinations of drug used to treat tuberculosis or leprosy. The related rifabutin has not only proved of value in the treatment of drug-resistant tuberculosis but is also useful in the prophylaxis and treatment of non-tuberculous mycobacterial infections such as *Mycobacterium avium* complex (MAC) infection, a common and troublesome opportunistic infection in AIDS patients.

AZOMYCIN ANALOGUES

Following the discovery that azomycin had trichomonacidal activity, researchers at the Rhône–Poulenc laboratories in Paris synthesised a variety of nitroimidazoles. One of these exhibited strong trichomonacidal activity and had a low level of toxicity.[60,61] It subsequently received the approved name of metronidazole.

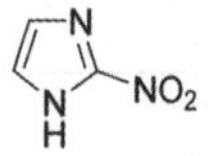

azomycin 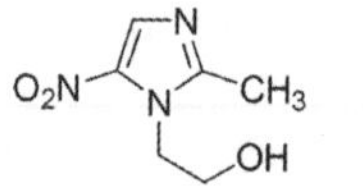   metronidazole 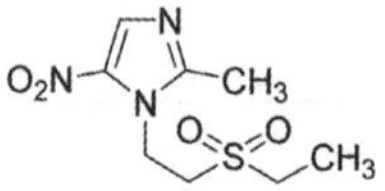   tinidazole

After oral administration, metronidazole was capable of eliminating *Trichomonas vaginalis* infections carried in semen and in the urine. It was the first effective drug for the treatment of trichomonal vaginitis and remains in use for this purpose. However, its antiprotozoal spectrum was wider than first anticipated. It extends to include *Entamoeba histolytica*, the causative organism of tropical dysentery, as a consequence of which metronidazole has become the standard oral treatment for invasive amoebic dysentery.

Metronidazole was also serendipitously found to be an effective antigiardial agent when patients who were infected with trichomoniasis and had diarrhoea due to *Giardia lambia* received treatment with it.[62] Hence it became the drug of choice for the treatment of giardiasis. Serendipity again played its part when it was observed that metronidazole alleviated ulcerative gingivitis, a bacterial infection of the gums.[63] This led to the realisation that metronidazole also had a wide spectrum of antibacterial activity, being especially effective against anaerobic infections. It is now widely used in the management of surgical and gynaecological sepsis. Tinidazole has similar activity to metronidazole.[64]

Serendipity was yet again involved when Janssen researchers were screening imidazoles for chemotherapeutic activity and found that one of them induced a profound hypnotic state in rats, whether injected or administered orally. Nearly 50 analogues were then synthesised and screened. It transpired that the best as an intravenous anaesthetic was etomidate, the ethyl ester analogue of the prototype methyl ester.[65] Preliminary studies found it to be an extremely potent, short-acting anaesthetic and it was introduced into anaesthetic practice in 1973. The duration of action was brief because esterase enzymes in the liver were responsible for its rapid metabolism to an inactive carboxylic acid, ensuring rapid postanaesthetic recovery.

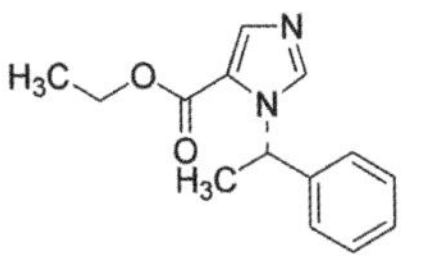

etomidate

After their unexpected success with etomidate, Janssen researchers developed miconazole as a broad-spectrum antifungal that was effective against dermatophyte, yeast and mould infections.[66] Since it was not well absorbed from the gut and was rapidly metabolised in the liver as it is a highly lipophilic molecule, miconazole was unsuitable for systemic medication other than by intravenous infusion in patients with severe infections such as aspergillosis, candidiasis or cryptococcosis. Such properties, however, were advantageous when miconazole was formulated for topical application to the mouth as a gel, swallowed as tablets in the treatment of intestinal fungal infections, or applied to the skin for ringworm infections such as athlete's foot.

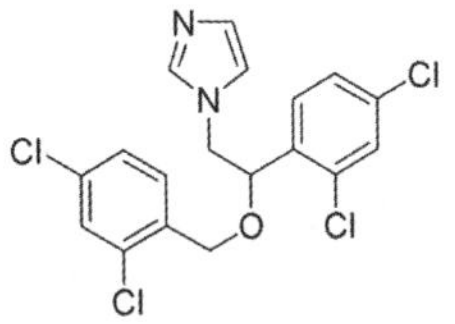

miconazole

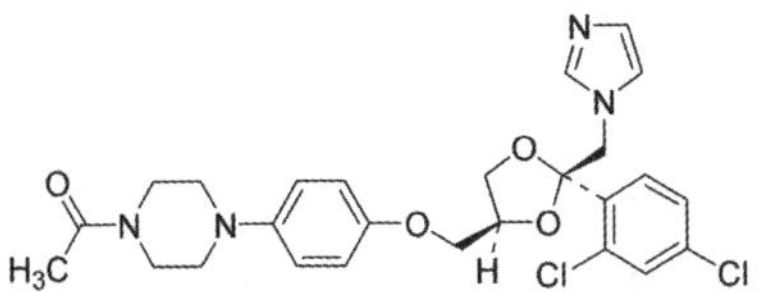

ketoconazole

Janssen researchers found that substituted imidazoles containing a dioxolane ring had some antifungal activity *in vitro*. By analogy with miconazole, potency was enhanced by adding a variety of aralkyl side chains while taking into account the spatial disposition of the aryl rings. Further refinement involved altering the alkyl portion of the side chain attached to the ketal by introducing a glycidyl ether fragment in its place. From the resulting compounds, the *cis* isomers were found to be more active than the *trans* isomers. Finally, ketoconazole emerged as a potent orally active antifungal agent.[67]

Ketoconazole was the first broad-spectrum imidazole suitable for the oral treatment of systemic mycoses.[68] Its enhanced resistance to metabolic inactivation resulted in higher plasma levels than with earlier imidazole antifungals. Nevertheless, ketoconazole was still extensively metabolised, less than 1% of an oral dose being excreted unchanged in the urine. In addition, it was not as lipophilic as the earlier drugs, hence less of it was protein bound. The greater proportion of unbound drug compensated for any shortcomings in the absorption from the gut and metabolism. Unfortunately, when given for systemic treatment, ketoconazole has sometimes proved to be a hazardous drug. Patients have died as a result of hepatoxicity produced by it.

When Janssen researchers replaced the imidazole ring of ketoconazole with a triazole ring that was less sensitive to nucleophilic attack, they obtained compounds that were more resistant to metabolic deactivation in the liver. Terconazole was the first of these triazoles to reach the clinic but, unfortunately, it caused photosensitivity reactions in some patients.[69] Itraconazole, an orally active analogue of terconazole, was free from this problem.[70] As it was highly lipophilic, itraconazole was highly tissue bound and persisted in the body to form a reservoir against infection within the skin and mucosa after dosing ceased. It is given by mouth for the treatment of candidiasis and dermatophytoses, but as there is still concern about hepatoxicity it is not administered to patients with a history of liver disease.

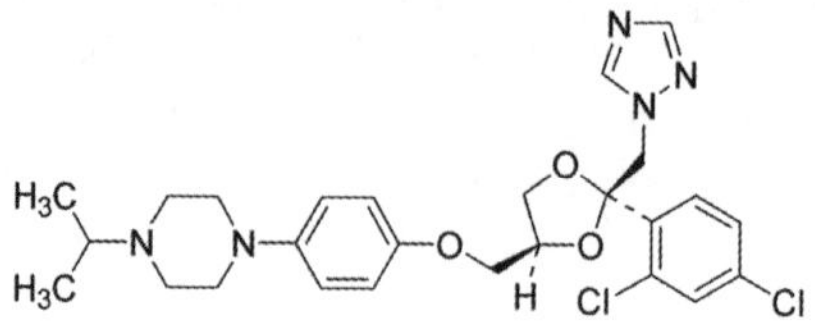

terconazole

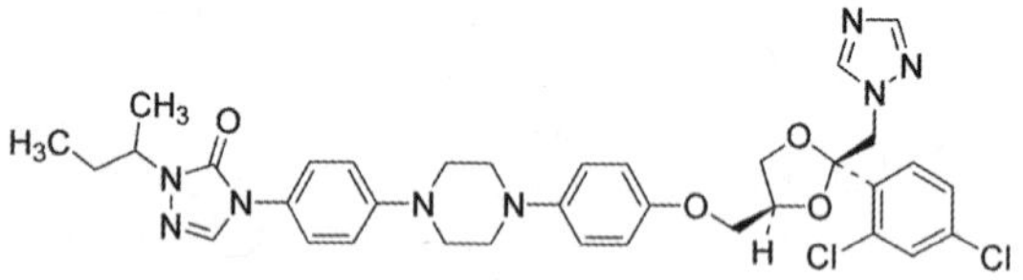

itraconazole

fluconazole

Pfizer researchers prepared hundreds of analogues of ketoconazole in an attempt to increase the oral bioavailability. Replacement of the imidazole ring by the metabolically more robust triazole ring, as had been done with the Janssen analogues, provided compounds that were more resistant to metabolism. When lipophilicity was lowered by replacing the dioxolane ring with a polar hydroxyl group, drug–protein binding was reduced and so antimicrobial potency was enhanced. Eventually, fluconazole was synthesised and found to be about 100 times as potent *in vitro* as ketoconazole, with good oral bioavailability.[71] It was also unique in possessing adequate water solubility for parenteral formulation. Due to its metabolic stability, fluconazole could be administered once daily, by mouth. It also penetrated the blood–brain barrier, making it a valuable therapeutic agent if infection had spread to the central nervous system.

ANALOGUES OF ANTHRACYCLINE ANTIBIOTICS

Many hundreds of analogues of doxorubicin have been prepared in the hope of finding less toxic compounds. Epirubicin, in which the 4′-hydroxyl of the sugar ring was epimerised, was found to have similar cytotoxic activity to doxorubicin, but its cardiotoxicity was reduced.[72] It has been of value in the treatment of leukaemia, lymphomas and some solid tumours. Idarubicin is an analogue of doxorubicin in which a methoxyl group has been removed from one of the aromatic rings. It has proved to be similar to doxorubicin in the clinic.[73]

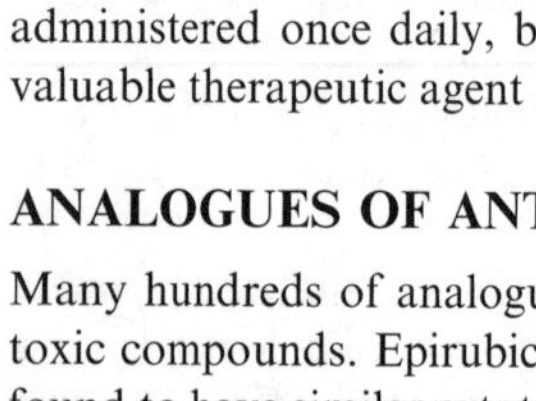

doxorubicin

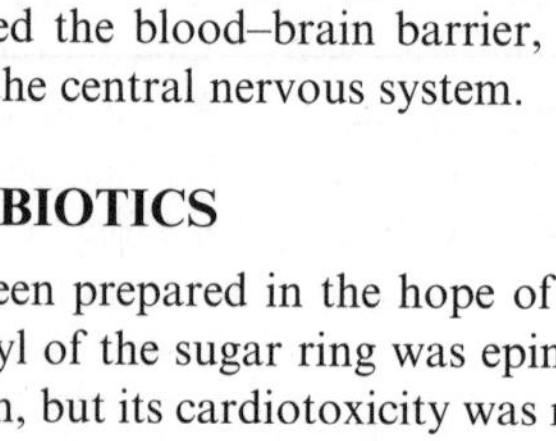

epirubicin

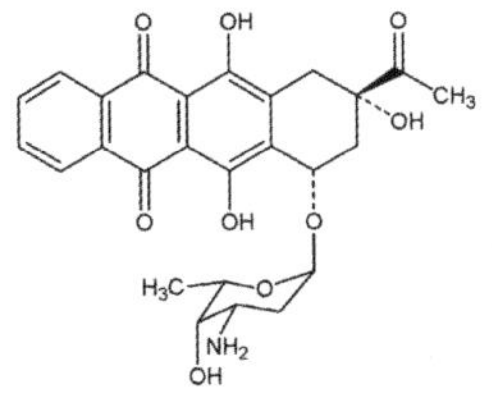

idarubicin

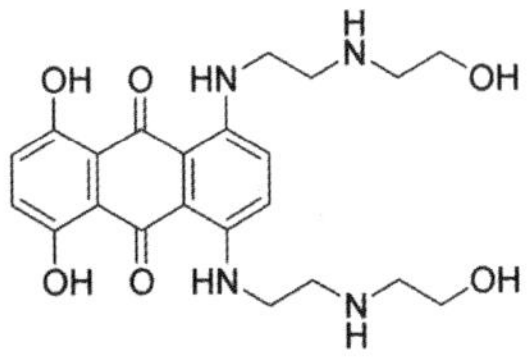

mitoxantrone

Lederle Laboratories introduced mitoxantrone (also known as 'mitozantrone'), a synthetic analogue of doxorubicin in which the sugar ring has been eliminated.[74] This was prepared in light of the evidence that the toxicity of the anthracycline antibiotics was probably due to the binding of their sugar rings to heart muscle. In place of the sugar ring in mitoxantrone was a moiety with a similar physicochemical nature. Dose-related cardiotoxicity still occurs, but is reversible – unlike that of the anthracyclines. Mitoxantrone is used in the treatment of breast cancer and leukaemia.

THIENAMYCIN ANALOGUE

The chemical instability of thienamycin was due to the basic character of its thioaminoethyl side chain. Merck researchers examined several hundred analogues before selecting imipenem, the _N_-formidimidoyl derivative, for clinical application.[75] Like thienamycin itself, it was an exceedingly potent, broad-spectrum antibiotic that was resistant to beta-lactamases.

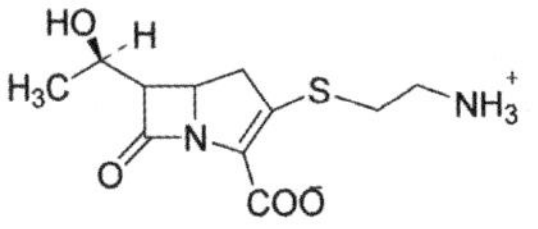

thienamycin

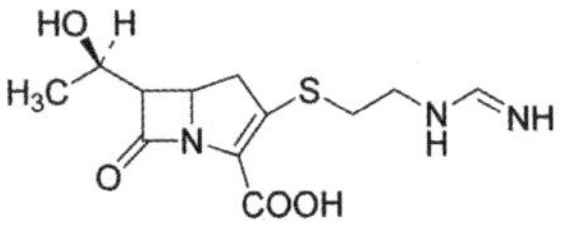

imipenem

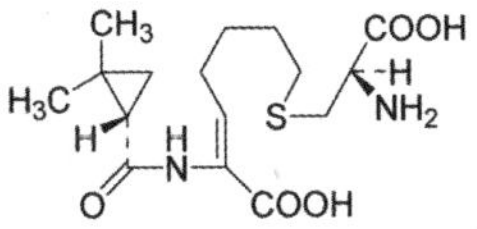

cilastatin

The main problems with imipenem were that it had to be administered parenterally and it was extensively degraded in the proximal tubules of the kidneys through cleavage of the beta-lactam ring by the action of the enzyme dehydropeptidase I. This rendered it useless in the treatment of urinary tract infections. However, the discovery that various synthetic acylamino-propenoates are inhibitors of dehydropeptidase I led to the marketing of a combination of imipenem with one of these inhibitors, cilastatin.[76]

ANALOGUES OF MONOBACTAM ANTIBIOTICS

Richard Sykes and his colleagues at the Squibb Institute developed a general synthesis of racemic analogues of SQ 26180 and proceeded to prepare non-methoxylated analogues as well as analogues exhibiting side chains previously found effective in the cephamycin series. A diverse range of compounds were prepared with the use of sophisticated synthetic approaches, leading to the conclusion that maximum activity was to be found in side chains featuring oxime and aminothiazole systems.[77] This culminated in the introduction of aztreonam, which contained a side chain seen previously in cefotaxime, a third-generation cephalosporin to which its activity can be compared.[78] The high degree of resistance to beta-lactamase was due to the presence of a 4-methyl group adjacent to the beta-lactam function, which sterically hindered enzymatic hydrolysis.

aztreonam

Aztreonam proved to be effective against a wide range of Gram-negative bacteria including *Pseudomonas aeruginosa*, *Neisseria meningitidis* and *Haemophilus influenzae*, but it had poor activity against Gram-positive organisms and anaerobes.

REFERENCES

1. J.C. Sheehan, K.R. Henery-Logan, The total synthesis of penicillin V. *J. Am. Chem. Soc.*, 1957; **79**: 1262–3.
2. F.R. Batchelor, E.P. Doyle, J.H. Nayler, G.N. Rolinson, Synthesis of penicillin: 6-aminopenicillanic acid in penicillin fermentations. *Nature*, 1959; **183**: 257–8.
3. J.C. Sheehan, *The Enchanted Ring. The Untold Story of Penicillin*. London: MIT Press; 1982.
4. E. Brandl, M. Giovanni, H. Hargreiter, [Studies on the acid stable, orally efficacious phenoxymethylpenicillin (penicillin V)]. *Med. Wochenschr.*, 1953; **103**: 602–7.
5. F.P. Doyle, K. Hardy, J.H.C. Nayler, *et al.*, Derivatives of 6-aminopenicillanic acid. Part III. 2,6-Dialkoxybenzoyl derivatives. *J. Chem. Soc.*, **1962**: 1453–8.
6. F.P. Doyle, J.H.C. Nayler, H. Smith, E.P. Stove, New penicillins stable towards both acid and penicillinase. *Nature*, 1961; **192**: 1183–4.
7. F.P. Doyle, J.H. Nayler, H. Smith, E.R. Stove, Some novel acid-stable penicillins. *Nature*, 1961; **191**: 1091–2.
8. W. Daehne, E. Frederiksen, E. Gundersen, *et al.*, Acyloxymethyl esters of ampicillin. *J. Med. Chem.*, 1970; **13**: 607–12.
9. P. Acred, D.M. Brown, E.T. Knudson, *et al.*, New semi-synthetic penicillin active against *Pseudomonas pyocyanea*. *Nature*, 1967; **215**: 25–30.
10. J.H.C. Nayler, in *Advances in Drug Research*, Vol.7, eds N.J. Harper, A.B. Simmonds. New York: Academic Press; 1973.
11. Fr. Pat. 1971: 2100682 (to Bayer).
12. H. Ferres, M.J. Basker, P.J. O'Hanlon, Beta-lactam antibiotics. I. Comparative structure–activity relationships of 6-acylaminopenicillanic acid derivatives and their 6-(D-alpha-acylaminophenylace-tamido) penicillanic acid analogues. *J. Antibiot.*, 1974; **27**: 922–30.
13. R.N.Jones, C. Thornsberry, A.L. Barry, *et al.*, Piperacillin (T-1220), a new semisynthetic penicillin: *in vitro* antimicrobial activity comparison with carbenicillin, ticarcillin, ampicillin, cephalothin, cefamandole and cefoxitin. *J Antibiot.*, 1977; **30**: 1107–14.
14. H.B. König, K.G. Metzger, H.A. Offe, W. Schrock, Mezlocillin. *Arzneimittel-Forsch.*, 1983; **33**: 88–90.
15. F.J. Lund, in *Chronicles of Drug Discovery*, Vol. 2, eds J.S. Bindra, D. Lednicer. New York: Wiley; 1983, p.149.
16. R.A. Volkmann, R.D. Carroll, R.B. Drolet, *et al.*, Efficient preparation of 6,6-dihalopenicillanic acids. Synthesis of penicillanic acid S,S-dioxide (sulbactam). *J. Org. Chem.*, 1982; **47**: 3344–5.
17. A.R. English, J.A. Retsema, A.E. Girard, *et al.*, CP-45,899, a beta-lactamase inhibitor that extends the antibacterial spectrum of beta-lactams: initial bacteriological characterization. *Antimicrob. Agents Chemother.*, 1978; **14**: 414–19.
18. S. Mehtar, R.J. Croft, A. Hilas, A non-comparative study of parenteral ampicillin and sulbactam in intra-thoracic and intra-abdominal infections. *J. Antimicrob. Chemother.*, 1986; **17**: 389–96.
19. S.C. Aronoff, M.R. Jacobs, S. Johenning, S. Yamabe, Comparative activities of the beta-lactamase inhibitors YTR 830, sodium clavulanate, and sulbactam combined with amoxicillin or ampicillin. *Antimicrob. Agents Chemother.*, 1984; **26**: 580–2.
20. L. Guttman, M.D. Kitzis, S. Yamabe, J.F. Acar, Comparative evaluation of a new beta-lactamase inhibitor, YTR 830, combined with different beta-lactam antibiotics against bacteria harboring known beta-lactamases. *Antimicrob. Agents Chemother.*, 1986; **29**: 955–7.
21. B. Loder, G.G.F. Newton, E.P. Abraham, The cephalosporin C nucleus (7-aminocephalosporanic acid) and some of its derivatives. *Biochem. J.*, 1961; **79**: 408–16.

22. R.R Chauvette, E.H. Flynn, B.G. Jackson, *et al.*, Chemistry of cephalosporin antibiotics. II. Preparation of a new class of antibiotics and the relation of structure to activity. *J. Am. Chem. Soc.*, 1962; **84**: 3401–2.

23. M.H. Richmond, S.Wotton, Comparative study of seven cephalosporins: susceptibility to beta-lactamases and ability to penetrate the surface layers of *Escherichia coli*. *Antimicrob. Agents Chemother.*, 1976; **10**: 219–22.

24. J.L. Spencer, E.H. Flynn, R.W. Roeske, *et al.*, Chemistry of cephalosporin antibiotics. VII. Synthesis of cephaloglycin and some homologs. *J. Med. Chem.*, 1966; **9**: 746–50.

25. C.W.Ryan, R.L. Simon, E.M. Van Heyningen, Chemistry of cephalosporin antibiotics. XIII. Deacetoxycephalosporins. Synthesis of cephalexin and some analogs. *J. Med. Chem.*, 1969; **12**: 310–13.

26. R.E. Buck, K.E. Price, Cefadroxil, a new broad-spectrum cephalosporin. *Antimicrob. Agents Chemother.*, 1977; **11**: 324–30.

27. C.W. Hale, G.G. Newton, E.P. Abraham, Derivatives of cephalosporin C formed with certain heterocyclic tertiary bases. The cephalosporin C–A family. *Biochem J.*, 1961; **79**: 403–8.

28. Fr. Pat. 1965: 1384197 (to Glaxo).

29. K. Kariyone, H. Harada, M. Kurita, T. Takano, Cefazolin, a new semisynthetic cephalosporin antibiotic. I. Synthesis and chemical properties of cefazolin. *J. Antibiot.*, 1970; **23**: 131–6.

30. D.B. Boyd, W.H.W. Lunn, Electronic structures of cephalosporins and penicillins. 9. Departure of a leaving group in cephalosporins. *J. Med. Chem.*, 1979; **22**: 778–84.

31. Ger. Pat. 1973: 2439880 (to Glaxo).

32. C.H. O'Callaghan, R.B. Sykes, A. Griffiths, *et al.*, Cefuroxime, a new cephalosporin antibiotic: activity *in vitro*. *Antimicrob. Agents Chemother.*, 1976; **9**: 511–19.

33. R.R. Chauvette, P.A. Pennington, Chemistry of cephalosporin antibiotics. 30. 3-Methoxy- and 3-halo-3-cephems. *J. Med. Chem.*, 1975; **18**: 403–8.

34. S.M. Harding, P. Williams, J. Ayrton, Pharmacology of cefuroxime as the 1-acetoxyethyl ester in volunteers. *Antimicrob. Agents Chemother.*, 1984; **25**: 78–82.

35. R. Wise, T. Rollason, M. Logan, *et al.*, HR 756, a highly active cephalosporin: comparison with cefazolin and carbenicillin. *Antimicrob. Agents Chemother.*, 1978; **14**: 807–11.

36. C.H. O'Callaghan, P. Acred, P.B. Harper, *et al.*, GR 20263, a new broad-spectrum cephalosporin with anti-pseudomonal activity. *Antimicrob. Agents Chemother.*, 1980; **17**: 876–83.

37. R. Reiner, U. Weiss, U. Brombacher, *et al.*, Ro 13-9904/001, a novel potent and long-acting parenteral cephalosporin. *J. Antibiot.*, 1980; **33**: 783–6.

38. T. Kamimura, H. Kojo, Y. Matsumoto, Y. Mine, *et al.*, *In vitro* and *in vivo* antibacterial properties of FK 027, a new orally active cephem antibiotic. *Antimicrob. Agents Chemother.*, 1984; **25**: 98–104.

39. Y. Utsui, M. Inoue, S. Mitsuhashi, *In vitro* and *in vivo* antibacterial activities of CS-807, a new oral cephalosporin. *Antimicrob. Agents Chemother.*, 1987; **31**: 1085–92.

40. R.E. Kessler, M. Bies, R.E. Buck, *et al.*, Comparison of a new cephalosporin, BMY 28142, with other broad-spectrum beta-lactam antibiotics. *Antimicrob. Agents Chemother.*, 1985; **27**: 207–16.

41. G. Seibert, N. Klesel, M. Limbert, *et al.*, HR 810, a new parenteral cephalosporin with a broad antibacterial spectrum. *Arzneimittel. Forsch.*, 1983; **33**: 1084–6.

42. S. Karady, S.H. Pines, L.M. Weinstock, *et al.*, Semisynthetic cephalosporins via a novel acyl exchange reaction. *J. Am. Chem. Soc.*, 1972; **94**: 1410–11.

43. H. Kawaguchi, T. Naito, S. Nakagawa, K.I. Fujisawa, BB-K 8, a new semisynthetic aminoglycoside antibiotic. *J. Antibiot.*, 1972; **25**: 695–708.

44. B.J. Magerlein, R.D. Birkenmeyer, F. Kagan, Chemical modification of lincomycin. *Antimicrob. Agents Chemother.*, **1966**: 727–36.

45. G.H. Miller, G. Arcieri, M.J. Weinstein, J. A. Waitz, Biological activity of netilmicin, a broad-spectrum semisynthetic aminoglycoside antibiotic. *Antimicrob. Agents Chemother.*, 1976; **10**: 827–36.

46. Ger. Pat. 1962: 1134071 (to Carlo Erba).

47. C.R. Stephens, K. Murai, H.H. Rennhard, *et al.*, Hydrogenolysis studies in the tetracycline series-6-deoxytetracyclines. *J. Am. Chem. Soc.*, 1958; **80**: 5324–5.

48. R.K. Blackwood, J.J. Beereboom, H.H. Rennhard, *et al.*, 6-Methylenetetracyclines. I. A new class of tetracycline antibiotics. *J. Am. Chem. Soc.*, 1961; **83**: 2773–5.

49. M.S. Wittenau, J.J. Beereboom, R.K. Blackwood, C.R. Stephens, 6-Deoxytetracyclines. III. Stereochemistry at C.6. *J. Am. Chem. Soc.*, 1962; **84**: 2645–7.

50. M.J. Martel, J.H. Boothe, The 6-deoxytetracyclines. VII. Alkylated aminotetracyclines possessing unique antibacterial activity. *J. Med. Chem.*, 1967; **10**: 44–6.

51. S. Morimoto, Y. Takahashi , Y. Watanabe, S. Omura, Chemical modification of erythromycins. I. Synthesis and antibacterial activity of 6-*O*-methylerythromycins A. *J. Antibiot.*, 1984; **37**: 187–9.

52. R.N. Jones, A.L. Barry, C. Thornsberry, *In vitro* evaluation of three new macrolide antimicrobial agents, RU28965, RU29065, and RU29702, and comparisons with other orally administered drugs. *Antimicrob. Agents Chemother.*, 1983; **24**: 209–15.

53. Belg. Pat. 1982: 892357 (to Sour Pliva).

54. G.M. Bright, A.A. Nagel, J. Bordner, *et al.*, Synthesis, *in vitro* and *in vivo* activity of novel 9-deoxo-9a-AZA-9a-homoerythromycin A derivatives; a new class of macrolide antibiotics, the azalides. *J. Antibiot.*, 1988; **41**: 1029.

55. J.C. Chabala, H. Mrozik, R.L. Tolman, *et al.*, Ivermectin, a new broad-spectrum antiparasitic agent. *J. Med. Chem.*, 1980; **23**: 1134–6.

56. N. Maggi, S. Furesz, P. Sensi, Influence of the carboxyl group on the antibacterial activity of rifamycins . *J. Med. Chem.*, 1968; **11**: 368–9.

57. P. Sensi, N. Maggi, R. Ballotta, *et al.*, Rifamycins. XXXV. Amides and hydrazides of rifamycin B. *J. Med. Chem.*, 1964; **7**: 596–602.

58. N. Maggi, V. Arioli, P. Sensi, Rifamycins. XLI. A new class of active semisynthetic rifamycins. *N*-Substituted aminomethyl derivatives of rifamycin SV. *J. Med. Chem.*, 1965; **8**: 790–3.

59. N. Maggi, C.R. Pasqualucci, R. Ballotta, P. Sensi, Rifampicin: a new orally active rifamycin. *Chemotherapia*, 1966; **11**: 285–92.

60. C. Cosar, L. Julou, Activité de l' (hydroxy-2-éthyl)-1-méthyl-2-nitro-5 imidazole (8,823 R.P.) vis-à-vis des infections expérimentales *Trichomonas vaginalis. Ann. Inst. Pasteur*, 1959; **96**: 238–41.

61. C. Cosar, C. Crisan, R. Horclois, *et al.*, Nitro-imidazoles – Préparation activité chimiothérapeutique. *Arzneimittel. Forsch.*, 1966; **16**: 23–9.

62. W. Fowler, *Br. J. Vener. Dis.*, 1960; **36**: 157.

63. D.L.S. Shinn, Metronidazole in acute ulcerative gingivitis. *Lancet*, 1962; *i*: 1191.

64. M.W. Miller, H.L. Howes Jr, R.V. Kasubick, A.R. English, Alkylation of 2-methyl-5-nitroimidazole. Some potent antiprotozoal agents. *J. Med. Chem.*, 1970; **13**: 849–52.

65. E.F. Godefroi, P.A.J. Janssen, C.A.M. Van der Eyken, *et al.*, DL-1-(1-Arylalkyl)imidazole-5-carboxylate esters. A novel type of hypnotic agents. *J. Med. Chem.*, 1965; **8**: 220–3.

66. E.F. Godfroi, J. Heeres, J. Van Cutsem, P.A.J. Janssen, Preparation and antimycotic properties of derivatives of 1-phenethylimidazole. *J. Med. Chem.*, 1969; **12**: 784–91.

67. J. Heeres, L.J.J. Backs, J.H. Mostmans, J. Van Cutsem, Antimycotic imidazoles. Part 4. Synthesis and antifungal activity of ketoconazole, a new potent orally active broad-spectrum antifungal agent. *J. Med. Chem.*, 1979; **22**: 1003–5.

68. D. Thienpont, J. Van Cutsem, F. Van Gerven, *et al.*, Ketoconazole – a new broad spectrum orally active antimycotic. *Experientia*, 1979; **35**: 606–7.

69. J. Heeres, R. Hendrickx, J. Van Cutsem, Antimycotic azoles. 6. Synthesis and antifungal properties of terconazole, a novel triazole ketal. *J. Med. Chem.*, 1983; **26**: 611–13.

70. J. Heeres, L.J.J. Backx, J. Van Cutsem, Antimycotic azoles. 7. Synthesis and antifungal properties of a series of novel triazol-3-ones. *J. Med. Chem.*, 1984; **27**: 894–900.

71. K. Richardson, K.W. Brammer, M.S. Marriott, P.F. Troke, Activity of UK-49,858, a bis-triazole derivative, against experimental infections with *Candida albicans* and *Trichophyton mentagrophytes. Antimicrob. Agents Chemother.*, 1985; **27**: 832–5.

72. F. Arcamone, S. Penco, A. Vigevani, *et al.*, Synthesis and antitumor properties of new glycosides of daunomycinone and adriamycinone. *J. Med. Chem.*, 1975; **18**: 703–7.

73. F. Arcamone, L. Bernardi, P. Giardino, *et al.*, Synthesis and antitumor activity of 4-demethoxydaunorubicin, 4-demethoxy-7,9-diepidaunorubicin, and their beta anomers. *Cancer Treat. Rep.*, 1976; **60**: 829–34.

74. R.K.Y. Zee-Cheng, C.C.Cheng, Antineoplastic agents. Structure–activity relationship study of bis(substituted aminoalkylamino)anthraquinones. *J. Med. Chem.*, 1978; **21**: 291–4.

75. W.J. Leanza, K.J. Wildonger, T.W. Miller, B.G. Christensen, *N*-Acetimidoyl- and *N*-formimidoyl-thienamycin derivatives: antipseudomonal beta-lactam antibiotics. *J. Med. Chem.*, 1979; **22**: 1435–6.

76. S.R. Norrby, K. Alestig, B. Bjornegard, *et al.*, Urinary recovery of *N*-formimidoyl thienamycin (MK0787) as affected by coadministration of *N*-formimidoyl thienamycin dehydropeptidase inhibitors. *Antimicrob. Agents Chemother.*, 1983; **23**: 300–7.

77. C.M. Cimarusti, in *Chronicles of Drug Discovery*, Vol. 3, ed. D. Lednicer. New York: American Chemical Society; 1993; p.239.

78. R.B. Sykes, D.P. Bonner, K. Bush, N.H. Georgopapadakou, Aztreonam (SQ 26,776), a synthetic monobactam specifically active against aerobic Gram-negative bacteria. *Antimicrob. Agents Chemother.*, 1982; **21**: 85–92.

24

Pharmacodynamic Agents
from Micro-organisms

Fungi have been the source of several drugs in addition to penicillin. During the Middle Ages, tens of thousands of people died after eating bread made from rye contaminated with a fungus known as 'ergot', *Claviceps purpurea* (Hypocreaceae).[1] Outbreaks of ergotism reached epidemic proportions in the rye-bread eating areas of France and Germany, death ensuing after gangrene of the limbs had set in following prolonged vasoconstriction produced by a toxic fungal metabolite. As this manifested itself as an overall blackening of the diseased extremity, superstitious minds attributed it to charring by the Holy Fire! Such was the severity of the problem at the end of the eleventh century that a religious order was established in southern France to care for the afflicted. The scourge came to be named after the patron saint of that order as 'St Anthony's Fire'. Not until the seventeenth century did the cause of the affliction become generally recognised, since when isolated outbreaks have occurred right up until modern times.

Early attempts to use ergot medicinally preceded the realisation that it was the cause of so many deaths. In the second edition of his *Kreuterbuch*, published in Frankfurt in 1582, Adam Lonicer mentioned that midwives employed Kornzapfen to hasten labour. Kornzapfen is believed to be ergot. It has been suggested that the introduction of ergot for this purpose may have come about from midwives observing that pregnant women miscarried during outbreaks of ergotism.[2] European midwives continued to administer their *pulvis ad partum* (birth powder), but orthodox practitioners did not seriously consider the therapeutic potential of ergot until after a letter appeared in 1808 in the *Medical Repository*, the first scientific journal to be published in the United States. It was written by the influential physician John Stearns of Saratoga County, New York State, who described how he had been using ergot to hasten prolonged labour for several years since being told about it by an old woman from Germany.[3]

On 2 June 1813, 100 members of the Massachusetts Medical Society heard Oliver Prescott deliver his *Dissertation on the Natural History and Medicinal Effects of the Secale cornutum, or Ergot* (note: *Secale cornutum* is the Latin name for ergot of rye). This was published later that year in pamphlet form, being reprinted in London and Philadelphia and then translated into French and German.[4] This finally ensured the general acceptance of the drug by the medical profession. At first, enthusiastic practitioners ignored Stearn's warning to avoid administering ergot if prolongation of labour was due to obstruction. Many stillbirths and maternal deaths were reported.

Another use for ergot was found when the English physician Edward Woakes introduced it for treating migraine in 1868.[5] It was one of many remedies for the relief of migraine proposed in the nineteenth century, but it was well received because there was a theoretical basis for its use, namely the postulated effect of its vasoconstrictive properties on swollen blood vessels in the brain.[6]

Drug Discovery. A History. W. Sneader.
©2005 John Wiley & Sons Ltd

Ergotamine

Ergot presented chemists with a formidable challenge. French pharmacist Charles Tanret isolated an alkaloid called ergotinine in 1875.[7] Disappointingly, it proved to be inert. The first pure alkaloid that was active was obtained after the Sandoz Company of Basle entered the field of pharmaceutical research in 1917 by appointing the eminent Swiss chemist Arthur Stoll as its director of research. By 1920, he had isolated ergotamine.[8] It was not until 1951 that he and Albert Hofmann were able to establish its full chemical structure.[9] Ten years later, Hofmann achieved its total synthesis.[10]

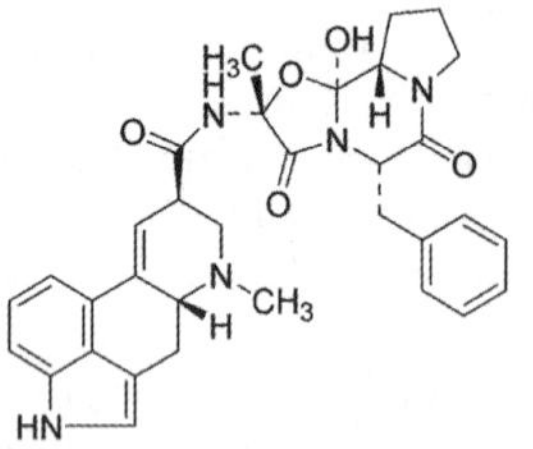

ergotamine

Ergotamine was initially marketed as a uterine stimulant, but was found to be inferior to ergot preparations. However, Sandoz pharmacologist Ernst Rothlin administered ergotamine to a patient with intractable migraine, in 1925.[11] Encouraged by the outcome, he persuaded the eminent Zurich psychiatrist Hans Maier to evaluate its use in the treatment of migraine.[12] This soon became the principal therapeutic application of ergotamine. However, it was a hazardous drug, with highly variable absorption after oral administration, increasing the risk of vasoconstriction and thus causing gangrene in the extremities. The maximum dose by mouth in an attack of migraine was established as 8 mg, with not more than 12 mg being taken in the course of one week. It could not be administered more than twice monthly, ruling out any possibility of giving ergotamine to prevent migraine.

Ergometrine (Ergonovine)

At University College Hospital, London, John Moir devised a sensitive method of measuring uterine contraction by inserting a small balloon in the uterus of patients undergoing routine pelvic examination one week after giving birth.[13] The balloon was connected to a manometer that indicated the slightest pressure change. Moir used this apparatus to investigate the controversial liquid extract of ergot that many authorities had decried on the grounds that its method of preparation precluded the presence of any alkaloids known to be active. Moir unexpectedly found that the response when the extract was given by mouth was unprecedented, both in the intensity of the contractions and the rapidity of their onset. Harold Dudley, the chief chemist at the National Institute for Medical Research, then joined him to isolate the active substance in the extract. They obtained pure crystals of ergometrine in 1935.[14] It was now clear that the traditional actions of ergot on the uterus had been due to ergometrine.

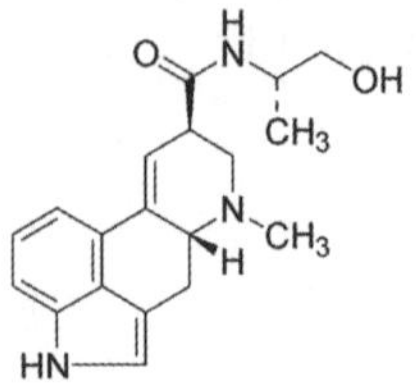

ergometrine

A few weeks before the publication of the British work, Morris Kharasch and his colleagues in Chicago reported their isolation of an active substance from ergot.[15] This was later shown to be ergometrine, although the Chicago workers had concluded that it was not an alkaloid. Shortly after, Marvin Thompson at the University of Maryland published details of his extraction method, the patent rights to which were acquired by Eli Lilly and Company.[16] The Sandoz group were also successful, with Stoll and Burckhardt publishing their results around the same time.[17] Because each of these groups gave a different name to the new alkaloid, the American Medical Association felt it necessary to adopt the name 'ergonovine', whereas the *British Pharmacopoeia* accepted the name 'ergometrine'.

Ergotoxine

At the Wellcome Research Laboratory in London in 1906, George Barger and Francis Carr isolated an amorphous powder from an ergot extract.[18] Henry Dale found that while it exhibited the actions of ergot, it was more toxic. This resulted in it being called 'ergotoxine'. After ergotamine had been introduced into the clinic, so too was ergotoxine because of the close similarity of its actions. Despite its lower cost of production, ergotoxine fell out of popularity because of the increased risk of causing gangrene.

For over 30 years, it was believed that ergotoxine was a pure alkaloid until Albert Hofmann in the Sandoz laboratories began refining it so that he could prepare lysergic acid from it for his synthetic work (discussed in the next chapter). He began to doubt that it was a single substance and soon established that it was a mixture consisting mainly of ergocornine, together with some ergocristine and ergocryptine.[19]

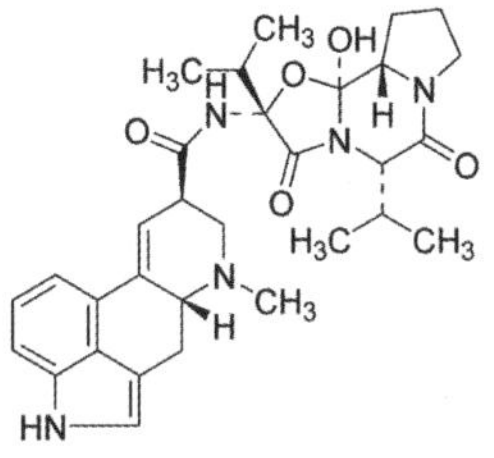

ergocornine

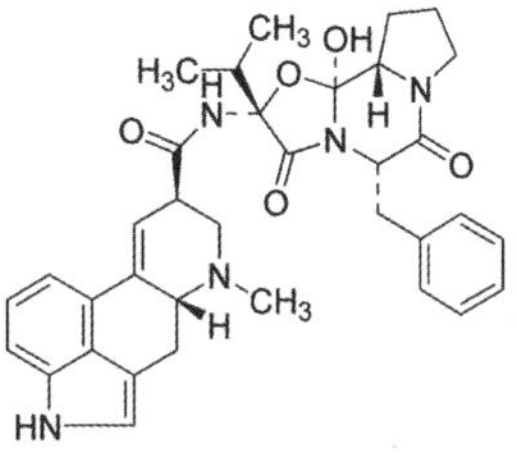

ergocristine

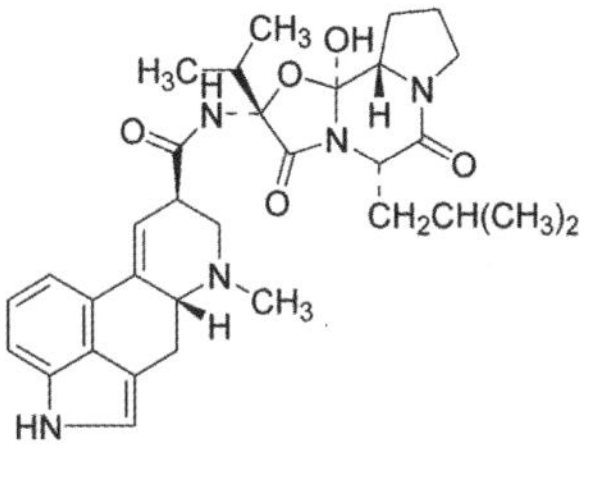

ergocryptine

The Statins

The ergot alkaloids were not the only important pharmacodynamic agents to be isolated from fungi. In 1976, two groups independently isolated mevastatin from fungi. Akira Endo and his colleagues at the Sankyo Company in Tokyo obtained it from *Penicillium citrinum* after having screened over 8000 microbial extracts for evidence of inhibition of sterol biosynthesis.[20,21] Beecham Research Laboratories in England isolated it from *Penicillium brevicompactum* and named it compactin.[22] Mevastatin was subsequently synthesised in the laboratory.[23]

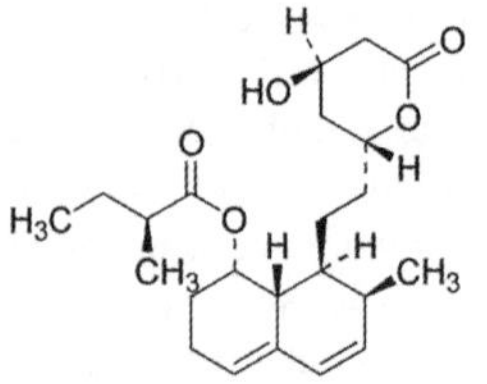

mevastatin

Pharmaceutical companies had begun to show interest in regulating sterol biosynthesis as a consequence of the findings from a long-term epidemiological study in which the health of the 5000 or so residents of Framingham, Massachusetts, has been closely monitored since the 1950s. The objective of this unique investigation established by the United States Public Health Service was to determine which biological and environmental factors may have been responsible for the rise in cardiovascular disease since the 1930s. One of the major outcomes of the Framingham study was the recognition of the relationship between high cholesterol values and cardiovascular disease.

Mevastatin was shown by Endo to inhibit the key enzyme that regulates hepatic synthesis of cholesterol, where at least 60% of the cholesterol in the body is biosynthesised from acetyl–coenzyme A (CoA). The enzyme is 3-hydroxy-3-methylglutaryl (HMG)–CoA reductase, which catalyses the rate-limiting conversion of 3-hydroxy-3-methylglutaryl–CoA to mevalonic acid.

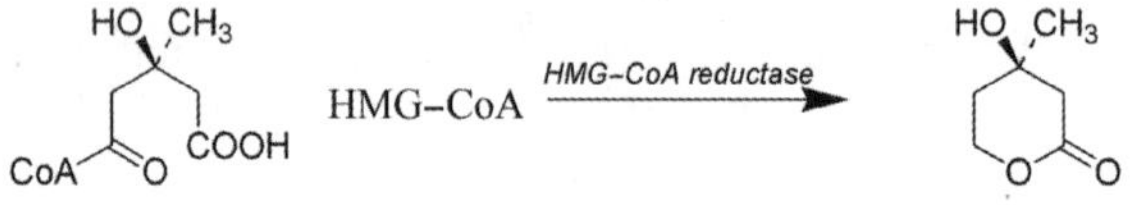

The administration of mevastatin was shown to lower cholesterol levels in the liver, causing more low-density lipoprotein (LDL) cholesterol receptors to be expressed. Circulating LDL cholesterol levels then fell due to the increased uptake of LDL cholesterol by the greater number of hepatic LDL cholesterol receptors.[24]

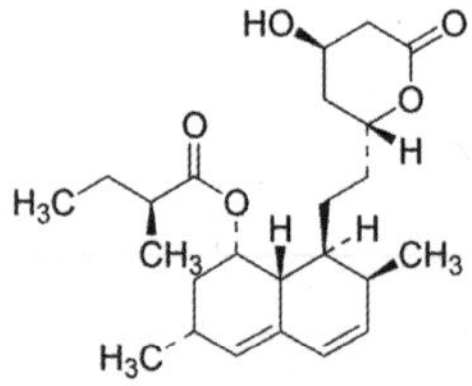

lovastatin

Endo also isolated lovastatin (which was originally called 'monacolin K') from the filamentous fungus *Monascus ruber*.[25] It was also obtained from the fungus *Aspergillus terreus* by researchers at Merck and Company (and named by them as 'mevinolin'), who confirmed that it was a potent inhibitor of HMG–CoA reductase.[26] It was more potent than mevastatin as an inhibitor of sterol biosynthesis and has become one of the most widely prescribed drugs in the world, as have other 'statins'. The natural and semi-synthetic statins now on the market have had a worldwide impact on heart disease, with the full implications of their use in people with elevated cholesterol levels only now becoming fully appreciated.

DRUGS FROM ACTINOMYCETES

During the 1950s, Ciba researchers led by Hans Bickel isolated from *Streptomyces pilosus* several iron-containing antibiotics known as 'ferrimycins'. Their antibiotic activity was disappointing as bacterial resistance quickly developed, but in addition the activity was inhibited by contaminants that contained iron. The major contaminant was sent to Victor Prelog at the Swiss Federal Institute of Technology in Zurich, where it was identified and named 'ferrioxamine B'.[27]

After animal tests had confirmed its safety, ferrioxamine B was supplied to Woehler at the University Hospital in Freiburg for evaluation in patients with iron deficiencies.[28] The first patient received an intravenous injection and was somewhat distressed when he shortly after passed urine coloured deep reddish brown. This was due to rapid and total elimination of unmetabolised ferrioxamine B by the kidneys. Woehler realised that was unique, all other known iron compounds breaking down and releasing their iron in the body. He then postulated that if iron could be removed from the ferrioxamine B, the residue might be a strong iron acceptor that was capable of removing excessive iron from the body in diseases involving iron overload.

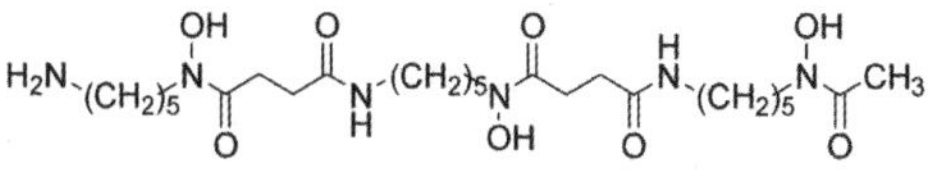

desferrioxamine

Bickel was able to prepare the compound Woehler wanted, namely desferrioxamine (also known as 'deferoxamine').[29] It was injected into animals intravenously and found to possess a very high affinity for ferric iron, as evidenced by the presence of iron in the urine. Further investigation revealed that iron in haemoglobin and in enzymes was not affected by the drug. This permitted its administration to a patient with severe haemochromatosis, a condition in which excessive amounts of iron are absorbed from dietary sources. Repeated injections were required, but that patient and several others were successfully treated during the next few months.[28] So, too, were two patients with thalassaemia, a genetic disorder of haemoglobin synthesis. Young patients given repeated blood transfusions frequently suffered life-threatening cardiac and renal complications from the consequent iron overload. Removal of excess iron by desferrioxamine radically transformed their lives. A subsequent clinical trial at the Hospital for Sick Children in London resulted in desferrioxamine being accepted as the standard chelating agent for treatment of iron overload in thalassaemia.[30]

Acarbose

Acarbose was isolated in 1975 from strains of *Actinoplanes* spp., actinomycetes found in cultures containing glucose and maltose. It was synthesised in the laboratory in 1988.[32]

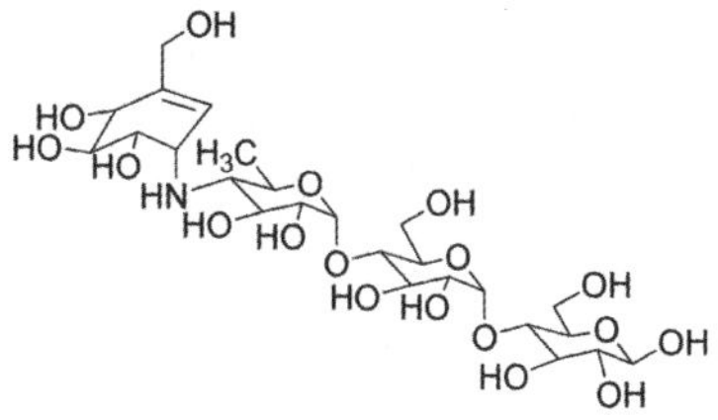

acarbose

Acarbose was found to be a reversible inhibitor of alpha-glucosidase, an enzyme that breaks down disaccharides, trisaccharides and oligosaccharides in the gut into glucose.[33] It was realised that such an alpha-glucosidase inhibitor might interfere with the digestion of dietary starch and sugar. A clinical trial in non-insulin-dependent diabetic patients found that it slowed the digestion of polysaccharides and sucrose and thereby reduced the rise in plasma glucose levels after meals.[34] Acarbose is currently used for this purpose, either on its own or with oral antidiabetic drugs.

DRUGS FROM BACTERIA

Streptokinase was the first fibrinolytic agent to be employed in the clinic. It was isolated from cultures of β-haemolytic streptococci in 1945 by researchers working under the guidance of William Tillett at New York University Medical School[35] and later purified in Lederle's laboratories. Its amino acid sequence was established in 1982.[36]

Fibrinolytic drugs act by increasing the amount of plasmin in the blood. Plasmin is present in blood to prevent unwanted clotting by catalysing the breakdown of the fibrin polymer that constitutes the framework of a blood clot. The plasmin is formed from plasminogen, a process that occurs once plasminogen has been activated by complexing with fibrin. The entire process is carefully balanced, but fibrinolytic drugs can tilt the balance to increase plasmin formation. Streptokinase administered intravenously reacts with uncomplexed plasminogen in the blood. The resulting streptokinase–plasminogen complex behaves in much the same way as the natural fibrin–plasminogen complex and converts uncomplexed plasminogen to plasmin. Thus the circulating levels of plasmin are increased and fibrinolysis occurs, dissolving intravascular blood clots.

The first major clinical application of streptokinase was in the dissolution of pulmonary embolisms. As more experience was acquired, 80–90% success rates in dissolving these clots were achieved. Streptokinase was next used to dissolve the occlusive clots in the coronary artery that are the cause of acute myocardial infarction, the commonest cause of death in industrialised countries. Large-scale clinical trials confirmed that streptokinase or other fibrinolytic drugs can reduce mortality by about one-quarter, thereby ensuring that this has become the initial means of treating such patients. Combination treatment with a fibrinolytic drug and low-dose aspirin doubled survival rates.[37]

Beecham laboratories prepared their semi-synthetic fibrinolytic agent anistreplase from human plasminogen and streptokinase. The site responsible for catalysing plasmin formation from free plasminogen was blocked with a 4-methoxybenzoyl group (anisoyl group), effectively converting the complex into a prodrug that slowly deanisoylated in the blood to release the active plasminogen–streptokinase complex.[38] The slow decomposition to the active complex meant that it did not need to be administered by continuous infusion for one to three days, as had been the case with streptokinase. Instead, it was intravenously injected over a period of five minutes.

Botulinum Toxin A

The toxin produced by *Clostridium botulinum*, an organism that causes fatal food poisoning, paralyses muscle by preventing the release of acetylcholine from presynaptic nerve terminals.[39,40] This relaxes muscles, including those in spasm.

Following a decade of clinical investigations, the US Food and Drug Administration in 1989 licensed a haemagglutinin complex of botulinum toxin A for treatment of certain spasmodic disorders such as hemi-facial muscle spasm or blepharospasm, a condition in which muscular spasm causes uncontrollable blinking and even total closure of the eyelids.[41] Local injections of nanogram quantities of the toxin are administered.

REFERENCES

1. F.J. Bove, *The Story of Ergot*. New York: Karger; 1970.
2. T.F. Baskett, A flux of the reds: evolution of active management of the third stage of labour. *J. Roy. Soc. Med.*, 2000; **93**: 489–93.

3. J. Stearns, Account of the pulvis parturiens, a remedy for quickening childbirth. *Med. Repos. New York*, 1808; **11**: 308–9.

4. O. Prescott, *Dissertation on the Natural History and Medicinal Effects of the Secale cornutum, or Ergot*. Andover: Cummings & Hilliard; 1813.

5. E. Woakes, On ergot of rye in the treatment of neuralgia. *Br. Med. J.*, 1868; **2**: 360–1.

6. P.J. Koehler, H. Isler, The early use of ergotamine in migraine. Edward Woakes' report of 1868, its theoretical and practical background and its international reception. *Cephalalgia*, 2002; **22**: 686–91.

7. C. Tanret, Sur la presence d'une nouvelle alkaloide, l'ergotine, dans seigle ergote. *C.R. Acad. Sci.*, 1875; **81**: 896–7.

8. A. Stoll, Zur Kenntnis der Mutterkornalkaloide. *Verhandl. Schweiz Naturf. Ges.*, 1920; **101**: 190–1.

9. A. Stoll, A. Hofmann, T. Petrzilka, Die Konstitution der Mutterkornalkaloide. Struktur des Peptidteils. III. *Helv. Chim. Acta.*, 1951; **34**: 1544–76.

10. A. Hofmann, A.J. Frey, H. Opt, [The total synthesis of ergotamine]. *Experientia*, 1961; **17**: 206–7.

11. E. Rothlin, Contribution á la méthode chimique d'exploration du système sympathique. *Rev. Neurol.*, 1926; **T1**: 1108–13.

12. H.W. Maier, L'ergotamine, inhibiteur du sympathique étudié en clinique, com moyen d'exploration et comme agent thérapeutique. *Rev. Neurol. Paris*, 1926; **33**: 1104–8.

13. J.C. Moir, Ergot: from 'St. Anthony's Fire' to the isolation of its active principle, ergometrine (ergonovine). *Am. J. Obstet. Gynecol.*, 1974; **120**: 291–6.

14. H.W. Dudley, J.C. Moir, The substance responsible for the traditional clinical effect of ergot. *Br. Med. J.*, 1935; **1**: 520–3.

15. M.E. Davis, F.L. Adair, G. Rogers, M.S. Kharasch, R.R. Legault, A new active principle in ergot and its effects on uterine motility. *Am. J. Obstet. Gynecol.*, 1935; **29**: 155–67.

16. M.R. Thompson, The active constituents of ergot. A pharmacological and chemical study. *J. Am. Pharm. Ass.*, 1935; **24**: 185–96.

17. A. Stoll, E. Burckhardt, *C.R. Acad. Sci.*, 1935; **200**: 1680.

18. G. Barger, F.H. Carr, An active alkaloid from ergot. *Br. Med. J.*, 1906; **2**: 1792.

19. A. Stoll, A. Hofmann, Partial Synthese von Alkaloiden vom Typus des Ergobasin. *Helv. Chim. Acta*, 1943; **26**: 944–65.

20. A. Endo, M. Kuroda, Y. Tsuyita, ML-236A, ML-236B, and ML-236C, new inhibitors of cholesterogenesis produced by *Penicillium citrinium*. *J. Antibiot.*, 1976; **29**: 1346–8.

21. A. Endo, Y. Tsuyita, M. Kuroda, K. Tanzawa, Inhibition of cholesterol synthesis *in vitro* and *in vivo* by ML-236A and ML-236B, competitive inhibitors of 3-hydroxy-3-methylglutaryl-coenzyme A reductase. *Eur. J. Biochem.*, 1977; **77**: 31–6.

22. A.G. Brown, T.C. Smale, T.J. King, *et al.*, Crystal and molecular structure of compactin, a new antifungal metabolite from *Penicillium brevicompactum*. *J. Chem. Soc. Perkin Trans.*, 1976; **1**: 1165–70.

23. N.Y. Wang, C.T. Hsu, C.J. Sih, Total synthesis of (+)-compactin (ML-236B). *J. Am. Chem. Soc.*, 1981; **103**: 6538–9.

24. A. Yamamoto, H. Sudo, A. Endo, Therapeutic effects of ML-236B in primary hypercholesterolemia. *Atherosclerosis*, 1980; **35**: 259–66.

25. A. Endo, Monacolin K, a new hypocholesterolemic agent produced by a *Monascus* species. *J. Antibiot.*, 1979; **32**: 852–4.

26. A.W. Alberts, J. Chen, G. Kuron, *et al.*, Mevinolin: a highly potent competitive inhibitor of hydroxymethylglutaryl–coenzyme A reductase and a cholesterol-lowering agent. *Proc. Natl Acad. Sci. USA*, 1980; **77**: 3957–61.

27. H. Bickel, E. Gäumann, W. Keller-Schierlein, *et al.*, Über eisenhaltige Wachstumsfaktoren, die Sideramine und ihre Antagonisten, die eisenhaltigen Antibiotika Sideromycine. *Experientia*, 1960; **16**: 128.

28. F. Woehler, Therapie von Hëmochromatosis. *Med. Klin.*, 1962; **57**: 1370–6.

29. H. Bickel, R. Bosshardt, E. Gäumann, *et al.*, Stoffwechsel-Produkte von Actinomyceten, über die Isolierung und Charakterisierung der Ferroxiamine A–F, neuer Wuchsstoff der Sideramin-gruppe. *Helv. Chim. Acta*, 1960; **43**: 2118–28.

30. R.S. Smith, Iron excretion in thalassaemia major after administration of chelating agents. *Br. Med. J.*, 1962; **2**: 1577–80.

31 Ger. Pat. 1975: 2347782 (to Bayer).

32. S. Ogawa, Y. Shibata, Synthesis of biologically active pseudo-trehalosamine: [(1S)-(1,2,4/3,5)-2,3,4-trihydroxy-5-hydroxymethyl-1-cyclohexyl] 2-amino-2-deoxy-alpha-D-glucopyranoside. *Carbohydr. Res.*, 1988; **176**: 309–15.

33. D.D. Schmidt, W. Frommer, B. Junge, et al., alpha-Glucosidase inhibitors. New complex oligosaccharides of microbial origin. *Naturwiss.*, 1977; **64**: 535–6.

34. D. Sailor, G. Rodger, *Arzneimittel.-Forsch.*, 1980; **30**: 2182.

35. L.R. Christensen, C.M. McLeod, A proteolytic enzyme of serum: characterization, activation and reaction with inhibitors. *J. Gen. Physiol.* 1945; **28**: 559–83.

36. K.W. Jackson, J. Tang, Complete amino acid sequence of streptokinase and its homology with serine proteases. *Biochemistry*, 1982: **21**: 6620–5.

37. ISIS-3: a randomised comparison of streptokinase vs tissue plasminogen activator vs anistreplase and of aspirin plus heparin vs aspirin alone among 41,299 cases of suspected acute myocardial infarction. ISIS-3 (Third International Study of Infarct Survival) Collaborative Group. *Lancet*, 1992; **339**: 753–70.

38. R.A.G. Smith, R.J. Dupe, P.D. English, J. Green, Fibrinolysis with acyl-enzymes: a new approach to thrombolytic therapy. *Nature*, 1981; **290**: 505–8.

39. I. Kao, D.B. Drachman, D.L. Price, Botulinum toxin: mechanism of presynaptic blockade. *Science*, 1976; **193**: 1256–8.

40. L.L. Simpson, The origin, structure, and pharmacological activity of botulinum toxin. *Pharmacol. Rev.*, 1981; **33**: 155–88.

41. E.J. Schantz, E.A. Johnson, Properties and use of botulinum toxin and other microbial neurotoxins in medicine. *Microbiol. Rev.*, 1992; **56**: 80–99.

CHAPTER

25

Analogues of Pharmacodynamic Agents from Fungi

Following his discovery that ergotoxine was a mixture consisting mainly of ergocornine, ergocristine and ergocryptine, Sandoz chemist Albert Hofmann prepared hydrogenated derivatives of each of these alkaloids, and also of ergotamine. Dihydroergotamine had a similar action to ergotamine and was introduced for the treatment of migraine as it had reduced peripheral vasoconstrictor activity, which arguably rendered it less likely to cause side effects.[1] After their pharmacological evaluation of dihydroergocornine, dihydroergocristine and dihydroergocriptine, Sandoz marketed Hydergine® in 1949. It was prepared by hydrogenating ergotoxine and now has the approved name of codergocrine. It was recommended as a coronary vasodilator to increase blood flow in senile patients with cerebral insufficiency and for many years was the most popular product Sandoz sold in continental Europe. It had fallen out of popularity by the 1980s, but recently it has been recognised as having clinical value and there is a need for further controlled trials to assess its proper role.[2]

Hofmann had chosen to work with ergot alkaloids even though Stoll abandoned this area once ergotamine had been introduced into the clinic. The introduction of ergometrine was one of two developments that convinced Hofmann that the moment was opportune. The other was the isolation of lysergic acid from an alkali digest of ergotinine by Walter Jacobs and Lyman Craig of the Rockefeller Institute in New York in 1934, and their subsequent recognition of it as the common component of all ergot alkaloids.[3] Dihydrolysergic acid was synthesised in 1945 by Frederick Uhle and Jacobs,[4] but it was not till 1954 that Robert Woodward and his colleagues synthesised lysergic acid.[5]

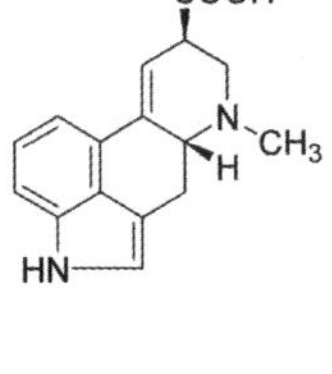

lysergic acid

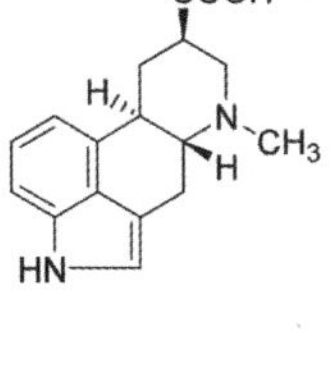

dihydrolysergic acid

Lysergic Acid Derivatives

Lysergic acid had no pharmacological activity, but Hofmann successfully exploited it for the preparation of several clinically important semi-synthetic analogues. He began by making ergometrine from it in 1937, thereby opening the way to an economic route to it via lysergic acid from ergotoxine. He next prepared a series of ergometrine analogues.[6] Among these were methylergometrine, which was at first thought to be superior to ergometrine for obstetric use, but is no longer available in the United Kingdom.

Drug Discovery. A History. W. Sneader.

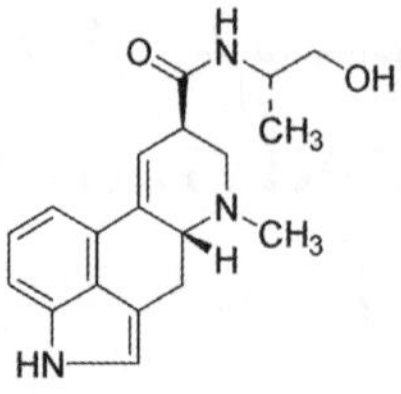

ergometrine

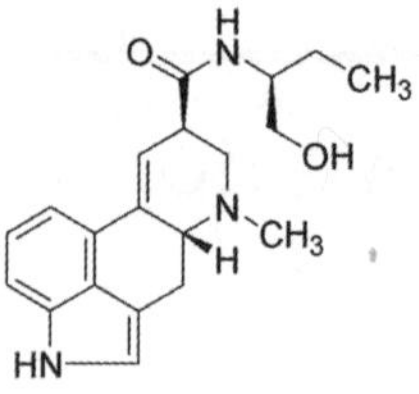

methylergometrine

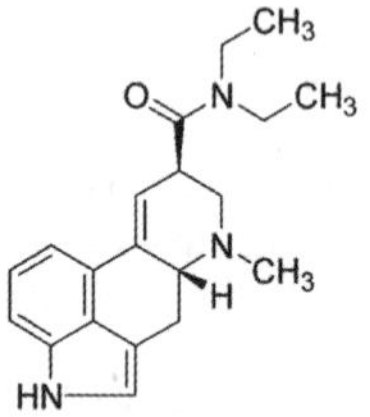

lysergide (LSD-25)

Another of the analogues prepared by Hofmann was one that was designed to enhance the activity of dihydroergotamine as a cerebral dilator for the elderly. In it, Hofmann incorporated molecular features found in nikethamide, a respiratory stimulant. The resulting molecule was code-named LSD-25 as it was the twenty-fifth lysergic acid derivative prepared by Hofmann.[6] It appeared unsuitable for clinical use as it caused excitement in some animals and a cataleptic condition in others, but Hofmann re-examined it in the spring of 1943 because it was a powerful uterine stimulant.[6] He was forced to stop work one afternoon because of dizziness and a peculiar restlessness. This was followed by extraordinary hallucinations for almost two hours. After recovering, Hofmann realised that the LSD-25 might somehow be responsible for his experience, even though he could only have ingested minute traces of it by accident. Subsequent self-experimentation confirmed that LDS-25 was an exceedingly potent psychotomimetic agent. A detailed study was then carried out at the psychiatric clinic of Zurich University, demonstrating that a few micrograms of LSD-25 taken by mouth could cause profound alterations in human perception.[7,8]

The subsequent discovery by John Gaddum at Oxford that LSD-25 was a potent, selective 5-HT antagonist stimulated worldwide research into the role of neuroamines in the brain.[9,10] Despite extensive investigations, however, no accepted role for LSD-25 in medical or psychiatric practice has yet been found, although it has been employed by military agencies for brain washing. Its illicit use since the 1960s as a recreational agent is well known. In medical circles it is now referred to by its approved name of 'lysergide'.

In 1957, Hofmann initiated further studies on psychotomimetic compounds related to lysergide after he received samples of mushrooms held sacred by Mexican Indians, who called them *teonanácatl* (flesh of God). These had been collected by Gordon Wasson, a retired financial journalist from New York. Together with his wife Valentina, he had made a life-long study of the role of mushrooms in human society. He had uncovered the story of the cult centred around 'magic mushrooms', which were categorised by the botanist Roger Heim as *Psilocybe mexicana*.[11] Hofmann isolated crystals of psilocybin from these mushrooms, which was an indole alkaloid – as was lysergide.[12] He swallowed it himself and found that it produced hallucinations.

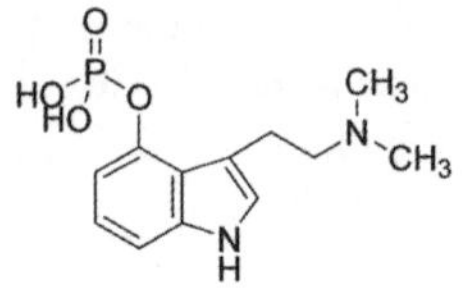

psilocybin

Hofmann next investigated *ololiuqui*, an Aztec magical plant still used by Mexican Indians in religious rituals. Wasson had also brought it to his attention. Hofmann identified it as the seeds of two types of morning glory, viz. *Rivea corymbosa* (L.) Hall and *Ipomoea violacea* L. On analysis in 1960, these were unexpectedly found to contain lysergide and alkaloids closely related to it, including ergometrine.[13]

In 1976, Wasson speculated that ergot might have been the hallucinogen that the Ancient Greeks employed in their celebration of the Mysteries at Eleusis. Hofmann knew that Ernst Chain and his colleagues in Rome had demonstrated that hallucinogenic lysergic acid derivatives were present in *Claviceps paspali*, a variety of ergot that often infected a wild grass growing around the Mediterranean basin.[14] Hofmann confirmed that the same alkaloids were present in ergot grown on wheat or barley as were to be found in that grown on rye – an important point so far as Wasson's speculation about the Eleusian Mysteries was concerned, for the Ancient Greeks did not cultivate rye. Hofmann realised that the only alkaloid readily extracted from ergot by water, as must have been the case in the Mysteries, was ergometrine. Once again, his self-experimentation revealed that it was hallucinogenic, but this raised the question of why this had not been previously discovered. His explanation for this was that its effects on the uterus were exerted by an oral dose of less than one quarter of a milligram, whereas the hallucinogenic dose was in the order of 1–2 mg. This lent support to Wasson's contention that ergot was used in the celebration of the Eleusian Mysteries, but it also explained the occurrence of madness in many victims of ergotism in the Middle Ages.[15] It is difficult to pass judgement on Wasson's assertions, although they have certainly been accepted in some quarters. The same is true of the claim that the bizarre events surrounding the Salem witchcraft trial in New England may have been associated with the consumption of contaminated rye.[16]

Methysergide

Apart from Hofmann's studies on hallucinogenic alkaloids, Sandoz investigators continued to exploit the medicinal potential of lysergic acid derivatives. They established that substitution on the 1- or 2-position of the indole ring enhanced potency as a 5-HT antagonist on the isolated rat uterus.[17] The placing of a methyl group on the ring nitrogen of ergometrine, for example, increased antagonist activity by a factor of 24.[18] Variation in the amide substituent led to the synthesis of methysergide, which was much more potent than lysergide as a 5-HT antagonist and hence was tested for its ability to prevent migraine.[19]

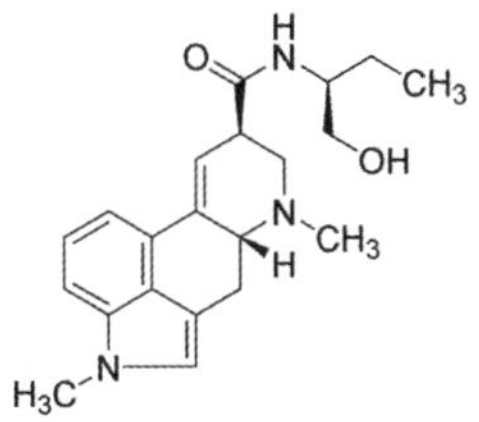

methysergide

Although an effective migraine remedy, methysergide was subsequently found to have dangerous side effects, including fibrosis of the heart valves and retroperitoneal fibrosis. Consequently, it has been reserved for the prophylaxis of migraine in patients who are severely incapacitated by migraine attacks.

Bromocriptine

At the Weizmann Institute of Science in Israel, Moses Shelesnyak began a search for a drug that could interfere with the uterine deciduoma reaction that was associated with ovum implantation in rats. In 1954, he reported that ergotoxine was active and attributed this to its inhibition of prolactin secretion.[20] This effect could be reversed by injections of either progesterone or prolactin, indicating that ergotoxine acted via the hypothalamus and pituitary to inhibit prolactin secretion.[21] This work by Shelesnyak opened up a new area of research in the field of the ergot alkaloids. The immediate outcome was that Sandoz researchers initiated a search for an ergot analogue that selectively inhibited prolactin secretion. This led to their development of the 2-bromo derivative of ergocriptine, namely bromocriptine.[22]

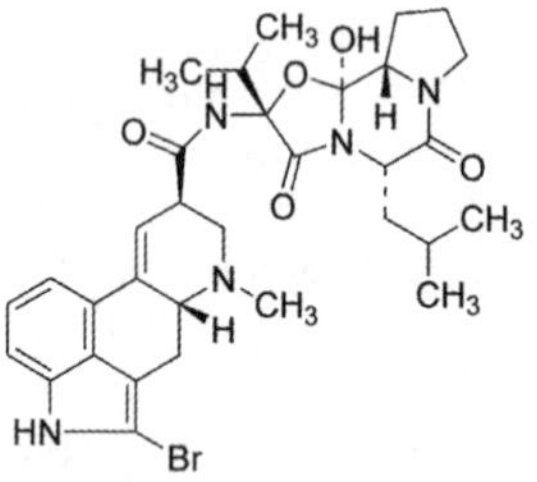

bromocriptine

While using bromocriptine to inhibit prolactin secretion during an investigation into the role of dopamine-secreting neurones in the hypothalamic–pituitary axis, Swedish investigators discovered that it had a dopamine-like action.[23] As deficiency of dopamine was associated with the occurrence of Parkinson's disease, it was examined as a supplement to levodopa treatment. Clinical trials confirmed its value in severe cases, but as a major side effect was involuntary movement, bromocriptine is reserved for patients who have ceased to respond to levodopa.[24]

Pergolide

In 1949, Hofmann reduced the methyl ester of lysergic acid with lithium aluminium hydride and obtained lysergol.[25] This compound was later isolated from *ololiuqui* and found to be one of its less psychoactive constituents insofar as oral doses of 8 mg merely induced slight sedation in volunteers.[26]

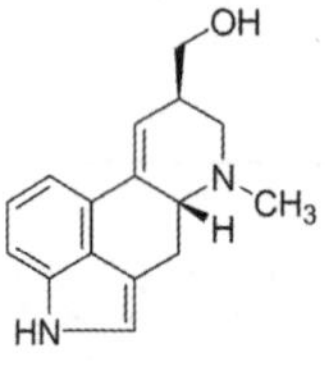

lysergol

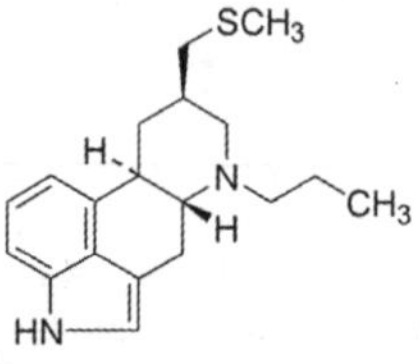

pergolide

Lysergol has been a valuable intermediate in the preparation of semi-synthetic analogues since its tosylate or mesylate heated with a thiol compound in a polar solvent readily underwent nucleophilic displacement, thereby introducing a sulfur-containing side chain. Pergolide was prepared in this manner by Lilly researchers and shown to be a potent dopamine receptor agonist.[27] It was then introduced as an adjunct to levodopa therapy in a similar manner to that of bromocriptine.

Quinagolide

Sandoz chemists Rene Nordmann and Trevor Petcher realised that, as both ergot alkaloids and apomorphine were dopamine agonists, it might be possible to postulate a common molecular moiety and base a simplified ergot alkaloid structure on this to obtain a novel dopamine agonist with high specificity of action.[28] They then synthesised a compound that retained key structural features of apomorphine and the dihydrogenated ergot alkaloid pergolide and also an experimental drug known as CQ 32-084.

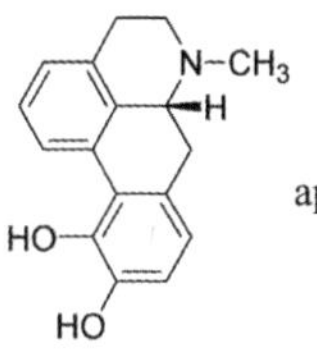

apomorphine

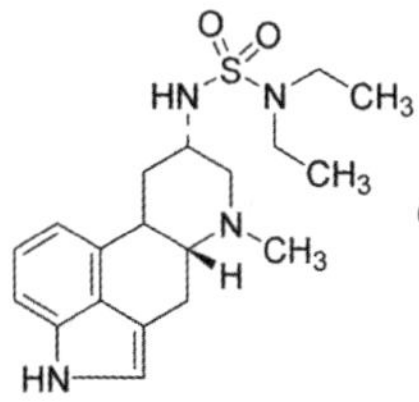

CQ 32-084

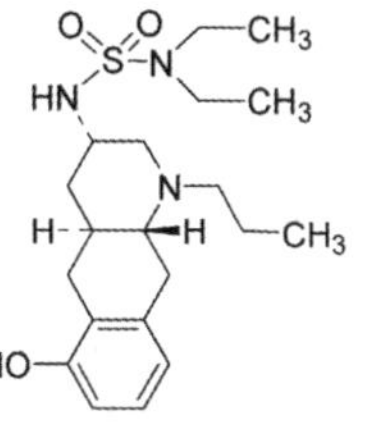

quinagolide

In order to avoid rapid metabolic deactivation of a catechol system, only a single phenolic group was included in the planned compound. This led to the development of racemic quinagolide in 1984. It combined the oral activity, high potency and long duration of action of the ergot alkaloids, with the higher specificity of action of apomorphine as a dopamine agonist. Further investigation revealed that the dopaminomimetic activity resided solely in the (−)-enantiomer,[29] which proved to be a highly selective dopamine D_2 receptor agonist.[30] In rats, quinagolide inhibited prolactin secretion without affecting levels of other pituitary hormones and it inhibited the growth of pituitary tumours.[31] In the clinic, it was effective in reducing prolactinaemia in patients with prolactin-producing pituitary tumours.[32]

SEMI-SYNTHETIC STATINS

Pravastatin is an active metabolite of mevastatin.[33] It was much less lipophilic than mevastatin and hence was less likely to penetrate extra-hepatic cells by passive diffusion. This meant that pravastatin was free of some of the side effects that arose with mevastatin. It could be given once daily by mouth in the management of hyperlipidaemia.

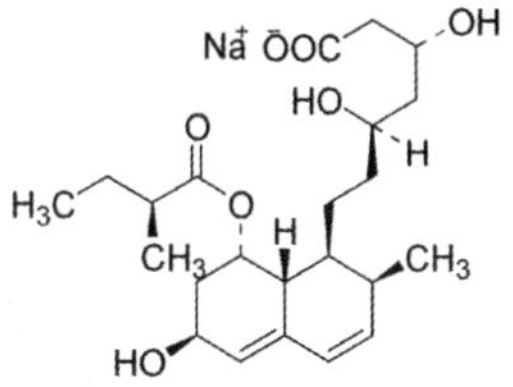

pravastatin

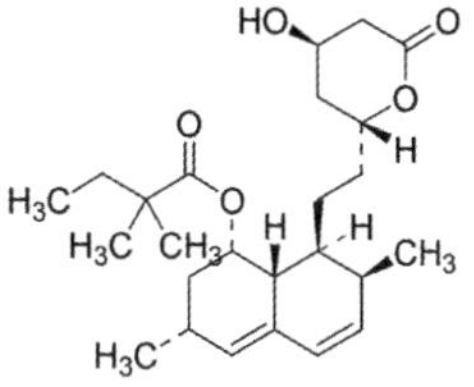

simvastatin

When Merck researchers introduced an additional methyl group adjacent to the ester function in the side chain of lovastatin to form simvastatin, they found that this more than doubled the potency as an inhibitor of HMG–CoA reductase.[34] Simvastatin was the first HMG–CoA reductase inhibitor for the treatment of severe hyperlipidaemia to be introduced into the clinic in the United Kingdom, shortly before pravastatin was marketed. There was little to choose between them. Several other statins have been introduced since then.

REFERENCES

1. A. Stoll, A. Hofmann, Die Dihydroderivate der natürlichen linksdrehenden Mutterkornalkaloide. _Helv. Chim. Acta_, 1943; **26**: 2070–81.
2. J. Olin, L. Schneider, A. Novit, S. Luczak, Hydergine for dementia (Cochrane Review), in _The Cochrane Library_, Issue 4. Chichester: Wiley; 2004.
3. W.A. Jacobs, L.C. Craig, The ergot alkaloids. II. The degradation of ergotinine with alkali. Lysergic acid. _J. Biol. Chem._, 1934; **104**: 547–51.
4. F.C. Uhle, W.A. Jacobs, The ergot alkaloids. XX. The synthesis of dihydro-_dl_-lysergic acid. A new synthesis of 3-substituted quinolines. _J. Org. Chem._, 1945; **10**: 76–86.

5. E.C. Kornfeld, E.J. Fornefeld, G.R. Kline, *et al.*, The total synthesis of lysergic acid and engrovine. *J. Am. Chem. Soc.*, 1954; **76**: 5256–7.
6. A. Stoll, A. Hofmann, Partial Synthese von Alkaloiden vom Typus des Ergobasin. *Helv. Chim. Acta*, 1943; **26**: 944–65.
7. A. Hofmann, Notes and documents concerning the discovery of LSD. *Agents Actions*, 1970; **1**: 148–50.
8. A. Hofmann, in *Discoveries in Biological Psychiatry*, eds F.J. Ayd, B. Blackwell, Philadelphia, Pennsylvania: Lippincott; 1970, pp. 91–106.
9. J.H. Gaddum, Antagonism between lysergic acid diethylamide and 5-hydroxytryptamine. *J. Pharm. Pharmacol.*, 1953; **121**: 15P.
10. J.H. Gaddum, K.A. Hameed, D.E. Hathway, F.F. Stephens, Quantitative studies of antagonists for 5-hydroxytryptamine. *Quart. J. Exp. Physiol.*, 1955; **40**: 49–74.
11. A. Hofmann, R. Heim, A. Brack, *et al.*, Psilocybin und Psilocin, zwei psychotrope Wirkstoffe aus mexikanischen Rauschpilzen. *Helv. Chim. Acta*, 1959; **42**: 1557–72.
12. A. Hofmann, R. Heim, A. Brack, H. Kobel, Psilocybin, ein psychotroper Wirkstoff aus dem mexikanischen Rauschpilz *Psilocybe mexicana* Heim. *Experientia*, 1958; **14**: 107–9.
13. A. Hofmann, H. Tscherter, Isolierung von Lysergsaure-Alkaloiden aus der mexikanischen Zauberdroge Ololiuqui (*Rivea corymbose* Hall.) *Experientia*, 1960; **16**: 414.
14. F. Arcamone, C. Bonino, E.B. Chain, *et al.*, Production of lysergic acid derivatives by a strain of *Claviceps paspali* Stevens and Hall in submerged culture. *Nature*, 1960; **187**: 238–9.
15. R.G. Wasson, A. Hofmann, C.A.P. Ruck, *The Road to Eleusis*. London: Harcourt Brace Jovanovich; 1978.
16. L.R. Caporael, Ergotism: the Satan loosed in Salem? *Science*, 1976; **192**: 21–6.
17. E. Rothlin, Lysergic acid diethylamide and related substances. *Ann. N.Y. Acad. Sci.*, 1957; **66**: 668–76.
18. A. Cerletti, W. Doepfner, Comparative study on the serotonin antagonism of amide derivatives of lysergic acid and of ergot alkaloids. *J. Pharm. Exp. Ther.*, 1958; **122**: 124–36.
19. A.P. Friedman, S. Losin, Evaluation of UML-491 in the treatment of vascular headaches. An analysis of the effects of 1-methyl-D-lysergic acid (plus) butanolamide bimaleate (methysergide). *Arch. Neurol.*, 1961; **4**: 241–5.
20. M.C. Shelesnyak, Ergotoxine inhibition of deciduoma formation and its reversal by progesterone. *Am. J. Physiol.*, 1954; **179**: 301–4.
21. M.C. Shelesnyak, Maintenance of gestation of ergotoxine-treated pregnant rats by exogenous prolactin. *Acta Endocrinol.*, 1958; **27**: 99–109.
22. E. Flückiger, H.R. Wagner, 2-Br-alpha-ergokryptin. *Experientia*, 1968; **24**: 1130–1.
23. H. Corrodi, K. Fuxe, T. Hökfelt, *et al.*, Effect of ergot drugs on central catecholamine neurons: evidence for a stimulation of central dopamine neurons. *J. Pharm. Pharmacol.*, 1973; **25**: 409–12.
24. D.B. Calne, P.F. Teychenne, L.E. Claveria, *et al.*, Bromocriptine in Parkinsonism. *Br. Med. J.*, 1974; **4**: 442–4.
25. A. Stoll, A. Hofmann, W. Schlientz, *Helv. Chim. Acta*, 1949; **32**: 1947.
26. E. Heim, H. Heimann, G. Lukacs, Die psychische Wirkung der mexikanischen Droge 'Ololiuqui' am Menschen. *Psychopharmacologia*, 1968; **13**: 35–48.
27. T.T. Yen, N.B. Stamm, J.A. Clemens, Pergolide: a potent dopaminergic antihypertensive. *Life Sci.*, 1979; **25**: 209–15.
28. R. Nordmann, T.J. Petcher, Octahydrobenzo[g]quinolines: potent dopamine agonists which show the relationship between ergolines and apomorphine. *J. Med. Chem.*, 1985; **28**: 367–75.
29. R. Nordmann, A. Widmer, Resolution and absolute configuration of the potent dopamine agonist *N*,*N*-diethyl-*N'*-[(3.alpha,4a.alpha,10a.beta)-1,2,3,4,4a,5,10,10a-octahydro-6-hydroxy-1-propyl-3-benzo[g]quinolinyl]sulfamide. *J. Med. Chem.*, 1985; **28**: 1540–2.
30. R.C. Gaillard, J. Brownell, Hormonal effects of CV 205-502, a novel octahydrobenzo [g] quinoline with potent dopamine agonist properties. *Life Sci.*, 1988; **43**: 1355–62.
31. J. Trouillas, P. Chevallier, B. Claustrat, *et al.*, Inhibitory effects of the dopamine agonists quinagolide (CV 205-502) and bromocriptine on prolactin secretion and growth of SMtTW pituitary tumors in the rat. *Endocrinology*, 1994; **134**: 401–10.
32. T. Nickelsen, E. Jungmann, P. Althoff, P.M. Schummdraeger, Treatment of macroprolactinoma with the new potent non-ergot D2-dopamine agonist quinagolide and effects on prolactin levels, pituitary function, and the renin–aldosterone system. Results of a clinical long-term study. *Arzneimittel.-Forsch.*, 1993; **43**: 421–5.
33. N. Serizawa, K. Nakagawa, K. Hamano, *et al.*, Microbial hydroxylation of ML-236B (compactin) and monacolin K (MB-530B). *J. Antibiot.*, 1983; **36**: 604–7.
34. W.F. Hoffman, A.W. Alberts, P.S. Anderson, *et al.*, 3-Hydroxy-3-methylglutaryl-coenzyme A reductase inhibitors. 4. Side-chain ester derivatives of mevinolin. *J. Med. Chem.*, 1986; **29**: 849–52.

26

The First Synthetic Drugs and their Analogues

August Wilhelm von Hofmann distilled aniline from coal tar in 1843 while working in Giessen as a research student with Justus Liebig. Two years later, he moved to the Royal College of Chemistry in London, where he demonstrated that benzene was present in coal tar. One of his students, Charles Mansfield, subsequently isolated it by fractional distillation of the tar. Nitration of benzene with nitric acid then provided the basis of a route to the industrial manufacture of aniline dyes and other important organic chemicals.

In 1856, another of Hofmann's students, William Perkin, oxidised the aniline derivative allytoluidine in an overly ambitious attempt to synthesise quinine. Instead he obtained a dark substance that turned fabrics purple. This was the first synthetic dyestuff, which Perkin initially called aniline purple but later changed to mauveine.[1] Realising the commercial value of the dye, Perkin established his own factory the following year. This marked the start of the synthetic dyestuffs industry that was to fuel a demand for organic chemists who could discover new products through the application of research. The high price of natural dyes was a matter of concern for the rapidly expanding textile industry, which was trying to match the demand from a growing population for cheap clothing. Perkin's mauvine remained expensive to produce, but within a decade several manufacturers were developing a range of affordable new dyes from aniline, toluidine and quinoline. Although the industry began in England, it was in Germany that it thrived. Two factors largely accounted for this. Once the German economy had recovered from the collapse of its stock market in 1873, industrialisation entered its second phase in which the chemical and electrical industries rapidly expanded to compete in importance with the existing coal, iron and steel industries. The heavy investment in the manufacture of synthetic dyes soon put Germany well ahead of all its competitors in this field. The second factor that made this possible was the willingness of German universities in the 1870s and onwards to meet the need of the moment. German chemists rapidly became the leaders in the emerging field of organic chemistry and remained so until the outbreak of the Second World War. They wrestled with the nature of the structures of the novel molecules they had synthesised, skilfully breaking them apart to identify known fragments, and then deducing how the atoms were assembled in the intact molecules. New synthetic reactions were also introduced, providing routes to a vast range of novel dyes and other commercially important organic compounds, including synthetic drugs. Success bred success.[2] In marked contrast to the situation in Germany, the failure of the United Kingdom to maintain the lead that Perkin had given it with mauvine was in no small measure due to the disdain with which its universities at the time viewed industrial contacts.

The German pharmaceutical industry developed directly out of the dyestuffs industry when leading manufacturers like F. Bayer & Company and Farbenfabriken Hoechst realised that their chemists could produce medicines as well as dyes. Initially, a few dyes served as drug prototypes, but during the twentieth century the industry became completely independent of its origins and instead concentrated on chemically modifying the structures of natural

Drug Discovery. A History. W. Sneader.
©2005 John Wiley & Sons Ltd

products from plant or biochemical sources. In Britain, France, the United States, Canada and to a much lesser extent Switzerland, the industry continued to focus on the extraction of alkaloids and glycosides from plants, with only a minimal effort being expended on the development of synthetic drugs. Such an approach was still capable of bringing immense benefits to the sick, as illustrated by the isolation of insulin in Canada and penicillin in the United Kingdom. However, a gradual change of direction in favour of synthetic drugs came about because of shortages during the two World Wars of essential medicines normally supplied by Germany.

The majority of natural products, be they from the plant or animal kingdoms, have been isolated in academic laboratories. However, the opposite is true of synthetic drugs. With the exception of a handful of hypnotics, the first synthetic drugs were all developed in industrial laboratories or research institutes where the *raison d'être* was the development of new medicines, such as the Institute for Experimental Therapy established by Paul Ehrlich in Frankfurt.

PHENOL

When the inventor William Murdock first used coal gas in 1794 to illuminate his home in Redruth, Cornwall, he could not have envisaged the full consequences of his actions. Within seven years, buildings in Birmingham were being lit by gas and before long the streets of other major British cities were no longer dark at night thanks to locally produced gas. In the United States, gas lighting had been installed in Baltimore by 1817. There was, however, one unwelcome by-product that arose from this exciting development, namely vast amounts of apparently worthless coal tar. One of the first to examine it was Friedlieb Runge, the chief chemist of a gas works at Oranienburg near Berlin, who steam-distilled a light oil from it.[3] A portion of this oil was acidic and so dissolved in milk of lime. Runge gave the name carbolic acid to the material he then recovered by acidifying the lime solution. Charles Gerhardt named it phenol in 1842.

Friedlieb Runge had been impressed by the ability of carbolic acid to prevent the decay of animal tissue and wood, but felt that it would be too expensive to market it as a preservative. In 1844, however, the French physician Henri-Louis Bayard incorporated coal tar in a clay-based powder for disinfecting manure to be used as a fertiliser. This won him a prize from the Société d'Encouragement for its contribution to hygiene.[4]

The first person to exploit the disinfectant properties of phenol was the industrial chemist Frederick Calvert, who had studied and worked in France from 1835 to 1846.[5] Returning to Manchester, he became a consultant chemist and introduced phenol for embalming. He became closely involved with the inventors of McDougall's Powder, a crude mixture of calcium salts and phenol patented in 1854 for purifying water and deodorising sewage. Calvert manufactured the powder and saw it become very popular as a disinfectant for stables, farmyards and any place where putrefying material was to be found. It was also applied to sores.[5]

Calvert became convinced that the disinfectant in coal tar was phenol and accordingly informed the Académie des Sciences in 1859.[6] This encouraged the Dresden physician Friedrich Küchenmeister to employ pure phenol as a wound dressing.[7] Meantime, a pharmacist from Bayonne, LeBeuf, asked Jules Lemaire to evaluate his emulsified coal tar.[8] It proved very successful in treating septic wounds and in April 1862 was authorised for wound disinfection in the civil hospitals of Paris. The following year, Lemaire's book entitled *De l'Acid Phénique* was published, followed by an enlarged 2nd edition in 1865. This established Lemaire as the leading advocate of the use of phenol in surgery at that time, although his work aroused little enthusiasm in Britain.

Gilbert Declat's lengthy volume entitled *Nouvelles Applications de l'Acide Phénique en Médecin et en Chirurgie* was also published in 1865. Declat referred to phenol as a 'parasiticide'. In contrast to Lemaire, he was fully cognisant with Pasteur's ideas. He expressed the hope that phenol would be used to prevent infection and even recommended washing of the walls and surroundings of the sick room with it.

British surgeons continued to ignore the developments in France. Fortunately, Calvert had convinced public authorities in Britain of the benefits of phenol for treating sewage and a newspaper report about its use in Carlisle was read by Joseph Lister, the Professor of Surgery at the University of Glasgow.[9] Though deeply concerned about the high incidence of lethal infections following surgery, reaching 40% after amputations, he had been unaware of the studies being carried out in France with phenol. After reading the newspaper article, he applied German creosote when operating on a patient with a compound fracture of the leg. The prognosis was poor since puncturing of the skin by the broken bone had resulted in infection and Lister was unable to save the patient. He went on to modify his technique by covering wounds with dressings soaked in solutions of pure phenol obtained from Calvert. On 12 August 1865, a boy run over by a cart was admitted to the Royal Infirmary with a compound fracture of the leg. This time, the dressing successfully prevented infection. Eleven more patients were treated, with only one death. The first of several papers by Lister on antiseptic surgery then appeared in the *Lancet* in 1867.[10] By basing his use of phenol on a clear understanding of Pasteur's researches, Lister was highly successful in preventing wound sepsis and he transformed surgical practice and rendered it safe. Antiseptic surgery was soon replaced by aseptic surgery, itself a logical development of Lister's approach. With the demise of antiseptic surgery, phenol became much less important. The main objection to its use was its corrosive nature, which permitted only low concentrations to be applied to the skin. Today phenol is found only in antiseptic creams and liquids such as mouth washes.

ANALOGUES OF PHENOL

Alternatives to phenol were sought as early as 1867 when Arthur Sansom at London's Royal Hospital for Diseases of the Chest administered sulfocarbolate of potash by mouth in the mistaken belief that it would slowly decompose in the body to release small amounts of phenol and thereby act as an internal antiseptic. The product he used was a mixture of the salts of *ortho*- and *para*-phenolsulfonic acids, apparently consisting largely of the former.[11,12] This was subsequently named solozic acid, and became available commercially as a one-in-three solution in water. It was widely used for treatment of diphtheria, scarlet fever and puerperal fever until Heinrich Bechhold and Paul Ehrlich revealed its inferiority to other phenolic compounds.[13]

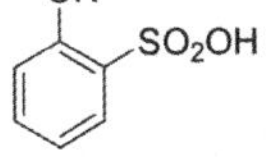

solozic acid

Salicylic Acid

Carl Thiersch, the Professor of Surgery at Lepizig, adopted a similar approach to Sansom in seeking a compound with less deleterious effects than phenol on tissues. As the first German surgeon to adopt Lister's methods, he had become well aware of its damaging effects. When he discussed the matter with Hermann Kolbe, the Professor of Chemistry and by now the leading chemist in Germany, the latter recalled how in 1860 he and Lautemann had treated phenol with carbon dioxide in the presence of sodium under pressure to form salicylic acid.[14] Kolbe knew that, on heating salicylic acid, carbon dioxide was liberated and the acid decomposed into phenol, so he now carried out some simple tests on salicylic acid and confirmed that it had antiseptic properties. His idea that it might release phenol was never realised, but salicylic acid did find a role as an antiseptic that was somewhat less corrosive than phenol. That it was still damaging to tissues is evident from its continued use to burn out warts. Convinced of the value of salicylic acid as a substitute for phenol, Kolbe modified his original synthesis so that the acid could be produced on an industrial scale. One of his former students then opened the Salicylsäurefabrik Dr F. von Heyden in Dresden, as a consequence of which salicylic acid became cheaply available in 1874.

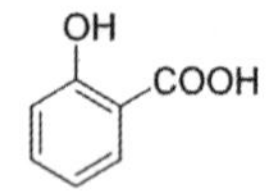

salicylic acid

Although it never replaced phenol in surgical practice, salicylic acid became popular as an internal antiseptic at a time when it was widely believed that many diseases arose from the presence of pathogenic bacteria in the gut. This led Carl Buss at St Gallen in Switzerland to administer salicylic acid by mouth to typhoid patients. When the course of their disease was routinely checked by thermometry, it became obvious that salicylic acid was an effective antipyretic. However, it did not lower body temperature by curing the typhoid infection. Widespread interest was aroused in 1875 when Buss published his observation that repeated doses of salicylic acid could control fevers without causing the side effects of quinine, at that time the standard antipyretic.[15] Salomon Stricker at the University of Vienna Medical School then tested salicylic acid for its ability to reduce the temperature of patients with rheumatic fever. To his surprise, it also proved to be of definite value as an antirheumatic drug.[16] A similar observation was made by the Scottish physician Thomas MacLagan,[17] while the French physician Germain Sée confirmed the specific value of salicylic acid in rheumatoid arthritis and gout.[18] Surprisingly, many physicians were unaware of these reports until the 1950s when, at last, there was universal recognition of the importance of salicylate therapy in rheumatoid arthritis.[19]

Phenyl Salicylate

Salicylic acid was normally prescribed as its sodium salt. Many patients complained about its unpalatibility and irritating effects on the stomach. An attempt to improve upon both it and phenol as internal antiseptics was made by the Polish chemist and physician Marceli Nencki in 1883, when he reacted the two drugs together to form phenyl salicylate.[20] After being swallowed, this passed unchanged through the stomach because it was highly insoluble. It was more soluble in the small intestine, where the portion that dissolved then decomposed to liberate small amounts of the parent drugs. After Hermann Sahli had tested phenyl salicylate in Berne, it was generally believed that some benefit was to be derived from these small amounts.[21]

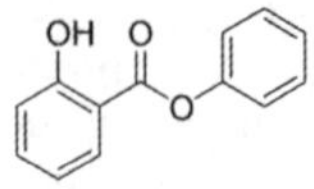

phenyl salicylate

Phenyl salicylate was marketed under the name 'Salol' and it was many years ι was general recognition that any advantage it had over salicylic acid was offʂ prolonged onset of activity and variability of therapeutic response. Until then, it was ɀ substitute for salicylic acid as an antipyretic and antirheumatic. It was an early exampⳇ gullibility of many when presented with a drug exhibiting chemical novelty unsuppoⳇed by reliable clinical proof of efficacy.

Aspirin

After being appointed in 1896 by F. Bayer & Company of Elberfeld to stimulate research so as to free the company from its dependence upon universities for the supply of new compounds, Arthur Eichengrün began to prepare esters of phenolic compounds that irritated the stomach. He expected that the masking of the phenol would protect the stomach, while the esters would decompose once they reached the more alkaline conditions of the gut and release the active drug for it to be absorbed into the circulation. Felix Hoffmann was given the task of preparing acetylsalicylic acid, a crude version of which may have been synthesised in 1853 by Charles von Gerhardt.[22] After acetylsalicylic acid had been prepared, it was tested in the spring of 1897 by the company pharmacologist Heinrich Dreser. He rejected it – despite the tests appearing to Eichengrün to show that it was superior to any other salicylate. Acting on his own initiative, Eichengrün tested the compound on himself, then arranged for it to be clandestinely evaluated by physicians in Berlin. The outcome of this was not only to confirm that acetylsalicylic acid was an effective substitute for salicylic acid, but also that it had unexpectedly relieved pain when a patient with toothache happened to be given a sample to consume. Once the analgesic properties had been confirmed in other patients, the colleague who had conducted the secret trials brought this to the attention of Bayer management. They responded by arranging for Kurt Witthauer of the Deaconess Hospital in Halle and Julius Wohlgemuth in Berlin to conduct independent clinical trials of the drug.[23,24] The outcome persuaded Bayer management to market acetylsalicylic acid under the proprietary name Aspirin®, which was coined by Eichengrün from 'a' for acetyl and 'spirin' from *Spirea ulmaria*, the now obsolete name of the plant from which salicin was obtained. To ensure the success of the new drug, F. Bayer & Company circularised more than 30 000 doctors in what was probably the first mass mailing of product information. Eichengrün was rewarded for his efforts by being promoted to Director of Pharmaceutical and Photographic Research, while Hoffmann became Director of Pharmaceutical Sales.

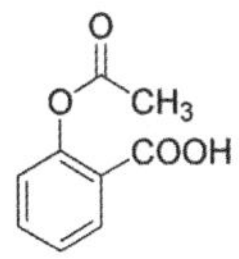

aspirin

Dreser was asked by the Bayer management to publish the results of his further examination of aspirin after its testing in Berlin, in order to lend scientific credibility to the new product.[25] His paper omitted any reference to either Eichengrün or Hoffmann and gave no indication of how aspirin came to be developed. The first account of this did not appear until a year after the Nazi party came to power in Germany in 1933. It was published in a history of chemical engineering as a short footnote that claimed to be based on a communication from Felix Hoffmann to the author.[26] This alleged that when Hoffmann had been asked by his rheumatic father to find an alternative to the foul-tasting sodium salicylate, he searched the literature and came across acetylsalicylic acid, then preparing it in pure form. On the fiftieth anniversary of the introduction of aspirin, Arthur Eichengrün published the only detailed account ever written by any of those directly involved in the development of aspirin.[27] In this, he implied that history had been rewritten by the Nazis to hide the fact that it was a Jew who was primarily responsible for the development of the most famous drug in history. This appears to have been as unpalatable to some as sodium salicylate was supposed to have been for Hoffmann's father, leaving the present writer to attempt to set the record straight on the centenary of the introduction of aspirin.[28]

How aspirin worked remained a mystery until 1971 when John Vane at the Institute of Basic Medical Sciences of the University of London and the Royal College of Surgeons of England demonstrated that it blocked prostaglandin synthesis.[29] The precise manner in which this occurred was subsequently shown to be through the permanent transfer of the acetyl group from aspirin on to the hydroxyl group of a serine residue located 70 amino acids from the C-terminal end of the cyclooxygenase (COX) enzyme that promotes the formation of prostaglandins.[30]

Another effect of aspirin that has been successfully exploited is its antiplatelet activity, which also arises from blocking of prostaglandin synthesis. In 1949, Gibson described his successful use of aspirin in a small group of patients with vascular problems.[31] Around the same time, Lawrence Craven in California realised that his tonsillectomy patients who had taken a chewable aspirin preparation for pain relief were more likely than others to bleed.[32] Craven went on to conduct an uncontrolled investigation on 8000 patients who regularly consumed aspirin and claimed that none suffered heart attacks.[33] None of his publications appeared in prominent journals and, when he died of a heart attack despite taking aspirin, any credibility his work might have carried was undermined. Fortunately, in New York in 1967, Harvey Weiss established that the prolongation of bleeding time caused by aspirin was due to an impairment of platelet aggregation.[34] He suggested that aspirin might be an antithrombotic drug and in 1971 was able to provide experimental evidence that this was the case. He urged that clinical trials be carried out. Three years later physicians in Wales published the results of the first randomised, controlled clinical trial of aspirin in patients who had experienced a previous heart attack.[35] Since then, it has taken many years for it to be generally accepted that low doses of aspirin reduce the risk of myocardial infarction in patients with cardiovascular disease.

Aspirin Analogues

Anthranilic acid (o-aminobenzoic acid), an analogue of salicylic acid in which the phenolic hydroxyl is replaced by an amino group, is inactive. Parke, Davis and Company developed a non-steroidal anti-inflammatory agent called mefenamic acid, by adding a second benzene ring that drastically reduced the basicity of the aromatic amino group in order to prevent zwitter ion formation. It was patented in 1961.[36] The researchers also confirmed that flufenamic acid, which had originally been synthesised in 1948, was a useful anti-inflammatory drug.[37] Geigy researchers subsequently developed diclofenac by taking into account the structural physicochemical characteristics of existing anti-inflammatory agents.[38]

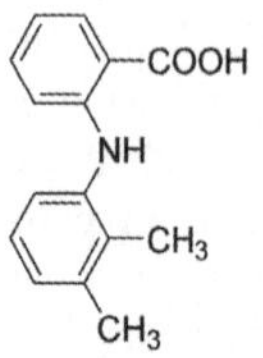

mefenamic acid 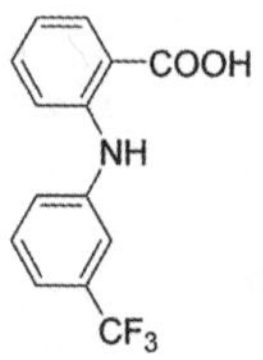   flufenamic acid 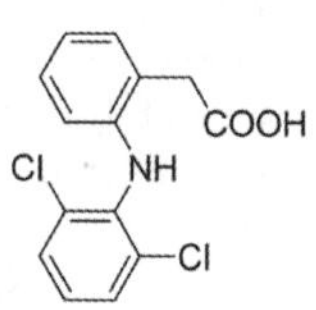   diclofenac

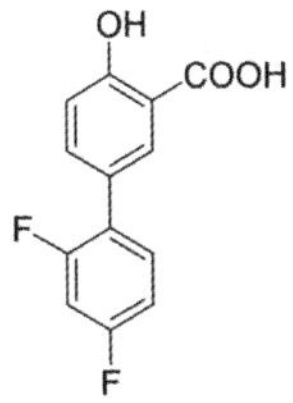

diflunisal

Diflunisal was introduced by Merck Sharp and Dohme after more than 500 compounds had been synthesised and evaluated in a 15 year long search for a longer-acting, safer analogue of aspirin.[39] It is similar in its therapeutic profile to arylpropionic acid-derived non-steroidal anti-inflammatory drugs.

p-Aminosalicylic Acid

While investigating the nutritional requirements of the causative organism of tuberculosis, *Mycobacterium tuberculosis*, in 1940 Frederick Bernheim, a biochemist at Duke University Medical School in North Carolina, discovered that benzoic and salicylic acids increased oxygen utilisation.[40] This indicated that these acids were serving as nutrients for the bacteria. Taking into consideration the recently announced antimetabolite theory, he went on to antagonise the effect of these acids with 2,3,5-triiodobenzoic acid.[41] In conjunction with Alfred Burger and others, Bernheim then examined a diverse range of halogenated aromatic acids and phenolic ethers as potential antimetabolites. Some of the latter were active, but unsuitable for clinical application because of side effects on the central nervous system.

Bernheim had communicated his findings to his friend Jorgen Lehmann at the Sahlgren's Hospital in Gothenburg. Reflecting on the past at the age of 83, Lehmann has written that in 1943 he was convinced that the positioning of the amino group in the sulfanilamide antagonist p-aminobenzoic acid was critical; hence he felt that a p-amino group should be introduced into salicylic acid to provide a tuberculostatic drug.[42]

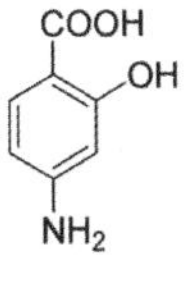

p-aminosalicylic acid

Lehmann asked the Ferrosan Company of Malmo (now incorporated into Kabi Pharmacia) to supply him with p-aminobenzoic acid as it had previously been prepared only in quantities insufficient for biological evaluation. As a result, he was able to test it in January 1944 and found it to be tuberculostatic in animals, with a wide margin of safety. In March of that year, a child with a severely infected wound was successfully treated by local application of the drug. By the end of the year, 20 patients had received the drug by mouth and results were most promising. Lehmann published the result of two years of clinical trials, confirming that p-aminosalicylic acid (PAS) could cure tuberculosis.[43]

It was later established that p-aminosalicylic acid was best used in combination with the much more potent drugs streptomycin and isoniazid. The combination of these drugs proved to be a major step in overcoming the problem of bacterial resistance towards streptomycin. On its own, p-aminosalicylic acid lacked sufficient potency for routine clinical application. There were two other shortcomings. It was rapidly excreted via the kidneys, resulting in the need for oral administration with quantities of 12 g daily in four or more divided doses. This compounded the second problem, which was that it caused distressing gastrointestinal disturbance. The problem could not be overcome either by formulating it differently or by the synthesis of analogues. Once a range of alternative drugs became available in the 1980s, *p*-aminosalicylic acid was no longer prescribed.

Cresols

In 1886, Oswald Schmiedeberg claimed that cresol was not only more potent but also less toxic than phenol.[44] As cresol consisted mainly of *m*-cresol, together with its *ortho* and *para* isomers, any reduced toxicity was probably due to the smaller amount that could dissolve in water. Kalle and Company of Frankfurt introduced chlorocresol as a bactericide in 1897. Many alkly, halo and haloalkylphenols have been introduced since then.

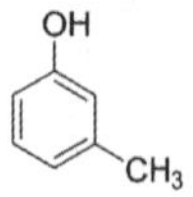

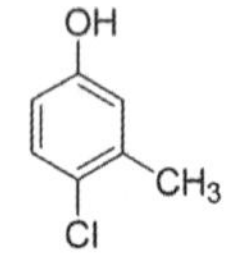

The findings of the first extensive investigation into their activity was reported in 1906 by Bechhold and Ehrlich, who found that although polyhalo compounds were more potent than monohalo compounds, they did not retain activity in the presence of serum.[13] Mono-halophenols were subsequently shown by Laubenheimer to be less affected by the presence of serum.[45] This was of considerable importance for compounds that were to be used in the clinic. Klarmann discovered that in higher molecular weight phenols the spectrum of antibacterial activity did not change uniformly with alteration to the chemical structure.[46] In several instances he found that structural modification enhanced activity against most organisms, yet removed all activity against specific organisms. This phenomenon was to be encountered repeatedly in the antibiotic era.

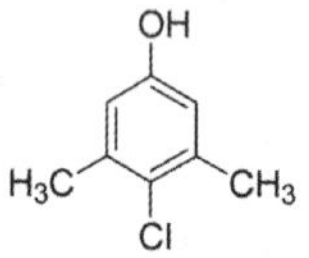

A well-equipped unit supported by the Medical Research Council, the Rockefeller Foundation and the Bernhard Baron Trustees was opened at Queen Charlotte's Maternity Hospital in London in 1931 with the objective of finding a solution to the problem of puerperal fever. This had been the cause of death in two or more out of every thousand women within days of giving birth. Leonard Colebrook, the bacteriologist in charge of the new unit, was particularly concerned about a form of the disease caused by haemolytic streptococci, in which there had been a mortality rate of over 25%. He collaborated with his cousin, a chemist who worked for the Reckitt company in Hull, in the development of a non-irritant antiseptic that could kill streptococci on the skin of the midwives' hands. He experimented on himself by smearing his hands with virulent bacterial cultures, a procedure that led to the development of chloroxylenol solution as a non-irritant antistreptococcal hand disinfectant, which greatly reduced the incidence of puerperal fever caused by streptococcal infection.[47] It remains the most important of the cresols to have been introduced into medicine.

Hexachlorophene

William Gump began an investigation of halogenated bisphenols in the laboratories of Givaudan–Delawanna in New York in 1937, which resulted in the development of hexachlorophene as a skin disinfecting and cleansing agent after the Second World War had ended.[48] It had the advantage of retaining activity in the presence of soap and hence was introduced in creams, soaps and cleansing lotions sold to the general public.

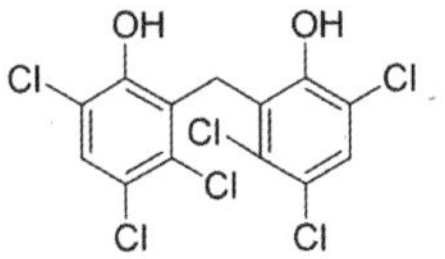

hexachlorophene

Tragically, a manufacturing blunder in France led to the sale of a baby powder containing 6% hexachlorophene, which resulted in the death of 20 children before the cause was discovered. It was subsequently confirmed that severe neurotoxicity had previously occurred in infants after absorption of hexachlorophene through the skin on repeated application.[49] This led the US Food and Drug Administration in 1972 to ban sales to the public of all formulations containing more than 0.1% hexachlorophene. Other drug authorities did likewise, with hexachlorophene being allowed to remain in use as a skin disinfectant for health care workers.

HYPNOTICS

Chloral hydrate was synthesised in 1832 by Justus Liebig, who also discovered that it decomposed into chloroform and formic acid when treated with alkali.[50] This caught the attention of two investigators, Rudolf Buchheim and Oskar Liebreich, who both then discovered the hypnotic action of chloral hydrate.[51]

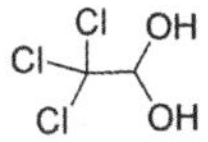

chloral hydrate

Buchheim had wondered whether excessive alkalinity of the blood, which was thought to be a complicating factor in some diseases, could be reduced by administration of chloral hydrate. The idea behind this was that as treatment of chloral hydrate with caustic alkali liberated chloroform and formic acid, then alkaline blood should do likewise. Buchheim believed that the chloroform thus released in the blood might even be converted into hydrochloric acid, thereby supplementing the alkali-neutralising action of the formic acid. However, on taking a draught of chloral hydrate to test his hypothesis he and several of his colleagues quickly fell asleep. Buchheim thought this proved that chloroform had been released, but had not then broken down into hydrochloric acid. His investigation was abandoned and not reported until 1872, three years after Liebreich had introduced chloral hydrate as a hypnotic drug.[52]

Liebreich, an assistant professor at Berlin University's Pathological Institute, also tried to use chloral hydrate to liberate chloroform in the blood. Unlike Buchheim, he actually hoped the chloroform would induce unconsciousness. He was therefore delighted when experiments on rabbits confirmed his expectations. The animals awakened unharmed several hours later. When 1.35 g of chloral hydrate was administered to a disturbed individual by subcutaneous injection, he slept for 5 hours. A subsequent dose of 3.5 g in water kept him asleep for 16 hours. Liebreich published his findings in August 1869.[53,54] Within a few months, chloral hydrate was in use all over the world as the first safe hypnotic, despite its unpleasant taste and the frequency with which it caused gastric irritation. An early shortage of supplies of chloral hydrate raised the price of a draught to three shillings and sixpence in the United Kingdom, or just under US $1, leading to the expression, 'A sleep costs a dollar!' The shortage ended after Schering built a factory in Berlin to produce it. Daily consumption in both Britain and the United States passed the 1 ton mark within a decade, which no other contemporary drug even remotely rivalled. Chloral hydrate remains in use around the world.

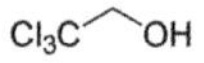

Cl₃C⌒OH   trichloroethanol   Cl₃C⌒O⌒P(=O)(OH)OH   triclofos

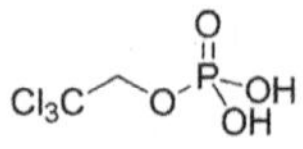

It soon became evident that if any chloroform at all was released in the blood after administration of chloral hydrate, it could only be trace amounts. It is nowadays realised that the alkalinity of blood is so slight as to be unable to induce decomposition of chloral at all, but in the 1860s there was no awareness of the subtleties of Sørenson's pH scale, which was not introduced until 1909. Joseph von Mering, a protégé of Schmiedeberg, correctly suggested that chloral hydrate was converted in the body into the active hypnotic trichloroethanol,[55] but it was not until 1948 that there was experimental proof of this.[56]

Trichloroethanol could not be administered as a drug because of its unpleasant taste, as well as a tendency to cause nausea. Its phosphate ester, triclofos sodium, was introduced by Glaxo in 1962.[57] This is rapidly hydrolysed in the gut to liberate trichloroethanol. There was no problem with palatability when triclofos was formulated in an elixir as its water-soluble sodium salt.

The discovery of the hypnotic properties of chloral hydrate brought home to many people the potential of synthetic drugs as therapeutic agents. It also set the scene for what was to become for many years the sole alternative to basing the structures of synthetic drugs on those of natural products, namely the idea of designing a new drug that would decompose to release a pharmacologically active agent. An early example of this approach is seen when Schmiedeberg selected urethane as a potential anaesthetic in small animals.[44] He thought it would break down in the body to release not only alcohol, a central nervous system depressant when large doses were consumed, but also ammonia and carbon dioxide, which were both known to be respiratory stimulants. The anaesthetic action of urethane that Schmiedeberg then observed was later shown to be due solely to the intact molecule, with neither carbon dioxide nor ammonia being released in the body. His hypothesis may have been wrong, but it resulted in the introduction of an anaesthetic that is still used in small animals.

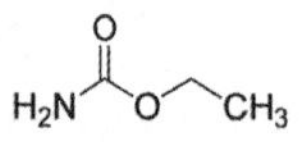

H₂N-C(=O)-O-CH₃   urethane 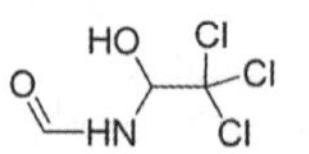   chloralformamide

The approach of designing drug molecules to liberate active substances can also be seen in analogues of chloral hydrate that were marketed in the 1880s. For example, Joseph von Mering patented chloralformamide as a hypnotic in 1889, more than 50 years after it had first been synthesised.[58] He also believed that ammonia and carbon dioxide would be liberated as respiratory stimulants, which would counteract any respiratory depression that occurred as a result of overdosing. The new drug turned out to be no safer than chloral hydrate, though it was less irritating to the stomach.

Sulfonmethane

In the summer of 1887, Eugen Baumann at the University of Freiburg asked his colleague Alfred Kast to see whether some novel sulfur compounds that he had prepared had any pharmacological activity. Kast began by injecting a suspension of 2 g of sulfonmethane into a dog.[59] Initially, there was no apparent reaction from the animal, but several hours later it staggered and fell unconscious. The dog did not awaken until several hours later. The experiment was repeated on other animals, confirming that sulfonmethane was a hypnotic.[60] It was marketed the following year by F. Bayer & Company.

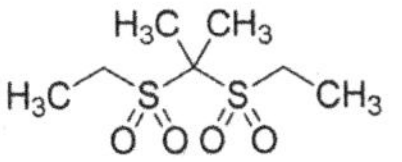

sulfonmethane

As a hypnotic that combined palatibility and absence of gastric irritancy with freedom from circulatory disturbance, sulfonmethane was to be one of Bayer's first profitable pharmaceutical products. It retained its popularity until the introduction of the more rapidly acting barbiturates rendered it obsolete.

The Barbiturates

Consideration of the chemical nature of the hypnotics discovered during the last two decades of the nineteenth century convinced von Mering that a key feature in their molecular structure was the presence of a carbon atom containing two ethyl groups. Knowing of work already carried out by others on urethane and urea derivatives, he and Emil Fischer investigated diethylacetylurea, finding it to be as potent a hypnotic as sulfonmethane.[61] Its bromo derivative, carbromal, was later marketed by Bayer as a hypnotic.[62]

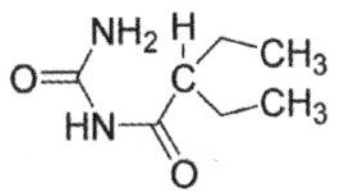

diethylacetylurea 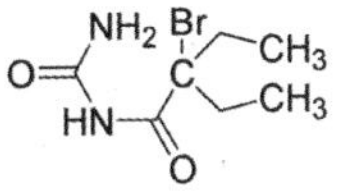   carbromal 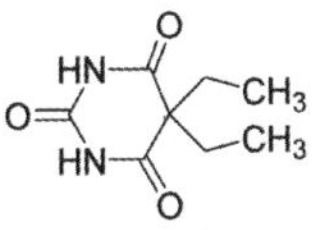   barbital

After finding diethylacetylurea to be a hypnotic, von Mering prepared 5,5-diethylbarbituric acid, unaware that it had already been made 20 years earlier.[63] The parent compound of this series, barbituric acid, had been synthesised by von Baeyer in 1864, and is variously said to have been so named after a young maiden with whom its discoverer was then in love, or, more prosaically, on account of its first preparation being on St Barbara's Day.[64] After von Mering established that 5,5-diethylbarbituric acid was a hypnotic in animals, he discussed the compound with Fischer. The latter doubted the reliability of the synthesis and instructed his nephew, Alfred Dilthey, to synthesise it and several related compounds. When tested on a dog, 5,5-diethylbarbituric acid proved to be the most potent of the 19 compounds that had been synthesised and was much more potent than von Mering's compound. This provoked Fischer to remark that he now had the true compound, which explains why it was given the proprietary name of Veronal® (Latin: *verus* = true) when it was marketed by F. Bayer & Company.[65] Fischer filed a patent on the new hypnotic at the end of January 1903 and a detailed report appeared the next year.[66] All previous hypnotics, with the possible exception of chloral hydrate, were now rendered obsolete.

When the United States entered the First World War in 1917, Congress passed the *Trading with the Enemy Act* to allow American firms to manufacture unobtainable German drugs covered by patents, such as Veronal® and Salvarsan®. Royalties were paid to the Alien Property Custodian for distribution to the American subsidiaries of German companies when the war ended. The Act required the American products to be given a new name approved by the American Medical Association (AMA). This practice of giving a drug an approved, or generic, name in addition to that chosen by its original manufacturer ultimately became standardised throughout the world. In the case of Veronal®, Roger Adams at the University of Illinois devised a manufacturing process for the Chicago-based Abbott Laboratories. The drug was then given the AMA approved name of barbital.

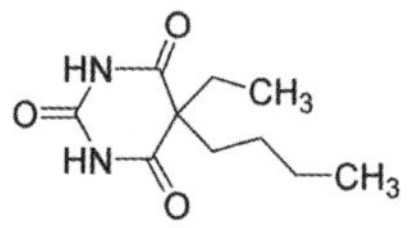

butobarbital

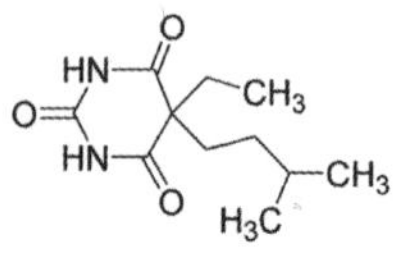

amobarbital

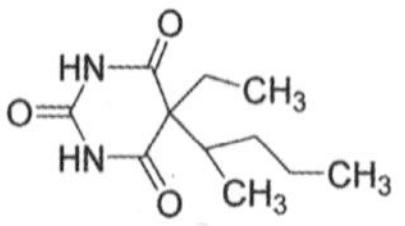

pentobarbitone

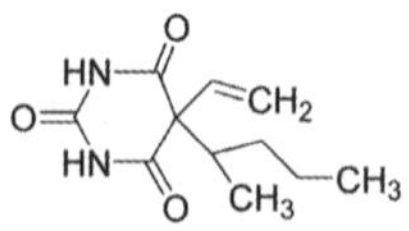

quinalbarbitone

During the First World War, Chaim Weizmann (who later became the first president of Israel) discovered that a bacterium known as *Clostridium acetobutylicum* could convert cheap starchy materials to acetone and *n*-butanol. This was of immense military significance as the United Kingdom was desperately short of acetone for the production of naval explosives. Once peace was restored, the Weizmann process resulted in a sudden drop in the price of *n*-butanol, previously an expensive chemical. Carl Marvel and Roger Adams at the Urbana campus of the University of Illinois synthesised 5-butyl-5-ethylmalonic ester in 1920.[67] This was to be the key intermediate in the synthesis of the butyl analogue of barbital, butobarbital (also known as 'butethal'), by Arthur Dox and Lester Yoder of Parke, Davis and Company.[68] This new barbiturate was about three times as potent as barbital, with a shorter duration of action, which minimised any drowsiness on awakening. The increased potency and more rapid metabolic destruction of butobarbital are both due to the enhanced lipophilicity caused by introduction of the longer butyl group. This favours entry into the brain, which is the site of action, and into the liver, which is the site of metabolic deactivation. Shonle and Moment of the Eli Lilly Company in Indianapolis announced their synthesis of amylobarbital (now known as 'amobarbital') a year after the introduction of butobarbital.[69] Both drugs were of similar potency, but the branched carbon atom on amobarbital rendered it more susceptible to metabolic deactivation, shortening the duration of action still more. It and similar barbiturates such as pentobarbitone and quinalbarbitone (secobarbitone) became highly popular hypnotics until the 1960s, when mounting concern about both their habit-forming properties and use in suicide led to their decline.

Drugs Structurally Related to Barbiturates

No hypnotics to challenge the barbiturates were developed until the 1950s, by which time there was concern about accidental overdosing by drowsy patients, as well as their use in suicide attempts. In 1952, Tagmann and his colleagues at Ciba in Basle announced that they had found a potent hypnotic among a series of dioxotetrahydropyridines structurally related to the barbiturates.[70] The new compound, glutethimide, was initially hailed as safer than the barbiturates – a claim that did not withstand the test of time.

glutethimide   aminoglutethimide   thalidomide

Aminoglutethimide was marketed for the treatment of epilepsy in 1960 after Ciba researchers found it was a stronger anticonvulsant but had weaker sedative–hypnotic properties than glutethimide.[71] In 1963, Ralph Cash, a paediatrician at the Sinai Hospital in Detroit, reported that it had induced the typical signs of Addison's disease (adrenal insufficiency) in a young girl who had been receiving it for five months to control her epilepsy. After similar reports from other doctors appeared, laboratory studies revealed that the drug had blocked steroid biosynthesis. It was withdrawn from the market in 1966. Cash demonstrated that aminoglutethimide inhibited the desmolase enzyme that removed the side chain from cholesterol to form pregnenolone, a prerequisite for steroid hormone synthesis.[72] Subsequently, it was administered to patients with Cushing's disease in the hope that they might benefit from its ability to inhibit overproduction of corticosteroids, but results were disappointing. In the 1970s physicians began administering aminoglutethimide to women with metastatic breast cancer, supplementing the drug with dexamethasone to compensate for diminished cortisone levels in the body.[73] The value of aminoglutethimide, especially in those women who had relapsed after initially responding to tamoxifen, is now established.

Another analogue of glutethimide was introduced by Chemie Grünenthal, a company established immediately after the Second World War by a soap and toiletries manufacturer keen to obtain a stake in the growing market for antibiotics, then in desperately short supply. Heinrich Mueckter, who qualified in medicine before the war, was appointed as research director on the basis of his wartime experience with the German army virus research group. In 1953, his assistant Wilhelm Kunz was given the task of preparing simple peptides required for antibiotic production. In the course of this he isolated a by-product that was recognised by a Chemie Grünenthal pharmacologist Herbert Keller to be a structural analogue of glutethimide. A series of related compounds were examined, from which one was examined in detail by Keller for its suitability as a hypnotic agent. Unusually, it did not abolish the righting reflex of animals – a standard laboratory test for hypnotic activity. Keller conducted a series of studies on the mobility of mice exposed to the drug, thalidomide, comparing it with several barbiturates and other central nervous system depressants.[74] After investigating its toxicity in mice, rats, guinea pigs and rabbits, he came to the conclusion that it was a remarkably safe sedative.

Chemie Grünenthal approached manufacturers throughout the world, with the outcome that several who were keen to enter the market for sedative–hypnotics marketed it with their own brand name. This was to lead to the greatest tragedy in the history of modern drugs, for the new sedative was a teratogen.

Thalidomide was introduced on the German market in 1956. In November 1961, Hamburg paediatrician Widukind Lenz reported a large increase in the number of infants with phocomelia attending ten clinics in North Germany. Instead of limbs, they had stumps. This had previously been one of the rarest malformations known, with no cases having been seen in these clinics in the decade prior to 1959. Yet there were 477 cases in 1961. Lenz attributed the increase to the taking of thalidomide by mothers during the first trimester of their pregnancies and notified Chemie Grünenthal and the authorities. The report of thalidomide teratogenicity was immediately picked up and publicised by a German newspaper, forcing the manufacturer to withdraw the drug. Like most others introduced up till then, it had never been tested for teratogenicity.

By the time thalidomide was withdrawn, 3000 deformed babies had already been born in Germany and at least twice that number elsewhere.[75] The United States was spared because Frances Kelsey at the Food and Drug Administration had not approved a new drug application, having been dissatisfied with the limited safety data that had been submitted. At that time, it was only the United States that required manufacturers to seek government approval before launching a new drug. Within a few years of the thalidomide disaster, countries around the world had quickly emulated the American system.

Barbiturate Anaesthetics

Intravenous medication did not become feasible until after the development of the hypodermic syringe by Alexander Wood of Edinburgh in 1853. The earliest attempt at intravenous anaesthesia was due to the work of Pierre Oré of Bordeaux, who reported to the Surgical Society of Paris in 1872 that he had injected a solution of chloral hydrate and achieved deep enough anaesthesia to remove a fingernail.[76] Oré published a detailed report of a further 36 operations in which he had used the technique with some success, but in one case the patient died.[77] It was not until 1905 that further development occurred when N.P. Krawkow of St Petersburg successfully administered a saline solution containing the Bayer Company's recently introduced urethane analogue Hedonal®. Fedoroff subsequently used this method in more than 500 operations.[78] The technique was taken up in Russia and some parts of Europe, where it stimulated others to seek more suitable drugs.

Daniel Bardet reported in 1921 that he had anaesthetised patients with injections of Somnifen®, a water-soluble formulation of barbital and allobarbital.[79] He found that recovery was too slow, and patients awoke with headaches. Amobarbital, butallylonal and pentobarbital were occasionally used in the latter half of the decade. Particularly disconcerting, however, was a tendency for anaesthesia to deepen alarmingly without warning. This was because the delay in its onset prevented the anaesthetist from knowing how much drug was required to render the patient unconscious. Not until I.G. Farben introduced hexobarbital in 1931 did a safe intravenous anaesthetic become available. With it, the onset of action was rapid and so the anaesthetist could control the level of anaesthesia by giving the injection slowly.

Hexobarbital was synthesised by the chemists Kropp and Traub at Elberfeld.[80] Its rapid onset of anaesthesia was due to the replacement of a hydrogen atom on one of the barbiturate ring nitrogen atoms by a methyl group. This rendered the molecule less water soluble and more lipophilic. Small though this molecular change may have been, it ensured rapid transposition of the drug from the blood into the brain cells. As a consequence, patients fell unconscious in the few seconds it took for the blood to carry the anaesthetic to the brain from the site of injection.[81] Hexobarbital was deservedly successful, and it is estimated that over the next 12 years some 10 million injections of it were administered.

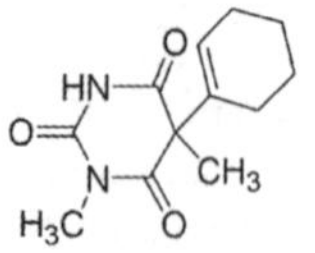

hexobarbital

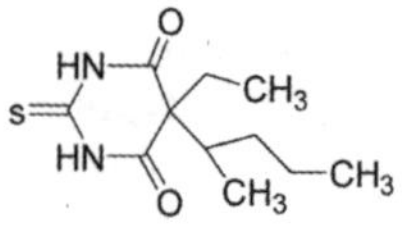

thiopental

Even before the first reports of the success of hexobarbital had begun to circulate, Donalee Tabern and Ernest Volwiler of Abbott Laboratories were on the trail of the drug that was ultimately to render it obsolete. Their work on pentobarbital encouraged them to seek very short-acting barbiturates, probably with a view to introducing them as hypnotics free from any tendency to produce a hangover. They followed up old reports stating that sulfur-containing thiobarbiturates were chemically less stable than the familiar oxobarbiturates. Thiobarbiturates had been among the earliest barbiturates examined in 1903 by Fischer and von Mering, but had been rejected after an oral dose of the sulfur analogue of barbital had killed a dog. Notwithstanding, Tabern and Volwiler pursued their idea that a chemically unstable thiobarbiturate might decompose fast enough in the body to ensure that its effects quickly wore off. By 1934 they were convinced that thiopental, the sulfur analogue of pentobarbital, was a promising agent.[82] Doubtlessly inspired by the recent success of hexobarbital, they arranged for thiopental to be investigated by Ralph Waters,[83] who had just completed his pioneering investigations into cyclopropane anaesthesia, at the University of Wisconsin Medical School, Madison, and by John Lundy[84] of the Mayo Clinic in Rochester, Minnesota. Both confirmed the superiority of thiopental over existing intravenous anaesthetics. It ultimately achieved recognition as the single most useful agent for the induction of anaesthesia prior to the administration of an inhalational anaesthetic. It was only in the 1990s that it was rivalled by propofol. It also became widely used as an intravenous anaesthetic for short operations.

Anticonvulsants

Phenobarbital was one of the compounds reported by Fischer and Dilthey in their paper of 1904. It was later found to be superior to barbital and was marketed by F. Bayer & Company under the proprietary name of Luminal®.[85] For half a century it was a commonly prescribed hypnotic and sedative, but it remains in use today principally on account of its anticonvulsant activity. This was discovered by chance shortly after its introduction into the clinic, when a young doctor, Alfred Hauptmann, supplied it to epileptic patients in his ward who kept awakening him at night due to their fits. He expected the hypnotic to keep them asleep during the night, but did not expect that the incidence of their fits would decline during the day, particularly in those with grand mal epilepsy.[86] At first, there were few who believed Hauptmann's claims that he had stumbled upon the first truly anti-epileptic drug that did not produce the severe sedation hitherto associated with the use of bromides. Only after the First World War was there general recognition of this valuable property of phenobarbital.

phenobarbital

phenytoin

Tracy Putnam, the Director of the Neurological Unit of the Boston City Hospital, initiated experiments in 1934 aimed at finding a less sedating anticonvulsant than phenobarbital. He and Frederick Gibbs established the first electroencephalographic laboratory in the world designed for routine clinical studies of brain waves. An important observation to emerge from the new laboratory was that epileptic seizures were accompanied by an electrical 'storm' in the brain. This led Putnam to conclude that it might be possible to induce convulsions in laboratory animals by applying an electrical current to the brain. Furthermore, it might also be possible to quantify the strength of current required, thereby affording a method of recognising whether a drug was able to give some degree of protection to the animal. Having then set up an improvised piece of apparatus, Putnam and Gibbs demonstrated that phenobarbital markedly raised the convulsive threshold in cats.

Putnam next sought a wide variety of phenyl compounds from several chemical manufacturers, believing that the phenyl group in phenobarbital was somehow responsible for its efficacy. Only Parke, Davis and Company responded. They provided 19 analogues of phenobarbital, all of which had been found to be inactive as hypnotics. Putnam screened these, as well as over a hundred other available chemicals. A few were active but, with one exception, were too toxic for clinical use. The exception was one of the Parke, Davis and Company compounds, phenytoin, which was more effective in protecting cats from electrically induced convulsions than even phenobarbital. As it was known to have no hypnotic or other untoward effects, this seemed to be just what Putnam had been seeking.

Putnam gave phenytoin to Houston Merritt for clinical evaluation in 1936. The first patient to receive the drug had suffered from seizures every day for many years, but as soon as his treatment began these ceased permanently. Subsequent studies confirmed that phenytoin was at least as effective as phenobarbital, with the added advantage of causing less sedation.[87] Paradoxically, this absence of marked sedation initially prejudiced many physicians against accepting the new drug!

phensuximide

methsuximide

ethosuximide

primidone

After the Second World War ended, Parke, Davis and Company initiated a major research project to find a less toxic drug to replace Abbott's troxidone in petit mal. This involved the synthesis and testing of over 1000 aliphatic and heterocyclic amides. This resulted in the discovery of three useful anticonvulsants, namely phensuximide,[88] methsuximide[88] and ethosuximide. The last of these was originally synthesised in 1927[89] and was put on the market in 1958. It remains in use for the treatment of petit mal absence seizures in children.

ICI scientists also tried to find improved anticonvulsants in the early 1950s. Herbert Carrington at the company's research laboratories in Manchester considered that the new hydantoins being developed by Parke, Davis and Company, although relatively free of sedating properties, produced too many side effects. Barbiturates, in contrast, were usually free from these, but instead caused sedation. However, as not all sedating barbiturates were anticonvulsants, Carrington concluded that it should be possible to find a barbiturate analogue in which this separation of activity was reversed. This led on to the development of primidone by Charles Vasey and William Booth.[90] It was given its first clinical trial in 1952 and results were satisfactory in grand mal epilepsy.[91] How much of its efficacy is due to phenobarbital formed from it and how much is due to unchanged primidone is uncertain, but primidone is prescribed when neither phenytoin nor carbamazepine is acceptable.

IMIDAZOLINES

Piperazine was synthesised at the University of Breslau in 1888 by Alfred Ladenburg.[92] When he discovered that it formed a soluble salt with uric acid, he suggested that this might dissolve the deposits of uric acid that caused much pain in patients with gout.[93] Piperazine was immediately marketed for this purpose and remained in use well into the twentieth century despite repeated criticism on the grounds that it was ineffective. The same situation arose with 2-methylimidazoline, which Ladenburg synthesised in 1894.[94] After clinical studies were conducted at his suggestion, it was marketed for the treatment of gout on the grounds that it dissolved uric acid.

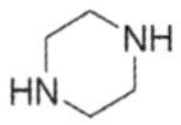

piperazine

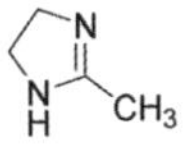

methylimidazoline

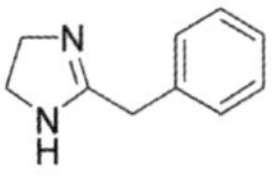

tolazoline

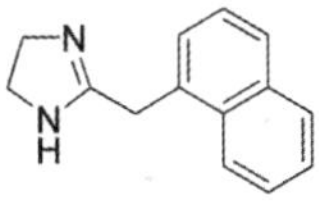

naphazoline

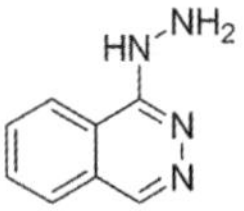

hydralazine

In 1935, Henry Chitwood and Emmet Reid at the Chemistry Department in Johns Hopkins University decided to reinvestigate 2-methylimidazoline and its homologues.[95] Only the methyl homologue had any effect on the acidity of the urine, a pointer to increased excretion of uric acid – which was the usual mode of action of drugs that relieved gout. Toxicity decreased as the methyl group was replaced with longer alkyl groups. As this was the reverse of the tendency usually found in a series of homologues, it prompted researchers at the Ciba laboratories in Basle to re-examine the series of compounds. They obtained the opposite effect so far as the influence of chain length on toxicity was concerned, contradicting the earlier claims. In the course of this investigation of the toxicity of imidazolines, a drop in blood pressure caused by dilation of peripheral blood vessels was observed. When Ciba chemists introduced cyclic substituents such as benzene and naphthalene rings at the 2-position of the imidazoline ring, the toxicity decreased. The most potent of these compounds, tolazoline,[96] was found to have weak adrenergic blocking activity and was introduced clinically.[97] Its use had to be limited to the treatment of Raynaud's disease and certain spastic vascular disorders since it stimulated the heart. Unexpectedly, the naphthyl analogue increased blood pressure by acting as a vasoconstrictor. It was introduced into clinical practice in the early 1940s as a long-acting nasal decongestant called 'naphazoline'.

In an expansion of their studies on imidazolines, Ciba researchers investigated other heterocyclic compounds containing two nitrogen atoms. A series of phthalazines were found to be active in screens for hypotensive activity, from which hydralazine emerged as a long-acting peripheral dilator.[98] It became the first orally active peripheral vasodilator to be introduced for the treatment of high blood pressure. With regular use, patients experienced side effects and became tolerant to it. As a result, other drugs superseded it. However, hydralazine became popular once again when it was found that tolerance was due to physiological compensatory mechanisms that could be overcome by combining it with a beta-blocker and a diuretic. Since the dose of hydralazine required in such a combination was smaller than that when used on its own, patients experienced fewer side effects. It continues to be widely used.

REFERENCES

1. S. Garfield, *Mauve: How One Man Invented a Colour that Changed the World.* London: Faber and Faber; 2000.

2. J.P. Swazey, *Chlorpromazine in Psychiatry: A Study of Therapeutic Innovation.* Cambridge, Massachusetts: The MIT Press; 1974, pp. 23–33.

3. F.F. Runge, Ueber einige Produkte der Steinkohlendestillation (*Poggendorff's*) *Ann. Phys. Chem.*, 1834; **31**: 65–78, and 1835; **32**: 308–33.

4. H.A. Kelly, Jules Lemaire: the first to recognise the true nature of wound infection, and the first to use carbolic acid in medicine and surgery. *J. Am. Med. Ass.*, 1901; **36**: 1083–8.

5. J.K. Crellin, The disinfectant studies by F. Crace Calvert and the introduction of phenol as a germicide. *Veroef. Int. Ges.Gesch. Pharm.*, 1966; **28**: 61.

6. F.C. Calvert, *C.R. Acad. Sci.*, 1859; **49**: 262.

7. F. Küchenmeister, *Deutsche Klinik am Eingange des zwanzigsten Jahrhunderts*, 1860; **12**: 123.

8. D.C. Schecter, H. Swan, Jules Lemaire: a forgotten hero of surgery. *Surgery*, 1991: **49**: 817–26.

9. R. Godlee, *Lord Lister.* London: Macmillan; 1917.

10. J. Lister, On a new method of treating compound fracture, abscess, etc. With observations on the conditions of suppuration. *Lancet*, 1867; **1**: 387–9.

11. A.E. Sansom, On the uses of septicidal agents in disease. *Retrospect. Med.*, 1868; **57**: 6–11.

12. A.E. Sansom, *The Antiseptic System: a Treatise on Carbolic Acid and its Compounds.* London: 1871, p. 324.

13. H. Bechhold, P. Ehrlich, Beziehungen zwischen chemischer Konstitution und Desinfektionswirkung – Ein Beitrag zum Studium der inneren Antisepsis. *Z. Physiol. Chem.*, 1906; **47**: 173.

14. H.J. Kolbe, E. Lautemann, Über die Constitution und Basicitat der Salicylsäure. *Ann. Chem.*, 1860; **115**: 157–9.

15. C.E. Buss, Über die Anwendung der Salicylsäure als Antipyreticum. *Deutsches Arch. Klin. Med.*, 1875; **15**: 457–501.

16. S. Stricker, Über die Resultate der Behandlung der Polyarthritis rheumatica mit Salicylsäure. *Berl. Klin. Wochenschr.*, 1876; **13**: 1–2, 15–16, 99–103.

17. T.J. MacLagan, The treatment of acute rheumatism by salicin and salicylic acid. *Lancet*, 1876; **1**: 342–3.

18. G. Sée, Études sur l'acide salicylique et les salicylates: traitement du rhumatisme aigu et chronique de la goutte, et de diverses affections du système nerveux sensitif par les salicylates. *Bull. Acad. Med. Paris*, 1877; **6**: 689, 897.

19. J.S. Goodwin, J.M. Goodwin, Failure to recognize efficacious treatments: a history of salicylate therapy in rheumatoid arthritis. *Persp. Biol. Med.*, 1981; **25**: 78–92.

20. M. Nencki, *Arch. Exp. Path. Pharmakol.*, 1886; **20**: 396.

21. H. Sahli, Ueber die Spaltung des Salols mit Rücksicht auf dessen therapeutische Verwerthung zu innerlichem und äusserlichem Gebrauch. *Therap. Monatsch.*, 1887; s. 333–9.

22. C.F. von Gerhardt, Untersuchungen über die wasserfreien organischen Säuren. *Annalen*, 1853; **87**: 149–79.

23. K. Witthauer, Aspirin, eine neues Salicylpräparat. *Die Heilkunde*, 1899; **3**: 396–8.

24. J. Wohlgemuth, Über Aspirin (Acetylsalicylsäure). *Therap. Monatshefte*, 1899; **3**: 276–8.

25. H. Dreser, Pharmacologisches über Aspirin (acetylsalicylsäure). *Pflüger's Archiv. Anat. Physiol.*, 1899; **76**: 306–18.

26. A. Schmidt, *Die industrielle Chemie in ihrer Bedeutung im Wetbild und Errinnerungen an ihren Aufban.* Berlin: De Gruyter; 1934, p. 775.

27. A. Eichengrün, 50 Jahre Aspirin. *Pharmazie*, 1949; **4**: 582–4.

28. W. Sneader, The discovery of aspirin: a reappraisal. *Br. Med. J.*, 2000; **321**: 1591–4.

29. J.R. Vane, Inhibition of prostaglandin synthesis as a mechanism of action for Aspirin-like drugs. *Nature New. Biol.*, 1971; **231**: 232–5.

30. G.J. Roth, P.W. Majerus, The mechanism of the effect of Aspirin on human platelets: 1. Acetylation of a particulate fraction protein. *J. Clin. Invest.*, 1975; **56**: 624–32.

31. P.C. Gibson, Aspirin in the treatment of vascular disease. *Lancet*, 1949; **2**: 1172–4.

32. L.L. Craven, Acetylsalicylic acid, possible preventive of coronary thrombosis. *Ann. West. Med. Surg.*, 1950; **4**: 95.

33. L.L. Craven, Experiences with aspirin (acetylsalicylic acid) in the non-specific prophylaxis of coronary thrombosis. *Miss. Valley Med. J.*, 1953; **75**: 38–44.

34. H.J. Weiss, L.M. Aledort, Impaired platelet-connective-tissue reaction in man after aspirin ingestion. *Lancet*, 1967; **2**: 495–7.

35. P.C. Elwood, A.L. Cochrane, M.L. Burr, *et al.*, A randomized controlled trial of acetyl salicylic acid in the secondary prevention of mortality from myocardial infarction. *Br. Med. J.*, 1974; **1**: 436–40.

36. C.V. Winder, J. Wax, L. Scott, *et al.*, Anti-inflammatory, antipyretic and antinociceptive properties of *N*-(2,3-xylyl)anthranilic acid (mefenamic acid). *J. Pharmacol. Exp. Ther.*, 1962; **138**: 405–13.

37. C.V. Winder, J. Wax, B.S. Querubin, _et al._, Anti-inflammatory and antipyretic properties of _N_-(alpha,alpha,alpha-trifluoro-_m_-tolyl) anthranilic acid (CI-440; flufenamic acid). _Arthritis Rheum.,_ 1963; **6**: 36–47.
38. A.R. Sallmann, The history of diclofenac. _Am. J. Med.,_ 1986; **80** (Suppl. 4B): 29–33.
39. J. Hannah, W.V. Ruyle, H. Jones, _et al.,_ Novel analgesic–antiinflammatory salicylates. _J. Med. Chem.,_ 1978; **21**: 1093–100.
40. F. Bernheim, The effect of salicylate on the oxygen uptake of the tubercle bacillus. _Science,_ 1940; **92**: 204.
41. A.K. Saz, F. Bernheim, Effect of 2,3,5-triiodobenzoate on growth of tubercle bacilli. _Science,_ 1941: **93**; 622–3.
42. Cited in F. Ryan, _Tuberculosis: The Greatest Story Never Told._ Bromsgrove: Swift; 1992, p. 244.
43. J. Lehmann, Para-aminosalicylic treatment of tuberculosis. _Lancet,_ 1946; **1**: 15–16.
44. O. Schmiedeberg, Über die pharmakoligischen Wirkungen und therapeutischen Anwendungen einiger Carbaminosäuren. _Arch. Exp. Path. Pharmakol.,_ 1886; **20**: 203–16.
45. K. Laubenheimer, _Phenol und seine Derivate als Desinfektionsmittel._ Berlin: Urban Schwartzenberg; 1909.
46. E.G. Klarmann, V.A. Shternov, L.W. Gates, _J. Lab. Clin. Med.,_ 1934; **19**: 835.
47. L. Colebrook, M. Kenny, Treatment of human puerperal infections, and of experimental infections in mice, with Prontosill. _Lancet,_ 1936; **1**: 1279–86.
48. US Pat. 1941: 2250480 (to Burton T. Bush).
49. H.C. Powell, O. Swarner, L. Gluck, P. Lampert, Hexachlorophene myelinopathy in premature infants. _J. Pediat.,_ 1973; **82**: 976–81.
50. J. Liebig, _Annalen,_ 1832; **1**: 182.
51. T.C. Butler, The introduction of chloral hydrate into medical practice. _Bull. Hist. Med.,_ 1970; **44**: 168–72.
52. R. Buchheim, _Arch. Path. Anat.,_ 1872; **56**: 1.
53. O. Liebreich, Das Chloral, ein neues Hypnotikum und Anästhetikum. _Berl. Klin. Wochenschr.,_ 1869; **6**: 325.
54. O. Liebreich, _Das Chloralhydrat, ein Neues Hypnotikum._ Berlin: G.F.O. Müller; 1869.
55. J.F. von Mering, F. Musculus, _Ber.,_ 1875; **8**: 662.
56. T.C. Butler, _J. Pharmacol. Exp. Ther.,_ 1948; **92**: 49.
57. B. Hems, R.M. Atkinson, M. Early, E.G. Tomich, Trichloroethyl phosphate. _Br. Med. J.,_ 1962; **1**: 1834.
58. J.F. von Mering, _Pharm. Central.,_ 1889; **484**: 494.
59. E. Baumann, _Ber.,_ 1885; **18**: 883.
60. A. Kast, Sulfonal, ein Neues Schlafmittel. _Berl. Klin. Wochenschr.,_ 1888; **25**: 309–14.
61. E. Fischer, J. von Mering, Über eine neue Klasse von Schlafmitteln. _Therap. Gegenw.,_ 1903; **44**: 97.
62. Ger. Pat. 1910: 225710 (to F. Bayer & Co.).
63. M. Conrad, M. Guthzeit, Über Barbitursäure-derivate. _Ber.,_ 1882; **15**: 2844.
64. G.B. Kauffman, Adolf von Baeyer and the naming of barbituric acid. _J. Chem. Educ.,_ 1980; **57**: 222.
65. G.W. Collins, P.N. Leech, The indispensable uses of narcotics. Chemistry of barbital and its derivatives. _J. Am. Med. Ass.,_ 1931; **96**: 1869.
66. E. Fischer, A. Dilthey, Ueber C-Dialkylbarbitursäuren und über die Ureide der Dialkylessigsäuren. _Annalen,_ 1904; **335**: 334–68.
67. R. Adams, C. Marvel, Organic chemical reagents. VI. Reagents from _n_-butyl alcohol. _J. Am. Chem. Soc.,_ 1920; **42**: 310–20.
68. A. Dox, L. Yoder, Some derivatives of normal-butyl-malonic acid. _J. Am. Chem. Soc.,_ 1922; **44**: 1578–81.
69. H.A. Shonle, A. Moment, Some new hypnotics of the barbituric acid series. _J. Am. Chem. Soc.,_ 1923; **45**: 243–9.
70. E. Tagmann, E. Sury, K. Hoffmann, _Helv. Chim. Acta,_ 1952; **35**: 1541.
71. US Pat. 1958: 2848455 (to Ciba).
72. R. Cash, A.J. Brough, M.N.P. Cohen, P.S. Satoh, Aminoglutethimide (Elipten–Ciba) as an inhibitor of adrenal steroidogenesis: mechanism of action and therapeutic trial. _J. Clin. Endocrin. Metab.,_ 1967; **27**: 1239–48.
73. C.T. Griffiths, T.C. Hall, Z. Saba, Preliminary trial of aminoglutethimide in breast cancer. _Cancer,_ 1973; **32**: 31–7.
74. W. Kunz, H. Keller, H. Muckter, _N_-Phthalyl-glutaminic imide. _Arzneimittel.-Forsch.,_ 1956; **6**: 426–30.
75. Sunday Times Insight Team, _Suffer the Children. The Story of Thalidomide._ London: Andre Deutsch; 1979.

76. P.C. Oré, Études cliniques sur l'anesthésie chirurgicale par la methode des injections de chloral dans les veines. *Bull. Soc. Chir. (Paris)*, 1872; **1**: 400.

77. P.C. Oré, *Études Cliniques sur l'Anesthésie Chirurgicale par la Methode des Injections de Chloral dans les veines.* Paris: J.B. Bailliè; 1875.

78. I. Kissin I, A.J. Wright, The introduction of Hedonal: a Russian contribution to intravenous anesthesia. *Anesthesiology*, 1988; **69**: 242–5.

79. D. Bardet, G. Bardet, Contribution l'étude des hypnotiques ureiques. Action et utilisation du diethyl-diallyl-barbiturate de diethylamine, *Bull. Gen. de Therap.*, 1921; **172**: 173.

80. Br. Pat. 1931: 401693 (to I.G. Farben).

81. H. Weese, W. Scharpff, Evipan, ein neuartiges Einschlafmittel. *Deut. Med. Wochenschr.*, 1932; **58**: 1205.

82. D.L. Tabern, E.H. Volwiler, Sulfur-containing barbiturate hypnotics. *J. Am. Chem. Soc.*, 1935; **57**: 1961–3.

83. T.W. Pratt, A.L. Tatum, H.R. Hathaway, R.M. Waters, Preliminary experimental and clinical study. *Am. J. Surg.*, 1936; **31**: 464.

84 J.S. Lundy, Intravenous anesthesia: preliminary report of the use of two new thiobarbiturates. *Proc. Staff Meet. Mayo Clinic*, 1935; **10**: 534–43.

85. Ger. Pat. 1911: 247952 (to F. Bayer & Co.).

86. A. Hauptmann, Luminal bei Epilepsie. *Munch. Med. Wochenschr.*, 1912; **59**: 1907.

87. T.J. Putnam, in *Discoveries in Biological Psychiatry*, eds F. Ayd and B. Blackwell, Philadelphia: Lippincott; 1970.

88. C.A. Miller, L. Anticonvulsants. I. An investigation of *N-R-α-R1-α*-phenylsuccinimides. *J. Am. Chem. Soc.*, 1951; **73**: 4895–8.

89. S.S.G. Sircar, The influence of groups and associated rings on the stability of certain heterocyclic systems. Part II. The substituted succinimides. *J. Chem. Soc.*, 1927: 1252.

90. Br. Pat. 1952: 666027 (to ICI).

91. J.Y. Bogue, H.C. Carrington, The evaluation of Mysoline – a new anticonvulsant drug. *Br. J. Pharmacol.*, 1953; **8**: 230–6.

92. A. Ladenburg, *Ber.*, 1888: **21**: 758.

93. F.S. Kipping, Ladenburg Memorial Lecture. *J. Chem. Soc.*, 1913: 1871–95.

94. A. Ladenburg, *Ber.*, 1894; **27**: 2952.

95. H. Chitwood, E. Reid, Some alkyl-glyoxalidines. *J. Am. Chem. Soc.*, 1935; **57**: 2424–6.

96. M. Hartmann, H. Isler, Chemische Konstitution und pharmakologische Wirksamkeit von in 2-Stellung substituierten Imidazolinen. *Arch. Exp. Path. Pharm.*, 1939; **192**: 141.

97. R. Meier, R. Müller, *Schweiz. Med. Wochenschr.*, 1939; **69**: 1271.

98. J. Druey, B.H. Ringier, Hydrazinederivate der Phthalazin- und Pyridazinreihe. *Helv. Chim. Acta*, 1951; **34**: 195–210.

Drugs Originating from the Screening of Dyes

In 1777, Wilhelm Friedrich von Gleichen-Russworm described how he had stained microbes with the natural dyes indigo and carmine so that they could be examined under the microscope.[1] His procedure was refined in 1869 when the botanist Hermann Hoffmann replaced indigo with the synthetic dye magenta, which had first been marketed ten years earlier. Carl Weigert at the University of Breslau then adapted the method for use in pathological investigations, employing a microtome to slice infected tissues so thinly that they could be stained and viewed under a microscope. He also selectively coloured different cellular structures in tissue slices with synthetic dyes. This caught the imagination of his younger cousin, Paul Ehrlich.

Ehrlich entered the University of Breslau in 1872, where Weigert persuaded him to study medicine. When Wilhelm Waldeyer moved from Breslau to the new University of Strassburg, Ehrlich went with him. As his tutor, the great anatomist had both welcomed Ehrlich into his home and encouraged him to continue with his microscopic investigations. Stimulated by reading Huebel's book on lead poisoning, which explained that chemical analysis of various organs had established that lead concentrated in the brain, Ehrlich examined slides of brain tissue in the hope of determining where the lead was stored. When this proved to be a futile exercise, it forced him to change his approach. Instead of lead salts, he injected dyes that could easily be detected in the cellular components where they concentrated.

In 1874, having completed his pre-clinical studies, Ehrlich returned to Breslau, where he avoided all unnecessary clinical involvement. He preferred to study in the laboratory of Julius Cohnheim, who encouraged him, Weigert and Robert Koch in their endeavours. The final part of Ehrlich's medical studies was completed at Leipzig. His doctoral dissertation, submitted in 1878, was entitled *Beiträge zur Theorie und Praxis der Histologischen Färbung* (*Contributions to the Theory and Practice of Histological Staining*). It was highly critical of histologists for failing to base their work on a theoretical understanding of how dyes bind to tissue components.

On taking up his first post at the Charité Hospital in Berlin, Ehrlich spent most of his time on histological studies, especially in the field of haematology. He continued to strive for an understanding of the factors influencing the uptake of dyes by cells, and came to the conclusion that the size of the molecule was critical. He soon became disillusioned with the limitations imposed by examination of tissues exposed to dyes only after the death of the animal from which they were taken. By 1885, he had developed a new approach in which he injected dyes into living animals, then left them to diffuse into the tissues before killing them. For the first time, it became possible to examine the disposition of chemical substances in living animals, a process that Ehrlich called 'vital staining'. The wide diversity of synthetic dyes that were available enabled Ehrlich to draw conclusions about the influence of chemical structure on the distribution in live animals of different types of coloured molecules.[2] These conclusions still influence drug design. For example, he observed that acidic dyes possessing the sulfonic acid function introduced by dye manufacturers to enhance water solubility were

Drug Discovery. A History. W. Sneader.
©2005 John Wiley & Sons Ltd

unable to penetrate into the brain or adipose (fat) tissue. In marked contrast, basic dyes such as methylene blue and neutral red readily stained these tissues. To explain this, Ehrlich drew an analogy between the extraction of fat-soluble, basic alkaloids from alkaline solution into ether and the transfer of basic dyes from the alkaline blood to the brain. Alkaloids could not be extracted into ether from acid solutions, where they existed in the form of their salts, which were insoluble in ether, just as acidic dyes existed as water-soluble salts in blood. This remarkably accurate assessment of the situation was made during the years 1886 and 1887.

methylene blue

In 1881 Ehrlich stained bacteria with methylene blue, which had been synthesised by Heinrich Caro five years earlier. Four years later, he found that this lipophilic dye had a strong affinity for nerve fibres, leaving other tissues unaffected. He described it as being neurotropic, for upon injecting it into a living frog all the nerve fibres were gradually tinted blue. Ehrlich then reasoned that as methylene blue stained nerves it might possibly interfere with nervous transmission and exert an analgesic action. In 1888, he and Arthur Leppmann gave it to patients suffering from a variety of severe neuritic and arthritic conditions. They found that it did relieve pain, but its tendency to damage the kidney discouraged its use as an analgesic.[3] Examination of Ehrlich's unpublished letters by Henry Dale revealed that he wrote to chemists in the dyestuffs industry asking advice about the possibility of obtaining analogues of methylene blue that might be more potent analgesics.[4]

In 1891, Ehrlich carried out further experiments with methylene blue after returning from Egypt, where he had gone to recuperate from tuberculosis. Knowing that it stained the plasmodia that caused malaria and that it could be administered to patients, he and Paul Guttmann administered daily five capsules each containing 100 mg of methylene blue to two patients who had been admitted to the Moabite Hospital in Berlin with malaria. Both recovered as a result of this treatment.[5] Although methylene blue was later found to be ineffective against the more severe manifestations of the disease experienced in the tropics, this cure of a mild form of malaria represented the first instance of a synthetic drug being used with success against a specific disease. In 1995, it was reported that methylene blue had exhibited high antimalarial activity in laboratory studies.[6]

Ehrlich could not pursue his work with methylene blue any further, for two reasons. Firstly, the inability to infect animals with malaria prevented testing of potential drugs in the laboratory, a prerequisite for the development of any chemotherapeutic agent for use in human or veterinary medicine. Secondly, he was working in Robert Koch's Institute for Infectious Diseases in Berlin, where his skills were fully deployed in transforming Emil von Behring's diphtheria antitoxin into a clinically effective preparation. This was later to result in his sharing the Nobel Prize for Medicine with Mechnikov in 1908.

MEDICINAL DYES

Following his early investigation of the action of the organic arsenical Atoxyl® against cultures of trypanosomes (see Chapter 7), Paul Ehrlich examined more than 100 synthetic dyes that were injected into mice infected with either *Trypanosoma equinum*, the organism that caused mal de Caderas in horses, or *T. brucei*, which caused nagana in cattle. The only dye to exhibit activity belonged to the benzopurpurin series. Ehrlich named it 'Nagana Red'. Like arsenious acid, it caused the disappearance of trypanosomes from the blood of the mice for a short time, the mice surviving five or six days instead of the usual three or four.

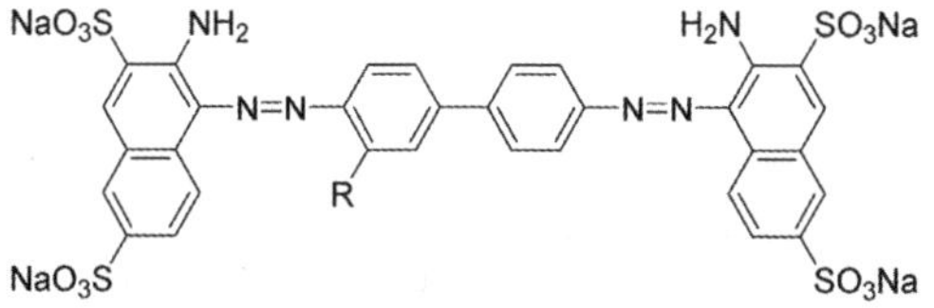

Nagana Red, R = H
Trypan Red, R = SO$_3$Na

Ehrlich believed that the low solubility of Nagana Red was the cause for it being poorly absorbed into the circulation from the site of its injection under the skin. He therefore contacted the Cassella Dye Works near Frankfurt, the longest established of several dye manufacturers in the area (it was soon to become affiliated with the larger Hoechst Dyeworks). Ehrlich asked for a derivative of the Nagana Red with an extra sulfonic acid function to enhance water solubility. Ludwig Benda was able to give him a dye first prepared in 1889, which was subsequently to be called 'Trypan Red'. Injections of it cured mice infected with *T. equinum* but not with other strains of trypanosomes.[7] The British Sleeping Sickness Commission requested supplies of the dye for trials in Uganda, but the results were disappointing since doses high enough to be effective against the disease were found to be likely to cause blindness and sometimes death.

Ehrlich obtained about a further 50 derivatives of Trypan Red from the Cassella Dye Works. The most active of these were more potent than Trypan Red itself. A supply of the 7-amino derivative of Trypan Red was tested in Africa during an expedition led by Koch in 1906, but it proved to be no better than the parent compound.

SURAMIN

Maurice Nicolle and Felix Mesnil at the Pasteur Institute examined analogues of Trypan Red supplied by Friedrich Bayer & Company, the leading manufacturer of acidic azo dyes. The French workers disclosed that a single injection of Trypan Blue could cause the disappearance of all trypanosomes from the blood.[8] Further trials showed that it had a mild action against several different strains of trypanosomes that caused disease in cattle, being more effective than either Atoxyl® or Trypan Red but still not acceptable for human use. After a demonstration that it could cure piroplasmosis (babesia), it was introduced into veterinary practice.[9]

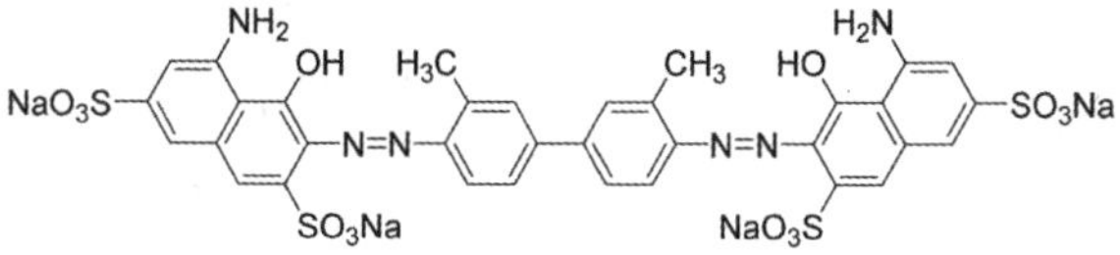

Trypan Blue

Wilhelm Roehl moved from Ehrlich's laboratory in Frankfurt, where he had been testing the azo dyes provided by the Cassella Company, to join the Bayer research group at Elberfeld in 1905. He found that none of the dyes prepared by the company for Nicolle and Mesnil were effective in his own infected mice. Nevertheless, Bernhard Heymann, who was director of the scientific laboratory at Elberfeld, asked Oskar Dressel and Richard Kothe to synthesise new analogues. Roehl requested that colourless compounds should be made since a drug that tinted the skin would not be acceptable to patients. Dressel and Kothe therefore synthesised analogues of one of the least coloured Bayer dyes that had been screened by Nicolle and Mesnil, namely Afridol Violet. Although this had only feeble activity, several red analogues exhibited stronger activity against trypanosomes. It was this advance that led to priority being given to the project in 1913. Superior analogues were developed, with the first patents being applied for just before the outbreak of the First World War. At this point there were arguments within the company over the wisdom of continuing with a line of research that had not delivered a truly outstanding drug. Only the insistence of Heymann prevented the project from being dropped. By the autumn of 1917, more than 1000 naphthalene ureas had been synthesised and tested. Only then did the long-sought agent emerge in the form of a colourless compound that had remarkable antitrypanosomal activity both in experimental animals and in humans. It had the code name of Bayer 205 and was later marketed by Bayer under the proprietary brand name of Germanin®. Its chemical structure was not revealed because F. Bayer & Company feared this would enable foreign manufacturers to develop similar products.[10]

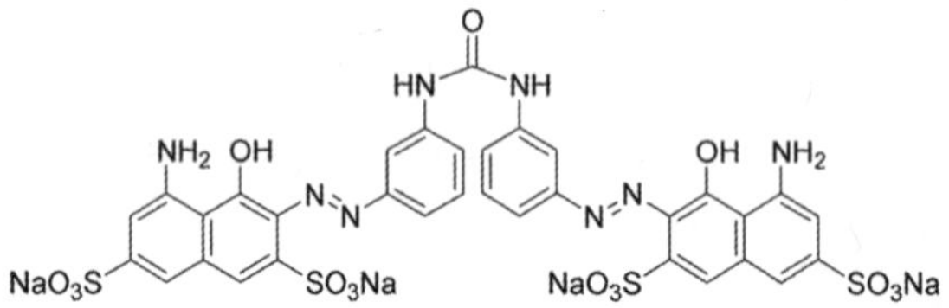

Afridol Violet

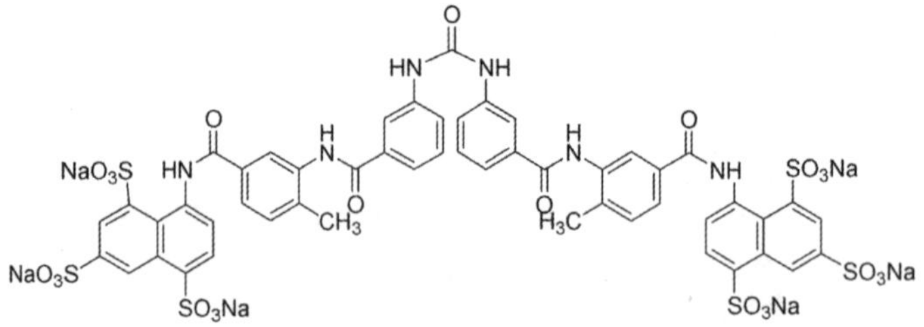

suramin

The first reports of the discovery of the new trypanocide began to circulate outside the Elberfeld laboratories towards the end of 1920. For a while the drug was only made available to German doctors and a few foreign investigators who undertook not to allow it to pass into the hands of anyone capable of determining its chemical structure. This sorely irked Ernest Fourneau, head of the medicinal chemistry laboratory at the Pasteur Institute. Originally trained in France as a pharmacist, he had then studied under leading German chemists, including Emil Fischer and Richard Willstätter. On returning to France, he had been anxious to lessen the dependence of his native country on drugs imported from Germany. He was now determined to establish the chemical structure of Bayer 205 by one means or another. To this end, he conducted a critical examination of 17 Bayer patents covering trypanocidal ureas derived from naphthalene sulfonic acids. His persistence enabled him to conclude that Bayer 205 must have been one of 25 possible structures. He then synthesised several of these and had them tested on infected mice. After one exhibited antitrypanosomal properties identical to those of Bayer 205, Fourneau published its structure in 1924, calling it Fourneau 309.[11,12] Because the structure had never been previously published, F. Bayer & Company could not claim that its patents were being infringed. For this reason, disclosure of structures thereafter became standard practice in pharmaceutical patents. It was not until 1928 that F. Bayer & Company finally admitted that Bayer 205 was identical to Fourneau 309. The approved name of the drug is suramin.

The success of suramin can be measured by the fact that three-quarters of a century after its discovery it remains one of the principal drugs for the prevention and treatment of trypanosomiasis. In keeping with Ehrlich's observation concerning dyes featuring the sulfonic acid function, it cannot enter the central nervous system, ruling out its use in advanced forms of sleeping sickness. However, of even more significance than its undoubted therapeutic value was the stimulus suramin gave to the subsequent development of chemotherapy in Germany. Within 12 years of its discovery, researchers at Elberfeld had developed the first effective synthetic antimalarials, the sulfonamides, and several other chemotherapeutic agents by an extension of the approach that had led to the introduction of suramin.

PAMAQUIN

Charles Louis Alphonse Laveran isolated the protozoal parasite that caused malaria as long ago as 1881, but it was not until 1924 that Wilhelm Roehl devised a technique for screening potential antimalarial drugs by administering them to canaries in the Bayer laboratories at Elberfeld. Compounds that appeared promising in the new screening programme were subsequently tested in syphilitic patients suffering from general paralysis of the insane who had been therapeutically inoculated with malarial parasites to produce fever. This approach had recently been introduced in Vienna by Julius Wagner-Jauregg, who claimed that at least one-third of his paralysed patients recovered after such treatment.

Once it had become possible to screen potential antimalarials, Werner Schulemann and his colleagues at Elberfeld, Fritz Schönhöfer and August Wingler, followed up the earlier suggestion by Paul Ehrlich that derivatives of methylene blue should be synthesised as potential antimalarials. They began by substituting a diethylaminoethyl side chain on one of the methyl groups in the blue dye. Roehl found this effective in canaries, with a therapeutic index of 8 (i.e. the ratio of the toxic dose to the therapeutic dose), but he was concerned that as the new compound was a strongly coloured dye there could be consumer resistance to its use. To avoid this problem, Schulemann switched his attention to quinolines, but he retained the basic side chain of the active methylene blue analogue as he was convinced that it was essential for antimalarial activity. After it was substituted on to 8-aminoquinoline, the resulting compound cured infected canaries. This quinoline subsequently served as the lead compound from which a diverse range of analogues were then synthesised and tested on canaries. The initial strategy was to examine the effect of altering the point of attachment of the side chain to the quinoline ring and then to investigate what happened when the side chain was varied in just about every conceivable manner. In order to increase the similarity to quinine, a methoxyl group was placed on the 6-position of the quinoline ring. As if all this were not enough, Schulemann and his colleagues in addition then introduced a variety of heterocyclic ring systems other than quinoline.[13] Literally hundreds, possibly even thousands, of compounds were prepared and tested by the small group of researchers at Elberfeld. In 1925, a promising compound with a therapeutic index of 30 was selected for clinical evaluation.[14] It was initially tried in patients who had been infected with the malarial parasite as part of the Wagner–Jauregg regimen. The new drug was effective, and Roehl confirmed that it was also able to cure patients with naturally acquired malaria. Clinical trials throughout the world followed, and then the drug was marketed as Plasmoquine®. It was given the approved name 'pamaquin', though its chemical structure was not disclosed until 1928, after the company had changed its policy on such matters.

pamaquin

The life cycle of the malarial parasites (sporozoites) after they enter the blood of a human bitten by a female anopheline mosquito is complex. Within an hour, liver cells are invaded and the sporozoites begin to divide, ultimately causing the cells to rupture. This releases merozoites into the blood, which then penetrate the red cells to initiate the erythrocytic phase of the disease. The merozoites multiply until the red cells rupture, causing the patient to experience chills, fever and sweating. The released merozoites then attack other blood cells to renew the cycle, accounting for the periodicity of malarial attacks. Quinine suppresses this erythrocytic stage of the disease. Roehl, however, discovered that the action of pamaquin was quite different, and it was clearly not simply a substitute for quinine. Large doses of pamaquin lowered the incidence of relapses among patients with benign tertian malaria caused by *Plasmodium vivax*. This is the commonest type of malaria and is so named as fewer deaths occur than with malignant tertian malaria, in which patients are infected with *P. falciparum*. Outright cures were even reported among those who could tolerate its inevitable side effects. However, better results were obtained when small doses of pamaquin were administered in conjunction with quinine. It was only some years later that it was established that pamaquin acted by destroying parasites that persisted in the liver. These were responsible for the characteristic relapsing fever associated particularly with *P. vivax* infection. The combination of pamaquin with quinine eradicated the infection in both the liver and the blood, thus producing outright cures.

Following the clinical introduction of pamaquin, the Joint Chemotherapy Committee of the Medical Research Council and the Department of Scientific and Industrial Research sponsored an ambitious programme of research in British universities, with the aim of developing new antimalarial drugs. In 1929, the first of many publications came jointly from Robert Robinson at University College in London and George Barger at the University of Edinburgh.[15] This described how they had been able to establish the chemical nature of pamaquin before its structure had been disclosed by what was by then no longer Bayer but I.G. Farbenindustrie. They then proceeded to make analogues of it. Although nothing superior to pamaquin arose out of either this collaboration or similar investigations by Fourneau in France, Magidson in Russia, Hegener, Shaw and Manwell in the United States or Brahmachari in India, the scene was set for a massive wartime effort that followed soon after. The I.G. Farbenindustrie group at Elberfeld, however, did succeed in discovering more effective quinoline antimalarials in the 1930s, showing how far ahead they were of their international competitors.

Mepacrine (Quinacrine)

The dichloro analogue of the triphenylmethane dye magenta was synthesised in 1909. Paul Ehrlich discovered not only that it had weak trypanocidal activity but also that this was due to its contamination by small amounts of acridines. He then asked Louis Benda at Farbwerke vorm. Meister, Lucius und Brüning to synthesise the yellow acridine dye, which was to become known as trypaflavine after Ehrlich found it to be the most potent trypanocide with which he had ever worked.[16] It was highly effective against virulent strains of *Trypanosoma brucei* in mice, yet proved to be worthless in larger animals or humans. However, Ehrlich's assistant Kiyoshi Shiga reported that, like magenta, it had bactericidal properties.[17] This was confirmed by Carl Browning, a former associate of Ehrlich who had returned to the University of Glasgow Medical School, where he tested a wide variety of dyes for antibacterial activity.[18] He discovered that trypaflavine was even effective against pathogenic organisms in the presence of serum. After war broke out the following year, the newly established UK Medical Research Committee asked Browning to continue his work on trypaflavine with a view to finding an antiseptic that could be used on deep wounds. Browning set up a laboratory in the Bland-Sutton Institute of Pathology at the Middlesex Hospital, London, where two dyes were selected for trial in casualty clearing stations at the front lines and in base hospitals.[19] These were brilliant green and trypaflavine, the latter now being renamed as acriflavine since it had no clinical value in trypanosomiasis. It was not until 1934 that it was realised that acriflavine was a mixture of two components, one of which was introduced during the Second World War under the approved name of proflavine.[20] Aminacrine, a non-staining analogue of acriflavine, was also made available at that time.

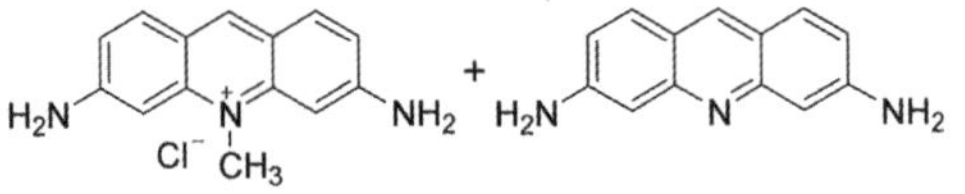

acriflavine

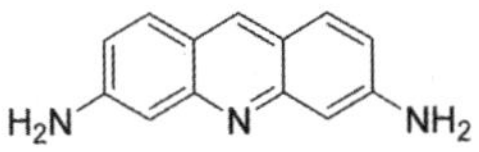

proflavine

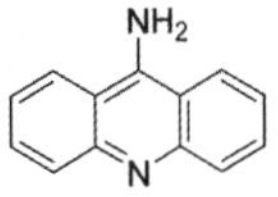

aminacrine

By 1926 Robert Schnitzer of I. G. Farbenindustrie was able to control acute streptococcal infections in mice by administering, either orally or by injection, large doses of 9-aminoacridines substituted with a nitro group at the 3-position.[21] One of these nitroacridines incorporated the same basic side chain that had conferred antimalarial activity on pamaquin. While effective against both trypanosomes and streptococci in animals, this particular nitroacridine was not potent enough to be considered for clinical use. An analogue of it prepared as a potential trypanocide by Schnitzer and Silberstein was shown to be more effective as an antibacterial agent. It was marketed as Entozon® (also known as Nitrokridin 3582), but it caused severe tissue irritation at the site of injection, as well as unpleasant side effects. A closely related compound in which the nitro group was replaced with a chlorine atom was the most promising among more than 12 000 compounds synthesised at Elberfeld that were entered in a screening programme that Walter Kikuth had established to discover a new antimalarial. The superiority of the chloroacridine was probably recognised in 1930, though the first report appeared in 1932.[22] It was initially called Plasmoquine E®, but to avoid confusion with pamaquin this was changed to Erion® and then to Atebrin®. It was later given the approved name of mepacrine; it is also known as quinacrine.

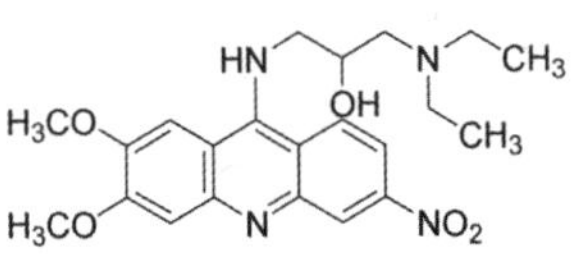

Entozon®

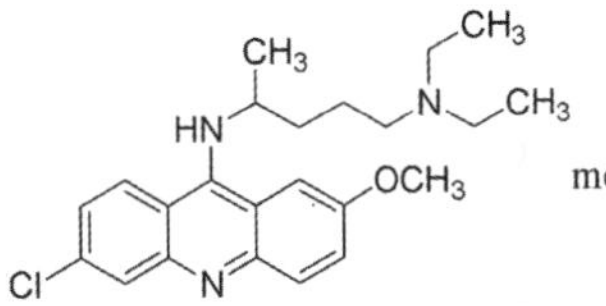

mepacrine (quinacrine)

Kikuth's original tests on mepacrine had convinced him that its action was similar to quinine insofar as, unlike pamaquin, it could kill merozoites in the erythrocytic phase of malaria. This meant that it could suppress the symptoms of malaria and cure those types of malaria in which the parasites did not persist in the liver cells. Although it was marketed throughout the world as a substitute for quinine, its full potential was not recognised until after the outbreak of the Second World War.

As the war clouds gathered, the importance of a quinine substitute was recognised since Germany had difficulty obtaining supplies of quinine during the First World War. Now the tables were likely to be turned if the Japanese took control of the East Indies, from whence came most of the world's supplies of cinchona bark. Both pamaquin and mepacrine were included in the Association of British Chemical Manufacturers' list of essential drugs that would need to be produced in the United Kingdom if war broke out. ICI were requested to devise suitable manufacturing processes, and by September 1939 mepacrine was being manufactured in a pilot plant. Shortly afterwards, full-scale production to meet the requirements of the armed forces had begun. In 1941, the American government also responded to the threat of war. One of its earliest moves involved the Winthrop Chemical Company, which had been set up after the First World War to distribute Bayer Pharmaceuticals in the United States following the purchase from the Custodian of Enemy Property, by Sterling Drug Inc., of the Bayer Company of New York. By a subsequent agreement concluded in 1926, the newly constituted I.G. Farbenindustrie, which had taken over control of Bayer in Germany, became half-owners of Winthrop. Three months before the bombing of Pearl Harbor brought the United States into the war, a government antitrust suit severed the ties between the Winthrop Chemical Company and I.G. Farbenindustrie, leaving it a wholly American owned company. When the Japanese moved into the East Indies, Winthrop was called upon to supply mepacrine for the US armed forces. Prior to this, the company had merely produced about 5 million of its Atabrine® brand tablets annually from six chemical intermediates imported from Germany. Winthrop responded by sublicensing 11 leading American manufacturers on a royalty-free basis, and the outcome was that in 1944 alone some 3500 million mepacrine tablets were produced in the United States. This and the wartime effort to produce large amounts of penicillin laid the foundations for the United States to become the biggest producer of pharmaceuticals in the post-war world.

When British and American production of mepacrine first began, the drug was considered to be nothing more than a synthetic substitute for quinine. As a result of its widespread use by American forces in the Far East, it became apparent that mepacrine was superior to quinine for the treatment and suppression of malaria. While the war lasted, great care was taken to keep this vital information from the enemy so that they would continue to use quinine, an inferior drug.

The chance observation that the condition of a patient suffering from the chronic autoimmune disease of the connective tissues known as lupus erythematosus improved dramatically while taking mepacrine led to a detailed study that confirmed its value in lupus.[23] Furthermore, this clinical trial also revealed that in patients with associated rheumatoid arthritis this also ameliorated as their skin condition improved. This led to further trials not only of mepacrine but also of other antimalarial drugs in patients with rheumatoid arthritis. Eventually the value of chloroquine and hydroxychloroquine for treating rheumatoid arthritis was generally recognised.

Chloroquine

During their North African campaign, German troops were equipped with supplies of sontoquine. This was a quinoline compound that could be considered as an analogue of mepacrine in which one of the benzene rings (i.e. that containing the methoxyl group) was absent. Samples of sontoquine were obtained from captured German prisoners of war and then sent to the United States for analysis. Particular attention was paid to the fact that sontoquine was an aminoquinoline substituted in the 4-position rather than the 8-position. Biological screening of a close analogue in which the methyl group on the 3-position of sontoquine was absent revealed outstanding antimalarial activity. This analogue, chloroquine, had been synthesised by Hans Andersag at Elberfeld in 1934 at the same time as sontoquine, and a German patent on it was awarded to the I.G. Farbenindustrie.[24] Wilhelm Roehl's successor, Walter Kikuth, had dismissed it as being toxic, preferring sontoquine instead. The American investigators, however, found chloroquine to have fewer side effects than mepacrine, as well as reducing malarial fevers more quickly. Another important advantage over mepacrine was that it did not colour the skin yellow. It was not possible to institute large-scale production of chloroquine before the war ended, but eventually it rendered mepacrine obsolete.

Resistance of most strains of *Plasmodium falciparum* towards chloroquine has developed during the years since the introduction of this drug. In benign tertian malaria caused by *P. vivax*, *P. ovale* or *P. malariae*, chloroquine can be an effective treatment. It is also widely used in the prophylaxis of malaria in many regions throughout the world, often in combination with another antimalarial agent.

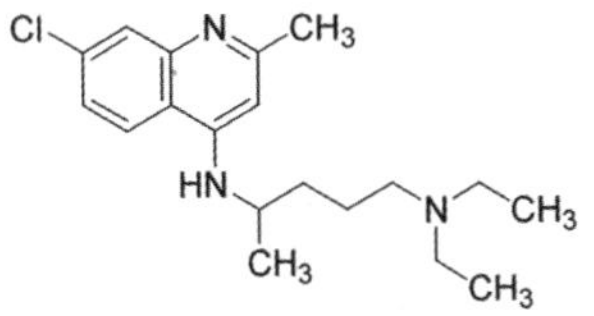

sontoquine

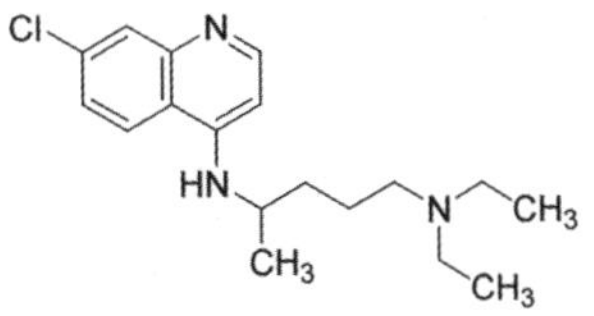

chloroquine

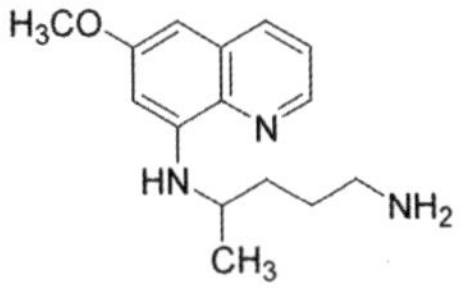

primaquine

Primaquine

In the United States, the government Office of Scientific Research and Development established a research programme to discover antimalarial drugs. From 1941 to 1945, over 14 000 compounds were screened, around one-third of these being new substances. Robert Elderfield and his colleagues at Columbia University in New York contributed to the programme by noting that among the vast range of substituted 8-aminoquinolines related to pamaquin, few primary or secondary amines had been reported. A variety of such compounds were then synthesised, resulting in the emergence of primaquine as a less-toxic analogue of pamaquin, which it immediately superseded as the drug of choice for eradication of benign tertian malaria.[25] No superior agent for this purpose has been found.

Impact of Research on Antimalarials

The research that was involved in developing synthetic antimalarial drugs from the mid-1920s to the mid-1940s in Germany and in the United States and the United Kingdom during the Second World War completely transformed the process of drug discovery from something that could readily be conducted in academic institutions or small commercial laboratories into a much more highly involved endeavour involving the preparation of large numbers of synthetic drugs by teams of chemists who received early information on their biological properties that could then be utilised in the selection of the next round of compounds to be prepared. The foundations for drug research in the second half of the twentieth century were firmly established by the end of the Second World War. The rate of discovery of new drugs since then has been unparalleled in any other period of history.

THE ANTIBACTERIAL SULFONAMIDES

At the University of Breslau in 1913, Philipp Eisenberg found that the azo dye known as chrysoidine possessed powerful antiseptic properties.[26] Twenty years later, phenazopyridine was introduced as a urinary antiseptic by Ivan Ostromislensky, a New York industrial chemist who had suspected it might have bacteriostatic activity.[27] After confirming this to be the case, he noticed that it imparted a red colour to the urine of animals. This observation was profitably exploited by marketing it as a relief for urinary tract pain. It is still prescribed even though it has questionable efficacy as a urinary antiseptic.

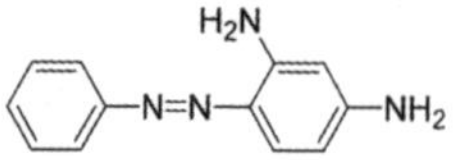

chrysoidine

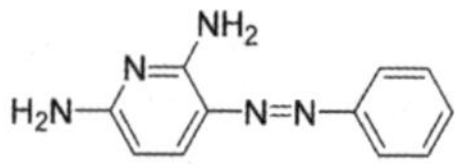

phenazopyridine

Sulfamidochrysoidine (Prontosil Rubrum®)

In 1927, I.G. Farbenindustrie opened an outstandingly well-equipped suite of new research laboratories at Elberfeld. Gerhard Domagk was appointed as Director of the Institute of Experimental Pathology, with the responsibility of continuing the search initiated by Robert Schnitzer at Hoechst for a drug effective against generalised bacterial infection. He introduced intensely rigorous test conditions that would eliminate all but the most effective compounds. This involved screening compounds on mice inoculated with a highly virulent strain of haemolytic streptococcus, i.e. *Streptococcus pyogenes*. The commonest diseases caused by this organism were tonsillitis and scarlet fever, from which most patients recovered uneventfully. However, the streptococci sometimes invaded the middle ear to cause otitis media, resulting in permanent deafness. Occasionally, fatal meningitis also resulted. Further complications of infection with the haemolytic streptococcus included rheumatic fever and acute nephritis, both of which could also be fatal. During the worldwide influenza epidemic of 1918–1919, pneumonia caused by this organism was a common cause of death. It was also responsible for many fatalities after wounding, both mild and severe, during the First World War. Burning and scalding were particularly likely to be followed by haemolytic streptococcal infection. Whatever the original cause of infection, the appearance of haemolytic streptococci in the blood of a patient, septicaemia, was an ominous sign.

The particular strain of haemolytic streptococci selected by Domagk was isolated from a patient who had died from septicaemia, and its virulence had been increased by repeatedly subculturing it in mice. This ensured that the test system was reliable and 100% of the mice consistently died within four days of inoculation. Only an exceptional drug could influence such an otherwise inevitable outcome.

Domagk began his investigations by testing three classes of substances that had been reported as having antibacterial properties, namely gold compounds, acridines and azo dyes. The first to exhibit activity in the mice were organic gold compounds. Domagk confirmed their activity and showed that they could even cure larger animals that were similarly infected. Unfortunately, kidney damage prevented the administration of the dosage required to cure streptococcal infections in patients. After this setback, Domagk turned his attention to dyes. The low toxicity of phenazopyridine encouraged Fritz Mietzsch and Josef Klarer to synthesise analogues of it, so they attached to chrysoidine a side-chain that had previously conferred antistreptococcal activity on acridines. Domagk found that this greatly enhanced the activity against cultures of streptococci, but negative results were still obtained in infected mice.

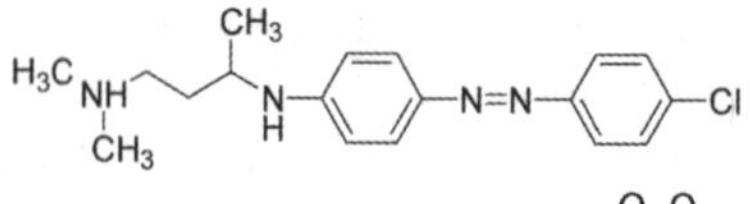

chrysoidine derivative

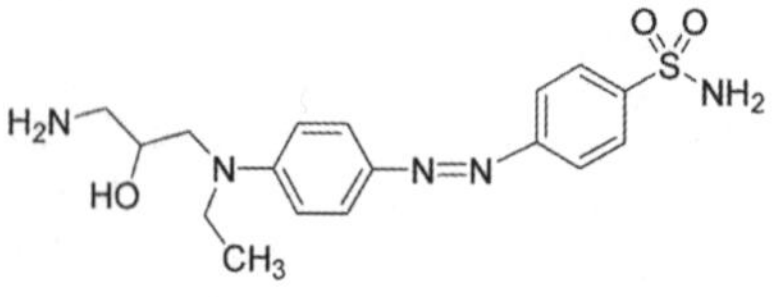

sulfonamide derivative

sulfamidochrysoidine (Prontosil Rubrum®)

Mietzsch and Klarer next took up an old idea originally tried in 1909 by Heinrich Hörlein, now Director of the Medical Division of I.G. Farbenindustrie. This involved introducing a sulfonamide function into azo dyes in order to enhance their ability to bind to wool. When Domagk tested the sulfonamide derivatives of the most promising of the dyes already examined, they turned out to be almost ineffective against cultures of streptococci. Undaunted by this, he then tested them on infected mice. For the first time in four years since the screening programme had begun, a genuine protective effect was apparent when high doses of one of the dyes were administered. A patent was applied for on 7 November 1931. During the next year a large range of sulfonamide dyes were synthesised and tested. Many of them not only cured the mice in an unprecedented manner, but were also non-toxic. Finally, on Christmas Day 1932, a patent application for another batch of sulfonamide dyes was submitted.[28] Among these was a red dye that was to make medical history.

Early the following year, the medical division of I.G. Farbenindustrie was asked whether there might be a drug that could help a ten month old boy dying from staphylococcal septicaemia. After being informed that the only drug available was intended for use in streptococcal infections, Dr Foerster, the physician treating the infant, was supplied with tablets of Streptozon®. Treatment began at once, the boy receiving half a tablet twice daily by mouth. To everyone's astonishment, he did not die. After four days his temperature gradually lowered to normal and his general condition improved markedly. Treatment was eventually stopped after three weeks and he was discharged from hospital. The case was reported to a meeting of the Düsseldorf Dermatological Society on 17 May 1933 by Foerster.[29]

Several other Rhineland physicians received supplies of Streptozon®, and three brief reports citing case histories appeared in German medical journals during 1934. No experimental details or chemical information appeared in print until Domagk's first publication on the subject appeared in the _Deutsche Medizinische Wochenschrift_ of 15 February 1935.[30] In this he described how small, non-toxic doses of a brick-red sulfonamide dye called Prontosil Rubrum® prevented every single mouse that received it by stomach tube from succumbing to an otherwise lethal inoculation of haemolytic streptococci. Thirteen out of fourteen untreated mice died within three to four days. Domagk also explained that Prontosil Rubrum® had been able to cure chronic streptococcal infections in rabbits and also alleviated those caused by staphylococci. Prontosil Rubrum® was a new name for Streptozon®, which was chosen in view of the broader spectrum of antibacterial activity that had been observed. Later, the drug was given the approved name of sulfamidochrysoidine, although this was infrequently used.

Accompanying Domagk's paper were three others with clinical reports of two years of investigations into the remarkable antistreptococcal action of Prontosil Rubrum®. Although the bacteriological work supporting these and the earlier clinical studies left much to be desired, the papers did testify to both the efficacy and safety of the dye, which made it the first truly effective chemotherapeutic agent for any generalised bacterial infection. Prontosil Rubrum® was put on the market shortly after the publication of these papers. Suprisingly, it did not create the sensation that might have been expected. There was a conviction in medical circles that chemotherapeutic agents could have little effect against generalised infections. Only clear-cut clinical results could be expected to allay any doubts. That such results quickly became available was principally due to the efforts of Leonard Colebrook at Queen Charlotte's Maternity Hospital in London. After listening to a lecture on Prontosil Rubrum® delivered by Hörlein[31] to the Royal Society of Medicine in London, Colebrook tried to obtain samples of it. With material eventually supplied from France, he confirmed Domagk's results in infected mice, although only after selecting a particularly virulent strain of haemolytic streptococci. In January 1936 he began to use the drug in patients. Shortly after, he received supplies of Prontosil Rubrum® and Prontosil Soluble® from Germany for use in a clinical trial on 38 dangerously ill women. The following June, Colebrook and Kenny reported in the *Lancet* that only three of these patients died.[32] This publication had considerable impact, not least among pharmaceutical companies. Colebrook and his colleagues continued their studies by treating a further 26 seriously ill women, none of whom died.

Domagk was awarded the Nobel Prize for Medicine in 1939 for his discovery of the antibacterial properties of Prontosil Rubrum®.[33] After acknowledging notification of the award he was detained by the Gestapo and persuaded to reject it. This was a direct consequence of Hitler's rage over the award of the 1936 Nobel Peace Prize to the pacifist journalist Carl von Ossietsky. It was not until two years after the war ended that Domagk was able to travel to Stockholm to receive his medal, by which time the prize money had reverted to the Nobel Foundation. For Domagk, however, the greatest reward must surely have been when, in February 1935, the life of his own daughter, Hildegarde, was saved by Prontosil Rubrum® after she developed a severe septicaemia caused by pricking her finger with a needle.

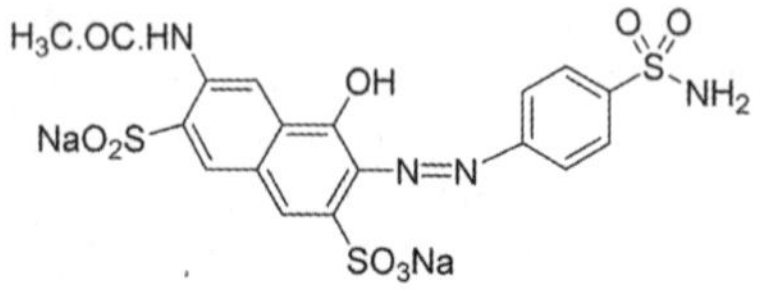

azosulfamide

Apart from the introduction in July 1935 of the now obsolete azosulfamide, an injectable sulfonamide known as Prontosil Soluble®, no other major development in this field came from the I.G. Farbenindustrie laboratories.[34] The initiative passed to French, British, American and Swiss workers – despite the fact that over 1000 sulfonamides had been synthesised at Elberfeld during the five years following the first recognition of antibacterial activity among such compounds.

Sulfanilamide

An important report was issued from Fourneau's laboratory in 1935 by Jacques and Therese Trefouel, Federico Nitti and Daniele Bovet, who suggested that the azo linkage of sulfamidochrysoidine was cleaved in the patient's body to form 4-aminobenzene sulfonamide, a colourless compound.[35] Fourneau promptly synthesised it as 1162 F, using a method described in the literature in 1908 by Paul Gelmo of Vienna, who had prepared it for his doctoral thesis.[36] Tests quickly revealed that it retained the activity of Prontosil Rubrum®; it was therefore called, for a while, Prontosil Album. The approved name given to this non-patentable compound was sulfanilamide.

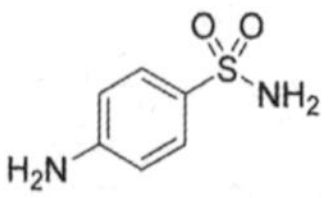

sulfanilamide

Sulfanilamide was then isolated from the urine of patients by Colebrook's assistant, A.T. Fuller.[37] Hörlein subsequently admitted that his company had already discovered that this was the active form of Prontosil Rubrum®, but had considered it possible that the unmetabolised Prontosil Rubrum® stimulated the immune system to fight off infection.[38] This is significant because there has been speculation that the two-year delay (unprecedented in those days!) by I.G. Farbenindustrie in bringing Prontosil Rubrum® on to the market could have been caused by its efforts to find some way of protecting their discovery from exploitation by rival manufacturers once it was known that the active species was the non-patentable sulfanilamide. The company's reply to this was that careful validation of their clinical results was necessary since they were both unprecedented and unlikely to be believed – as was indeed the case. Hörlein claimed they were also trying to establish whether the intact drug was an immuno-stimulant. This could explain the concentration of effort by I.G. Farbenindustrie on the development of analogues of Prontosil Rubrum® rather than analogues of sulfanilamide and tends to support Hörlein's contention.

In 1937 an elixir of sulfanilamide was put on sale in the United States. It contained 10% of diethylene glycol as a solvent to render the sulfonamide soluble. During the two months that this preparation was on sale, 107 people died from severe damage to the liver and kidneys caused by the solvent.[39] The chemist who devised the formulation committed suicide. As a consequence of this episode, the US Congress enacted a Food and Drug Act in an attempt to prevent such a state of affairs ever arising again. That the United States was spared the thalidomide tragedy 20 years later was one direct consequence of this legislation.

The revelation that sulfanilamide was a systemically active antibacterial agent caught the attention of several companies. The first of its analogues was marketed in 1938 by Schering–Kahlbaum of Berlin, namely sulfacetamide.[40] It was a stronger acid (pK_a 5.4) than most other sulfonamide drugs because of resonance stabilisation of its anion. This meant that a greater proportion was ionised in the glomerular filtrate, causing it to be rapidly excreted by the kidneys rather than reabsorbed through the renal tubules. While this was undesirable for systemic therapy, it did guarantee the high urinary levels that for many years ensured its use in the treatment of urinary tract infections.

sulfacetamide

Another consequence of the enhanced acid strength of sulfacetamide was that solutions of its sodium salt were not as alkaline as those of other sulfonamides, rendering them ideal for application to the eye, a purpose for which it was widely employed for many years despite lack of any evidence of efficacy. This use has been abandoned.

Heterocyclic Sulfonamides

The most important early development following the introduction of sulfanilamide was the synthesis of sulfapyridine.[41] This was carried out in 1937 at the suggestion of Arthur Ewins, the Director of Research at the May and Baker laboratories in Dagenham, London, who wanted to study the action of sulfonamides substituted with a heterocyclic ring on the sulfonamide nitrogen. Sulfapyridine proved to be not only more potent than sulfanilamide but it also had a wider spectrum of antibacterial activity, being effective against pneumococci, meningococci, gonococci and other organisms. As a result, it was supplied to Lionel Whitby at the Middlesex Hospital in London, where he first established that it had unprecedented efficacy against mice inoculated with pneumococci, despite being less toxic than sulfanilamide. He then organised a clinical trial in which sulfapyridine fully lived up to its early promise by reducing the mortality rate among patients with lobar pneumonia from 1 in 4 to only 1 in 25.[42] At a stroke, this dreaded disease was no longer to be one of the commonest causes of death among otherwise healthy adults. One whose life was saved was Winston Churchill, when he contracted pneumonia during his wartime visit to North Africa in December 1943. This event had as great an impact on the popular imagination as had the anaesthetising of Queen Victoria with chloroform some 90 years earlier. The proprietary name of **M & B 693**® at once became world famous, though few realised that the life of the British Prime Minister had been saved through research initiated in Germany. However, increasing bacterial resistance caused by extensive overprescribing of sulfapyridine and other sulfonamides eventually led to their replacement by antibiotics, most of which had fewer side effects. Typical side effects of the sulfonamides included rashes, renal damage owing to their insolubility and blood dyscrasias. The principal use of sulfonamides at present is in the treatment of urinary tract infections.

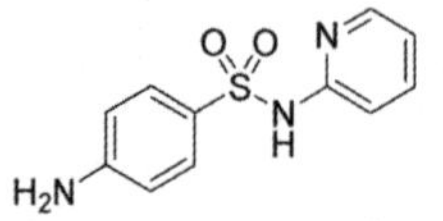

sulfapyridine

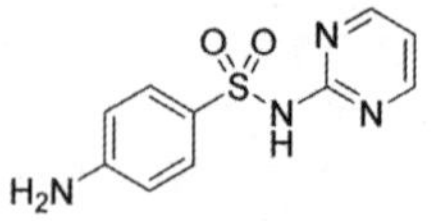

sulfadiazine

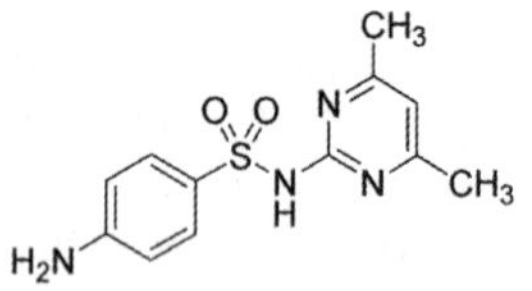

sulfadimidine

Following the introduction of sulfapyridine, varieties of heterocyclic sulfonamides were developed in different laboratories during the early 1940s, often involving patent conflicts between rival companies rushing to market their products. Few of them offered any advantage over sulfapyridine other than perhaps a reduced tendency to deposit crystals in the kidneys. This nasty side effect was due not only to the inherent insolubility of the sulfonamides but also to that of their N^4-acetyl metabolites (i.e. where the acetyl substituent is attached to the aromatic nitrogen atom). This insolubility was attributable to intermolecular H-bonding. The N^4-acetyl metabolites, which were usually the major ones, were devoid of antibacterial activity.

The risk of crystals depositing in the kidneys was diminished by encouraging patients to drink plenty of fluids or by raising the urinary pH to increase the proportion of water-soluble ionised species in the glomerular filtrate. This was achieved by giving potassium citrate or sodium bicarbonate by mouth until the urinary pH reached about 7.

The problem of crystals depositing in the kidneys was finally overcome by the introduction of sulfonamides that were more highly ionised in the urine because they were stronger acids. There was a limit to the degree of acidity that could be considered since compounds with a pK_a value as low as 5–6 were very rapidly excreted by the kidneys, as in the case of sulfacetamide. The ideal sulfonamides for treating systemic infections required a pK_a value in the range of 6.5–7.5 in order to balance the risk of kidney damage against rapid excretion. Two drugs, in particular, came within this range and both remain in use for the parenteral treatment of meningococcal meningitis, viz. sulfadiazine (pK_a 6.52) and sulfadimidine (pK_a 7.4). Sulfadiazine, prepared in 1940 by Richard Roblin and his colleagues at the Stamford Research Laboratories of the American Cyanamid Company, was not only more potent than sulfapyridine but also less toxic.[43] Furthermore, it had a wider spectrum of activity than any previous sulfonamide, hence it was used extensively during the Second World War. Sulfadimidine is also known as sulfamezathine and was first synthesized at Temple University in Philadelphia by William Caldwell and two of his Masters students.[44] It had even greater solubility in urine than sulfadiazine, though it was less potent.

DRUGS DERIVED FROM THE ANTIBACTERIAL SULFONAMIDES

The relative ease with which sulfonamides could be synthesised from commercially available 4-aminobenzensulfonyl chloride encouraged many companies to venture into this area, with the result that several classes of novel chemotherapeutic agents were discovered. However, this was not the sole stimulus to drug research that evolved from the introduction of the sulfonamides. Their side effects caused many problems in the clinic, but were successfully exploited to provide oral antidiabetic drugs and valuable diuretics that have benefited many patients with cardiovascular disease.

Antileprotic drugs

Arthur Buttle and his colleagues at the Wellcome Research Laboratories in London investigated 4:4′-diaminodiphenylsulfone as a potential analogue of sulfanilamide.[45] It had been synthesised in 1908 at the University of Freiburg and was later to become known as dapsone.[46] Although it was found to be 30 times more potent than sulfanilamide when tested on mice infected with streptococci, it was 15 times as toxic.

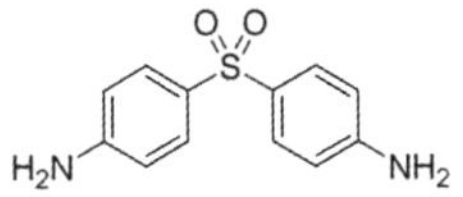

dapsone

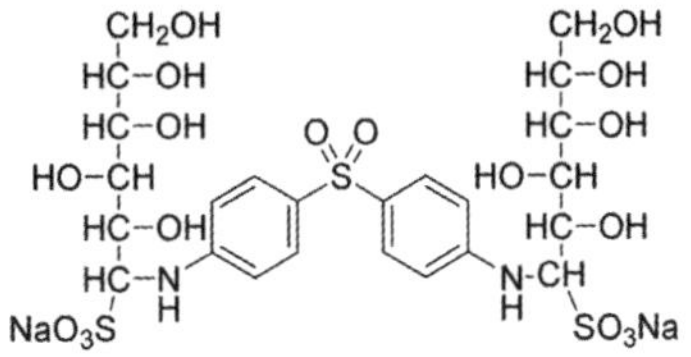

glucosulfone

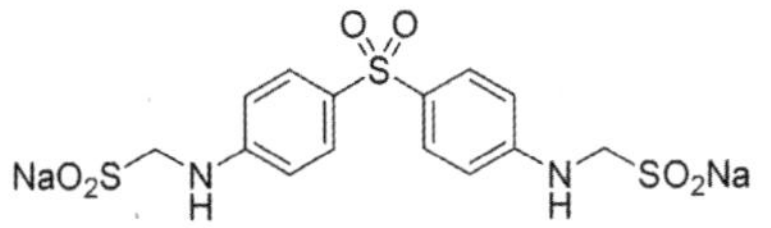

sulfoxone sodium

Several derivatives of dapsone were made in different laboratories in the hope of finding safer sulfones. Among them was a water-soluble analogue, glucosulfone, synthesised in the laboratories of Parke, Davis and Company. It initially appeared to be safer than dapsone, so samples were sent to the Mayo Clinic where it was found to be active against *Mycobacterium tuberculosis* in guinea pigs.[47] Guy Faget of the US National Leprosarium in Carville, Louisiana, was given details about this work after he contacted Parke, Davis for information about their new drug. His interest in it arose from his view that any drug effective against tuberculosis could also have value in leprosy, as both diseases were caused by a mycobacterium. Parke, Davis also put Faget in touch with Edmund Cowdry at Washington University School of Medicine in St Louis, Missouri, who was evaluating glucosulfone in rats infected with *M. leprae*, the causative organism of leprosy. Cowdry informed Faget that glucosulfone not only reduced the size of the lesions in rats but also brought about an improvement in their general physical condition. The results were published in 1941.[48] By this time Cowdry's work had convinced Faget that studies on humans should begin. With the support of Parke, Davis, he tested glucosulfone on six volunteers at the National Leprosarium. Early signs that the drug was effective led to the setting up of a controlled clinical trial of glucosulfone and sulfoxone sodium, a sulfone synthesised by Hugo Bauer of Abbott Laboratories.[49] This concluded that both drugs were effective in the treatment of leprosy, although severe side effects could occur.[50] Further trials around the world confirmed that this was the long-awaited breakthrough in treating leprosy.[51]

After the clinical trials on glucosulfone sodium had been reported, investigations into the value of dapsone revealed that it was as effective as any of the sulfones derived from it for treating leprosy.[52] Despite problems with resistance, it remains in use today as the standard antileprotic sulfone, while its analogues have been largely abandoned.

Diuretics

At the University of Cambridge in 1940, Thaddeus Mann and David Keilin carried out an experiment to determine whether the fall in carbon dioxide binding power of the blood caused by some of the recently discovered antibacterial sulfonamides could be accounted for by inhibition of carbonic anhydrase.[53] This enzyme, which they had isolated in a pure state a year before, was known to play an important role in the output of carbon dioxide by the lungs. The experiment confirmed their suspicions. However, only those sulfonamides in which both hydrogen atoms on the sulfonamide function were unsubstituted were enzyme inhibitors. These included sulfanilamide and seven sulfonamides devoid of antibacterial activity. Horace Davenport[54] at the Harvard Medical School then discovered large amounts of carbonic anhydrase in the kidneys, leading Rudolf Höber to suggest that the alkaline diuresis in patients who had been given massive doses of sulfanilamide might be accounted for by increased excretion of sodium bicarbonate caused by carbonic anhydrase inhibition.[55] It had been shown, shortly before this, that resorption of water from the tubules of the kidney depended principally on the absorption of sodium ions from the lumen. Subsequently, it was established that carbonic anhydrase promoted the exchange of sodium for hydrogen ions in the distal portion of the renal tubules.[56] When the enzyme was inhibited, sodium ions were excreted in the urine because the process responsible for their reabsorption was blocked.

Davenport next sought a more potent inhibitor of carbonic anhydrase from Richard Roblin at the Lederle Division of the American Cyanamid Company. Thiophen-2-sulfonamide was provided in the belief that it would be more acidic than conventional sulfonamides and that this would enhance its ability to compete with carbon dioxide for the active site on the enzyme. When tested, it proved to be about 40 times more potent an inhibitor than sulfanilamide, as a consequence of which Davenport recommended it as more reliable for investigating the role of carbonic anhydrase in the tissues.[57]

thiophen-2-sulfonamide   acetazolamide

In 1949, Boston physician William Schwartz administered large doses of sulfanilamide by mouth to obtain a diuretic effect in three patients with congestive heart failure, but he had to abandon this approach because of toxic side effects.[58] However, his report rekindled Roblin's interest in carbonic anhydrase inhibitors. He and James Clapp set about synthesising some 20 heterocyclic sulfonamides and, within a year, acetazolamide was found to be around 330 times more potent than sulfanilamide as an inhibitor of the enzyme.[59] It was introduced clinically as an orally active diuretic in 1952, but inhibition of carbonic anhydrase throughout the body led to a variety of complications. The only acceptable way to use acetazolamide as a diuretic was on an intermittent schedule. Happily, the inhibition of carbonic anhydrase in other parts of the body was turned to advantage in the treatment of glaucoma. By acting on the aqueous humour of the eye in much the same way as in the kidneys, namely to reduce bicarbonate levels and the water secreted with it, the build-up of pressure from excess fluid was overcome. Acetazolamide remains in use for this purpose.

Leading a new project for Merck, Sharp and Dohme, Karl Beyer considered that the problem with sulfanilamide as a diuretic for clinical use was essentially that it inhibited carbonic anhydrase at the distal end of the renal tubules, rather than solely at the proximal end. This, he believed, accounted for the increased excretion of bicarbonate. He sought a carbonic anhydrase inhibitor that acted in the proximal portion, as indicated by increased excretion of chloride in the form of sodium chloride. Such a drug might, he hoped, have the added bonus of being a useful antihypertensive agent, for clinicians were beginning to believe that low salt diets were an effective means of controlling high blood pressure. To identify a saluretic drug of this type, Beyer carried out salt assays on urine collected from specially trained dogs. These assays were done almost instantaneously by means of flame photometry.

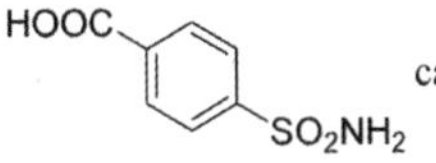

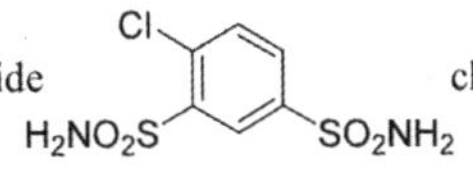

The first carbonic anhydrase inhibitor that Beyer found to increase chloride excretion was 4-sulfonamidobenzoic acid, which received the approved name of carzenide.[60] It was a cheap by-product from the synthesis of saccharin, which had been shown to be a weak carbonic anhydrase inhibitor by Hans Krebs.[61] Carzenide still increased bicarbonate excretion, indicating a lack of specificity of action within the kidneys. In humans, it was poorly absorbed from the gut and had weak diuretic activity. Nevertheless, James Sprague and Frederick Novello were encouraged by it to synthesise more aromatic sulfonamides for Beyer and his associates to test. They then found that when a second sulfonamido group was introduced, chloride ion excretion was markedly increased.[62] Further enhancement of activity followed on the introduction of a chlorine atom into the benzene ring. The outcome was the discovery that clofenamide was a potent carbonic anhydrase inhibitor. As it was a known compound,[63] clofenamide could not be patented, but that did not apply to the dichlorphenamide.[64] Unlike acetazolamide, it produced an increase in chloride secretion in man.[65]

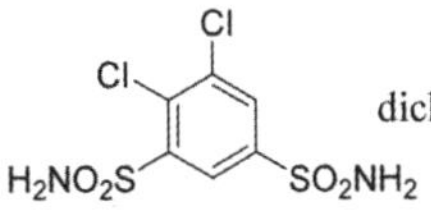

Merck researchers found that when an amino group was attached to the benzene ring of dichlorphenamide, there was a reduction in carbonic anhydrase inhibitory potency. Surprisingly, there was no corresponding reduction in chloride ion excretion. This proved to be a major breakthrough towards the goal of obtaining a saluretic agent. Novello proceeded to synthesise analogues with substituents on the amino group. In the course of this, he tried to make the *N*-formyl analogue with formic acid. This resulted in an unplanned ring closure to form a benzothiadiazide. As a matter of routine, this novel compound was entered in the screening programme. One can well imagine the surprise and delight when it was found to be a potent diuretic which did not increase bicarbonate excretion. Clinical tests confirmed that it was a safe, orally active diuretic with marked saluretic activity. The first reports appeared in 1957, and it was given the name 'chlorothiazide'.[66] It had a short duration of action of 6–12 hours. Chlorothiazide remains in use because of its low price.

Chlorothiazide was the first of many thiazide diuretics. Literally overnight, it rendered mercurial diuretics obsolete for the treatment of cardiac oedema associated with congestive heart failure. However, that was not all. Beyer was correct in his long-held belief that a safe diuretic that could increase sodium chloride excretion would be of value in the treatment of hypertension. Thiazide diuretics and related compounds are still used for this purpose.

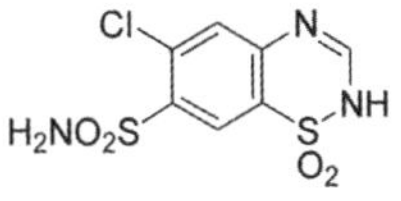

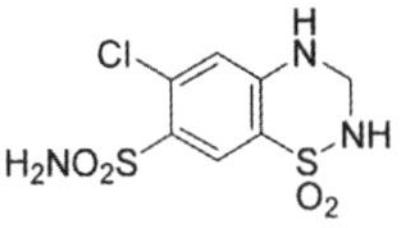

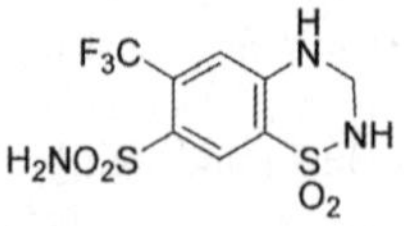

hydroflumethiazide

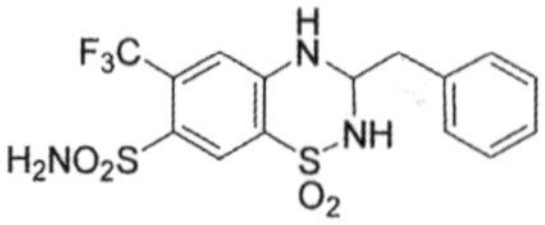

bendrofluazide

Ciba scientists led by George De Stevens replaced the formic acid used to produce chlorothiazide with formaldehyde and thereby obtained hydrochlorothiazide, which was ten times as potent as chlorothiazide.[67] At least four American companies then synthesised hydroflumethiazide, which had similar properties to hydrochlorothiazide.[68-71] Bendrofluazide was synthesised at the same time as hydroflumethiazide.[68] Its action lasted 18–24 hours. As one of the cheapest diuretics on the market, it remains widely used in patients with either mild heart failure or hypertension. Many other thiazides have been developed.

The realisation that the second acidic group in dichlorphenamide may be replaced with a carboxyl group, so long as an appropriate substituent is present on the amino group, led Hoechst to introduce frusemide (also known as 'furosemide') in 1962.[72] It had a quicker onset of activity, within one hour of oral administration, which was more intense and of shorter duration than that of other diuretics.[73]

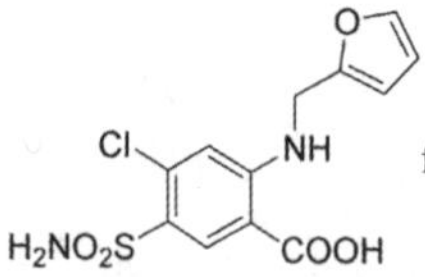

frusemide 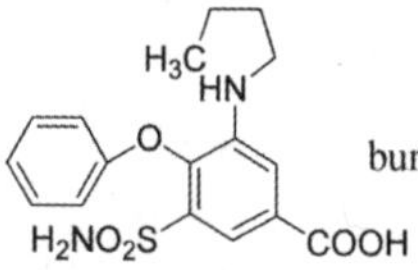 bumetanide 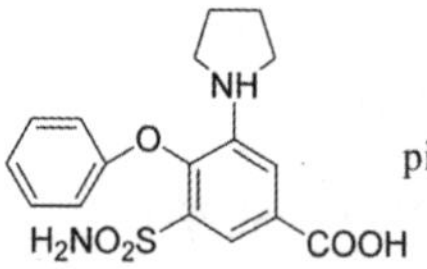 piretanide

Frusemide had a different site of action within the kidney tubule and became known as a loop diuretic because it acted in the region known as the loop of Henle. Loop diuretics were valuable in patients with pulmonary oedema arising from left ventricular failure. Despite thiazides being indicated for most patients requiring a diuretic, frusemide is widely prescribed.

Bumetanide is a more potent loop diuretic introduced by Leo researchers ten years after frusemide.[74] Hoechst introduced its analogue known as piretanide when their patent on frusemide expired.[75]

Antidiabetic Sulfonylureas

In March 1942, Marcel Janbon arranged a clinical trial at Montpellier University of 2254 RP, an experimental sulfonamide, in patients with typhoid.[76] Some of them became very ill, and a few died. The survivors made a startling recovery after receiving intravenous glucose, which led to suspicion that the drug had been producing severe hypoglycaemia. When this was confirmed, Janbon asked August Loubatieres to conduct a full investigation into the effects of the drug on animals for his doctoral studies. Although Loubatieres concluded that 2254 RP could be of value in diabetes, this suggestion was ignored.[77]

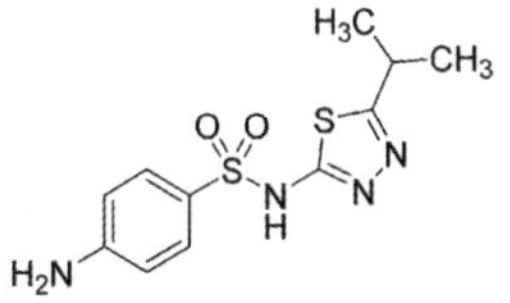

2254 RP

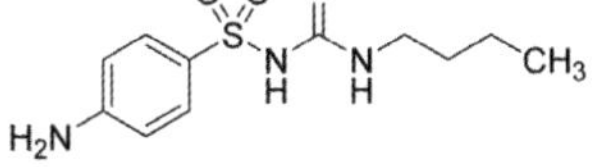

carbutamide

When a sulfonylurea developed by the C.H. Boehringer Company as a long-acting sulfonamide was put on clinical trial at the Auguste Viktoria Hospital in Berlin in 1954, it produced severe toxic effects. On testing the drug on himself, Joachim Fuchs found that it produced the symptoms of severe hypoglycaemia. The chief of his clinic, Hans Franke, conducted further investigations that led to the introduction of the drug as an oral hypoglycaemic agent with the approved name of carbutamide.[78] When the Eli Lilly Company arranged extensive clinical trials of it in the United States, the incidence of side effects was unacceptable, even though the drug was already being used in Europe. Upjohn meantime arranged a trial of Hoechst's closely related tolbutamide,[79] involving 20 000 patients and 3000 doctors. This time the drug received approval from the Food and Drug Administration for use in type 2 (non-insulin-dependent) diabetes. Unlike carbutamide, it did not possess antibacterial properties, so there was no likelihood of inducing resistant bacteria. It had to be taken three times a day because the methyl group was rapidly metabolised to a carboxylic acid.

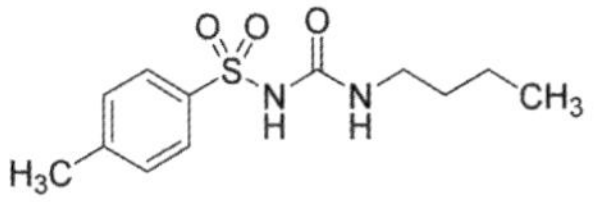

tolbutamide

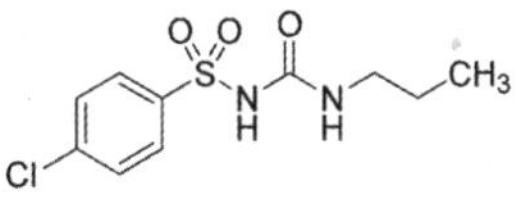

chlorpropamide

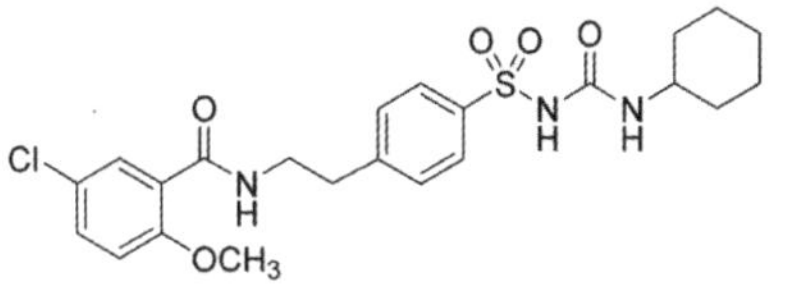

glibenclamide

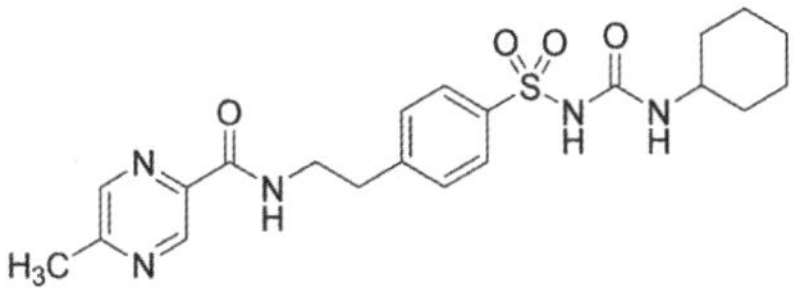

glipizide

Soon after the introduction of tolbutamide, Pfizer marketed chlorpropamide.[80] As it did not feature the metabolically sensitive methyl group of tolbutamide, it was about twice as potent as tolbutamide and could be taken once daily. While initially welcomed, this feature has been found to be disadvantageous in elderly patients in whom the drug can accumulate and thereby produce hypoglycaemia. Other longer-acting sulfonylureas have been marketed, including highly potent agents such as glibenclamide (glyburide)[81] and glipizide.[82]

Proguanil

On the outbreak of the Second World War, a high-priority rating was given to a scheme established by the British Medical Research Council to develop new antimalarial drugs. This resulted in a harmonious collaboration between 20 leading academics and a similar number of industrial chemists, involving the synthesis of around 1700 novel compounds, of which one-third exhibited antimalarial activity against experimental infections. One of the companies involved was ICI, where Francis Rose followed up evidence that sulfadimidine had weak activity against malaria. Attributing this to the presence of the pyrimidine ring, he and his colleagues synthesised a range of pyrimidines incorporating features present in mepacrine.

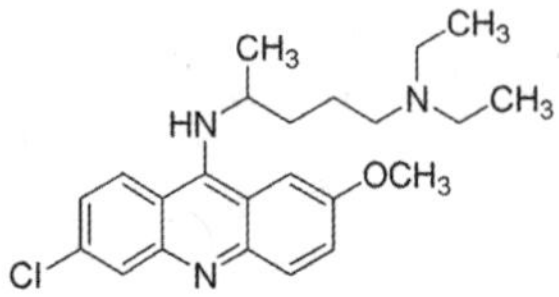

mepacrine

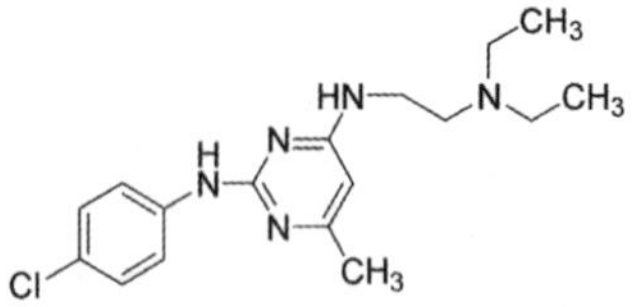

2666

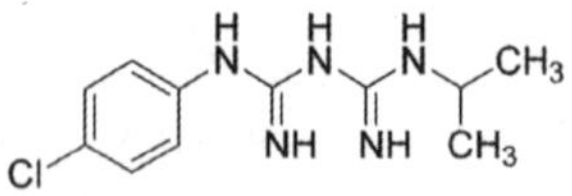

proguanil

Compound 2666, which contained a basic side chain and a chlorophenyl moiety, was active in chickens infected with *Plasmodium gallinaceum*. The most effective of its analogues were biguanides in which the pyrimidine ring was split open. The first of the biguanides to be screened had been devoid of antimalarial activity, but it was realised that this could have been due to the presence of too many basic nitrogenous side chains. Replacement of one of these by an isopropyl group led to the reappearance of strong antimalarial activity. Around 200 biguanides were synthesised and tested before proguanil emerged as being superior to mepacrine.[83]

Clinical trials at the Liverpool School of Tropical Medicine confirmed that proguanil was a first-line drug for the treatment of the erythrocytic phase of malaria, although it is now used mainly for the prophylaxis of malaria in those parts of the world where the parasites have not yet developed resistance.

Quinolones

In 1946, Alexander Surrey and H.F. Hammer of the Sterling–Winthrop Research Institute in Rensselaer, New York, devised a novel synthesis of chloroquine which produced a by-product, viz. 7-chloro-1,4-dihydro-1-ethyl-4-oxoquinoline-3-carboxylic acid. Some years later, this by-product was included in a screening programme and was found to be effective against fowl coccidiosis. When analogues were prepared by George Lescher and his colleagues, nalidixic acid emerged as a potent antibacterial agent effective against Gram-negative rods, with no cross-resistance to other antibiotics then in routine use.[84] Being a polar compound it was rapidly excreted by the kidneys, hence adequate tissue levels of drug could not be achieved. This, however, was ingeniously exploited when nalidixic acid was introduced as a urinary antiseptic in 1962. It was initially believed to bind to the A subunit of bacterial DNA gyrase, thereby interfering with the supercoiling of chromosomes required for them to be packed sufficiently tightly for cell replication to proceed. However, it is now known that nalidixic acid and its quinolone analogues bind to topoisomerase IV.[85]

Analogues of nalidixic acid were investigated in the hope of widening the spectrum of action. Early ones have been described as 'second-generation quinolones'. While the 3-carboxyl and 4-oxo groups were essential for activity, it was possible to alter the heterocyclic ring. One of the first compounds to exhibit superior activity was Warner–Lambert's oxolinic acid, which was found to be more potent than nalidixic acid.[86] It was rapidly excreted in the urine, once again limiting its use to the treatment of urinary tract infections. Lescher synthesised acrosoxacin, which was about 10 times as potent as nalidixic acid and had a similar spectrum of activity. Its principal value was in the treatment of gonorrhoea in patients who were either allergic to penicillins or infected by strains resistant to other antibiotics.

nalidixic acid

oxolinic acid

acrosoxacin

norfloxacin

enoxacin

ofloxacin

ciprofloxacin

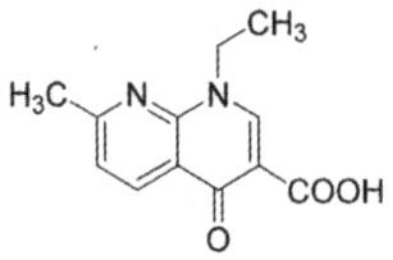

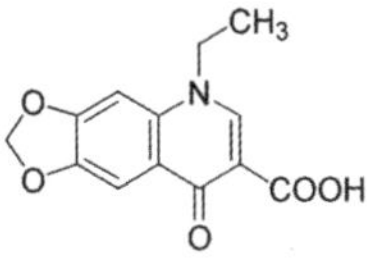

Several quinolones were developed in Japan. Norfloxacin was prepared in 1978 by the Kyorin Pharmaceutical Company in Tokyo.[87] The combination of the piperazine ring at the 7-position of the quinolone ring and the fluorine atom at the 6-position had the effect of radically altering the spectrum of activity, which was widened to include in a quinolone for the first time a low level of activity against Gram-positive organisms. Unfortunately, norfloxacin was poorly absorbed from the gut, with only about one-third of a dose entering the circulation. It then underwent both hepatic oxidation and renal elimination, hence it could not be considered for anything other than treatment of gastrointestinal or urinary tract infections. However, it served as the lead for further development of the fluoroquinolones.

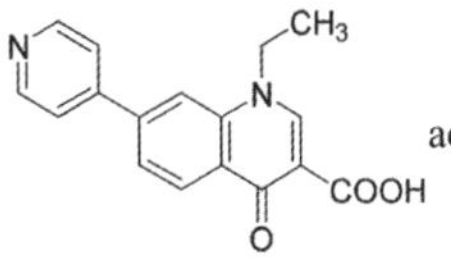

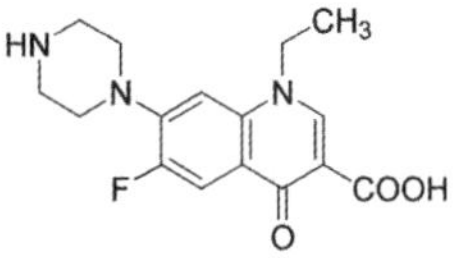

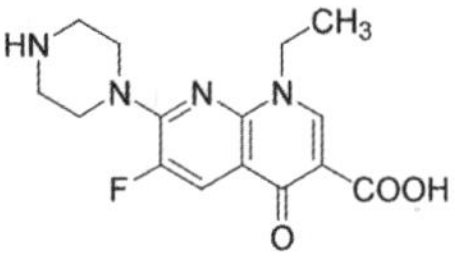

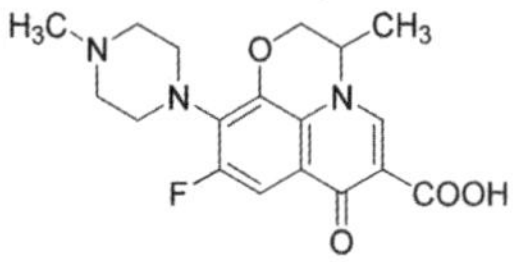

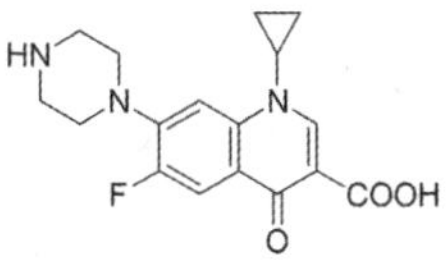

Researchers at the Dainippon Pharmaceutical Company in Osaka introduced an extra hetero-nitrogen atom to produce enoxacin, which by good fortune provided better oral bioavailability than that found with norfloxacin.[88] Ofloxacin is another quinolone developed in Japan, having been synthesised by the Daiichi Seiyaku Company in Tokyo.[89] It had a spectrum of activity and absorption profile comparable to that of enoxacin, but was resistant to metabolic oxidation in the liver. It is mainly used in urinary tract infections, gonorrhoea and respiratory infections. The *S*(−) stereoisomer is also available as levofloxacin. The still more potent quinolone ciprofloxacin was introduced by Bayer AG in 1987.[90] Several more fluoroquinolones have been developed since then, including grepafloxacin, alatrofloxacin, sparfloxacin and trovafloxacin.

Antituberculous Drugs

In 1938, Arnold Rich and Richard Follis at Johns Hopkins Hospital reported that sulfanilamide had weak activity in animals infected with *Mycobacterium tuberculosis*.[91] The following year, Gerhardt Domagk found that sulfathiazole and the related sulfadithiazole were much more effective. Domagk also screened the thiosemicarbazides used in the synthesis of the sulfadithiazoles. To his surprise, they were more active than the sulfonamides, benzaldehyde thiosemicarbazone being particularly so. A series of thiosemicarbazones were then prepared[92] for Domagk to test, from which thiacetazone emerged as a potential therapeutic agent.[93] Clinical trials took place in Germany at the end of the war during an epidemic of tuberculosis, but were inadequately organised and led to unrealistic claims being made. American physicians subsequently discovered that thiacetazone was too toxic to the liver for routine therapeutic application.

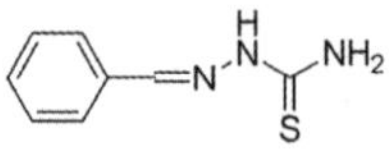

benzaldehyde thiosemicarbazone 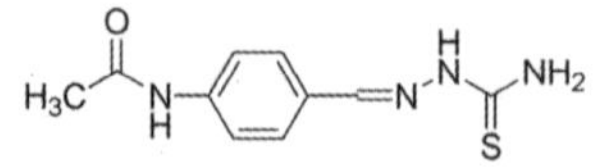   thiacetazone

When researchers at Lederle Laboratories reported in 1948 that nicotinamide had mild tuberculostatic activity,[94] it caught the attention of both Domagk in Germany as well as Robert Schnitzer of Hoffmann–La Roche laboratories in Nutley, New Jersey, where Hyman Fox had made several pyridine derivatives with antituberculous activity. Both men immediately recognized the possibility of combining a nicotinamide residue with a thiosemicarbazone to form isonicotinaldehyde thiosemicarbazone. Each of them subsequently discovered that the chemical intermediate used in the synthesis of the thiosemicarbazone was itself a highly potent antituberculous drug.[95,96] It was tested in New York hospitals and quickly became established as the most valuable antituberculous drug ever discovered. It received the approved name of isoniazid and remains in use as an essential component of the combinations of drugs used to treat this deadly disease, which has reappeared with renewed virulence in recent years.

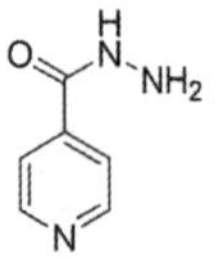

isoniazid 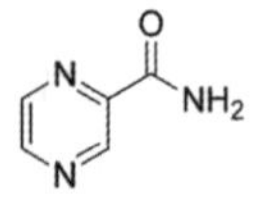   pyrazinamide 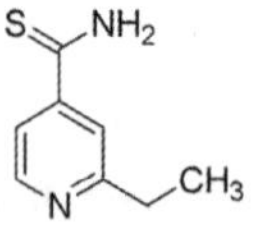   ethionamide

Shortly after the announcement of the discovery of the activity of isoniazid, both Kushner[97] of Lederle Laboratories and Solotorovsky[98] of Merck and Company simultaneously reported the antituberculosis properties of the nicotinamide analogue known as pyrazinamide. This remains an important drug because of its efficiency in meningeal penetration in cases of meningeal tuberculosis. The Theraplix company in Paris subsequently introduced ethionamide, but it is now rarely used.[99]

Antidepressant Drugs

When a clinical trial on an analogue of isoniazid developed by Hoffmann–LaRoche was conducted at Sea View Hospital on Statten Island, there was concern about its side effects, particularly central nervous system stimulation.[100] This particular property of the new drug, iproniazid, was then pursued elsewhere. At a meeting of the American Psychiatric Association in Syracuse, New York, in April 1957 there were several reports of the value of iproniazid in depression, including one from a group of psychiatrists led by Nathan Kline of Rockland State Hospital, Orangeburg, New York. He presented the meeting with results that revealed iproniazid to have been the first drug of value in chronically depressed psychotic patients.[101] Kline had begun to study the effects of iproniazid after being shown the results of animal experiments carried out by Charles Scott of the Warner–Lambert Research Laboratories in New Jersey. At that time it was known that reserpine caused brain cells to liberate 5-HT and noradrenaline. It was suspected that the tranquillising effect might be due to the release of the former, so Scott administered iproniazid to inhibit the enzymatic destruction of this substance. That iproniazid could inhibit the enzyme, monoamine oxidase, had been known since 1952.[102]

To Scott's surprise, pre-medication of animals with iproniazid before administration of reserpine caused stimulation, rather than the expected tranquillisation. When Kline saw the results of the animal studies, he carried out similar studies on humans. He then found that iproniazid on its own could stimulate depressed patients.

Since iproniazid had already been marketed as an antituberculous drug, psychiatrists were able to obtain supplies as soon as they heard of its antidepressant properties. More than 400 000 patients received it for depression during the year after the Syracuse conference, but Hoffmann–LaRoche withdrew iproniazid from the American market after a number of cases of jaundice had been reported. They replaced it with isocarboxazid, which was a more potent monoamine oxidase inhibitor.[103] This remains in use for treatment of patients who have not responded to other antidepressants. As it inhibits all types of monoamine oxidases, isocarboxazid may produce life-threatening hypertension after eating foods and wines rich in tyramine, a pressor amine. Warner–Lambert's phenelzine had similar properties to isocarboxazid.[104]

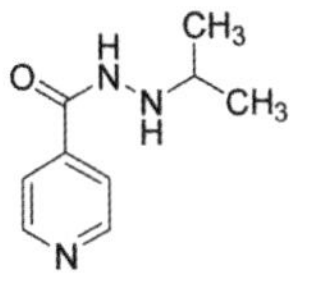

iproniazid 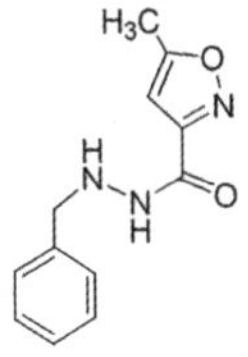   isocarboxazid 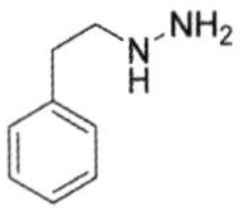   phenelzine

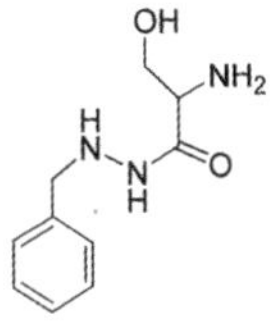

benserazide 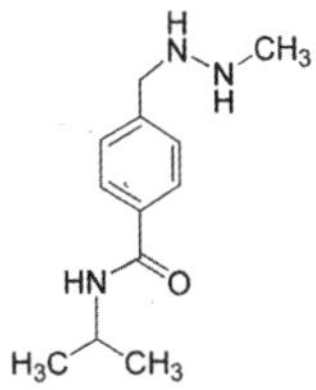   procarbazine

When the first attempts to treat Parkinson's disease with levodopa were made, it was realised that much of the drug was metabolised before it could reach the brain. Alfred Pletscher and his colleagues at the Hoffmann–LaRoche laboratories in Switzerland then investigated the possibility of finding an inhibitor of the enzyme responsible, DOPA decarboxylase.[105] They found that benserazide, a compound synthesised as a potential monoamine oxidase inhibitor and which did not enter the brain, was capable of inhibiting extra-cerebral DOPA decarboxylase to bring about a large reduction in the dose of levodopa required by patients. A combined formulation of both drugs was then marketed.

The liver toxicity following prolonged therapy with iproniazid ensured that Hoffmann–LaRoche investigators in Basle would examine its potential successors very thoroughly. This led to the discovery that 1-methyl-2-benzylhydrazine had a pronounced tumour inhibitory effect. Screening of several hundred of its analogues that had been synthesised as potential antidepressants revealed that 40 methylhydrazines were active antitumour agents. Two of them were selected for extended biological and clinical trials. In 1963, these revealed the value of procarbazine.[106] It was subsequently used in combination with mustine, vincristine and prednisolone in the 'MOPP' regimen (i.e. mustine + vincristine [Oncovin®] + procarbazine + prednisone), which transformed the prospects for survival of patients with advanced Hodgkin's disease.

Inhibitors of Gastric Acid Release

At the Hassle division of Astra, an analogue of ethionamide known as 'pyridylthioacetamide' was found to inhibit gastric acid secretion when it was routinely included in a screening test using dogs with chronic gastric fistulas.

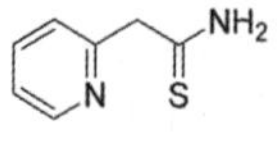

pyridylthioacetamide

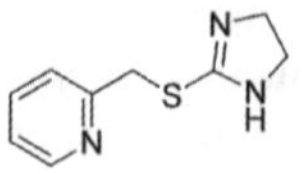

H 77/67

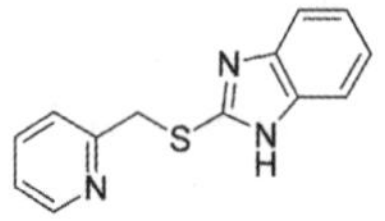

H 124/26

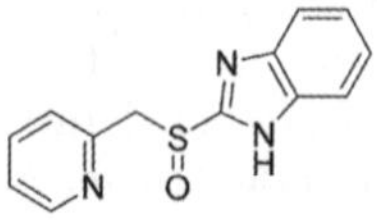

timoprazole

Because of the toxicity known to be associated with a thioacetamide function, other sulfur compounds were examined. This led to the discovery of antisecretory activity in H 77/67. Structural variants of this were prepared, culminating in the discovery of a potent antisecretory benzimidazole, compound H 124/26, just over one year later. This underwent metabolism to the more potent sulfoxide, subsequently called timoprazole. When this was submitted to chronic toxicity testing it was found to prevent uptake of iodine by the thyroid, so further analogues had to be prepared and tested for both antisecretory and antithyroid activity. Picoprazole was synthesised in 1977 and found to be safe and effective. More potent analogues were then obtained by increasing the pK_a of the pyridine ring by placing electron-donating groups on to it. One of the promising results of this was H 159/69, but this ester proved too unstable for clinical use. Modification of it led to omeprazole, in 1979.[107,108] This proved to be a safe, potent inhibitor of gastric acid secretion and soon rivalled cimetidine and ranitidine in the marketplace.

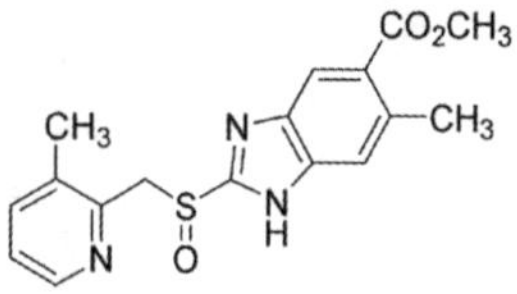

picoprazole

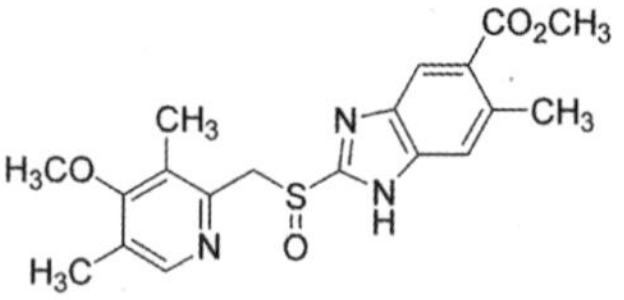

H 159/69

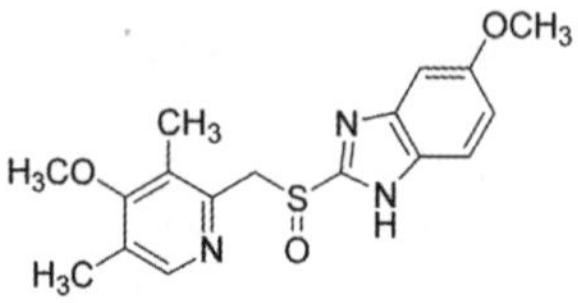

omeprazole

Aminosalicylates for Bowel Disorders

At the Karolinska Institute in Stockholm, Nanna Svartz began experimenting on the treatment of rheumatoid arthritis in 1938. Following the introduction of sulfapyridine, he experimented with it to eradicate a postulated diplostreptococcal infection which he believed was a causative factor in both rheumatoid arthritis and ulcerative colitis. When no improvement occurred in his patients, he tried to combine sulfapyridine with salicylic acid to form a drug that would possess the ability of the latter to target the connective tissue while possessing antistreptococcal activity. Failing to achieve a successful synthesis, he turned to the Pharmacia Company for assistance. The company arranged for Willstedt, their chemical consultant from the University of Stockholm, to prepare four test compounds for Svartz. One of these was sulfasalazine, which was administered to a patient suffering from ulcerative colitis.[109] Within days, the patient became symptom free and the diarrhoea ceased. More patients received the treatment and those with ulcerative colitis responded well, but only some of the arthritic patients improved.[110] Subsequent studies confirmed that sulfasalazine concentrated in connective tissue and in the lumen of the intestine. Bacterial azoreductase enzymes in the colon then metabolised it to release mesalazine, the active anti-inflammatory agent.[111] Sulfasalazine became the drug of choice for treatment and prophylaxis of ulcerative colitis, but newer analogues have proved to be safer by avoiding side effects from the sulfapyridine moiety, which has no role to play as streptococci are not involved in the aetiology of ulcerative colitis. One of these is mesalazine, the active metabolite of sulfasalazine.[112]

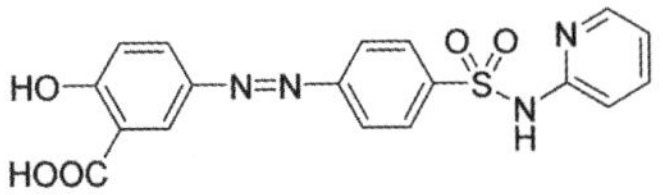

Olsalazine is a dimer of mesalazine that was once used as a dye called 'Mordant Yellow'. Pharmacia patented it in 1981 for use in ulcerative colitis. It is activated in the colon in the same manner as sulfasalazine.[113]

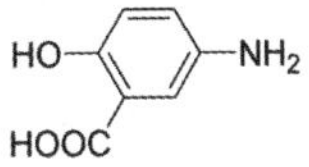

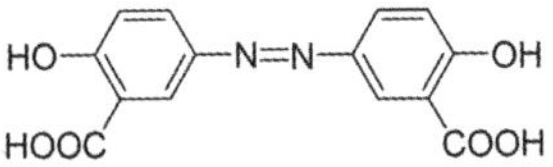

REFERENCES

1. W.F. Gleichen-Russworm, *Auserlesene mikroskopische Entdeckungen bey den Pflanzen, Blumen und Blüthen, Insekten und andern Merkwürdigkeiten.* Nurnberg: Winterschmidt; 1777.
2. P. Ehrlich, *Das Sauerstoff-Bedurfniss des Organismus: eine farbenanalytische Studie.* Berlin: A. Hirschwald; 1885.
3. P. Ehrlich, A. Leppmann, Ueber schmerzstillende Wirkung des Methylenblau. *Deut. Med. Wochenschr.*, 1890; **16**: 493–4.
4. H.H. Dale, in *The Collected Papers of Paul Ehrlich*, Vol. 3, ed. F. Himmelweit, London: Pergamon; 1960, p. 1.
5. P. Ehrlich, P. Guttman, Ueber die Wirkung des Methylenblau bei Malaria. *Berl. Klin. Wochenschr.*, 1891; **28**: 953–6.
6. J.L. Vennerstrom, M.T. Makler, C.K. Angerhofer, J.A. Williams, Antimalarial dyes revisited: xanthenes, azines, oxazines, and thiazines. *Antimicrob. Agents Chemother.*, 1995; **39**: 2671–7.
7. P. Ehrlich, Chemotherapeutische Trypansomen-Studien. *Berl. Klin. Wochenschr.*, 1907; **44**: 233–6.
8. M. Nicolle, F. Mesnil, Traitement des trypanosomiasies par les couleurs de benzidine. *Ann. Inst. Pasteur*, 1906; **20**: 417.
9. G.H.F. Nuttal, S. Hadwen, *Parasitol.*, 1909; **2**: 156, 236.
10. J. Dressel, R.E. Oesper, The discovery of Germanin by Oskar Dressel and Richard Kothe. *J. Chem. Educ.*, 1961; **38**: 620–1.
11. E. Fourneau, Représentation cinématographique d'une synthèse chimique (série du '205' Bayer). *Bull. Acad. Nat. Méd. (Paris)*, 1924; **91**: 458–61.
12. E. Fourneau, J. Tréfouel, J. Valée, Recherches de chimiothérapie. *Ann. Inst. Pasteur*, 1924; **38**: 81–4.
13. W. Schulemann, Synthetic anti-malarial preparations. *Proc. Roy. Soc. Med.*, 1932; **25**: 897–905.
14. W. Schulemann, F. Schönhöfer, A. Wingler, *Klin. Wochenschr.*, 1932; **11**: 381.
15. G.M. Robinson, Attempts to find new anti-malarials. Introduction by George Barger and Robert Robinson. Part I. Some pyrroloquinoline derivatives. *J. Chem. Soc.*, 1929; 2947–51.
16. L. Benda, Ueber das 3,6-Diamino-acridin. *Ber.*, 1912; **45**: 1787–99.
17. K. Shiga, *Z. Immun. Forsch.*, 1913; **18**: 144.
18. C.H. Browning, Exerpts from the history of chemotherapy. *Scot. Med. J.*, 1967; **12**: 310–13.
19. C.H. Browning, R. Gulbransen, E.L. Kennaway, L.H.D. Thornton, Flavine and brilliant green. Powerful antiseptics with low toxicity to the tissues: their use in infected wounds. *Br. Med. J.*, 1917; **1**: 73–6.
20. M. Gaillot, *Quart. J. Pharm. Pharmacol.*, 1934; **7**: 63.
21. R. Schnitzer, Chemotherapy of bacterial infections. *Ann. NY Acad. Sci.*, 1954; **59**: 227–42.
22. W. Kikuth, [Further development of synthetic anti-malarial drugs. I. The chemotherapeutic action of Atebrin]. *Deut. Med. Wochenschr.*, 1932; **58**: 520, 530–1.

23. F. Page, Treatment of lupus erythematosus with mepacrine. *Lancet*, 1951; *ii*: 755–8.
24. Ger. Pat. 1939: 683692 (to I.G. Farbenendustrie).
25. R.C. Eldenfield, W.J. Gensler, J.D. Head, Alkylaminoalkyl derivatives of 8-aminoquinoline. *Am. Chem. Soc.*, 1946, **68**: 1524–9.
26. P. Eisenberg, *Zbl. Bakteriol., Parasiten. u. Infektkr.*, 1913; **71**:, 420.
27. I. Ostromislensky, Note on bacteriostatic azo compounds. *J. Am. Chem. Soc.*, 1934; **56**: 1713–14.
28. Ger. Pat. 1935: 607537 (to I.G. Farbenindustrie).
29. R. Foerster, *Zentrallbl. Haut Geschlechtskr.*, 1933; **45**: 549.
30. G. Domagk, Ein Beitrag zur Chemotherapie der Bakteriellen Infektionen. *Deut. Med. Wochenschr.*, 1935; **61**: 250–8.
31. H. Hörlein, *Proc. Roy. Soc. Med.*, 1936; **29**: 313.
32. L. Colebrook, M. Kenny, Treatment of human puerperal infections and of experimental infections due to haemolytic streptococci in mice, with Prontosil. *Lancet*, 1936; *ii*: 1310.
33. L. Colebrook, Gerhard Domagk, 1895–1964. *Biographical Memoirs of Fellows of the Royal Society*, 1964; **10**: 39.
34. Ger. Pat., 1936: 638701 (to I.G. Farbenindustrie).
35. J. Trefouel, T. Trefouel, F. Nitti, D. Bovet, Activité du *p*-aminophénylsulfonamide sur les infections steptococciques expérimentales de la souris et du lapin. *C.R. Soc. Biol.*, 1935; **120**: 756–8.
36. P. Gelmo, *J. Prakt. Chem.*, 1908; **77**: 369.
37. A.T. Fuller, Is *p*-Amino benzene sulfonamide the active agent in Prontosil therapy? *Lancet*, 1937; *i*: 194.
38. H. Hörlein, *Practioner*, 1937; **139**: 635.
39. E.M.K. Geiling, P.R. Cannon, Pathological effects of elixir of sulphanilamide (diethylene glycol) poisoning. *J.Am. Med. Assoc.*, 1938; **111**: 919–26.
40. M. Dohrn, P. Diedrich, Albucid, a new sulfanilic acid derivative. *Münch. Med. Wochenschr.*, 1938; **85**: 2017–18.
41. Br. Pat. 1939: 512145 (to May & Baker).
42. L.E.H. Whitby, Chemotherapy of pneumococcal and other infections with 2-(*p*-aminobenzenesulphonamido)pyridine. *Lancet*, 1938; *i*: 1210–12.
43. R.O. Roblin, J.H. Williams, P.S. Winner, J.P. English, Chemotherapy. II. Some sulfanilamido heterocycles. *J. Am. Chem. Soc.*, 1940; **62**: 2002–5.
44. W.T. Caldwell, E.C. Kornfeld, C.K. Donnelly, Substituted 2-sulfanilamidopyrimidines. *J. Am. Chem. Soc.*, 1941; **63**: 2188–90.
45. G.A. Buttle, D. Stephenson, Smith, *et al.*, The treatment of streptococcal infections in mice with 4:4'diaminodiphenylsulfone. *Lancet*, 1937; *i*: 1331–4.
46. E. Fromm, J. Wittmann, Derivate des *p*-nitrophenols. *Ber.*, 1908; **41**: 2264–73.
47. W.H. Feldman, H.C. Hinshaw, H.E. Moses, Effect of promin (sodium salt of *p,p*-diaminodiphenyl-sulfone-*N*-didextrose sulfate) on experimental tuberculosis: preliminary report. *Proc. Staff Meet. Mayo Clin.*, 1940; **15**: 695–9.
48. E.V. Cowdry, C. Ruangsiri, Influence of promin, starch and heptaldehyde on experimental leprosy in rats. *Arch. Pathol.*, 1941; **32**: 632–40.
49. H. Bauer, Organic compounds in chemotherapy. II. The preparation of formaldehyde sulfoxylate derivatives of sulfanilamide and of amino compounds. *J. Am. Chem. Soc.*, 1939; **61**: 617–18.
50. G.H. Faget, R.C. Pogge, F.A. Johansen, *et al.*, The promin treatment of leprosy. *US Public Health Rep.*, 1943; **58**: 1729–41.
51. G. Wozel, The story of sulfones in tropical medicine and dermatology. *Int. J. Dermatol.*, 1989; **28**: 17–21.
52. R.G. Cochrane, The chemotherapy of leprosy. *Br. Med. J.*, 1952; **2**: 1220–3.
53. T. Mann, D. Keilin, Sulphanilamide as a specific inhibitor of carbonic anhydrase. *Nature*, 1940; **146**: 164–8.
54. H.W. Davenport, A.E. Wilhelmi, *Proc. Soc. Exp. Biol. Med.*, 1941; **48**:, 53.
55. R. Höber, Effect of some sulfonamides on renal excretion. *Proc. Soc. Exp. Biol. Med.*, 1942; **49**: 87–90.
56. R.F. Pitts, R.S. Alexander, The nature of the renal tubular mechanism for acidifying the urine. *Am. J. Physiol.*, 1945; **144**: 239–42.
57. H. Davenport, The inhibition of carbonic anhydrase by thiophene-2-sulfonamide and sulfanilamide. *J. Biol. Chem.*, 1945; **158**: 567–71.
58. W.B. Schwartz, The effect of sulphanilamide on salt and water excretion in congestive heart failure. *N. Engl. J. Med.*, 1949; **240**: 173.
59. R.O. Roblin, J.W. Clapp, The preparation of heterocyclic sulfonamides. *J. Am. Chem. Soc.*, 1950; **72**: 4890–2.

60. K.H. Beyer, Factors basic to the development of useful inhibitors of renal transport mechanisms. *Arch. Int. Pharmacodyn.*, 1954; **98**: 97–117.

61. H.A. Krebs, Inhibition of carbonic anhydrase by sulfonamides. *Biochem. J.*, 1948; **43**: 525–8.

62. J.M. Sprague, The chemistry of diuretics. *Ann. N.Y. Acad. Sci.*, 1958; **71**: 328–43.

63. S.J.C. Olivier, *Rec. Trav. Chim.*, 1918; **37**: 307.

64. US Pat. 1958: 2835702 (to Merck & Co.).

65. A.P. Crosley, R.E. Cullen, *Fed. Proc.*, 1957; **16**: 289.

66. F.C. Novello, J.M. Sprague, Benzothiadiazine dioxides as novel diuretics. *J. Am. Chem. Soc.*, 1957; **79**: 2028–9.

67. G. De Stevens, L.H. Werner, A. Holamandaris, S. Ricca, Dihydrobenzothiadiazine dioxides with potent diuretic effect. *Experientia*, 1958; **14**: 463.

68. C.T. Holdrege, R.B. Babel, L.C. Cheney, Synthesis of trifluoromethylated compounds possessing diuretic activity. *J. Am. Chem. Soc.*, 1959; **81**: 4807–10.

69. W.J. Close, L.R. Swett, L.E. Brady, J.H. Short, Synthesis of potential diuretic agents. I. Derivatives of 7-sulfamyl-3,4-dihydro-1,2,4-benzothiadiazine 1,1-dioxide. *J. Am. Chem. Soc.*, 1960; **82**: 1132–5.

70. H.L. Yale, K. Losee, J. Bernstein, 6-(Trifluoromethyl)-1,2,4-benzothiadiazine-7-sulfonamide-1,1-dioxide and related compounds. *J. Am. Chem. Soc.*, 1960; **82**: 2042–6.

71. F.C. Novello, S.C. Bell, E.L.A. Abrams, *et al.*, Diuretics: 1,2,4-benzothiadiazine-1,1-dioxides. *J. Org. Chem.*, 1960; **25**: 970–81.

72. US Pat. 1962: 3 058 882 (to Hoechst).

73. A.O. Robson, D.N. Kerr, R. Ashcroft, G. Teasdale, The diuretic response to frusemide. *Lancet*, 1964; **ii**: 1085–8.

74. P.W. Feit, Aminobenzoic acid diuretics. 2. 4-Substituted-3-amino-5-sulfamoylbenzoic acid derivatives. *J. Med. Chem.*, 1971; **14**: 432–9.

75. W. Merkl, D. Bormann, D. Mania, *et al.*, *Eur. J. Med. Chem.*, 1976; **11**: 399.

76. M. Janbon, Accidents hypoglycémiques graves par un sulfamidothiodiazol (le VK 57 ou 2254 RP). *Montpellier Méd.*, 1942; **441**: 21–2.

77. A. Loubatieres, Analyse du mechanism de l'action hypoglycémiante du *p*-aminobenzène-sulfamido-isopropyl-thiodazol. *C.R. Acad. Sci.*, 1944; **138**: 766.

78. H. Franke, J. Fuchs, [A new anti-diabetes principle; results of clinical research]. *Deut. Med. Wochenschr.*, 1955; **80**: 1449–52.

79. G. Erhart, *Naturwissen.*, 1956; **43**: 93.

80. F.J. Marshall, M.V. Sigal Jr, Notes – some *N*-arylsulfonyl-*N'*-alkylureas. *J. Org. Chem.*, 1958; **23**: 927–9.

81. W. Aumuller, A. Boender, R. Heerdt, *et al.*, [A new highly-active oral antidiabetic]. *Arzneimittel.-Forsch.*, 1966; **16**: 1640–1.

82. V. Ambrogi, K. Bloch, S. Daturi, *et al.*, New oral antidiabetic drugs. *Arzneimittel.-Forsch.*, 1971; **21**: 200.

83. F.H.S. Curd, F. Rose, Synthetic antimalarials. Part X. Some aryl-diguanide ('biguanide') derivatives. *J. Chem. Soc.*, 1946: 729–37.

84. G.Y. Lescher, E.D. Froelich, M.D. Gruet, *et al.*, 1,8-Naphthyridine derivatives. A new class of chemotherapeutic agents. *J. Med. Pharm. Chem.*, 1962; **5**: 1063–8.

85. A.M. Emmerson, A.M. Jones, The quinolones: decades of development and use. *J. Antimicrob. Chemother.*, 2003; **51** (Suppl S1): 13–20.

86. D. Kaminsky, R.I. Meltzer, Quinoline antibacterial agents. Oxolinic acid and related compounds. *J. Med. Chem.*, 1968; **11**: 160–3.

87. H. Koga, A. Itoh, S. Murayama, *et al.*, Structure activity relationships of antibacterial 6,7- and 7,8-disubstituted 1-alkyl-1,4-dihydro-4-oxoquinoline-3-carboxylic acids. *J. Med. Chem.*, 1980; **23**: 1358–63.

88. J. Matsumoto, T. Miyamoto, A. Minamida, *et al.*, Pyridonecarboxylic acids as antibacterial agents. 2. Synthesis and structure–activity relationships of 1,6,7-trisubstituted 1,4-dihydro-4-oxo-1,8-naphthyridine-3-carboxylic acids, including enoxacin, a new antibacterial agent. *J. Med. Chem.*, 1984; **27**: 292–301.

89. K. Sato, Y. Matsuura, M. Inoue, *et al.*, *In vitro* and *in vivo* activity of DL-8280, a new oxazine derivative. *Antimicrob. Agents Chemother.*, 1982; **22**: 548–53.

90. Ger. Pat. 1982: 3033157 (to Bayer).

91. A.R. Rich, R.H. Follis, The inhibitory effect of sulphonamide on the development of experimental tuberculosis of the guinea pig. *Bull. Johns Hopkins Hosp.*, 1938; **62**: 77–84.

92. R. Behnisch, F. Mieztsch, H. Schmidt, *et al.*, Chemical studies on thiosemicarbazones with particular reference to antituberculosis activity. *Am. Rev. Tuberculosis*, 1950; **61**: 1–7.

93. G. Domagk, R. Behnisch, F. Mieztsch, H. Schmidt, Ueber eine neue, gegen Tuberkelbazillen *in vitro* wirksame Verbindungsklasse. *Naturwissen.*, 1946; **33**: 315.

94. D. McKenzie, L. Malone, S. Kushner, J.J. Oleson, Y. Subbarow, The effect of nicotinic acid amide on experimental tuberculosis of white mice. *J. Lab. Clin. Med.*, 1948; **33**: 1249–53.
95. H.H.Fox, The chemical approach to the control of tuberculosis. *Science*, 1952; **116**: 129–34.
96. H.A. Offe, W. Siekfen, G. Domagk, *Z. Naturforschr.*, 1952; **7B**: 446.
97. S. Kushner, H. Dalalian, J.N. Sanjurio, *et al.*,Chemotherapy of tuberculosis. II. The synthesis of pyrazinamides and related compounds. *J. Am. Chem. Soc.*, 1952; **74**: 3617–21.
98. E.F. Rogers, W.J. Leanza, H.J. Becker, *et al.*, *Science*, 1952; **116**: 253.
99. D. Liebermann, M. Moyeux, N. Rist, F. Grumbach, *C.R. Acad. Sci.*, 1956; **242**: 2409.
100. H.H.Fox, J. Gibas, Synthetic tuberculostats. VII. Monoalkyl derivatives of isonicotinyhydrazine. *J. Org. Chem.*, 1953; **18**: 994–1002.
101. N.S. Kline, in *Discoveries in Biological Psychiatry*, eds F.J. Ayd, B. Blackwell. Philadelphia: Lippincott; 1970, p. 194.
102. E.A. Zeller, A.J. Barsky, *In vivo* inhibition of liver and brain monoamine oxidase by 1-isonicotinyl-2-isopropyl hydrazine. *Proc. Soc. Exp. Biol. Med.*, 1952; **81**: 459–61.
103. T.S. Gardner, E. Wenis, J. Lee, Monoamine oxidase inhibitors – I. 1-Alkyl and 1-aralkyl-2-(picolinoyl and 5-methyl-3-isoxazolylcarbonyl)hydrazines. *J. Med. Pharm. Chem.*, 1960; **2**: 133–45.
104. J.H. Biel, A.E. Drukker, T.F. Mitchell, *et al.*, Central stimulants. Chemistry and structure–activity relationship of aralkyl hydrazines. *J. Am. Chem. Soc.*, 1959; **81**: 2805–13.
105. W.P. Burkard, K.F. Gey, A. Pletscher, A new inhibitor of decarboxylase of aromatic amino acids. *Experientia*, 1962; **18**: 411–12.
106. Belg. Pat. 1962: 618638 (to Hoffmann–La Roche).
107. A. Brändström, P. Lindberg, U. Junggren, Structure activity relationships of substituted benzimidazoles. *Scand. J. Gastroenterol. Suppl.*, 1985; **108**: 15–22.
108. P. Lindberg, A. Brändström, B. Wallmark, *et al.*, Omeprazole: the first proton pump inhibitor. *Med. Res. Rev.*, 1990; **10**: 1–54.
109. N. Svartz, Salazopyrin: a new sulphanilamide preparation. *Acta Med. Scand.*, 1942; **110**: 577–96.
110. N. Svartz, The treatment of 124 cases of ulcerative colitis with salazopyrine and attempts at desensitization in cases of hypersensitivity to sulfa. *Acta Med. Scand.*, 1948; **139** (Suppl 206): 465–72.
111. N. Svartz, Sulfasalazine: II. Some notes on the discovery and development of salazopyrin. *Am. J. Gastroenterol.*, 1988; **83**: 497–503.
112. A.K. Azad Khan, J. Piris, S.C. Truelove, An experiment to determine the active therapeutic moiety of sulphasalazine. *Lancet*, 1977; *ii*: 892–5.
113. H. Sandberg-Gertzen, J. Kjellander, B. Sundberg-Gilla, G. Jarnerot, *In vitro* effects of sulphasalazine, azodisal sodium, and their metabolites on *Clostridium difficile* and some other faecal bacteria. *Scand. J. Gastroenterol.*, 1985; **20**: 607–12.

Drugs Originating from the Screening of Organic Chemicals

The previous chapter dealt with a variety of drugs that can trace their origins to the screening of synthetic dyes for chemotherapeutic activity. Even more drugs have been discovered through the screening of other synthetic compounds.

Screening has been used both to discover drug prototypes and to find improved analogues of other compounds that exhibit useful activity but require enhancement in some manner. Examples of the latter process can be found throughout this book, but the present chapter is concerned solely with the discovery of drug prototypes through screening and their subsequent development to provide therapeutic agents.

CLASSICAL ANTIHISTAMINES

After the presence of histamine in the body had been established there was considerable interest in its physiological role. Daniel Bovet at the Pasteur Institute realised that antagonists of acetylcholine and adrenaline, e.g. atropine and ergotamine, had made it possible for physiologists to investigate and understand their actions. Since no antagonist of histamine existed, Bovet screened compounds that had previously been synthesised at the Institute and found that certain adrenaline analogues and antagonists diminished the action of histamine on the guinea pig intestine.[1] However, when the most promising of these, piperoxan, was injected into live guinea pigs it failed to protect them against the lethal effects of histamine administered by the intrajugular route.

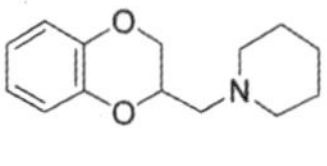

piperoxan (933F) 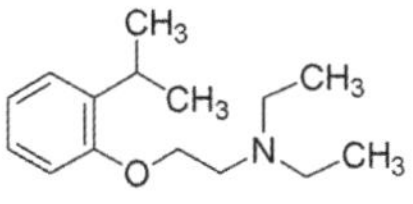   929F

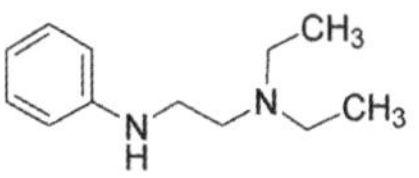

1167F 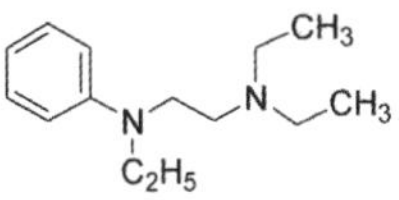   1571F

On examining similar compounds prepared at the Institute, Bovet and Anne-Marie Staub found Ernest Fourneau's compound 929F to be the most potent histamine antagonist yet tested on the guinea pig intestine.[2] When it was administered to guinea pigs, it consistently protected them against the otherwise lethal bronchoconstrictive action of histamine. Compound 1167F, in which the oxygen atom was replaced by nitrogen, was also effective.[3]

Staub investigated several more compounds and proceeded to define the molecular requirements for antihistaminic activity, with remarkable accuracy.[4] Unfortunately, none of the compounds examined were safe enough for human administration. It was not until 1941 that phenbenzamine was found suitable for clinical use after Mosnier had synthesised 24 analogues of 1571F at the Rhône–Poulenc laboratories in Paris.[5]

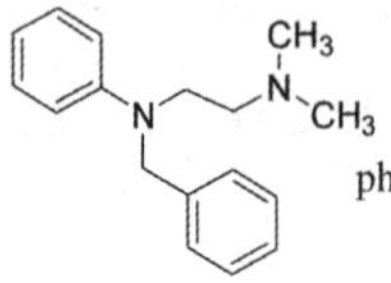

phenbenzamine

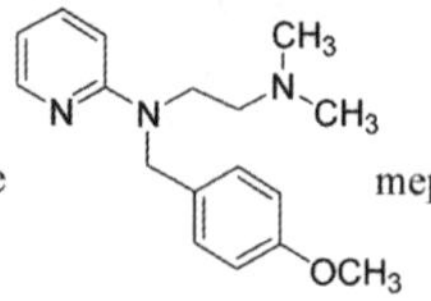

mepyramine

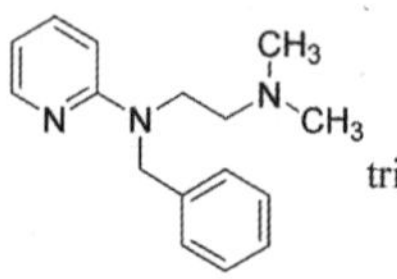

tripelennamine

Two years after the introduction of phenbenzamine, Bovet and his colleagues published their studies on the closely related mepyramine, in which a pyridine ring replaced one of the benzene rings.[6] This alteration was probably introduced because of the experience Rhône–Poulenc had acquired from their British subsidiary, May and Baker, who had developed sulfapyridine. Researchers at Ciba Pharmaceuticals in Summit, New Jersey, synthesised tripelennamine, which differed only in the absence of the methoxyl group attached to the benzene ring.[7]

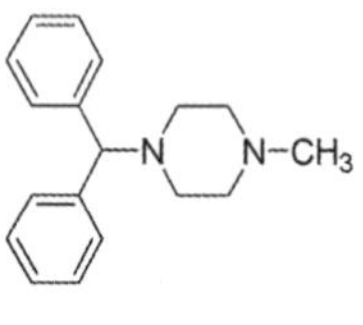

cyclizine

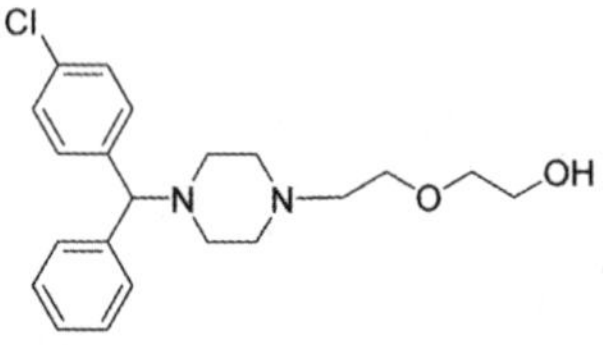

hyroxyzine

The American division of Burroughs Wellcome developed cyclizine, a long-acting antihistamine in which the amino group was derived from piperazine instead of dimethylamine.[8] In the clinic it proved to be a useful anti-emetic drug, achieving the notable distinction of being selected by the US National Aeronautic and Space Agency for use as a space sickness remedy on the first manned flight to the moon. The related hydroxyzine is an antihistamine that has also been used as a minor tranquilliser.[9]

During the Second World War, Rhône–Poulenc followed up Ehrlich's demonstration of antimalarial activity with methylene blue by investigating phenothiazines. This line of research was abandoned after negative results were obtained, but Bernard Halpern and René Ducrot realised that one of the phenothiazines synthesised by Paul Charpentier was an analogue of phenbenzamine in which its two benzene rings were bridged by a sulfur atom. When tested, it proved to be an antihistamine and was given the name 'fenethazine'.

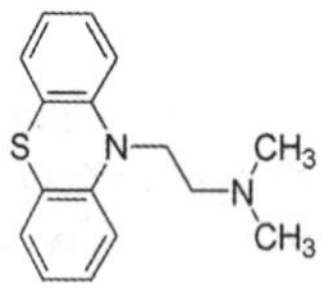

fenethazine

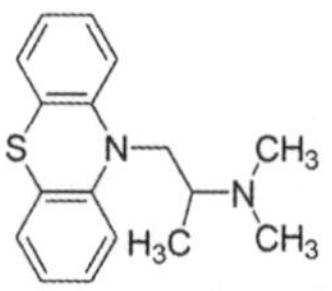

promethazine

Promethazine, a derivative of fenethazine synthesised in 1946, had an extra methyl group on the dimethylaminoethyl side chain and turned out to be highly potent and a very long-acting antihistamine.[10] Charpentier had intended to place the methyl group on the adjacent carbon atom. Had he succeeded, the compound would not have been as successful. It was marketed as an antihistamine, but its ability to cause prolonged central depression also led to its use as a non-prescription hypnotic.

Aminoalkyl Ether Antihistamines

Diphenhydramine was one of several compounds designed to be antispasmodics by George Rieveschl, an assistant professor at the University of Cincinnati. It was synthesised in 1943 by Fred Huber, one of his research students. Parke, Davis and Company tested the new compounds on the guinea pig ileum and found diphenhydramine to be a highly potent antispasmodic. Extensive testing revealed not only that it had an exceptionally high safety margin but also that it was a potent antihistamine.[11] Parke, Davis bought the patent rights from Rieveschl, granting him a 5% royalty on all sales for the next 17 years while the patent lasted. Rieveschl joined Parke, Davis and became Director of Research in 1947, in which role he was responsible for the development of the very similar antihistamine known as 'orphenadrine'.[12] Both compounds had atropine-like anticholinergic effects, which were more marked in orphenadrine and resulted in its use in the treatment of Parkinson's disease. In an attempt to develop an analogue of orphenadrine with reduced side effects, the Riker Company synthesised nefopam in 1966.[13] At first, this was thought to be a centrally acting muscle relaxant, but was later shown to be an analgesic. It is used in the management of moderate pain.

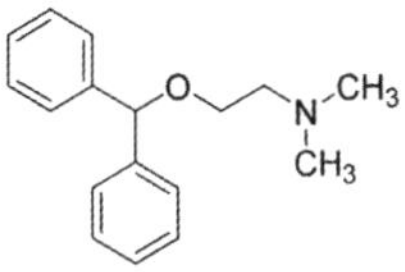

diphenhydramine 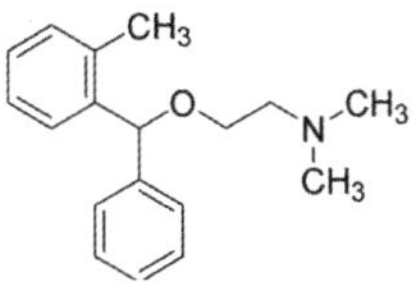   orphenadrine 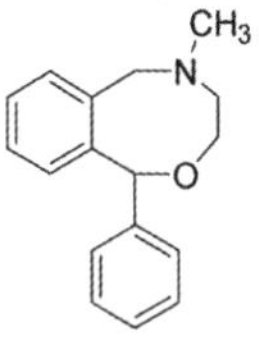   nefopam

Antihistamines became immensely popular during the late 1940s, being hailed in some quarters as miracle drugs! Although their principal use was in the control of certain conditions such as hay fever or urticaria, they were initially believed to be of value for a wide range of ailments, including the common cold. Leading pharmaceutical manufacturers competed vigorously to develop new antihistamines, resulting in the introduction of a plethora of new drugs with little to choose between them, none being free of the tendency to cause drowsiness.

G.D. Searle and Company, a family-owned Chicago pharmaceutical distributor, broke new ground by introducing a formulation of diphenhydramine designed to minimise drowsiness by formulating it with a mild stimulant, namely 8-chlorotheophylline.[14] The resulting salt, dimenhydrinate, did not prevent drowsiness, but it became one of the most profitable antihistamines on the market after it was found to have an unexpected therapeutic action. Samples had been sent to the allergy clinic at Johns Hopkins University in Baltimore for evaluation by Leslie Gay and Paul Carliner.[15] They administered dimenhydrinate to a patient suffering from urticaria. She then discovered that when travelling on a streetcar after having swallowed the drug, she was not car sick – for the first time in years! Tests on other patients who suffered from travel sickness confirmed the apparent value of dimenhydrinate. The matter was reported to Searle, who organised an ambitious clinical trial. On 27 November 1947, the *General Ballou* sailed from New York to Bremerhaven in Germany. The crossing was particularly rough, yet only 4% of the troops on board who received the drug were sick, in contrast to a quarter of the others who received a placebo. Furthermore, all but 17 of 389 sick soldiers recovered within a couple of hours of receiving dimenhydrinate. Unlike earlier remedies such as hyoscine, the only significant side effect was drowsiness. Searle quickly exploited this before their rivals began to discover that other antihistamines were also effective against motion sickness.

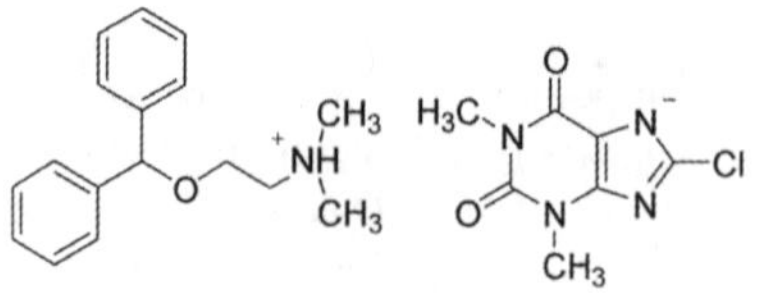

dimenhydrinate

Monoaminopropyl Antihistamines

Research workers at the Schering Corporation realised that all the potent antihistamines contained two aromatic rings joined to either a nitrogen or an oxygen atom. They therefore synthesised a new series in which these rings were instead attached to a carbon atom as this had roughly similar atomic dimensions, being an example of isosteric replacement of an atom. This led to the development of the long-acting antihistamines pheniramine, brompheniramine and chlorpheniramine in 1948, the latter two being more potent.[16]

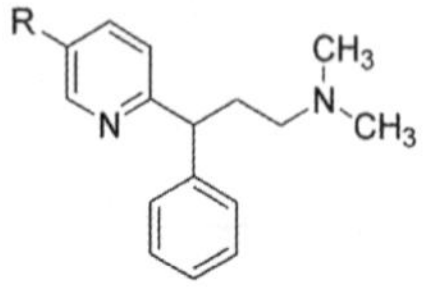

pheniramine, R = H
brompheniramine, R = Br
chlorpheniramine, R = Cl

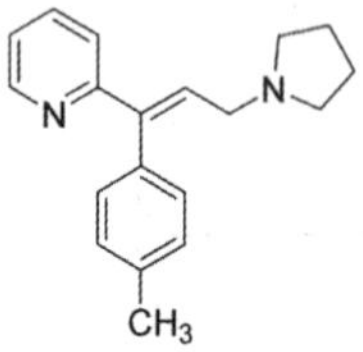

triprolidine

A similar approach was adopted around the same time, at the Wellcome Research Laboratories in England, although a few years passed before triprolidine was marketed.[17]

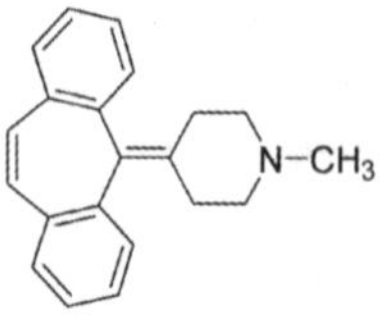

cyproheptadine 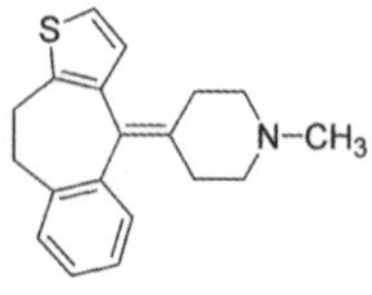   pizotifen 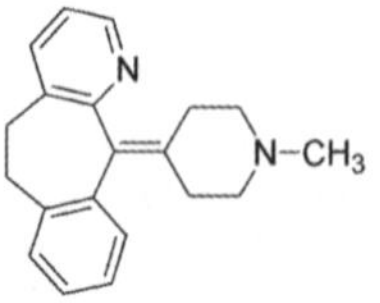   azatadine

Merck introduced a cyclic analogue of pheniramine known as cyproheptadine.[18] It had similar properties to chlorpheniramine, but was also of some value in the prophylaxis of migraine due to its ability to act as a 5-HT$_2$ antagonist. It is now reserved for refractory cases of migraine. Pizotifen proved to be better than cyproheptadine in the prophylactic treatment of migraine.[19,20]

The Schering–Plough Corporation developed azatadine as a potential non-sedating antihistamine.[21] A pre-clinical behavioural test on cats indicated an absence of central activity, but azatadine turned out to be a typical potent sedating antihistamine when administered to human volunteers.

Non-sedating Antihistamines

Richardson–Merrell chemists synthesised terfenadine in 1973 as a potential tranquilliser, but found it to be inactive as it did not enter the central nervous system. Pharmacologist Richard Kinsolving noticed that it had a resemblance to diphenhydramine and it was tested and then found to be an antihistamine that did not cause sedation.[22] Clinical trials confirmed that terfenadine was the first non-sedating antihistamine to have been discovered.

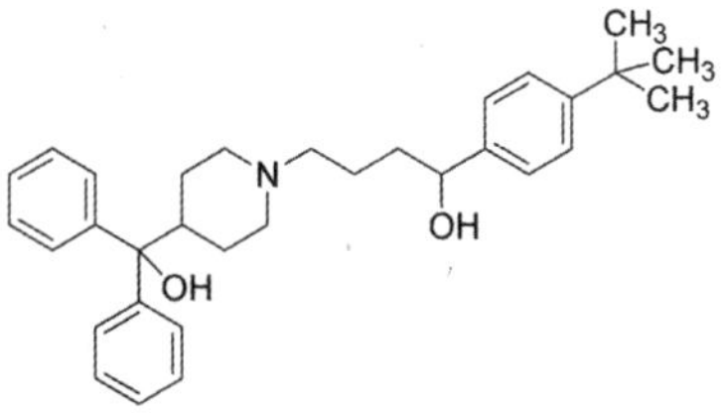

terfenadine

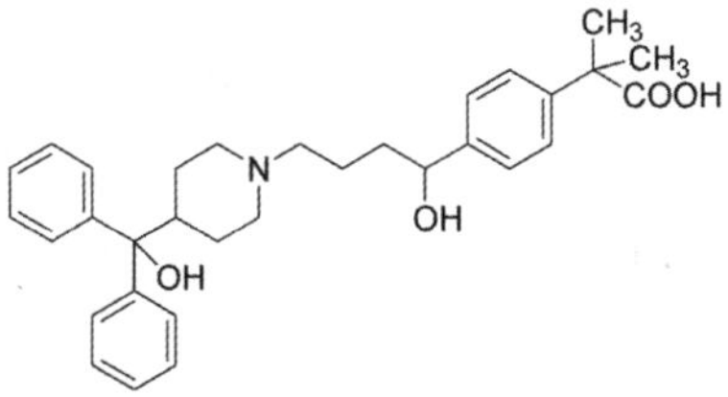

fexofenadine

In 1992, the US Food and Drug Administration issued a warning that some patients who took terfenadine might develop a life-threatening ventricular arrhythmia called 'torsades de pointes'. Use of the drug was ruled out in patients with liver disease as it was not being efficiently metabolised. This problem was overcome when the active metabolite, fexofenadine, was introduced.[23]

Despite the fact that it had taken almost 40 years to discover a non-sedating antihistamine, several more were introduced soon after the launch of terfenadine. Wellcome marketed a derivative of triprolidine known as 'acrivastine' which was developed as a non-sedating antihistamine by incorporating an ionisable side chain to reduce central nervous system penetration.[24]

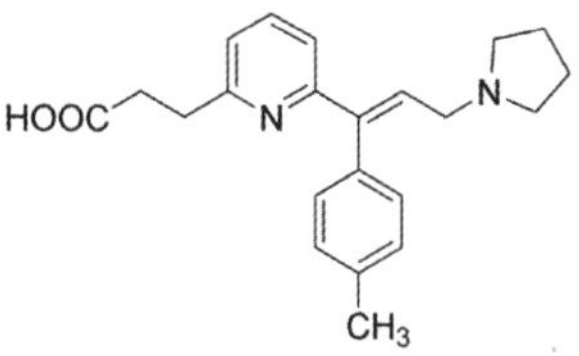

acrivastine

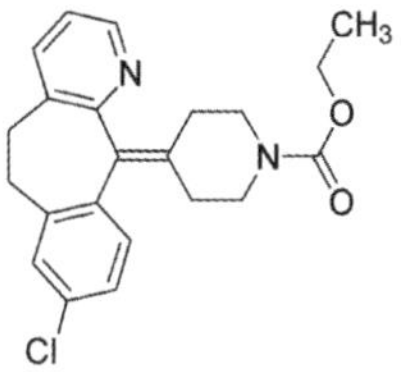

loratadine

Frank Villani at Schering–Plough synthesised potential antihistamines designed to antagonise both histamine H_1 and H_2 receptors. He hoped that these might have useful anti-ulcer properties. He began by making analogues of azatadine in which the basicity of the piperidine ring nitrogen was reduced by the formation of urea, sulfonamide and carbamate derivatives. The resulting compounds failed to exhibit histamine H_2 blocking activity. However, when the ethyl carbamate ester was later screened it had no effect on the central nervous system. Further investigation confirmed that it was a non-sedating antihistamine.[25] Attempts were then made to find a longer-acting analogue. When a chlorine atom was placed at the 8-position of the ring system to reduce oxidative metabolism, it increased the duration of action in human volunteers from under 8 hours to permit once-daily medication. Unexpectedly, this modification also increased potency by a factor of four.[26] This new 8-chloro analogue was also active when given by mouth. It received the approved name of 'loratadine', and was marketed as a non-sedating antihistamine.

Antipsychotic Agents

The phenothiazine tranquillisers were developed as a consequence of studies initiated at the Sidi Abdallah Hospital near Bizerte, Tunisia, in April 1949 by Henri Laborit, a French Navy surgeon who had been one of the first to use antihistamines to pre-medicate patients undergoing surgery. His concern about the traumatic effects of surgical shock led him to consider the possibility that antihistamines might prevent the capillary hyperpermeability caused by histamine release in patients in shock. He therefore incorporated mepyramine and promethazine in the mixture of drugs he was administering.[27] Over an 8 month period it then became apparent to Laborit that the antihistamines had unusual central actions which were contributing to the antishock action.[28] The mood of his patients had improved and, particularly in the case of promethazine, they were less anxious and required less morphine.[29] An army psychiatrist confirmed this effect of the drug, but matters rested there for the time being.

On being transferred to the Val-de-Grâce Military Hospital in Paris, Laborit took the opportunity to investigate the central effects of antihistamines in more detail. In collaboration with the anaesthetist Pierre Huguenard, he was able to show that they lowered body temperature and so reduced basal metabolism to produce a reduction in the amount of anaesthetic required during operations. This, in turn, lowered the risk of shock, so once more Laborit turned his attention to the effects of pre-medication on shock. He now experimented with a 'lytic cocktail' of drugs to cool the bodies of patients wrapped in ice bags even further.[30]

Laborit visited the manufacturer of promethazine, the Specia Laboratories of Rhône–Poulenc at Vitry-sur-Seine, near Paris, and described his work. In the autumn of 1950 they began a search for a drug that would have an action on the central nervous system that met Laborit's requirements.[31] Simone Courvoisier screened phenothiazines that Paul Charpentier had synthesised as potential antihistamines, investigating those that were previously rejected because of sedating effects. When the fenethazine analogue now known as promazine proved most interesting despite its low level of antihistaminic activity, Charpentier synthesised analogues of it.[32] One of them was chlorpromazine, a chlorinated derivative prepared in December 1950. It was passed to Courvoisier who identified its outstanding activity and low toxicity.[33]

chlorpromazine

In the spring of 1951, samples of chlorpromazine were given to Laborit. He confirmed that it was indeed the agent he had long sought. After completing appropriate animal tests, he incorporated the new drug into a 'lytic cocktail' in combination with promethazine and the fenethazine analogue known as ethazine for use on patients undergoing surgery. Before long, he observed that not only did they fare better both during and after their operations, due to the antishock action, but they also seemed relaxed and unconcerned with what was happening to them during the normally stressful pre-operative period. The significance of this was not lost on Laborit.[34] He persuaded his psychiatric colleagues at the Val-de-Grâce Hospital to test the mixture on psychotic patients. On 19 January 1952 Joseph Hamon, the Director of the Neuropsychiatric Service, assisted by Jean Paraire and Jean Velluz, began to treat a manic patient who was decidedly agitated until he was given his first injection.[35] At once, he became calm and remained so for several hours. It was recognised that the mixture of drugs was palliative rather than curative, but this did not stop the release of the patient from hospital three weeks later. However, the psychiatrists at the Val-de-Grâce Hospital did not observe the full effects of chlorpromazine as it was only one component of a mixture, with the dose duly modified to take account of the other two central depressants, promethazine and pethidine. Consequently, they soon abandoned the mixture and returned to using electroshock therapy on their patients.

On learning of the effects of the mixture containing chlorpromazine, Pierre Deniker of the Sainte Anne Hospital in Paris requested samples of chlorpromazine from Rhône–Poulenc for a detailed study of its psychopharmacological action when administered without other drugs.

He and his senior colleague Jean Delay conducted a clinical trial on 38 patients, soon confirming its outstanding value as a tranquilliser for manic, agitated and psychotic patients.[36] Chlorpromazine was found to be a relatively sedating antipsychotic drug, but unlike the then popular sedatives (i.e. central depressant drugs such as hypnotics administered in subhypnotic doses) it did not aggravate disorders of wakeful consciousness. Indeed, mental confusion was alleviated. In schizophrenic patients, chlorpromazine produced a diminution in aggressiveness, agitation and delusion. Particularly characteristic of chlorpromazine was its effect on the central control of movement whereby a type of akinesia, or psychomotor indifference, was induced in patients. This led Delay and Deniker in 1955 to introduce the term 'neuroleptic' to describe any antipsychotic drug with this effect.[37]

Chlorpromazine was marketed in France by Rhône–Poulenc in the autumn of 1952. The early observations by French psychiatrists and others became generally accepted, with the result that psychiatry was transformed and psychotic patients were released from the restraints of straight-jackets and locked wards. They were not cured, but the phenothiazines controlled their behaviour. Some critics believe that physical restraint had simply been substituted by chemical restraint. That view remains a minority one. As with all other drugs, problems occurred when insufficient care was taken with their administration. A discussion of these here is inappropriate, but considering that millions of patients have received these drugs, their record is impressive.

The remarkable success of chlorpromazine stimulated rival manufacturers to introduce analogues of it. Many of these compounds had a different substituent incorporated in place of the chlorine atom attached at position 2 of the phenothiazine ring. This was motivated not merely by a desire to circumvent the Rhône–Poulenc patents, but also by a belief that potency was influenced by the electron-withdrawing power of the substituent, a view that has not been upheld. Typical variants included acetyl, methoxyl, nitrile, trifluoromethyl, thioalkyl and dialkylsulfonamide groups. The differences between the scores of phenothiazine tranquillisers that have been introduced into the clinic are less significant than the variability in patient response to any single drug.

The Rhône–Poulenc researchers found that more potent analogues could be obtained by replacing the dimethylamine function on the side chain of chlorpromazine with a piperazine group. This increased side effects involving dopaminergic extrapyramidal pathways in the nervous system, leading to a Parkinson's disease-like tremor in some patients. This may have been a direct consequence of the reduction of anticholinergic activity arising from the use of smaller doses than for chlorpromazine. Anticholinergic drugs are actually used to treat phenothiazine-induced extrapyramidal tremor. There is, however, a benefit from the diminished anticholinergic response in patients treated with the piperazine compounds, namely that they are less sedating. For this reason they are preferred to other phenothiazines for the prevention and treatment of nausea. The first of the piperazine compounds to be marketed was prochlorperazine.[38]

prochlorperazine

fluphenazine

thioridazine

chlorprothixene

zuclopenthixol

Fluphenazine had a very similar activity to prochlorperazine, but as it had an alcohol function in the side chain, it was also formulated as either the enanthate (i.e. heptanoate) or decanoate ester in an oily depot injection.[39] This was injected every 14–28 days for the long-term control of psychotic behaviour. In contrast to fluphenazine, thioridazine had a low potency. However, its anticholinergic activity helped to counter extrapyramidal tremor and was an important advantage for elderly patients. This was to some extent offset by an increased risk of hypotension.

In 1958, Petersen and his colleagues, working with the Danish firm H. Lundbeck, published their first report on the thioxanthenes,[40,41] a new series of tranquillisers in which the nitrogen of the phenothiazine ring had been isosterically replaced by a carbon atom. As these tricyclic compounds had strong chemical similarities to the phenothiazine tranquillisers, it is hardly surprising that their therapeutic activity proved to be similar. The first member of the series to be introduced underwent a clinical trial with 70 patients in 1958 and was marketed the following year with the approved name of chlorprothixene.[42] Lundbeck introduced clopenthixol three years later after clinical evaluation had shown it to be a better antipsychotic agent than chlorprothixene. It was a mixture of *cis* and *trans* isomers. The active *cis* isomer was introduced into the clinic as zuclopenthixol.[43] The acetate ester was also developed for depot medication.

Reports of persistent abnormal facial movements among patients who had taken chlorpromazine began to be published within five years of its introduction.[44] The condition was termed 'tardive dyskinesia'. Lawsuits were brought against companies that marketed the drug. Once it was realised that related drugs could also cause tardive dyskinesia, companies were discouraged from developing new tranquillisers.

Anxiolytic Drugs

Early in 1954 at the laboratories of Hoffmann–LaRoche in Nutley, New Jersey, Leo Sternbach decided to reinvestigate some tricyclic compounds he had synthesised about 20 years earlier at the University of Cracow as part of his post-doctoral studies on dyestuffs. He had in mind the tricyclic nature of chlorpromazine, which had just been discovered, and he believed that the introduction of a basic side chain into his own compounds might create derivatives with a degree of overall similarity to it. He prepared around 40 new compounds by reacting his key intermediate, an alkyl halide, with a variety of secondary amines selected to confer structural analogy with the tricyclics then being patented. When these compounds were submitted to Lowell Randall for screening for muscle relaxant, sedative and anticonvulsant properties, they were all found to be inactive. Renewed chemical studies then revealed that the tricyclic system of the key synthetic intermediate was not that of a benzheptoxdiazine, as had been believed, but was instead a quinazoline-3-oxide. This seemed to account for the lack of biological activity in the derivatives synthesised from this intermediate. The last compound Sternbach had prepared remained untested until a year and a half later, when a colleague who was tidying up the laboratory suggested it should be sent for screening. Sternbach agreed, and a few days later Randall informed him that his compound appeared to approach the activity of chlorpromazine as a tranquilliser. Furthermore, it had a low level of acute toxicity and was free from significant side effects. This report engendered considerable excitement and raised the obvious question of why only this single compound was active. The answer was soon found when Sternbach reinvestigated its chemistry. It became clear that by using the primary amine methylamine in the last stage of the synthesis, the reaction had followed a different pathway (ring enlargement) from that undergone when secondary amines had been employed. The product formed was a benzodiazepine.[45,46] Sternbach filed a US patent application for this new tranquilliser, chlordiazepoxide, in May 1958. The initial clinical studies were conducted on 16 000 patients before it was granted approval by the US Food and Drug Administration in 1960. Thousands of benzodiazepines have been synthesised since then, of which several are still used throughout the world as anti-anxiety agents and hypnotics. While they can be of value in patients whose anxiety interferes with their work, leisure and personal relationships, the benzodiazepines have been widely misused in the treatment of the most trivial symptoms of stress. Dependence and tolerance occur after prolonged use. This has recently resulted in litigation brought by patients who have suffered from dependence on benzodiazepines.

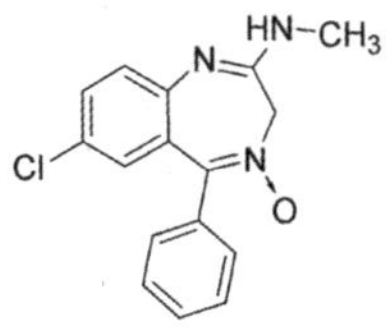

chlordiazepoxide

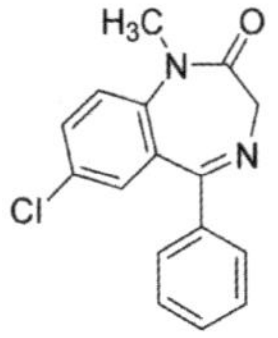

diazepam

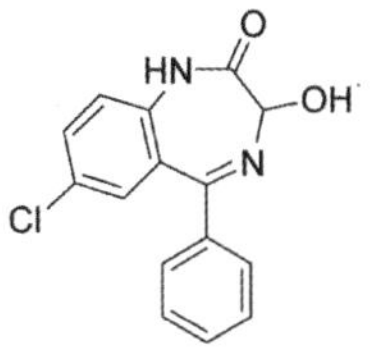

oxazepam

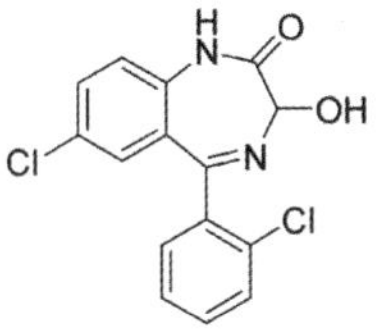

lorazepam

Because chlordiazepoxide was not designed to be a benzodiazepine, certain features of its chemical structure were superfluous, notably the basic side chain and the *N*-oxide function. Simpler analogues were found, the first of these being synthesised in 1959 and marketed four years later as diazepam.[47] It had more pronounced muscle relaxant properties than chlordiazepoxide and a half-life of one to two days as it was slowly cleared from the body.

Benzodiazepines were used for treating chronic anxiety states. Some of them, including diazepam, formed an active metabolite such as nordiazepam or something similar, which was responsible for their effects. Unfortunately, nordiazepam took from two to five days before being cleared from the body, hence its concentration gradually built up as more doses were taken. Recognition of this led to the introduction of benzodiazepines that did not form this type of active metabolite or which were rapidly eliminated. Examples of this include oxazepam,[48] itself a metabolite of diazepam, and lorazepam.[49] As these are alcohols, they are glucuronidated in the liver and quickly eliminated by the kidneys.

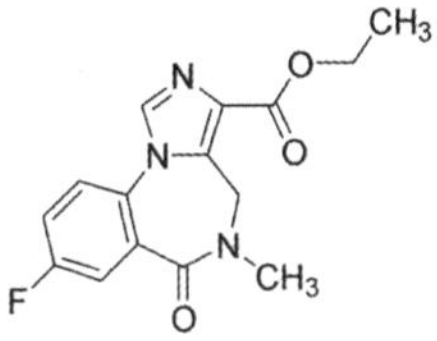

flumazenil

Flumazenil was synthesised in 1979.[50] The replacement of a phenyl group by a carbonyl removed most of the typical sedative and anxiolytic activity, but as it still fitted the benzodiazepine receptor flumazenil acted as an antagonist when a moderate dose was administered. It has been used to reverse the sedation produced by benzodiazepines, either when given as medication during short surgical procedures or in overdosage by drug abusers.

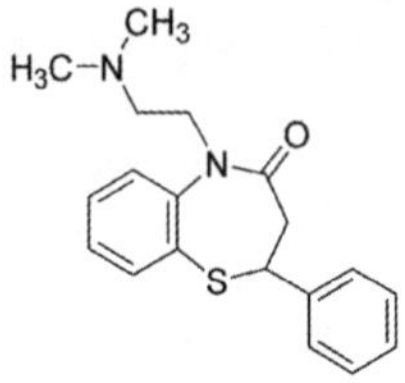

thiazesim

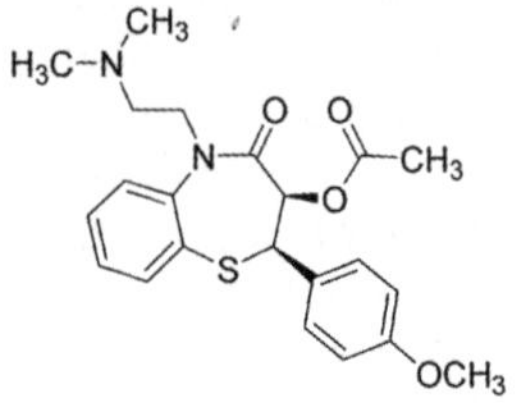

diltiazem

Seeking a product with which to enter the growing market in psychotropic drugs, the Tanabe Seiyaku Company of Japan prepared analogues of an antidepressant called thiazesim, in which a hydroxyl group or an *O*-acyl group was introduced at the 3-position of the benzothiazepine ring system. This manoeuvre had been instituted in the knowledge that the presence of a hydroxyl group in the equivalent position of diazepam had produced a more potent drug that did not accumulate in the body, e.g. oxazepam and lorazepam. When fully evaluated, these 3-substituted 1,5-benzothiazepines lacked sufficient novelty for them to be marketed. However, routine screening revealed that the 3-*O*-acyl benzothiazepines exerted a strong coronary vasodilator effect in the anaesthetised dog at dose levels that produced minimal central effects. Analogues were synthesised and it was established that introduction of a methyl or methoxy group at the 4-position of the benzene ring enhanced potency. Diltiazem was then found to have good oral absorption coupled with high efficacy and low toxicity.[51] As it was a racemic mixture, its isomers were examined. As the dextro isomer possessed all the vasodilating activity, it was selected for clinical evaluation after it was fully evaluated by pharmacologists who demonstrated that it had a papaverine-like vasodilating action on the coronary artery and antagonised calcium ion flow across cardiac muscle membrane stores.[52] The novelty of this action resulted in diltiazem being the first coronary vasodilator to be described as a 'calcium antagonist'. The anti-arrythmic activity also observed was due to the fact that there are fast and slow calcium channels, and diltiazem not only blocks calcium transport through the slow channels but also delays their recovery.

Diltiazem is now used in the treatment of angina and a slow-acting formulation is available for patients with hypertension who respond poorly to beta-blockers.

Hypnotic Benzodiazepines

Among the earliest chemical modifications effected on the benzodiazepine nucleus was the introduction of the nitro group, as this offered chemists an opportunity of subsequent structural variation. Several nitro compounds were prepared by Sternbach and his colleagues, of which nitrazepam proved to be much more potent than chlordiazepoxide in both mice and cats.[53] Subsequent investigations showed that sleep could be induced by larger doses that were well below the toxic threshold. Indeed, so wide was the margin of safety that self-poisoning with nitrazepam was most unlikely to occur. This safety factor alone ensured worldwide acceptance of this new hypnotic. However, nitrazepam had a half-life of around 26 hours and so persisted in the body long enough to cause a hangover effect when patients awakened.

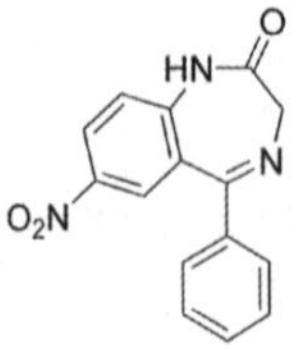

nitrazepam

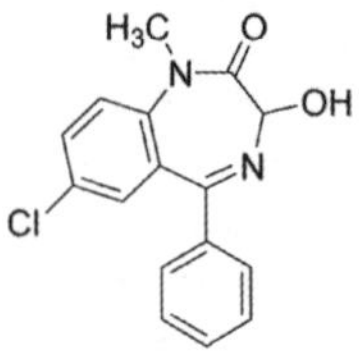

temazepam

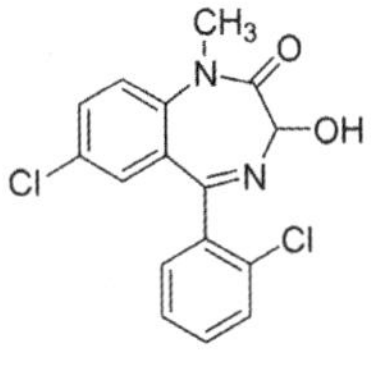

lormetazepam

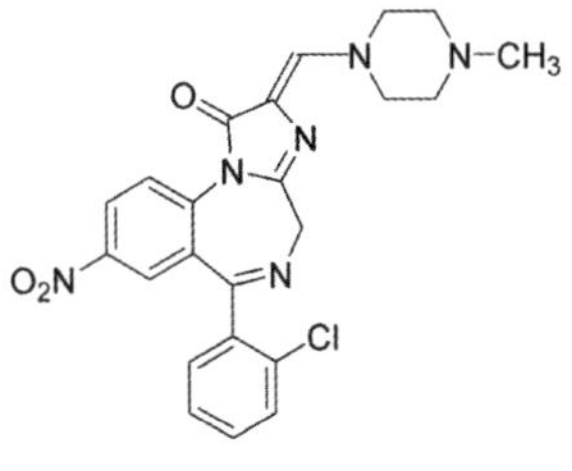

loprazolam

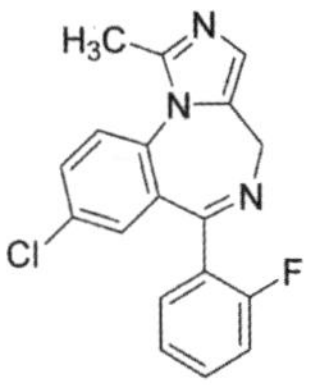

midazolam

Temazepam was developed by Wyeth Laboratories in Radnor, Pennsylvania, and became a very popular hypnotic because there was no hangover effect.[49] It was more susceptible than nitrazepam to metabolic deactivation in the liver, with a half-life of 8–10 hours, and no active metabolite formed. Unfortunately, temazepam was widely abused as an illicit recreational drug. Other short-acting hypnotics that were introduced include lormetazepam[54] and loprazolam.[55]

Midazolam was a short-acting anxiolytic agent developed by Hoffmann–LaRoche.[56] Formulated as the hydrochloride salt, it is the only water-soluble benzodiazepine available for injection. When injected it has a half-life of about 1–3 hours and produced amnesia for a period of about 10 minutes after administration. It is also given continuously by the intravenous route in order to sedate patients undergoing intensive care.

Tricyclic Antidepressants

The recognition of the tranquillising properties of chlorpromazine in the mid-1950s led psychiatrists to test it and its analogues in a variety of clinical conditions. Roland Kuhn of the Cantonal Psychiatric Clinic, Munsterlingen, Switzerland, noticed that chlorpromazine produced effects that reminded him of those he had observed when testing an antihistamine that had been sent to him by Geigy for testing as a hypnotic. On that occasion, Kuhn had suggested further studies would be worth while, but this suggestion was ignored. This time, a long letter he wrote to Geigy was taken seriously, especially as the antihistamine had a striking structural resemblance to chlorpromazine. He received further samples of the antihistamine, code-named G22150. While it was soon found to have interesting properties, it had too many side effects. Geigy then sent Kuhn samples of imipramine, an analogue of G22150 with a side chain identical to that of chlorpromazine.[57,58]

Kuhn thoroughly evaluated imipramine in a variety of psychiatric conditions. Early in 1956, it was administered to several patients suffering from endogenous depressions. After only three patients had been treated, it became clear that this new tricyclic compound had unique properties. A letter sent to Geigy at the beginning of February that year referred to the pronounced antidepressant activity of the new drug. At the Second International Congress of Psychiatry, held in Zurich 7 months later, an audience of a dozen people heard the first public disclosure of this major advance. A subsequent publication caught the attention of a wider audience.[59,60] Since then, imipramine has been administered to millions of patients with impressive results.

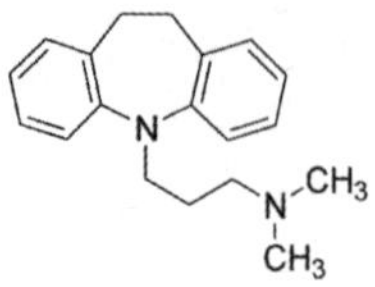

imipramine

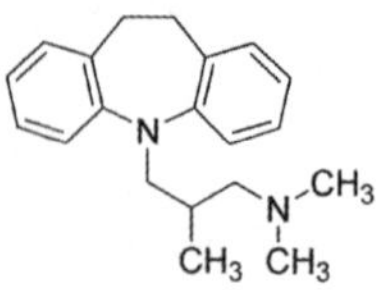

trimipramine

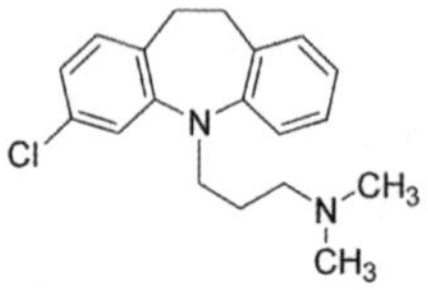

clomipramine

Rival companies responded by introducing their own antidepressants, which had similar activity to imipramine. For example, trimipramine was synthesised as an analogue of trimeprazine at the Rhône–Poulenc laboratories,[61] while clomipramine was introduced by Smith, Kline and French.[62]

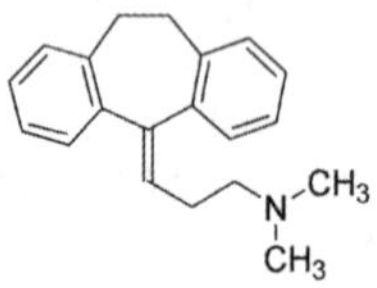

amitriptyline

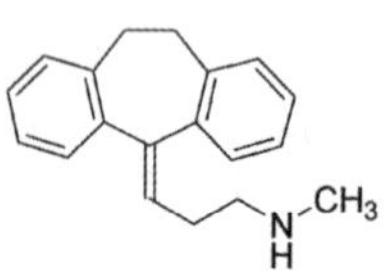

nortriptyline

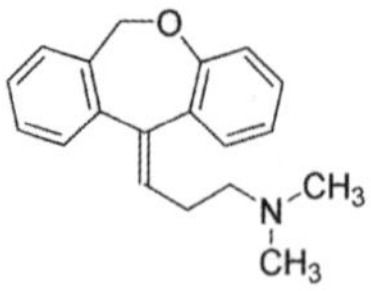

doxepin

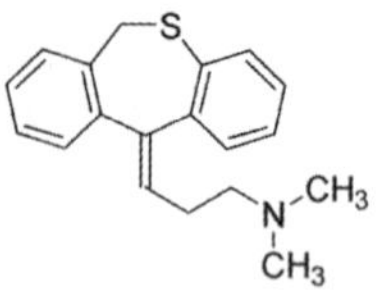

dosulepin

The recognition of the antidepressant action of imipramine revealed that minor structural alterations in the central ring of phenothiazine tranquillisers could radically change their pharmacological profile. This stimulated medicinal chemists to synthesise novel tricyclic compounds. As has been seen, replacement of the nitrogen atom in the central ring of chlorpromazine led to the introduction of the thioxanthenes as tranquillisers in 1958. Similarly, replacement of the sulfur atom in the thioxanthene system resulted in the first of the dibenzocycloheptadienes, namely amitriptyline, which was synthesised by several groups in 1960. One of the first was Merck Sharp & Dohme Research Laboratories, who also prepared nortriptyline.[63] Both resembled imipramine insofar as they were antidepressants rather than tranquillisers, but were noticeably less stimulating. This has made them more suitable than imipramine for treating agitated, anxious patients who were also depressed.

Analogues of amitriptyline include doxepin, in which a carbon atom in the central ring is replaced by oxygen.[64] It has a similar clinical profile to amitriptyline, but may be somewhat less cardiotoxic in overdosage. Dosulepin is similar in its activity.[65]

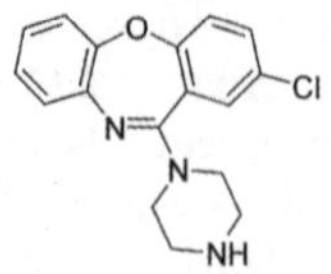

amoxapine

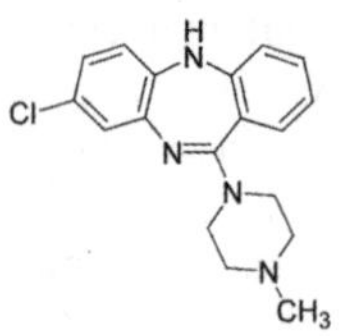

clozapine

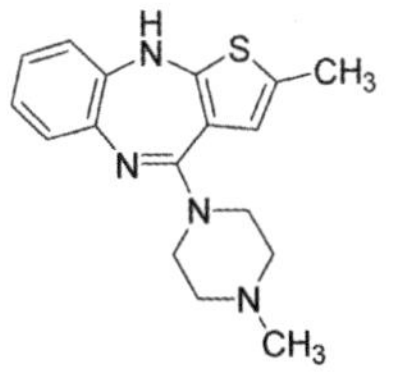

olanzapine

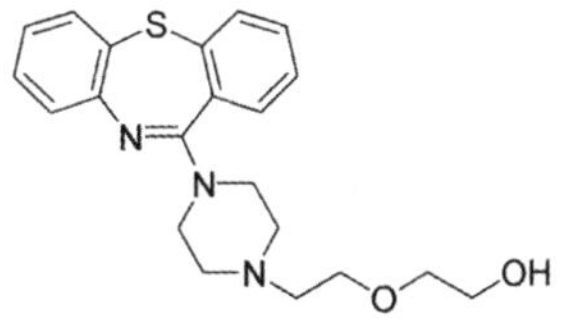

quetiapine

In 1958, researchers at the Wander Research Institute in Basle synthesised analogues of imipramine in which one or more heteroatoms replaced carbon atoms in the central ring. Particularly interesting from a pharmacological point of view were several amidines, including the antidepressants amoxapine and clozapine. Amoxapine[66] had very similar antidepressant properties to imipramine, but clozapine proved to be an antipsychotic drug. However, it was atypical insofar as it did not produce extrapyramidal side effects and was of value in patients who had failed to respond to treatment with other antipsychotic drugs.[67] It was marketed in Switzerland and Austria in 1972, but a high incidence of agranulocytosis was observed during a clinical trial in Finland three years later.[68] This resulted in the use of clozapine being severely restricted. However, in 1988 a major study by John Kane of Hillside Hospital in Glen Oaks, New York, revealed the outstanding activity of clozapine in the treatment of schizophrenics who had not responded to therapy with conventional antipsychotic drugs.[69] It became more frequently prescribed thereafter.

When Lilly researchers examined the effect of replacing either of the benzene rings in clozapine with a thiophen ring, they found four compounds worthy of further study. Only one of these proved to be safe enough for human studies. It was marketed under the name 'olanzapine' in 1996 as a safer alternative to clozapine.[70] Quetiapine is a similar atypical antipsychotic drug.[71]

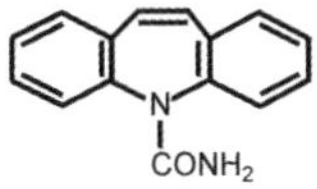

carbamazepine

Carbamazepine was synthesised by Walter Schindler at the Geigy laboratories in Basle in 1953 when the company was investigating analogues of chlorpromazine.[72] It was only some years later that its anticonvulsant properties were recognised. The first clinical study was not carried out until 1963, and it seems to have taken longer than most anticonvulsants to become established in clinical practice. Carbamazepine is now considered to be as effective as phenytoin in the control of partial and tonic–clonic seizures.

Selective Serotonin Reuptake Inhibitors

During the 1960s, Swiss psychiatrist Paul Kielholz differentiated tricyclic antipressants for clinical application on the basis of whether they possessed the ability to sedate, stimulate drive or improve the mood of patients. At that time there was a broad consensus that the tricyclic antidepressants acted by inhibiting the reuptake of norepinephrine back into the neurones from which it was released, thereby elevating the level of the hormone. By the end of the decade, however, there was mounting evidence that the reuptake of 5-HT was also blocked. The first to associate the thinking of Kielholz with the biochemical advances were the Russians Izyaslav Lapin and Gregory Oxenkrug of Bekhterev's Psychoneurological Research Institute in Leningrad. In 1969, they suggested that increased serotonergic activity in the brain involving tryptophan and its metabolites, including 5-HT, accounted for the mood-elevation effect of antidepressants, while increased noradrenergic activity was responsible for the motor and energising effects.[73] Several groups of researchers followed this up by seeking selective serotonin reuptake inhibitors (SSRIs).

At the Karolinska Institutute, Arvid Carlsson examined the effects of antihistamines on both 5-HT and norepinephrine uptake in tissues. Although most had mixed activity, diphenhydramine affected only 5-HT uptake.[74] In collaboration with Peder Berntsson and Hans Corrodi of Aktiebolaget Hassle, based in Gothenburg and part of Astra, Carlsson quickly developed a pheniramine analogue as a potent SSRI in the spring of 1971.[75] Several years later it was marketed in Europe as zimelidine in 1982, but was withdrawn by Astra the following year because ten cases of the Guillaine–Barré syndrome had been reported out of 200 000 prescriptions. This neurological disorder was characterised by progressive muscular weakness. Fortunately, a slow recovery over a period of months occurred in all patients once medication had ceased.

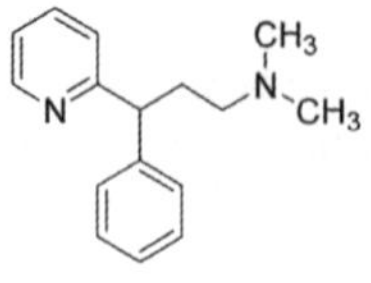

pheniramine

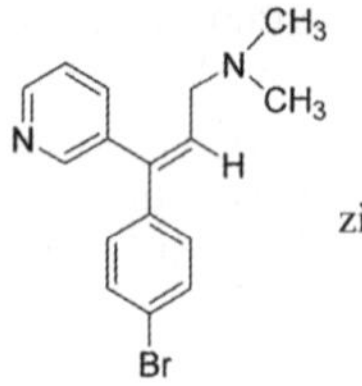

zimelidine

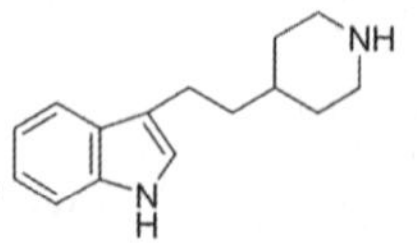

indalpine

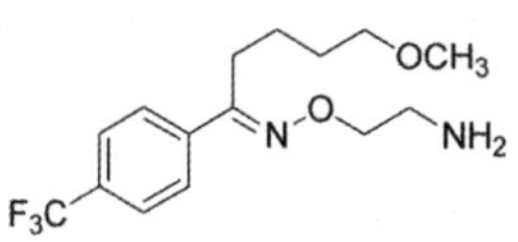

fluvoxamine

The second SSRI to be marketed in Europe also had to be withdrawn shortly after its introduction, this time because it produced agranulocytopenia in a few patients. The drug was an antihistamine analogue called indalpine, developed by Gerard Le Fur and his colleagues at Fournier Frères, a company that became part of Rhône–Poulenc.[76]

The next SSRI to be introduced was fluvoxamine. It remained on the market without encountering the problems faced by its predecessors.[77] However, during early trials concern was expressed over the number of patients who committed suicide before the drug had exerted its beneficial action. Similar concerns have been raised about other SSRIs, despite their main advantage over tricylic antidepressants being their enhanced safety margin when deliberate overdoses are consumed. This issue is highly contentious and is being examined in the law courts.

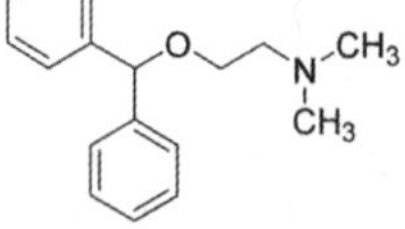

diphenhydramine

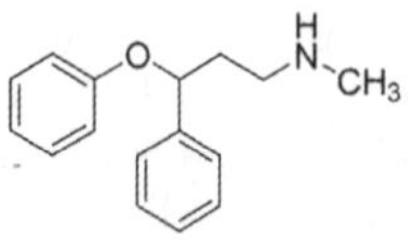

N-methyl-phenoxyphenylpropylamine

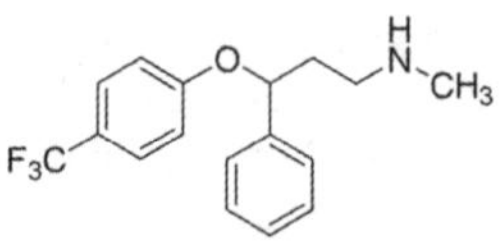

fluoxetine

Several other SSRIs were subsequently marketed, including one that has become a household name. Bryan Molloy of Eli Lilly made analogues of diphenhydramine for Robert Rathbun and Richard Kattau to screen for inhibition of norepinephrine and 5-HT uptake.[78]

They found *N*-methyl-phenoxyphenylpropylamine to be twice as potent at inhibiting uptake of 5-HT as it was at inhibiting norepinephrine uptake, so a series of analogues of it were synthesised. This resulted in the discovery of fluoxetine as an SSRI. Eli Lilly marketed it in 1988. Since then, it has become the most frequently prescribed antidepressant drug. It was especially popular in the United States under its proprietary name of Prozac®.

ANTIFIBRINOLYTIC DRUGS

S. Okamoto set up a screening programme to find antifibrinolytic drugs that could be used to stop haemorrhage. Among the 400 or so compounds he tested were basic amino acids that were found to have some activity, the most effective being lysine.

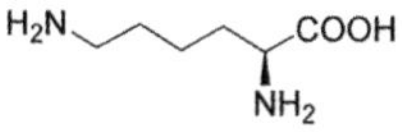

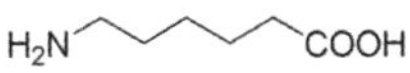

Since lysine had inadequate potency for clinical applications, analogues of it were examined. This led to the discovery in 1957 that removal of the α-amino group greatly enhanced activity, ε-aminocaproic acid being ten times as potent as lysine.[79] Further investigation showed that it was an inhibitor of plasminogen activation. It was introduced into the clinic as a haemostatic agent, but was superseded by tranexamic acid.

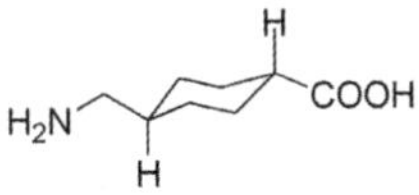

When Okamoto and his colleagues screened a large number of analogues of ε-aminocaproic acid they discovered that the distance between the carboxylic and amino groups was critical, as was the nature of the linkage between them. After finding that a benzene ring could be used as a linkage, a cyclohexane ring was shown to be even more potent. The most potent compound was tranexamic acid, so named because it was a *trans* isomer.[80] The *cis* isomer was inactive as the amino and carboxylic groups were positioned too close to each other. Tranexamic was introduced to stop haemorrhage during surgery.

NON-STEROIDAL ANTI-INFLAMMATORY DRUGS

During the 1950s, it became widely recognised that the long-term use of corticosteroids in rheumatoid arthritis caused serious problems that were inherent in the nature of the medication. In 1955, Stewart Adams at the Boots Pure Drug Company laboratories in Nottingham established a screen to find a safe, orally active anti-inflammatory agent. This involved ultraviolet (UV) irradiation of the backs of guinea pigs 30 minutes after they had received a test compound by mouth. Adams had established that this procedure reliably indicated the ability of aspirin to reduce inflammation and could be used to test alternatives to it in the search for a more potent compound with fewer side effects.

The chemist working with Adams was John Nicholson. Both were convinced that the presence of a carboxylic acid group was responsible for the anti-inflammatory activity of aspirin and some of its analogues. Adams screened phenylacetic and phenoxyacetic acids previously made by the company as potential herbicides. After 2-(4'-ethylphenoxy)propionic acid proved to be several times as potent as aspirin, more than 200 aryloxyalkanoic acids were synthesised and tested.[81] Eventually, 2-(4'-phenylphenoxy)propionic acid emerged as a candidate for clinical investigation. Its ethyl ester was preferred since it was expected to cause less gastric irritancy. However, when put on clinical trial in 1960 it turned out to be inactive in patients with rheumatoid arthritis. Significantly, its analgesic and antipyretic effects were feeble.[82] This led to a decision that all active test compounds should in future be tested for analgesic and antipyretic as well as anti-inflammatory activity.

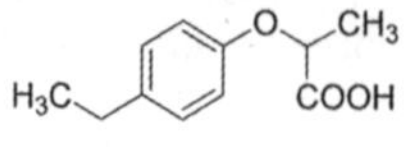

2-(4'-ethylphenoxy)propionic acid

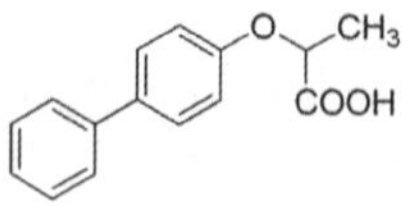

2-(4'-phenylphenoxy)propionic acid

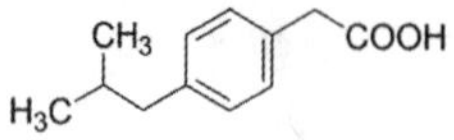

ibufenac

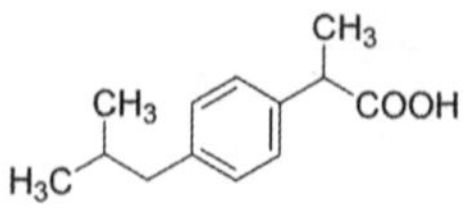

ibuprofen

Attention was now switched to 4-alkylphenyl and biphenylalkanoic acids. Approximately 450 more compounds were synthesised and screened. Two of these were examined in the clinic, but produced rashes in patients. Finally, ibufenac was found to exhibit more than four times the potency of aspirin.[83] After an early clinic trial gave encouraging results, it was marketed in 1966.[84] However, it had to be withdrawn soon after when evidence of an unacceptable incidence of jaundice appeared.

Ibuprofen was selected as an acceptable alternative to ibufenac after animal studies confirmed that it did not accumulate in the liver or produce ulcers in dogs, as had some similar compounds.[85] It proved to be a safe, effective anti-inflammatory agent with analgesic and antipyretic properties and was marketed in 1969. Ibuprofen became widely prescribed throughout the world in the wake of increasing concern about the hazard of gastric bleeding caused by aspirin. Such was its relative safety that in 1983 it became available in the United Kingdom as a non-prescription analgesic on account of its having the lowest overall rate of reporting of suspected adverse reactions among the non-steroidal anti-inflammatory agents, some 20 million prescriptions having been issued over the preceeding 15 years.

Among the 600 compounds screened by Adams before the introduction of ibuprofen were several 4-biphenylalkanoic acids. These proved to be highly potent anti-inflammatory agents, but were abandoned in favour of the less-potent phenylalkanoic acids that were at that time believed to be less toxic. After the introduction of ibuprofen these acids were again investigated, resulting in the introduction of flurbiprofen. Although it turned out to be 5–10 times as potent as ibuprofen, this did not confer any significant therapeutic advantage. The indications for its use are identical to those for ibuprofen.

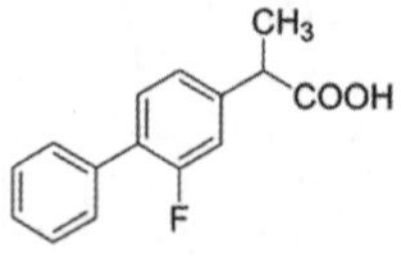

flurbiprofen

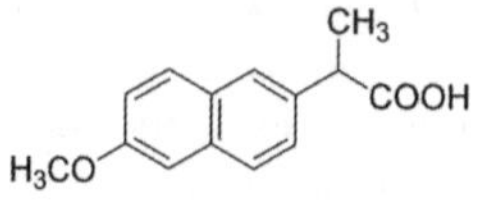

naproxen

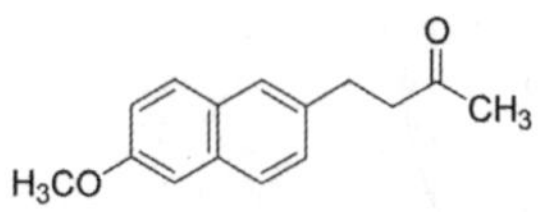

nabumetone

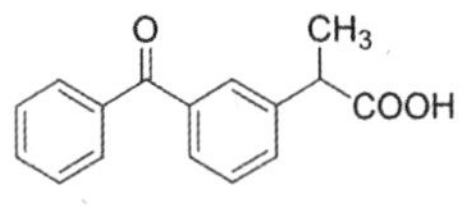

ketoprofen

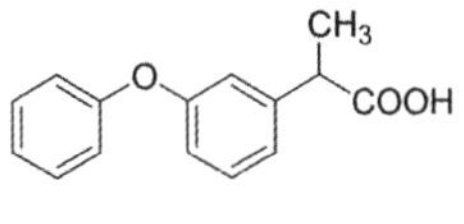

fenoprofen

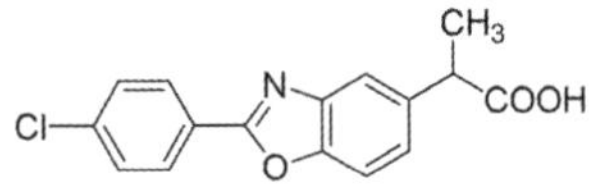

benoxaprofen

Rival manufacturers quickly developed analogues of ibuprofen. Syntex introduced naproxen, which was twice as potent as ibuprofen and had a longer duration of action, allowing twice-daily dosing.[86] Beecham Pharmaceuticals developed the related nabumetone as a prodrug that rapidly underwent metabolic activation in the liver to form the acid.[87] Their intention had been to minimise the inhibition of prostaglandin synthesis in the stomach and so reduce gastric irritation, but this would only have been relevant if that inhibition had been a local rather than a systemic effect. Rhône–Poulenc marketed ketoprofen, which was about ten times as potent as ibuprofen and had a longer duration of action.[88] Fenoprofen was introduced by Lilly.[89] It had a similar duration of action to ibuprofen but was 2–3 times as potent.

A number of other non-steroidal anti-inflammatory acids were marketed in the 1970s and 1980s. Among them was benoxaprofen, which was developed at the Lilly Research Centre in Surrey, England.[90] When it was launched in the United Kingdom in March 1980, benoxaprofen was promoted as an anti-arthritic agent that could be taken as a single daily dose because of its resistance to metabolic degradation. At the time, this was considered helpful in ensuring good patient compliance. There were also anecdotal reports of dramatic improvements in the condition of seriously crippled patients.

In February 1982, Hugh Taggart at Queen's University in Belfast reported the deaths of five elderly patients who had received benoxaprofen.[91] Urgent enquiries ensued and the UK Committee on Safety of Medicines withdrew the Product Licence in August of that year, amid intense media coverage. By that time, 83 fatalities had occurred among three-quarters of a million patients in the United Kingdom who had taken the drug. Most had died from renal or hepatic failure.

The 15th International Congress of Rheumatology held in Paris in June 1981 had been told that benoxaprofen was slowly excreted, its biological half-life being as long as four days in elderly patients. Earlier reports had indicated that the half-life was 33 hours in humans, justifying the convenience of once-daily dosage. The implications of a more prolonged half-life in elderly patients had not then been obvious. It is with the benefit of hindsight that they certainly are now. The issue relates to the question of toxicity, for it is only if a drug is relatively toxic that a problem arises from its accumulation in patients with poor renal function. Although reports of photosensitivity and nail damage had been received, benoxaprofen was not at that time thought to be any more toxic than other non-steroidal anti-inflammatory agents.

The outcome of this tragic affair would have been different if more information about the effects of benoxaprofen in elderly people had been available. The manufacturer had mentioned the need for dosage reduction in a pamphlet issued three months after the Paris Symposium, but this seems to have been generally overlooked. Only 52 patients over the age of 65 had received the drug in clinical trials, but this was not exceptional since it was not expected that elderly patients should be recruited specifically for such trials. Lessons were learned from this affair, not least being the importance of an exemplary level of vigilance required from both manufacturers and licensing authorities. Furthermore, there is now recognition that elderly patients should not be prescribed long-acting drugs when alternatives are available.

Selective COX-2 Inhibitors

In 1991, Dan Simmons and his colleagues at Brigham Young University in Utah discovered that there was a second type of cyclooxygenase enzyme inhibited by aspirin and the non-steroidal anti-inflammatory agents.[92] This COX-2 was principally involved in producing prostaglandins during inflammation, whereas COX-1 was involved in routine physiological processes such as platelet aggregation. Pharmaceutical companies immediately recognised the implications and began seeking selective COX-2 inhibitors that would be free from the side effects of existing anti-inflammatory drugs that all lacked selectivity.

G.D. Searle launched a screening programme that was to test over 2500 compounds. These were at first screened against cloned COX-2 enzyme, but it was found that an assay in rodents was more reliable. Around 10% of the compounds were selected for further screening, from which seven emerged as potential drug candidates and were examined in several species of animals. A compound from the company's agrochemical library of compounds turned out to be both a selective COX-2 inhibitor and an anti-inflammatory agent. It was marketed in 1999 with the approved name of 'celecoxib'.[93] It was hoped that it would produce a lower incidence of side effects than other anti-inflammatory drugs.

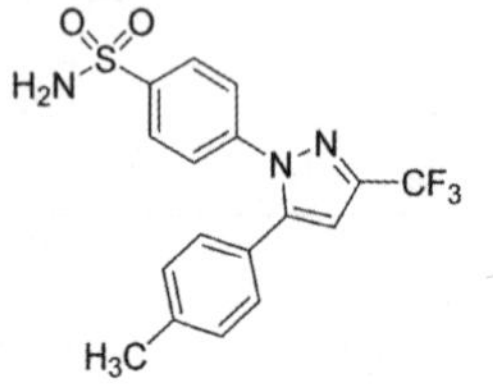

celecoxib

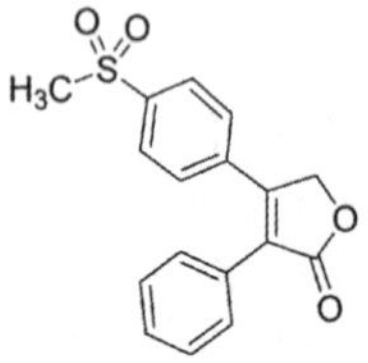

rofecoxib

Rival companies introduced several more selective COX-2 inhibitors during the next few years. One of these, rofecoxib, was suddenly withdrawn in 2004 after Merck had conducted a study that revealed that patients taking their product faced a higher risk of heart attacks and stroke than those on a placebo.

ETHAMBUTOL

In the course of an extensive screening programme, researchers at Lederle Laboratories discovered that N,N'-diisopropylethylenediamine had antituberculosis activity comparable with that of isoniazid. The sole drawback was its greater toxicity.

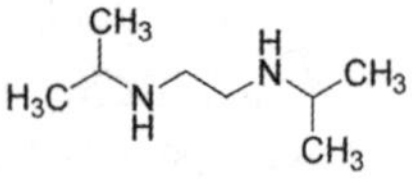

N,N'diisopropylethylenediamine

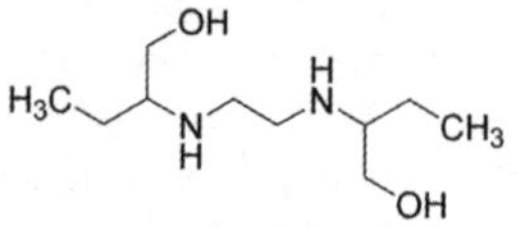

ethambutol

An extensive series of analogues was synthesised, culminating in the development of ethambutol, which was reported in 1961 to be a particularly promising drug.[94] This early promise was fulfilled and ethambutol is still used in combination with other agents. The main problem with it is that visual side effects occur and thus patients require to be regularly monitored while receiving treatment.

LEVAMISOLE

The Janssen Research Laboratory at Beerse in Belgium initiated an extensive screening programme in which 2721 novel heterocyclic compounds were tested for anthelmintic activity against three types of parasitic worms before an aminothiazole derivative, R6438, was found to be effective in chickens and sheep.

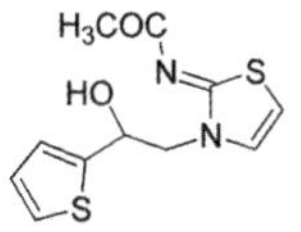

Its failure in mice and rats pointed to the possibility that it had to undergo metabolic conversion to an active drug that was only formed in some animals. All the metabolites were then isolated and synthesised. R8141 was the only active one, but was difficult to produce and, in addition, was unstable in water. A large series of its analogues was synthesised, of which one met all the requirements for possible clinical application. This was given the approved name of 'tetramisole',[95] but it was its laevo isomer that was selected for medicinal use since it was several times more potent, yet no more toxic.[96,97] It is employed under the name 'levamisole' as an ascaricide to eliminate the common roundworm.

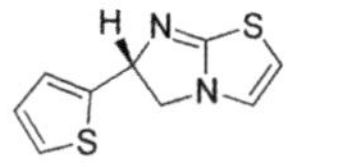

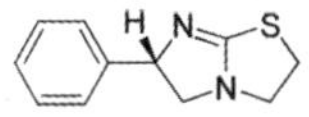

PYRANTEL

In the mid-1950s, Pfizer researchers at Groton, Connecticut, established a screening programme to find new anthelmintic agents. In order to widen the score of the screens, laboratory mice were inoculated with three different organisms, namely the tapeworm *Hymenolepsis nana*, the nematode *Nematospiroides dubius* and the pinworm *Syphacia obvelata*. Out of a large number of compounds submitted for screening, only compound I emerged with any evidence of activity.[98] When administered by mouth to sheep, it had little activity, probably because it hydrolysed to the compounds from which it had been synthesised, namely 2-thenylthiol and 2-imidazolidone.

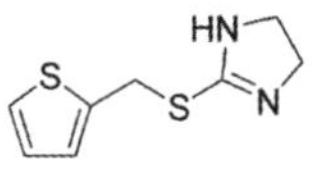

Analogues of compound I designed to resist hydrolysis were synthesised at the Pfizer research centre in Sandwich, England. An early advance came when a methylene group replaced the sulfur atom in the bridge between the two rings to give a compound that was active against a variety of roundworms that infested sheep. As this compound was toxic, analogues were synthesised. Optimal activity against a variety of nematodes was obtained by enlarging the imidazoline to a tetrahydropyrimidine ring, as in pyrantel.[99] This was an orally active broad-spectrum anthelmintic that was effective against roundworms, hookworms and threadworms in both humans and animals.

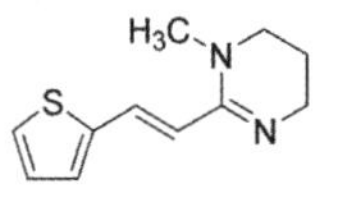

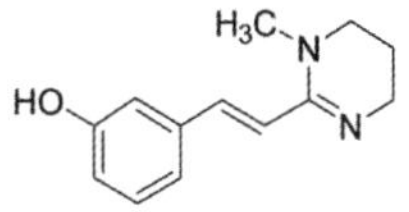

Oxantel was one of several analogues of pyrantel that were prepared in order to examine the relationship between aromatic ring substitution and anthelmintic potency. It had only one-tenth of the activity of pyrantel in the mouse screen against *N. dubius*, but was active against the tapeworm *H. nana*, unlike pyrantel. When tested in dogs with whipworm infestation, it was also active.[100] This activity against whipworm infection compensated for its narrow spectrum of anthelmintic activity, and became the principal clinical application.[101]

NIFEDIPINE

Because 1,4-dihydropyridines played an important role in biochemical processes yet had never been investigated pharmacologically, the medicinal chemistry group at the Bayer laboratories submitted a variety of these for screening. After some of these compounds were found to exhibit a measurable effect on cardiac output, more than 2000 analogues were synthesised and screened.[102] In 1967, Friedrich Bossert and Wulf Vater applied for a South African patent in which they claimed that nifedipine possessed marked coronary vasodilating activity.[103]

Further investigations revealed that nifedipine selectively blocked calcium channels in the conductive cells of the heart and vascular smooth muscle, thereby inhibiting the entry of ionised calcium and its release from intracellular stores. Since calcium was required for membrane depolarisation and muscle contraction, nifedipine relaxed smooth muscle both in the myocardium and in the walls of blood vessels. The outcome of this was dilation of the coronary vessels and a fall in vascular resistance, with a consequent reduction of cardiac afterload, work and oxygen consumption. As this was of major importance in the treatment of vascular disorders such as angina and hypertension, nifedipine was marketed in 1975.

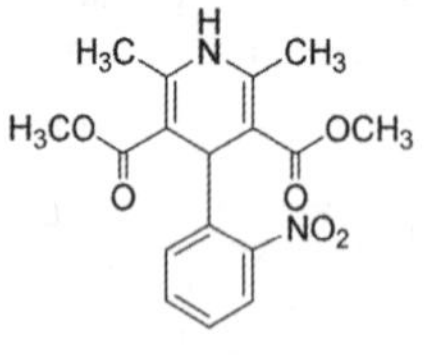

nifedipine

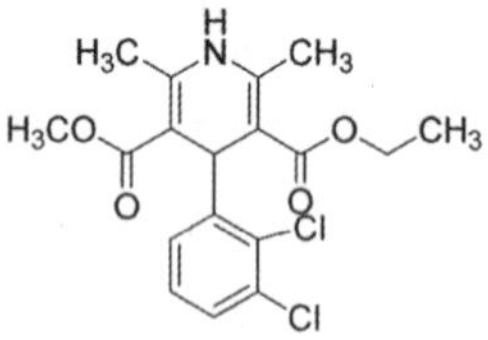

felodipine

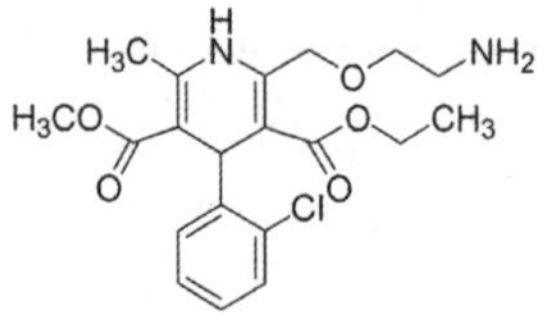

amlodipine

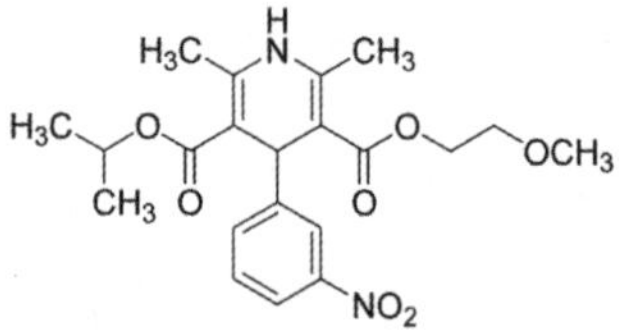

nimodipine

Several analogues of nifedipine have been developed which have a longer duration of action and hence are less likely to cause fluctuations in blood pressure and reflex tachycardia, e.g. felodipine[104] and amlodipine.[105] Nimodipine had a high specificity for calcium channels in cerebral blood vessels and hence was able to increase cerebral blood flow without decreasing blood pressure.[106] This made it valuable in the prevention of cerebral arterial spasm after subarachnoid haemorrhage.

CARMUSTINE

In 1955, following a decade of remarkable progress in the sphere of cancer chemotherapy in the United States, Congress allocated large sums of money to set up a screening programme run by the Cancer Chemotherapy National Service Center (CCNSC), which was part of the National Cancer Institute. The first contracts were awarded to four screening centres that confidentially tested large numbers of compounds submitted by academic and industrial researchers. By the end of the decade, these centres were testing around a thousand chemicals each month against animal tumours.

In 1959, researchers at the Wisconsin Alumni Research Foundation discovered that 1-methyl-1-nitroso-3-nitroguanidine (MNNG), an intermediate used by organic chemists to prepare the unstable alkylating agent diazomethane, had antileukaemic activity in mice. Unfortunately, human trials were disappointing. Nevertheless, when Thomas Johnston, George McCaleb and John Montgomery at the Southern Research Institute in Birmingham, Alabama, were notified by the CCNSC of the activity of MNNG, they immediately began an evaluation of related compounds as there was considerable concern that half of the long-term survivors among children who received combination chemotherapy for acute leukaemia were dying from meningeal leukaemia. They found that an alternative compound used in the synthesis of diazomethane, namely 1-methyl-1-nitrosourea (MNU), was more active than MNNG. As it was also more lipophilic, they believed it would penetrate the central nervous system and so be effective in meningeal leukaemia. After preliminary tests in animals injected intracerebrally with leukaemic cells, MNU was investigated further by the National Cancer Institute.[107] While this was taking place, analogues of MNU were synthesised and tested at the Southern Research Institute. The researchers there believed that the antileukaemic activity was due to MNU decomposing into diazomethane hydroxide, a powerful alkylating agent. Attempts were therefore made to find a more active biological alkylating agent by replacing the N-methyl group with similar groups that would produce diazoalkyl hydroxides with different activity profiles. This led to the synthesis of carmustine.[108] When it was tested in mice injected with L-1210 leukaemic cells, not only was it the most active of 23 compounds submitted for evaluation and far superior to MNU but it was also the first to cure such mice.[109] An early clinical trial confirmed the value of carmustine.[110]

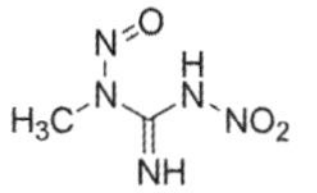

1-methyl-1-nitroso-3-nitroguanidine

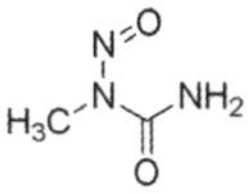

1-methyl-1-nitrosourea

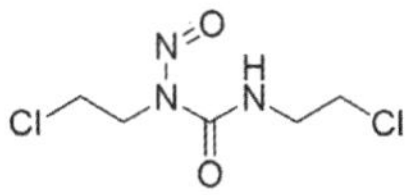

carmustine

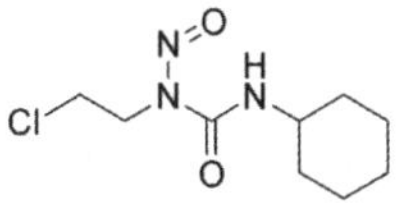

lomustine

As expected, carmustine penetrated into the central nervous system because of its high lipid solubility. It controlled meningeal leukaemia in some children, but as this soon became preventable by irradiating the cranium at the outset of acute lymphocytic leukaemia treatment, this application of carmustine became redundant. It is instead used to treat brain tumours, advanced Hodgkin's disease, lymphomas and myelomas. Because of its alkylating activity, local tissue damage would be caused if it were taken by mouth or injected intramuscularly. Hence carmustine is given intravenously so that immediate dilution by blood occurs. Lomustine, which was also developed at the Southern Research Institute, can be taken by mouth.[111]

PRAZIQUANTEL

In 1972, a screening programme at the Bayer Institute for Chemotherapy in Wuppertal revealed anthelmintic activity in novel pyrazinoisoquinolines that had been synthesised in the laboratories of E. Merck and Company of Darmstadt in the course of a joint project between these two German companies. Praziquantel was selected from over 400 compounds for further investigation. It exhibited outstanding efficacy against all known intestinal cestode infections in humans, as well as a great many in animals.[112]

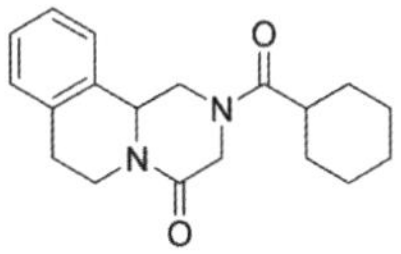

praziquantel

Further studies revealed not only that a single oral dose was capable of eradicating these infections but also that praziquantel was highly effective against schistosomiasis and a variety of other parasitic infections.[113] It became the first drug to receive World Health Organization approval for use in mass eradication programmes aimed at eliminating a broad range of parasitic infections.

MILRINONE

At the Sterling–Winthrop Research Institute in Rensselaer, New York, a screening programme was established in order to find inotropic compounds with cardiotonic activity similar to that of the cardiac glycosides. Amrinone was discovered to be among the most active of the compounds capable of increasing the contractility of cardiac muscle.[114] In addition, it has a useful vasodilating action. These properties were found to be due to its ability to act as a selective phosphodiesterase inhibitor in cardiac and vascular muscle, raising intracellular levels of cAMP.[115] This resulted in an increase in calcium levels, which accounted for the increased force of cardiac contraction. As it had little effect on heart rate, amrinone was suitable for use as a cardiac muscle stimulant in congestive heart failure. However, the Sterling–Winthrop researchers found milrinone to be many times more potent, so it was preferred for clinical use.[116]

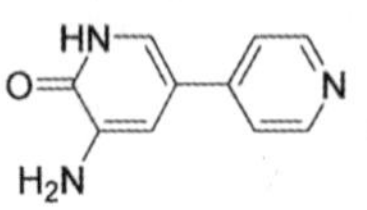

amrinone 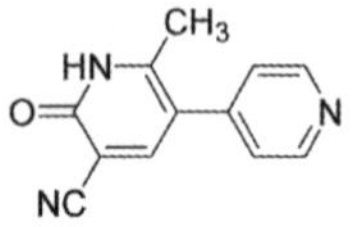 milrinone 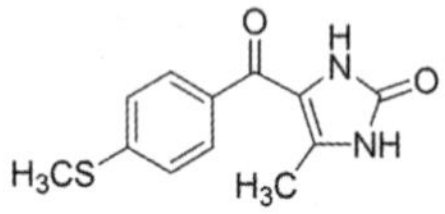 enoximone

Several other companies sought selective phosphodiesterase inhibitors, enoximone being developed at the Merrell Dow Research Center in Cincinnati after it was discovered that imidazole had some activity.[117] It was less potent than milrinone.

NAFTIFINE

Naftifine was found to be an antifungal agent during screening at the Sandoz Research Institute in Vienna.[118,119] It was shown to have a novel mode of action which involved blocking the synthesis of ergosterol by inhibiting squalene oxidase, but it was only suitable for topical use.

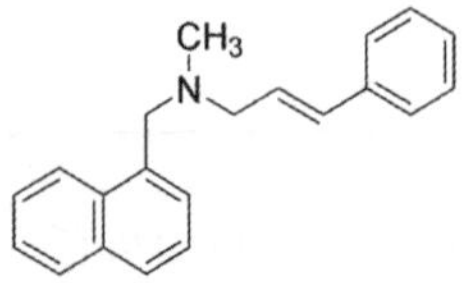

naftifine 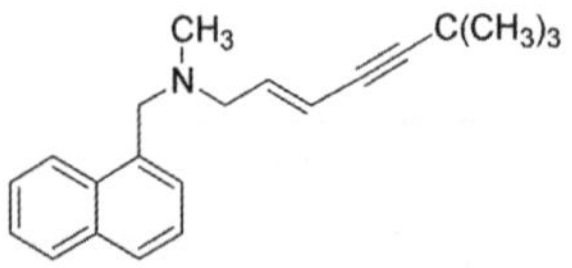 terbinafine

Over a thousand analogues of naftifine were prepared in order to establish the requirements for optimal antifungal activity.[120] From this, acetylenic allylamines were selected for special attention, resulting in the development of terbinafine as an orally active antifungal agent with significantly greater activity than naftifine.[121,122]

PROPOFOL

At the Alderley Edge laboratories of ICI (now AstraZeneca), screening in mice for potential anaesthetics revealed the anaesthetic activity of 2,6-diethylphenol.

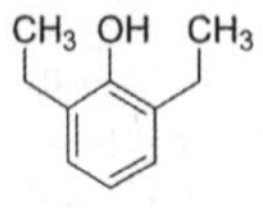

2,6-diethylphenol 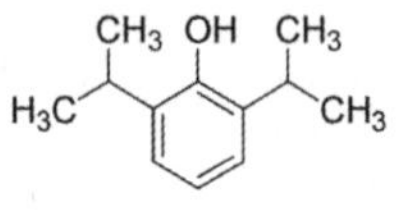 propofol

Roger James and J.B. Glen then prepared and examined further alkyl-substituted sterically hindered phenols in order to optimise activity in this series.[123] They found that it was necessary to strike a balance between the minimum steric hindrance providing adequate potency and excessive steric crowding, which caused loss of anaesthetic activity. In addition, lipophilicity had to be limited in order to avoid slower kinetics through binding to plasma proteins. The most active compounds were di-*sec*-alkyl substituted and had 6 to 8 carbon atoms in the side chains. Propofol was the only compound among those studied that emerged with a satisfactory profile when evaluated as an intravenous anaesthetic. It is now widely used.

ANGIOTENSIN II ANTAGONISTS

Angiotensin II is a powerful hypertensive agent because it binds to receptors known as AT_1 and AT_2 to cause vasoconstriction. Using these receptors as a target for high throughput screening (HTS), the Japanese company Takeda tested a vast number of compounds and obtained a 'hit'. Analogues of this were then prepared by DuPont chemists in order to find a compound that was selective for the AT_1 receptor. This was given the approved name of 'losartan' and was introduced into the clinic in 1993.

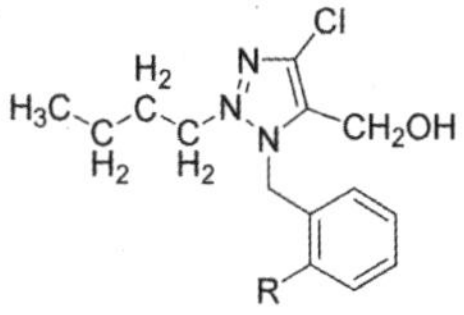

lead compound from HTS

R= Cl or NO$_2$

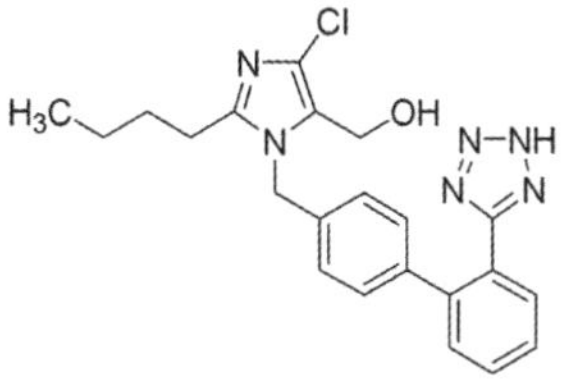

losartan

Unlike the ACE inhibitors, losartan and its analogues do not cause the breakdown of bradykinin, hence patients do not experience the unwelcome dry cough caused by ACE inhibitors. In general, the clinical value of angiotensin II antagonists matches that of the ACE inhibitors.

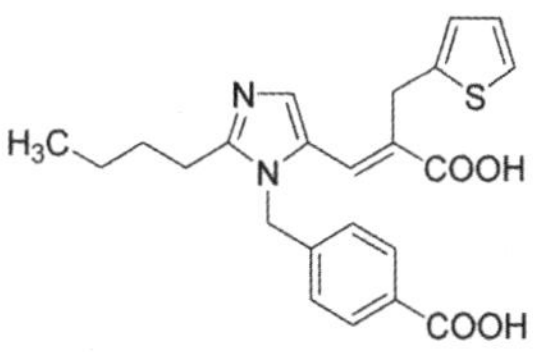

eprosartan

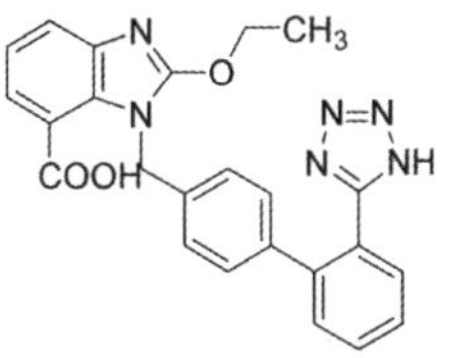

candesartan

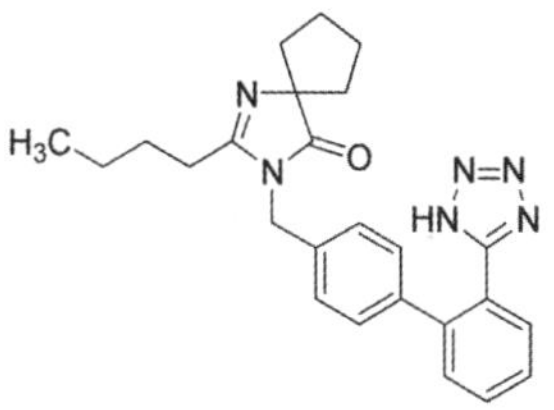

irbesartan

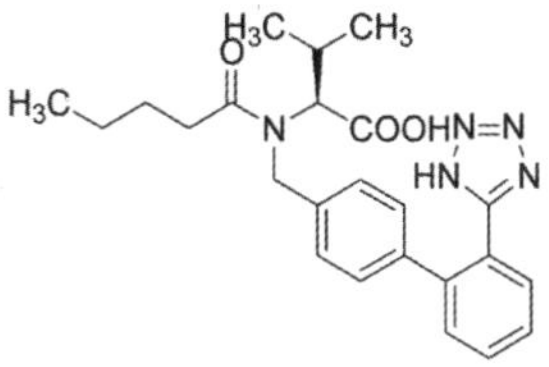

valsartan

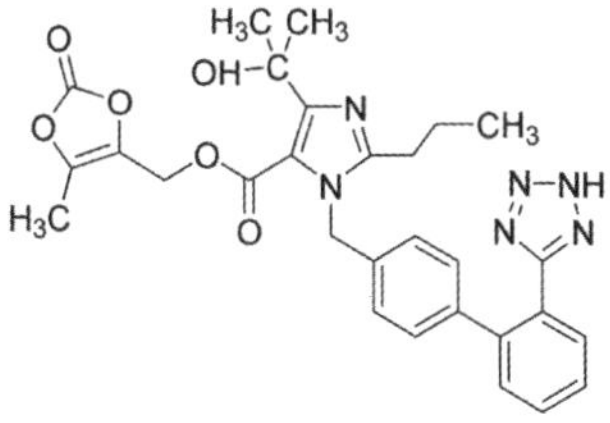

olmesartan

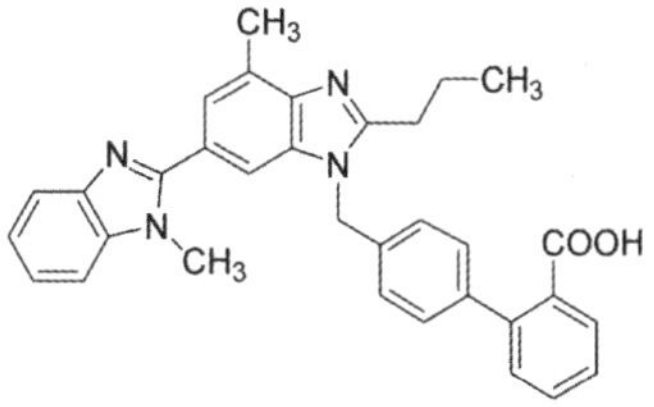

telmisartan

Several other angiotensin II antagonists have been introduced, including eprosartan,[124] candesartan,[125] irbesartan,[126] valsartan,[127] olmesartan[128] and telmisartan.[129] Even a superficial glance at their chemical structures shows that they are derived from either losartan or the lead compound from HTS that led to its development. This reveals the pattern of drug development by rival companies that will be seen when future drugs derived from HTS are introduced.

IMATINIB

In 1973 at the University of Chicago, Janet Rowley established that the missing portion of chromosome 22, the 'Philadelphia chromosome', had translocated to chromosome 9 in patients with chronic myelogenous leukaemia (CML). During the next decade it was confirmed that one gene from each chromosome (Bcr in chromosome 22, Abl in chromosome 9) fused to form a new gene, designated Bcr–Abl. This gene produced the rogue protein that caused CML. David Baltimore and his colleagues at the Whitehead Institute for Biomedical Research in Cambridge, Massachusetts, reported that this protein was a tyrosine kinase enzyme. In the early 1990s, the Swiss company Ciba-Geigy became interested in the possibility of designing an inhibitor of the enzyme. Using high throughput screening of the company's small molecule libraries, a 2-phenylaminopyrimidine emerged as a weak inhibitor of several protein kinases that could serve as a lead for further development. Introduction of a methyl at the 6-position and of a benzamide on the phenyl ring enhanced inhibitory activity towards Abl. A promising inhibitor ultimately emerged from this work, but it lacked water solubility and oral bioavailability was poor. These difficulties were overcome by attaching an N-methylpiperazine to form imatinib.[130]

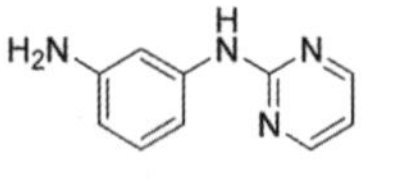

2-phenylaminopyrimidine

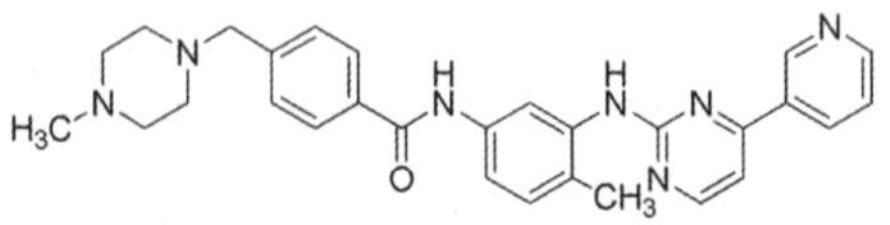

imatinib

Imitinab appears to be giving encouraging results as a highly selective chemotherapeutic agent for the treatment of CML.[131] Among untreated patients who receive imitinab, around 60% respond to treatment and experience few side effects. There have been no long-term studies yet, but it is clear that this is the first drug that can treat cancer by targeting a protein that causes the disease. Several companies are currently investigating other protein kinase inhibitors.

REFERENCES

1. G. Ungar, J.L. Parrot, D. Bovet, Inhibition des effets de l'histamine sur l'intestine isolé du cobaye par quelques substances sympathomimétiques et sympathicolytiques. *C.R. Soc. Biol.*, 1937; **124**: 445–6.
2. D. Bovet, A.-M. Staub, Action protectrice des éthers phenoliques au cours de l'intoxication histaminique. *C.R. Soc. Biol.*, 1937; **124**: 547–9.
3. D. Bovet, A.-M. Staub, Action protectrice des éthers phenoliques au cours de l'intoxication histaminique. *C.R. Soc. Biol.*, 1937; **124**: 547–9.
4. A.-M. Staub, Recherches sur quelques bases synthétique antagonistes de l'histamine. *Ann. Inst. Pasteur*, 1939; **63**: 400–36, 485–524.
5. B.N. Halpern, Les antihistaminiques de synthése: essais de chimothérapie de états allergiques. *Arch. Int. Pharmacodyn.*, 1942; **68**: 339–408.

6. D. Bovet, R. Horclois, F. Walthert, Antihistamine properties of 2-[(p-methyoxybenzyl)(2-dimethylaminoethyl)amino]-pyridine (RP2786). *C.R. Soc. Biol.*, 1944; **138**: 99–100.
7. C.P. Huttrer, C. Djerassi, W.L. Beears, *et al.*, Heterocyclic amines with antihistaminic activity *J. Am. Chem. Soc.*, 1946; **68**: 1999–2002.
8. R. Baltzly, S. DuBreuil, W.S. Ide, E. Lorzl, Unsymmetrically disubstituted piperazines (III). *N*-Me-*N'*-benzhydrylpiperazine as histamine antagonists. *J. Org. Chem.*, 1949; **14**: 775–82.
9. US Pat. 1959: 2899436 (to UCB).
10. P. Charpentier, The constitution of 10-(dimethylaminopropyl)phenothiazine. *C. R. Acad. Sci.*, 1947; **225**: 306–8.
11. E.R. Loewe, M.E. Kaiser, V.Moore, Synthetic benzhydryl alkamine ethers effective in preventing experimental asthma in guinea pigs exposed to atomised histamine. *J. Pharm. Exp. Ther.*, 1945; **83**: 120–9.
12. A.F. Harms, W.Th. Nauta, The effects of alkyl substitution in drugs – I. Substituted dimethylaminoethyl benzhydryl ethers. *J. Med. Chem.*, 1960; **2**: 57–77.
13. Neth. Pat. Appl. 1966: 6606390 (to Rexall).
14. J.W. Cusic, Note on the chemistry of Dramamine. *Science*, 1949; **109**: 574.
15. L.N. Gay, P.E. Carliner, The prevention and treatment of motion sickness. *Science*, 1949; **109**: 359.
16. US Pat. 1951: 2567245 (to Schering).
17. D.W. Adamson, Aminoalkyl tertiary carbinols and derived products. Part I. 3-Amino-1:1-diphenylpropan-1-ols. *J. Chem. Soc.*, **1949**: S144–S155.
18. E.L. Englehardt, H.C. Zell, W.S. Saari, *et al.*, Structure–activity relationships in the cyproheptadine series. *J. Med. Chem.*, 1965; **8**: 829–35.
19. J.-M. Bastian, A. Ebnöther, E. Jucker, *et al.*, 4H-Benzo[4,5]cyclohepta[1,2-b]thiophene. *Helv. Chim. Acta*, 1966; **49**: 214–34.
20. F. Sicuteri, G. Franchi, P.L. Del Bianco, An antaminic drug, BC 105, in the prophylaxis of migraine. Pharmacological, clinical, and therapeutic experiences. *Int. Arch. Allergy Appl. Immunol.*, 1967; **31**: 78–93.
21. F.J. Villani, P.J.L. Daniels, C.A. Ellis, *et al.*, Derivatives of 10,11-dihydro-5H-dibenzo[a,d]cyclo-heptene and related compounds. 6. Aminoalkyl derivatives of the aza isosteres. *J. Med. Chem.*, 1972; **15**: 750–4.
22. C.R. Kinsolving, N.L. Munro, A.A. Carr, Separation of the CNS and H$_1$ receptor effects of antihistamine agents. *Pharmacologist*, 1973; **15**: 221.
23. S.H. Kawai, R.J. Hambalek, G. Just, A facile synthesis of an oxidation product of terfenadine. *J. Org. Chem.*, 1994; **59**: 2620–2.
24. A.F. Cohen, M.J. Hamilton, S.H. Liao, *et al.*, Pharmacodynamic and pharmacokinetics of BW 825C: a new antihistamine. *Eur. J. Clin. Pharmacol.*, 1985; **28**: 197–204.
25. A. Barnett, M.J. Green, Loratadine, in *Chronicles of Drug Discovery*, Vol. 3, ed. D. Lednicer. Washington: American Chemical Society; 1993, pp. 83–99.
26. F.J. Villani, C.V. Magatti, D.B. Vashi, *et al.*, *N*-substituted 11-(4-piperidylene)-5,6-dihydro-11H-benzo-[5,6]cyclohepta[1,2-b]pyridines – antihistamines with no sedating liability. *Arzneimittel.-Forsch.*, 1986; **36**: 1311–4.
27. H. Laborit, Etude expérimentale du syndrome d'irritation et application clinique à la maladie post-traumatique. *Thérapie*, 1949; **4**: 126–39.
28. J.P. Swazey, *Chlorpromazine in Psychiatry: A Study of Therapeutic Innovation*. Cambridge, Massachusetts: MIT Press; 1974, pp. 82–5.
29. A.E. Caldwell, *Origins of Psychopharmacology. From CPZ to LSD*. Springfield, Illinois: Charles C Thomas; 1970.
30. H. Laborit, Le phénomède potentialisation des anesthétiques généraux. *Presse Méd.*, 1950; **58**: 416.
31. P. Viaud, Les amines derives de la phenothiazines. *J. Pharm. Pharmacol.*, 1954; **6**: 361–89.
32. P. Charpentier, P. Gaillot, R. Jacob, J. Gaudechon, P. Buisson, Recherches sur les diméthylami-nopropyl-*N*-phénothiazines substitueés. *C. R.*, 1952; **235**: 59–60.
33. S. Courvoisier, J. Fournel, R. Ducrot, *et al.*, Propriétés pharmacodynamiques du chlorhydrate de chloro-(3-dimethylamino-3'propyl)-10 phénothiazine (4560 R.P.). *Arch. Int. Pharmacodyn.*, 1953; **92**: 305.
34. H. Laborit, P. Huguenard, R. Alluaume, Un nouveau stabilisateur végétative (le 4560 RP). *Presse Méd.*, 1952; **60**: 206–8.
35. J. Hamon, J. Paraire, J. Velluz, Remarques sur l'action du 4560 RP sur l'agitation maniaque. *Ann. Medicopsychol. (Paris)*, 1952; **110**: 331–5.
36. J. Delay, P. Deniker, J.M. Harl, Utilisation en thérapeutique psychiatrique d'une phénothiazine d'action centrale elective (4560 RP). *Ann. Medicopsychol. (Paris)*, 1952; **110**: 112–31.
37. W. Sneader, The 50th anniversary of chlorpromazine. *Drug News Perspect.*, 2002; **15**: 466–71.
38. S. Courvoisier, R. Ducrot, J. Fournel, *et al.*, *C.R. Soc. Biol.*, 1958; **152**: 1371–5.

39. H.L. Yale, F. Sowinski, 4-{3-[10-(2-Trifluoromethyl)-phenothiazinyl]-propyl}-1-piperazine-ethanol l and related compounds. II. *J. Am. Chem. Soc.*, 1960; **82**: 2039–42.

40. P.V. Petersen, N.L. Lassen, T. Holm, *et al.*, Chemical structure and pharmacological effects of the thiaxanthene analogs of chlorpromazine, promazine and mepazine. *Arzneimittel.-Forschung.*, 1958; **8**: 395–7.

41. J. Ravn, The history of thioxanthenes, in *Discoveries in Biological Psychiatry*, eds F.J. Ayd and B. Blackwell, Philadelphia, Pennsylvania: Lippincott; 1970, p. 180.

42. G.E. Bonvicino, H.G. Arlt Jr, K.M. Pearson, R.A. Hardy Jr, Tranquilizing agents. Xanthen- and thioxanthen-$\Delta^{9,\gamma}$-propylamines and related compounds. *J. Org. Chem.*, 1961; **26**: 2383–92.

43. C.L. Cazzullo, M.L. Andreola, [Central nervous system observations of a thioxanthenic derivative: N 746. (Semeiologic, anatomo-pathologic, experimental study)]. *Acta Neurol.* (*Napoli*), 1965; **20**: 162–84.

44. M. Schonecker, Ein eigentumliches Syndrom im oralen Bereich bei Megaphen Applikation. *Nervenarzt*, 1957; **28**: 35.

45. L.H. Sternbach, E. Reeder, Quinazolines and 1,4-benzodiazepines. II.1 The rearrangement of 6-chloro-2-chloromethyl-4-phenylquinazoline 3-oxide into 2-amino derivatives of 7-chloro-5-phenyl-3H-1,4-benzodiazepine 4-oxide. *J. Org. Chem.*, 1961; **26**: 1111–8.

46. L.H. Sternbach, The discovery of librium. *Agents Actions*, 1972; **2**: 193–6.

47. L.H. Sternbach, E. Reeder, Quinazolines and 1,4-benzodiazepines. IV. Transformations of 7-chloro-2-methylamino-5-phenyl-3H-1,4-benzodiazepine 4-oxide. *J. Org. Chem.*, 1961; **26**: 4936–41.

48. S.C. Bell, S.J. Childress, A rearrangement of 5-aryl-1,3-dihydro-2H-1,4-benzodiazepine-2-one 4-oxides. *J. Org. Chem.*, 1962; **27**: 1691–5.

49. S.J. Childress, M.I. Gluckman, 1,4-Benzodiazepines. *J. Pharm. Sci.*, 1964; **53**: 577–90.

50. W. Hunkeler, H. Mohler, L. Pieri, *et al.*, Selective antagonists of benzodiazepines. *Nature*, 1981; **290**: 514–6.

51. M. Sato, T. Nagao, I. Yamaguchi, *et al.*, Pharmacological studies on a new l,5-benzothiazepine derivative (CRD-401). *Arzneimittel.-Forsch.*, 1971; **21**: 1338–43.

52. T. Nagao, M. Sato, Y. Iwasawa, *et al.*, Studies on a new 1,5-benzothiazepine derivative (CRD-401). 3. Effects of optical isomers of CRD-401 on smooth muscle and other pharmacological properties. *Jap. J. Pharmacol.*, 1972; **22**: 467–78.

53. L.H. Sternbach, R.I. Fryer, O. Keller, *et al.*, Quinazolines and 1,4-benzodiazepines. X. Nitro-substituted 5-phenyl-1,4-benzodiazepine derivatives. *J. Med. Chem.*, 1963; **6**: 261–5.

54. S.C. Bell, R.J. McCaully, C. Gochman, *et al.*, 3-Substituted 1,4-benzodiazepin-2-ones. *J. Med. Chem.*, 1968; **11**: 457–61.

55. I.R. Ager, G.W. Danswan, D.R. Harrison, *et al.*, Central nervous system activity of a novel class of annelated 1,4-benzodiazepines, aminomethylene-2,4-dihydro-1H-imidazo[1,2-a][1,4]benzodiazepin-1-ones. *J. Med. Chem.*, 1977; **20**: 1035–41.

56. A. Walser, L. E. Benjamin, T. Flynn, Quinazolines and 1,4-benzodiazepines. 84. Synthesis and reactions of imidazo[1,5-a][1,4]benzodiazepines. *J. Org. Chem.*, 1978; **43**: 936–44.

57. US Pat. 1951: 2554736 (to Geigy).

58. W. Schindler, F. Haefliger, *Helv. Chim. Acta*, 1954; **37**: 472.

59. R. Kuhn, [Treatment of depressive states with an iminodibenzyl derivative (G 22355).] *Schweiz. Med. Wochenschr.*, 1957; **87**: 1135–40.

60. R. Kuhn, The imipramine story, in *Discoveries in Biological Psychiatry*, eds F.J. Ayd and B. Blackwell, Philadelphia: Lippincott; 1970, p. 205.

61. R.M. Jacob, M.N. Messer, Préparation de la (diméthylamino-3′-méthyl-2′-propyl)-5-dihydro-10,11-dibenz-[b,f]-azépine racémique et de ses isomères optiques. *C. R. Acad. Sci.*, 1961; **252**: 2117–8.

62. P.N. Craig, B.M. Lester, A.J. Saggiomo, *et al.*, Analogs of phenothiazines. I. 5H-dibenz[b,f]azepine and derivatives. A new isostere of phenothiazine. *J. Org. Chem.*, 1961; **26**: 135–8.

63. R.D. Hoffsomer, D. Taub, N.L. Wendler, The homoallylic rearrangement in the synthesis of amitriptyline and related systems. *J. Org. Chem.*, 1962; **27**: 4134–7.

64. K. Stach, F. Bickelhaupt, *Monatsh.*, 1962; **93**: 896.

65. M. Protiva, M. Rajner, V. Seidlová, E. Alderová, Z.J. Vejdêlek, *Experientia*, 1962; **18**: 326.

66. J. Schmutz, F. Künzle, F. Hunziker, R. Gauch, Über 11-Stellung amino-substituierte Dibenzo[b,f]-1,4-thiazapine und -oxazepine. *Helv. Chim. Acta*, 1967; **50**: 245.

67. J. Schmutz, E. Eichenberger, Clozapine, in *Chronicles of Drug Discovery*, Vol. 1, eds J.S. Bindra, D. Lednicer. New York: Wiley; 1982, pp. 39–59.

68. H.A. Amsler, L. Teerenhovi, K. Barth, *et al.*, Agranulocytosis in patients treated with clozapine. A study of the Finnish epidemic. *Acta Psychiat. Scand.*, 1977; **56**: 241–8.

69. J. Kane, G. Honigfeld, J. Singer, H. Meltzer, Clozapine for the treatment-resistant schizophrenic. A double-blind comparison with chlorpromazine. *Arch. Gen. Psychiatry.*, 1988; **45**: 789–96.

70. N.A. Moore, N.C. Tye, M.S. Axton, F.C. Risius, The behavioral pharmacology of olanzapine, a novel 'atypical' antipsychotic agent. *J. Pharmacol. Exp. Ther.*, 1992; **262**: 545–51.

71. C.F. Saller, A.I. Salama, Seroquel: biochemical profile of a potential atypical antipsychotic. *Psychopharmacology (Berl.)*, 1993; **112**: 285–92.

72. US Pat. 1960: 2948718 (to Geigy).

73. I.P. Lapin, G.F. Oxenkrug, Intensification of the central serotoninergic processes as a possible determinant of the thymoleptic effect. *Lancet*, 1969; **1**: 132–6.

74. A. Carlsson, M. Lindqvist, Central and peripheral monoaminergic membrane-pump blockade by some addictive analgesics and antihistamines. *J. Pharm. Pharmacol.*, 1969; **21**: 460–4.

75. S.B. Ross, S.O. Ogren, A.L. Renyi, (Z)-dimethylamino-1-(4-bromophenyl)-1-(3-pyridyl) propene (h 102/09), a new selective inhibitor of the neuronal 5-hydroxytryptamine uptake. *Acta Pharmacol. Toxicol. (Copenhagen)*, 1976; **39**: 152–66.

76. C. Gueremy, F. Audiau, A. Champseix, *et al.*, 3-(4-Piperidinylalkyl)indoles, selective inhibitors of neuronal 5-hydroxytryptamine uptake. *J. Med. Chem.*, 1980; **23**(12): 1306–10.

77. V. Claasen, J.E. Davies, G. Hertting, P. Placheta, Fluvoxamine, a specific 5-hydroxytryptamine uptake inhibitor. *Br. J. Pharmacol.*, 1977; **60**: 505–16.

78. D.T. Wong, F.P. Bymaster, E.A. Engleman, Prozac (fluoxetine, Lilly 110140), the first selective serotonin uptake inhibitor and an antidepressant drug: twenty years since its first publication. *Life Sci.*, 1995; **57**: 411–41.

79. S. Okamoto, A. Hijikata, Rational approach to proteinase inhibitors. *Drug Des.*, 1975; **6**: 143–69.

80. S. Okamoto, S. Oshiba, H. Mihara, U. Okamoto, Synthetic inhibitors of fibrinolysis: *in vitro* and *in vivo* mode of action. *Ann. N.Y. Acad. Sci.*, 1968; **146**: 414–29.

81. B.J. Northover, B. Verghese, The action of aryloxyaliphatic acids on the permeability of blood vessels. *J. Pharm. Pharmacol.*, 1962; **14**: 615–6.

82. S.S. Adams, R. Cobb, in *Salicylates*, eds A. St. J. Dickson, B.K. Martin, M.J. H. Smith, P.H.N. Wood. London: Churchill; 1963, p. 127.

83. S.S. Adams, E.E. Cliffe, B. Lessel, J.S. Nicholson, Some biological properties of 'ibufenac', a new anti-rheumatic drug. *Nature*, 1963; **200**: 271–2.

84. M. Thompson, P. Stephenson, J.S. Percy, Ibufenac in the treatment of arthritis. *Ann. Rheum. Dis.*, 1964; **23**: 397–404.

85. J. Nicholson, Ibruprofen, in *Chronicles of Drug Discovery*, eds J.S. Bindra, D. Lednicer. John Chichester: Wiley; 1982, pp. 149–72.

86. I.T. Harrison, B. Lewis, P. Nelson, *et al.*, Nonsteroidal antiinflammatory agents. I. 6-Substituted 2-naphthylacetic acids. *J. Med. Chem.*, 1970; **13**: 203–5.

87. A.C. Goudie, L.M. Gaster, A.W. Lake, *et al.*, 4-(6-Methoxy-2-naphthyl)butan-2-one and related analogs, a novel structural class of antiinflammatory compounds. *J. Med. Chem.*, 1978; **21**: 1260–4.

88. S. Afr. Pat. 1968: 6800524 (to Rhône–Poulenc).

89. A. Rubin, B.E. Rodda, P. Warrick, *et al.*, Physiological disposition of fenoprofen in man. I. Pharmacokinetic comparison of calcium and sodium salts administered orally. *J. Pharm. Sci.*, 1971; **60**: 1797–1801.

90. D.W. Dunwell, 2-Aryl-5-benzoxazolealkanoic acid derivatives with notable antiinflammatory activity. *J. Med. Chem.*, 1975; **18**: 53–8.

91. H.M. Taggart, Fatal cholestatic jaundice in elderly patients taking benoxaprofen. *Br. Med. J.*, 1982; **284**: 1372.

92. W. Xie, J.G. Chipman, D.L. Robertson, *et al.*, Expression of a mitogen-responsive gene encoding prostaglandin synthase is regulated by mRNA splicing. *Proc. Natl Acad. Sci USA*, 1991; **88**: 2692–6.

93. T.D. Penning, J. Talley, S.R. Bertenshaw, *et al.*, Synthesis and biological evaluation of the 1,5-diarylpyrazole class of cyclooxygenase-2 inhibitors: identification of 4-[5-(4-methylphenyl)-3-(trifluoromethyl)-1H-pyrazol-1-yl]benzenesulfonamide (SC-58635, Celecoxib). *J. Med. Chem.*, 1997; **40**: 1347–65.

94. R.G. Wilkinson, M.B. Cantrall, R.G. Shepherd, Antituberculous agents. III. (+)-2,2 -(Ethylene-diimino)-di-1-butanol and some analogs. *J. Med. Chem.*, 1962; **5**: 835–45.

95. A.H.M. Raeymaekers, F.T.N. Allewijn, J. Vanderbek, *et al.*, Novel broad-spectrum anthelmintics. Tetramisole and related derivatives of 6-arylimidazo[2,1-b]thiazole. *J. Med. Chem.*, 1966; **9**: 545–51.

96. M.W. Bullock, J.J. Hand, E. Waletzky, Resolution and racemization of *dl*-tetramisole, *dl*-6-phenyl-2,3,5,6-tetrahydroimidazo-[2,1-b]thiazole. *J. Med. Chem.*, 1968; **11**: 169–71.

97. P.A.J. Janssen, The levamisole story. *Prog. Drug Res.*, 1976; **20**: 347–83.

98. J.E. Lynch, B. Nelson, Preliminary anthelmintic studies with *Nematospiroides dubius* in mice. *J. Parasitol.*, 1959; **45**: 659–62.

99.   J.W. McFarland, L.H. Conover, H.L. Howes Jr, *et al.*, Novel anthelmintic agents. II. Pyrantel and other cyclic amidines. *J. Med. Chem.*, 1969; **12**: 1066–79.

100. J.W. McFarland, in *Chronicles of Drug Discovery*, Vol. 2, eds J.S. Bindra, D. Lednicer. Chichester: Wiley; 1983, p. 87.

101. E.L. Lee, N. Iyngkaran, A.W. Grieve, *et al.*, Therapeutic evaluation of oxantel pamoate (1, 4, 5, 6-tetrahydro-1-methyl-2-[trans-3-hydroxystyryl] pyrimidine pamoate) in severe *Trichuris trichiura* infection. *Am. J. Trop. Med. Hyg.*, 1976; **25**: 563–7.

102. F. Bossert, W. Vater, [Dihydropyridines, a new group of strongly effective coronary therapeutic agents.] *Naturwiss.*, 1971; **58**: 578.

103. S. Afr. Pat. 1968: 6801482.

104. Eur. Pat. Appl. 1980: 7293 (to AB Hassle).

105   J.E. Arrowsmith, S.F. Campbell, P.E. Cross, Long-acting dihydropyridine calcium antagonists. 1. 2-Alkoxymethyl derivatives incorporating basic substituents. *J. Med. Chem.*, 1986; **29**: 1696–702.

106. R. Towart, S. Kazda, The cellular mechanism of action of nimodipine (BAY e 9736), a new calcium antagonist. *Br. J. Pharmacol.*, 1979; **67**: 409P–410P.

107. S.A. Schepartz, Early history and development of the nitrosoureas. *Cancer Treat. Rep.*, 1976; **60**: 647–9.

108. T.P. Johnston, G.S. McCaleb, J.A. Montgomery, The synthesis of antineoplastic agents. XXXII. *N*-Nitrosoureas. *J. Med. Chem.*, 1963; **6**: 669–81.

109. F.M. Schabel Jr, T.P. Johnston, G.S. McCaleb, *et al.*, Experimental evaluation of potential anticancer agents. VIII. Effects of certain nitrosoureas on intracerebral L1210 leukemia. *Cancer Res.*, 1963; **23**: 725–33.

110. S.K. Carter, F.M. Schabel Jr, L.E. Broder, T.P. Johnston, 1,3-bis(2-Chloroethyl)-1-nitrosourea (BCNU) and other nitrosoureas in cancer treatment: a review. *Adv. Cancer Res.*, 1972; **16**: 273–332

111. T.P. Johnston, G.S. McCaleb, J.A. Montgomery, The synthesis of potential anticancer agents. XXXVI. *N*-Nitrosoureas.1 II. Haloalkyl derivatives. *J. Med. Chem.*, 1966; **9**: 892–911.

112. J. Seubert, R. Pohlke, F. Loebich, Synthesis and properties of praziquantel, a novel broad spectrum anthelmintic with excellent activity against schistosomes and cestodes. *Experientia*, 1977; **33**: 1036–7.

113. P. Andrews, H. Thomas, R. Pohlke, J. Seubert, Praziquantel. *Med. Res. Rev.*, 1983; **3**: 147–200.

114. A.E. Farah, A.A. Alousi, New cardiotonic agents: a search for digitalis substitute. *Life Sci.*, 1978; **22**: 1139–47.

115. S.D. Levine, M. Jacoby, J.A. Satriano, D. Schlondorff, The effects of amrinone on transport and cyclic AMP metabolism in toad urinary bladder. *J. Pharmacol. Exp. Ther.*, 1981; **216**: 220–4.

116. A.A. Alousi, J.M. Canter, M.J. Montenaro, *et al.*, Cardiotonic activity of milrinone, a new and potent cardiac bipyridine, on the normal and failing heart of experimental animals. *J. Cardiovasc. Pharmacol.*, 1983; **5**: 792–803.

117. R.A. Schnettler, R.C. Dage, J.M. Grisar, 4-Aroyl-1,3-dihydro-2H-imidazol-2-ones, a new class of cardiotonic agents. *J. Med. Chem.*, 1982; **25**: 1477–81.

118. H. Loibner, A. Pruckner, A. Stutz, Reduktive Methylierung primarer und sekundarer Amine mit Hilfe von Formaldehyd und Salzen der phosphorigen Saure. *Tetrahedron Lett.*, 1984; **25**: 2535–6.

119. A. Georgopoulos, G. Petranyi, H. Mieth, J. Drews, *In vitro* activity of Naftine®. A new antifungal agent. *Antimicrob. Ag. Chemother.*, 1981; **19**: 386.

120. A. Stütz, A. Georgopoulos, W. Granitzer, *et al.*, Synthesis and structure–activity relationships of naftifine-related allylamine antimycotics. *J. Med. Chem.*, 1986; **29**: 112–5.

121. A. Stütz, G. Petranyi, Synthesis and antifungal activity of (*E*)-*N*-(6,6-dimethyl-2-hepten-4-ynyl)-*N*-methyl-1-naphthalenemethanamine (SF 86-327) and related allylamine derivatives with enhanced oral activity. *J. Med. Chem.*, 1984; **27**: 1539–43.

122. G. Petranyi, A. Stütz, N.S. Ryder, J.G. Meingassner, H. Mieth, in *Recent Trends in the Discovery, Development and Evaluation of Antifungal Agents*, ed. R.A. Fromtling. Barcelona: J.R. Prous Science Publishers; 1987, p. 441.

123. R. James, J.B. Glen, Synthesis, biological evaluation, and preliminary structure–activity considerations of a series of alkylphenols as intravenous anesthetic agents. *J. Med. Chem.*, 1980; **23**: 1350–7.

124. J. Weinstock, R.M. Keenan, J. Samanen, *et al.*, 1-(Carboxybenzyl)imidazole-5-acrylic acids: potent and selective angiotensin II receptor antagonists. *J. Med. Chem.*, 1991; **34**: 1514–7.

125. K. Kubo, Y. Kohara, E. Imamiya, *et al.*, Nonpeptide angiotensin II receptor antagonists. Synthesis and biological activity of benzimidazolecarboxylic acids. *J. Med. Chem.*, 1993; **36**: 2182–95.

126. C.A. Bernhart, P.M. Perreaut, B.P. Ferrari, *et al.*, A new series of imidazolones: highly specific and potent nonpeptide AT1 angiotensin II receptor antagonists. *J. Med. Chem.*, 1993; **36**: 3371–80.

127. P. Bühlmayer, *et al.* Valsartan, a potent orally active angiotensin II antagonist developed from the structurally new amino acid series. *Bioorg. Med. Chem. Lett.*, 1994; **4**: 29–34.

128.   H. Yanagisawa, Y. Amemiya, T. Kanazaki, *et al.*, Nonpeptide angiotensin II receptor antagonists: synthesis, biological activities, and structure–activity relationships of imidazole-5-carboxylic acids bearing alkyl, alkenyl, and hydroxyalkyl substituents at the 4-position and their related compounds. *J. Med. Chem.*, 1996; **39**: 323–38.

129.   U.J. Ries, G. Mihm, B. Narr, *et al.*, 6-Substituted benzimidazoles as new nonpeptide angiotensin II receptor antagonists: synthesis, biological activity, and structure–activity relationships. *J. Med. Chem.*, 1993; **36**: 4040–51.

130.   J. Zimmermann, E. Buchdunger, H. Mett, *et al.*, Phenylamino-pyrimidine (PAP) derivatives: a new class of potent and highly selective PDGF receptor autophosphorylation inhibitors. *Bioorg. Med. Chem. Lett.*, 1996; **6**: 1221–6.

131.   M.W. Deininger, B.J. Druker, Specific targeted therapy of chronic myelogenous leukemia with imatinib. *Pharmacol Rev.*, 2003; **55**: 401–23.

Drugs Discovered through Serendipitous Observations Involving Humans

Horace Walpole introduced the word 'serendipity' over 200 years ago in a letter he wrote to a friend after reading a poem about three princes of Serendip (Sri Lanka) who had repeatedly made discoveries they were not seeking. Walpole explained that his new word referred to chance discoveries that were exploited with sagacity. It is a word that repeatedly reappears in the literature of drug research, not so much with regard to the discovery of new drug prototypes but rather with regard to discovering new applications of existing drugs and their analogues. This chapter and the next, however, deal with serendipitous discovery of several prototypes and the drugs that evolved from them.

Some drug prototypes have been discovered as a consequence of their effects being noticed by those exposed to them. While the actions of certain chemicals on the nervous system were obvious, their relevance to clinical practice could easily be overlooked unless an individual had the sagacity to exploit them. This has also been the case in the laboratory, when experiments sometimes resulted in unexpected outcomes, the most celebrated instance being the discovery by Fleming of penicillin.

NITRATES

The product generally known as amyl nitrite actually consists mainly of isoamyl nitrite. It was first synthesised at the Sorbonne in 1844 by Antoine Balard, who reported that its vapour had given him a severe headache.[1] Frederick Guthrie of Owens' College, in Manchester, experimented with amyl nitrite and found that its most significant action was upon the heart.[2] He found that it produced intense throbbing of the carotid artery, flushing of the face and an increase in heart rate.

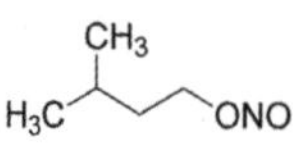

isoamyl nitrite

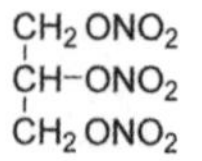

glyceryl trinitrate

$$CH_2\,ONO_2$$
$$CH\text{–}ONO_2$$
$$CH_2\,ONO_2$$

Thomas Lauder Brunton, a newly qualified house surgeon at the Edinburgh Royal Infirmary, pioneered the clinical application of the sphygmograph in 1867 by employing it to monitor the rise in blood pressure that accompanied attacks of angina pectoris in his patients. Faced with one particular patient who nightly experienced paroxysmal attacks of angina, he followed the fashion of the time by removing blood through either cupping or venesection. This appeared helpful, leaving Brunton convinced that the relief of pain was due to the lowering of arterial pressure. He then speculated that amyl nitrite might help his patient, having seen his friend Arthur Gamgee use it in animals to lower blood pressure.

Drug Discovery. A History. W. Sneader.
©2005 John Wiley & Sons Ltd: ISBN 0 471 89979 8 (HB); 0 471 89980 1 (PB)

Brunton obtained a sample of amyl nitrite from Gamgee and poured some on to a cloth for his patients to inhale. Within a minute, their agonising chest pains had disappeared, several remaining free of pain for hours. The success of the drug was assured, and it was universally adopted after Brunton reported his observations.[3,4]

Over the next few years, Brunton established that other nitrites had similar effects. He also examined nitroglycerin, which had become readily available following Alfred Nobel's discovery of its value as an explosive. Finding that when he and a colleague tried nitroglycerin it gave them both a severe headache, Brunton did not pursue its use further. His observations were mentioned in the *St Bartholomew's Reports* in 1876, but matters did not rest there as William Murrell at the Westminster Hospital decided to resolve the conflicting reports in the literature over whether or not nitroglycerin caused severe headache. Its discoverer, the Turin chemist Ascanio Sobrero,[5] had reported in 1847 that he experienced an intense headache after merely tasting a drop of it placed on his finger. Others had a similar experience, including Arthur Field, a Brighton dentist, who claimed that he had alleviated toothache and neuralgia by applying a drop or two of a dilute alcoholic solution of nitroglycerin to the tongue.[6] Were it not for the ensuing headache, so he claimed, nitroglycerin could be a valuable remedy. In an ensuing correspondence, some writers confirmed Field's observations, while others claimed nitroglycerin produced no effects whatsoever despite their swallowing large amounts. Murrell suspected that this confusion could have arisen from variation in the susceptibility of individuals to nitroglycerin.

In the course of one of his clinics, Murrell casually licked the moist cork of a bottle of nitroglycerin solution that was in his pocket. Within a few moments he began to experience throbbing in the neck and head, accompanied by pounding of his heart, the very effects that Field had so vividly described. Such was his discomfiture that Murrell could not continue with his examination of a patient. After five minutes, he had recovered sufficiently to resume his duties, but his severe headache lasted all afternoon. He subsequently tested nitroglycerin on himself on a further 30 or 40 occasions before persuading friends and volunteers to take part in a trial of its effects. This brought home to him the similarity between its action and that of amyl nitrite, but there was one important difference. Charting the changes in blood pressure of the volunteers, Murrell established that although it took 2 or 3 minutes for nitroglycerin to produce its effects, as opposed to only 10 seconds or so for amyl nitrite, they persisted for about half an hour. This was in marked contrast to amyl nitrite, the effects of which wore off after 5 minutes. This persuaded Murrell that nitroglycerin might be superior in the treatment of angina pectoris. He also confirmed that the headache was due to overdosage. The correct dose turned out to be in the order of 0.5–1 mg when the drug was formulated in tablets that were allowed to dissolve slowly under the tongue, whereas around 100–300 mg of amyl nitrite had to be inhaled to produce similar effects. If the nitroglycerin were swallowed, a larger dose was required since absorption from the gut was less efficient than from the mouth, or even through the skin.

Murrell began treating patients with nitroglycerin in 1878, and the following year published his first report in the *Lancet*.[7] This led to the general adoption of the drug into clinical practice. British doctors took steps to avoid any unnecessary alarm that might result if patients discovered they were receiving the same explosive as was in dynamite by renaming it as glyceryl trinitrate or trinitrin.

Tablets of glyceryl trinitrate taken sublingually are still commonly used to deal with attacks of angina caused by exertion, but any prophylactic effect is of short duration. In recent years, an old form of percutaneous treatment with ointment containing glyceryl trinitrate has been adapted by the use of much more reliable transdermal patches that release the active constituent in a controlled manner over a prolonged period during which there is effective prophylaxis against anginal attacks.

As nitroglycerin was an ester prepared by the nitration of a polyhydroxylic alcohol, analogues of it were synthesised by the nitration of sugars in attempts to obtain longer-acting drugs for prophylactic use. The first was pentaerythritol tetranitrate, which was introduced in 1896, although its synthesis was not reported until 1901. There has been controversy as to whether it and other longer-acting nitrate esters such as isosorbide dinitrate are of any real value in preventing attacks of angina as the rate of their metabolism in the liver is such that little unchanged drug enters the circulation.

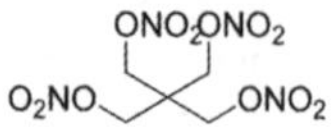

pentaerythritol tetranitrate

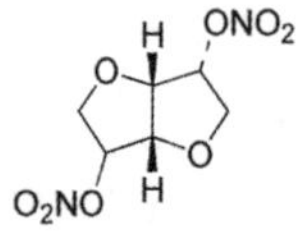

isosorbide dinitrate

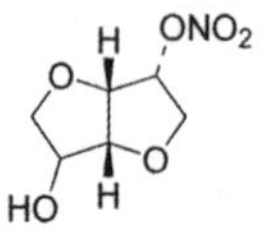

isosorbide mononitrate

The active metabolite of isosorbide dinitrate is the mononitrate, which is twice as potent and has a less complicated pharmacokinetic profile that makes the response more predictable.[8]

LIDOCAINE

In the course of a pioneering chemical plant taxonomy study on a chlorophyll-deficient mutant of barley, Hans von Euler at the University of Stockholm isolated an alkaloid called 'gramine'.[9] In order to confirm that it was 2-(dimethylaminoethyl)-indole, von Euler's assistant Holger Erdtman synthesised this compound in 1935. The synthetic product turned out to be an isomer of gramine and so was named 'isogramine'.

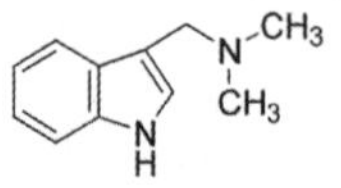

gramine

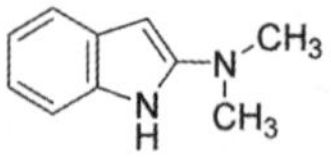

isogramine

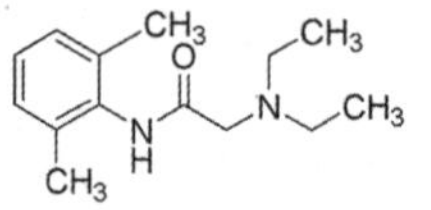

lidocaine

On tasting a trace of isogramine, Erdtman noted that it numbed his tongue. Further investigation revealed that a similar local anaesthetic activity existed in its open-chain synthetic precursor. This persuaded Erdtman and his research student Nils Löfgren to seek less-irritant analogues for possible clinical use. The task proved daunting, as most of the active analogues were irritant. Löfgren's persistence finally paid off after 57 compounds had been synthesised over a 7 year period. After his colleague Bengt Lundqvist had tested compound LL 30 on himself, he suggested it should be pharmacologically evaluated at the Karolinska Institute. The results were encouraging and clinical trials were then arranged. Löfgren then approached a small Swedish company called Astra, which resulted in lidocaine being marketed in Sweden in 1948.[10] Over the next few years, as a result of its rapid onset of action, relative safety and freedom from irritancy, it attained a pre-eminence over all other local anaesthetics. It was also administered intravenously as an anti-arrhythmic agent in coronary care units, being preferred to procaine as it had a faster onset of activity. In 1960, Löfgren synthesised prilocaine, which had a wider safety margin.[11]

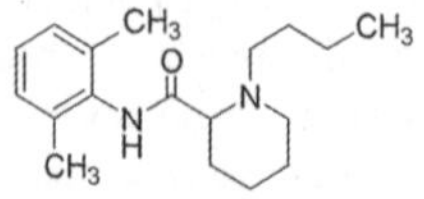

bupivacaine

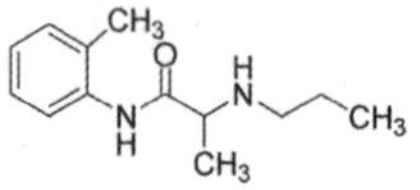

prilocaine

In 1957, scientists from the Swedish pharmaceutical company AB Bofors investigated the pharmacological activity of a series of lidocaine analogues in which the side chain had been partially incorporated into a cyclic system.[12] This had been done to establish whether a cyclic analogue offered any advantages. One of the resulting compounds, bupivacaine, turned out to be longer-acting, producing nerve blocks for up to 8 hours. It became widely used for continuous epidural anaesthesia during childbirth.

ANTICHOLINESTERASES

At the University of Berlin, in 1932, Willy Lange and his student Gerda von Kreuger took advantage of the recent ready availability of fluorine to prepare the first phosphorus–fluorine compounds. In the course of their work they experienced marked pressure in the larynx, followed by breathlessness, clouding of consciousness and blurring of vision. These effects were similar to those produced by nicotine, which was used as an insecticide, killing insects by relentlessly mimicking acetylcholine to disrupt nervous transmission. Lange mentioned this at the end of his paper.[13] He left Germany shortly after.

At the Leverkusen laboratories of I.G. Farbenindustrie, Gerhard Schrader followed up this observation and prepared more than 2000 organophosphorus compounds as potential insecticides. It was shown (although not divulged) by Eberhard Gross at the company's Elberfeld laboratories that these were anticholinesterases, some of which were highly poisonous to laboratory animals and thus of little value as insecticides. Unlike the natural anticholinesterase physostigmine, these organophosphorus compounds formed such a strong bond with the enzymes that their action was permanent. Under these circumstances, acetylcholine could not be broken down and nervous transmission was disrupted. This inevitably resulted in death.

The lethal action of the organophosphorus compounds was ruthlessly exploited in the development of war gases with the terrifying potential to kill entire populations. Their value as chemical warfare agents was fully appreciated by the German authorities, who stockpiled them for military use. The first agent to be so employed was tabun, synthesised by Schrader in 1936.[14] Described as a 'war gas', tabun was actually a liquid that could be deployed in a fine dispersion. A plant disguised as a soap-making factory was opened at Dyhernfurth-am-Oder, near Breslau, near the Polish border in 1942 for its production. By the end of the war, 12 000 tons had been manufactured, and field forces were equipped with tabun-filled shells.[15] Tabun was allegedly used for the first time in 1980 during the Iran–Iraq war.[16]

The far more toxic sarin was developed by Schrader and Otto Ambros in 1938, being named after them and Rudriger and van der Linde of the chemical warfare division of the Wehrmacht; a mere 1 mg of this was capable of killing an adult within minutes after being absorbed through the skin.[14] In the early 1950s, the United States supplied its chemical warfare units around the world with sarin, listed under the code name of GB.[15] Later in the decade, production began of the even more toxic VX agent.[17]

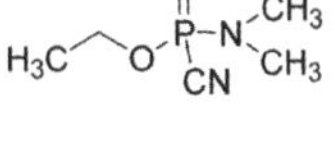

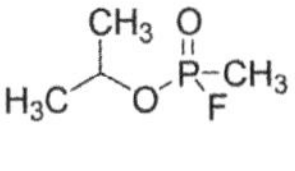

The British Ministry of Supply arranged for the potential of the fluorophosphonates to be studied at the physiology laboratory in Cambridge University by Edgar Adrian and his colleagues. They found, in 1941, dyflos to be the most toxic of the compounds originally prepared by Lange and von Krueger.[18] Its prolonged miotic action on the eye convinced them that it was an anticholinesterase, and direct evidence for this was obtained. When the war ended, dyflos found occasional clinical application in the treatment of glaucoma.

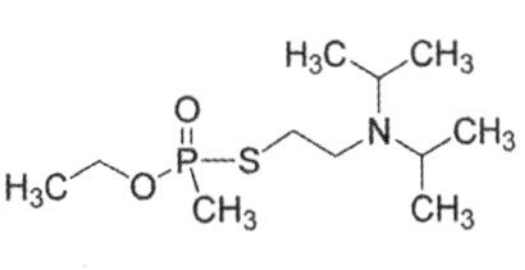

American Cyanamid introduced malathion as an insecticide in 1951.[19] It represented a major advance from the point of view of safety for both those who handled insecticides and also farm animals as it incorporated two features that reduced mammalian toxicity. Firstly, it was metabolised in insects to form the active insecticide malaoxon. The second inherent safety feature was that if it was absorbed it rapidly underwent esterase hydrolysis to form an acid that was rapidly excreted. This did not happen in insects, hence the enhanced safety of malathion which permits its topical application to treat scabies, head lice and crab lice.

In 1952, Schrader and his colleagues at the Bayer laboratories in Elberfeld synthesised metrifonate.[20] It had a sufficiently wide margin of safety for it to be administered as a systemic insecticide in domestic animals. Though itself not an anticholinesterase, metrifonate spontaneously rearranged in aqueous solution to form the active agent dichlorvos.[21]

Jacques Cerf, a physician working in the Belgian Congo (now Democratic Republic of the Congo), tested ten organophosphorus insecticides to see whether any of them could destroy samples of _Ascaris lumbricoides_ that he had cultured in Ringer's solution. After finding metrifonate to be active, he arranged for the commercial powder to be formulated in tablets at a local pharmacy. Cerf then experimented on himself to establish a safe dose and gave this amount to 15 volunteers whom he carefully monitored for toxic effects.[22] Cerf went on to conduct a trial on 2000 patients, most of whom were infected with _Ascaris_ or _Anchylostoma duodenale_. From this, it became clear that metrifonate was a highly effective anthelmintic agent.[23] Trials in Egypt subsequently established that metrifonate could also eradicate bilharzia caused by _Schistosoma haematobium_. It became the drug of choice for treating bilharzia until praziquantel was introduced.[24]

REFERENCES

1. A.J. Balard, _C.R. Acad. Sci._, 1844; **19**, 634.
2. F. Guthrie, Contributions to the knowledge of the amyl group. 1. Nitryl of amyl and its derivatives. _J. Chem. Soc._, 1859; **11**: 245–52.
3. T.L. Brunton, On the use of nitrite of amyl in angina pectoris. _Lancet_, 1867; **2**: 97–8.
4. W.B. Fye, Nitroglycerin: Lauder Brunton and amyl nitrite: a Victorian vasodilator. _Circulation_, 1986; **74**: 222–9.
5. A. Sobrero, Sur plusiers composes detonants produit avec l'acide nitrique et le sucre, la dextrine, la lacitine, la mannite et la glycerine. _C.R. Acad. Sci._, 1847; **24**: 247–8.
6. A. Field, On the toxicol and medicinal properties of nitrate of oxide of glycl. _Med. Times Gaz._,1858; **37**: 291; 1859: **39**: 340.
7. W. Murrell, Nitro-glycerine as a remedy for angina pectoris. _Lancet_, 1879; **1**: 80–1.
8. U. Elkayam, W.S. Aronow, Glyceryl trinitrate (nitroglycerin) ointment and isosorbide dinitrate: a review of their pharmacological properties and therapeutic use. _Drugs_, 1982; **23**: 165–94.
9. H. von Euler, _et al._, _Z. Physiol. Chem._, 1933; **217**: 23.
10. N. Löfgren, B. Lundqvist, Studies on local anaesthetics II. _Svensk. Chem. Tidskr._, 1946; **58**: 206–17.
11. N. Löfgren, C. Tegner, _Acta Chem. Scand._, 1960; **14**: 486.
12. B. Ekenstam, B. Egner, G. Pettersson, _N_-alkyl pyrolidine and _N_-alkyl piperidine carboxylic acid amides. _Acta Chem. Scand._, 1957; **11**: 1183–90.
13. W. Lange, G. von Kreuger, Ueber ester der Monofluorphosphorsaure. _Ber._, 1932; **65**: 1598–601.
14. B. Holmstedt, Synthesis and pharmacology of dimethyl-amidoethoxy-phosphoryl cyanide (Tabun) together with a description of some allied anticholinesterase compounds containing the N–P bond. _Acta Physiol. Scand._, 1951; **25** (Suppl.): 90.
15. S.M. Hersh, _Chemical and Biological Warfare. America's Hidden Arsenal_. London: MacGibbon and Kee; 1968.

16. J.P. Robinson, J. Goldblat, *Sipri Fact Sheet. Chemical Weapons I*. Stockholm: International Peace Research Institute; 1984.
17. Br. Pat. 1974: 1346409 (to UK Secretary of State for Defence).
18. E.D. Adrian, W. Feldberg, B.A. Kilby, *Nature*, 1946; **158**: 625.
19. U.S. Pat. 1951: 2578652 (to American Cyanamid).
20. W. Lorenz, A. Henglein, G. Schrader, The new insecticide *O,O*-dimethyl 2,2,2-trichloro-1-hydroxyethylphosphonate. *J. Am. Chem. Soc.*, 1955; **77**: 2554–6.
21. I. Nordgren, M. Bergström, B. Holmstedt, M. Sandoz, Transformation and action of metrifonate. *Arch. Toxicol.*, 1978; **41**: 31–41.
22. A. Lebrun, J. Cerf, Noté préliminaire sur la toxicité pour l'homme d'un insecticide organo-phosphoré (Dipterex). *Bull. World Health Org.*, 1960; **22**: 579.
23. J. Cerf, A. Lebrun, A new approach to helminthiasis control: the use of an organophosphorous compound. *Am. J. Trop. Med. Hyg.*, 1962; **11**: 514–17.
24. J.M. Jesbury, M.J. Cooke, M.C. Weber, Field trial of metrifonate in the treatment and prevention of schistosomiasis infection in man. *Ann. Trop. Med. Parasit.*, 1977; **71**: 67–83.

Drugs Discovered through Serendipity in the Laboratory

The previous chapter was concerned with the serendipitous discovery of drug prototypes after observations were made on people or patients exposed to chemical compounds. However, the most celebrated serendipitous discovery of all – the discovery of penicillin – was made in the laboratory. It was, however, by no means the sole discovery of that type. Others are now discussed.

ACETANILIDE

The department of internal medicine at the University of Strassburg in the 1880s was noted for its investigations into intestinal worms. Adolf Kussmaul, the director, asked two young assistants, Arnold Cahn and Paul Hepp, to treat patients with naphthalene as it had been used elsewhere as an internal antiseptic. The young doctors were disappointed with the initial results, but Hepp persevered with the naphthalene treatment in a patient suffering from a variety of complaints besides worms. Surprisingly, the fever chart revealed a pronounced antipyretic effect from this treatment. This had not been observed before, but further investigation revealed that Hepp had wrongly been supplied by Kopp's Pharmacy in Strassburg with acetanilide instead of naphthalene! Cahn and Hepp lost no time in publishing a report on their discovery of a new antipyretic.[1] This appeared in August 1886, and a small factory, Kalle and Company, situated outside Frankfurt set up in competition with Hoechst's Antipyrin® (i.e. phenazone), mischievously calling their product Antifebrin®. In 1908, however, the Farbenfabriken Hoechst obtained control of Kalle and Company, which had grown considerably in size largely due to the success of their antipyretic. As acetanilide was cheaper to manufacture than other antipyretics, it remained in use for many years, despite the fact that it inactivated some of the haemoglobin in red blood cells, a medical condition known as methaemoglobinaemia. Sometimes acetanilide was used illicitly as a cheap adulterant of other antipyretics.

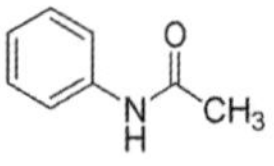

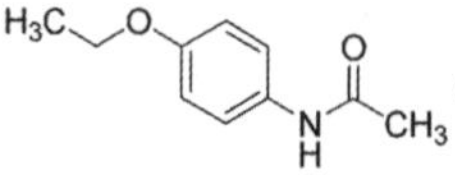

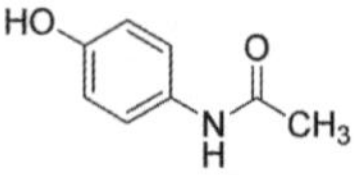

Immediately after the publication of the report of the antipyretic activity of acetanilide, Carl Duisberg, chief research chemist at F. Bayer & Company in Elberfeld, decided that its 4-methoxy and 4-ethoxy derivatives should be prepared. He assigned the task to Otto Hinsberg, a lecturer at the University of Freiburg who was working at Elberfeld during the summer vacation. When the task was completed, Hinsberg gave the two new compounds to the Professor of Pharmacology at Freiburg, Alfred Kast. Kast then demonstrated that both

Drug Discovery. A History. W. Sneader.
©2005 John Wiley & Sons Ltd

were antipyretics, noting that the ethyl ether was less toxic than acetanilide.[2] It was promptly put on the market as Phenacetin®, a proprietary name that suffered a similar fate to Aspirin® at the end of the First World War. In countries where this name continued to be recognised as the property of the Bayer Company, the approved name became acetophenitidin. It became a highly successful product, establishing F. Bayer & Company as a leading pharmaceutical manufacturer. Acetophenetidin remained popular for about 90 years until mounting concern about kidney damage in chronic users led to restrictions on its supply.

Many attempts were made to find an antipyretic superior to phenacetin. The eminent clinical pharmacologist Joseph von Mering collaborated with the Bayer Company in a trial of paracetamol in 1893. He found it to be an effective antipyretic and analgesic, but claimed that it had a slight tendency to produce methaemoglobinaemia. This could conceivably have been caused by the contamination of his paracetamol with the 4-aminophenol from which it was synthesised. Such was the reputation of von Mering that nobody challenged his observations on paracetamol until half a century later, when Lester and Greenberg[3] at Yale and then Flinn and Brodie[4] at Columbia University, New York, confirmed that paracetamol was formed in humans as a metabolite of phenacetin.

In 1953, paracetamol was marketed by the Sterling–Winthrop Company. It was promoted as preferable to aspirin since it was safer in children and anyone with an ulcer. Time has shown that it is not without its disadvantages, for it is far more difficult to treat paracetamol poisoning than that caused by aspirin.

It is fair to say that if an attempt were to be made today to introduce either paracetamol or aspirin into medicine, they might be denied a license. Nevertheless, when used in moderation they remain a boon to mankind.

ANTICONVULSANTS

During the late 1930s there was some interest in a class of sedative-hypnotics known as oxazolidine-2,4-diones. One of these, propazone, proved to be an anticonvulsant, but its potent sedative activity ruled out any clinical application. At the Abbott Laboratories in Chicago, tests confirmed that none of the known oxazolidine-2,4-diones had any analgesic activity. However, Marvin Spielman found that when the lipophilicity of these compounds was increased by substituting a methyl group on the nitrogen atom, analgesics comparable with aspirin were obtained.[5]

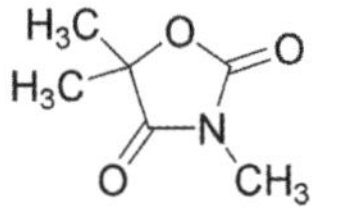

troxidone

In the course of trials to establish its clinical value, the most promising of the new analgesics, troxidone, was combined with a novel antispasmodic drug, amolanone. Toxicological studies in mice revealed that large doses of the antispasmodic that would normally induce convulsions did not do so when troxidone was concurrently administered. The obvious conclusion was drawn, namely that troxidone was an anticonvulsant. This discovery was made in 1943, and after extensive animal investigations the new anticonvulsant was administered the following year to children at the Cook County Hospital in Chicago. The trial established that, unlike phenytoin or phenobarbital, troxidone was capable of controlling petit mal absence seizures. It was the first drug ever to do this, but it caused many side effects. Nevertheless, troxidone was a turning point in the development of anticonvulsant drugs because it showed that these could be selective in their spectrum of activity. Henceforth, a battery of animal tests were set up for screening potential new anticonvulsants.

ALKYLATING DRUGS

Early in 1942, Yale University entered into a contract with the US Office of Scientific Research and Development, whereby Louis Goodman and Alfred Gilman agreed to investigate the pharmacological action of the recently developed nitrogen mustard chemical warfare agents. They were surprised to discover that the damage done to an animal after a nitrogen mustard was absorbed through its skin into the circulation was of greater consequence than the blistering action on initial skin contact. The toxicity was most extensive in rapidly dividing cells, notably the blood-forming elements in the bone marrow, lymphoid tissue and the epithelial linings of the gastrointestinal tract. The consistency of this phenomenon persuaded Goodman and Gilman to invite their colleague Thomas Dougherty to examine the influence of nitrogen mustards on transplanted lymphoid tumours in mice.

After making preliminary checks to establish a non-lethal dose range in normal mice, Dougherty administered a nitrogen mustard to a single mouse bearing a transplanted lymphoma that was expected to kill the animal within three weeks of transplantation. After only two injections had been administered, the tumour began to soften and regress, subsequently becoming unpalpable. On cessation of treatment, there was no sign of its return until a month had passed, whereupon it gradually reappeared. A second course of injections afforded a shorter respite than before; the lymphoma ultimately killed the mouse 84 days after transplantation. Such an unprecedented prolongation of life was never matched in subsequent studies on a large group of mice bearing a variety of transplanted tumours, although good remissions were frequently obtained. This was correctly seen by Goodman and Gilman to indicate a varying susceptibility of different tumours to specific chemotherapeutic agents, a view that did not accord with the perceived wisdom of the time. There was, however, no doubting the therapeutic implication of their results on animals, and in August 1942 treatment of a patient in New Haven Hospital began, under the supervision of an assistant professor of surgery at Yale. A 48 year old silversmith was dying from a radiation-resistant lymphoma that had spread over his chest and face, preventing chewing or swallowing, and causing considerable pain and distress. A nitrogen mustard, code-named HN3 (viz. 2,2',2"-trichlorotriethylamine), was administered at a dose level corresponding to that previously used for mice. Belatedly, this was found to be somewhat high, resulting in severe bone marrow damage. Nevertheless, the patient survived a full ten-day course of injections. Despite his apparently hopeless condition at the onset of therapy, he responded as dramatically as the first mouse had. An improvement was detected within two days, when the tumour masses began to shrink. On the fourth day the patient could once again swallow, while after two weeks there were no signs of any tumour masses. Bone marrow cells began to regenerate over a period of weeks, but so too did tumour masses. A brief second course of injections was of some value, but a third course failed to prevent the lethal progress of the disease.[6]

A further six terminally ill patients with a variety of neoplastic diseases were treated at New Haven before the nitrogen mustard group at Yale was disbanded in July 1943. Earlier that year, Charles Spurr, Leon Jacobson, Taylor Smith and Guzman Barron of the Department of Medicine at the University of Chicago began a full-scale clinical trial of another nitrogen mustard, then known as HN2 but later given the approved name of 'mustine', which has since been changed to chlormethine. They examined its effects on 59 patients with various blood dyscrasias and obtained spectacular remissions in patients with Hodgkin's disease, among whom were several who had ceased to respond to X-ray therapy. When Cornelius Rhoads, chief of the the Army Chemical Warfare Service based at Wedgewood Arsenal in Maryland, was informed of the results of this trial in August 1943, he arranged for further secret clinical trials in various American hospitals. Rhoads was on wartime leave of absence from his post of Director of the Memorial Hospital in New York, a leading cancer treatment centre. When the results of all the secret trials had been collated, a contract was awarded for David Karnovsky and his colleagues to organise a major clinical trial at the Memorial Hospital in order to establish the relative merits of HN2 and HN3 in patients with leukaemia, Hodgkin's disease and brain tumours.

Because of the strict wartime secrecy surrounding all work on nitrogen mustards, no information about any of the trials was released until 1946. It was then revealed that HN2 and HN3 had produced useful results in patients with Hodkgin's disease, lymphomas or chronic leukaemias, although there was doubt concerning whether they had any superiority over X-rays properly applied to solid tumours.[7] Poor results had been obtained in patients with acute leukaemia, although partial remissions occurred in some cases. Those with other neoplastic diseases failed to respond. On balance, mustine seemed a better drug than HN3.

During the war, information on nitrogen mustards had been freely exchanged between British and American investigators holding government contracts. Chemists George Hartley, Herbert Powell and Henry Rydon at Oxford University had obtained experimental proof that the chloroethyl side chain in these compounds cyclised to form highly reactive aziridine ions that could rapidly alkylate vital tissue components.[8] It is now known that inhibition of cell division occurs because DNA is alkylated by nitrogen mustards on guanine at N-7 and on adenine at N-3, with cross-linking from guanine to guanine or guanine to adenine then occurring.[9] Consequently, the generic term for anticancer drugs that act in this manner is 'alkylating drugs'.

chlormethine, R = CH$_3$
HN3, R = H

It was assumed that only compounds capable of forming aziridine ions could be effective alkylating drugs. That this was not necessarily so first became apparent in 1948 when Alexander Haddow, George Kon and Walter Ross at the Chester Beatty Institute (the research division of the Royal Cancer Hospital in London) discovered that aromatic nitrogen mustards were effective cytotoxic agents, despite being unable to form aziridine ions.[10] The following year, Reginald Goldacre, Anthony Loveless and Ross published a paper suggesting that the cytotoxic action of both aliphatic and aromatic nitrogen mustards might be due to their ability to cross-link cellular components such as the nucleic acids, and that it was not essential for aziridine ions to be formed for this to occur.[11] All that was necessary was the presence of two chemically reactive functional groups. The publication of this paper stimulated the development of several useful drugs.

Walter Ross decided to investigate the action of diepoxides that might conceivably act as cross-linking agents in a manner akin to that of the nitrogen mustards. Before he had an opportunity to put this to the test, John Speakman at the Department of Textile Industries at the University of Leeds suggested to Haddow that the biological properties of cross-linking agents used in textile technology should be examined, especially those of the diepoxides. Speakman had become interested in the work at the Chester Beatty Institute after examining the ability of aromatic nitrogen mustards to cross-link keratin fibres. A series of diepoxides subsequently prepared by James Everett and Kon turned out to act almost identically to the nitrogen mustards.[12] However, Haddow's group were not alone in discovering this. Francis Rose and James Hendry at the ICI research laboratories in Manchester had examined polymethylolamides used as cross-linking agents in paper and textile technology, fields in which their company had considerable expertise. The only compound with worthwhile cytotoxic activity turned out to be the product of condensing formaldehyde with melamine. The ICI researchers then tested epoxides and ethylene imines, also used to cross-link textile fibres. Of the former class, one of the most active compounds was diepoxybutane, which consisted of a mixture of isomers. This was one of the compounds that Ross at the Chester Beatty had examined, but it had not been considered suitable for clinical application. It was not until 1960 that Walpole found a diepoxide that was suitable, namely etoglucid.[13] It was subsequently marketed and used for a number of years in the treatment of bladder cancer.

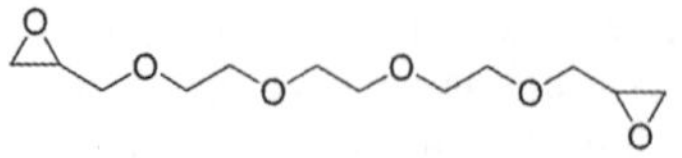

etoglucid

The most potent alkylating agent investigated by ICI was tretamine, a compound used in the textile industry to cross-link cellulose fibres. It had three aziridine rings attached to a triazine ring. On investigation, it turned out to be suitable for treating lymphatic and myeloid leukaemias, as well as Hodgkin's disease.[14] As it lacked the extremely high chemical reactivity of the nitrogen mustards, tretamine became the first alkylating drug that could be given by mouth.

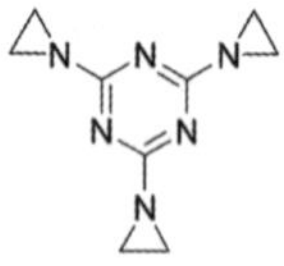

tretamine

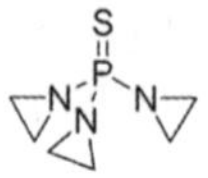

thiotepa

In 1950, after the ICI group had prepared, but not yet published, their first paper reporting their work on tretamine,[15] Joseph Burchenal and Chester Stock of the Sloane–Kettering Institute for Cancer Research in New York, in conjunction with Moses Crossley, the chief chemist at the Bound Brook laboratories of the American Cyanamid Company, published a prior report describing the action of tretamine against experimental tumours.[16] The following year, the Sloane–Kettering researchers introduced thiotepa, an alkylating drug synthesised by Crossley and his colleagues.[17] It is still used in the treatment of malignant effusions, as well as ovarian and bladder cancer.

Three frequently prescribed alkylating drugs were synthesised and evaluated at the Chester Beatty laboratories and marketed by Burroughs Wellcome. The first was busulphan, prepared in 1950 by Geoffrey Timmis, a former employee of Burroughs Wellcome.[18] It was the most potent of a series of sulfonic acid esters in which alkylsulfonyl functions could undergo nucleophilic displacement to act as alkylating agents. Clinical studies completed in 1953 confirmed that, when given by mouth, busulphan had a selective action on the blood-forming cells of the bone marrow and was effective in the treatment of chronic myeloid leukaemia.

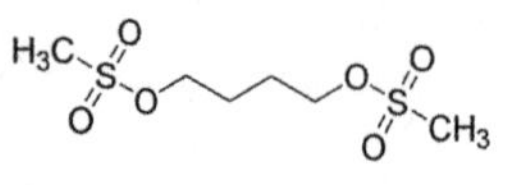

busulphan

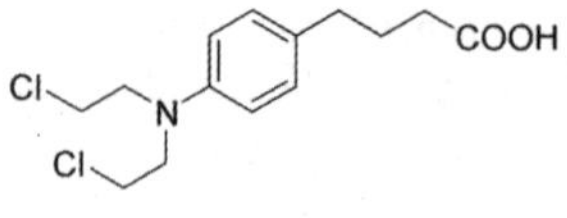

chlorambucil

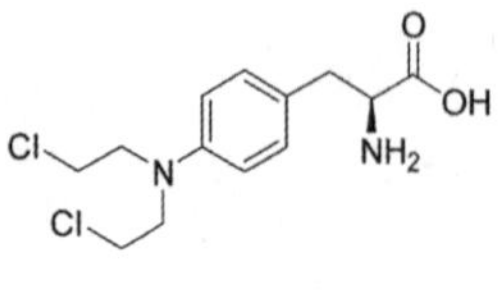

melphalan

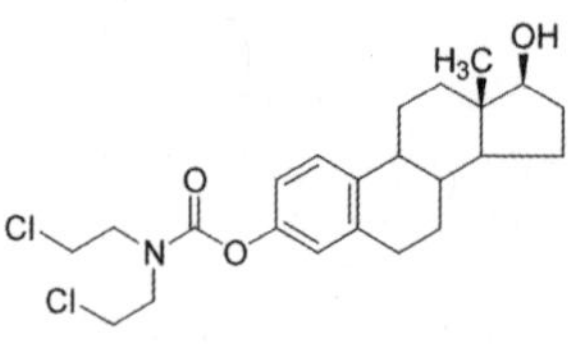

estramustine

The next important cross-linking agent developed at the Chester Beatty laboratories was chlorambucil.[19] Recognising that the clinical value of aromatic nitrogen mustards previously prepared was circumscribed by lack of specificity of action, Ross exploited a concept enunciated half a century earlier by Paul Ehrlich, who had pointed out that introduction of acidic or basic groups into dyes and other molecules greatly influenced their ability to penetrate tissues. By adding acidic side chains to aromatic mustards he hoped to obtain a less toxic drug. This objective was achieved when chlorambucil was shown, in 1952, to be less toxic to the bone marrow than was chlormethine. Since then it has been widely used in the treatment of chronic lymphocytic leukaemia and ovarian cancer.

The third alkylating drug developed at the Chester Beatty laboratories was melphalan, the synthesis of which was reported by Franz Bergel and John Stock in 1954.[20] This represented a further refinement of the design approach that led to the synthesis of chlorambucil. This time, the extra moiety incorporated in an attempt to improve tissue selectivity was alanine, a natural amino acid. It was chosen to render the polar drug similar to phenylalanine in the hope that it would enter target cells through the active transport pathway for phenylalanine. The approach succeeded and melphalan has proved to be a valuable drug in the treatment of multiple myeloma.

The success of the approach taken at the Chester Beatty laboratories encouraged researchers throughout the world to synthesise a vast range of nitrogen mustard analogues incorporating different biological carriers. Despite the expenditure of much effort, little was gained from this. The only drug of note to be developed was estramustine, which contained a nitrogen mustard function attached to oestradiol. It was developed by Niculescu-Duvaz, Cambani and Tarnauceanu at the Oncological Institute in Bucharest in 1966, and is used in the treatment of prostatic cancer.[21] The elaboration of another nitrogen mustard for treatment of prostatic cancer was based on a different, but flawed, approach. This was cyclophosphamide, developed in 1956 by Herbert Arnold, Friedrich Bourseaux and Norbert Brock of Asta-Werke AG in Brackwede, Germany.[22] The idea behind its synthesis was the same as that which had previously led Arnold to obtain a patent on the use of fosfestrol, the diphosphate ester of stilboestrol. He believed that this was devoid of hormonal activity until it decomposed in the presence of acid phosphatase, an enzyme present in prostatic tumours. Since large amounts of this enzyme were released into the circulation in patients with prostatic cancer, it is hardly surprising that there is little evidence to support the contention that fosfestrol is superior to stilboestrol. In cyclophosphamide, a nitrogen mustard function was combined with a phosphoramide residue in the hope that there would be no significant alkylating activity until enzymic action decomposed the drug. Paradoxically, cyclophosphamide turned out to have good activity against a wide variety of malignancies and chronic lymphocytic leukaemia, but not against prostatic tumours. Ten years after its introduction, Brock found that this was because rather than being decomposed by acid phosphatase, as originally hypothesised, the drug was metabolised by liver enzymes and thereby converted to an active species.[23]

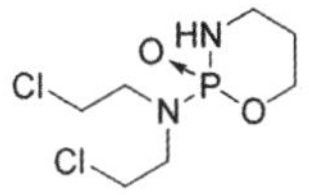

cyclophosphamide 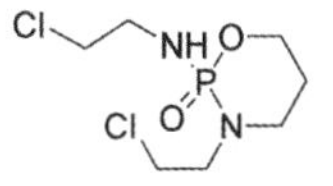   ifosfamide

Asta-Werke introduced an isomer of cyclophosphamide known as 'ifosfamide' in 1967.[24] It has similar therapeutic activity.

VALPROIC ACID

Pierre Eymard, a research student at the University of Lyon, synthesised a series of derivatives of khellin as part of his doctoral studies. After completing his thesis he arranged to have his new compounds evaluated, but when he tried to prepare a solution of the first compound to be tested he could not get it to dissolve. He then sought advice from Hélène Meunier of the Laboratoire Berthier in Grenoble. She suggested that valproic acid might be a suitable solvent as she had used it in the past to dissolve bismuth compounds for clinical evaluation.

valproic acid

The valproic acid did dissolve Eymard's compound and subsequent tests showed the khellin derivative to have anticonvulsant activity. Shortly after this, Meunier used valproic acid to dissolve a coumarin compound unrelated to Eymard's compound. When it also proved to have anticonvulsant properties, she suspected this was not mere coincidence. She immediately tested the valproic acid and discovered that it was an anticonvulsant. After detailed investigations, valproic acid was subjected to extensive clinical evaluation before its sodium salt was marketed in 1967 for the control of epileptic seizures.[25]

REFERENCES

1. A. Cahn, P. Hepp, Das Antifebrin, eine neues Fiebermittel. *Centralb. Klin. Med.*, 1886; **7**: 561.
2. A. Eichengrün, 25 Jahre Arzneimittelsynthese. *Z. Angew. Chemie*, 1913; **26**: 49–56.
3. D. Lester, L.A. Greenberg, Metabolic fate of acetanilide and other aniline derivatives. II. Major metabolites of acetanilide in the blood. *J. Pharm. Exp. Ther.*, 1947; **90**: 68.
4. F.B. Flinn, B.B. Brodie, Effect on pain threshold of *N*-acetyl-*p*-aminophenol, a product derived in the body from acetanilide. *J. Pharm. Exp. Ther.*, 1948; **94**: 76.
5. M. Spielman, Some analgesic agents derived from oxazolidine-2,4-dione. *J. Am. Chem. Soc.*, 1944; **66**: 1244–5.
6. A. Gilman, The initial clinical trial of nitrogen mustard. *Am. J. Surg.*, 1963; **105**: 574–8.
7. L.S. Goodman, M.W. Winetrobe, M.T. McLennan, W. Dameshek, M. Goodman, A. Gilman, in *Approaches to Tumour Chemotherapy*, ed. F.R. Moulton. Washington: American Association for the Advancement of Science; 1947, p. 338.
8. C. Golumbic, J.S. Fruton, M. Bergmann, Chemical reactions of the nitrogen mustard gases. I. The transformations of methyl-bis(β-chloroethyl)amine in water. *J. Org. Chem.*, 1946; **11**: 518–35.
9. M.R. Osborne, D.E.V. Wilman, P.D. Lawley, Alkylation of DNA by the nitrogen mustard bis(2-chloroethyl)methylamine. *Chem. Res. Toxicol.*, 1995; **8**: 316–20.
10. A. Haddow, G.A.R. Kon, W.C.J. Ross, Effect upon tumours of various haloalkylarylamines. *Nature*, 1948; **162**: 824–5.
11. R.J. Goldacre, A. Loveless, W.C.J. Ross, Mode of production of chromosome abnormalities by the nitrogen mustards. *Nature*, 1949; **163**: 667–9.
12. J.L. Everett, G.A.R. Kon, The preparation of some cytotoxic epoxides. *J. Chem. Soc.*, 1950; 3131–5.
13. Br. Pat. 1962: 901876 (to ICI).
14. D.A. Karnofsky, J.H. Burchenal, G.C. Armistead Jr, *et al.*, Triethylene melamine in the treatment of neoplastic disease; a compound with nitrogen-mustardlike activity suitable for oral and intravenous use. *Arch. Int. Med.*, 1951; **87**: 477–516.
15. F.L. Rose, J.A. Hendry, A.L. Walpole, New cytotoxic agents with tumour-inhibitory activity. *Nature*, 1950; **165**: 993–6.
16. J.H. Burchenal, M.L. Crossley, C.C. Stock, *et al.*, The action of certain ethyleneimine (aziridine) derivatives on mouse leukaemia. *Arch. Biochem.*, 1950; **26**: 321.
17. J.H. Burchenal, S.C. Johnstone, R.P. Parker, M.L. Crossley, E. Kuh, D.R. Seeger, Effects of N-ethylene substituted phosphoramides on transplantable mouse leukaemia. *Cancer Res.*, 1952; **12**: 251–2.
18. A. Haddow, G.M. Timmis, Myleran in chronic myeloid leukaemia. *Lancet*, 1953; **1**: 207–8.

19. J. Everett, J.J. Robertson, W.C. Ross, Aryl-2-halogenalkylamines. Part XII. Some carboxylic derivatives of *N,N*-di-2-chloroethylamines. *J. Chem. Soc.*, 1953; 2386–90.
20. F. Bergel, J.A. Stock, Cyto-active amino-acid and peptide derivatives. Part I. Substituted phenylalanines. *J. Chem. Soc.*, 1954; 2409–17.
21. I. Niculescu-Duvaz, A. Cambani, E. Tarnauceanu, Potential anticancer agents. II Urethane-type nitrogen mustards of some natural sex hormones. *J. Med. Chem.*, 1967; **10**: 172–4.
22. H. Arnold, F. Bourseaux, N. Brock, Chemotherapeutic action of a cyclic nitrogen mustard phosphoramide ester (B 518-ASTA) in experimental tumours of the rat. *Nature*, 1958; **181**: 931.
23. N. Brock, H.J. Hohorst, Metabolism of cyclophosphamide. *Cancer*, 1967; **20**: 900–4.
24. US Pat. 1973: 3732340 (to Asta).
25. H. Meunier, G. Carraz, Y. Meunier, *et al.*, Propriétés pharmacodynamiques de l'acide *n*-dipropylacetique. *Thérapie*, 1963; **18**: 435–8.

Concluding Remarks

Experimental pharmacology is an area of science that could only develop once pure chemicals of consistent quality had been isolated. Pharmacologists used live animals and isolated tissues to evaluate the alkaloids and glycosides isolated early in the nineteenth century and the synthetic drugs that followed them. In the field of chemotherapy, Ehrlich infected animals in order to screen for antitrypanosomal drugs, an approach that was also to prove successful when applied to the development of the synthetic antimalarials and sulfonamides. However, by the end of the twentieth century, many companies were deserting the traditional approaches involved in screening compounds. Instead, the search for novel drugs now embraces all that modern science can offer, as seen with the dual application of combinatorial chemistry and high-throughput screening. This harnesses the power of the computer to organise massive programmes in which hundreds of thousands of chemicals prepared by combinatorial chemistry, a form of robotic unitary construction involving the attachment of variant chemicals on to molecular templates, are routinely screened for activity in multiple arrays of protein targets identified by DNA technology and the like. Hundreds of thousands of different molecules can now be prepared and screened in a matter of weeks, but success in the field of drug research is not measured by the degree of its sophistication. The first decade of this new methodology has failed to live up to its promise and fewer novel drugs have been discovered during this period than in the other decades since the 1950s. In fairness to those who are labouring to overcome this disappointment, it has to be said that the prime reason for it arises from a decline in the number of novel drug prototypes from botanical and biochemical sources. Drug research in the twentieth century was driven by the development of analogues of alkaloids, hormones and similar natural substances, only a few of which have been isolated during the last 50 years. It is to be hoped that the conclusion of the human genome project will remedy this situation in years to come if novel enzyme targets for drug design evolve from it. The prospects for gene therapy, on the other hand, have been seriously damaged by the inability of pharmaceutical scientists to develop safe methods for its delivery.

In the interim period before new biochemical target molecules are discovered, there remains a serious threat to the development of new drugs that could have unforeseeable consequences if it is not addressed. Reflection on what has happened over the last two centuries reveals that it is relatively easy to discover a new drug, but exceedingly difficult to discover one that is safe enough to be administered as a medicine to heal the sick. The previous pages have described many successes of modern drug research. It would have required several more volumes if consideration had also been given to the far greater number of projects that failed.

The reality of the situation is that when any foreign substance is introduced into the body there will always be a risk of some unanticipated reaction occurring that existing safety tests cannot detect. The demands made upon the pharmaceutical industry to ensure that its products are safe are quite understandable, but the level of sophistication of current pre-launch chronic safety testing and post-marketing surveillance of patients now means that every new product faces the risk of being withdrawn shortly before or after its launch. Hundreds of millions of pounds or dollars will have been spent by this time, and much greater

Drug Discovery. A History. W. Sneader.
©2005 John Wiley & Sons Ltd: ISBN 0 471 89979 8 (HB); 0 471 89980 1 (PB)

expense may also be incurred to meet claims for compensation if a drug has to be withdrawn after it has been marketed. It is inevitable that all these costs will continue to rise and a time may soon come in which the commercial risks of developing new drugs will be considered too great when measured against the potential financial returns. Unless steps can be taken to resolve this issue, the prospects for drug discovery in the future could be bleak.

The public needs to be better educated about the nature of drug therapy. The message that must be put across is that no matter how carefully and conscientiously a pharmaceutical company designs, selects and evaluates drugs in order to introduce one that fulfils its intended medicinal role by affecting one specific target, the human body is so complex that there will always be the possibility of an unintended target also being hit. Even when the best practice in conducting safety tests on animals, volunteers and those patients involved in early trials has been followed to the point of perfection, the potential for disaster will still persist through no fault of those involved. By agreeing to accept any medication that has been exhaustively tested and correctly administered in the light of existing knowledge, patients should be considered to have accepted this minimal risk. If they are to be compensated for any damage to their health that arises, the settlement should take acceptance of that risk into account. Whether or not the government under whose authority the drug received a license indicating that it had been thoroughly tested should pay compensation could determine the likelihood of future drugs ever reaching the market. A levy on the sale of all drugs, including generics no longer covered by patent, would cover the cost of compensating the few who are damaged by the drugs that help restore the good health of so many.

Index

AB Bofors, 434
AB Hässle, 193
abacavir, 262
abamectin, 307, 332–3
Abbott, G., 80
Abbott Laboratories, 66, 111, 238, 264, 293, 296, 366, 368, 370, 389, 439
Abd-er-Rahaman III, 26
Abel, J., 155–6, 160, 167–8
abortion, 204
Abraham, E., 292, 296–7, 324–5
absinth, 14
acacia, 27
Académie des Sciences, 90
acarbose, 345–6
acebutolol, 194–5
acemetacin, 215
acetaldehyde, 144
acetanilide, 438–9
acetarsol, 55
acetazolamide, 390–1
acetic acid, 168, 274
acetone, 82, 366
acetophenetidin, 438–9
17-acetoxyprogesterone, 204
acetyl-coenzyme A (CoA), 344
acetylcholine, 5, 96, 98–9, 138–9, 160, 217, 252, 346, 435
  antagonists, 304
  chloride, 160
acetylcodeine, 115
  hydrochloride salts of, 116
acetylenic allylamines, 424
N-acetylneuraminic acid, 264
acetylsalicylic acid, 359
Achard, F., 88
aciclovir, 259–60
  triphosphate, 259
De l'Acid Phénique (Lemaire), 356
Ackerman, D., 159
aclacinomycins, 313
aclarubicin, 313
acne, 239
Acomatol®, 164
aconite, 38
Aconitum napellus L., 38
An Acount of the Foxglove, and Some of its Medical Uses: with Practical Remarks on Dropsy, and Other Diseases (Withering), 39
acquired immunodeficiency syndrome see AIDS
Acragas, 19
acridine yellow, 380
acridines, 380, 384
acriflavine, 381
acrivastine, 407
acrodermatitis enteropathica, 62
acromegaly, 168

Actinomyces, 299
  antibioticus, 300
actinomycetes, 299–300, 303–4, 310
  drugs from, 344–6
actinomycetin, 300
actinomycin A, 300, 311
actinomycin C, 311, 313
actinomycin CD₁₋₃, 311
actinomycin D, 311
Actinoplanes, 345
  teichomycetius, 310
active principles, isolation of, 3–4
activity
  antisecretory, 122
  antispasmodic, 122
  cholinergic, 131
  mydriatic, 121
  spasmolytic, 121
Actuarius of Constantinople, 92
Adams, R., 295, 366
Adams, S., 417–18
Adaouste, France, 16
Addison, T., 151
Addison's disease, 151, 179–80, 182, 207, 367
adenine, 252–4, 441
  arabinoside, 258–9
adenocarcinoma, 197
adenosine, 135
adhan, 28
adiphenine, 122
Admirand, W., 273
adrenal cortex, 180
  hormones, 179–84
adrenal glands, 155
Adrenalin®, 155
adrenaline (epinephrine), 5, 82, 101, 129, 155–6, 179, 212, 271
  acid, 156
  analogues, 188–92
  antagonists, 192–5, 403
  orcinol isomer of, 190
adrenaline, 156, 188
adrenoceptors, 158, 189, 192–3, 195
adrenocorticotrophic hormone (ACTH), 170
adrenosterone, 180
Adrian, E., 435
Die Aetiologie, der Begriff und die Prophylaxis des Kindbettfiebers (Semmelweis), 66
Africa, 49–50, 57, 333
Afridol Violet, 377–8
aglycones, 107, 143
agonists, selective 5-HT, 216–19
agranulocytopenia, 416
agranulocytosis, 214, 415
Ahlquist, R., 189, 192

AIDS (acquired immuno-deficiency syndrome), 260–2, 264, 272, 309, 334
ajmaline, 101–2
Akiebolaget Hassle, 416
akinesia, 409
alanine, 282, 443
alanylproline, 280–2
alatrofloxacin, 395
Albrecht, W., 303
Albright, F., 173
Albucasis see al-Zahrawi, Abu al-Qasim
albumin, 46
albuterol, 191–2
alchemists, 2
Alcmaeon, 20
Alcock, T., 65
alcohol, 9–10, 81
  first preparation of, 42
aldosterone, 183, 211, 280
  analogues, 211
Aldrich, T., 156, 167
Alexander the Great, 21
Alexander of Tralles, 39, 90
Alexandria, 19, 21–2, 24
alfalfa, 243
alfentanil, 124
alkaloids, 4–5, 18, 40, 88–105, 356
  ergot, 343, 349, 352–3
  mydriatic, 96
  opium, 89–92, 115
  rauwolfia, 101–2
  vinca, 102–3
  xanthine, 95
alkanoic acids, 275
alkyl iodides, 115
alkylphenols, 362
4-alkylphenyl acid, 418
Allen & Hanbury, 183, 191, 195
Allen, E., 173–4
Allen, W., 175–6, 200
allergy, 182
Alles, G., 130
allobarbital, 368
Allocrysine®, 60
allopathy, 77
allopurinol, 254
allyltoluidine, 125, 355
Almadén, Spain, 44
almond oil, 22
almonds, 28
Almquist, H.J., 243
aloes, 12, 27–8
alpha-glucosidase, 345
alprenolol, 193
alprostadil, 185–6, 220
Alt, 53
alteplase, 270
Altounyan, R., 142
Altschul, R., 231
Aludrin®, 189

alum, 14, 27
  water, 46
aluminium, sulfate, 27
Amanita muscaria, 15
amaurosis, 45
amber, 27, 34
Ambros, O., 435
amdinocillin, 323
amenorrhoea, 45, 175, 200, 202
America, 57, 65
American Birth Control League, 201
American Cyanamid Company, 294, 330, 388, 390, 435, 442
  Bound Brook Laboratories, 251
American Indians, 9
American Journal of Physiology, 201
American Medical Association, 82, 343, 366
  Council of Pharmacy and Chemistry, 101
American Physiological Society, 160
American Psychiatric Association, 396
American Society for Clinical Investigation, 293
Amgen, 272
amides, 370
amidines, 415
Amidon®, 123
amidopyrine, 126–7
amidoximes, 277
amikacin, 302, 328–9
aminacrine, 381
amines, 228
  sympathomimetic, 188, 195
aminitrozole, 309
amino acids, 417, 443
  analogues, 219–20
amino alcohols, 122, 129
9-aminoacridines, 381
4-aminobenzene sulfonamide, 386
4-aminobenzensulfonyl chloride, 388
4-aminobenzoic acid, 250, 252
o-aminobenzoic acid, 360
p-aminobenzoic acid, 361
aminobenzoic acids, 129
4'-aminobenzylpenicillin, 319
γ-aminobutyric acid see GABA
ε-aminocaproic acid, 417
7-aminocephalosporanic acid, 324
aminoglutethimide, 367
aminoglycosides, 300–2, 323
  analogues, 328–30
5-aminoimidazole ribotide, 256
5-aminoimidazole-4-carb-oxamide ribotide, 256–7